Understanding China

The series will provide you with in-depth information on China's social, cultural and economic aspects. It covers a broad variety of topics, from economics and history to law, philosophy, cultural geography and regional politics, and offers a wealth of materials for researchers, doctoral students, and experienced practitioners.

Wenheng Yang · Qinye Yang

A History of Ancient Chinese Geoscience

ZHEJIANG UNIVERSITY PRESS
浙江大学出版社
GEP 广西教育出版社
Guangxi Education Publishing House
Springer

Author
See next page

ISSN 2196-3134 ISSN 2196-3142 (electronic)
Understanding China
ISBN 978-981-97-4459-6 ISBN 978-981-97-4457-2 (eBook)
https://doi.org/10.1007/978-981-97-4457-2

Jointly published with Zhejiang University Press
The print edition is not for sale in China (Mainland). Customers from China (Mainland) please order the print book from: Zhejiang University Press.
ISBN of the Co-Publisher's edition: 978-981-97-4456-5

This book is supported by Chinese Fund for the Humanities and Social Sciences (20WZSB033).

Translation from the Chinese language edition: "中国地学史 (古代卷)" by Wenheng Yang et al., © Guangxi Education Publishing House 2014. Published by Guangxi Education Publishing House. All Rights Reserved.

This Springer imprint is published by the registered company Springer Nature Singapore Pte Ltd.
The registered company address is: 152 Beach Road, #21-01/04 Gateway East, Singapore 189721, Singapore

Wenheng Yang · Qinye Yang

A History of Ancient Chinese Geoscience

ZHEJIANG UNIVERSITY PRESS
浙江大学出版社
GEP 广西教育出版社
Guangxi Education Publishing House
Springer

Author
See next page

ISSN 2196-3134 ISSN 2196-3142 (electronic)
Understanding China
ISBN 978-981-97-4459-6 ISBN 978-981-97-4457-2 (eBook)
https://doi.org/10.1007/978-981-97-4457-2

Jointly published with Zhejiang University Press
The print edition is not for sale in China (Mainland). Customers from China (Mainland) please order the print book from: Zhejiang University Press.
ISBN of the Co-Publisher's edition: 978-981-97-4456-5

This book is supported by Chinese Fund for the Humanities and Social Sciences (20WZSB033).

This Springer imprint is published by the registered company Springer Nature Singapore Pte Ltd.
The registered company address is: 152 Beach Road, #21-01/04 Gateway East, Singapore 189721, Singapore

Wenheng Yang
Institute of History of Natural Sciences
Chinese Academy of Sciences
Beijing, China

Qinye Yang
Institute of Geographic Sciences
and Natural Resources Research
Chinese Academy of Sciences
Beijing, China

Translated by
Jiqing Dong
School of English Studies, Institute
of Language Science
Shanghai International Studies University
Shanghai, China

Co-Translated by
Christine WL Wilson
Center for Translating and Interpreting
in Scotland
Heriot-Watt University
Edinburgh, Scotland, UK

With Contributions by
Jie Fei
Fudan University
Shanghai, China

Xiuying Wen
Tianjin University of Finance
and Economics
Tianjin, China

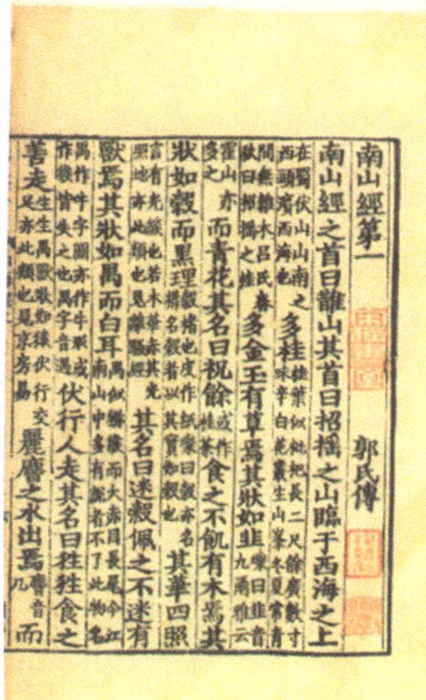

南山經第一 郭氏傳

南山經之首曰䧿山其首曰招搖之山臨于西海之上

THE CLASSIC OF MOUNTAINS AND SEAS/SHANHAI JING 山海经 [18 volumes] annotated by Guo Pu 郭璞 from the Eastern Jin dynasty. A facsimile copy produced by the Chiyangjun Workshop in the 7th year of the Chunxi reign of the Song dynasty (1180). Collection of the National Library of China

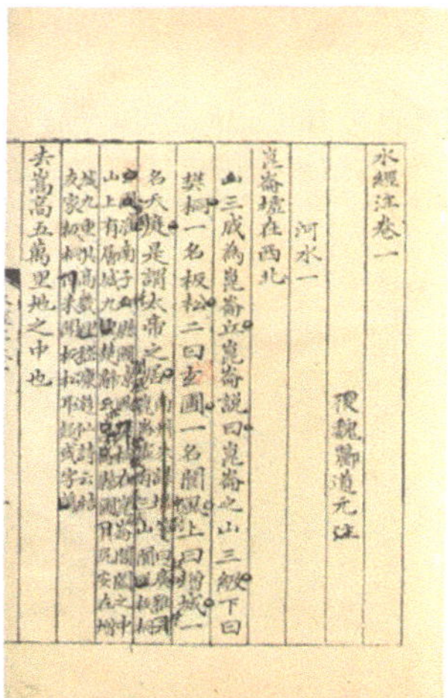

水經注卷一

河水一

崑崙墟在西北

COMMENTARY ON THE WATER CLASSIC/SHUIJING ZHU 水经注 [40 volumes] authored by Li Daoyuan 郦道元 from the Northern Wei dynasty. A copy collated by Quan Zuwang 全祖望 from the Qing dynasty conserved in the Xiaoshan Studio. Tianjin Library collection

FANGMATAN MAP. This map drawn on a wood board was excavated from Fangmatan No. 1 Qin tomb in Gansu province. *Zhongguo gudai dituji* 中国古代地图集 (1990) [Ancient Chinese atlas]. Cultural Relics Press 文物出版社, Image 4

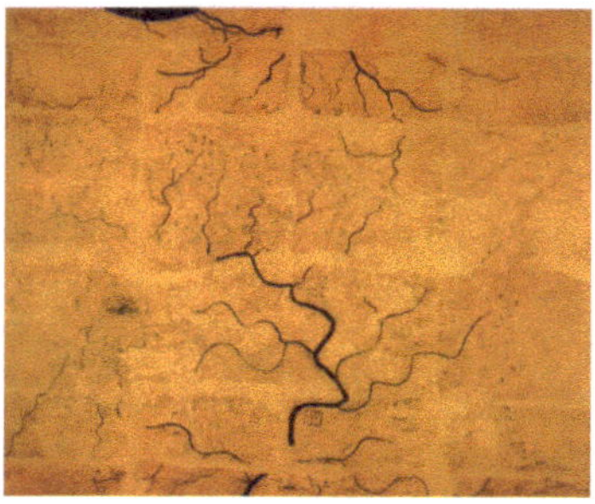

MAWANGDUI TOPOGRAPHIC MAP. This map on silk was excavated from Mawangdui No. 3 Han tomb in Hunan province. *Zhongguo gudai dituji (zhanguo-yuan)* 中国古代地图集 (战国-元) (1990) [Ancient Chinese atlas: from the Warring States to the Yuan dynasty]. Cultural Relics Press 文物出版社, Image 20

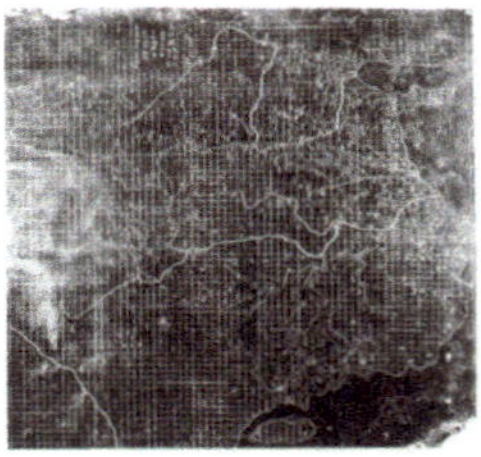

A rubbing of MAP OF TRACES OF YU/YUJI TU 禹迹图 dating from the Northern Song dynasty. *Zhongguo gudai dituji (zhanguo-yuan)* 中国古代地图集 (战国-元) (1990) [Ancient Chinese atlas: from the Warring States to the Yuan dynasty]. Cultural Relics Press 文物出版社, Image 55

A rubbing of THE CITY MAP OF PINGJIANG/PINGJIANG TU 平江图 created during the Southern Song dynasty. *Zhongguo gudai dituji (zhanguo-yuan)* 中国古代地图集 (战国-元) (1990) [Ancient Chinese atlas: from the Warring States to the Yuan dynasty]. Cultural Relics Press 文物出版社, Image 80

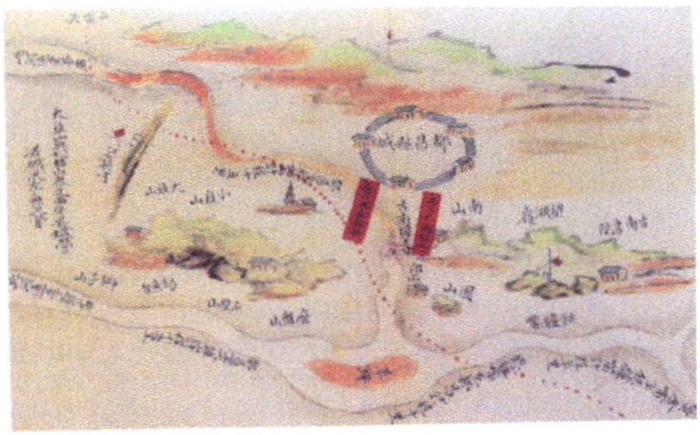

WATER POLICE MAP/JIANGFANG TU 江防图. This map represents the jurisdiction of the water police station at Tongzhou near the mouth of the Yangzi River during the Qing dynasty. The red strips on the map give patrol instructions, distance information, and information about tributaries. The *Staatsbibliothek zu Berlin—Preussischer Kulturbesitz* is the owner of the original map

Part of the NINE GARRISON DISTRICTS MAP/JIUBIAN TU 九边图 from the Ming dynasty. *Zhongguo gudai dituji (ming dai)* 中国古代地图集 (明代) (1994) [Ancient Chinese atlas: Ming dynasty]. Cultural Relics Press 文物出版社, Image 81

Part of THE GENERAL MAP OF THE RIVER DEFENCE FORTIFICATIONS/HEFANG YILAN TU 河防一览图 from the Ming dynasty. *Zhongguo gudai dituji (ming dai)* 中国古代地图集 (明代) (1994) [Ancient Chinese atlas: Ming dynasty]. Cultural Relics Press 文物出版社, Image 34

THE GREAT UNIVERSAL GEOGRAPHIC MAP/KUNYU WANGUO QUANTU 坤舆万国全图 (1602) drawn by Italian missionary Matteo Ricci. *Zhongguo gudai dituji (ming dai)* 中国古代地图集 (明代) (1994) [Ancient Chinese atlas: the Ming dynasty]. Cultural Relics Press 文物出版社, Image 80

A wooden tellurion made in China during the Ming dynasty (currently conserved in the British Museum). *Zhongguo gudai dituji (ming dai)* 中国古代地图集 (明代) (1994) [Ancient Chinese atlas: the Ming dynasty]. Cultural Relics Press文物出版社, Image 92

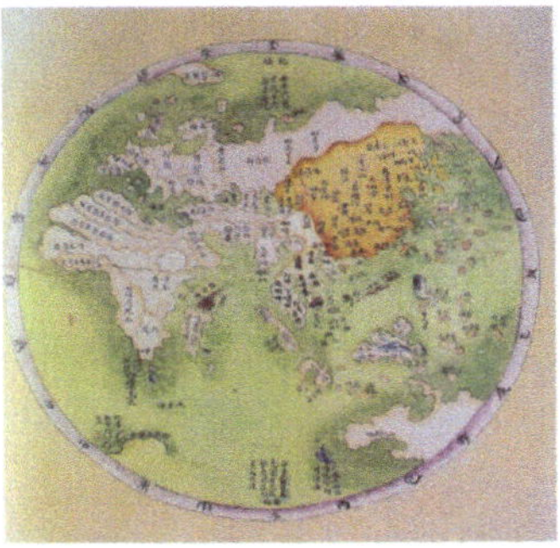

CHINESE MAP OF THE EASTERN HEMISPHERE, 1790. This map was drawn at the beginning of a scroll devoted mainly to the maritime territory requiring coastal defences during the mid-Qing dynasty. The *Staatsbibliothek zu Berlin—Preussischer Kulturbesitz* is the owner of the original map

A mariner's compass from the Qing dynasty (photographed by Liang Qing 梁擎)

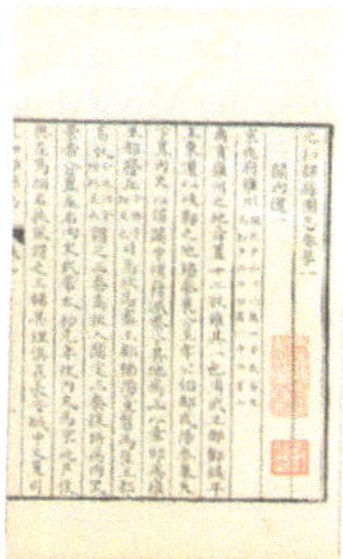

The ILLUSTRATED GAZETTEER OF THE YUANHE PREFECTURES AND COUNTIES 元和郡县图志 [40 volumes] by Li Jifu 李吉甫 from the Tang dynasty (a copy from the early Qing dynasty). Collection of the National Library of China

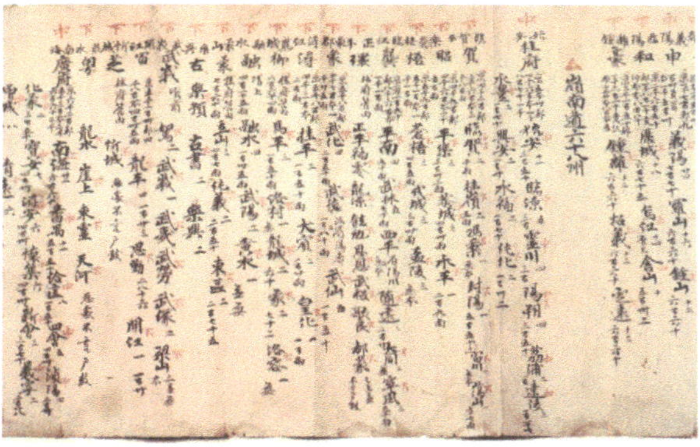

A handwritten excerpt of a LOCAL CHRONICLE 地志 on a scroll from the Tang dynasty. Dunhuang Museum collection. The POLARIS METEORIC MAP/ZIWEI GONGXING TU 紫微宫星图 and the BOOK OF PREDICTING WEATHER BY CLOUDS/ZHAN YUNQI SHU 占云气书 are found on the back of the scroll

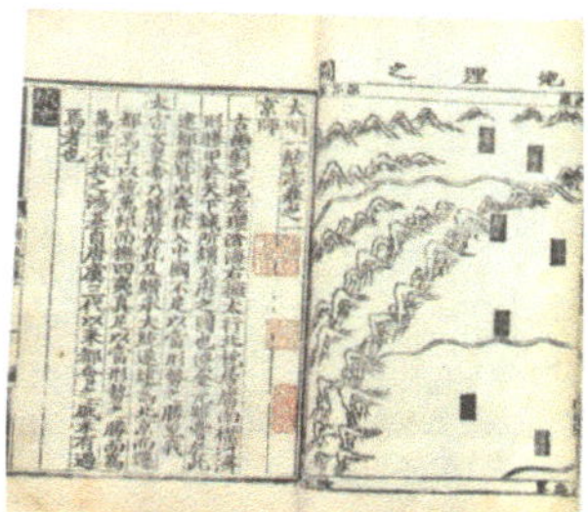

The Ming imperial block-printed edition of the RECORDS OF THE UNITY OF THE GREAT MING/DAMING YITONG ZHI [90 volumes] 大明一统志 authored by Li Xian 李贤 and Wan An 万安 in the 5th year of the Tianshun reign of the Ming dynasty (1461). Sichuan University Library collection

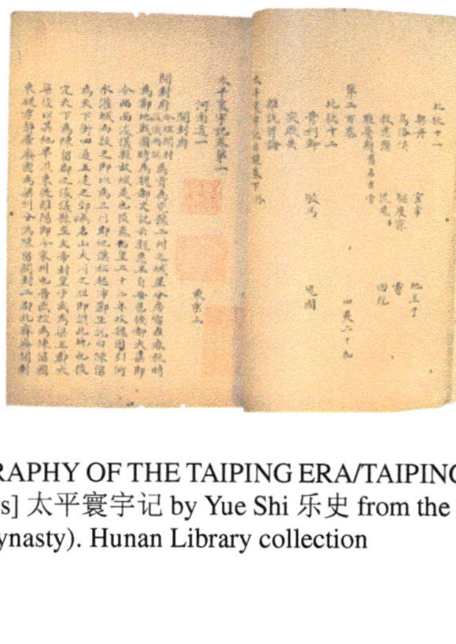

THE UNIVERSAL GEOGRAPHY OF THE TAIPING ERA/TAIPING HUANYU JI [200 volumes with 2 volumes of catalogues] 太平寰宇记 by Yue Shi 乐史 from the Song dynasty (a handwritten copy from the Early Qing dynasty). Hunan Library collection

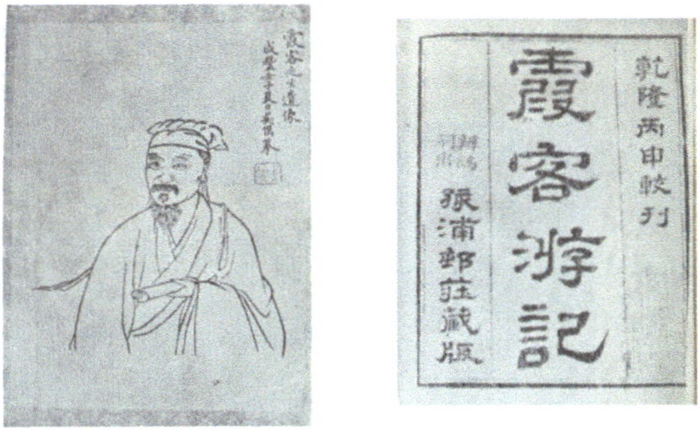

A portrait of Xu Xiake and the Qianlong Emperor edition of XU XIAKE'S TRAVELS. Tang Xiren 唐锡仁, Yang Wenheng 杨文衡 (1987). *Xuxiake jiqi youji yanjiu* 徐霞客及其游记研究 [Research on Xu Xiake and his Travelogues]. China Social Sciences Press 中国社会科学出版社

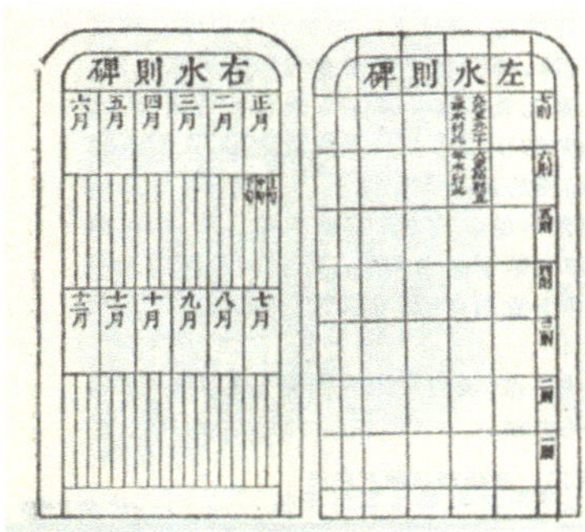

Inscriptions of flood level observations—Wu River levels recorded in the form of a carving in the INVESTIGATIONS OF WATER CONSERVANCY OF THE WU RIVER/WUJIANG SHUILI KAO 吴江水利考 from the Ming dynasty. *Shuiwen, shamo, huoshan kaogu* 水文、沙漠、火山考古 [Archaeological research on water, deserts and volcanoes] (1977). Cultural Relics Press 文物出版社, p. 39

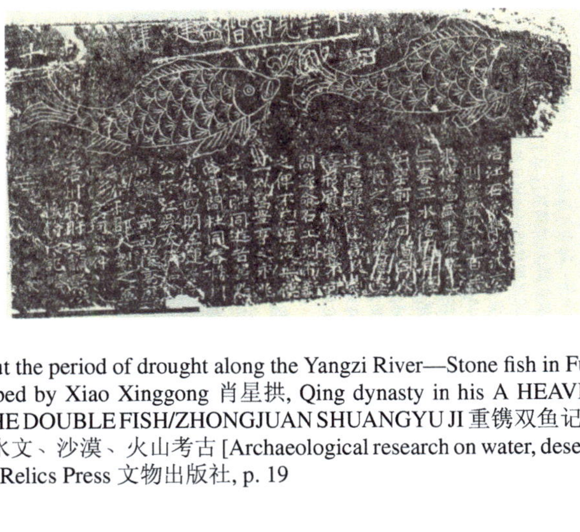

Inscriptions about the period of drought along the Yangzi River—Stone fish in Fuling, Chongqing, originally inscribed by Xiao Xinggong 肖星拱, Qing dynasty in his A HEAVILY ENGRAVED RECORD OF THE DOUBLE FISH/ZHONGJUAN SHUANGYU JI 重镌双鱼记. *Shuiwen, shamo, huoshan kaogu* 水文、沙漠、火山考古 [Archaeological research on water, deserts and volcanoes] (1977). Cultural Relics Press 文物出版社, p. 19

Foreword

Within the framework of modern science, mathematics, physics, chemistry, astronomy, geoscience and biology are often regarded as the six pillars of natural science. Geoscience, also known as Earth science, includes all fields of study on the Earth: geology, geophysics, geography, atmospheric science, marine science, to name but a few. Geoscience is the foundation subject that is most critical to mankind's knowledge of the Earth. Research on the planet Earth and its multi-sphere composition, structure and physical properties—as well as on how these sub-systems emerge, evolve and interact with each other—provides scientific evidence to address issues vital to the environment and crucial to the resources and sustainable development of every human society.

Ancient geoscience traces its origins to China, during the Spring and Autumn and the Warring States periods, and to Europe, during the time of ancient Greek civilisation. As a rule, early scholars began gathering information and describing geographic phenomena, with little intention of differentiating geoscience from other disciplines. Due to environmental differences and communication barriers, ancient geoscientific accounts displayed marked regional characteristics in terms of the topics and stylistic rules. Modern geoscience, in contrast, grew out of the industrial revolutions. From around the first half of the 1800s onwards, regional geographic observations and geological surveys were increasingly assisted by the scientific and technological advances. As a result, modern geographers and geologists founded their respective academic disciplines, and proposed a series of theories and models. The 1950s witnessed a rapid development of geoscience. Methodological innovation was exemplified in the wider use of scientific expeditions and fieldwork, as well as in greater modernisation of specific project designs and experiments. This not only fostered the cross-fertilisation of ideas between neighbouring disciplines but also accelerated the integration of sub-domain knowledge and a shift in research paradigms. Consequently, modern geography and geology took a substantial leap forward: a relatively complete academic system had been established, and major breakthroughs had been achieved in its multiple branches of study.

In the late twentieth century, a team of science historians worked with Guangxi Education Publishing House and conceived a plan to compile a book series entitled *Zhongguo kexueshi* ("A History of Chinese Science"). In this series, there are two volumes focusing on the history of geoscience, i.e. *Zhongguo dixueshi gudaijuan* ("A History of Ancient Chinese Geoscience") and *Zhongguo dixueshi xiandaijuan* ("A History of Modern Chinese Geoscience"). As a result of reviewing and summarising the chronicles of Chinese geoscience, we became acquainted with the upward trajectory of geoscience development in China and its distinct patterns. Building on an understanding of the current status quo as regards scholarship, we hope to further advance academic progress in the field of modern geoscience in China.

As expounded in the *Zhongguo dixueshi gudaijuan*, China was one of the first civilisations in the world to establish geoscientific principles. Ancient scholars often combined the term "*tianwen*" (天文, astronomy) with "*dili*" (地理, geography), believing that "*tianwen dili*" constituted the entire body of knowledge of the universe. Research indicates that the beginning of Chinese geoscientific thought may date from the prehistoric period. In the Spring and Autumn and the Warring States periods, the Chinese character writing system reached maturity, and the practice of inscribing topographical information and surface features together with graphic symbols on objects enabled a fast expansion and accumulation of geographic knowledge. Subsequently, a succession of historical works related to geoscience came into being in the Warring States period, including the *Shanhai jing* ("The Classic of Mountains and Seas"), the chapter "Yugong" (or "Tribute of Yu") in the *Shangshu* ("Book of Documents"), and *Guanzi*. Diverse schools of thought on the Earth were also taking shape during this period, with discussions ranging from the theory of regional differences and the philosophy of the human-nature relationship to the ecological propositions needed for human survival. From mediaeval to late imperial China, myriads of accounts and narratives were written in the domains of chronicles, the geography of historical administrations, physical geography, cartography, meteorology and climatology, geology and palaeontology, petrology and mineralogy. Important scholarly contributions included the *Datang xiyu ji* ("Great Tang Records on the Western Territories"), *Shuijing zhu* ("Commentary on the Water Classic"), *Mengxi bitan* ("The Dream Pool Essays"), *Xu Xiake's Travels*, *Nautical Chart of Zheng He* ("Mao Kun Map") and *Bencao gangmu* ("The Compendium of Materia Medica"). However, a great deal of ancient geographic text was descriptive in nature and lacked any holistic examination of the Earth's surface systems and their operating logic. Modern geoscience was largely impeded by the disregard for science and technology in feudal society, especially due to the isolationist foreign policy of the late imperial dynasties.

The *Zhongguo dixueshi xiandaijuan* outlines how geography and geology have evolved, as two disciplines, from around the period of the Opium Wars to the contemporary era. In the early twentieth century, a collection of modern Western geoscience works was introduced and developed by a group of senior Chinese geoscientists, including Zhang Xiangwen, Zhu Kezhen (Chu Kochen), Li Siguang, Weng Wenhao and Ding Wenjiang. It was through their industrious work and great perseverance that a solid foundation for the development of modern geoscience was successfully established in China. Meanwhile, the drive to improve the discipline was further stimulated by the founding of the Geographic Society of China, in 1909, and other research institutes related to Earth science. With more advanced models and methods being transmitted to China, many universities started setting up geoscience faculties in order to provide specialised training for young talents. All this progress represents an important milestone in the development of Chinese geoscience.

Since the mid-twentieth century, Li Siguang and Zhu Kezhen's call to establish more geoscience research centres has been addressed: the Institute of Geology and Geophysics and the Institute of Natural Resources, both at the Chinese Academy of Sciences, have been founded. More students have been enrolled on geoscience programmes to help disseminate knowledge. In response to the government's need for experts on the physical environment, land resources, and the economic and social conditions of all the regions in China, geoscientists have conducted a considerable amount of research. This has included integrated surveys of regional resources, land planning research, geological surveys, drilling and sampling at mining sites, fixed point observation and indoor testing, geoscience simulation experiments, remote sensing and geo-information system research. Thanks to the unswerving efforts of these colleagues, modern geoscience has made great strides in recent years.

This volume summarises the development pathway, major research orientations and findings of modern Chinese geoscience since the early twentieth century. Key representative figures and their academic achievements during this period are fully discussed. While focusing on the extensive progress in the secondary disciplines and major branches of modern geography and modern geology, the authors also expound on some of the key research trends and chart a course forward for this field.

As time passes, a diminishing number of people seem to be familiar with the early days of modern geography and geology. A couple of the authors of the second volume, *Zhongguo dixueshi xiandaijuan*, personally experienced the vicissitudes of the past and learned from the masters who made groundbreaking contributions to our field. I anticipate readers will gain an in-depth understanding of the various stages of development in the field of study after reading this geoscience chronicle. I believe this book will also be a good companion to future scientists and young scholars. With this guide, you can follow in the footsteps of our forefathers and take inspiration from their hard work.

We call on all academic colleagues of geoscience in China to work together to promote interdisciplinary and cross-disciplinary research. As we continuously improve the quality of research and education, I am convinced that modern geoscience in China can reach new heights and become internationally recognised. I hope we can all play a positive role in strengthening the foundation of Earth system science, promoting the revival and sustainability of the Chinese nation, and building a beautiful future for our planet home.

May 2011

Zheng Du
Academician of Chinese
Academy of Sciences
Institute of Geographic Sciences
and Natural Resources Research
Chinese Academy of Sciences (CAS)
Beijing, China

Contents

Chapter 1
Geoscience in Embryo: Exploring the Earth in Primitive Societies

1.1 An Overview of Primitive Societies

Primitive society is the first of the five stages of economic development of human society. Archaeologists and historians refer to the extended period of primitive society in which humans used stone tools as a means of production as the Stone Age. Different methods of making tools (i.e. chipped stone implements and ground stone implements) further divide the Stone Age into the Palaeolithic period and the Neolithic period respectively. The geological history of the Stone Age relies entirely on archaeological and ethnological sources.

Lasting from about two to three million years ago to ten thousand years ago, the Palaeolithic period was the earliest and longest stage in the development of human society. Humans had just evolved from *Homo erectus* and started to create stone tools, learned to use fire and shifted away from eating raw food, gradually improving their physical strength. The Middle and Lower Palaeolithic periods are marked by the "primitive horde" in Chinese history, when Renzidong Man in Fanchang, Anhui province, and Dingcun Man in Xiangfen, Shanxi province, were active. They survived by collecting fruits and hunting animals with basic stone tools. There were no laws, rulers or class divisions within these societies, and members had sexual intercourse solely for procreation and only within their community.

The Middle Palaeolithic period witnesses a transition from humans living in primitive hordes to the formation of tribes. By the late Palaeolithic period, a clan society based on blood ties emerged, leading to a giant leap in general productivity and skill development, e.g. stone spears and slings, horn implements, fur clothing, etc. The mastery of the controlled use of fire in this period enabled mankind to freely produce light and heat. As the clan system was bound by blood ties, there was a stable structure and leadership. The prohibition of within-clan marriage was a significant step forward and played a key role in improving people's physique. With the gradual establishment of communal cemeteries, clan councils and primitive religions, humans became not too dissimilar from their modern descendants.

W. Yang and Q. Yang, *A History of Ancient Chinese Geoscience*, Understanding China,
https://doi.org/10.1007/978-981-97-4457-2_1

Marked by a matrilineal clan society, the Mesolithic period in Chinese history is a transitional period between the Palaeolithic Age and the Neolithic Age. The bow and arrow were invented as a tool for hunting, which enormously improved the effectiveness of hunting. Overall, there are ten major aspects that characterise the New Stone Age in Chinese history:

1. an agricultural economy was taking shape and life conditions were significantly enhanced due to the agricultural breakthroughs;
2. the skill of making stone tools improved, with carving and grinding being applied to stones, bones, shells and jade wares;
3. pottery was invented and cookware was created for various purposes;
4. permanent living spaces were evolving, including many large and small settlements, and fortified and communal structures appeared in the late New Stone Age;
5. livestock farming further developed, providing additional resources to sustain life and grow agriculture;
6. textile products were used for clothing and fishing nets;
7. the practice of religions and the worship of deities was part of everyday life;
8. a variety of fine decorative artwork and religious artefacts appeared;
9. the matriarchy ended and the patrilineal clan society began;
10. productivity experienced another peak due to the introduction of metals, giving birth to the system of private ownership. As primitive society was collapsing, Chinese society was entering a new era of civilisation.

1.2 Preliminary Observations of the Earth in the Palaeolithic Period

1.2.1 Production Activities and the Understanding of Geoscience

(1) Geoscience reflected in fishing and hunting

As *Homo erectus*, men and women were in charge of hunting and collecting respectively, and this division of labour was mainly determined by their physiological differences [1].

Lantian Man hunted birds, frogs, turtles, lizards, snakes, mice, rabbits, porcupines and other small animals for food, 1.1 million years ago [2]. They must have been familiar with the morphology and ecology of these animals and their geographical distribution.

More than 500,000 years ago, Peking Man had a good understanding of the morphology, ecology and geographical distribution of the animals in the surrounding areas. Therefore, these hominins mainly targeted small creatures such as deer, but they also occasionally brought down tigers, leopards, wild boar, antelopes, horses and other medium to large animals.

About 70,000 years ago, Dingcun Man hunted not only the animals in the mountains and forests, but also the fish in the water and a large number of molluscs. It is evident that Dingcun Man was keenly aware of the multiple habitats of different animals, rather than focusing on a limited range of species.

Dating from about 28,000 years ago, at the Zhiyu archaeological site in Shuo County, Shanxi province, the largest number of fossils found are of wild horses and donkeys. Also known as "horse hunting man" [3], these were Zhiyu Man's main targets.

Dating from over 20,000 to 10,000 years ago, at the Xiaonanhai archaeological site in Anyang, Henan province, 22% of the animal fossils belong to extinct species, and the rest to species still in existence. Some of these were large herbivores, such as wild donkeys and woolly rhinos, and they were mostly juvenile and elderly animals, indicating that people at that time often chose prey that was easy to catch during hunting [4].

(2) Geoscience and gathering

Women in primitive society encountered a large number of plant species, accumulating a rich body of experience in the morphology, ecology and geographical distribution of this aspect of flora and fauna. They could mostly recognise poisonous food that was not suitable for eating.

At the cave sites inhabited by Peking Man, a large number of burnt elm tree seeds have been found, and it is assumed that these seeds were eaten as food by Peking Man.

The work of identifying edible plants requires persistent efforts and sometimes the price paid was life itself. The description of "the Divine Farmer Shennong who tastes hundreds of herbs" in a Chinese legend proves that medicine has its origins in practice. Similarly, deliberate cultivation and agriculture also developed from gathering practice. Early agriculture began in the Palaeolithic period and boomed in the Neolithic period.

1.2.2 *Living Environment and the Understanding of Geoscience*

As *Homo erectus*, our ancestors began to build rudimentary shelters, in addition to using caves and basic structures as dwellings. Several factors were considered when selecting an ideal location.

First, it should be close to rivers or water sources, and there should be sufficient raw materials to make stone tools.

Second, a relatively rich fauna for humans to hunt was necessary.

Third, there should be relatively abundant resources of flora for humans to gather. The supply of food through hunting is not as stable as by gathering. For many modern hunter-gatherer communities, 60–70% of their food is plant-based and provided by women, in addition to fish and aquatic shellfish [5].

Fourth, temporary dwellings should be able to provide shelter from the wind and rain, fire and wild beasts, so caves and rock shelters were the principal temporary dwellings for *Homo erectus.*

Fifth, there should be easy and safe access to the encampment, so that people could travel easily and the safety of children and the elderly would not be threatened. Overall, the living quarters should not be located in an area which was too steep area or too close to the waterside.

The above choices involved not only geomorphological knowledge, but also mineralogical, animal and plant, and hydrological understanding, which could be seen as an initial exploration of geoscience in ancient times.

For instance, Peking Man resided in the Zhoukoudian caves for more than 300,000 years, suggesting that this place had a good geographical environment. It is surrounded by high mountains to the north, low mountains to the west and southwest, and vast plains to the southeast. There is a river to the east of the caves, and there are also lakes or swamps. Not far from Zhoukoudian, there may have been a fairly wide area of dry grassland, and even desert areas. Similarly, Dingcun Man chose to live in the Fen River Basin because of the dense forests, rich animal and plant resources, and a large quantity of boulders by the riverside for making stone tools [6]. Xujiayao Man of Yanggao, Shanxi province, was found in the Datong Basin. There was a large lake here, during the Palaeolithic period, covering more than 9,000 km^2. The northern shore of the lake was flat and a stream flowed into the lake. To the north of this plain lay low hills and savannah. Lastly, the inhabitants of the Xiaonanhai site in Anyang, Henan province, lived a cave life. The cave is located in a gorge surrounded by mountains, close to the Huan River and Lake Xiaonanhai. East of the gorge was a vast grassland, with forests, swamps and rivers nearby.

1.2.3 Mineral-Petrological Evidence in Stone Artefacts

Around 2.4 million to two million years ago, Fanchang Man, in Anhui province, was able to make stone tools [7]. It can be said that, ever since they first appeared, human beings have formed an indissoluble bond with minerals and rocks.

Palaeolithic humans made stone tools using rocks or minerals as the raw material, but not all rocks or minerals were useful. Initially, they randomly picked up some stones to make tools. In the course of their experience over the long term, most stone tools stood out as easy to use and durable, although some could become blunt, or less durable. Experience taught them that it was necessary to choose a high-performing, extremely hard and tough stone. This was the earliest understanding of the properties of rocks and minerals shown by Chinese ancestors.

In addition to hardness, the material used for stone tools must have a certain degree of toughness and brittleness, so that it is relatively easy to break. However, stones that are too friable may produce fractures when used as tools, thus reducing their lifespan.

Consisting of cryptocrystalline quartz (SiO_2), flint is the most ideal material for stone tools. It has a value of 7 on the Mohs hardness scale and is tough and brittle. Once broken into a conchoidal fracture, flint can be chipped with a hammerstone into knife-like sharp edges. The distribution range of flint in China is extremely small, and the output was rather limited, so the Palaeolithic tools here were mostly made of other rocks such as quartz and quartzite.

Quartz is a metamorphic rock formed by the recrystallization of sandstone or chemical siliceous rock. Quartz (often as crystals or vein quartz) has a value of 7 on the hardness scale. It is generally light or white in colour and hard in texture, but its particles often form dense lumps. Quartz is widely distributed in China and is very abundant. For example, 88.8% of the stone tools of Peking Man were made of quartz [8]. The stone tools of the Kehe culture, except for a very few using vein quartz, were mostly made of quartzite [9].

Silbene limestone and horn shale were also good materials for stone tools, but, because of their limited availability, they are only found in large quantities at a few Palaeolithic sites. For example, at the Guanyin Cave site, siliceous limestone was used for 65% of the stone tools [10]. At the Dingcun cave, 95% of the stone tools were made of horn shale [11].

In addition, nearly 60 types of rocks, such as iron ore [12], sandstone, agate, basalt, andesite and diorite were used to make Palaeolithic tools. The best place to find the materials was the river bank, where gravel, commonly referred to as pebbles, can be found among all kinds of rocks and minerals [13].

Based on the author's statistics, the total number of rocks and minerals used by the inhabitants of the Palaeolithic period amounts to 37, of which nine were minerals, namely quartz, flint, chalcedony, crystal, hematite, graphite, opal, agate, jasper, and the remaining 28 were rocks. Could they recognise and differentiate between these 37 rocks and minerals? Unfortunately, the answer is no. The above-mentioned rocks and minerals were only occasionally encountered by chance, and then picked up and made into stone tools. However, these materials must have been known to the ancients as they were used regularly and especially to make stone tools or ornaments. Our ancestors in the Palaeolithic period could identify about 17 kinds of rocks and minerals, which were distinguished by colour, hardness, transparency and thickness.

Another piece of evidence which proves that Palaeolithic humans knew certain rocks and minerals is that the stones from which they made tools and ornaments were not all locally sourced. Some raw materials are not available in the local area but were collected from other places. For example, flint is scattered over the Sarausu site in Inner Mongolia, but only fine sand occurs locally, without any gravel layers [14].

Similarly, dating from more than 22,000 years ago, one of the chopping tools made of quartzite was not locally produced, because there is no rock of this kind at the Yanjiagang site in Harbin, Heilongjiang province, and it was apparently brought from elsewhere [15].

Quartz crystals were not local to the Peking Man site but were selected from granite-weathered hillsides about two kilometres north of the site, or from riverside terraces. The hematite used by Shandingdong Man was also obtained from hundreds of li away at Xuanhua [16].

1.2.4 Geoscientific Data from Human Migrations and Communications

So far, more than 260 Palaeolithic archaeological sites [17] have been discovered in China and these are widely distributed cross North China, Northeast China, the Yangzi River Basin, South China and the Qinghai-Tibet area. Due to the vast size of the territory and the different geographical environments, there are also regional differences in Palaeolithic culture.

First of all, the small stone tool tradition—flake tools—dominated the early Palaeolithic period. Influenced by this tradition, there are different types in the north and south, and each type has several cultural variations.

Secondly, there are two traditions of Palaeolithic culture in North China: the Kehe-Ding Village style, that is, the tradition of using hammerstones for chopping and crushing purposes; whilst the other was the Zhoukoudian-Zhiyu style, that is, the rounded and pointed scraper-carver tradition. These reflect differences in economic life.

Thirdly, from a global perspective, there were two major systems in the early Palaeolithic period: the West was marked by a hand-axe system, while the East features a system of hammerstone tools [18]. Such differences in cultural traditions show that hominins were adapting to their respective geographical environments in different ways, which lays the foundation for understanding regional differences in later human society.

In the early Palaeolithic period, our ancestors led a nomadic existence and there was little interaction between groups. By the Middle Palaeolithic period, the types of stone tools increased in number and their functions became further differentiated, making regional characteristics all the more obvious. With the growth in productivity, population size and the expansion of people's activities, more and more new families appeared. The marriage system had progressed from the bonds formed within the group to marriage outside the clans, which further promoted cultural exchange and integration.

At the Lingjing archaeological site in Xuchang, Henan province, many fine stone tools have been found, such as cone-shaped stone cores, pencil-shaped stone cores, long and narrow flint flakes, etc., which are very similar to those found at Mesolithic sites such as Hailar on the northeast border and Shayuan in Dali, Shaanxi province. In particular, the two small, thick-bladed axe-shaped tools found in Lingjing are the same as those from the Shayuan area, but they are more delicately crafted. These findings suggest that the Mesolithic culture spread to the surrounding area and that there were cultural exchanges with other regions. As Huang Weiwen commented: "The time when the fine stone tools from North China spread southwards to Xiqiaoshan Mountain was earlier than the time when fine stone tools from the Mediterranean spread eastwards to southern Australia." [19]

The tradition of fine stone tool craftsmanship in the Mesolithic period was widely distributed over Yunnan. For example, fine stone tools were found at eight locations in Danawu Village, in Yuanmou County. They were also discovered in Yangyi Xinzhai, Baoshan city, and in several regions of Tibet, including Naqu city, Nielamu County, Shenzha County and Shuanghu area, with exactly the same craft traditions and styles as those found in the Yellow River Basin [20]. Fine stone tools were also excavated at the La Yihai site in Guinan County, Qinghai province, as well as in Fenghuangling and Qingfengling in Linyi, Shandong province, and at Heilongtan in Tancheng County, and at Daxianzhuang in Donghai County, Jiangsu province. The stone tools from the Xueguan site in Puxian County, Shanxi province, display almost all the characteristics of Chinese Mesolithic stone toolmaking [20].

According to An Zhimin, the craft tradition represented by fine stone tools in eastern Asia and northern America must have originated in North China. During the Mesolithic period, hunter-gatherers gradually expanded their areas of activity throughout northern China and turned westward, through the Weihe River Valley, to Gansu, Qinghai, Xinjiang and Tibet; they followed the Yellow River east down to Hebei, Henan, Shandong and the north of Jiangsu province; they crossed over the Qinling Mountains to the south and reached the Yunnan-Guizhou Plateau and the shore of the South China Sea; they travelled to Siberia, via the Mongolian Plateau, and then through the Bering Strait to Alaska, in North America. It is improbable that this communication was linear or one way. It was linked with the Mesolithic people's migrations or fleeing from floods, plagues, earthquakes, forest fires and other natural disasters, or their trips in search of fauna that was easy to catch and a more suitable living environment. Intergroup encounters led to extensive contact, but such exchanges were clearly unbalanced, resulting in regional differences in cultural development [20].

1.3 Enhanced Knowledge of the Earth in the Neolithic Period

1.3.1 Enhanced Knowledge of the Earth from Production Activities

There are three main Neolithic production activities: agriculture, hunting-gathering and arts and crafts. They did not necessarily prosper in every region, but all feature distinctly in certain areas.

For example, the Neolithic culture of the Yellow River Basin is characterised by agriculture and pottery, while fishing, hunting and gathering appear to have been in decline. On the southeast coast of China and in the desert steppe areas of northern China, fishing, hunting and gathering were still fairly developed, while pottery lagged behind. The regional differences in the Neolithic period are more pronounced and complex than in the Palaeolithic period.

(1) Agriculture

The Emaokou toolmaking area in Huairen, Shanxi province, is a pre-pottery Neolithic site, where stone axes, stone hoes and other agricultural tools with slightly sharpened blades have been found, suggesting that primitive agriculture had emerged there. At the Shayuan cultural site in Dali County, Shaanxi province, a small number of worked stones similar to Neolithic ground stone axes were excavated. Among the fine stone wares, the stone axe, the conical blade core and the wedge-shaped core were all similar in appearance to those from the northern desert steppe area, part of the pre-pottery Neolithic sites [21].

Early Neolithic agriculture was a type of "slash and burn" fire cleared agriculture which required a certain degree of knowledge regarding the soil, climate and biogeography. If the soil was too dry or too barren, the crops would not grow well. Farmers were also expected to know the best time of the year to sow seeds. The emergence of dryland farming and rice farming reflected their grasp of meteorological and climatic knowledge, without which they could not have developed dry farming in the north and rice farming in the south adapted to local conditions.

In the middle of the Neolithic period, fire-based farming was replaced by hoe farming, and farmers knew how to turn the soil and cultivate. At that time, millet was widely grown in the Yellow River Basin, while rice was common in the Yangzi River Basin. For example, rice husks, stalks and leaves have been identified in the early strata of the Hemudu cultural site in Yuyao, in Zhejiang province, with the thickness of the layer reaching 780 cm. There is also a lot of chaff in the terracotta tiles. Rice fields are not the same as land where dry crops are grown. A level field is required so that the irrigation water can flow to every corner. An uneven field is hostile to rice growing conditions; in the higher sections the rice cannot reach the water, and in the lower sections it will drown. At the Hemudu cultural site, wells dug in the low-lying area appeared. These are the earliest remains of a well found in China so far, from about 6,000 years ago. The well is covered with a simple well pavilion to keep the water source clean [22]. The emergence of wells shows that the Hemudu people had a certain level of hydrological knowledge and had environmental awareness of the need to protect clean water quality.

At the Liangzhu cultural site, many ploughshares, adzes and field tools (*yun*) were found from the late Neolithic period [23]. A stone plough fixed on a frame is the oldest one discovered in China. The Liangzhu cultural site is densely surrounded by water systems and swamps, which are often overgrown with reeds and grasses. As a result, the various kinds of buried roots are entangled and difficult to turn over. The invention of earthbreakers and ploughs certainly improved productivity and the quality of the land, making the use of animal power a possible solution.

Many archaeological remains of wells have been found in the excavations at the Liangzhu cultural site. In Xingang Port in Jiashan, Zhejiang province, there is a wooden well whose bottom layer full of river clams and shells (to a depth of 10 cm) fulfilled the role of filtering and purifying the well water [24].

(2) Hunting and gathering

In the Neolithic period, due to the different geographical environments in various places, three vast economic and cultural zones were formed, namely the rice agricultural areas in central and southern China, the dryland millet agricultural areas in northern and southeastern China, and the hunter-gatherer economic zones in northeastern China, the Inner Mongolia Plateau, Xinjiang and the Qinghai-Tibet Plateau.

Hominins at the Xinglongwa site of Aohan Banner, in Inner Mongolia, which dates back more than 7000 years, mainly hunted or raised deer and wild boar. In the case of Xinglongwa Village, two kinds of bone fish darts were discovered, indicating the local people's strong focus on fishing. Some walnut kernels have also been unearthed at the site, evidence which shows that the Xinglongwa people often gathered the fruits of this plant to eat. They relied on hunting, fishing and the gathering of wild fruits, as their main means of subsistence.

Dating from more than 4,300 years ago, during the first phase of the Xiaozhushan cultural relic on Guanglu Island and Dachangshan Island in Changhai County, Liaoning province, stone weights for fishing nets have been unearthed, indicating that ancients here had adopted a maritime lifestyle [25].

Although primitive agriculture and poultry breeding were practised in the early Neolithic period in southern China, people there still primarily survived by gathering, fishing and hunting, and planting was only regarded as a supplement to economic activity. The hominins of the Beiqiu site lived by gathering fruit and fishing for molluscs [26].

During the Yangshao culture period, when hoe farming and agriculture were relatively developed, gathering, poultry-farming, fishing and hunting still prospered. Hazelnuts, chestnuts, pine nuts, simple trees, plant roots and snails were collected at that time. Raising livestock had become an important part of the production economy at that time.

Dating to more than 4,700 years ago, the Liangzhu cultural site in Qianshanyang, Zhejiang province, revealed silk fragments, silk threads and ribbons, suggesting that sericulture and silk weaving had begun by that time.

(3) Neolithic arts and crafts

There were three Neolithic crafts: making pottery, stone tools and ornaments. The geological knowledge acquired through the crafting of these productions in mineralogical and petrological knowledge.

① Mineralogical knowledge reflected in pottery-making

The emergence of pottery was a key aspect of the Neolithic period. The development of primitive agriculture and the planting of grain motivated the making of pottery.

The early Neolithic site of Chenqiao at Chao'an, Guangdong province, produced pottery containing one type of sand. The pottery wares are made of coarse sand or shell powder, heated to a lower temperature. Their surface was smoothed, and some of the mouths and necks of the wares were painted with red ochre.

The site discovered at the Dushan Zengpiyan Cave in Guilin, Guangxi, dates from the early Neolithic period. Pottery wares found there are made of red and grey clays that contain both fine and coarse sand. They are more sophisticated than those made using simple pure sand [27].

Coarse, red sand pottery was excavated at the early Neolithic site in Xianren Cave, Wannian County, Jiangxi province. Its clay was heated to a low temperature and the colour of the ceramic is mottled with an uneven inner wall. The crafting skill remains rudimentary. A few containers are painted with vermilion decoration in swirl and wing motifs [28].

The early Hemudu culture pottery included black pottery and grey pottery made with charcoal and sand respectively, with charcoal black pottery being the most common (see Fig. 1.1, Hemudu pottery pot). The charcoal tempered pottery is formed of organic matter, such as clay, plant stems, leaves and chaff, which was fired in a reducing flame in an oxygen-starved environment. In this way, the organic matter only reached the stage of carbonisation, making the shape of the objects rather simple and irregular, often with uneven thicknesses and colours. The high quality Hemudu pottery pots are coated with a layer of fine white clay, and the surface is well polished in bright and shiny colours.

At the early Neolithic site of Jiahu Lake (7,500–8,500 years ago) in Wuyang, Henan province, a small amount of reddish-brown pottery has been found. It is made of a mixture of grass dust, mussel chips, mica flakes and powdered talc [29].

In the middle and late Neolithic periods, the pottery industry flourished, and the level of craftsmanship made significant progress. For example, the pottery from the Dahe Village site in Zhengzhou, Henan province, is mainly red pottery, but there is also grey pottery, an extremely small amount of white pottery and a certain amount of painted pottery. Some faience ware was first coated with white glaze, and there were two common coating colours: white and light yellow. The subsequent coloured painting was usually in black and brown.

Fig. 1.1 Hemudu black pottery pot (photographed by Liang Qing 梁擎)

Fig. 1.2 Painted pottery of the Yangshao culture (photographed by Liang Qing 梁擎)

The painted pottery wares, painted mostly in black and brown, were mainly found among the cultural relics of the Qinwangzhai in the middle of Henan province, which is part of the third and fourth phases of the Dahe Village site. The colours of the painted decorations are red, brown, grey, black and sometimes red mixed with black.

The pottery from the Yangshao culture is mainly made from fine, red clay. Grey pottery is rare, and black pottery is even scarcer, but a small amount of white pottery has been found. The ornaments are polished, hand-patted and painted. Painted pottery artwork is one of the remarkable achievements of the Yangshao culture, a symbol of China's prehistoric culture and a historical treasure of the world.

Faience is an artistic creation that combines painting and moulding. Painted in black, or black in combination with red, on the outer walls of unfired utensils (very rarely also on the inner walls), the colour of the utensil will not change easily nor be rubbed off after firing. To distinguish this "pre-firing painting" method from the later "post-firing painting" method, archaeologists specifically called it faience (see Fig. 1.2, faience of the Yangshao culture).

The chemical composition of the pigment used in faience can be learned from spectral analysis: the main colouring component of red ochre is iron, and in black they are iron and manganese. Except for a small amount of iron, white pigment basically has no colourants. Based on this, it can be deduced that the red ochre colour may be ochre (hematite Fe_2O_3), the black paint may be a kind of red clay with a high percentage of iron, and the white material may be a type of porcelain clay mixed with solvents [30].

By the late period of the Yangshao culture, there was white pottery and, by the time of the Dawenkou culture and the Longshan culture, white pottery became much more common. The chemical composition of the white pottery closely resembles that of kaolin. It is characterised by a much lower iron oxide than clay, so it appears white after firing, hence the name white pottery. The emergence of white pottery shows that China was the first country in the world to use kaolin, which was key in bringing porcelain to the fore, since kaolin was one of the main raw materials required for the manufacture of porcelain.

The pottery of the Longshan culture in Shandong province was mainly black, and there was not much grey. A small amount of red, yellow and white pottery has been found. There were three ways of making black pottery, using fine clay, common clay and sand inclusion; of these the production of fine clay, thin-walled black pottery required the highest level of skill. After rounds of fine washing and moulding, the wall of the pottery could be minimised to between only 0.5–1 mm, with its surface appearing pitch-black, known as "eggshell black pottery".

This is the pottery art most representative of the Longshan culture. Its raw materials include laterite, sedimentary deposits, chernozem, clay and kaolin [30].

② The knowledge of minerals and rocks reflected in stone tool making

The knowledge of minerals and rocks reflected in stone tool making reached its peak in the Neolithic period, and both the types of tools and the materials used greatly exceeded those of the Palaeolithic period. The author lists the mineral rocks used or recognised by people in the Neolithic period, based on statistics from 84 cultural sites in the early, middle and late Neolithic periods (see Table 1.1).

Table 1.1 shows the following sets of figures:

a. The total number of mineral rocks used or recognised in the Neolithic period was 90, an increase of 53 species compared with 37 in the Palaeolithic period, and a growth rate of 143%. This growth rate is very high.
b. In the early, middle and late Neolithic periods, the number of minerals and rocks used or recognised was different: there were fewer in the early period, only 9 species; the medium-term increase was relatively fast, with an increase of 52 species, bringing the total to 61; in the late period, 10 species were added, bringing the total to 71.
c. The mineral rocks most used by people in the Neolithic period are sandstone (47), flint (32), shale (28), quartzite (25), slate (22), limestone (21), granite (16), jade (13), basalt (10) and agate (10), accounting for the top ten on the list. Further down the list are quartz (8), opal (8), fine sandstone (8), diabase (8), igneous rock (7), crystal (7), chalcedony (7), metamorphic rock (6), gabbroite (6), turquoise (6), horn shale (6), talc (6), rhyolite (6), hematite (5), andesite (5), slate (5), jasper (5), argillite (4), siltstone (4), tuff (4), coarse sandstone (4), metamorphic sandstone (4), limestone (4), quartz sandstone (4), pyrophyllite (3), quartz feldspar sandstone (3), metamorphic sandstone (3), gneiss (3), diorite (3), serpentinite (3), shale (3), marble (3), cinnabar (3), etc. Those featuring less than twice may be largely incidental and not intentionally sought mineral rocks. For example, copper (2): in the Neolithic period, people mainly used natural copper or red copper. Although it appeared in the late Yangshao culture, that is, around 5,500 years ago, it was not easy to find due to the small amount of natural copper production, so it was not possible to consciously look for copper ore, let alone mine copper ore and smelt copper. Because the melting point of copper is 1,084 °C, it was not easy to achieve this temperature at that time, so for a long time after the discovery of copper, only small pieces of copper ware could be handed down. The late Yangshao culture to the Longshan culture (5,000–4,000 years ago) as well as the Qijia culture (4,200–3,900 years ago) are periods

Table 1.1 Mineral rocks used or understood by people in the Neolithic Period

Name		Quartzite	Sandstone	Hematite	Flint	Quartz	Metamorphic rocks	Shale	Crystal	Jade	Fluorite	Jasper	Opal
Amount	Early period	3	3	1									
	Middle period	12	18	3	17	4	3	7	2	3	1	2	3
	Late period	10	26	1	15	4	3	21	5	10		3	5
Total		25	47	5	32	8	6	28	7	13	1	5	8

Name		Agate	Shale	Porphyry	Chlorite slate	Marlstone	Basalt	Tuff	Limestone	Gneiss	Gabbro	Igneous rock	Fine sandstone	Granite
Amount	Early period								1					1
	Middle period	5	1	1	1	2	4	2	7	3	4	3	3	5
	Late period	5	4				6	2	13		2	4	5	10
Total		10	5	1	1	2	10	4	21	3	6	7	8	16

Name		Siltstone	Bertholite	Chlirite schist	Diabase	Schist	Marl	Gritstone	Slate	Siliceous metamorphic rock	Argillaceous mudstone	Andesite	Sedimentary rock	Graphite
Amount	Early period													
	Middle period	2	1	1	3	1	2	1	4	1	1	2	1	1
	Late period	2			5	1	2	3	18			3	1	
Total		4	1	1	8	2	4	4	22	1	1	5	2	1

(continued)

Table 1.1 (continued)

Name		Jet	Argillaceous sandstone	Quartzose arkose	Phyllite	Jadeite	Argillaceous shale	Volcanic breccia	Quartz sandstone	Hornfels	Pyrophyllite	Talc
Amount	Early period								2			
	Middle period	1		1						1		4
	Late period		1	2	2	1	1	1	3	5	3	2
Total		1	1	3	2	1	1	1	5	6	3	6

Name		Siliceous slate	Metasandstone	Orthophyre	Monzonite	Rhyolite	Tuffaceous sandstone	Quartz porphyry	Ferricrete	Biotite quartz schist	Granodiorite	Sandshale	Pelite	Felsite
Amount	Early period											1	1	1
	Middle period		1			2						1	1	
	Late period	1	3	1	1	4	1	1	1	1	1	1	2	1
Total		1	4	1	1	6	1	1	1	1	1	3	4	2

Name		Diorite	Sandy lime	Serpentinite	Tuff clastic rock	Chalcedony	Hyalopsite	Siliceous rock	Breccia	Sandy slate	Marble	Hornstone	Chlorite	Cinnabar
Amount	Early period													

(continued)

Table 1.1 (continued)

Name		Diorite	Sandy lime	Serpentinite	Tuff clastic rock	Chalcedony	Hyalopsite	Siliceous rock	Breccia	Sandy slate	Marble	Hornstone	Chlorite	Cinnabar
	Middle period	2	1	1	1	1					1		3	1
	Late period	1		2		6	1	2	2	2	2	1	3	2
Total		3	1	3	1	7	1	2	2	2	3	1	6	3

Name		Dolomite	Oil Shale	Argillaceous limestone	Sillitite	Black jade (Moyu)	Hornblende	Metamorphic limestone	Gypsum	Copper	Silliceous hornstone	Argillite	Tremolite	Burnt jade
Amount	Early period													
	Middle period	1	1	1	1	1	1	1						
	Late period			1					1	2	1	1	1	1
Total		1	1	2	1	1	1	1	1	2	1	1	1	1

Name		Albinite	Kaolin Jade	Total mineral and rock species		90
Amount	Early period			Total	9	
	Middle period			Total	61	
	Late period	1	1	Total	71	
Total		1	1		141	90

of bronze and stone; bronze making at that time required both casting and forging processes, which were reliant on mining and the smelting of copper ore.

d. Among the 90 kinds of mineral rocks used or recognised in the Neolithic period, there are 23 kinds of minerals. They are turquoise [$CuAl_6(PO_4)_4(OH)_8 \cdot 5H_2O$], pyrophyllite [$Al_2(Si_4O_{10})(OH)_2$], talc [$Mg_3(Si_4O_{10})(OH)_2$], tremolite [$Ca_2Mg_5(Si_4O11)_2(OH)_2$], copper (Cu), flint ($SiO_2$), agate ($SiO_2$, a type of chalcedony), chalcedony (SiO_2), opal ($SiO_2 \cdot nH_2O$), crystal (SiO_2, transparent crystalline quartz), quartz (SiO_2), jasper (SiO_2, also known as iron-banded jasper), hematite (Fe_2O_3, also known as ochre), graphite (C), coal essence (C), fluorite (CaF_2), jadeite (a complex composition with many types, the reference here is to nephrite), gypsum ($CaSO_4 \cdot 2H_2O$), cinnabar (HgS), ink jade, burnt jade, kaolin and kaolinite. The remaining 67 species are rocks by nature. Compared with the Palaeolithic period, the Neolithic period used or recognised 14 additional minerals, which are: turquoise, ink jade (black nephrite), pyrophyllite, talc, tremolite, copper, coal essence, fluorite, jade, gypsum, cinnabar, fire jade, white stone resin, and kaolin jade. The number of rock species should increase to 39, but, in fact, the rocks used in the Palaeolithic period did not all continue to be used in the Neolithic period; rocks such as siliceous limestone, peridotite, silicified rock, and silicified volcaniclastic rocks were used in the Palaeolithic period, but no evidence of their continued use has been found in the Neolithic period. Thus, the Neolithic used 43 more rocks than the Palaeolithic. They are: metamorphic rock, mud slate, green mud slate, marl, granite, fine sandstone, siltstone, greenschist, schist, limestone, coarse sandstone, siliceous metamorphic rock, argillaceous mudstone, sedimentary rock, argillaceous sandstone, quartz feldspar sandstone, millerite, emeraldite, volcanic breccia, siliceous slate, syenite, monzonite, tuff sandstone, quartz porphyry, ferrous sandstone, biotite quartz schist, granodiorite, sandshale, argillite, fine-grained stone, sandy limestone, serpentinite, tuff clastic, sandy slate, marble, dolomite, oil shale, siliconite, hornblende, metamorphic limestone, siliceous hornstone, clay slate and quartz sandstone.

e. In terms of using the same mineral rocks, the early, middle and late Neolithic periods show a trend: the number of uses of sandstone in the early, middle and late periods is 3, 18 and 26 respectively, shale is 0, 7 and 21, and granite is 1, 5 and 10. This reflects a trend of increasing frequency of use as time passes. In particular, the increase of jade from 3 to 10 highlights its dramatic development in the late Neolithic period.

f. Among the 90 kinds of mineral rocks used or recognised in the Neolithic period, some are only used occasionally, some are used more often, and others may have special features, such as colour, hardness, transparency, crystal shape, etc., which were easier to distinguish and recognise. It is not possible to rely on a strict nomenclature (in order) to define the rocks, as in modern geoscience, because there may only be a few matches because of historical naming traditions.

③ Mineralogical knowledge reflected in the crafts and ornaments

Based on the author's statistics, the raw materials of relevant crafts and ornaments in the Neolithic period are listed in Table 1.2.

It can be seen from Table 1.2 that:

First, there are 25 kinds of raw materials used in Neolithic craft ornaments. Excluding the pottery, the remaining 24 are mineral rocks. Among them, there are 15 kinds of minerals including turquoise, pyrophyllite, talc, jade, false jade, copper, coal essence, fluorite, chalcedony, dolomite [$CaMg(CO_3)_2$], agate, gypsum, crystal, quartz and cinnabar. The rocks include volcanic limestone, marble, limestone, shale, sandstone and rock. There are seven types of quartz rocks and coral rocks. Pearls and ivory are organic gemstones, members of the family of organic minerals.

Second, there are differences in the types of decoration materials used in the early, middle and late Neolithic periods. The steadily increasing use of these materials indicates a rapid development of skills and knowledge. In the early days, there were only four kinds of minerals, including turquoise, jade and talc. In the middle period, the total number increased to 11 species, adding minerals such as fluorite, coal essence, dolomite, ivory, false jade, cinnabar, as well as volcanic limestone and sandstone. In the late period, it rose to 19 kinds, adding minerals such as chalcedony, gypsum, agate, copper, pyrophyllite, crystal, quartz, and pearls, as well as five kinds of rocks such as quartzite, marble, shale, coral rock and limestone.

Third, as far as the individual mineral type is concerned, this also reflects an upward development trajectory. For example, turquoise has a score of only four in the early period, and five in the middle period, with an increase to 10 in the late period. Jade was used once in the early Neolithic period, but it reached 14 times in the middle period and 26 times in the late period, a fast rate of development.

Fourth, Neolithic ornaments were developed from their Palaeolithic counterparts. The Palaeolithic cave dwellers had more than 100 carved and processed ornaments, including drilled stone beads dyed red with hematite powder and yellow-green oval bored gravel. Dating from the early Neolithic period, black talc carvings in the shape of monkeys and stone owls were used at the Shangzhai site in Pinggu, Beijing. Animal and humanoid talc carvings have been unearthed in the lower layer of the Houwa site on the Liaodong Peninsula. In the middle Neolithic period, there were jade and stone carvings, such as the two-sided stone carved human face and human face stone toys excavated at the Daxi site of Wushan, Sichuan province; the human-face jade carvings excavated at the Dawenkou cultural site in Shandong province; and the bird-shaped and fish-shaped jade ornaments of the Songze culture. In the late Neolithic period, there were many types of jade and stone carvings, and the level of craftsmanship was very high. There were professionals specialising in jade and stone carving; jade and stone processing workshops were up and running, such as at the Huangniangniangtai relic site in Wuwei, Gansu province.

Table 1.2 Statistical table of the materials for craft ornaments in the Neolithic Age

Name		Calaite	Pottery	Jade	Fluorite	Jet	Cinerite
Amount	Early period	4	1				
	Middle period	5	14	1	1	1	1
	Late period	10	26				
Total		19	41	1	1	1	1
Name		Chalcedony	Dolomite	Quartzite	Marble	Talc	Gypsum
Amount	Early period					1	
	Middle period		1			1	
	Late period	2		1	2	1	1
Total		2	1	1	2	3	1
Name		Ivory	Limestone	Agate	Copper	Pyrophyllite	Shale
Amount	Early period						
	Middle period	2					
	Late period	3	1	3	1	1	1
Total		5	1	3	1	1	1
Name		Crystal	Fake jade	Quartz	Vermilion	Sandstone	Pearl
Amount	Early period						
	Middle period		1		1	1	
	Late period	2	2	1			1
Total		2	3	1	1	1	1
Name		Coral rock	Total types	25			
Amount	Early period		Total number	7			
	Middle period		Total number	29			
	Late period	1	Total number	61			
Total		1					

1.3.2 *Enhanced Knowledge of the Earth from Living Environments*

The geological knowledge reflected in the Neolithic living environment have four features in common as outlined below.

(1) Choice of place of residence (site selection). In the Neolithic period, after the emergence of agriculture, people lived a settled life and had a fixed place of residence; these settlements had four features in common as outlined below:

First, the places of residence in riverside areas were chiefly located on terraces or river benches, and most of them were in the curve of a river or in a delta area where two rivers meet. This kind of location was close to the water source, and the water needed for production activities and life was easily accessible. It did not suffer from flooding, when the river levels rose during the rainy season.

① The places of residence are in riverside areas. For example, the locations selected for the sites in Yangshao are mostly on the Loess Plateau or in the canyons separated by watercouses, and some are on river terraces. In the desert steppe areas of Inner Mongolia and Xinjiang, the locations of the fine stone tool sites are mostly on the loess platform near the river, such as some of the sites in Baotou, in Qingshuihe County, and in Junwangqi. The site in Jiaojiazhuang, Lanzhou, is located on the north bank of the Yellow River, which is about 30 m above the riverbed. One hundred metres west of the Neolithic site in Huaide County, Jilin province, is the Dongliao River, and 800 m to the east is the Immortal Cave Mountain, which is fertile and beautiful, an ideal place for humans to live. Neolithic sites in Liuzhou, Guangxi, are distributed on first-level terraces along both sides of the river, backed by mountains, and are mostly located at the confluence of small rivers, small ravines and the Liu River.

② The place of residence is located at the bend of a river. In the case of the Peiligang ruins at Xinzheng, where the Shuangji River flows from north to south, and then from west to east around the ruins, the ruins are in the bend of the Shuangji River, about 25 m above the riverbed. The Cishan Mountain site at Wu'an is located at the foot of Gushan Hill, part of the Taihang Mountains, south of the Ming River, on a platform 25 m above the riverbed. The Ming River flows from west to east, south of the ruins, and then flows north, with the ruins right in the curve of the river. The Baijia Village site at Lintong, Shaanxi province, is located on the north bank of the Wei River, where the Wei River curves from west to east, in the south of the site, and flows north, the site is on the shore of the ground nearest to the bend of the river. The Neolithic site of Tanwan in Chenxi County, in the middle and lower reaches of the Yuan River, in Hunan province, is located on the north bank of the Chen River, on a platform 135 m above the riverbed. The water flows from west to east, curving around the south of the site. A small stream flows from north to south, through the east of the site, into the Chen River. A hill about 100 m high cascades down to the northwest of the site, creating a favourable terrain surrounded by water on three sides, backed by mountains on one side, and facing the sun.

③ The place of residence is located at the junction of two rivers. For example, the Gugou site in Mi County, Henan, dating from 7,000–8,000 years ago, is in the triangle where the rivers Suishui and Weishui meet, and the site is 70 m above the riverbed. There are large tracts of arable land around. Gugou Beigang is a ridge stretching for more than ten li. The ground is relatively flat, and it is close to the river, making it a good place for farming. The site is surrounded by mountains and forests, making it a good place for fishing and hunting. Such a geographical environment made it a very suitable place to settle. At the site of Wuyang Jiahu Lake in Henan province, the Hui River flows into the Sha River three kilometres northeast of the site, and the site is about ten kilometres northwest of the confluence of the present-day Sha River and the Ru River.

The two sites of Peiligang and Gugou are in the hillfoots of the Song Mountains, and the Jia Lake site is located on the alluvial plain at the eastern foot of the Funiu Mountains. Although these Neolithic ancients had moved to the river, they were still inseparable from the hills and mountains.

Second, the Neolithic settlements in the lower reaches of the Yangzi River were mostly established on mounds, where the terrain was higher than the plain and not easily flooded. These mounds could be five to ten metres higher than the nearby water level, mostly next to a lake or river, and such a site is called a mound site. Mound sites are the most common sites identified in the Liangzhu culture, accounting for about 60% of the Liangzhu cultural site. Typical examples include Caoxie Mountain and Zhangling Mountain in Wu County, Jiangsu province, and temple sites in Wujin County. These sites are often distributed across the flat ground, surrounded by crisscrossing rivers and hills. They account for about 30% of the entire number of the Liangzhu cultural sites, and typical examples include Shuitianfan in Hangzhou, Zhejiang province; Qianshanyang in Wuxing; and Guangfulin in Songjiang, Shanghai, etc. [31].

Third, the Neolithic settlements in the coastal areas were mostly established on Gaofugang hills, which are recognised as *beiqiu* (shell mound) sites today. Mainly distributed in Shanghai, Shandong Peninsula, Guangxi and Guangdong provinces, e.g. the Maqiao ruins in Shanghai, the Beiqiu ruins at Penglai, Yantai, Weihai and Rongcheng in Shandong province, their shared features include the following. From one to three sides are adjacent to mountains or hills, and the other sides face the valley plain or low-lying land. Their centres are generally located on a higher platform, 20–30 m above sea level. They are less than six kilometres from the coastline. For example, Neolithic shell mound sites in Nanning, Guangxi province, are often situated at the bends of large rivers, or in the delta area where large and small rivers meet, generally looking over the river with mountains behind, and with relatively open flat land nearby, 3–20 m above the water level. The stacked shells are also varied and can be divided into four types: oysters, clams, corbicula and mud cockles. Shellfish are very sensitive to their living environment, and these types have adapted to live in seawater at different temperatures and levels of salinity [32].

Fourth, the invention of artificial ponds and the digging of cisterns created conditions which enabled Neolithic settlers to choose plains, plateaux, hills and mountains far from rivers. At that time, it was not possible to drill very deep wells, and only artificial ponds and cisterns could be relied on to store rainwater. The bottles with small mouths and pointed bottoms found at the Yangshao site were the main vessels for drawing water from cisterns [33].

(2) Settlement planning

The rise of primitive agriculture brought about settlements. The design of settlements was very diverse, reflecting people's thinking about how to plan their living space. For example, at Xi'an Banpo Yangshao cultural site, an area of 50,000 m^2, houses and cellars were designed for the storage of goods and as enclosures for raising livestock. They are all concentrated in the centre of the site, accounting for about 30,000 m^2. There is a 5–6 m deep ditch around the residential area. Beyond Daweigou, to the north of the site, there are mainly public cemeteries, and there are also a small number of cellars. To the east of the site is the pottery kiln area. The 300-m-long ditch is a defensive structure. This kind of orderly and systematic planning arranged the locations of the houses, tombs and craft workshops in the residential area. There is also a large hall intended for collective activities, with the door facing east. The clan leader and some old and young people regularly lived inside. Clan meetings and religious activities also took place here.

The site of Jiangzhai in Lintong, Shaanxi province, covers an area of about 20,000 m^2. In the residential area, there are ditches dug on the north, east and south sides of the central square, and the river in the west acts as a natural ditch. The central square measures about 4,000 m^2 and contains two roads, parts of which are paved with calcium stones or clinkers. Outside the ditch is the clan cemetery area and craft workshops. The planned layout is similar to that of the Banpo site.

The ruins of E'gou Beigang in Mi County, Henan province, also have a planned layout: at the centre is the house which is partially surrounded by or next to the cellar area or pottery kiln area. To the north-west of the settlement is the clan's public cemetery. Basically, it encapsulates a planning policy of division into three main districts.

The Xinglongwa Village site in Baoguotu, Ao Han Banner, Chifeng, Inner Mongolia, dates from about 7,000 years ago. The houses are neatly arranged and organised in an orderly fasion. The site lies along a northwest-southeast axis. There is a ditch around the settlement, nearly circular, with a diameter of 160–183 m. The ditch is two metres wide and the depth is currently one metre, which had a certain defensive value and served to demarcate the boundary [34].

The late Neolithic period site at Zhaobaogou in Aohan Banner, Inner Mongolia, is located on a gentle slope, and covers an area of about 90,000 m^2. The structures on the site are distributed in rows lying along a northeast-southwest axis. There are three types of houses: large, medium and small, and there are already obvious social and hierarchical differences. The evolution of the site has led to several groups of houses, each of which is roughly divided into three levels: (1) first-class or second-class houses are in the middle; (2) third-class houses are located on both sides;

(3) fourth-class houses are on the outermost fringes. The central settlement and the surrounding small and medium-sized settlements constitute a larger settlement group [35].

At the Hongshan cultural site of Niuheliang, Lingyuan, Liaoning province, a large-scale religious centre called the "Goddess Temple" was discovered. The layout of the entire temple respects the rules of symmetry on a grand scale. The principal and the secondary zones are clearly marked. A large number of stone cairns are scattered over the summit of a hill about one kilometre south of the temple. In Kazuo Dongshanzui, 50 km away from Niuheliang, there is also a Hongshan altar foundation site. The whole layout consists of the centre, two wings and the front and rear ends, arranged symmetrically along a north–south axis, reflecting the concept of "the earth is square whilst the sky is round" [35].

(3) The birth of cities

At present, the earliest known ancient city in China is at the Chengtoushan site in Nanyue Village, Chexi Town, Li County, Hunan province. The site has been built on four times, the first time being in the early days of the Daxi culture, about 6,000 years ago. After that, in the third or fourth phase of the Daxi culture about 5,500 years ago, and in the early and middle stages of the Qujialing culture, 4,600–4,800 years ago, the ancient city site of Chengtoushan experienced another three rounds of construction, during which the city wall was widened and raised, and the 35-m wide moat outside the city was added to realise a grand vision [36].

The second relatively early city is at the Yinxiang site in Jingzhou, Hubei province. The city was already a large-scale settlement site during the Daxi culture period. In the early days of the Qujialing culture, the building of a large-scale city wall began, with a radius of tens of li. The Yinxiang site was founded during the Qujialing culture period. At that time, the cross-section of the city wall was trapezoidal, consisting of walls and slope protection, built with mounds of earth forming a slope. The wall was 8 m high. The tall walls and the wide moats were ideal as defences and as protection against attack [37].

The third most ancient city is at the middle-phase Longshan cultural site on the east side of Mengzhuang Town, Huixian, Henan province. It is surrounded by the Taihang Mountains in the north and the ancient highway of the Yellow River in the south, laid out in a square shape with a city wall around, and about 400 m long and wide, covering an area of about 160,000 m^2. There is a moat outside the city wall, 5.7 m deep, which was a large defensive fortification at that time [38].

The fourth ancient city is Jingyanggang Longshan Cultural City in Yanggu County, Shandong province. It is rectangular with rounded corners, about 1,150 m long and 300–400 m wide, including the city wall, with a total area of about 380,000 m^2. This is the largest of the Longshan cultural city sites discovered in the Yellow River basin, and it is also the first Longshan cultural city site scientifically excavated in northwest Shangdong province. Two sets of foundations have been found within the circle of the city, of one large structure and one small one. The layout is the first of its kind in any excavated Longsan cultural city site [39].

The fifth ancient city is Pingliangtai Longshan Cultural City in Huaiyang, Jiangsu province. Although not large in scale, with a total area of only over 50,000 m^2, it was sturdily built and distinctive. Its main features included:

① Orderly planning. The city is square in shape, orientated towards the north and the south, in a direction which virtually coincides with the meridian line. The south gate is larger, with the main entrance located in the middle of the southern wall; the north gate, which is very small and slightly to the west, serves as a back door. The idea of square symmetry influenced the development of ancient Chinese cities for thousands of years and has become a major feature of Chinese cities.
② Facilities for the strict control of security. In addition to the walls being the largest defensive structures, the city also set up dedicated gatekeeper rooms, where the guards were responsible for managing the gates. Facing east and west, these guardrooms were made of adobe, and the passage between the two rooms was only 1.7 m wide, which facilitated keeping guard.
③ A public sewer system. Cities enclosed by walls must solve the problem of water supply and drainage. The city's water supply is presumed to have come from wells. The sewage system which has been found is more than five metres long, although the entire length and network are unknown. Even so, there is sufficient evidence to say that there were public sewage facilities at that time. This section of the sewer passed through the South City Gate and was buried 0.3 m deep underground. The waterway is made of pottery pipes, each of which is 35–45 cm long, 23–26 cm in diameter, or 27–32 cm at the thicker end. In the drain, the three pipes lay parallel to each other, which increased the drainage. The waterway slopes from the inside of the city to the outside, so that the wastewater in the city could run away easily.
④ Buildings with advanced structures. Some houses constructed on rammed earth foundations and with internal corridors were more sophisticated than the houses in ordinary villages. It is evident that the inhabitants of the city were mainly nobles and rulers.
⑤ There were craft facilities. Pottery kilns have been found in the southeast, northeast, and southwest of the city; copper slag was found in ash pit No 15 in the southeast corner, indicating that there was a copper smelting and a pottery craft industry actively working at that time, proving that it was not simply a hillfort.
⑥ There are the remains of religious activities. For example, traces of animal sacrifice can be identied in the southwest corner of the city.

The above examples show that the Pingliangtai site already had the basic elements of a proper city, and it may have been the political, economic and religious centre. This kind of fortified settlement was obviously not the earthen enclosure of an ordinary village, but rather the prototype of a city.

In addition to the above five city sites, some ancient city sites from the Longshan culture period have also been found. For example, the layout of the Chengziya ruins in Zhangqiu, Shandong province, is virtually square, and the east, south and west walls are straighter, with the north wall bulging outwards and with rounded corners.

The city is more than 430 m wide from east to west, 530 m across from north to south, and covers an area of more than 200,000 m^2, making it the only large-scale city site constructed during the Longshan culture period [40].

In short, it is a fact that fortified settlements and the initial cities appeared during the Longshan culture and even the Daxi and Qujialing culture periods. This reflects the development of a city from its embryonic state to maturity.

⑦ A growing awareness of enviromental protection.

At the Neolithic sites excavated in various places, settlements, public cemeteries and craft workshops were built separately from each other. and the enclosures for raising domestic animals were also separated from the residential houses, which is most evident at the Banpo and Jiangzhai sites. At some Neolithic sites in the Taihu Lake area, basic pens enclosed by tree branches or bamboo poles for raising livestock have been found. A pottery model of a livestock house was unearthed at the Shizihang Longshan cultural site in Weixian County, Shandong province. At some sites, there are also ash pits, gutters and cellars where waste was dumped. Underground pipelines and a sewage system were found at the Longshan cultural relic site at Pingliangtai in Huaiyang. The well at Hemudu was protected by timber piles and logs forming the walls and roof. The inner wall of an ancient well of the Liangzhu culture at the Xingang site, Jiashan, was protected by a hollowed out piece of timber, and its base was covered with river clam shells to filter and purify the groundwater [41].

(4) Well-digging techniques and geohydrological understanding

The invention of hand dug wells not only created the conditions for Neolithic settlers to expand their production and living areas, but also showed that people had a certain understanding of geohydrology at that time. Wells do not supply water unless they are located in a plain area fed by rivers and lakes. Common water-storing landforms include sloping plains at the foot of the mountains, intermountain basins, river terraces, alluvial plains, and karst caves, etc. Neolithic inhabitants could not have had this knowledge of the landscape, but life experience would have taught them that wells had to be drilled in the foothills, hollows, and alluvial plains, etc. In the south, the Hemudu site, dating back more than 7,000 years, had a relatively high number of wells. Speculation suggests that the digging began earlier than the Hemudu period. Quite a few wells have also been found at the Liangzhu site. At the Longshan site, wells have been found in Jiangou, Handan, Hebei province; and also in Baiying, Tangyin, and in Cuoli, Luoyang, both in Henan province.

(5) Choice of direction in house-building

Neolithic settlers chose the direction when building houses and, according to the author's statistics, there were about seven kinds: ① to the south; ② towards the centre of the settlement; ③ to the southeast; ④ to the east, southeast or south; ⑤ to the east; ⑥ facing in three directions: east, west and south; ⑦ to the southwest. Of the houses, 55.5% are south-facing, 16.6% are southeast-facing, 16.6% are east-facing, 5.5% are west-facing, and 5.5% are southwest-facing. There are no north-facing or north-west facing examples. This shows that the orientation of the house was closely related to the need for sunshine, warmth and shelter from the cold wind from the north, a sign of the geographical knowledge acquired by Neolithic people regarding the relationship between temperature, sunlight and direction.

1.3.3 Geoscientific Implications of Burial Sites

(1) Knowledge of minerals and rocks reflected in tombs

Since the late Palaeolithic period, Shandingdong Man had consciously buried his dead in the earth, hence the creation of tombs. Funerary objects have been found in the Peiligang tombs from the early Neolithic period. These objects, in addition to pottery, are stone tools and ornaments, reflecting the ancestors' knowledge of minerals and rocks at that time. Moreover, different burial customs in different places also reflect the Neolithic people's awareness of certain minerals. According to the author's statistics, Neolithic tombs mostly contained hematite powder and cinnabar, each accounting for more than 38%, the total of the two combined amounting to 76%. Red ore, flower stones and turquoise each account for 7%, the sum of the three amounting to 21%. It can be seen that the minerals used in Neolithic burials as part of the burial rituals were mainly hematite and cinnabar. Why were these two used? Some believed that they were linked to worship of the sun or fire, others believed they had religious significance, and yet others thought they had the function of protecting the body [42].

(2) Geographical knowledge applied to the selection of tombs

From the time tombs were built in the late Palaeolithic period, the burial sites chosen were not located far from the residential areas. In the Neolithic period, cemeteries were also near residential areas.

For example, the Chengzishan Cemetery at Sanguan Dianzi Village, Lingyuan County, in Liaoning province, part of the Hongshan culture, mostly spreads out on high platforms or hillsides, and some graves are even on the top of the mountain. The cemetery at Liangzhu is located on higher ground, especially for some clans and nobles. The Qijia cemeteries, like the residential sites, are mostly located on terraces on both sides of the upper reaches of the Yellow River and its tributaries, with the second level terraces as the preferred locations.

(3) Directions

Based on the author's statistics, tombs are mainly orientated in eight directions. Among them, the most frequent is westward, accounting for 21%, followed by towards the south, east and northwest, accounting for 15.8% each; the next directions are to the northeast and southwest, accounting for 10.5% each; the least common directions are to the southeast and to the north, accounting for 5% each. The reason for emphasising the orientation of burials is that this was related to the social customs or the beliefs of an ethnic group. Different ethnic groups followed different practices. The Yao ethnic group believed that wherever people originated from, their head should point towards that place. The Bulang people believed that the head should be orientated towards the setting sun [43]. They compared a person's life to the rising sun and one's death to the sunset in the west. The different head-orientation in the tombs reflects the ideological or religious beliefs of the people at that time [44]. Another concept worth considering is that death is like life. When a man dies, his soul lives on, and it will live like a living person. Consequently, burials were accompanied by production tools, daily necessities, ornaments, and even livestock and human beings. Therefore, the orientation of a burial should be the same as the orientation of the house the person had lived in. This view later developed into the main theories of Feng Shui.

(4) Burial customs and geographical understanding

① "Round heaven and square earth"

More than 6,000 years ago, No 45 Xishuipo Tomb, Puyang, in Henan province, placed the owner's head towards the south and the feet towards the north, reflecting the ancient Chinese tradition that where the head and the south are should be the sky, and where the feet and the north are should be the earth. The plan of the tomb clearly shows that its southern part (also the skull) is round, and that the northern part (the foot) is square, symbolising the "round heaven and squre earth" [44].

At the Dongshanzui site of the Hongshan culture, Kazuo County, in Liaoning province, there is a large altar whose layout follows "the round south and square north" principle. At the Yaoshan site, Yuhang, in Zhejiang province, part of the Liangzhu culture, the altar used to worship the heavens and the earth also reflects this idea [45]. Further evidence of this is provided by the jade sacrificial vessel *cong*. It has a square outer surface and a rounded interior [46].

② The idea of the Chinese dragon and tiger

Two Chinese dragon motifs moulded on the reddish-brown earthen ground with pure yellow clay were found at the Neolithic site of Yangjiawa, Tashan Village, Huludao Island, in Liaoning province, dating from 8,000 years ago. This was the first time such an ancient technique had been discovered in China [47].

In Tomb No 45 of Xishuipo, Puyang, in Henan province, a dragon and tiger pattern composed of mussel shells was found: the dragon is in the east and the tiger is in the west. A second group of mussel-sculpted patterns show dragons, tigers, deer and spiders; in a third group there are people riding dragons and tigers.

At the Dongshanzui site of Kazuo, Liaoning province, part of the Hongshan culture, a single jade ornament in the shape of double-headed dragon was unearthed. At the Sanxingtala Village site, part of the Hongshan culture, in Inner Mongolia, a large piece of jade in the form of a dragon was unearthed, dark green in colour, 26 cm high, and intact [48].

This is the image of a dragon appearing as a sculpted work of art. In the Longshan tombs of Tao Temple at Xiangfen, Shanxi province, painted pottery plates with curled-up dragons were discovered [49].

The image of the dragon appears as a painted work of art in the Yongsan culture. This shows that the idea of the dragon and tiger was shaped in the Neolithic period, creating a conceptual foundation for the dragon veins and the dragon-tiger mountain ranges (*longhu sha*) in Feng Shui theories.

1.3.4 Geographical Data from Human Migrations and Communications

(1) Geographical data from human migrations

① Road transport

In the Palaeolithic period, the concept of roads did not really exist and there were only natural pathways regularly travelled along by people. During the Yangshao culture period, man-made roads appeared, such as at the site of Jiangzhai in Lintong, Shaanxi province. This road was paved with *jiangshi* (calcium stones) or clinkers [50]. Roads are a symbol of inter-group communication.

② Water transport

In the Neolithic period, the Yangshao people were already able to build canoes. At the Yangshao site in Baoji, Shaanxi province, a boat-shaped pottery pot was unearthed, on which was painted a fishnet pattern. Apart from catching fish, canoes were also used to travel on the rivers [51]. In the Fanshan tomb of the Liangzhu culture in Yuyao, Zhejiang province, a huge canoe was used as the inner coffin [52]. Six wooden oars were unearthed at the Hemudu site in Yuyao, Zhejiang province. From this we learn that, at least six or seven thousand years ago, there were canoes and water transportation in China. More wooden oars and wooden paddles were unearthed from the Liangzhu culture period,; there are wide (up to 26 cm) and narrow blades, with a large waterfront surface that could achieve a strong reverse thrust. Most of the paddles had long wooden handles attached, which would make them quick and easy to use. This shows that, during the Liangzhu culture period, the canoe was much more advanced than in the Hemudu period [53].

③ Maritime transport

The Hemudu people used both canoes and rafts. They ventured out to sea and explored coastal islands. Pottery kettles have been found at prehistoric sites on islands in Changdao County, Shandong province. These kettles are essentially identical to those found in the first layer of Hemudu and the middle layer of the Songze site indicating that, 5,000 years ago, the aboriginal peoples of the Yangzi river area seem to have had the ability to sail along the coast of the sea [54]. There is a very short distance between Hemudu and the sea, so residents could reach the sea by going east or north, and the geographical conditions were extremely favourable. The fact that canoes can cross the sea has already been proved by modern explorers through simulation tests.

Therefore, it was possible for the Hemudu people to travel between the mainland and the coastal islands. Archaeological excavations in recent years have shown that the Hemudu culture gradually expanded to the Zhoushan Islands after more than 1,000 years of development. At least 5,000 years ago, the first inhabitants settled on the Baiquan Cross Road on the main island of Zhoushan, four kilometres from the sea. Around 4,000 years ago, there were more residents on larger islands such as Dinghai, Daishan, and Shengsi. These original inhabitants were directly influenced by the Hemudu culture and were an important part of the Hemudu culture [55].

In the north, the original inhabitants also travelled across the sea. Penglai in Shandong province is across the sea from Changdao County, which has been continuously inhabited since at least the end of 5,000 BC. This county is an archipelago of islands connected with each other in a chain. From Penglai Port all the way to the northernmost Huangcheng Island, there are many intermediate islands, with no more than ten kilometres of distance between any two islands. People could use canoes to travel between islands on a sunny day. The islands in the north have been more influenced by the primitive culture of the area east of the Liao River, while the islands in the south have been shaped more by the primitive culture of Shandong. Maritime communication between the islands and the mainland was frequent and extensive [56].

(2) Geographical data from communication

① Exchange of goods

As craft work became independent of agriculture, production activities for trading purposes started to expand. During the Dawenkou culture period, all the movables may have become trading targets, such as painted pottery, thin-walled black pottery cups, white pottery, bone carvings, turquoise-inlaid bone carvings and all kinds of exquisite jade which were precious commodities that people craved. These precious works of art were produced by a small number of talented craftsmen with rarer materials, and they must have circulated across large areas. Some of the jade materials in the Dawenkou culture may have been produced locally in Shandong, such as ink crystals from Jimo and jade from Taishan, Zouxian and Laiyang.

However, the jade shovels and chisels excavated from the Dawenkou cemetery cannot have been sourced in Shangdong, because they are made of hard jade which is abundant in Yunnan and Myanmar. Charcoal white stone, which has a hardness of 11, has also been uncovered at the Dawenkou cemetery, so it cannot have been produced locally either. Turquoise was produced on the border between northwestern Hubei and southwestern Henan. During the Dawenkou period, travel conditions were inadequate, and it was impossible for merchants to travel long distances to trade goods. In most cases they were obtained thanks to exchanges between private individuals.

Trading in commodities requires a dedicated location: a market. The sea shells and stone spinning wheels excavated from the tombs of the Qijia culture in Liuwan, Qinghai province, were not produced locally but were imported from the south sea regions, and from Fujian and Zhejiang provinces. The tombs of the Liangzhu culture on Fuquan Mountain contain a type of carved-on-the-back jars which were introduced from the Dawenkou culture of Shandong province, reflecting the long-distance trade taking place during the Liangzhu period [57].

② Cultural exchange

Cultural exchange encompasses both the material (commodity) and various production techniques, experiences and certain ideas. A bowl known to archaeologists as the "red top bowl" first appeared in the central plains of Shaanxi; with the development of the clan society, it quickly spread in all directions. It reached the north bank of the Yangzi river in Hubei province in the south, the grasslands of Inner Mongolia and the upper reaches of the Liao River in the north, the foot of Mount Tai in the east, and the southeast of Gansu in the west. The Dahe Village clan, Liulin Town, Zhengzhou province, was a new tribe founded by migrants who travelled through various places along the banks of the Yellow River 5,000 years ago.

The early remains from the Xiawanggang site in Chunchuan, Henan Province, contain obvious elements of the Daxi culture, yet its fundamentals were still consistent with the Yangshao culture. This is a typical example of the blending of two cultures. The Qujialing culture evolved from the late Daxi culture and the original culture of the neighbouring areas, and was already in the early stage of becoming a patrilineal clan society.

The North Yinyangying culture soon lost the style of its motherland, the Huai River Valley, and became a blend of the exchange between and integration of Qingliangang, Qujialing, Majiabang and Songze cultures, becoming a new type of primitive culture.

Viewed through a geographical lens, two obvious conclusions can be drawn from the cultural exchanges during the Neolithic period: first, in the early and middle Neolithic periods, due to the different geographical environments of various places, there were only two obvious cultural sites, namely the Yellow River Basin and the Yangzi River Basin. Second, the exchange between and integration of the two gradually weakened regional differences and gradually eliminated regional boundaries, thereby producing a Chinese civilisation dominated by the Central Plains [58].

1.3.5 Geographical Facts in Ancient Inscriptions

Dating from the Neolithic period, tortoise shells, bones and stone tools excavated from the site of Wuyang Jiahu, in Henan province, have inscribed symbols, which were probably related to primitive writing. Among them, the  shaped symbol is very similar to the character "目" (pinyin: mù, "eye") appearing in the oracle inscriptions found at the Yinxu site, Anyang [59].

On the pottery excavated from the Yangshao cultural site, carved symbols are often found, including at Banpo, in Xi'an; Jiangzhai, Lingkou, Yuantou, in Lintong; Wulou, in Chang'an; Xinye, in Yunyang; Lijiagou, in Tongchuan; Beishouling, in Baoji, etc. Among them, on the 113 pieces excavated at the Banpo site, there are 27 kinds of carved symbols, and on the 129 pieces unearthed in Jiangzhai, there are

Fig. 1.3 Pictograms on pottery of the Yangshao culture (photographed by Liang Qing 梁擎)

27 kinds of carved symbols. A total of 52 types of symbols were uncovered in the central Shaanxi plain area (see Fig. 1.3).

Some of the 52 symbols mentioned above, such as 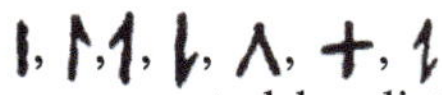 have appeared at sites such as Banpo, Jiangzhai, and Lijiagou, separated by distances of up to 100 km. The use of the same symbols on such a large scale shows that they held the same meaning for different tribes, and can be regarded as a form of writing [60]. In the late Dawenkou culture, there were already more complex runic inscriptions on pottery. Some scholars regard them as relatively mature runes, while others disagree. To date, there have been 17 such graphic symbols found. Some have two different carved symbols on the same pottery jar, and some sites have found several jars with the same symbols, and among different sites several of the same symbols have appeared. The structure of these graphic symbols looks much more complex and they carry more profound meaning than those that have been discovered at the Yangshao cultural sites. Most scholars believe that the graphic symbols in the Dawenkou culture constitute writing, indicating that writing was already being used by the Shaohao clan in the Haidai region. For example, "炟" (pinyin: dá, "a person's name") is depicted in a symbol as "", "南" (pinyin: nán, "south") is depicted in a symbol as "", and "山" (pinyin: shān, "mountain") as "".

References

1. Zhang Z张之恒, Wu J吴健民 (1991) *Zhongguo jiushiqi shidai wenhua*中国旧石器时代文化 [Chinese Palaeolithic culture]. Nanjing University Press南京大学出版社, Nanjing, p 142
2. Huang W黄慰文 (1983) Lantian ren蓝田人 [The Lantian Man]. In: Huang W黄尉文, Jia L贾兰坡, An Z安志敏 (eds) *Zhongguo lishi de tongnian* 中国历史的童年 [The infancy of Chinese history]. Zhonghua Book Company中华书局, Beijing, p 33
3. Zhang Z张之恒, Wu J吴健民 (1991) *Zhongguo jiushiqi shidai wenhua* 中国旧石器时代文化 [Chinese Palaeolithic culture]. Nanjing University Press南京大学出版社, Nanjing, p 256

4. Zhang Z张之恒, Wu J吴健民 (1991) *Zhongguo jiushiqi shidai wenhua* 中国旧石器时代文化 [Chinese Palaeolithic culture]. Nanjing University Press南京大学出版社, Nanjing, p 262
5. Zhang Z张之恒, Wu J吴健民 (1991) *Zhongguo jiushiqi shidai wenhua* 中国旧石器时代文化 [Chinese Palaeolithic culture]. Nanjing University Press南京大学出版社, Nanjing, p 143
6. Zhang Z张之恒, Wu J吴健民 (1991) *Zhongguo jiushiqi shidai wenhua* 中国旧石器时代文化 [Chinese Palaeolithic culture].Nanjing University Press南京大学出版社, Nanjing, p 214
7. *Beijing wanbao*北京晚报 [Beijing evening], 3 June 1999
8. Pei W裴文中, Zhang S张森水 (1985) *Zhongguo yuanren shiqi yanjiu* 中国猿人石器研究 [A study on Chinese ape-man stone tools]. Science Press科学出版社, Beijing, p 219
9. Jia L贾兰坡, Wang Z王择义, Wang J王建 (1962) *Kehe* 匼河 [Ke River]. Science Press科学出版社., Beijing
10. Li Y李炎贤, Wen B文本亨 (1986) *Guanyin dong*观音洞 [Guanyin cave]. Cultural Relics Press 文物出版社, Beijing, p 30
11. Pei, W裴文中, Wu R吴汝康, Jia L贾兰坡 (1958) *Shanxi xiangfenxian dingcun jiushiqi shidai yizhi fajue baogao*山西襄汾县丁村旧石器时代遗址发掘报告 [The excavation report of the Palaeolithic site Ding Villiage, Xiangfen County, Shanxi province]. Science Press科学出版社., Beijing
12. *Beijing wanbao*北京晚报 [Beijing evening] 3 June 1999
13. Zhang Z张之恒, Wu J吴健民 (1991) *Zhongguo jiushiqi shidai wenhua* 中国旧石器时代文化 [Chinese Palaeolithic culture]. Nanjing University Press南京大学出版社, Nanjing, p 66
14. Jia L贾兰坡 (1957) *Jiushiqi shidai wenhua*旧石器时代文化 [Culture in the Palaeolithic age]. Science Press科学出版社, Beijing, p 40
15. Cultural Relics Administrative Commission of Heilongjiang Province黑龙江省文物管理委员会, Harbin Culture Bureau哈尔滨市文化局, The Northeast Expedition Team of the Institute of Vertebrate Palaeontology and Palaeoanthropology, and the Chinese Academy of Sciences中国科学院古脊椎动物与古人类研究所东北考察队 (1987) *Yanjia gang*阎家岗 [The hillock of the Yan family]. Cultural Relics Press文物出版社, Beijing, p 2
16. Zhang Z张之恒, Wu J吴健民 (1991) *Zhongguo jiushiqi shidai wenhua* 中国旧石器时代文化 [Chinese Palaeolithic culture]. Nanjing University Press南京大学出版社, Nanjing, p 175
17. Zhang Z张之恒, Wu J吴健民 (1991) *Zhongguo jiushiqi shidai wenhua* 中国旧石器时代文化 [Chinese Palaeolithic culture]. Nanjing University Press南京大学出版社, Nanjing, p 19
18. Shi X石兴邦 (1992) Zongxu总序 [Preface]. In: Enzhi C陈恩志 (ed) *Zhongguo huashi gurenlei he jiushiqi wenhua kaogu faxian yu yanjiu*中国化石古人类和旧石器文化考古发现与研究 [Archaeological discovery and research on fossil hominids and Palaeolithic culture in China]. Shaanxi Science and Technology Press陕西科学技术出版社, X'ian
19. Huang W黄慰文,, Li C李春初, Wang H王鸿寿 (1979) Guangdong nanhai xiqiaoshan yizhi de fucha广东南海西樵山遗址的复查 [Revisiting the Xiqiao Mountain ruins in Nanhai, Guangdong province].*Kaogu* 考古 [Archaeology] (4)
20. Ma H马洪路 (1989) *Yuangu zhi lü - zhongguo yuanshi wenhua de jiaorong*远古之旅——中国原始文化的交融 [Travelling to an ancient civilisation—The blending of Chinese primitive cultures]. Shaanxi People Press陕西人民出版社, Xi'an, pp 95–114
21. Zhang Z张之恒 (1995) *Zhongguo kaoguxue tonglun*中国考古学通论 [Conspectus of Chinese archaeology]. Nanjing University Press南京大学出版社, Nanjing, p 99.
22. Institute of Archaeology, Chinese Academy of Social Science 中国社会科学院考古研究所 (1984) *Xin zhongguo de kaogu faxian yu yanjiu*新中国的考古发现和研究 [Archaeological discovery and research in New China]. Cultural Relics Press文物出版社, Beijing, p 148.
23. Mu Y牟永抗, Song Z宋兆麟 (1981) Jinagzhe de shili he potuqi-shilun woguo ligeng de qiyuan 江浙的石犁和破土器—试论我国犁耕的起源 [Stone ploughs and earthbreakers of Jiangsu and Zhejiang—The origin of ploughing in China]. *Nongye kaogu* 农业考古 [Agricultural archaeology] (2)
24. Lu Y陆耀华, Zhu R朱瑞明 (1984) Zhejiang jiashan xingang faxian liangzhuwenhua mutong shuijin浙江嘉善新港发现良渚文化木筒水井 [Discovery of the wooden well from the Liangzhu culture at Xingang, Jiashan, Zhejiang province]. *Wenwu*文物 [Relics] (2):94

25. Bai, S白寿彝, Su B苏秉琦 (1994) *Zhongguo tongshi: di'er juan* 中国通史: 第二卷 [A general history of China, vol 3]. Shanghai People's Publishing House上海人民出版社, Shanghai, p 375
26. Dai G戴国华 (1985) Huanan diqu xinshiqi shidai zaoqi wenhua de dongwu kaoguxue yanjiu 华南地区新石器时代早期文化的动物考古学研究 [A Zooarchaeological study of the early Neolithic culture in South China]. *Shiqian yanjiu*史前研究 [Prehistorical studies] (2):95
27. Institute of Archaeology, Chinese Academy of Social Science中国社会科学院考古研究所. (1984). *Xinzhongguo de kaogu faxian he yanjiu* 新中国的考古发现和研究 [Archaeological discovery and research in New China]. Cultural Relics Press文物出版社, Beijing, pp 167–168
28. Institute of Archaeology, Chinese Academy of Social Science中国社会科学院考古研究所 (1984) *Xinzhongguo de kaogu faxian he yanjiu* 新中国的考古发现和研究 [Archaeological discovery and research in New China]. Cultural Relics Press文物出版社, Beijing, p 139
29. Feng Y冯沂 (1989) Henan wuyang jiahu xinshiqi shidai yizhi dier zhi liuci fajue jianbao河南舞阳贾湖新石器时代遗址第二至六次发掘简报 [A brief report on excavations two to six of the Neolithic ruins in Jiahu Wuyang, Henan province]. *Wenwu*文物 [Relics] (1):1
30. Zhou R周仁, Zhang F张福康, Zheng Y郑永圃 (1964) Woguo huanghe liuyu xinshiqi shidai he Yinzhou shidai zhitaogongyi de kexue zongjie我国黄河流域新石器时代和殷周时代制陶工艺的科学总结 [A scientific report of pottery making in the Neolithic Age and the Yin Zhou Age in the Yellow River Basin]. *Kaogu xuebao*考古学报 [Acta Archaeologica]
31. Zhang Z张之恒 (1992) *Zhongguo xinshiqi shidai wenhua* 中国新石器时代文化 [Chinese Palaeolithic culture]. Nanjing University Press南京大学出版社, Nanjing, p 224
32. *Kaogu* 考古 [Archaeology] (1997) 5(4):25–32
33. X, S许顺湛 (1993) *Huanghe wenming de shuguang*黄河文明的曙光 [The dawn of Yellow River civilisation]. Zhongzhou Ancient Texts Press中州古籍出版社, Zhengzhou, p 193.
34. *Zhongguo wenwu bao*中国文物报 [Chinese cultural relics news] 13 Dec 1992
35. *Wenwu* 文物 [Relics] (1997) 8:48–56
36. *Zhongguo wenwu bao*中国文物报 [Chinese cultural relics news] 10 Aug 1997
37. *Kaogu* 考古 [Archaeology] (1997) 5:1–10
38. *Guangming ribao* 光明日报 [Guangming daily] 4 Dec 1992
39. *Kaogu* 考古 [Archaeology] (1997) 5:11–24
40. Bai S白寿彝, Su B苏秉琦 (eds) (1994) *Zhongguo tongshi: dier juan*中国通史: 第二卷 [A general history of China, vol 2]. Shanghai People's Publishing House上海人民出版社, Shanghai, pp 318–322
41. *Wenwu* 文物 [Relics] (1984) 2:94
42. Tao S 石陶 (1961) Huanghe shangyou de fushishize shehui-qijiawenhua shehui jingji xingtai de tansuo黄河上游的父系氏族社会—齐家文化社会经济形态的探索 [Patrilineal clan society in the upper reaches of the Yellow River—An exploration of the social and economic form of Qijia culture] *Kaogu*考古 [Archaeology] (1):3.
43. *Kaogu* 考古 [Archaeology] (1986) 2:147
44. Xu S许顺湛 (1993) *Huanghe wenming de shuguang*黄河文明的曙光 [The dawn of Yellow River civilisation]. Zhongzhou Ancient Texts Press中州古籍出版社, Zhengzhou, p 227
45. *Wenwu* 文物 [Relics] (1988) 1:32
46. *Wenwu* 文物 [Relics] (1987) 10:1
47. *Zhongguo wenwu bao*中国文物报 [Chinese cultural relics news] 8 June 1997
48. *Wenwu* 文物 [Relics] (1984) 6
49. *Kaogu* 考古 [Archaeology] (1983) 1:30
50. Xu S许顺湛 (1993) *Huanghe wenming de shuguang*黄河文明的曙光 [The dawn of Yellow River civilisation]. Zhongzhou Ancient Texts Press中州古籍出版社, Zhengzhou, p 76
51. *Kaogu* 考古 [Archaeology] (1961) 1:62
52. Zhang Z张之恒 (1995) *Guangming ribao* 光明日报 [Guangming daily] 16 Oct 1995
53. Lin H林华东 (1992) *Hemudu wenhua chutan*河姆渡文化初探 [A preliminary study of Hemudu culture]. Zhejiang People's Press浙江人民出版社, Hangzhou, p 144
54. Lin, H林华东 (1992) *Hemudu wenhua chutan*河姆渡文化初探 [A preliminary study of Hemudu Culture]. Zhejiang People Press浙江人民出版社, Hangzhou, p 140

55. Yuxian W吴玉贤 (1983) Cong kaogu shidai tan Ningbo yanhai diqu yuanshijumin de haishang jiaotong从考古发现谈宁波沿海地区原始居民的海上交通 [A study of the aboriginals' maritime transportation in the coastal area of Ningbo through archaeological discoveries]. *Shiqian yanjiu*史前研究 [Prehistory studies] 1
56. Beijing University Archaeology Intern Team北京大学考古实习队, Yantai Cultural Management Association烟台地区文管会, and Changdao County Museum长岛县博物馆 (1983) Shandong changdao shiqian yizhi山东长岛县史前遗址 [Prehistoric site in Changdao County, Shandong province]. *Shiqian yanjiu*史前研究 [Prehistory Stud] 1:128–130
57. Xu S许顺湛 (1993) *Huanghe wenming de shuguang*黄河文明的曙光 [The dawn of Yellow River civilisation]. Zhongzhou Ancient Texts Press中州古籍出版社, Zhengzhou, pp 502 & 520
58. Ma H马洪路 (1989) *Yuangu zhi lu - zhongguo yuanshi wenhua de jiaorong*远古之旅——中国原始文化的交融 [Travelling to an ancient civilisation—The blending of Chinese primitive cultures]. Shaanxi People Press陕西人民出版社, Xi'an, pp 157–174
59. *Wenwu* 文物 [Relics] (1989) 1:1
60. Zhang Z张之恒 (1992) *Zhongguo xinshiqi shidai wenhua* 中国新石器时代文化 [Chinese Palaeolithic Culture]. Nanjing University Press南京大学出版社, Nanjing, p 47

Chapter 2
Expanding Knowledge of the World in Slave Societies: The Xia, Shang and Zhou Dynasties

2.1 The Impact of the Social Environment on Geoscience

The legendary Xia dynasty in China was a period of transition from primitive societies to slave societies. From the reign of Yu the Great to King Shaokang, a hereditary system marked by military democracy was adopted by the six rulers of the Xia dynasty. Following Shaokang and his son Yu, there was no evidence of an election or any mechanism for dismissal, suggesting that the union of tribes, a form of social structure, had been replaced by state institutions, hence a class society [1], also called a slave society, had emerged. Slave societies existed in the Xia, Shang and Western Zhou dynasties. The Xia dynasty marked the end of primitive societies and the beginning of class societies in Chinese history.

The slave societies saw a rapid development in productivity. As Engels commented in his work *Anti-Dühring*, only the slavery system made a larger division of labour between agriculture and manufacturing possible, thus paving the way for cultural prosperity, that is, Greek culture. Similarly, with the slavery system, Chinese culture thrived, making China one of the world's remarkable ancient civilisations. The impact of the social environment on the development of geoscience during this period is mainly reflected in the following six aspects.

First, the improvement of crafting techniques led China into the Bronze Age and created a splendid bronze culture. The lacquerware manufacturing sector emerged due to the invention of lacquerware. In the *Hanfeizi*: *Shiguo*, "Yu" is recorded as a sacrificial vessel with ink on the outside, and cinnabar on the inside. The development of crafting skills promoted the expansion of geoscientific knowledge.

Second, the crafting of bronzewares boosted the mining industry and contributed to the improvement of mining technologies.

Third, the agricultural sector, comprising planting, forestry, livestock farming, and subsidiary production and fisheries, had essentially taken shape. This type of agriculture required a relatively rich knowledge of geology, which in turn promoted

W. Yang and Q. Yang, *A History of Ancient Chinese Geoscience*, Understanding China,
https://doi.org/10.1007/978-981-97-4457-2_2

the development of geoscience, including meteorology, climatology, phenology and water conservancy.

Fourth, thanks to the invention of writing, scientific and cultural knowledge had been rapidly accumulated and disseminated. Knowledge in the field of geoscience was no exception.

Fifth, the improvement in productivity caused the slave state to continue to develop. With diverse ethnic groups further integrating into the main population, the territory had been constantly expanding and the horizon broadening, which had stimulated huge interest amongst people in understanding geoscientific phenomena. According to the *Shiji: Sunzi wuqi liezhuan* ("Records of the Grand Historian: Biographies of Sunzi and Wuqi"), the territory during the Xia dynasty stretched east to the junction of the Yellow River and the Ji River, west to the Hua Mountain in southern Huayin, Shaanxi, south to Longmen, over ten kilometres south of Luoyang, Henan, and north to the less accessible sides of the Taihang Mountains in Shanxi.

Sixth, the legendary tale of Yu the Great tackling floods and the great project of water control reflected the level of geoscientific knowledge at that time.

2.2 Developing Understanding of the Earth During the Xia Dynasty

2.2.1 Geoscientific Evidence in Ancient Capitals

Based on research into defining the periods of the Xia, Shang and Zhou dynasties, it is presumed that the dividing point between the Xia and Shang dynasties is around 1500–1600 BC [2]. In the chapter "Chronicles of the Law" of the *Hanshu: Lüli zhi*, which quotes *Emperor Families*: "people were the descendants of the Xia clan. Seventeen emperors succeeded to the throne, which was four hundred and thirty-two years old." Therefore, the period of the Xia dynasty is estimated to be 2032–1932 BC, about 4,000 years ago. Consequently, the urban ruins dated to 3,500–4,000 years ago and excavated in the territory of the Xia are certainly the cities of the Xia.

According to historical accounts, there are two main areas where the Xia people were active: one was the upper reaches of Ying River in western Henan and the lower reaches of the Yi River and Luo River near Luoyang. The second was the lower reaches of the Fen River and the Su Water regions in the south of Shanxi province. Most of the major events that occurred in the capital of the Xia Kingdom and during the Xia dynasty in historical legends are related to these two areas. For example, the Xia ruins in western Henan or southern Shanxi, the Erlitou sites in Henan and other places, and the lower layer of the Dongxiafeng cultural site in Xia County, Shanxi province, are all parts of the Xia dynasty cultural sites.

The Erlitou ruins contain dozens of palace building sites. Among them, the No 1 and No 2 palaces that have been excavated belong to the third phase of the Erlitou culture, and the two sites where foundations of constructions have been discovered in recent years belong to the early stage of the second phase and are also palace buildings. There are sacrificial pits in the courtyard of the No 1 palace and under the eaves of the pillars of the hall, proving that the foundation site of this palace forms part of the temple buildings. An ancient temple was not only a place for rulers to worship their ancestors, but also an important administrative location and a symbol of the regime. Therefore, the Erlitou site is surmised to be a city of the Xia state. According to Zheng Jiexiang's findings in the *Xiashi chutan* ("A Preliminary Study of the History of the Xia"), it is believed that the ruins of the large palaces in Erlitou may be from the capital during five different periods of Xia emperors, namely Taikang, Zhongkang, Shaokang, Kongjia and Jie [3].

Although the Erlitou site is a large-scale capital city site, no defensive facilities, such as city walls or moats, have been found so far. In the lower cultural layer of the Dongxiafeng site in Xia County, Shanxi, two rings of trenches have been found, around a trapezoidal plane. The inner trench is about 130 m long, and the outer trench is about 150 m long. The two trenches are 5.5–12.3 m apart, and the existing ditch is about three metres deep. Zheng believes that this may be a pre-Shang cultural site from the same period as the Erlitou culture, while other scholars believe that it belongs to the Xia dynasty culture.

The Wangchenggang Fortress site in Kaifeng, Henan, comprises two small square fortresses juxtaposed at the east and west. Their walls are made of rammed earth; the west city wall is 94.8 m long, and the south city wall is 97.6 m long. Its age coincides with the Xia era [4]. Zheng holds that this may be the location of "the Yang Capital ruled by Yu" (featured in the *Shiben*, or *Generational Records*).

On the terraces along both sides of the Qinglong River, Dongxiafeng Village, Xia County, Shanxi province, a small-scale fortress-city was found. Its southern wall is about 400 m long, 8 m wide, with the surviving height amounting to 1.8 m. The wall was built of rammed earth, about 4000 years ago. The city's age seems to coincide with the Xia dynasty and its location is right within the area of the so-called Xia relic site [5].

In addition, ancient works also document some of the cities of the Xia dynasty. According to the *Zuozhuan: Zhaogong sinian*, Qi's ruling centre was located in Xiayi, formerly known as Yangzhai, present-day Yu County, Henan. In the *Zhushu jinian* ("Bamboo Annals"), King Xiang succeeded to the throne and lived in Shangqiu, which is the present-day Puyang County, Henan, according to annotions in the fourth volume of the *Zizhi tongjian* ("Comprehensive Mirror to Aid in Government"). The *Zhushu jinian* also mentioned that King Xiang lived in Zhenguan (not far from Shangqiu), the present-day Guancheng County, Shandong province.

The modern text of the *Zhushu jinian* mentions that "Shaokang relocated to Yuan", and Yuan is the present-day Yuan Village in Jiyuan, Henan.

The ancient text of the *Zhushu jinian* also mentioned that "King Ning, also called Yu, lived in Yuan, and from Yuan he relocated to Laoqiu", the present-day Kaifeng, Henan.

The ancient text of the *Zhushu jinian* mentions that Qin, also called Yinjia, took power and lived in Xihe (West River). Its present-day location may be in the north-eastern Henan, or Fenyang in Shanxi, or in the Luoyang-Huayin of the western Henan and eastern Shaanxi area.

The political centres of the above-mentioned Xia kings were the cities of that time, and they were distributed over the area from the east of Huashan, in the west, to the eastern Yudong Plain, indicating that the territory of the Xia extended around these political centres, and could not have reached too far from them.

The choice of location for a city reflected the geographical knowledge of the time. For example, the terrain was generally flat, high and dry, and preferably close to a river or the sea, but not too low-lying. The prime function of the city was not only to be a political and cultural centre, but also a centre for promoting the exchange of materials and information.

2.2.2 Significance of Bronzewares for Understanding Geoscience

Bronze is an alloy of copper and tin (or lead), the production of which must be based on the industries that produce these elements. Judging from the materials excavated, the Xia dynasty had entered the Bronze Age.

The site of a bronze smelting workshop dating from the second phase of the cultural remains at the Erlitou site in Yanshi, Henan, has been found, and a considerable number of molten copper crucible fragments were unearthed nearby. A collection of bronze utensils, such as copper knives, copper scythes, copper bells and turquoise-inlaid bronze plaques, were also found. The bronze casting sites found in the cultural remains of the third phase of Erlitou are larger than those of the second phase of Erlitou. A large number of bronze smelting crucible fragments and some pottery moulds were discovered nearby, indicating that bronzes could be produced in batches. According to chemical analysis, the composition of the bronzeware from this period contains 91.6% copper, 7.03% tin and 1.23% lead. One of the two wine vessels contains 92% copper and 7% tin and the other contains 91.89% copper, 2.62% tin and 2.34% lead. A common feature of the three specimens is that they contain low levels of tin and lead, with an average copper content of 91.85%, tin 5.55% and lead 1.19%, which means they can be categorised as bronzeware, but compared to those from the middle and late Shang dynasty, they were still primitive in some ways [6].

Apart from being used as production tools, bronzes in this period also included weapons, such as the *ge*, *qi*, and scythe, as well as items for daily use, such as vessels, bells, copper ornaments, etc. Historical texts such as the *Yuejue shu* ("The Glory of Yue"), *Zuozhuan* and *Mozi* also mention multiple copper wares and their production processes.

Five categories of geoscientific knowledge can be identified in relation to the bronzes.

(1) Sourcing copper, tin and lead ore was a prerequisite of bronze production; these were generally traced and found by relying on natural copper and malachite outcrops. This was the earliest and most primitive prospecting method.
(2) The quarrying technology of the Neolithic period developed into a mining technology in the Xia dynasty which ensured the supply of copper, tin and lead ore.
(3) The mining and smelting of copper, tin and lead enriched people's understanding of these elements and their geographical distribution.
(4) From collecting the raw materials to trading the finished products, people's expertise on the subject of transportation developed rapidly. Ores were generally located in distant mountainous areas, while smelters were often located in the central areas where people lived, so the organisation of complex bronze production offered the potential to enhance people's understanding of their living environment.
(5) The copper, tin and lead mineral resources in an area were limited, perhaps insufficient, and needed to be exchanged with neighbours, which promoted economic and cultural exchanges between tribes.

Apart from the metallic minerals involved in bronzes, the Xia people had other knowledge of mineral rocks. According to the archaeological data processed by the author, other mineral rocks used in the Xia dynasty included: turquoise, nephrite, cinnabar, mica, *wuhuatu* (tesselated soil), mud slate, limestone, sandy shale, diorite, gabbro, flint, quartz, hematite, gold, silver, sandstone, malachite, laterite, clay, kaolin, sedimentary soil, black soil, ochre, etc. Of these, the ones most used were turquoise, jade, cinnabar and *wuhuatu*. Using cinnabar and *wuhuatu* in tombs was a popular burial custom at that time, and some huge ones needed to use 308–586.5 kg. As gemstones, the turquoise and jade were used to make ornaments and ceremonial vessels. Many exquisite jade artefacts were unearthed in the third phase of the cultural relic at Erlitou in Henan, including jade shovels, a jade dagger, a jade battle-axe, jade handle-shaped tools and sacrificial utensils. These were new types of artefacts that appeared in the Erlitou culture and were important ceremonial objects at that time. The main use of hematite was as a pigment. Argillite and limestone were mainly used to make stone resonators. The gold and silver wares were precious metal objects first found in the Yumen Huoshaoogou cemetery, Gansu province, dating back to the Xia dynasty, and which included gold earrings, gold rings and silver rings [7].

2.2.3 A Legendary Map

Legend has it that Xia Yu once cast nine *ding* [ritual vessel on legs] from copper, and their bodies were cast with patterns of mountains, rivers, roads, birds and beasts, plants and trees, which were the original maps of ancient times. The nine *ding* were passed down to the late Shang dynasty and transferred to Luo city after the establishment of the Western Zhou dynasty [8]. By the Qin dynasty, the nine *ding* had been destroyed, but the patterns on these *ding* had been copied and painted into a map called the *Shanhai tu* [9]. Later, it was proposed that the *Shanhai jing* ("Classic of Mountains and Seas") and the *Shanhai tu* had both originated from the nine *ding*. The maps appended to the *Shanhai jing* only contain gods and spiritual figures without any geographical maps. Therefore, the claim that the original map was cast on the nine *ding* remains a legend today and has not been confirmed.

2.2.4 The Chinese Flood Myth and Its Geoscientific Narrative

According to legend, there had been serious floods before the Xia dynasty [10]. Since Yu's father Gun had been unsuccessful in controlling the water, Shun sent Yu to control the water. Yu was under pressure as he faced punishment if he failed and he learned from his father's mistakes [11], deciding to change the approach by both blocking and dredging the water rather than by purely relying on blocking it. First of all, he personally surveyed the mountains and rivers to obtain a clear picture of the distribution of terrains and rivers [11]. According to *Shiji*, he travelled across the fields, swamps, mountains and rivers of the entire territory; with ropes, rulers and compasses, he marked which places were to be dug away, the places on which embankments were to be built, and which sections of the rivers were to be channelled or drained. He then organised teams to draw up water control plans grounded in collective wisdom [12]. In this way, he understood the nature of the water and dredged the rivers, allowing the water to flow unimpeded, from small rivers to the large rivers, and to the sea. At the same time, some natural lakes, swamps and low-lying lands were used to store water, thus performing the function of dispersing floodwaters and irrigating farmland. In the meantime, he also dug ditches in the fields to improve irrigation [13]. For thirteen years, Yu eliminated water damage in the middle and lower reaches of the Yellow River, enabling people to live and work in peace and contentment [14]. For eight years, he worked so hard to benefit his people that he refused to enter his home even if he passed by three times [10]. There are still many records of Yu's deeds in ancient texts, suggesting that there is (some) truth behind the legend of Yu. Yu's success was universally praised and revered. The geological knowledge reflected in the legend concerns not only topography and the geographical distribution of mountains and water, but also surveying, hydrology and water conservancy.

2.2.5 Phenological Changes in the Xia Xiaozheng Calendar

The agricultural development in the Xia dynasty deepened people's understanding of the seasons and climates. The *Xia Xiaozheng Calender,* which has been handed down to this day, was considered by Confucius, in the Spring and Autumn period, to be a document from the Xia dynasty. The celestial phenomena and climate it records are very close to those of the Central Plains region. The phenological knowledge recorded in the book is fairly rich, and it is the earliest extant phenological monograph in China. Phenological data for the first month noted the long inflorescence of willow, plum, apricot and mountain peach blossom, one after another; voles coming out to move around, pheasants chirping, males and females mating; fish rising from the river bottom to be close to the icy surface; and the farmland pest—the mole cricket—also singing loudly. Month-by-month records of this kind formed a complete phenological calendar throughout the year. According to Wang's statistics [15], there are 68 phenological phenomena related to birds, animals, fish, insects and plants, seven related to meteorological conidtions, and eleven related to agriculture and animal husbandry in this calendar.

2.2.6 Maritime and Land Routes in the Xia Dynasty

Near the foundations of the third phase of Erlitou, Shiyong Road (District 3) and Shizi Road (District 6) were found. The fragment of Shiyong Road is 0.35–0.60 m wide, with its western end paved with stone slabs, and the east made of cobblestones. The road surface is smooth. The use of stones to build roads indicates that the people lived in close proximity to each other and that frequent exchanges took place, hence it was necessary to build a relatively solid road to facilitate traffic. The road construction skills of that period were much more advanced than during the Yangshao culture period.

The *Shiji: Xiaben ji* ("Records of the Grand Historian: History of the Xia") mentions that when Yu dealt with the flood "he travelled across the land by carriage and crossed the water by boat". This indicates that land transportation was far more developed than during the Yangshao culture period. The wars during the Xia dynasty, as well as the exchange of goods and supplies, were also reliant on roads. It is possible that the Xia state may have had some degree of a reasonably developed transportation network. During the Taikang reign, Xiangtu, the ancestor of merchants, led an army to conquer the tribes on the coast and on the islands, which led to him being highly admired by later generations [16].

From the accounts in the *Yugong*, we learn that the people living in the Liaodong Peninsula and the Shandong Peninsula, as well as in the coastal areas of Jianghuai, all landed on the mainland after sailing a certain distance by sea, in order to pay tribute to the Xia dynasty in the Central Plains. For example, for residents of Jizhou, dressed in fur clothes, to travel to the Central Plains, they first sailed from east to west across the northern Bohai Sea, with the Jieshi Mountain on their right side. On entering the Yellow River, they could reach the Central Plains by following the river. When people from the Shandong Peninsula came to pay tribute, they would first sail around the Peninsula. For residents wearing straw clothes from the southeast sea of

Yangzhou, they would sail north from the mouth of the Yangzi River, would enter the Huai River, and then would turn into the Si River [17].

2.3 Developing Understanding of the Earth During the Shang Dynasty

Historians studying the Xia, Shang and Zhou dynasties believe that the phase separating the Shang from the Zhou was between 1050–020 BC, about 3,060 years ago. The Shang dynasty existed between 1500 BC (or 1600 BC) and 1020 BC (or 1050 BC) and lasted approximately 480 (or 550) years. The bronze culture of the Shang dynasty was more mature than that of the Xia dynasty. The economic and cultural boom of the Shang dynasty inspired the development of a writing system with inscriptions appearing on pottery, jade, bones and bronzes, etc. and the writing on the oracle bones in the late Shang dynasty became the dominant form. The use of a large number of oracle bone scripts was the most distinctive feature of this period.

2.3.1 Geoscientific Facts Found in Oracle-Bone Inscriptions

Information recorded on the oracle bones related to celestial phenomena, calendars, numbers, meteorology, fauna and flora, floods and droughts, farming and animal husbandry, field hunting, disease, procreation, ghosts and gods, dreams, sacrifices, conquests, institutions, human activities, the genealogy of the Shang emperors and so on: a truly historical encyclopedia of the Shang dynasty [18]. Information related to geoscience can be summarised under the seven following headings [19].

(1) Meteorology

① Sunshine conditions

- Sunny; further divisions include various degrees of clarity and cloudiness
- Cessation of rainfall and appearance of sun
- Overcast
- Cloudy
- Thick clouds
- Blinding sunlight
- Sun halo

② Humidity conditions

- Rainbow
- Clouds; further divisions include dark clouds, stagnant clouds, endless clouds, etc.
- Rain, including the direction of movement and the magnitude, e.g. heavy rain, sharp showers, rainstorm, continuous rain, light rain, drizzle, residual rain, steady rain, timely rain, etc.

- Snow
- Hail
- Fog, haze and mist

③ Lightning conditions

- Thunder; further descriptions include fierce thunder, gentle thunder, and loud thunder.

④ Atmospheric motion

The understanding of wind, including wind direction, the naming of the winds (the eastern Xie wind, southern Ji wind, western Yi wind and northern wind) and the magnitude of the wind (gale, hurricane, wild wind, breeze, etc.).

⑤ Weather forecasts

Most of the weather forecasts on oracle bones are divination with elements of superstition, but a few were used to observe the routines of animals as a sign of changes in the weather, which was relatively objective.

⑥ Meteorological records

The meteorological records are quite detailed, including the meteorological changes within one day, a decade, and even continuous records throughout a number of decades. The oracle inscription capturing meteorological knowledge is the earliest achievement in the history of Chinese meteorology and a treasure in the world history of meteorology.

(2) Surveying

The Shang people were familiar with the method of measuring angles of inclination with water. Specifically, they first dug two ditches perpendicular to each other, and then made four smaller channels perpendicular to each other, at each end of the ditch, to form a "卐" shape. Water was then poured into the ditches so that the level of the ground could be measured. This method has been confirmed at the archaeological site of Yinxu, as a water trench like this was discovered in the ruins of the Yinxu palace chamber [20]. The Yinxu people also knew how to confirm vertical direction with a hanging rope, which can be seen from the pictographic character "直" (pinyin: zhí, straight). Moreover, a set of well-digging measurement tools–a wooden beacon and two wooden plummets–was unearthed at the end of 1993 in Chenjiadun Village, De'an, Jiangxi. The plummet looks like a teetotum, with a round top and a pointed tip, and a small hole on top in the centre, which is in line with the tip. This was a fixed-point location measurement tool used for digging wells at that time (see Fig. 2.1). Once the position of the well mouth had been determined and, after digging down to a certain depth, a support structure would be placed over the mouth. Next, the plummet was tied to a rope and dropped vertically into the hole. After adjusting the beacon to ensure that it was centred on the teetotum, the digging could continue without the plummet in the way. Later, the the beacon would be used again to recalibrate the centre and define the circumference, until the mouth of a spring was found. A well

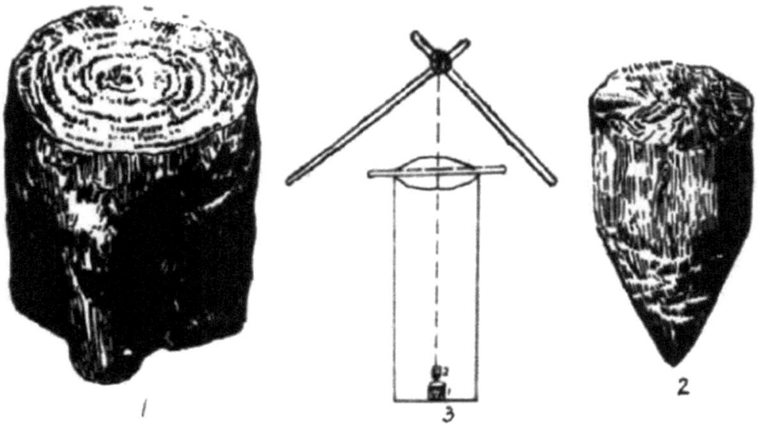

Fig. 2.1 The measuring instruments of the Shang dynasty. 1—Wooden beacon; 2—Wooden plummets; 3—Sketch map

dug by this method appears to be a straight cylinder without any deflection [21]. The pictographic character “规” (pinyin: guī, ruler) appears on the oracle bones [22]. Its arrow shape signifies direction (“准望”, pinyin: zhǔn wàng), line detection and measurement (“准绳”, pinyin: zhǔn shéng) [23]. The signs that record the directions on oracle bones signify: north, south, east, west, and middle [24], and the established signs for the four directions are [illegible](north), [illegible](south), 析 (east) and [illegible](west) [25].

(3) Mineralogy

The character “金” (pinyin: jīn) appeared in the oracle bone text, but it did not refer to real gold; people called bronze “gold” at that time. In addition, they could also recognise real gold, lead, iron and tin [26].

The oracle sign “丹” (pinyin: dān) depicts the scene of mining ore. There is also a divination that records the mining process [27].

(4) Borderland geography

According to the *Shijing* (“Book of Poetry”), this was the territory during the Shang period: “When the tribal chief, Chengtang, established the Shang dynasty, no one in the Qiang ethnic group from afar dared not come to pay tribute to him. The Emperor is the father of all people.” During this time, the Shang not only firmly controlled the middle and lower reaches of the Yellow River, but also integrated ethnic minorities such as the Qiang in the West, forming a slavery-based state that “has thousands of miles of frontiers where people build their homes and the territory is everywhere”. Its territory included all or most of present-day Henan, Shandong, Hebei, Liaoning, Shanxi, Shaanxi, Anhui, and Hubei provinces [28]. So far, archaeological excavations have shown that the distribution of Shang culture extended from Kazuo in Liaoning and Keshketeng in Inner Mongolia in the north, to Shaanxi and Gansu in the west,

Sichuan in the southwest, Shimen and Ningxiang in Hunan in the south, Qingjiang in Jiangxi in the southeast and Haiyang in Shandong in the east [29].

The Shang territory was surrounded by many tribal states (*-fang*). While some were subordinate to the Shang power, others were its enemies. These included Guifang, during the Wuding period, in present-day Jinnan; Tufang, the Tangdu state in the Western Zhou dynasty, predecessors of the Shiwei state in the Shang dynasty; Shefang in the northwest region of present-day Taihang Mountain; Genfang, ten kilometres west of present-day Yuanqu County; Qiangfang, in the west of present-day Dali County, Shaanxi; Longfang, close to Qiang; Yufang, the Rong clan; Mafang, near Qiangfang; Infang, approximately in the southern Jinnan region; Shifang, in the west; Lifang, adjacent to the Qiangfang; Jifang, in present-day Hejin, Shanxi; Jingfang, probably the Geng feudatory state and also in Hejin; Jifang, the Ji feudatory state in Guancheng; Dafang, near Qinyang; and, in addition, there were Hufang, Xingfang and Pangfang. In the area of southern Jin, there were also Zhou, Ling, Quan, Chuan, Guo, Shu, Zhi and other states; in Yuxi, there were Zhi, Que and other states. After King Wuding's reign, the main belligerents were: Zhafang, adjacent to the Qiangfang; Suifang, close to Zhafang; Xifang, probably also known as Manfang; Beifang, which is related to the Bei kingdom in the early Western Zhou dynasty; Weifang, between present-day Yongcheng and Su County, around the junction of present-day Anhui and Jiangsu; Renfang, which was conquered in the late Shang dynasty, that is, Yifang which belonged to the northern territory of the Zheng, Zhou, Song and Chu states in the Spring and Autumn period; and Yufang, near present-day Sui County, Henan [30]. The total number of *fang* is more than 40 [31].

(5) Toponymy

There are more than 500 place names recorded on the oracle bones [32]. Apart from over 40 names of *fang* and 150 river names, there are also names of mountains, hills, roads, ferry ports and settlements [33].

There are also geographical terms found on the oracle bones, serving as the names for entities such as *bang*, land, fields, stones, wells, springs, water bodies, rivers, prefectures, cities, mountain foothills, towns, countries, etc., reflecting the hydrography, topography, territorial concepts and land planning knowledge of the time.

(6) Water conservancy

The pictographic sign "汫" (pinyin: jǐng) found on the oracle bones symbolises the process of drilling wells ("井") for irrigation. The sign "录" (pinyin: lù) symbolises the installation of a shadoof on the well to draw water. The sign "洗" (pinyin: xǐ) means "pouring water on the ground", suggesting the irrigation process. "洗田" (pinyin: xǐ tián) in oracle inscriptions means "channelling water to the farmland" [34].

There are also irrigation water conservancy projects signified by "畎" (pinyin: quǎn) and "浍" (pinyin:kuài). The sign "甽" (pinyin: quǎn) shows that the Yin people knew how to use natural standing water to irrigate fields. The sign "畴" (pinyin:

chóu) represents ditches in the field. "作沑" (pinyin: zuō niǔ) means to build an embankment to store water. "方商" (pinyin: fāng shāng) means to protect the capital from flood by building dams. "正河新圩" (pinyin: zhèng hé xīn wéi) means to control the river and repair the polder [34].

2.3.2 Geoscientific Implications of Artefacts

The Shang excelled in bronze handicraft and produced high-quality vessels with fascinating abstract designs, such as the Ya Chou bronze battle-axe unearthed in Yidu, Shandong, and the Pig Zun (Fig. 2.2) unearthed in Ningxiang, Hunan province. So far, we have seen tens of thousands of bronze objects from the Shang dynasty, up to as many as 40 to 50 different kinds [35], including agricultural production tools, handicraft production tools, utensils for everyday use, weapons, ritual vessels, musical instruments, and ornaments, etc.

The larger ones, such as the Rectangular *Ding* with Animal Mask Motif unearthed in Zhengzhou, and the Simuwu *Ding* unearthed in Anyang are heavy vessels of about 800–900 kg. To manufacture such a large number of bronzes, the amount of materials required, such as copper, tin and lead, is also astounding, and was only possible with a relatively developed mining industry and large-scale workshops for casting these objects. The earliest large-scale copper mining site found in China, dating from the mid-Shang, is the Ruichang Tongling site in Jiangxi province, with an area of about 250,000 m^2. Within the excavated area of 300 m^2, there are 24 shafts, three horizontal drifts, one open pit and one concentration tank. This shows that the combination of the mining technology of shafts, horizontal drifts and pits adopted at that time was fairly mature [36].

Worksites for smelting and casting copper during the Shang dynasty have been discovered in multiple locations, including a large one in Huangpi, Hubei province, two in Zhengzhou (one is outside Nanguan and the other is north of Zijin Mountain) and one with an area of more than 10,000 m^2 in Xiaotun, Anyang, Henan province [37].

Fig. 2.2 The Pig Zun (photographed by Liang Qing 梁擎)

Regarding the raw materials for bronzes in the Shang dynasty, Guo Moruo, in his work *Bronze Age,* argued that that copper came from the Wuyue area in the south, but Jai Bozan believes that it originated from the upper reaches of the Yangzi River. In fact, within two or three hundred kilometres of Anyang, Henan, there were many places that produced copper, such as Jiyuan in Henan, Shixian and Xingtai in Hebei, Quwo, Yicheng, Jiangxian, Yuanqu, Xia County, Wenxi in Shanxi, the southwest of Taihang Mountain, and Laiwu in Shandong, which all produced copper that could be mined nearby. However, to date, the scale and quality of no other mining site can match those of the Tongling site in Jiangxi during the Shang dynasty, which somewhat supports Guo's comment.

There are generally three types of copper ore: natural copper (i.e. red copper), copper sulfide and copper oxide (i.e. malachite). Judging from the raw materials left by the remains of copper metallurgy, malachite was the main material used to make bronze. At the Yinxu ruins in Anyang, Henan, a piece of malachite weighing 18.8 kg was unearthed, which may have been obtained from a copper mining area northwest of Anyang.

There were also tin mining areas near Yinxu, such as Tangyin, Wu'an, Yangcheng and Qinshui in Shanxi. Judging from the unearthed artefacts, tin and lead utensils already existed in the Shang dynasty, indicating that metal tin and lead could be extracted from tin and lead ore at that time.

To give a few examples, at Wanggang and Zhaoji in Yingshang, Anhui province, 12 pieces of lead from the Shang dynasty have been found [38]. Excavation at the copper-casting site at Miaopu Beidi of Yinxu has unearthed crucibles, copper blocks, tin blocks and charcoal [39]. Rock resonators, gold leaves, turquoise, bronze, lead *ge* (4 pieces) and jadewares have been excavated from the Yin tombs in Hougang, Anyang [40].

Judging from the existing relics, the amount of gold consumed by the Shang people greatly exceeded that of the Xia. For instance, two pieces of gold armlets, one pair of gold earrings and one example of a gold hairpin (85% gold) were unearthed in the mid-Shang tomb of Liujiahe at Pinggu, Beijing, and other items mainly contained silver and copper, with no impurities [41]. At the Xuejiazhuang site of Anyang, Henan, three examples of gold leaves were unearthed [42]. A piece of unworked gold weighing more than 50 g, probably cooled molten natural gold, was unearthed at the Yinxu ruins. Ring-shaped gold items were unearthed from the large tombs at Wuguan Village [43]. Two bow-shaped gold artefacts of 95% pure gold were unearthed at Baode, Shanxi; these are 95% pure gold and the craftsmanship is exceptional too.

The expertise in crafting jade was highly developed at that time. Merchants had an extraordinary love of jade and the aristocrats regarded it as a treasure that could be exchanged for money.

Not only did the living love jade and use jade, but the dead, ghosts and gods also craved fine jade. Slave masters and nobles owned a large amount of exquisite jade for burial and sacrificial rituals. For example, at the Fuhao cemetery at Anyang more than 750 jade objects have been unearthed, including *huang*, *jue* and *bi*, and *gui* ritual vessels. Jadewares for the guard of honour included *ge*, spears, *qi*, *yue*, choppers and axes. Jade tools included shovels, *ben* (short axes), chisels, sickles and various knives; jade objects for daily use included *gui* (a food vessel), plates, mortars, pestles, combs and hairpins; and *gongs*, jade musical instruments. There were also countless jade ornaments and crafted carvings, all demonstrating a high-level of craftsmanship [44].

The Sanxingdui site at Guanghan, Sichuan, dates back to 2,800–4,800 years ago. Thousands of temple utensils were discovered in the first and second sacrificial pits, such as golden staffs and masks; nearly 100 bronze figures, heads, masks, sacred trees, etc.; groups of stone decorations and beautifully carved jade objects, including *gui*, *zhang*, *bi*, *yuan*, *chong*, rings, beads, pipes, *pei*, *ge*, spears, axes, short axes, chisels, etc. Their raw materials were found locally from nearby mines in Longmen Mountain and Qionglai Mountain. The gold artefacts unearthed at Sanxingdui represent the earliest examples of high level gold craftsmanship in the Shang dynasty [45].

There were nearly twenty kinds of jade varieties used in the Shang dynasty. The main source of materials was from nearby mining, such as from the Du Mountain at Nanyang, Mi County, Weichuan, Laiyang, and Lantian, etc. However, Hetian suet jade, greenish-white jade, topaz jade and *moyu* (black jade) products were unearthed in the Yinxu ruins, suggesting that Hetian jade mining existed from no later than the Yin period and had been introduced into the Central Plains [46]. Of the jadewares unearthed from the Fuhao cemetery at the Yinxu ruins, more than 700 pieces were mostly green and white. Jade artefacts in greenish-white were fewer, and the number of topaz or ink jade pieces was very limited. Most of these types of jade were produced in Xinjiang.

According to the author's statistics on archaeological data, the mineral rocks recognised and used by the Shang people were: gold, copper, tin, lead, iron, cinnabar, malachite, jade, turquoise, agate, sandstone, kaolin, *wuhuatu* (tesselated soil), flint, granite, basalt, slate shale, marble, limestone, diorite, gabbro, siliceous rock, hornblende, shale, andesite, slate, serpentine, fluorite and salt, a total of 29 types. Among them, the seven types most used were: jade, turquoise, copper, cinnabar, gold, agate and sandstone.

2.3.3 Significance of Ancient Capitals to Geoscientific Understanding

According to the documentation, there were five capitals in the Shang dynasty, of which Yanshi Shangcheng was the first capital of the Shang. In addition, the ruins of Zhengzhou Shangcheng and Yinxu at Anyang have also been uncovered. The *Shiji* records that the Tang lived in Bo. Historically there were three places called Bo, "Meng is the northern Bo; Gushu is the southern Bo, and Yanshi is the western Bo." It can be seen that Yanshi Shangcheng is what the historical text refers to as the western Bo.

Located south of Dahuaishu Village, Xichengguan Town, Yanshi, in Henan province, Yanshi Shangcheng was enclosed by Mangshan Mountain in the north and the Luo River in the south, with flat terrain and fertile soil. The city site is more than 1,700 m long from north to south, and different lengths from east to west. It was divided into three sections, 1,215 m from east to west in the north, 1,120 m in the middle, and 740 m in the south. The city is surrounded by rammed earth walls, with seven gates and eleven roads. Inside there are palace ruins, and three stone-built drainage ditches have been found in the northeast, southeast and south of the palace site [47]. The ruins of the royal garden there are 130 m long and 20 m wide in total. On both the east and west sides of the city there is a large stone-built water channel. This was the earliest water conservancy facility in the palace area, which was mainly used to supply water to the people in the palaces, and may also have served to enhance the environment and entertain people [48].

The ruins of Zhengzhou Shangcheng are located in the Zhengxian County and Beiguan area in the east of Zhengzhou. The plan of the city looks roughly rectangular in shape, and its circumference is approximately 6,960 m: the northern side is about 1690 m in length, the southern and eastern sides are each about 1,700 m long, and the side to the west measures about 1,870 m. The city is surrounded by walls, with the palace area located in the northeast of the city. There is a copper casting workshop site at both the north and the south side of the city, a pottery workshop site 1,300 m to the west of the city centre, a bone work site in the north of Zijinshan Mountain in the north of the city, and a bronzeware storage facility in both the western city wall and the southeast corner of the city. Outside the city, the northeast, west and south are burial areas. This represents the vision of a city with orderly planning.

Yinxu was the capital of the Shang dynasty after King Pangeng, and was about six kilometres long and four kilometres wide, with a total area of about 24 square kilometres. It was bordered by the Taihang Mountains to the west, the Yellow River to the east, and rolling hills and vast plains to the south and north. Near the capital city is an area of terrain with fertile soil. The climate is mild and the produce is abundant. The Huan River divides the site into two parts, the palace area in Henan and the cemetery area in Hebei. Around the palace there are handicraft workshops, residential areas and the tombs of the ordinary people. Around the palaces there is a large man-made ditch as a defensive facility. The division of the cemeteries reflects the ethnic burial system in the Shang dynasty, that is, each tomb belonged to a different "clan". The choice of this place as the capital reveals their grasp of some geographical knowledge.

The Panlong City site at Huangpi County, Wuhan, Hubei province, has an approximately square plan. It is about 290 m long from north to south and about 260 m wide from east to west. The city is surrounded by a rammed earth wall, which is encircled by a moat on the outside with a suspension bridge for access. There are palace ruins facing south, situated on the northeastern highlands of the city. Its layout suggests that there were only palaces in the city, that by nature it was a palace city, whilst residential and workshop areas lay outside the city: an early design for a city at that time. It might be the remains of a tribal state sitting on the banks of the Yangzi River during the Shang dynasty [49].

2.3.4 Early Communication Systems and the Understanding of the Earth

The Shang imperial government had close ties with the feudal states within its territory and the outlying tribal states, thanks to some vital communication arteries of the time. The round-trip route to conquer Ren (Yi) fang appearing on oracle bones may also show a road leading to the present-day Xu-Huai area. The places that this route follows are related to some sections of the Longhai Railway (Xuzhou to Zhengzhou) and the Jinpu Railway (Xuzhou to the northern Huai River) [50].

From evidence that involves the Marquis Zhu mentioned on the oracle bone, the Guzhu state appearing in the ancient text and the anecdotal relationship with the Sushen clan, it can be inferred that there was a traffic artery connecting the northeast of the Shang capital to present-day Lulong, reaching Chaoyang in Liaoning and beyond. Several Shang dynasty sites uncovered along or near this path, including Handan, Xingtai, Shijiazhuang, Gaocheng, Beijing and Kazuo, are likely to have been essential nodes that travellers had to pass through.

Heading towards the south from the Shang capital, there were obvious traffic arteries connected to the major clans based in what are present-day Hubei, Hunan and Jiangxi. Sites and relics found in these regions over the years fully support this view.

A traffic route was also built to the south from the Shang capital to connect to the Zhou capital and other outlying tribal states. It is this route that King Wu followed to go to battle against the Shang. At that time, soldiers and horses from different places joined forces in Mengjin, with chariots wheeled directly to Mu—an outlying district of the Shang. A route that vehicles could pass through was definitely not a small track. It probably led from Fenghao, the capital of Zhou, along the Wei River to the east, out of Shaanxi into Henan, and northeast to Qi County (i.e. Chaoge) and to Anyang.

There were also key roads from the capital of the Shang to Gupugu on the shore of the Bohai Sea (probably the present-day Yidu area of Shandong). The archaeological discovery that is the Shang dynasty Subutun ruins was probably a large capital at that time. It is assumed that the journey first necessitated travelling along a waterway, down the ancient Yellow River or the Ji River, and then transferring onto the land route. As for the means of transportation, the Shang dynasty already had chariots and boats, as evidenced by the Yinxu site and oracle bone writing. The oracle bone script has a pictogram for vehicle and, although inconsistently written in both complex and simple ways, it always foregrounds the characteristics of two wheels and one axis.

In the Yinxu excavation, more than 30 vehicles were found, some were for three people pulled by four horses, some carried three people with two horses, and others were for one person with two horses. Based on the remains found, the structure of a chariot was the same as what appears on the oracle, as both have a single shaft and two wheels. In addition to animal-powered chariots, there were also rickshaws; these are called *ji* and the pictogram can be found in the inscriptions of Shang dynasty bronze objects.

Regarding waterway vessels, canoes already existed in the Neolithic period. Boats in the Shang dynasty were made of multiple wooden planks. The word *zhou* on the oracle clearly does not resemble the shape of a canoe. Apart from the types of boats with oars, there were also boats pulled by ropes, as shown by the evidence on oracle bones [51]. Notably, according to Yu Shengwu's study, the Shang dynasty had established an early form of a courier system with multiple courier stations [52].

2.4 Developing Understanding of the Earth in the Western Zhou Dynasty

In the second half of the eleventh century BC, King Wu of Zhou conquered the Shang and established the Zhou dynasty, with Hao as its capital (present-day Chang'an district, Xi'an, Shaanxi). The period from King Wu to King Ping, who moved the capital to Luoyang (770 BC), is called the Western Zhou. Since King Mu, the Zhou had extended its territory to Wei, Tai, Rui, Qi, and Bi in the west, Pugu (around present-day Zibo, Shandong) and Shangyan (around present-day Qufu, Shandong) in the east, Ba, Pu, Chu and Deng in the south, and Sushen (present-day Liaodong), Yan and Bo in the north [53], becoming a powerful state without precedent based on the slavery system. The understanding of geoscience in the Western Zhou dynasty can be summarised according to the five following categories.

2.4.1 Earthquake Narratives

The earliest earthquake recorded in the Zhou dynasty is in the *Lü'shi chunqiu* ("Spring and Autumn of Master Lü") which says that "in June of the eighth year since the reign of King Wen, King Wen was ill and slept for five days, during which the earth moved east, west, south and north but did not have any impact outside the suburbs". This description suggests that an earthquake was taking place. The author who wrote this was from the end of the Warring States period, hence he could not have witnessed it in person, but rather he was recounting information handed down through time.

In the second year of King You of Zhou (780 BC), the three rivers within the country were shaken. A man named Bo Yangfu explained that "when the yang is so concealed underground that it cannot come out, and the yin is so suppressed that it cannot transpire, an earthquake will strike" [12]. This is the earliest theory of earthquakes in China, and Bo Yangfu was the first scholar to apply yin and yang to explain the cause of earthquakes.

2.4.2 Early References to Maps

When the Duke of Zhou assisted the government, he obeyed the will of King Wu and started to build the eastern capital, Luoyi. The Duke of Shao was appointed to survey and plan the potential area chosen by King Wu. After the completion of the new capital Luoyi, it became known as Chengzhou. This event was mentioned in the *Shangshu* ("Book of Documents") in that "the investigator came to the King with a map". This was the plan drawn for the building of the capital Luoyi [54].

A long inscription cast on the base of the "Yi Hou Ze" bronze *gui* records King Kang of Zhou (1020–996 BC) changing the territory granted to a duke named Ze from Yu to Yi. The inscription also mentions that King Kang reviewed the maps of Shang and the eastern states that were subdued by King Wu and King Chen in their battles [55].

The first volume of the *Mutianzi zhuan* ("Biography of King Mu") mentioned that King Mu of Zhou checked the map and documents when he travelled to the west approaching a mountain.

In the chapter "Praises to Zhou" of the *Shijing,* a sacrifice made by King Wu as an offering to the mountains is recorded. There the King consulted the map where the mountains at the confluence of the nine rivers were marked and chose to visit the more significant places first.

During the reign of King Li, records of maps were also found in the inscription on the San Family Plate. When the Shi state paid compensation to the San state in the form of farmland, the land was drawn in a map presented to the San family [56].

In short, the application of maps in the Western Zhou dynasty was not rare, indicating that people could make maps at that time.

2.4.3 Minerals and Stoneworking

Bronzeware reached its peak in the Western Zhou dynasty and people's understanding of copper, tin and lead deepened all the more as a consequence. So far, more copper mining sites dating from the Western Zhou have been found than from the Shang dynasty, including the copper mine site of Huangshi in Hubei; the Dajing site in Linxi County, Chifeng, Inner Mongolia; the Gangxia sites in Yangxin County, Hubei; and the Tongling, Nanling and Fangyang sites in Anhui. Copper ores included natural copper, malachite, copper sulfide, arsenic-bearing copperite, chalcopyrite, bornite and the symbiotic sulfide ore involving copper, tin and arsenic. During the Western Zhou, magnetite and hematite were used to grind heavy sword points. Siliceous rock and hornblende [57] were also used in grinding.

The lacquerware also reflects the use of mineral rocks at that time. There were three main colours used on an object's surface: brown, vermilion and black. Vermilion was made by adding cinnabar to the translucent paint; the black colour was adjusted by adding iron hydroxide to the lacquer and mixing and stirring these elements evenly.

The lacquerware of the Western Zhou was gilded with gold leaf, as shown in a piece with three gold leaf stickers on its body found in the Liuli River in Beijing. This suggests that the processing and application of gold leaf had become very advanced at that time. The technique of applying inlaid turquoise has also been found in Western Zhou lacquerware [58].

In the early Western Zhou dynasty, tin-plating technology was widely applied to chariots and horse tools, weapons, everyday utensils, etc. The specific method employed is called amalgam coating [59].

Pure lead tools already existed in the Western Zhou dynasty, which shows that pure lead could be extracted from the lead ore. Remnants of *ge* made of lead were unearthed in the Baifu tomb in Changping, Beijing [60]. Two examples of lead ware were also unearthed at the Beiyao Village site, Luoyang, one is *jue* and the other is *zhi* (both are wine vessels) [61].

The craft of jade carving was also well developed in this period. Jade artefacts can mainly be divided into two categories: used as decorative items and wearables, mostly animal-shaped figurines (e.g. fish, birds, silkworms, deer, tigers, beasts); used as ritual vessels (e.g. *bi*, rings, *yuan, huang, gui, zhang, chong, jue*) and models of weapons (e.g. *ge*, axes, *qi*, knives and others). The raw materials used were jade, stone, carbon essence, agate, turquoise, and glass, etc. [62].

There were also new developments in the ceramic craft industry. In addition to pottery for daily use, architectural pottery began to appear, such as plate tiles, tube tiles and *wadang* (eaves tiles). There were also primitive celadon pieces, such as the piece of porcelain unearthed from the Western Zhou tomb in Xiang County, Henan. Its base is made of greyish-white kaolin, with the cyan glaze applied outside, which is similar to the porcelain jar unearthed from the Western Zhou tomb at Pangjiagou, Luoyang [63].

In the late Western Zhou tomb at Yiwu, Zhejiang, 100 pieces of primitive porcelain were unearthed, including cups, bowls, plates, beans, bowls, lids, etc. They were glazed on both the inside and the outside. The wares were made of greyish-white kaolin clay. Most were covered in green glazes, but a small quantity had greenish-yellow glazes [64]. This proves that the Western Zhou people recognised the nature of kaolin and knew how to make porcelain ware.

The mineral pigments used in the Western Zhou dynasty included cinnabar, mineral yellow, hematite and so on [65]. Lime appeared in this period too, which was a leap forward in people's understanding of the nature of limestone at that time. In 1976, walls with lime were found at the ruins of one Western Zhou construction. Lime was obtained by heating limestone to produce quicklime, which was mixed with water to become hydrated lime and could finally be used to coat the walls.

According to the author's analysis of some archaeological data, there were 40 types of mineral rocks known or used during this period, such as gold, copper (malachite, natural copper, copper sulfide, copper symbiotic minerals with silver, tin, lead and zinc), cassiterite, lead, jade (nephrite, such as Hetian jade, Dushan jade, Lantian jade, Laiyang jade, emerald), carbon essence, cinnabar, sandstone, mica, *wuhuatu* (tesselated soil), kaolin, clay, mineral yellow, agate, hematite, magnetite, turquoise, talc, basalt, limestone, marble, diorite, diabase, granite, slate, andesite, porphyry, phyllite rocks, siliceous rocks, flint, quartzite, amber and salt.

2.4.4 Geoscientific Phenomena in Literature

(1) Geoscience in the *Zhouyi*

The ancient scripture of the *Zhouyi* ("Book of Changes") was compiled around the late Shang dynasty and involves historical accounts up to the reigns of King Wen and King Wu of Zhou. A supplementary section of "commentaries" is believed to have been written during the Spring and Autumn and Warring States periods. Overall, multiple authors contributed to the work over many years [66]. It extensively documents the social life of the Western Zhou people, the rise and fall of the Zhou nation, as well as the development of scientific knowledge. Below, we mainly focus on the part relating to geoscience.

The basic Eight Trigrams (*Ba-gua*) in the *Zhouyi* are: *qian*, the heavens; *kun,* the earth, *zhen*, thunder; *xun*, wind; *kan*, water; *li*, fire, *gen*, mountains, and *dui*, rivers. All eight elements are core to people's understanding of the earth. However, the text in the hexagram (*guagxiang*, i.e. the six-line diagram), the *yaoci* (the explanation of each line of the hexagram), and the *xiangshu* (the number or regularity included in a certain phenomenon) is not very consistent, sometimes it is explicit whilst at others it remains obscure [67]. For example, the symbol of *qian-gua* are the heavens, but its hexagram and *yaoci* do not relate to the heavens; the hexagram and *yaoci* of the *xun-gua* do not concern wind, nor do those of the *li-gua* discuss fire. *Gen* and *dui* are the same, and although *kun-gua* is related to the earth, it talks about more than that, which is similar to *kan-gua.* Only *zhen* is completely focused on thunder. To link *ba-gua* to the eight substances of nature is a sign of people's efforts in searching for the truth at that time, but to attribute many diverse phenomena to limited symbols seems unfounded, hence the inconsistences in the examples mentioned above.

Furthermore, there are aspects of the trigrams and *yaoci* which also reveal geoscientific knowledge. For example, core to the hexagram *kan* (the perilous pit) and its *yaoci* is that as people's lives shifted from fishing and hunting to agricultural production, the pits and traps originally set up for fishing or hunting had become a hindrance to the farmers and so they needed to be filled or removed. These were changes in the local landscapes which were due to the needs of humans.

The hexagram *jing* (a well) and its *yaoci* reflect drinking water hygiene among the Zhou people. "Water with mud in the well cannot be drunk." Then what method can be used to treat it? "(It) can be drawn." This means that the pollution could be removed by drawing the water up and purifying it. To maintain its cleanliness without weakening the well wall, tiles were used to reinforce the well, which is called a *zhou.* Archaeological excavations have confirmed that the Zhou people used flat tiles to strengthen wells in Shaanxi. After this remediation, the well water was clean, fresh, and became the best drinking water. To protect the water source, railings or well covers were added, not only for the safety of people and animals, but more importantly to protect the wells against mud, dust and other impurities that might fall into the water [68].

The hexagram *shihe* (biting through) mentioned those who "eat dry meat and find *huangjin* in the meat". Some believe that *huangjin* refers to gold grains [69], and others think they are copper arrow heads [70]. If the former meaning was intended, then this was the earliest proper noun for "gold" in Chinese history. In addition to the word "gold", there are more than 30 terms related to geoscience in the *Zhouyi*, among these the geographical terms include cropland, field, suburb, country, city, etc. Regarding geomorphology, there are the terms large river, high hill, cave, terrain, shore, chasm, mountain, valley, rock, land, spring, etc. In terms of meteorology and climate, there are dense clouds without rain, frost, ice, rain, humidity, rain loss, etc. The directions are east, south, west, northeast and southwest, and stone, mud and sand appear describing the quality of the earth.

(2) Geoscience in the *Shijing*

The *Shijing* ("Classic of Poetry") is an anthology of 305 poems that are classified as ballads (*feng*, "wind"), courtly songs (*ya*, "elegant") and eulogies (*song*). Dating from the beginning of the Western Zhou to the Spring and Autumn periods, the collection comprehensively covers aspects of social life, as well as people's understanding of nature and knowledge of geoscience.

① Hydrological knowledge

The *Shijing* documents more than 30 rivers, mainly the three major water systems of the Yellow River, the Yangzi River and the Huai River and their tributaries, with multiple names of sub-divisions called river, large river, ditch of stagnant water, bay, spring, stream, pond, swamp, abyss, river with islets, sea, and other water types. Relevant terms related to water phenomena such as waves and ripples, winter springs and fountains, swamps and damp low-lying lands appeared in the collection.

There are lines that describe the erosive effect of living water: "Amidst the gurgling waters, the white rocks stand up grandly"; the motion of the water: "Look at the water of that spring, sometimes clear, sometimes muddy"; and the sedimentation caused by water: "The Jing water turns muddy because of the Wei water, but the water remains crystal clear around the islets" [71]. These were the earliest descriptions of the impact of water on the geographical surroundings.

② Topographic knowledge

The *Shijing* mentions over 20 mountains, such as Mount Tai, Mount Song, Mount Shouyang, Mount Zhongnan and Mount Qishan. At least 60 geomorphological terms are used, such as mountain, hummock, hill, mound, plain, wetland, islet in a river, slope, valley, stone, big mound, land, dune, rolling hill, etc. Landforms near flowing water are written as shore, tributary, margin, water's edge, estuary, water bank, riverside, water front, mountain close to water, islet, dyke and pool.

According to the explanation in the *Erya*, a classic work of Chinese literature assembled in the Qin or early Han dynasty, mountains covered with vegetation are divided into two types: those with grass and trees are called 岵 (pinyin: hu, "luxuriant hill") and those without are called 峐 (pinyin: gai, variant of 屺, "bare hill"). Moreover, these definitions match the interpretations in the *Shuowen jiezi* ("An Explication of Written Characters"). The two characters "岵" and "屺" both appeared in the *Shijing*, with the verse that "I ascended to a luxuriant hilltop, thinking of my father from afar; I ascended to a barren hilltop, searching for memories of my mother".

In Mao Chang's commentary (often known as *Maoshi*, "Mao Poetry") on the *Shijing*, the following names of landforms are explained: *cuiwei*—an earth hill with stones; *ju*—a stone hill with earth; *fu*—the largest earth mountain; *ling*—the largest *fu*; *yan*—a mountain summit; *gang* –a relatively low and flat ridge; *wanqiu*—a basin; *dunqiu*—a single mound; *eqiu*—a relatively high mound; *jing*—a very high mound; *e*—a large loess hill; *qiu'e*—the slope of a mound; *yu*—the corner of a mound; *qiu'ce*—the side of mound.

In the section entitled "Xiaoya" of the *Shijing*, there is an account of some topographic changes caused by a large-scale landslide after a major thunderstorm in the sixth year of the King of Zhou's reign (776 BC): "Hundreds of rivers boil, mountains and cairns collapse, high banks sink into valleys, and deep valleys turn into hills." This is the first description of topographic changes in a short period of time in Chinese literature and it also reflects the naïve dialectical thinking of the Western Zhou people.

(3) Meteorological and climate knowledge

Twenty-five meteorological terms appear in the *Shijing*, such as wind, thunder, rain, haze, snow, thunderbolt, frost, ice, cloud, mist, drizzle, etc. There is a weather forecasting proverb suggesting that when a rainbow appears in the west, it will rain heavily in the morning. Another one foretells that when the crescent moon does not occult Aldebaran, the rainy season is coming and it will rain heavily.

Rain can be understood by its scale and type, such as cloudy and rainy, windy and rainy, scattered rain, rainfall, rain after a long drought, rain, timely rain, rain and snow.

The names of types of winds include mild wind, headwind, strong wind, valley wind, north wind, storm, gale, etc.

In terms of climate, there are names for the four seasons and many phenological records. In particular, the phenological poem "The Wind at Bin in July" describes the climate features for several months. For example, "in April, grass has grown green; in May, cicadas are singing loudly, and grasshoppers are rubbing their legs; in June, katydids are flapping their wings; in July, crickets are in the fields; in August, they huddle under the roof; in September, they enter my home; and in October, they slip under my bed.

2.4.5 Networks of Exchange and Geoscience

The east–west traffic artery of the Western Zhou extended from Zhouyuan to Luoyi via Fenghao, and from Luoyi eastwards to the Qi and the Lu states. Some sections of the old road were built during the Shang dynasty, starting from Luoyi to Chaoge via Mengjin, and further north to the Xing state (present-day Xingtai, Hebei). At the beginning of the Zhou dynasty, one could travel north to the Huo state (present-day Huo County, Shanxi) across the Fen River and the Shu River; from Luoyi, this road went south through Fangcheng to the Shen state (present-day Nanyang, Henan), and then south to the Han River.

From Luoyi, one could go east to the Song state (present-day Shangqiu, Henan), and then southeast to the Xu state, that is, the Huaiyi region.

This was the main road at that time, but there were other pathways. For example, the route taken by the king of Zhou to patrol and worship the mountains and seas was also an important traffic artery at that time, yet it cannot be compared with the one mentioned earlier, because that was the avenue through which the king of Zhou controlled the surrounding states.

In the "Xiaoya" section in *Shijing*, it is noted that "the road of Zhou is as smooth as gravel, as straight as an arrow. Such roads are only used by the ruling class, and ordinary people can only look at it from the sidelines". The text also mentions that "the roads of Zhou are very long", which illustrates the level of road building and management skills at that time. There were already various regulations in place, such as "no roadworks in the rain; resume when the rain stops". Trees should be planted along the roads, and dwelling places should be established at a distance so that roadworkers and pedestrians can be accommodated [72].

References

1. Bai S白寿彝, Xu X徐喜辰, Si W斯维至 (eds) (1994) *Zhongguo tongshi: disanjuan shangce* 中国通史: 第三卷上册 [A general history of China, vol 3, I]. Shanghai People's Publishing House 上海人民出版社, Shanghai, pp 200 & 207
2. *Beijing wanbao*北京晚报 [Beijing evening], 8 Jan 2000
3. Zheng J郑杰祥 (1988) *Xiashi chutan*夏史初探 [A preliminary study of Xia history]. Zhongzhou Ancient Books Press中州古籍出版社, Zhengzhou, pp 78–84
4. Institute of Archaeology, Chinese Academy of Social Sciences中国社会科学院考古研究所 (1984) *Xinzhongguo de kaogu faxian he yanjiu*新中国的考古发现和研究 [Archaeological discovery and research in New China]. Cultural Relics Press文物出版社, Beijing, p 84.
5. Liu D刘德岑 (1986) *Gudu pian*古都篇 [Texts on ancient capitals]. Xinan Normal University Press西南师范大学出版社, Chongqing, p 41
6. Zheng J郑杰祥 (1988) *Xiashi chutan* 夏史初探 [A preliminary study of Xia history]. Zhongzhou Ancient Books Press中州古籍出版社, Zhengzhou, p 215
7. *Wenwu*文物 [Relics] (1984) 10:88
8. *Zuozhuan: Henggong ernian*左传·桓公二年 [The chronicle of Zuo: the second year of Duke Huan]

9. *Tao Yuanming quanji juansi: Liuguan shanhai tu*陶渊明全集·卷四 "流观山海图" [The complete collection of Tao Yuanming's works, vol 4: views of the *Map of Mountains and Seas*]
10. *Mengzi*: *Tengwengong shang*孟子·滕文公上 [Mencius: Duke Wen of Teng I]
11. *Shiji: Xia benji*史记·夏本纪 [Records of the grand historian: history of the Xia]
12. *Guoyu: Zhouyu xia*国语·周语下 [Discourses of the states: Zhou II]
13. *Lunyu*: *Taibo*论语·泰伯 [The analects of Confucius: Taibo]
14. *Zhuangzi: Tianxia* 庄子·天下 [Zhuangzi: land under heaven]
15. Geoscience History Group, Institute of History of Natural Sciences, Chinese Academy of Sciences中国科学院自然科学史研究所地学史组 (1984) *Zhongguo gudai dili xueshi*中国古代地理学史 [A history of ancient Chinese geography]. Science Press科学出版社, Beijing, p 82
16. *Shijing: Shangsong: Changfa* 诗经·商颂·长发 [Book of songs: Ode to the Shang: eternal prosperity].
17. Zhang X章巽 (1986) *Yugong* pianzhong suojian de haishang jiaotong《禹贡》篇中所见的海上交通 [The sea traffic recorded in the *Tribute of Yu*]. In: *Zhangxun wenji*章巽文集 [Collection of Zhang Xun's articles]. Ocean Press海洋出版社, Beijing, pp 38–40
18. Meng S孟世凯 (1986) *Xiashang shihua*夏商史话 [History of the Xia and Shang dynasties]. China Youth Press中国青年出版社, Beijing, p 119
19. Wen S温少峰,, Yuan T袁庭栋 (1983) *Yinxu buci yanjiu: kexue jishu pian*殷墟卜辞研究: 科学技术篇 [A study of the inscriptions in ruins of the Yin era: on science and technology]. Sichuan Academy of Social Sciences Press四川省社会科学院出版社., Chengdu
20. Li Y李亚农 (1962) Yindai shehui shenghuo殷代社会生活 [Social life in the Yin era]. In: Xinranzhai shilun ji欣然斋史论集 [Collection of essays from the Xinran study]. Shanghai People's Publishing House上海人民出版社, Shanghai, pp 548–549
21. *Zhongguo wenwu bao* 中国文物报 [Chinese cultural relics news]. 17 Apr 1994
22. Li X李孝定 (1970) *Jiagu wenzi jishi: disanjuan*甲骨文字集释: 第三卷 [A collection of inscriptions on oracle bones, vol 3]. Institute of History and Language, Academia Sinica (台湾) 中央研究院历史语言研究所, Taibei
23. Wang K王克陵 (1992) Zhongguo xianqinshiqi de dixing celiang gongju——'guiyi'中国先秦时期的地形测量工具——"规仪" [Topographic survey tools from the pre-Qin period in China—The Guiyi]. *Ziran kexueshi yanjiu*自然科学史研究 [Studies on the history of natural science] (3)
24. Wu H吴浩坤,, Pan Y潘悠 (1985) *Zhonguo jiagu xueshi*中国甲骨学史 [A history of the oracle bone studies in China]. Shanghai People's Publishing House上海人民出版社, Shanghai, p 105
25. Wen S温少峰,, Yuan T袁庭栋 (1983) *Yinxu buci yanjiu: kexue jishu pian*殷墟卜辞研究: 科学技术篇 [A study of the inscriptions in ruins of the Yin era: on science and technology]. Sichuan Academy of Social Sciences Press四川省社会科学院出版社, Chengdu, p 156
26. Wang Y王宇信 (1982) *Jianguo yilai jiaguwen yanjiu*建国以来甲骨文研究 [Studies of inscriptions on oracle bones since the founding of New China]. China Social Science Press中国社会科学出版社, Beijing, p 157
27. Wen S温少峰, Yuan T袁庭栋 (1983) *Yinxu buci yanjiu: kexue jishu pian*殷墟卜辞研究 科学技术篇 [A study of the inscriptions in ruins of the Yin era: On science and technology]. Sichuan Academy of Social Sciences Press四川省社会科学院出版社, Chengdu, p 354
28. Wang Y王宇信 (1982) *Jianguo yilai jiaguwen yanjiu*建国以来甲骨文研究 [Studies of inscriptions on oracle bones since the founding of New China]. China Social Science Press中国社会科学出版社, Beijing, p 141
29. Wang Y王宇信 (1982) *Jianguo yilai jiaguwen yanjiu*建国以来甲骨文研究 [Studies of inscriptions on oracle bones since the founding of New China]. China Social Science Press中国社会科学出版社, Beijing, p 138
30. Wang, Y王宇信 (1982) *Jianguo yilai jiaguwen yanjiu*建国以来甲骨文研究 [Studies of inscriptions on oracle bones since the founding of New China]. China Social Science Press中国社会科学出版社, Beijing, pp 136–137

31. Wu H吴浩坤, Pan Y潘悠 (1985) *Zhonguo jiagu xueshi*中国甲骨学史 [A history of the oracle bone studies in China]. Shanghai People's Publishing House上海人民出版社, Shanghai, p 265
32. Chen M陈梦家 (1956) *Yinxu buci zongshu*殷墟卜辞综述 [An overview of the inscriptions of the Yin Ruins]. Science Press科学出版社, Beijing, p 249
33. Wu H吴浩坤, Pan Y潘悠 (1985) *Zhonguo jiagu xueshi*中国甲骨学史 [A history of the oracle bone studies in China].Shanghai People's Publishing House上海人民出版社, Shanghai, p 267
34. Wen S温少峰, Yuan T袁庭栋 (1983) *Yinxu buci yanjiu: kexue jishu pian*殷墟卜辞研究: 科学技术篇 [A study of the inscriptions in ruins of the Yin era: on science and technology]. Sichuan Academy of Social Sciences Press四川省社会科学院出版社, Chengdu, pp 200–206
35. Peng, B彭邦炯 (1988) S*hangshi weitan*商史探微 [A preliminary exploration of the history of the Shang dynasty]. Chongqing Press重庆出版社, Chongqing, p 240
36. *Zhongguo wenwu bao*中国文物报 [Chinese cultural relics news]. 27 Jan 1989
37. Peng B彭邦炯 (1988) S*hangshi weitan*商史探微 [A preliminary exploration of the history of the Shang dynasty]. Chongqing Press重庆出版社, Chongqing, p 242
38. *Wenwu*文物 [Relics] (1985) 10:36
39. *Kaogu*考古 [Archaeology] (1961) 2:63
40. *Kaogu*考古 [Archaeology] (1961) 3:14
41. *Wenwu*文物 [Relics] (1977) 11:2–4
42. *Kaogu*考古 [Archaeology] (1958) 8:23
43. Guo B郭宝钧 (1963) *Zhongguo de qingtong shidai* 中国的青铜时代 [The Bronze age of China]. SDX Joint Publishing Company北京三联书店, Beijing, pp 48–49
44. Peng B彭邦炯 (1988) S*hangshi weitan*商史探微 [A preliminary exploration of the history of the Shang dynasty]. Chongqing Press重庆出版社, Chongqing, p 254
45. Chen D陈德安, Wei X魏学峰, Li W李伟纲 (1988) *Sanxingdui - changjiang zhongshangyou wenming zhongxin tansuo*三星堆——长江上游文明中心探索 [Exploration of the Sanxingdui: a centre of civilisation in the upper reaches of the Yangzi River]. Sichuan People's Press四川人民出版社, Chengdu
46. *Wenwu*文物 [Relics] (1987) 4:53
47. Zhang Z张之恒, Zhou Y周裕兴 (1995) *Xiashangzhou kaogu*夏商周考古 [Archaeology of the Xia, Shang and Zhou dynasties]. Nanjing University Press南京大学出版社, Nanjing, p 52
48. *Beijing wanbao* 北京晚报 [Beijing evening], 24 May 1999
49. Zhang Z张之恒, Zhou Y周裕兴 (1995) *Xiashangzhou kaogu*夏商周考古 [Archaeology of the Xia, Shang and Zhou dynasties]. Nanjing University Press南京大学出版社, Nanjing, pp 48–65
50. Chen, M陈梦家 (1956) *Yinxu buci zongshu*殷墟卜辞综述 [An overview of the inscriptions of the Yin Ruins]. Science Press科学出版社, Beijing, pp 301–310
51. Peng B彭邦炯 (1988) S*hangshi weitan*商史探微 [A preliminary exploration of the history of the Shang dynasty]. Chongqing Press重庆出版社, Chongqing, pp 268–273
52. Yu X于省吾 (1995) Yindai de jiaotong gongju he yichuan zhidu殷代的交通工具和驿传制度 [The systems of transportation and correspondence in the Shang dynasty]. *Dongbei renmin daxue xuebao*东北人民大学学报 [J Northeast Renmin Univ] (2)
53. *Zuozhuan: Zhaogong jiunian*左传·昭公九年 [The chronicle of Zuo: The ninth year of Duke Zhao]
54. Hou R侯仁之 (1962) *Zhongguo gudai dilixue jianzhi*中国古代地理学简史 [A brief history of ancient Chinese geography]. Science Press科学出版社, Beijing, p 3
55. Tang L唐兰 (1977) Xizhou shidai zuizao de yijian tongqi lihe mingwen jieshi西周时代最早的一件铜器利簋铭文解释 [An explanation of the inscriptions on the first bronzeware named Ligui]. *Wenwu*文物 [Relics] (8):10
56. Guo M郭沫若 (1961) *Shi pipanshu*十批判书 [Ten critiques]. People's Literary Press人民文学出版社, Beijing, p 49
57. *Kaogu* 考古 [Archaeology] (1983) 2:133
58. *Kaogu* 考古 [Archaeology] (1983) 8:734
59. *Wenwu* 文物 [Relics] (1988) 3:77

60. *Kaogu* 考古 [Archaeology] (1976) 4:246
61. *Wenwu* 文物 [Relics] (1981) 7:52
62. Zhang Z张之恒, Zhou Y周裕兴 (1995) *Xiashangzhou kaogu*夏商周考古 [Archaeology of the Xia, Shang and Zhou dynasties]. Nanjing University Press南京大学出版社, Nanjing, p 266
63. *Wenwu* 文物 [Relics] (1977) 8:13
64. *Kaogu* 考古 [Archaeology] (1985) 7:608
65. *Wenwu* 文物 [Relics] (1976) 4:34
66. Huang S黄寿祺, Zhang S张善文 (1989) *Zhouyi yizhu*周易译注 [The translation and notes to the *Book of Changes*]. Shanghai Ancient Books Press上海古籍出版社, Shanghai, p 9
67. Gao H高亨 (1984) *Zhouyi gujing jinzhu*周易古经今注 [Modern notes to the ancient text *Book of Changes*]. Zhonghua Book Company中华书局, Beijing, p 12
68. Li J李镜池 (1982) *Zhouyi tanyuan*周易探源 [A probe into the origin of the *Book of Changes*]. Zhonghua Book Company中华书局, Beijing, pp 214–218
69. Gao H高亨 (1984) *Zhouyi gujing jinzhu*周易古经今注 [Modern notes to the ancient text *Book of Changes*]. Zhonghua Book Company中华书局, Beijing, p 223
70. Li J李镜池 (1982) *Zhouyi tanyuan*周易探源 [A probe into the origin of the *Book of Changes*]. Zhonghua Book Company中华书局, Beijing, p 45
71. All quoted from the *Book of Songs: wind*: *Fatan/Kaifeng/Fenjuyun/Yangzhishui/Gufeng*. Gathering Shu/April, Xiaoya
72. Bai S白寿彝, Xu X徐喜辰, Si W斯维至 (eds) (1994) *Zhongguo tongshi: disanjuan shangce* 中国通史: 第三卷上册 [A general history of China, vol 3, I]. Shanghai People's Publishing House上海人民出版社, Shanghai, pp 709–715

Chapter 3
Geoscience in the Spring and Autumn and the Warring States Periods

3.1 The Impact of the Social Environment on Geoscience

The periods from 770 BC, when King Ping of Zhou moved eastwards to Luoyang, to 221 BC, when Qin Shi Huang unified the whole country, are called the Spring and Autumn period (770–476 BC, 295 years in total) and the Warring States period (475-221 BC, 254 years in total), altogether 549 years. During these periods, Chinese society was undergoing a transition from a slave society to a feudal society, which was a time of great social changes. It was also a period when the instruments of production shifted from bronze to iron, enabling productivity to grow significantly. The hydraulic engineering, mining, metallurgy and foundry sectors became the backbone of social production. Various ideas, thoughts, and the collision between different minds were embraced and welcomed by society. It was against such a social background that geoscience developed rapidly and many ancient Chinese geographical classics were produced, forming the start of geographic studies. Therefore, it can be said that geoscience in ancient China reached its first peak during these periods, and thus laid a solid foundation for the traditional geoscience system.

3.2 Literature Review

The Spring and Autumn and Warring States periods were the "axis" of Chinese civilisation, a time when many classic works were written. Subsequent generations of scholars needed to build on these prototypical works to develop or justify their own claims. In fact, substantial theories in the history of Chinese philosophy were published in the form of commentaries on those original texts.

Many of the classics of that time were rich in geoscientific content, and some of them were geoscientific works. One could even argue that the traditions of the discipline of geoscience were preliminarily shaped at that time.

W. Yang and Q. Yang, *A History of Ancient Chinese Geoscience*, Understanding China,
https://doi.org/10.1007/978-981-97-4457-2_3

It has been proved that the Chinese term *dili* (geography) was invented in the classics of these periods. In the *Zhouyi: Xici*, there is the statement: "looking up, to observe the *tianwen* (heavenly phenomena) and looking down, to investigate the *dili* (earth's law)". Here *dili* correlates with *tianwen*, and the explanation of *dili* can be found in the later work *Huainanzi*: "reviewing the geographical conditions to formulate measurement systems, to identify waters in the regions of hills and plains and to examine the fertility and level of the ground, so that production can be arranged, wealth created and famine removed." Wang Chong, one of the most original thinkers and literary critics in the Eastern Han dynasty, proposed in his work *Lunheng* that "the universe has the sun, moon and stars, hence called *tianwen* (astronomy); the earth has the mountains, rivers, hills and valleys, hence called *dili*". The Tang dynasty philosopher Kong Yingda, annotator of the *Zhouyi*, further explained that "The universe has various phenomena that can be traced, and the earth has diverse topographies that are distributed in an orderly manner". Based on these accounts, Li Jiuchang believes that the concept of *dili*, during the pre-Qin periods, not only referred to the observation of geographical and hydrologic conditions, but also included the investigation of the patterns of natural phenomena, i.e. the laws of nature, in order to adapt to different geographical conditions and sustain life [1]. Just as argued by Guan Zhong (c. 645 BC): "a wise ruler will not go against the heavens or waste the land, hence heaven rewards him with favourable times and the land generates wealth for him; a fatuous one does the contrary and deserves nothing." In short, the word *dili* referred to the geographical environment on which human beings depended, and this concept was built around the lives of human beings and the functioning of human society [2].

In the same period, the concept of geography also appeared in ancient Greece. Hecataeus (c. 550-476 BC) wrote the *Description of the Earth*, in which the expression "new geography" was used in a subheading on one of its surviving pages [3]. This is the earliest record of the term "geography" in Western literature. The expression "new geography" indicates that the word "geography" had already existed before, but unfortunately only fragments of this book remain, so it is difficult to comprehend its exact meaning. The ancient Greek geographer Eratosthenes (275-195 BC) used "Geography" as the title of his work. The word "geography" is composed of "geo" ('earth' in Greek) and "graphy" ('to describe' in Greek), meaning "to describe the earth" [4]. This shows that there is a great difference between the word "geography" of ancient Greece and "*dili*" of ancient China, which also reflects the disparity between these two civilisations in terms of the technology, traditions and way of thinking [5].

The term *dizhi* (geology) emerged relatively late in China [6]. According to Li E'rong, it was first used in the *Zhouyi zhu*, written by Wang Bi (226–249 AD), during the Empire of Cao-Wei, yet *dizhi* in its modern sense only appeared in Chinese literature in 1853, although popular opinion holds that the term was first introduced by the writer Lu Xun from the article "A Brief Discussion of Chinese Geology" written in the Japanese language in 1903. Many classics of the Spring and Autumn and Warring States periods had already documented geological knowledge, including the *Shanhai jing* ("Classic of Mountains and Seas"), *Yugong* ("Tribute of Yu") and *Guanzi* ("Book of Master Guan").

The term *giologia* was first used in the early seventeenth century in the will of Ulysses Aldrovandus (1522–1605), an Italian natural philosopher and assiduous collector of plants, animals and fossils. The will set out what should happen to his library and museum which exhibited knowledge of palaeontology, minerals and rocks. Of course, there was already a wealth of geological knowledge in the West as early as the ancient Greek period, and the most representative work is *On Rocks* by Theophrastus (371-285 BC).

A number of geoscientific works were written during the Spring and Autumn and Warring States periods, including both comprehensive and specialised types. Those that have been passed down to this day include: *Shanhai jing*, *Yugong*, *Mutianzi zhuan* ("Biography of King Mu"), as well as some parts of *Zhouli* ("Rites of Zhou") and *Guanzi*, several chapters of *Lüshi chunqiu* ("Spring and Autumn of Master Lü"), *Sunzi bingfa* ("Art of War by Sunzi"), and *Sunbin bingfa* ("Art of War by Sunbin"). Below is an overview of each work.

3.2.1 Comprehensive Studies on Shanhai jing *and* Yugong

A discussion of traditional Chinese geography generally begins with the review of these two books. They are believed to have been written in the Eastern Zhou dynasty. Two reasons may explain why they were produced at that time. First, during the Spring and Autumn and the Warring States periods, the interaction between China and the outside world had expanded. The extent of such interaction may have gone beyond the barriers of mountains and deserts, as Fang Hao commented: "After China entered the era of recorded history, that is, between the ninth and sixth centuries BC, no trace of Sino-foreign interaction has been found ... According to the artefacts that have been discovered, since the sixth century BC, that is, the Spring and Autumn and the Warring States periods, the relationship between China and the West can be traced, and the routes of such interactions can be divided into two: the North Route, where the artefacts were conveyed directly from West Asia to the tribes in the north, and then to China; the South Route, where the artifacts were transported from West Asia to Central Asia and India, and then to China [7]." The large amount of geographic information brought by foreign exchanges required people to organise and review it. Second, during this period, the world was divided and wars were frequent; in order to mitigate the scourge of war and fight against barbarians, unification became the common desire of the people, and the idea of *dayitong* (great unity) was formed and became the guiding ideology for compiling geographical information. As Zhou Zhenhe pointed out: "Studies since the May Fourth Movement have proved that *Yugong* was a work of the Warring States period, and that the concept of *jiuzhou* (nine regions of ancient China) was a product of unification awareness and the geographical knowledge of the nations of that time [8]." Ancient Chinese geography had become established by then.

The *Shanhai jing* and *Yugong* were the earliest works to provide a systematic description of the people's perception of the world at that time, covering nature, the economy, society and geography, and were the most representative geoscientific works of that time, both in terms of the geographical area covered and the issues discussed. The first priority when describing the geography of a large area is to organise the text in a proper way. To achieve this, the large area should first be divided into regions, and then be described accordingly. Such a method of describing certain regions was already used in the comprehensive geoscientific works of the Spring and Autumn and Warring States periods. In the development of geoscience, regional description was one of the oldest methods of geoscientific research.

(1) *Shanhai jing*

The *Shanhai jing* pioneered the method of regional description in ancient Chinese geoscience. It describes a wide area, including not only the vast area of later China, but also parts of Central Asia and East Asia. The age of the book has been much debated, but it is widely believed that the book had more than one author, and was not written at a certain period time, but was composed from the late Spring and Autumn period to the early Warring States period, and was later revised in the Qin and Han dynasties [9].

A total of 18 volumes of the book's transcript have survived and circulated until the present day, consisting of 31,000 characters in three parts: the "Mountains" (the five-volume Wuzang shanjing, or Classic of Mountains in Five Directions), the "Seas" (eight volumes) and the "Great Wilds" (five volumes). As many of the records in the book are believed to be groundless and even absurd, the book has been considered "strange" and "weird" since ancient times, and people's opinions regarding the book have been diverse throughout the ages, and it is difficult for historians of later dynasties to determine which category of books or bibliographies it belongs to. Until the modern day, it has been deemed by many scholars to be a work of mythology, or even a book written by witch doctors. For example, Lu Xun in *A Brief History of Chinese Fiction* described the *Shanhai jing* as "the ancient book of witchcraft". However, an increasing number of scholars have noticed the geographical and geological value of the book. Hou Renzhi emphasises that the "Mountains" is "the first work on geography that has been handed down to this day in China" and is "a summary of all the geographic knowledge that our ancestors acquired in their struggle to live since ancient times" [10].

Yet the *Shanhai jing* held a unique place in the academic life of the traditional Chinese literati. In a poem *Understanding Shanhai jing* by Tao Yuanming, there is a verse "to browse the *Biography of the King of Zhou* and to appreciate the *Map of Shanhai*". The "King of Zhou" refers to King Mu, Son of Heaven, and the "Map" is the maps and images in the *Shanhai jing*. Mr Ge Guangzhao cited the verse and commented: "Many of the ancient scholars had a similar reading experience and it was from these mind-blowing books that they gained surreal and non-canonical knowledge … Often Tao's poems provided them with the legitimacy of gaining such knowledge [11]." As such, the main significance of this type of book is that it created a place for geoscience to develop within the general framework of traditional Chinese

Confucianism, and it continued to stimulate the interest of literati and scholars in geography and exotic places [12].

Abundant evidence indicates that there were maps in the book, but the origin of these diagrams, their original appearance, and their relationship with the content of the book remain inconclusive. According to Zhu Xi, the celebrated philosopher in the Southern Song dynasty, the map existed before the written content, and the original book was not a picture collection of mysterious objects and beasts as it is in modern times, but a landscape map of mountains and rivers marked with legendary objects and beasts [13]. According to today's studies, instead of illustrating the mountains and rivers, the *Shanhai jing* shifted to the depiction of strange objects and beasts in the Song dynasty [14], and the original book gradually fell into oblivion.

The extant version of the *Shanhai jing* is centred around geographical records and is of great geoscientific value in terms of its content: it covers the geography, history, ethnicity, religion, flora and fauna, water conditions, mythology, witchcraft, etc. from ancient times to the Zhou dynasty. And of all the sections, the "Mountains" in particular contains the most natural geographic content and has the greatest geological value [15]. The "Seas" and the "Great Wilds" also cover a lot of geographical knowledge.

The "Mountains" divides the country into five regions, and each region is described by the coordinates of mountains (Fig. 3.1). It considers the southwestern part of Shanxi and the western part of Henan as the "Central Mountains" and divides the country into four parts in the east, west, south, and north, which are described respectively in the "Eastern Mountains", "Western Mountains", "Southern Mountains" and "Northern Mountains". Tan Qichang analysed the 140 mountains recorded in the seven articles of the "Mountains" that could be located with certainty, and based on this, he examined the accuracy of the direction and distance of each mountain recorded in the *Mountains*, and then examined the area that those recorded mountains cover. He expected that the total range would be much larger than that recorded in *Yugong*, yet it turned out to be much smaller than the map of China today [16].

In the description of each region, the "Mountains" still adopts the mountain system as the outline, and divides the region into several subparts of the mountains. The descriptions of mountains in each subpart are clearly organised and follow a certain pattern, including the mountains, rivers, plants, animals, water systems, minerals and other geological aspects. The accounts are quite extensive, and although they are still primitive, rough and contain a lot of inaccurate speculations, nevertheless they are the oldest comprehensive geological work covering the entire nation. Moreover, the geological knowledge introduced in the book is inclusive. For instance, for mountains, the location, direction, distance, height and slope are included; for rivers, the origin, flow direction and the seasonal changes of water flow are included; for plants and animals, their characteristics and medical effects are described; for minerals, their hardness, colour, lustre and identification methods are described.

The "Mountains" documents the names of more than 5,370 mountains and 300 rivers, 130 species of plants and 260 species of animals, as well as 70 to 80 species of minerals [17]. As the "Mountains" covers a wide variety of records, opinion differs

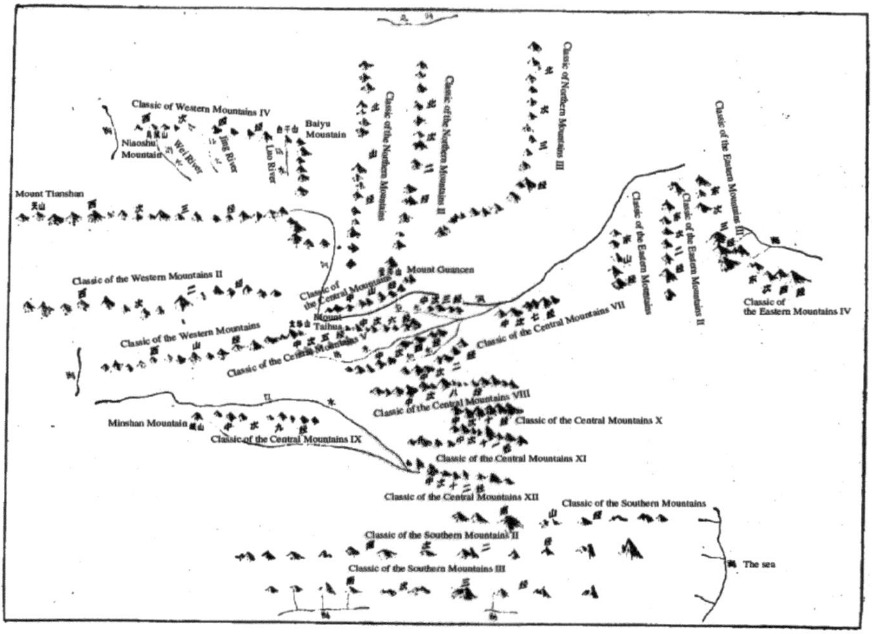

Fig. 3.1 The sketch map of the "Mountains". Wang [32]

regarding the reliability of some segments and how to categorise them, thus the statistics remain inconsistent. The "Mountains" concludes:

> Yu the Great said that he had travelled to 5,370 of the world's most famous mountains, with a length of 64,056 li, and that these mountains were located in the east, west, north, south, and centre of the earth. There are countless other small mountains, so it is impossible to list them all. They stretch 28,000 li from the east to the west, and 26,000 li from the south to the north. There are mountains where rivers flow out and where rivers flow through, both with a length of 8,000 li, among which there are 467 copper-producing mountains and 3,690 iron-producing mountains. [18]

Although the data here are far from accurate, they still reflect the overall style of the book.

The author(s) of the *Shanhai jing* tried to depict the whole world they knew at that time, and such efforts deserve recognition. At that time, civilisation was rather primitive, and there were very few written materials to draw on, so there were only two possible ways to gather facts: from the oral information passed down from generation to generation; and from the information obtained through fieldwork. Judging from the limited means of transportation and geographical surveying at that time, the reliability of the accounts is dubious. As the book also features bizarre landscapes, exotic animals and unfamiliar clans, it is conceivable that the authors often incorporated absurdity and novelty to attract readers. On this point, some scholars comment that "the interest is not in the stories themselves, but in checking how the natural phenomenon is interpreted and comprehended by people" [19].

In addition, the style of description applied in the book was successful and had influence on many later geological works, including the *Yugong*. As for the "Seas" and the "Great Wilds", since the accounts are more distant from the areas where the authors lived, more myths and speculations appear in these chapters, so they are of less geological value than the "Mountains". But they are still valuable in terms of the style and thoughts, especially the descriptions of various ethics, which could be described as the incipient form of cultural geography.

(2) *Yugong*

The "Yugong" is a chapter of the *Book of History*, which is the oldest historical book in China, and is also the earliest and most systematic comprehensive geographic work, known as "the most scientific geographic account in ancient China" [20].

It is said that the *Yugong* was written during the Xia dynasty and regarded by scholars as the classic that "remain unchanged for all ages" [21] and as "the forefather of ancient and modern geography" [22], after it was included in the *Book of History* by Confucius. It has had an unprecedented influence on later dynasties, to the extent that nearly all local chronicles were edited based on it and the names of today's regions, such as Taiyuan in Shanxi province and Jiujiang in Sichuan province were derived from this chapter [23]. In the past, the *Yugong* was a must-read for scholars, and was traditionally regarded as a credible history book. It divided ancient China into nine regions, or *jiuzhou*. However, later studies have proved that it was written during the Warring States period, and the concept of *jiuzhou* was the product of people's awareness of unity and geographical knowledge [24].

The *Yugong* abandoned the narrative style of the *Shanhai jing*, which features enumeration of geographical existence, and provided a systematic account of the available materials in a very concise text. The text has only 1,191 characters in total, but it offers a realistic and comprehensive geographic description of the whole country, without any mythological elements, set against a very primitive social background, which is rare and fascinating, showing that people at that time had already acquired more geoscienctific knowledge than before. The *Yugong* follows the writing style of the "Mountains", which introduces geographical conditions by regions, but it does not divide the region by the mountain systems, as the "Mountains" does.

In the *Yugong*, geographical divisions were made according to two systems: the Jiuzhou System and the Wufu System. The Wufu System was an idealisation of the feudal system: it took the capital as the centre, and divided its peripheral places into the Tianfu, Houfu, Suifu, Yaofu and Huangfu, based on their distance to the capital, with 500 li in between. The Wufu System was more like a political ideal rather than a geographical division [25].

The main part of the *Yugong* outlines the Jiuzhou System, which divided the then known world, or *tianxia*, into nine regions, or *jiuzhou* [26], according to natural elements of geographical division, including major mountains, climate, soil and vegetation conditions (Fig. 3.2). The *Yugong* adopted a consistent style to systematically present the mountains, lakes, soil, tribute and products of each state, and thus has a higher geoscientific value [27].

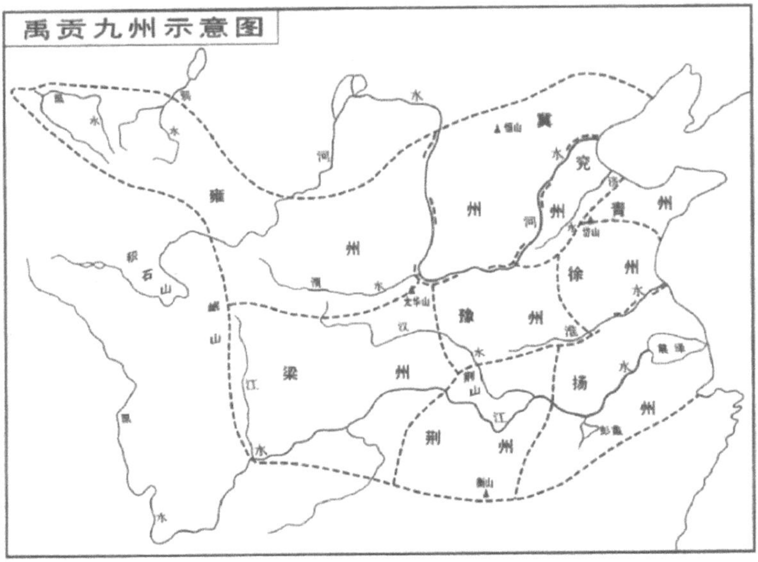

Fig. 3.2 The sketch map of the Nine Regions (*Jiuzhou*): Jizhou冀州, Yanzhou兖州, Zhou Q青州, Xuzhou徐州, Yangzhou扬州, Jingzhou荆州, Liangzhou梁州, Yongzhou雍州, Yuzhou豫州. Li C李长傅 (2007) *Lichangbo wenji*李长傅文集 [Lichangbo collected works]. Kaifeng: Henan University Press河南大学出版社, p 269

In addition to the geographical division, the *Yugong* also contains the two sections "Daoshui (the guide to waters)" and "Daoshan (the guide to mountains)". Whilst the former is concerned with the hydrographical conditions, the latter addresses the distribution of mountains in the nine regions.

The "Daoshui" documents the water systems in the *jiuzhou* and their distribution patterns, which was the first attempt by the ancient Chinese people to understand their national water systems. It describes the rivers in both northern and southern China, and provides an intuitive description of the distribution of the water system by including not only the origins of the rivers and their inlets, but also the main course and the tributaries of the rivers.

Although the "Daoshan" only notes the names of the mountains, it already delineated the pattern of mountain distributions and discovered that the general topography features a high west and a low east. The concept of mountain ranges is not defined, but their directions are implied in the text: the first range is north of the Yellow River, from present-day Shaanxi through Shanxi to Henan, with a total of 12 mountains; the second range is on the south bank of the Yellow River, starting from Shaanxi, through Henan to Shandong, with a total of eight mountains; the third range is in the Han River basin, from Shaanxi to Hubei, Anhui and other provinces, with a total of four mountains; the fourth one is on the north bank of the Yangzi River, starting from Sichuan, through Hubei to Jiangxi, for a total of three mountains.

In the comparison between the *Shanhai jing* and *Yugong*, early scholars tended to value the latter more than the former. More recent researchers, however, highlight the respective contributions of the *Shanhai jing to* natural geography and of the *Yugong* to human geography [28]. Moreover, the focus has been on the "Mountains" in the studies of the *Shanhai jing,* leaving the "Seas" and the "Great Wilds" largely underexplored, even though a growing number of scholars have noticed the geographic value of these two sections. Although the elements of human geography in the *Shanhai jing* are not as systematic and clear as those in the *Yugong*, they still offer interesting aspects of human geography to be explored beyond the mythologies and fantasies. To summarise, they are both seminal and comprehensive classics in the development of geoscience in China (Table 3.1).

In the same period, Western descriptive geography was also in its prime. The most representative works were *The Histories* by Herodotus and *The Geographica* by Strabo. Herodotus attached great importance to the characteristics of ethnicity

Table 3.1 Geoscience reflected in the *Shanhai jing* and *Yugong*

		Shanhai jing	*Yugong*
Physical geography	Mountain ranges	Characteristics, locations, heights and shapes of 26 mountain ranges in five regions	Generally describing the mountain ranges as "three rows and four columns"
	Rivers	The mountain ranges where the rivers originate, their direction of flow, estuaries, and water volumes resulting from seasonal changes	Main river network and waterways, covering water sources, flow directions, areas traversed, tributaries and estuaries; the phenomenon of "reverse flow" (i.e. the tidal reach); approximately 30 rivers, including the "Si Du (Four Estuaries)", namely, the Jiang River, Huai River, He River and Ji River
	Natural resources	Flora, fauna and mineral resources	Soil classification, characteristics of soils and vegetation
Economic Geography	Production features	Customs of distant ethnic groups such as hunting and fishing	Types of production including agriculture, handicrafts and mining
	Regional specialities	Special local products in the "Shanjing"	Special local products, land taxes, and other related information
Cultural Geography	Administrative divisions	Five administrative districts	The *jiuzhou* and the *wufu* systems
	Ethnic groups	Over 150 ancient states and tribes, and their racial characteristics, customs and religions	Attire, production, lifestyle, customs of certain ethnic groups

and the general features of natural landscape. Yet as his works were voluminous and largely unstructured, he had limited influence on later studies. In contrast, the descriptive approach in Strabo's *The Geographica* is clear and practical and, therefore, itbecame a classic reference for future Western scholars. However, among the descriptive geographic works of ancient Greece, no all-encompassing geoscience work comparable to the *Shanhai jing and Yugong* was produced.

3.2.2 Thematic Studies on Geoscience: Mutianzi zhuan, Guanzi *and Others*

The *Shanhai jing* and *Yugong*, as examples of encyclopaedic writing on geoscience, present a broad view of the known world composed of different regions. There are other authors, however, who focus only on certain regions or even on a certain geoscience problem, producing thematic works on geoscience. During the Spring and Autumn and Warring States periods, examples of this kind of study gradually emerged as well, reflecting more advanced explorations into specialised areas of geoscience. Whilst they may seem quite preliminary, the thematic studies cover a wide range of topics and established a basic framework for the later development of Chinese geoscience.

(1) *Mutianzi zhuan*: the origin of the Chinese travelogue genre

As an innate desire within man, expeditions allow the brave to conquer the world and achieve self-fulfilment, and have been a popular topic in numerous ancient and modern works of literature. In the West, one of the most marvellous expeditions is represented in the *Odyssey*, a journey that has fuelled the inspiration of Western literati, philosophers, anthropologists and historians throughout many generations. Although there is no such comparable example in China, accounts of travels and voyages abound, such as the legendary Kua Fu who pursued the sun and the regional inspections or hunting trips undertaken by emperors of every dynasty.

Among them, the most notable is the westward journey of Mu Tianzi ("King Mu of Zhou"), as it was the first geoscience work in the travelogue genre. Also known as the *Zhouwang zhuan* ("Biographies of King of Zhou"), the book was one of the pre-Qin ancient books discovered in the royal tombs of the Warring States during the second year of the Taikang reign (281 AD) in the Western Jin dynasty. It is now commonly accepted that the book was written in the late Warring States period.

As regards when the *Mutianzi zhuan* was completed, modern scholars mostly accept Gu Jiegang's view that it was written in the late Warring States period [29]. As for the nature and cataloguing of this book, just as with the *Shanhai jing*, there have been different views: the *Book of the Sui: Jingji zhi*, *New Book of the Tang: Yiwen zhi,* and *Comprehensive Treatise: Yiwen lue* put it under "history", while in the *Yuhai: Yiwen zhi*, it was seen as a biography; *Songshi: Yiwen zhi* ("History of the Song") considered it a specialised history, yet in the *Siku quanshu zongmu* ("Annotated Catalogue of the Complete Library of the Four Treasuries") it appeared as a fictional work. In short, it tended to be increasingly marginalised with the changing times. The evaluation of this book by modern scholars also varies greatly, but in the history of geography, it is mostly studied and discussed as a scholarly work. Wang Yong regarded its genre as a travelogue with absurd facts documented [30], while Chen Zhengxiang believes that "it is a realistic travelogue but written in a surprisingly mysterious and abbreviated manner" [31]. Wang Chengzu treats it as a mythical travelogue [32], and Shi Weile considers its text to be "quaint and authentic, displaying the style of a pure geographical work" [33].

Whether or not it is true that King Mu of Zhou travelled west in history [34], *Mutianzi zhuan* was undoubtedly the first geoscience text in China to be written in the style of a travelogue. It shows people's analysis and understanding of the surroundings during their trip, as well as the physical and geographical conditions of the northwest of China and its interaction with the Central Plains. The book is rich and extensive in its content, and is of high geographic and historical value, especially in the current situation where very few pre-Qin works are preserved.

Collated by Xun Xu in the Jin dynasty and annotated by Guo Pu, the *Mutianzi zhuan* records the westward trip of King Mu of Zhou pulled by eight horses. Along the way the group observed the landscapes, exotic fauna and flora, resources and products, and initiated links with the western nations and experienced the local folk customs. The King started from Luoyang, crossed the Yellow River and reached the area that is the present-day north of Hetao Plain in Inner Mongolia, and then went westwards through present-day Ningxia, Gansu, Qinghai, Xinjiang and other provincial administrative regions to further areas. Finally, he passed through Shanxi province and returned to Luoyang. Wang Shouchun [35] suggests that he travelled to a region with vast deserts, grasslands, and an abundance of horses, cattle, sheep, and dogs, and such a natural landscape and cultural characteristics can only be matched to the grassland and desert regions in the northwest of China, instead of as interpreted by Ding Qian and Gu Shi—that the King had reached Central Asia, Southwest Asia, or even Warsaw in Eastern Europe [36].

Travelogues are usually situated in the natural geography of mountains and rivers: this book is no exception. It notes a large number of mountain and river names, and its descriptions of mountain and river distributions are similar to those in the "Mountains". However, unlike the "Mountains", it provides a detailed account of the ethical groups, as well as the flora and fauna, mineral resources and the weather conditions along the way. The notes of wind direction, precipitation and temperature and also indicate the author's awareness of the changes in meteorological conditions. The book also records the animals, such as tigers, leopards, elk, deer, foxes, raccoons, horses, and pigs, as well as the names of birds and plants, as well as the details of resources such as the places of origin and the geographic distribution of jade.

As the first Chinese work in the travelogue style, *Mutianzi zhuan* is of great geoscientific value. Together with *Shanhai jing*, it enabled the conservative literati to acquire a great deal of geoscience knowledge beyond the classics and even reality, which was conducive to the expansion of traditional knowledge and the development of geoscience [37].

(2) Geoscience in *Guanzi*

Since the beginning of the farming era, the land had become an indispensable part of the lives of the Chinese people, so knowledge of farming and the land naturally accumulated. As an agricultural country, ancient China attached great importance to the study of soil. During the Spring and Autumn and Warring States periods, many geological texts discussed the nature and classification of soil. Not only did a comprehensive geoscientific work like the *Yugong* involve soil classification, but other works also included chapters on the study of soils, such as the *Guanzi*: *Diyuan*, *Zhouli: Caoren*, and *Lüshi chunqiu: Rendi.*

In the *Yugong,* the soil was classified into different types based on its colour (black, yellow, red, white, green) and characteristics (grave earth, fine clay, loessial soil, mud); based on the soil types, the local production and the growth of plants, the amount of local tribute could be estimated, which was reasonable in some ways, but such a classification was not detailed enough.

Guanzi: *Diyuan* presents a comprehensive and in-depth study of soil. It discusses the relationship between the terrain, soil, plants and groundwater, and proposes a soil classification system. The whole text can be divided into two parts: the first half addresses the relationship between soil and plants in different terrains, and the second half classifies various soils across the *jiuzhou* (nine regions of ancient China). The entire book is structured around the framework of soil classification.

The first half section of the book elaborates the idea of understanding different types of soil: farm fields, grave mounds, hills, and mountains. In addition to grave mounds, each other type is divided into several subtypes: the farm fields and mountains have five subtypes, while there are 15 subtypes of hills. For every subtype of farm fields, the book describes in detail for this type of land the crops suitable for cultivation, the plants that can grow, the depth and quality of the groundwater, and the population and their corresponding dialects.

The author had a deeper understanding of the soil conditions in the plains, which was due to the developed nature of agriculture and the large size of population in such areas. In contrast, the division and description of the 15 subtypes of hills are less detailed: these are simply divided according to the altitude of the terrain and the depth of the water springs, with certain increases in altitude indicated as new land types. Such division is perfunctory and of less significance. The five subtypes of mountains are arranged according to the altitude of the terrain, the vegetation, and the depth of the groundwater, as well as the vertical variation of the vegetation being recorded. At the end of the first half, the author explains the basis for soil classification, that is, "for different types of soil, there are plants that grow accordingly", which reveals the close relationship between the vegetation and the soil.

The second half section of the book focus on the soils in *jiuzhou.* It first classifies the soil into 18 categories according to their properties, and then further divides each category into five subcategories according to the colour of the soil. Also, according to the soil's texture, structure, pore space, organic matter, salinity and alkalinity tolerance and fertility, and combined with topography, hydrology, vegetation and other natural conditions, the 18 types of soil are divided into "upper soil", "middle soil" and "lower soil": in each class there are six types of soil. Among the types of "upper soil", the *xi*, *wo* and *wei* soils are the best: these are collectively termed the "three types of soils" and used as the standard for measuring the quality of other soils.

The book describes the "three types of soils" in the most detailed manner, which not only includes the colour, particle size, viscosity, sand content and other characteristics of the soils, but also covers the plants suitable for cultivation on the soils. Other soil types are also described in terms of their properties and suitable crops, and their agricultural productivity is compared with that of the "three soils".

In conclusion, the *Guanzi: Diyuan* describes in detail the various kinds of soils and their relationship with plants and attempts to summarise some laws and patterns derived from them. Although its classification and descriptions are somewhat mechanical in some respects, the book generally reflects the enrichment and improvement of people's knowledge about the soil during the Spring and Autumn and Warring States periods, and it is a significant work on soil geography.

Apart from the section "Diyuan", in the *Zhouli: Diguan situ* (The Earth officials), there is a special article on the position of "Caoren (the official of the soil and vegetation)" [38]. According to Xia Weiying's research, the main duty of the Caoren was to "improve the soil, identify the properties of the land, and identify the varieties of crops that were suitable for each kind of land" [39]. In the article "Caoren", there are nine types of soil, and this classification is also determined by the properties of the soil, including the colour, texture, structure, salinity and alkalinity.

Among the studies of various natural elements during the Spring and Autumn and Warring States periods, the most detailed research was conducted on soil, which reflected the outstanding characteristics of China as an agricultural country. The idea of soil classification established the foundations for the in-depth study of soil geography thereafter.

(3) *Lüshi Chunqiu: Shi'er ji* ("Twelve Records") and others

During the Spring and Autumn and Warring States periods, the political and economic centres of China were situated in the middle and lower reaches of the Yellow River, which is located in the temperate zone and the largest monsoon climate zone in the world with four distinct seasons, facilitating a smoother mastery of phenological patterns. The most important thing in agrarian society is to follow the farming calendar. Therefore, in order to properly manage farming, people from high and low are all conscious of the need to comply with the routines of farming. Phenology is the study of the changes in natural phenomena and their impact on agricultural production; it originated very early in China and reached maturity by the Spring and Autumn and the Warring States periods. The *Xia xiaozheng* ("Small Calendar of the Xia") was the first Chinese phenological calendar. The *Shijing* also recorded many natural events [40]. By the Spring and Autumn period, there was already a tradition of marking the natural changes in each solar term.

The *Lüshi chunqiu* ("Master Lü's Spring and Autumn Annals") is a collection of miscellaneous works by the guests of Lü Buwei, the prime minister of the state of Qin during the late Warring States. Within the collection, the "Twelve Records" focuses on climate and its effects. The text is divided into four chapters: "Spring", "Summer", "Autumn" and "Winter" with each chapter starting with the "Monthly Order". Each season is then divided into three phrases: *meng*, *zhong* and *ji*. The year is thus divided into twelve segments, corresponding to the twelve months of the solar calendar.

Changes in weather are directly reflected in the change of plants and animals. By observing plants, animals and various natural phenomena, the ancient Chinese people could tell the change of climate and season. The "Twelve Records" describes both the normal and abnormal climate phenomena of each month, and the relationship between climate change and human activities, rendering it a comprehensive work on both nature and humanity. However, the whole text is centred around climate and how its changes affect social activities, so it can be considered to be a thematic work on phenology. Overall, thematic studies of geoscience during the Spring and Autumn and Warring States periods were quite extensive, such as those on hydrography, geology, geomorphology, military geography and maps. The development of these aspects of geoscience will be discussed in later sections.

3.2.3 Cultural Geography in the Ancient Classics

The ancient classics of the Spring and Autumn and Warring States periods are rich in cultural geography content. While some of them are geographical works themselves, others can confirm certain perspectives of cultural geography even if they do not directly address it. This section attempts to review the perspectives of cultural geography in the classics from three aspects: settlement geography, regional cultural geography, and the development of cultural geography within the Chinese context.

(1) On the geographical conditions of settlements

Ancient Chinese scholars attached great importance to the study of human-earth relations, thus contributing to the accumulation of cultural geography knowledge. One of the key issues in the study of cultural geography is the spatial combination of human culture and its interaction with the geographical environment. The settlement is the centre of human activities; it is where people live, rest, conduct various social activities, produce and exchange products. Therefore, the understanding of the geographical conditions around the settlements is one of the main tasks of cultural geographers. In the pre-Qin literature, the geographical conditions suitable for the development of settlements were often mentioned. The *Book of Documents* records that the locations of Wangcheng city and Chengzhou city of the Zhou dynasty were selected through geographical examination and divination in the field, as well as the use of maps. This kind of fieldwork chiefly included the evaluation of the natural environment, such as the terrain and hydrologic conditions.

The sections "Classic of Regions within the Seas" and "Beyond the Seas" in the *Shanhai jing* document the geographical conditions of remote areas. Although the descriptions are mixed with many magical and bizarre elements, they still constitute valuable evidence for cultural geography. In the *Seas*, there are many descriptions about the environment around the human settlements, reflecting people's understanding of the geographic conditions of settlements at that time. For example, the "Classic of Regions Beyond the Seas: The West" mentions that "there is a place called Woye ('fertile land'), where the inhabitants eat the eggs laid by the phoenix and drink the manna that comes down from the sky" [41], which depicts an ideal human settlement with fertile fields, sufficient food and abundant precipitation. The "Classic of Regions Within the Seas: The West" says that "the country Liu Huang Feng (present-day Ordos Plateau) has a territory with an area of 300 square li. There is a road leading to the west and a big mountain in the middle of the country" [41], which indicates a relationship between the siting of the settlement and road transportation.

Many ancient works also touched upon the relationship between settlement sites and the landscape. The authors of the *Huainanzi* mentioned that, in ancient times, when flooding occurred, people "chose hills to live on" [42]. Once the flood was under control, the *Yugong* records that "the land has been able to grow mulberry trees and raise silkworms, so people have moved down from the small hills to live on flat land" [43]. There is a similar account in the *Mencius: Tengwengong II*: the flood control of Yu the Great made "the waters pursue their course through the country...and after this, men found the plains available to them and occupied them" [44]. Hence, the settlement sites moved from hilly areas to fertile plains, which indicates that human habitation altered according to changes in the geographical environment. When flooding occurred, the ideal place to live was on high ground to avoid flooding, and after the flooding was controlled, people moved to the plains, which were more suited to agricultural activities and the life of the people. This also reflects the merits of Yu's water control, which enabled people to live and work in peace and happiness.

The *Guanzi* first outlined the geographical requirements for the siting of state capitals: "the state capital is usually built at the foot of mountains or next to rivers. The site should not be too high, which is subject to drought, and have access to sufficient water sources; nor should it be too low, so as to save the cost of digging ditches. The building of city walls should take advantage of the natural terrain, so the walls do not have to be square and round, nor the roads to be straight [45]." It also highlighted that "the capital where the Shengren (Son of Heaven) is located must be in a firm location with fertile land. It can be on both the left and right sides of the mountain, and close to a river or lake. Ditches are to be built inside, and the drainage will flow into the natural river. This will allow the use of natural resources and agricultural products to feed the people and breed livestock. And all the people on the earth will turn to his virtuous government to receive his kindness ...The Shengren resides in the centre (of the vassal states), so that the natural resources and wealth of the whole country can be concentrated there. Walls should be built to form the inner and outer city, with trenches around the capital. If the terrain around the city is elevated, ditches should be dug, and if it is low, dykes should be built. And that can be called a solid city" [46]. The *Guanzi* took many geographical elements into account, such as the terrain, landform, soil, hydrologic conditions and the position of the city, which reflects the fact that the people at that time already had some knowledge about the siting of cites. In the *Lüshi Chunqiu: Shen fenlan: Shenshi* (Views on Clarifying Duties and Responsibilities: On Being Cautious with the Power of the Throne), the conditions for choosing the location of the state capital were discussed: "The ancient kings who unified the world had all chosen the centres of the world to build their palaces; and they had all chosen the centres of their palaces to build their ancestral temples as well. The territory of the world is vast, but they established their capitals on a land of only about one thousand square li, because their kingdoms could be arranged in the best possible order this way." The capital was located at the centre of the world with convenient transportation, which was conducive to gathering in the wealth of the whole country and facilitating the rule of the monarch. Therefore, the central position of the state capital was one of the important geographical conditions for the siting of the human settlement.

The ancient scholars not only highlighted the surrounding environment and the position of the settlement, but also considered the internal structure and pattern of a settlement. In the *Zhouli*, the concept of *jiufu*, or "nine rings" for the city layout was mentioned, which divided the royal domain into nine concentric rings (or squares) based on the distance to the royal capital. The capital area was called *wangji*, and for every 500 li away from the capital, the other eight rings (*-fu*) were the *houfu*, *dianfu*, *nanfu*, *caifu*, *weifu*, *manfu*, *yifu*, *zhenfu* and *fanfu* respectively [47]. The *Yugong* also explains a similar concept, which is the *wufu* or Wufu System. Settlement geography is one of the main elements of cultural geography. Although aspects of the geographical conditions of settlements were scattered across many works, it is obvious that people at that time already had some knowledge about the environment around human settlements.

(2) On regional cultural geography

In the course of utilising natural resources, human beings are also changing the natural environment. Hence, wherever there is human activity, there are traces of human modification of nature. As a result, these human landscapes become an integral part of the natural environment and infuse the latter with the characteristics of cultural geography. Hence, the cultural geography becomes an indispensable element in the description of the region. In the chapter of the "Seas" and the "Great Wilds" in the *Shanhai jing*, more than 150 ancient countries and tribes are recorded [48]. In the description of these countries and tribes, cultural geography elements such as the racial characteristics, ethnic customs, practices and religions are covered. In the "Classic of Regions Beyond the Seas: The South", there are many records about the various races, for example: "In the country of feather people, the citizens have long necks and feathers all over their bodies [49]." In the "Classic of Regions Beyond the Seas: The North", it is written that "in the country of Shenmu (deep eyes), the citizens have deep eyes" [50]. However, most descriptions of the races and their customs in the book are bizarre and grotesque. In terms of religion, the "Classic of Regions

Beyond the Seas: the West" records that "there are two witches called Ji and Qi … Qi holds a small wine cup, while Ji holds a *zu* (container)"; "there is a mountain called Dengbao, which is the passage through which witches travel between heaven and earth" [51]. There is a description here of how the witches perform the sacrificial ceremony, as well as its purpose, which is to "declare the divine decree to the earth, and convey the people's needs to heaven" [52]. The tribes and ancient countries noted in the *Shanhai jing* are mostly located in remote areas, and the accounts are mostly based on hearsay and lack actual investigation. Nevertheless, the book still pioneered the development of regional cultural geography.

The *Yugong* divides the known world into nine regions, or *jiuzhou*, and offers a rather detailed description of the ethnic groups, handicrafts and land tax in each *zhou*. For example, in Jizhou, "people on the island give leather clothes as tribute"; in Yangzhou, "people along the southeast coast give hemp clothes as tribute". In terms of the local customs, in Qinzhou, "the Lai people are used to take their animals to graze", and in Xuzhou, "the Huai people collected pearls and fish as tributes". The descriptions of tributes from each regions indicate that the craftsmanship at that time was of quite a high level.

The *Zhouli: Zhifang* also divides the world into nine regions. In addition to the types of crops and livestock suitable for agriculture in each state, the book documents the male–female ratio of the population in each state, such as "five males and three females" in Jizhou and "two males and three females" in Bingzhou, which indicates that people had come to realise the importance of population in regional geography. During the Warring States period, some vassal states attached great importance to their population. For example, Shang Yang of the state of Qin realised that in order to rule the country, it was necessary to know the size of and changes in the population. He once conducted a population census within the state of Qin [53].

(3) The agricultural background in cultural geography

The emergence and development of cultural geography knowledge largely resulted from the merging of human society and the natural landscape. In ancient China, this was mainly evident in agriculture. Hence, the body of knowledge in cultural geography in ancient China grew along with the emergence and development of agricultural society and culture.

The Spring and Autumn and Warring States periods witnessed the beginning of traditional Chinese agriculture. The use of iron farming tools and ox ploughing greatly facilitated the development of agriculture, while the division of labour in society became increasingly specialised during this period. In the *Rites of Zhou: Offices of the Heaven (on general governance)*, it can be read that "the citizens are appointed to nine types of occupations. The first one is farming, which is practised in three kinds of terrains to produce various kinds of grains. The second one is gardening, cultivating all kinds of fruits and vegetables. The third one is forestry, exploiting the resources found in the mountains, forests, rivers and lakes. The fourth is animal husbandry, breeding all kinds of livestock. The fifth is the handicraft industry, processing raw materials into various crafted items. The sixth is commerce, promoting the prosperity of the market and the distribution of goods. The seventh is the job of the wife: weaving and sewing, and the eighth is the job of concubines: gathering plants and fruits. The last type is for people who only have temporary jobs and have to change employers frequently to make a living [54]." This reflects the economic situation of society at that time. Among these occupations, the first four types are related to agriculture, which shows how much importance people attached to agriculture. At that time, not only was the natural environment clearly marked by human activities, but the domestication of wild animals and plants also took place. There were workers with expertise in horse breeding and officials who evaluated the price of horses. With the development of animal breeding, veterinarians also appeared. Some wild species were domesticated, becoming captive breeding species.

The use of iron tools made it possible to build large water conservancy and irrigation projects to meet the needs of agricultural production. A number of water conservancy facilities were built, including Dujiangyan, Lingqu, etc. A water conservancy project is in essence a cultural landscape, and its construction also changes the ecological environment. In particular, it mitigates the salinisation of soil. Human production activities have left obvious traces of human culture in agricultural areas, both in terms of the landscape and soil properties, making these areas completely different from the natural landscape.

The Spring and Autumn and Warring States periods were the formative stage of agriculturally orientated thinking. Many works at that time, such as the *Laozi*, *Book of Rites*, *Mencius*, *Zhuangzi*, *Xunzi*, *Hanfeizi*, and *Lüshi chunqiu*, all covered the thinking and guidance on agriculture. At the same time, the theory of the "trinity of heaven, earth and human" emerged: "As for the crops, human beings grow them, Earth generates them and Heaven nurtures them [55]." On the one hand, this way of thinking stressed the innate connection between heaven, the earth and human beings in agricultural production, and on the other hand, it emphasised the importance of man in agriculture. This centrality of human beings undoubtedly promoted the emergence of cultural geography ideas. The background to the development of cultural geography at that time was twofold: it was rooted in both the flourishing of thoughts on agriculture, as well as in the agricultural culture, which was based on the natural geographical environment. This duality was also the characteristic of cultural geography in ancient China.

3.3 Development of Geoscientific Thought

3.3.1 The Yin-Yang and Wuxing Theories and Geoscience

During the Spring and Autumn and the Warring States periods, people's understanding of nature was no longer confined to their sensory and superficial contact with the physical world. Along with the empirical progress made in the field, geoscience was set fair to build its conceptual framework out of perceptual representations. As suggested by Cang: "When people's cognitive ability improves to the extent that they are no longer susceptible to mythology, and when science offers no clear answers regarding the matters being studied, they then turn to philosophy for an explanation [56]". The yin-yang and *wuxing* (five phases) theories exerted a strong influence over Chinese geoscientific thought.

(1) The geoscientific origins of yin-yang and wuxing

The origins of the Chinese characters for yin and yang are truly ancient. The symbol for yang was found in inscriptions on oracle bones, and the combined use of yin-yang also appeared in the *jinwen* (the inscriptions on ritual bronzes such as bells and vessels) [57]. As a concept, however, the earliest known reference to yin-yang is in the *Shijing* ("Book of Songs"). One poem tells the story of Duke Liu, who led the Zhou tribe's migration from Tai and inspected the hilly terrains in Bin to plan dwellings and barracks so that his people could live and work in peace. The poem recounts [58]:

He surveyed a fertile plain ... He came up a mountain-crest, and descended again...He saw a hundred springs, and surveyed the plain wide, by the southern hillside. He founded a new capital, wide enough for all his people ... Ascending the ridges, he surveyed the shade and the shine, also viewing the course of the streams and springs.

It is important to note that the words yin (translated as "shade" in the text quoted) and yang ("shine") above do not yet indicate any Daoist philosophy. Similar to what was noted in the *Shuowen* dictionary, an early second century Chinese dictionary from the Han dynasty [59], "yin, the shade, the south of a river, the north of a hill" and "yang, the height and the light", these were merely geographical expressions for locations and directions. The philosophical principle of yin and yang traces its origins to the *Zhouyi* ("Book of Changes"). Well-known quotes include: "That which lets the dark and light alternate is the Way" [60]; "The yin and yang, two vital energies transform each other ... After heaven and earth interact, all beings are born" [61]. Evidence can also be found in the *Zhuangzi*: "All changes in the universe are interlinked in harmony with yin and yang". The yin-yang theory may have its roots in people's activities related to production, living and religion, resulting from their consistent observation and exploration of the various changes in nature. In many texts written during the Spring and Autumn and the Warring States periods, the concept of yin-yang had already been adopted as a philosophical principle [62]. It held that yin and yang are two opposite forces that exist in all things and phenomena in the cosmos, as pointed out by Laozi in the *Dao De Jing*: "Everything has a bright (strong, hard) side and a dark (weak, soft) side, co-existing in harmony" [63].

Meanwhile, another philosophical school—Wuxing, or the Five Phases—began to take shape. The two branches of thought subsequently interacted and integrated, providing fundamental frameworks for the history of ancient Chinese philosophy, and ultimately becoming the building blocks of Chinese culture. Just as was the case with the notion of yin-yang, *wuxing* also emerged from people's understanding of nature. It postulated that the different kinds of matter in the physical world, though varied in shape, size and weight, could all be reduced down to a system of five phases or processes, namely, metal, wood, water, fire and earth. *Wuxing* proponents attempted to use several common elements and relationships to explain the structure of everything in the cosmos. The oldest written account of the *wuxing* principle has been found in the chapter entitled *Hongfan*, in the *Shangshu* ("Book of Documents"):

> ...all the soldiers cheer and sing:
>
> Water and fire, make us the food to survive;
>
> Metal and wood, build us the tool to thrive;
>
> Earth is the mother of all, and of great use to all.
>
> The text further clarifies [64]:

Of the five phases: the first is water; the second is fire; the third, wood; the fourth, metal; and the fifth, earth. [The nature of] water is to soak and descend; of fire, to blaze and ascend; of wood, to be crooked and straight; of metal, to yield and change; while [that of] earth is seen in the sowing of seeds and harvesting. That which soaks and descends becomes salt; that which blazes and ascends becomes bitter; that which is crooked and straight becomes sour; that which yields and changes becomes acrid; and from the sowing of seeds and harvesting comes sweetness.

The *wuxing* cycle in its entirety stemmed from people's direct experience of the world and was closely linked to their everyday lives. According to the *Guoyu* ("Discourses of the States"), Shibo discussed the five phases: "The former emperor mixed earth with metal, wood, water and fire, so that all things were created" [65]. In Wei Zhao's commentary on the *Guoy*u, it is written that "a mixture is a blend of things. All kinds of things were made through acts like casting, smelting, melting deep-frying and boiling".

One of the early references that combined the idea of yin-yang and *wuxing* is in the chapter entitled "Four Seasons" in the *Guanzi*:

So the yin-yang theory underpins the principle of both Heaven and Earth. And the four seasons are the most important rule of the yin and yang. Punishment and acts of benevolence should correspond to the four seasons. If they are in harmony with the four seasons, they will lead to good fortune. If not, they will lead to misfortune...

The east is referred to as *xing* (stars) and the corresponding season is spring. Its qi (or *chi*, vital energy) is wind. Under the influence of wind, woods and bones can develop. The character of spring is the growth of things, so all the plants thrive and all creatures grow well in this season...

The south is referred to as *ri* (sun) and the corresponding season is summer. Its qi is Yang. Under the influence of Yang, fire and gas can develop. The virtue of the summer is to provide benefits to others and enjoyment of leisure activities.

The centre is referred to as *tu* (earth). The virtue of *tu* is to assist the four seasons so that wind and rain can come at the right time and the fertility of the soil can be enhanced.

The west is referred to as *chen* (celestial bodies) and the corresponding season is autumn. Its qi is Yin. Under the influence of Yin, metal and weapons can develop.

The north is referred to as *yue* (moon) and the corresponding season is winter. Its qi is the cold. Under the influence of the cold, water and blood can develop.

Here the concept of yin-yang is integrated with the four seasons and *wuxing*. In fact, the theories of yin-yang and *wuxing* exemplify a naive materialism and a dialectical outlook. Since they were grounded in careful scrutiny of the fundamental physical matter of the universe, these theories played a very positive role in accounting for the changes in nature—a primary field of enquiry in geoscience—and they have proved to be extremely influential.

(2) Explanations of natural phenomena using the yin-yang and *wuxing* theories

The principles of yin-yang and *wuxing* were often referenced when discussing the physical environment during this era and were especially prevalent during the Warring States period.

① Explaining the meteorological phenomena

Ancient scholars believed that "yang qi is warm whilst yin is cold" [66]. It was argued that changes in temperature represent the cyclical movement of yin and yang energies. The two forces alternate once a year, with the Winter Solstice at the start of the cycle. At this point, the yang air germinates underground, and gradually percolates up to reach the Earth's surface and above. This also marks the beginning of the yearly four-seasons cycle. According to the Wuxing school, the seasonal dynamics were triggered by the various phases of vital energy. In spring, "it resides in the Wood"; in summer, "it stays with the Fire"; in autumn and winter, the qi goes to Metal and Water respectively [67]. Furthermore, weather variation was often caused by a yin-yang imbalance. For example, a summertime plagued by rainstorms and hail was because of the "lurking cold" [68], and an iceless spring was due to the seasons being out of kilter—"the yin principle cannot overcome that of the yang" [69], etc. As for the more common weather conditions, yin-yang theory offered an explanation: "Yin and Yang confront each other… and the conflict results in fog. If Yang triumphs over Yin, it will disperse and thus produce rain and dew. If Yin triumphs over Yang, it will coagulate and thus bring frost and snow" [70]. As the theory expanded, the yin-yang principle was further blended with the concepts of *wuxing*; *wufang* (five directions), east, south, west, north, centre; and *wuqi* (five airs) wind, heat, dryness, chill, dampness. This reflected a holistic climate tendency across China. As is recorded in *Huangdi neijing* ("The Inner Canon of the Yellow Emperor"):

The east produces wind, the wind promotes [the growth] of trees … the south produces heat, the heat produces fire … the central region produces dampness, the dampness produces soil … the west produces dryness, the dryness produces metal … the north produces cold, the cold produces water …

This description corresponds to the overall profile of the climate features of China's heartland: while there are frequent sea breezes in the east due to the long coastline, inhabitants in the south suffer from the heat in the summer. Meanwhile, in contrast to the Yellow River and Yangzi River region in central China, where warm and humid air circulates, the western land is rainless and dry, and the north is typically cold.

② Explaining the spatial distribution of the soil

Yugong documented that "the tribute that is the soil is in five colours". According to the *Shiming* (or *Erya*, a dictionary compiled at the end of the Later Han period), "the tribute of the soil in Xuzhou has five colours, namely, cyan, yellow, red, white and black." This is in line with the five-colour and five-phase pair mentioned in *wuxing* theory.

In his description, Wan suggested [71]:

Shaanxi and Henan, home to the ancient capitals of many ruling dynasties, are rich in yellow earth. It has often been observed that there is red soil in the south of China, black and podzolic in the north, and salinised (whose surface turns white when it is dry) in the west. Towards the northwest side, a large area of sand dunes, sierozem and dessert soil is all light in colour, and when the air turns moist, the eastern soil appears caesious, which barely makes sense. Therefore, it is plausible that the structure of the positions of the five elements might have been based on the colour of the soil in different locations.

③ Explaining earthquakes

In the second year of the reign of King You of Zhou (780 BC), during the Western Zhou dynasty, an earthquake rocked the valleys of the Jing, Wei and Luo rivers. The grand scribe Bo Yangfu believed that "the vital energies of Heaven and Earth function in their own order. When the yang is so concealed underground that it cannot come out, and the yin is so suppressed that it cannot transpire, an earthquake will strike. The three rivers felt an earthquake today because the yang lost its position and yielded to yin" [72].

④ Explaining astronomy and other natural phenomena

Guoyu documented the twenty-fourth year of the rule of Duke Zhao (518 BC): "On the first day of May there will be a solar eclipse." Zi Shen replied: "It is going to rain." According to Du Shi's commentary: "It rains when yin energy overtakes yang." But Zhaozi thought differently: "There will be a drought. The sun has passed the equinox and the yang influence has not overcome the yin yet. When it does so, it will be on a very massive scale, and drought is the certain consequence [73]· In the text *Tianlun* ("Discourse on Nature"), Xunzi also emphasised that "the four seasons present themselves in succession" and explained that "the falling of stars is due to a modification in the relations of Heaven and Earth or a transmutation of Yin and Yang".

⑤ Explaining the balance of nature

Using the yin-yang theory to explain the universe instead of following the Mandate of Heaven can be viewed as ideological progress in the field of geoscience. Mother Nature may have her laws, but she does not always comply with them. In ancient societies, these abnormal occurrences were often proof of the existence of an almighty power that men could only submit to. Yet, there were already scholars in the Spring and Autumn and the Warring States periods applying the yin-yang relationship to explain all the normal and abnormal phenomena of the cosmos. From around the late Western Zhou dynasty, yin-yang began to be imagined as qi, or specifically part of the Six Qi from Heaven, together with wind, rain, dark and light. All natural phenomena, whether common or otherwise, tended to be interpreted according to the yin-yang theory. A large number of intellectuals in those periods proposed similar ideas. Apart from Laozi's maxim that "everything comprises yin and yang" [74], Xunzi recounted that "when Heaven and Earth conjoin, a myriad of things are conceived; when the yin and yang principles combine, transformations and transmutations are produced" [75]. The author of the *Huangdi neijing* called *wuxing* by the name of *wuyun* (five motions) and argued that "the *wuyun* and yin-yang are the Dao (laws) of the heavens and earth, the fundamental principle of all things, the parents of change, the beginning of birth and death and the home of divinities" [76]. The balance of nature is maintained by the proportion of yin to yang. Once the balance is tipped, yin-yang will suffer from chaos. Although this explanation sounds unduly ambiguous, it was recognised as a relatively scientific way of understanding the universe. In light of this theory, many natural phenomena were no longer viewed as acts of divine providence, but rather as comprehensible to human beings.

3.3.2 Early Views of the Shape of the Earth

When exploring their living environment, human beings also conjecture about the shape of the land they inhabit. Constrained by limited technology, the ancient people's awareness of the Earth's appearance was largely intuitive. As such, the flat-Earth theory was then considered highly realistic due to the horizontal arc of human beings' visual field. This was a commonly held belief across all ancient civilisations. Old Babylonians, for instance, portrayed the universe as a vault with a hemispherical dome and a flat disc. Alternatively, in ancient Egypt, the world was imagined to be a rectangular box with a lower-lying bottom. The ancient Indians held a similar view that earth was flat.

In ancient China, speculation about the Earth's form had existed for a long time. The ancestral homeland was bordered to the north by the vast expanse of the Gobi Desert, sandy deserts and wild steppes that extended to the Siberian rainforest and mountain tundra; to the west and south by some of the world's tallest mountain peaks and largest plateaux; and to the east by the immense oceanic divide. Living within a relatively enclosed geographical space, it is not surprising that the ancient Chinese believed in a flat-Earth cosmography.

(1) The inception of the "Four Poles" concept

The Chinese *ji* (pole) refers to the farthest known point of the universe. The same concept can also be conveyed by *yu* (border) or *zou* (corner). In the *Shanhai jing*, the section "Seas" opens with [49]:

All the living creatures on earth and all that is within the whole universe and the four seas are illuminated by the sun and the moon and governed by the movement of

Jupiter, the Master of the year. All beings that were born from different deities have different shapes and figures. Some die young while others enjoy a long life. Only the sage knows the secret.

This paragraph shows that the regions described in the *Shanhai jing* involved the entire world inhabited by humans. In particular, the authors of "Seas" documented the remote areas clockwise: starting from the South Pole towards the west, and then passing through the northern end until they reached the East Pole. Indeed, scholars initially thought that the Earth had poles at its outer corners, but as their vision broadened, they also realised that the east, west, south and north poles extended further away from the centre. Since some places mentioned in "Seas" were so distant that they went beyond the reach of humans, it is conceivable that a sizeable portion of it might have been based on imagination and mythology, especially the unclear locations of the four poles. Nevertheless, the authors affirmed the efforts made to map the entire planet. As noted in the *Shanhai jing*: "The God of Heaven orders a god called Shuhai to pace out the distance from the East Pole to the West Pole. It amounts to 500,109,800 steps [77]." Although this estimate lacks accuracy, it is evident that the concept of "four poles" was already established in this book.

The earliest account of the locations of the four poles was found in the chapter "Wan Zhang", in the *Mencius*: "Shun banished Gonggong to Youzhou, banished Huandou to Chongshan Mountain, exiled the chieftain of Sanmiao to Sanwei, and imprisoned and killed Gun on Yushan Mountain. When these four offenders were punished, the whole kingdom submitted to him [78] ..." Tong commented that "the four locations were very far away". They were in Youzhou, in what is now the northern part of Hubei province; Chongshan, which is around Hunan province; Sanwei, which is currently in the region of Shaanxi and Gansu provinces; and Yushan, which is in Jiangsu province [79]. Therefore, from today's perspective, the area marked out by the four poles is actually quite small.

The four cardinal points were described differently in the *Lüshi chunqiu*: "From Da Xia in the north to Bei Hu in the south, from San Wei in the west to Fu Sang in the east." According to Tong, Da Xia is in the northern part of today's Shanxi (unconfirmed); Bei Hu is in central Vietnam, based on older sources; and Fu Sang is Fu Mu in the east of Liaoning (unconfirmed). The four poles in this quotation were obviously located further outwards. Surrounding the land was ocean that also had its limits, and together they constituted the entire universe known by the inhabitants of that time. The scale of the Earth was also discussed in the *Lüshi chunqiu*: "Within the Four Poles, the world spans five hundred million and ninety-seven thousand li from east to west, and it spans five hundred million and ninety-seven thousand li from north to south as well." In contrast, the size of the land was described as: "Within the Four Seas, the total land spans twenty-eight thousand li from east to west and twenty-six thousand li from north to south [80]." The concept of the Four Poles suggests that during the Spring and Autumn and the Warring States periods, the Chinese ancestors viewed the Earth as a flat surface, with the central land encircled by the ocean.

(2) The flat Earth model

As debates on geoscience gained momentum, perspectives on the shape of the Earth were no longer restricted by assumptions based on the four poles and a land entirely enclosed. People started to reflect on their living space and called such a space "six folds", i.e., east, west, south, north, up and down. This spatial awareness was further theorised and served as the foundation for the ancient Chinese cosmological models about the figure of the Earth. Amongst multiple schools of thought, one of the most influential models handed down through the generations was the Gai Tian model. Originating in the late Shang dynasty (1600-1046 BC) and early Zhou dynasty (1046-771 BC), this model became more sophisticated in the Spring and Autumn and the Warring States periods and reached its peak around 100 BC when the book *Zhoubi suanjing* (a mathematical text dedicated to astronomical calculations) first appeared.

Based on practical experience, proponents of the Gai Tian (or Canopy Heaven) model believed that our vast and flat Earth had a "canopy" suspended above and that the expansive land stretched away to meet with the sky. This formed the premise for early views on the cosmological structure: "while the Heaven is round like an umbrella, the Earth is flat and square like a chessboard [81]." However, this notion did not withstand closer scrutiny. In the sixth century BC, Zeng Shen (505 BC-unconfirmed), a disciple of Confucius, asked the question: "If the sky is circular and the land is square, then its four corners cannot be fully covered [82]." To address this problem, ancient scholars revised the model and termed it the Second Gai Tian model saying that "heaven is like a bamboo hat covering over a bowl placed upside down", which indicated that although the shape of heaven remained unchanged, the Earth was not flat but arched. Zeng Shen explained that "both Heaven and Earth are high in the middle, low at the sides; beneath the North Pole lies the centre of the universe and the highest point of the Earth. The sun, moon and stars now appear, now disappear, and that is how the day and night alternate". Furthermore, the *Zhoubi suanjing* provided the distance between heaven and the Earth: "From the North Pole to where people live is as high as 60,000 li; its four sides sloping down, the middle of heaven is 60,000 li above the four sides." "The distance between heaven and the Earth is 80,000 li [83]." He said that wherever the Earth was, it was lower than heaven. Although the Earth had changed from a completely flat plane to a curved surface, this claim, together with the Gai Tian model at that time, still subscribed to a flat Earth cosmography, as did other ancient civilisations.

The flat Earth view also prevailed in ancient Greece. The people in Homer's epics, for example, believed that the Earth was a circular surface surrounded by ocean. Ionian historians, on the other hand, depicted the Earth as a disk floating on the water. In the world map they imagined, Greece remained at the centre of the planet. The inhabited continent was round but circumscribed by oceans. This could partly be explained by the Pythagorean hypothesis, in the sixth century BC, that only circles or spheres were the perfect shapes, since they symbolised a harmonious universe for mankind. Consequently, they posited that the Earth must be spherical and geocentric. A hundred years later, Aristotle (384-322 BC) accepted this hypothesis and found proof. From that time onwards, the concept of a spherical Earth gained more credence and became predominant in ancient Greece.

However, there have been conflicting accounts amongst historians regarding whether Chinese astronomers were aware of the sphericity of the Earth during the same period of time. The philosopher Hui Shi (370-310 BC), during the Spring and Autumn and the Warring States periods, proposed that "the south is endless but at the same time it has an end"; "I know the centre of the world; I can say it is to the north of the Yan state or to the south of the Yue state [84]". He also suggested that two people will meet up eventually, even if they walk towards south and north respectively. These claims were often cited as proof of the spherical nature of the Earth in ancient China. Admittedly, it is not impossible for Chinese people to have developed this idea, as the era of the Hundred Schools of Thought was indeed full of active minds and free discussions. However, Hui Shi, as a Sophist philosopher, tended to put forward proposals underpinned by ancient dialectics. For instance, he deemed "everything is relative, and has no substantial difference [85]". Moreover, his works are now lost, and only a few words are documented in the *Tianxia* by Zhuangzi, including these two vague and elusive lines. It is very difficult to verify their scientific implications and to associate him with the spherical Earth idea without more sources of evidence.

There was a wealth of theories related to the figure of the Earth in this period. Apart from the dominant school of Gai Tian, Zou Yan (305-240 BC), from the state of Qi, proposed the theory of the Great Nine Continents by drawing upon and extending the contemporary nine-region division of the Central Plains.

Located in the Shangdong Peninsula, the state of Qi, during the Warring States period, benefited significantly from its prosperous maritime navigation routes and the people's broad view of the world. Based on earlier expeditions to the three sacred mountains [86], Zou Yan pointed out:

The Middle Kingdom, known as the Divine Red Land, comprises nine regions...Outside the Middle Kingdom there are nine continents as large as the Divine Red Land, and these are the true Nine Continents. These are each surrounded by small seas, which separate the people and beasts on one from those on the others, and each one forms a region which makes up one continent. There are nine such continents, surrounded by a great ocean at the boundary between Earth and Heaven [87].

Zou Yan expanded the world much further. The Divine Red Land, where the Middle Kingdom was situated, was only one of the eighty-nine (9×9) parts of the world, and it was not necessarily the centre. The theory of Great Nine Continents is an example of a "non-traditional open-marine perspective of the Earth in ancient China" [88]. It viewed the Earth as eighty-one separate "continental islands" surrounded by the ocean, and the farthest reaches of the ocean met the edges of the dome-shaped sky. It should be noted that this was still a flat Earth outlook because of the influence of the traditional models of the shape of the Earth in ancient China.

3.3.3 Theories of Regional Differences

There are broad and narrow definitions of the term "region" in geographical research. In its broader sense, a geographical region refers to the global distribution of the continents and oceans and the latitudinal/longitudinal variations in zonality. Narrowly speaking, it is merely viewed as the demarcation of land borders in light of differences in temperature, hydrology, biology and other natural conditions. Based on their existing knowledge of the world, the ancient scholars acknowledged that, owing to different geographical factors or their combination, discrepancies in spatial distribution did exist and contributed to the forming of different regions. Nevertheless, they also realised that these factors remained relatively identical within the same geographical region. During the long-standing development of geoscience, regional geography was once the most time-honoured core subject. As Liu elucidated: "The reason why it plays such an important role in the history of geographical thought is that it recorded how the unknown universe had been transformed into a known one, and it is this process that becomes the history of geographical thought [89]." It should be noted that historians' analyses of regional differences were entirely based on practical experience, and the region itself was subject to the individual's own definition. This is the reason why ancient thought on regional geography truly reflects how much knowledge of nature humans actually possessed.

(1) Ancient thought on latitudinal zonality

Latitudinal zonality is a key structural component of the natural geographical environment. Although the way latitudinal zones are determined is man-made, they do conform to real-world conditions. It is only when one has a full and accurate grasp of the distribution patterns of the varied local environments that the divisions selected can truly reflect nature. Thus, discussions of zonality were an indispensable step towards a fully rounded knowledge of the natural environment, as the latitudinal zonality of heat on the surface of the Earth is the fundamental cause of the differences found in the world's geographic regions.

China is mainly located in the temperate region of the Northern Hemisphere, with most human activities taking place in the middle latitude. Here, the regional differences between the south and north are relatively evident, so people have long realised that the north is colder whilst the south is warmer, and that this difference is related to the height of the sun. To cite from the *Zhouli*: "On hot days, the shadow is shorter when the sun is in the south; on cold days, the shadow is longer when the sun is in the north [90]." Temperature differences result in diverse natural landscapes, and biological zonality is an excellent case in point. Not only were the zonal features distinctly outlined in the *Shanhai jing*, but the patterns in vegetational distribution tendencies across the southern and northern zones were also very clearly depicted in the *Yugong*. For example, Yanzhou (now north of Henan, south of Hebei and west of Shandong provinces) in the northern part of China had "lush growth of grass and trees that are tall but often sparsely planted". Towards the south, in Xuzhou (now south of Shandong, north of Jiangsu and north of Anhui provinces), "grass and trees grow thickly". As for Yangzhou in the south (now southern Jiangsu, southern Anhui, eastern Jiangxi, eastern Henan and eastern Hubei provinces), "the land is clothed with the most luxuriant woods and bushes". This reflected the north–south change in the vegetation of the eastern plains area. Moreover, Chinese scholars also noticed the boundaries between biological zones. According to the *Kaogong ji* ("Record of Trades"), "*ju* (orange) trees, if transferred to the north of the Huai River, will be called *zhi* (trifoliate orange) trees; a *quyu* (plover) will not fly over the Ji River; a *diao* (raccoon dog) will not survive if it crosses over the Wen River" [91]. This is all because of the variety in the geographical conditions. In particular, the Huai River is an important regional boundary in China. "As an ancient zoogeographic borderline, the Ji River and the Wen River match the current south borderline of the North China Region, located within the Palaearctic region" [92]. Marked by the Qinling Mountains and the Huai River, this southern border was a significant division between geographic regions in China. Since ancient geographers were already aware of it, it is natural to infer that they had developed an understanding of the geographical zones in the south and north regions during the Spring and Autumn and the Warring States periods.

Concurrently in the West, the ancient Greek seafarers were developing geographic thinking on global climate zones informed by the concept of a spherical Earth. The Greek philosopher Parmenides (c. 515 BC-unknown) pioneered the method of projecting a celestial sphere onto the Earth, which inspired him to divide the Earth's surface into five zones: a torrid tropical zone in the middle, two temperate zones in the mid-latitudes, and two frigid zones beyond that. Aristotle provided an explanation regarding this: due to variation in the angle of the sunlight reaching the spherical surface, heat spreads differently in the south and the north. This is why there is a climate classification scheme [93].

(2) Division into natural geographic regions

A region is characterised by its duality. "On the one hand, it is an established condition that was formed historically by Nature and that humans have to accept; on the other hand, it represents a historical, cultural space–it is through human activity that regional demarcations become meaningful [94]." When describing the geographic conditions of a large area, Chinese scholars tend first of all to divide it into several regions. Faced with numerous and complicated geographical elements, the first challenge is often to identify a criterion that can be applied to classify those regions. Undoubtedly, geofactors which are linear in appearance, e.g. mountain ranges and rivers, are the most distinguishable. Consequently, they were often adopted by ancient geographers to mark regional boundaries. In the "Mountains" chapter, regions were simply identified according to their direction: east, west, south, north and centre. Regions with different topographical features were mixed together, instead of being grouped together according to shared characteristics and set apart by clear demarcations. However, as the first attempt to collate such a wealth of material in a logical format, the regional references recorded in "Mountains" had a far-reaching impact. Its wide geographic coverage included social customs, folk cultures, shamans, powerful gods, mysterious monsters and other natural conditions–locations, hydrology, flora and fauna, mineral resources and local produce in particular. In comparison, the *Yugong* was a leap forward in terms of regional knowledge. It set aside the traditional method of classification by orientation and concentrated instead on selecting certain geofactors as the benchmark. With a clearer boundary, it made it possible not only to learn what the similarities were within regions, but also to find out how one region's natural and geographic conditions differed from another's.

In ancient Greece, one classification method gained popularity early on, i.e. that the north was Europa and the south was Asia. Subsequently, new progress was made in the understanding of regional differences during the period of Alexander the Great (356-323 BC), when geofactors such as climate and landscape were used as key indicators of regional divergence. The Greek geographer Strabo wrote a summary of the regional classification theory proposed by the descriptive geography school. Typical of region-specific thinking, the author compared regional classification to "dissecting a body at its joints rather than by random parts" in his work *Geographica,* which consists of 17 volumes. He added that "to identify the location of joints is to respect the peculiarity of the natural structure, so that one can handle it based on the appropriate joints and defined contours". He further pointed out that "in geography, when we need to classify a region for a more detailed description, we must follow the 'joints', and avoid casually selecting parts. Only the former can determine an obvious form and accurate boundary [95]." Strabo emphasised objectivity in regional classification and alerted us to the existence of some fundamental differences (joints) between regions when drawing the lines. This core guiding principle had a massive impact on thinking regarding regional division in the West.

(3) Comparative studies of regional differences

One of the major objectives in regional geography is to ascertain the geographic features of a region. For ancient historians, this target was mainly achieved by comparing regional differences before grouping similar ones together. Many texts from the Spring and Autumn and the Warring States periods, such as the "Mountains", the *Yugong* and the *Zhouli: The Map Officer,* documented a general awareness of such differences. In the "Mountains", for example, accounts of regional features often recorded inconsistencies in directions and distances between mountains and in the distribution of minerals and flora and fauna across mountainous areas. Although the different traits were listed clearly, they were not compared and contrasted systematically. Both the *Yugong* and *Zhouli* divided China into nine regions, or *jiuzhou.* However, the names and profiles of the nine regions in these texts were not entirely compatible with each other, although the system of classification developed was based on one understanding of regional differences [96]. It is worth noting that the *Zhouli* provided a much richer and better organised description of each region's landscape, natural produce, family structure and the livestock and grains suited to a particular region. The differences among the nine regions could be compared easily in the text—taking Yangzhou as an example [97]:

A region called Yangzhou is located in the southeast. Its mountain town is named Kuaiji. Its lakes and ponds are called Juqu. Its waterways have three branches. Its hydraulic resources are five lakes, and its local products are gold, tin, bamboo and arrows. Its ratio of men to women is two to five. Its livestock usually includes birds and dogs, and its grain is usually rice.

Such information helped readers to quickly grasp the differences between these regions, hence making more progress than the "Mountains". Unfortunately, facts were only reported sequentially, and there was still a lack of consistent criteria used for comparison.

The *Yugong* is not just an example of regional geographic classics, but it is also generally acknowledged to be an inventory of geographic comparisons between regions. With its emphasis on dissimilar natural conditions, the text analysed and compared each region's soil conditions, the grading of farmlands, types of farming produce, natural resources and tribute goods. First, the author classified the soil in the nine regions, and then he established an agricultural land grading scheme for tax purposes, thus creating a unified standard across China. Specifically, agricultural land was graded into top, middle and bottom bands, and each band was sub-divided into another top, middle and bottom sub-classification, leading to nine bands in total. In addition, attention was also paid to the greenery and local produce in the different regions. These comparisons were relatively direct and accurate, which effectively improved the users' understanding of the regional variations. In short, the idea of comparing regional differences in *Yugong* represented the state-of-the-art theory of regional classification at that time.

3.3.4 Theories on the Human-Earth Relationship

The human-Earth relationship can be defined as how humans and the natural environment interact and their mutual influence on each other. Views on this relationship occupy a critical position in the history of geographical thought, and even of human civilisation as a whole. The central idea varies depending on the mode of production in different eras. According to Tang: "The earliest thoughts on the human-Earth relationship appeared in the pre-Qin period...evolving perspectives in later generations nearly all have their roots in this period [98]." There are a variety of theories addressing human and nature dynamics, and four models are proposed here based on how men perceived their own status in relation to Nature.

(1) The interactionist view

The beginnings of human society are marked by extremely low production power. Even if humans managed to change their physical environment, they found it difficult to explain their achievements. It was even more difficult to grasp the intricacies of the world when they encountered failure. Therefore, our ancestors tended to attribute any anomalies to Heaven, or God, hence the origins of the interactionist view of the heaven-human relationship. This was further strengthened during the Spring and Autumn and the Warring States periods.

The interactionist standpoint remained mythological in that heaven was not defined in a cosmological or geographic sense but was perceived to be a divine ruler, or God, who could intervene consciously in human affairs. Human behaviour could also be sensed by Heaven, which would respond with natural disasters to condemn individuals or with good fortune to reward them.

Ancient thought on communicating with Heaven had long been prevalent in China. Proof was found in the *Shijing*: "the eclipse of the sun will make the world wild run" [99]. According to Kong's interpretation, the eclipse was an ill omen warning that "the Emperor, or the 'Son of Heaven', has not chosen a kind person for the job" [100]. In the Spring and Autumn and the Warring States periods, the *Zuozhuan* ("Commentary of Zuo") documented a case that associated natural catastrophes with men's indecent behaviour: "When the seasons of Heaven are reversed, we have prodigious phenomena. When the virtues of men are reversed, there appear disorders. It is those disorders which give rise to calamities and prodigious phenomena" [101]. Another example in the *Mozi* says that "the cold and heat would arrive untimely, so would snow, frost, rain and dew. The grains would not ripen and domestic animals would not mature. Diseases, epidemics and pestilences would spread far and wide and storms and bitter rains would repeatedly afflict the earth. All these are the punishments that Heaven flings upon men" [102].

The Greek historian Herodotus provided abundant evidence of the interactionist school of thought in his ground-breaking work, *The Histories*. In his description of the war between Greece and Persia, every time the Persians faced defeat, a particular kind of natural disaster would occur. Herodotus linked this to the divination of omens, typically in the form of thunderstorms, windstorms, earthquakes and eclipses, etc. [103].

Due to humans' limited knowledge of nature and the impact of fatalism, the interactionist perspective on the heaven-human relationship emerged in both the East and the West, albeit with differences in detail. In ancient China, while the Mandate of Heaven may have revealed a kingdom's rise and fall, the prehistoric Greeks used hints from Nature to forecast the results of battles.

(2) The determinist view

In the late Western Zhou period (1045-771 BC), the Emperor was losing power to his nominal vassals, resulting in a change in his status as "Son of Heaven". The frequent overthrowing of rulers and their subjects and the continual annexation of different states threatened people's belief in the Mandate of Heaven. As an agrarian society, water and soil were the two elements Chinese people were most intimately acquainted with in the entire physical realm. It is conceivable that ancient thought on physical determinism, especially the power of water and earth, was very diverse. The *Liji* ("Book of Rites") pointed out that: "those who were born in deep valleys and live on riversides look different from those in other areas, and so do their manners and customs [104]."

As noted in the *Guanzi* [105].

Earth is the foundation of everything in the world. All things are dependent on it. All people, kind or devilish, sensible or unworthy, foolish or talented, live on it... what is water? Water is the foundation of everything in the world. All kinds of lives depend on it to survive. And the quality of all people, kind or devilish, sensible or unworthy, is affected by it. How could I know this? The waters of the state of Qi are turbulent, so the people of that state are rapacious, impertinent and intrepid. The waters of the state of Chu are soft and clear, so the people of that state are nimble and decisive. The waters of the state of Qin are dense, stagnant, thick and motley, so the people of that state are greedy, cruel and bellicose. The waters of the state of Jin are almost dried up, turbid, silted and motley, so the people of that state are all deceitful, shrewd and covetous. The waters of the state of Yan are deep, weak, stagnant and motley, so the people of that state are foolish, chaste, hot-tempered and not afraid of death. The waters of the state of Song are light, but powerful and clear, so the people of that state are simple, easy-going and disinterested.

The text quoted emphasised that it was the water and land that shaped men's virtues or vices, wisdom or stupidity. Similar ideas can also be found in the *Lüshi chunqiu*. The *Dadai liji* ("Records of Ritual Matters by Dai the Elder") presented a focused discussion on the impact of the soil: "People who live on the hard soil are portly, on the loose soil are burly…on the sandy soil are thin…on the fertile soil are handsome, on the infertile soil are plain" [106]. Notably, some of the arguments in *Huainanzi* ("Book of the Master of Huainan") were similar. *Zhouli—Officers of Earth* described a deterministic view of landscapes: "those who live in mountain forests have long hair and sturdy figures, near rivers and ponds look tanned and sweat often, in hills are round and tall, near lakesides and marsh look thin and have fair skin, on wet plains are chubby and short" [107]. Although these interpretations seem over-generalised, they suggest that ancient scholars were aware of the impact of the natural environment on humans.

Whilst the view of physical determinism existed in both the East and the West, in China more attention was paid to the effects of water, soil and other topographical features which were closely related to agricultural work. In contrast, the West was more focused on the impact of the climate and oceans due to people's reliance on maritime activities. For example, the subject of how human character may be determined by the climate was discussed by the Greek scholar Hippocrates (480-400 BC) in his work on *Airs, Waters and Places*, as well as by Aristotle in *Politics* where he argued that northerners are brave and southerners are smart. Alternatively, Plato proposed an ocean determinist view in *The Laws* stating that "oceans help citizens to develop a businessman's temperament and fill their souls with selfishness" [108].

(3) Man the Conqueror

As humans increasingly improved their productivity, they refused to submit to Nature's power. Men demonstrated their skill with a number of large-scale hydraulic engineering works such as the twelve irrigation canals along the Zhang River, the Han Gou canal and the Dujiangyan irrigation system. One of the main supporters of the Conqueror view was Xunzi (313-238 BC) during the Spring and Autumn and the Warring States periods. He elucidated that the laws of nature could be acquired: "the course of Heaven is constant; if you respond to the constancy of Heaven's course with good government, there will be good fortune; if you respond to it with disorder, there will be misfortune" [109]. This explained the physical nature of the universe, suggesting that people could use and conquer nature if they observed the natural order and followed reasonable measures of governance. A crucial idea in Xunzi's work, the *Tianlun* ("Discussion of Heaven"), was that Man has dominion over the natural world. In other words, Man is the active agent, and Nature is the passive object. He adamantly contested: "How can glorifying Heaven and contemplating it be as good as tending its creatures and regulating them? How can obeying Heaven and singing it hymns of praise be better than regulating what Heaven has mandated and using it?" As a core concept in Xunzi's Conqueror Theory, "regulating Heaven" served humans' ultimate purpose of knowing nature's laws.

(4) Heaven and Mankind as One

What distinguished the Heaven-Mankind unity perspective from the interactionist and the deterministic models was that, instead of debating which side takes precedence, it laid stress on harmony between man and nature.

The "Heaven and Mankind United as One" was one of the key concepts in the heaven-human relationship theory in ancient Chinese philosophy. During the Spring and Autumn and the Warring States periods, this concept emphasised that man is an integral part of nature. According to Mozi (468-376 BC), one of the harmonious ways of living with nature was that man should source everything needed from nature, as much as possible. But he also implied that humans' demands should be proportional to nature's supply, as explained by Zhou: "Humans should only ask for what nature can offer; neither too much nor too little is appropriate [110]." A suitable estimate should be made based on nature's capacity. Guanzi advanced this argument further and claimed that [111].

Those who are capable of acclimatising to the outside world and controlling its changes can be revered as deities. Those who are capable of adjusting to the ever-changing situation of things and controlling their changes can be called wise men…Those who are capable of undivided attention and maintaining it constantly can be leaders of everything in the world. Gentlemen can make good use of the external things but will not be used by these things, because they know how to adhere to the rule of maintaining undivided attention.

This signified that men should make a sustained effort to explore nature; once they comprehended its regular patterns, they could proactively make good use of it. Environmental protection was a hotly debated topic during the Spring and Autumn and the Warring States periods, and many philosophers subscribed to Guanzi's view that an excessive consumption of natural resources would disrupt the ecological balance. In that context, a belief in the harmonious coexistence of man and nature essentially encouraged people to work in a manner appropriate to local conditions and, therefore, improved public awareness of environmental protection. Meanwhile, a series of environmental laws were formulated and strict measures were also introduced to preserve nature.

3.3.5 Thought on Environmental Protection

During the Spring and Autumn and the Warring States periods, the use of iron tools enabled people to exploit more natural resources. As a consequence of deforestation from agricultural reclamation, war damage, massive civil engineering projects, animal husbandry, logging, etc., the forest area had been greatly reduced. Mencius noticed that [112]:

> The trees of Niu Mountain were once beautiful. However, situated on the border of a large state, they were hewn down with axes and billhooks, so how could they retain their beauty?... then came the cattle and goats which browsed upon them. The bare and stripped appearance of the mountain is due to these things.

The shrinking of the forest area, together with excessive hunting activity, had greatly harmed animal resources. Hence, during this period the resources of natural flora and fauna had been seriously damaged.

At the same time, with the deepening understanding of the human-Earth relationship, the initiatives that human beings could take in the natural world had been noticed. People had come to recognise, on the one hand, the damage that human production exerts upon nature, and on the other hand, the ability of humans to change and protect the natural environment though their own efforts.

(1) Environmental policies

The idea of maintaining the natural ecological balance had already been stressed during the Spring and Autumn and Warring States periods. Guanzi suggested that "if a king does not know how to protect his mountains, marshes and vegetation from being damaged, then he cannot be upheld as the king of the world" [113]. He also recognised the importance of soil and water conservation in national development and in sustaining people's livelihoods and argued that "people need clothing and food to live, and the production of food depends on water and soil" [114]. Xunzi also emphasised that the protection of natural resources was one of "the duties of a sage king" [115]. Ancient scholars not only stressed the need for environmental protection, but also proposed many specific measures. One of the earliest environmental protection decrees in China was the *Fachong ling* ("Decree on Felling") issued by the Western Zhou dynasty in the eleventh century BC. The decree clearly stated "do not destroy houses or wells, do not cut down trees, and do not injure the livestock. Those who do not comply with the decree will not be spared death." [116]

Many written texts from the Spring and Autumn and Warring States periods, such as the *Zuozhuan*, *Guanzi*, *Mencius*, *Xunzi* and *Zhouli*, are very rich in discussion of specific measures to protect the environment.

The damage to natural resources at that time was mainly evident in the over-exploitation and serious destruction of animal and plant resources. Regarding this issue, the *Lüshi chunqiu* recorded "the injunctions for the four seasons: do not fell trees in the mountain forests until the right time comes; do not set fire to grass in swampy areas; do not take nets for catching birds or animals beyond the gates of the capital; do not cast nets to catch fish in any water; do not travel by boat for any reason except for officials in charge of boats. These activities might impede farm work" [117]. This indicates that logging, cutting and burning grass, hunting and fishing were banned except at the prescribed times. Activities in each month were also specified in the *Lüshi chunqiu: Shi'er ji.*

> January (*mengchun*: the first month of the spring season): during this month, do not fell trees; do not harm birds' nests; do not kill young birds or animals; do not shoot at deer or destroy birds' eggs.
>
> February (*zhongchun*: the second month of the spring season): during this month, do not drain the water from rivers, lakes or ponds during this month; do not set fire to mountain forests.
>
> March (*jichun*: the third month of the spring season): during this month, make sure that nothing used for hunting, such as nets or the poison used to catch animals, is taken out of the Capital, and forbid people from felling mulberry trees or chopping down silkworm thorns.
>
> April (*mengxia*: the first month of the summer season): during this month, avoid large-scale hunting. Do not chop down big trees.
>
> May (*zhongxia*: the second month of the summer season): during this month, the common people are ordered not to collect indigo plants to dye fabrics nor to make charcoal.
>
> June (*jixia*: the third month of the summer season: during this month, trees are thriving and the Son of Heaven orders officials in charge of mountain forests to patrol the forests all over the country to stop logging. No large-scale construction should be undertaken.
>
> July (*mengqiu*: the first month of the autumn season): during this month, dykes should be maintained and repaired, and places where silt has accumulated should be dredged to prevent flooding.
>
> August (*zhongqiu*: the second month of the autumn season): during this month, officials in charge of ceremonies are ordered to inspect the livestock used as offerings at festivals and to check their breeding to ensure they are healthy, in good shape and of the right colour. They should be examined carefully and compared with other animals of their kind. Their measurements should also be taken to make certain that they are in accordance with the set standards. After these details are confirmed, they can be offered to God.
>
> September (*jiqiu*: the third month of the autumn season): plants wither during this month, and it is the right time to collect firewood to make charcoal. Dormant insects and animals stay in their lairs and cover themselves with earth.
>
> October (*mengdong*: the first month of the winter season): during this month, officials in charge of mountain forests and waters are ordered to collect the taxes on ponds, lakes and other bodies of water.
>
> November (*zhongdong*: the second month of the winter season): during this month, all construction should be stopped. If mountain lakes and forests can provide any kind of food or meat, officials in charge of these areas should show people how to find it. If anyone dares to plunder others, he will be punished without mercy. On the day of the Winter Solstice, bamboos and trees should be used to make arrows.
>
> December (*jidong*: the third month of the winter season): during this month, wild geese fly towards the north. Fishing is ordered to begin. Officials notify farmers to get the seeds for crops ready.

These policies not only specified the resources that could not be exploited each month, but also instructed people as to which resources were available at which time and how to use these resources. The above text shows that ancient scholars attached great importance to protecting forest resources, and similar ideas are also found in other texts from the same period. To summarise, the specific measures proposed in these writings fall into three categories: no logging or lighting of fires on mountains, and the protection of animal and plant resources. No logging means the prohibition of logging during the growing period of plants to protect the forest resource. Such a practice was widely accepted at that time. For example, in *Mencius*, it is written that "if the axes and billhooks only enter the hills and forests at the proper time, there will be more wood than can be used" [118]. In the *Guanzi*, "though the mountains and forests are vast, and the trees and grass there grow exuberantly, logging should still be restricted at certain times." [119] In *Xunzi*, "the use of the mountains, forests, marshes and weirs is permitted or forbidden according to the season." "Not undertaking cutting and pruning, growing and planting outside the appropriate season is the reason why the mountain forests are not denuded and the Hundred Clans have more than enough timber." [115] The *Zuozhuan* also covers a similar idea. Those who violated the rules and damaged the forest resources would be severely punished [120]:

> If resources are found on the mountain, the king of the country should strictly seal off the mountain and forbid people from entering. Those who enter the closed mountain will be punished by death. Any violator entering with his left foot will have his left foot cut off, and if he enters with his right foot, his right foot will be cut off.

Along with strict bans, it was only with active enforcement and clear instructions on logging that the forest could really be offered protection. So which seasons were suitable for logging and which ones were not? In the *Zhouli*, it was suggested that "trees in the south of the mountain can be cut down in mid-winter, and those in the north can be cut down in mid-summer" [121]. In the *Guanzi*, "trees should not be logged during the spring, summer and autumn. And it is only after the tree has grown that it can be decided which weapons can be made from it." [122] In the *Xunzi*, "if it is the season when the grasses and trees are in the splendour of their flowering and sprouting new leaves, axes and halberds are not permitted in the mountain forest to avoid ending their lives prematurely or interrupting their maturation."

Apart from avoiding illegal logging, the protection of the forest also included preventing accidents, especially forest fires. In the *Guanzi*, it was particularly emphasised "do not start a great fire in the forest, and do not cut down large trees" and "if a forest fire cannot be extinguished, and the plants cannot grow abundantly, then a nation will be impoverished; while if the forest is kept safe from fire, and plants can grow well, then a nation will be prosperous". Hence, it further stressed that "decrees should be made to prohibit piling up dry grass near the mountain forests and swamps; and the exploitation of these natural resources should be opened and closed at proper times" [123]. In the *Xunzi*, a similar idea was proposed in that "it is the duty of the Yu Shi (official in charge of forest resources) to issue a ban on lighting fires in mountainous areas to conserve the plant and animal resources, and to open and close the exploitation of resources at proper times, so that the state will not run short of natural resources".

The ancient scholars also stressed the protection of animal resources. In the aforementioned *Lüshi chunqiu: Shi'er ji*, there were bans on illegal and indiscriminate hunting. In the *Xunzi*, some specific measures were proposed that "if the Six Traditional Domestic Animals are nurtured and bred according to the seasons, they will multiply and increase; if it is the season when the giant sea turtles, water monitors, fish, freshwater turtles, loach and eels are depositing their eggs, nets and poisons are not permitted in the marshes to avoid prematurely ending their lives or interrupting their maturation; ... Ensuring that ponds, lakes, pools, streams, and marshes are strictly closed during the proper season is the reason why fish and turtles are in plentiful abundance and the Hundred Clans have a surplus for other uses". In the *Mencius*, it was also argued that "if close-mesh nets are not allowed into the pools and ponds, there will be more fish and turtles than can be consumed" [118]. In the *Liji*, it was also suggested that "during the spring hunts, the ruler of a state will not surround a marshy thicket, nor will high-ranking officers try to surprise a whole herd, nor will (other) officers take young animals or eggs" [124].

Ancient scholars also recognised the importance of soil and water conservation. The *Guanzi* suggested that "people rely on clothing and food to survive; and the production of food depends on water and soil" [114]. As recorded in the *Guoyu*, "the ancient rulers did not destroy hills, fill swamps, block rivers, or dredge lakes [125]." Ancient scholars considered the ability to protect soil and water resources as a fundamental requisite of a sage ruler and an important matter for the country's development, which illustrates the emphasis on environmental production at that time.

During the Spring and Autumn and Warring States periods, the environmental protection decrees were comprehensive in their coverage. After unifying China, the state of Qin built on these decrees and issued the environmental protection law, *Tianlü* ("Laws on fields and farmland"). The *Tianlü* is a part of the *Qinlü* ("Laws of the Qin"), which was revised and supplemented based on the original laws of the Qin state after the unification of China by Emperor Qin Shi Huang in 221 BC. Therefore, the ideas in the *Tianlü* can be viewed as a further development of the environmental protection ideas in the Spring and Autumn and Warring States periods. The regulations in the *Tianlü* were mainly prohibitions related to the exploitation of natural resources.

(2) Establishment of institutions

During the Spring and Autumn and Warring States periods, not only were there clear regulations and systems for environmental protection, but institutions and officials responsible for protecting the environment had also been established. As early as the time of Emperor Shun, the Shanling Ju (Mountain and Forest Bureau) already existed. [126]. According to records in the *Zuozhuan*, during the Spring and Autumn period, the officials responsible for environmental protection included the Heng Lu, Zhou Jiao, Yu Hou, and Qi Wang, etc. Their main responsibilities were [127]:

> The Heng Lu guards the mountains and forests; the Zhou Jiao guards the reeds in the marshes; the Yu Hou guards the firewood and the crops in the fields; the Qi Wang guards the clams in the sea.

The *Zhouli: Diguan situ* provides the most detailed description of the institutional infrastructure for environmental protection. One of the responsibilities of the Grand

Si Tu (officials) was to "understand the size and geography of the national territory based on maps of the land, and to identify the names of the various geographic features in each region, such as mountains, forests, rivers, swamps, hills, mounds, plains, plateaux and marshes" [128]. The duties of the officials under the Grand Si Tu's command were all related to environmental protection, such as the Shan Yu, Lin Heng, Chuan Heng, Ze Yu, Ji Ren, and Kuang Ren, etc., all of whom were responsible for protecting the environment and the natural resources. The Shan Yu was responsible for "issuing regulations related to mountains and forests, managing the logging activity, and authorising the number of people participating in the logging and the dates for the logging...They are also in charge of punishing illegal felling operations". The Lin Heng was responsible for "patrolling the forests and enforcing the relevant prohibitions". The Chuan Heng was responsible for "patrolling rivers and swamps and enforcing the relevant prohibitions". The Ze Yu was responsible for "issuing the regulations related to lakes and swamps in the state, and for setting up boundaries and prohibitions to guard them". Therefore, the main responsibilities of the Shan Yu, Lin Heng, Chuan Heng and Ze Yu were to protect the natural resources in the forests, rivers and marshes. As for the Ji Ren, their main responsibility was to protect the wildlife resources: "the Ji Ren was in charge of issuing regulations relating to the hunting and farming land in the state. He set up boundaries and prohibitions to safeguard these lands, and anyone who engaged in hunting or farming had to obey his orders. The hunting of young animals, the collection of birds' eggs and the use of poisoned arrows to shoot birds and animals were prohibited." The main responsibility of the Kuang Ren (official of mining) was to "manage the production sites of gold, jade, tin, and stone", that is, to protect mineral resources.

Ancient China was based on agriculture and the development of the state relied on the exploitation of natural resources. Therefore, in order to ensure the sustainable use of these resources, it was necessary to protect the natural ecological environment from destruction by issuing corresponding laws, some of which were very strict.

3.4 Cartographic Drawing

"In the early days, geographical knowledge was probably expressed in maps rather than in written text." [129] According to Zhu Xi, the *Shanhai jing*, one of the earliest geoscience works in China, was a set of maps at first, and was then followed by the text. Maps, which reflect human understanding of the world, are effective tools in human production and life and are one of the oldest forms of human communication. Before the creation of writing, our ancestors used drawing to record their surroundings and discoveries, and it was from such drawings that early maps emerged.

The development of Chinese maps has a long history, and there are many ancient mythological legends about maps. However, the earliest reliable records of the use of maps can be found in inscriptions on the Yi Hou Shi Gui (a food utensil from the early Western Zhou) and the San Shi Pan (a bronze washbasin from the late Western

Zhou). There are also records of maps in the *Shijing: Zhou Song* and *Shangshu: Luohao*, which record that maps had already been made and employed before the Spring and Autumn and Warring States periods.

There are additional records relating to maps and providing more details in the texts which date from the Spring and Autumn and Warring States periods. Furthermore, the level of cartography in the Warring States period can be better understood from the maps drawn in the late Warring States period that have been handed down to the present day.

3.4.1 Literature and Classification Systems

During the Spring and Autumn and Warring States periods, various maps were widely used, and state rulers all attached great importance to maps. They had become the symbol of the territorial sovereignty of a nation. When a country surrendered, the map of its territory was often offered to the conqueror as a sign of dedication. As Han Fei said [130]:

> There may not be any practical benefit to be gained from pleasing the great powers, but it is necessary to offer the map and the seal of the country, so that the plea for military assistance may be considered by the great power. Yet offering the map will result in a shrinkage of the country, and submitting the seal will reduce the authority of the king. Losing territories will weaken the country, while cheapening the reputation will cause political chaos.

According to the *Zhanguo ce: Yance* ("Annals of the Warring States: the Yan Annal"*)*, in order to assassinate the king of the Qin state, Jing Ke used the map of Yan as a decoy to be able to meet the King of Qin. All of the rulers highly valued maps, and after the Qin had conquered the six other kingdoms, their maps were stored in Qin's capital, Xianyang. After the fall of Qin, Xiao He, the prime minister of the Han dynasty at that time, came to Xianyang, and the first thing he did was to securely store those maps. The maps were not only a symbol of power, but more importantly, provided the information required to govern the world.

During the Spring and Autumn and Warring States periods, each vassal state had special officials in charge of managing maps, and they were respected by the people. In the *Lunyu* ("Analects of Confucius"), it is recorded that, when meeting these map officials, even Confucius himself had to dismount from his carriage to pay his respects [131].

The writings of this period also covered the content of these ancient maps. The *Guanzi: Ditu* was an early monograph dedicated to the geographical content of maps. In discussing the relationship between maps and military operations, it highlights the nature and main components of the maps of that era [132]:

> Leaders of armies must know maps well. They should be aware of all the conditions of the convoluted paths of deep waters that might inundate chariots as well as those of famous mountains, huge valleys, big rivers, tablelands, hills, etc. They should also be aware of the places where grass, trees, cattails or reeds thrive, of where all the roads lead to, of the sizes

> of all the cities and towns, and of the locations of all the famous cities, towns, the abandoned cities and ruins. And they should be familiar with all the geographical conditions of these places as well. Thus, they can march through these places successfully to attack enemy cities and organise the order of all the military operations according to the advantages provided by favourable geographical conditions. That is the value of geographical maps.

Although the quotation above focuses more on the general principles of using maps for military purposes, it indicates that the maps of that period were very rich in content. There are similar records in the *Zhouli* [107]:

> Based on maps of the land, it is possible to understand the size and geography of the national territory, identify the names of various geographic features such as the mountains, forests, rivers, swamps, hills, mounds, plains, plateaux, and marshes in each region, and clearly distinguish the number of fiefs in vassal states and the royal capital, and to then delineate the boundaries of each state's territories by digging ditches and building walls.

At that time, maps not only included geographical elements such as mountains, rivers, terrain, vegetation and cities, but were also drawn to a certain scale. Only maps drawn in proportion could precisely show the distance between cities and the size of a city, as well as the size of *jiuzhou* (nine regions of ancient China). This is reflected in the records of the *Zhanguo ce* ("Stratagems of the Warring States") when Su Qin said to the king of the Zhao state: "I secretly measured the territory of *tianxia* (under the heaven), and the territories of the vassal states are five times larger than Qin [133]." Only maps drawn to scale could have helped analyse the actual size of those vassal states.

During the Spring and Autumn and Warring States periods, maps were widely used and some focused on specific geographical elements, giving rise to thematic maps.

(1) National maps

According to legend, in the Xia dynasty, nine *ding* (ceremonial tripod cauldrons) were cast with motifs of mountains, rivers, and strange creatures. And these relief patterns were later called the *Jiuding tu* (nine *ding* map). There are many records of this in pre-Qin literature, for example, the *Zhanguo ce* records that "the state of Qin sent troops to Zhou to seek the nine *ding*" [134]. There has been much speculation about the specific content of this map. According to the research of Bi Yuan, a scholar of the Qing dynasty, "there were the names of countries, mountains and rivers as well as strange and supernatural creatures on the map" [135], which indicates that the *Nine Ding Map* was a relatively primitive national map.

According to Yang Shen, a scholar of the Ming dynasty, the maps in the *Shanhai jing* evolved from the *Jiuding tu* [136]. Although the original maps are now lost, we can still examine the explanatory texts in the *Shanhai jing* that recorded their content: according to the explanatory texts, the maps not only recorded the names of mountains, rivers and lakes, but also the relative positions of the mountains to each other, and the natural products, mineral resources and strange creatures in the mountains. This was more detailed content than in the *Jiuding tu*. Nonetheless, due to people's underdeveloped understanding of the world, the maps contained many fictional and legendary elements, and hence remained rather primitive.

During the Spring and Autumn and the Warring States periods, all the vassal states wanted to dominate the Central Plains, and lobbyists from various countries often used maps as the basis for their discussions of world affairs. It is certain that all countries at that time had managed to draw "the map of the world". According to the *Shuyi ji* ("Tales of Unusual Matters"), during the Spring and Autumn period, Lu Ban, a native of the state of Lu, had created the earliest rock map, the *Map of Jiuzhou* (under the heaven), on a rockface on Shibao Mountain near Luoyang [137]. The *Zhouli* records that "the Zhi Fang Shi (map officers) are in charge of the map of the world, so that they can manage the land of the world, as well as the people, wealth, food crop production and livestock of each vassal state, royal fief, and ethnic group, and know about their positive and negative aspects" [138]. A map of this kind contained the national and regional boundaries and the territories of the surrounding ethnic groups, which reflected the people's understanding of *jiuzhou.* Also, in the *Zhouli*, it is recorded that the "Si Xian (official) is in charge of the map of *jiuzhou*, so that he can be aware of any danger within the mountains, forests, rivers and ponds in each state, and can build roads between them" [139]. From the records in the *Zhouli*, it can be seen that the former focused on human geographic factors such as boundaries, people and products, while the latter emphasised factors such as mountains and rivers and road traffic. Hence, national maps fell into different categories depending on the practical requirements.

(2) Land maps

Land maps are also called "cadastral maps". In the *Zhouli: Diguan situ*, it is recorded that "the duties of the Grand Si Tu were to manage and draw land maps of the realm and to record the number of the people, in order to assist the king"; while "the duties of the Minor Si Tu were .... to solve land disputes with official land maps". The development of agricultural production, the expansion of cultivated land and the need to divide the land made land maps indispensable to the ruler in order to establish the state and manage the land. Such maps were important proof of land ownership and hence were managed by special officials.

(3) Maps of mineral resources and natural products

The *Zhouli: Diguan situ* also records the Kuang Ren (official in charge of mineral resources) and his duties were to "manage the production of gold, jade, tin and stone, and to guard the sites by establishing boundaries and issuing bans. Also, to establish when the proper time is to do the mining, and to survey and map the mining area, and give the map to the miners" [128]. Therefore, the Kuang Ren was responsible for surveying the mining areas and managing the maps of mineral resources to support the mining industry.

Another occupation recorded in *Zhouli: Diguan* is the Tu Xun (official of the land), and his duties were "to explain the maps, show the king the proper agricultural practices for different terrains, and list the evil things that are harmful to people as well as the natural products that people benefit from in each place, so that the king knows when the products of a certain place can be obtained when he needs them. If the king is out on a tour, his duty is to accompany the king by chariot and offer advice." The maps that the Tu Xun was in charge of mainly related to terrain and natural products [140].

(4) Military maps

During the Spring and Autumn and the Warring States periods, wars were frequent and military maps were widely used. In the *Guanzi*, the use of maps for military purposes was discussed: "When it comes to directing military forces, it is necessary to use a map and clearly comprehend all the geographical conditions." [141]

In the chapter "Ditu", the value of maps in war was further discussed, as mentioned in the above section [142].

Therefore, maps played an important role in the deployment and movement of troops, in avoiding danger and in taking advantage of the terrain.

According to the *Hanshu: Yiwen zhi*, major military texts in the Spring and Autumn and the Warring States periods were accompanied by maps, amongst which the *Sunzi bingfa* (Sunzi's "Art of War") comprised nine volumes with maps, and the *Sunbin bingfa* (Sun Bin's "Art of War") comprised four volumes with maps. These maps most likely formed part of the major military maps of that time. Although these maps, along with the notes marked on them, have been lost, the contents of these two military works (which we will discuss in the next section) show that military maps at that time were highly focused on topographic features that related to the deployment, garrisoning and combat of military forces, as well as on geographic features such as rivers, lakes and transportation points.

During the Spring and Autumn and the Warring States periods, various kinds of maps were produced, and apart from the those mentioned above, there were also thematic maps, such as maps of mausoleums, cities and transportation routes.

3.4.2 Cartographic Relics

Compared to books, the preservation of ancient maps is more challenging [143]. The earliest extant cartographic relics in China are the *Zhaoyu tu* ("Map of the Area of the Mausoleum"), preserved from the Warring States period, and the maps excavated at Fangmatan, which are from the Qin dynasty. These map antiques provide evidence which merits further research on cartographic development during the Spring and Autumn and the Warring States periods, and they are of great significance in the history of Chinese cartography.

(1) *Zhaoyu tu*

During the Spring and Autumn and the Warring States periods, many maps were produced for the planning and construction of mausoleums. During the Western Zhou dynasty, there were already special officials responsible for drawing and managing such kinds of maps, who were known as the Zhong Ren and Mu Dafu. While the former's duties were "to manage the royal mausoleum and map out the surrounding area" [144], the latter's were "to manage the public cemeteries and map them"

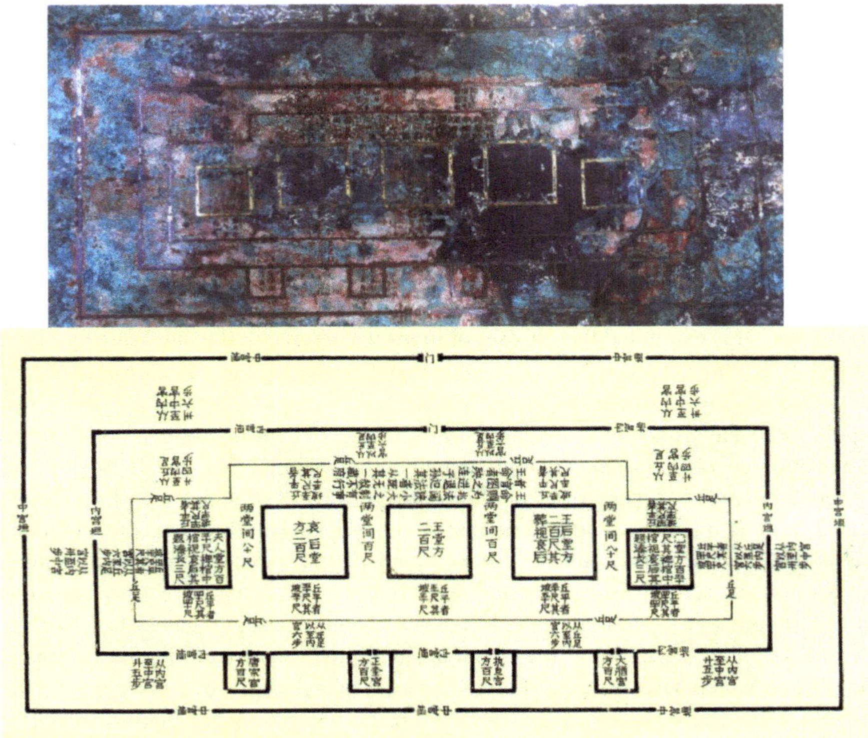

Fig. 3.3 *Zhaoyu tu* on a copper plate. Sohu [215]

[145]. "Zhaoyu" means the area or room where the coffins were placed in a royal mausoleum, therefore, the maps of mausoleum planning are known as the *Zhaoyu tu* (Fig. 3.3).

From November 1974 to June 1978, archaeologists were excavating the site dating from the Warring States period at Zhongshan, in Pingshan, Hebei province, when a map of the mausoleum's layout was found. This was engraved on a copper plate and is estimated to have been painted around 310 BC. It is the earliest planimetric map still in existence today. It is estimated that this copperplate map from the Warring States period dates from at least 2,200 years ago. The copper plate is 94 cm long, 48 cm wide, and about 1 cm thick. The map consists of three parts: line symbols, numerical notes and textual descriptions. The data indicate the distance between the lines and the actual measurements of the mausoleum. The map is drawn to scale, which is calculated to be about 1:500. Moreover, the south lies at the top (edge) of the map and the north is at the bottom. Although there are no indications of directions (visible) on this map, the charting of the *Zhaoyu tu* did require directionality to be taken into consideration. It can be said that this map contains the basic elements of cartography, and its drawing had reached a certain level of technical specificity.

The map was also annotated with text, including an imperial decree issued by the King of Zhongshan, and it has probably been preserved to the present day because of this imperial decree. The decree says that "there are two copies of this map; one is to accompany the burial, and the other is to be archived in the imperial storehouse" [146].

(2) Maps excavated in Fangmatan

In 1986, four pine boards with maps of different sizes were excavated from the No 1 Qin tomb in Fangmatan, southeast of Tianshui, Gansu province, three of which had maps on both sides and one had a map on only one side. In total seven maps were discovered. It is estimated that this set of maps was drawn in the late Warring States period in the third century BC. The map's name, scale, legend and cartographer are not noted on any of the seven maps. Modern scholars have divided the maps into topographic maps, and maps of administrative areas, of natural products and of forest distribution according to the contents drawn on the original maps (Fig. 3.4).

There are over 80 annotations on these maps [147], 28 of which are place names, plus 2 mountain names, 8 stream names, 4 valley names and 6 names of mountain passes [148]. The maps mainly centre around rivers, depicting mountain ranges, passes, forests, settlements, roads and other features. The rivers are represented by single sinuous lines, and the names of the rivers and places are generally written in the order of their direction of flow, from upstream to downstream. Roads, mountain ranges and watersheds are also mostly depicted using single sinuous lines. Passes are represented by symmetrical symbols, while the annotations for forest distribution are relatively detailed, with the types of trees and logging activities indicated. Some settlements are indicated by a square drawn around them. The maps were drawn with the north at the top, and some places also indicated the distance in li. The orientations and locations of the river systems depicted are largely accurate, indicating that the

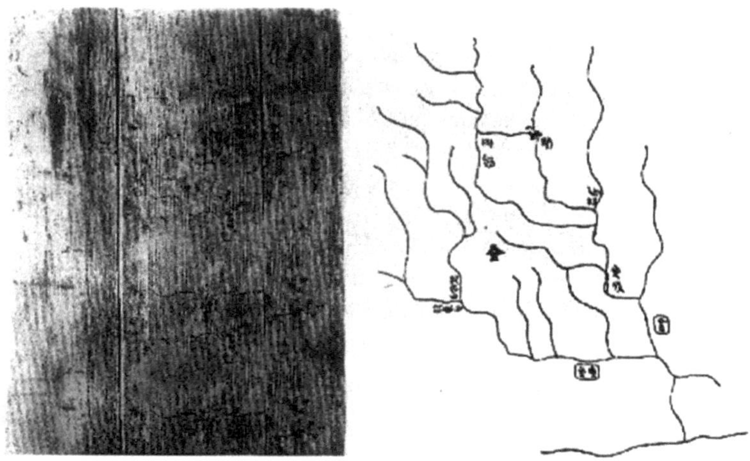

Fig. 3.4 The Qin maps discovered at Fangmatan

maps were drawn to a certain scale. Researchers hold different views regarding the regions covered by the seven maps and the relationship between them. Some believe that the eastern edge sketches up to 20 km away from the west of present-day Baoji, Shaanxi province, and that the northern edge stops at Qingsui County, Tianshui, Gansu province. Also, that the western edge extends to Tianshui Village, Tianshui city, and the southern edge reaches the north of Liangdang and Wei counties [149]. Others argue that the maps present the upstream section of the western Han River and the area of the Yongning River at the upper reaches of the Jialing River [150]. What can be confirmed is that they provide a map of the small regions of the Fangmatan site.

The maps excavated at Fangmatan once again prove that there was a rather advanced level of cartography during the Spring and Autumn and Warring States periods.

3.5 Applications of Geoscience

During the Spring and Autumn and Warring States periods when wars were frequent, almost all major countries focused on agriculture and military affairs and their state policies centred around having "sufficient food supplies and armaments", which are both closely related to geoscience. The deployment of troops depends on military geography and agricultural production relies on hydraulic engineering, which are the two most important practical applications of geoscience knowledge. This section will give a brief description of their development in the Spring and Autumn and Warring States periods.

3.5.1 Military Geography

In ancient China, the Yellow Emperor and Chi You had fought a battle in Zhuo Lu, Hebei province. The Yellow Emperor knew that Chi You and his tribe came from the south and were unfamiliar with the environment of the north. Therefore, taking advantage of a familiar geographical environment and favourable weather conditions, the Emperor had managed to defeat Chi You [151]. Hence, in ancient warfare, the idea of military geography had emerged, which studied the interrelationship between the geographic environment and human warfare activities.

More detailed treatises on the relationship between war and geography were produced mainly during the Spring and Autumn and Warring States periods, especially in the Warring States period. At the beginning of the Western Han dynasty, Zhang Liang and Han Xin collected and compiled a total of 182 military books, most of which had been written during the Warring States period. [152] Among the various schools during the Spring and Autumn and Warring States periods, the Bing Jia (School of Military) was very active in producing such works. The main works included *Sunzi bingfa*, *Wuzi bingfa*, *Sima fa* ("Methods of the Sima"), *Sunbin bingfa* ("Art of War by Sun Bin"), *Wei Liaozi* ("Master Wei Liaozi"), and *Liutao* ("Six Secret Teachings"), etc., among which *Sunzi bingfa* contains the most detailed discussion of the relationship between the military and the geographical environment and had a great influence on later generations. In addition, many ancient pre-Qin texts also cover military geography, but the military books are richest in such content and represent the extent of the development of military geography in these periods.

During the Spring and Autumn and Warring States periods, annexation wars between vassal states were frequent. From these wars people accumulated a great deal of experience of using the geographical environment, which facilitated the development of military geography theory at that time. Given the conditions of military technology at that time, the constraints and influence of the geographical environment on warfare were significant. In particular, geographic factors such as geographic location, topography, hydrology and meteorology are most closely connected to warfare activities. Meanwhile, elements of human geography, which is a part of the geographical environment, also play a role in war. However, given the socio-historical context at the time, the weapons used in warfare were mainly swords, spears and bows, and their use was subject to the terrain and the weather conditions for the warfare. Thus, the influence of natural geographic factors on warfare activities was more prominent.

Also, at that time, larger scale naval battles were already taking place along the eastern coast of China. The state of Qi, Wu, Chu and Yue had massive naval forces. Naval warfare is more constrained by natural conditions, such as wind direction, currents, underwater topography, storms, etc. However, in ancient Chinese military theory, more effort was dedicated to the study of land warfare, whilst naval warfare was less discussed. When General Wu Zixu gave a briefing on the training method of the navy to King Wu, he merely referred to the method used for the land force:

> There are different types of naval vessels, such as the *dayi* (attack battleship), *xiaoyi* (defence battleship), *tumao* (cruiser), *louchuan* (floating fortress) and *qiaochuan* (destroyer). We can compare the navy to the ground army and use a similar method. While the *dayi* resembles the attack chariot, the *xiaoyi* represents the light chariot; the function of the *tumao* is to initiate an assault, the *louchuan* is comparable to a *louche* (chariot with a high tower for reconnaissance), and the *qiaochuan* acts as a well-equipped horseman [153].

Few of the major military books of the time addressed naval warfare. In short, the theory of military geography focused more on terrestrial geography, and the application of military geography was mainly informed by the physical geography.

(1) Concept of *di* and its significance in war

In the military books of the Spring and Autumn and Warring States periods, the study of *di* takes a considerable proportion of the text. *Di* often refers to the natural geographical environment. The strategic significance of *di* is emphasised in military books, which advocate the full use of favourable natural geographical conditions in warfare.

Sunzi, a celebrated military strategist at the end of the Spring and Autumn period, suggested that "those who know both heaven and earth will achieve victory" [154]. In the *Art of War*, Sunzi defined *di* as "the distance, whether it is great or small; the terrain, whether it is treacherous or secure; the land, whether it is open or constricted; and the place, whether it portends life or death" [155]. He also suggested that when deploying troops, the following five steps could help achieve victory: "the measurement of the space refers to the difference in the size of the territories of the opposing parties; from that an estimation of quantity can be derived, which refers to the difference in resources; from that, a calculation of the numbers can be estimated, which refers to the difference in the size of their troops; from that there can be a comparison of the relative strengths of the armies and, finally, an assessment of the material basis for the chances of victory." [156] Hence, Sunzi had already concluded that the territory and population of a nation were the factors that would determine the outcome of a war. Sun Bin in the middle of the Warring States period also considered the geographical environment as being one of the main factors in warfare tactics: "He who has mastered this art knows the way of heaven and earth, has the support of the populace and is fully aware of the enemy's situation." [157]

This idea was popular in the military works at that time. In *Wuzi bingfa*, which is said to have been written by Wu Qi of the Warring States period, it is argued that "there are four principles in deploying military forces: the first is to master the morale, the second is to make use of the terrain, the third is to apply strategies, and the fourth is to enhance the military strength". The "use of terrain" here refers to a place that is "narrow, dangerous, with great mountains as blockades, which can be guarded by ten men but cannot be passed by a thousand men" [158]. It also refers to the natural geographical environment. Written in the Warring States period, the *Sima fa* stressed that "when the former kings ruled the world, they obeyed the laws of heaven and adapted to the geographical conditions on earth" [159]. When listing the five factors that influence victory or defeat in war, the factor *li di* (adaptation to the natural environment) was also included [160].

The *Liutao* recorded that Jiang Taigong proposed to King Wu that an army leader should have 72 assistants, with three of them specialised in "the advantages of terrain". They were also responsible for "analysing the movement and disposition of the army, the advantages and disadvantages of the terrain, the distance and traffic capacity on the roads, the water and dry land, and the dangerous and difficult mountains in order to avoid losing the advantages of the terrain" [161]. The "advantages of the terrain" here refers to the typological conditions through which an army passes. The work also argued that "when we venture deep into enemy territory, we must investigate the terrain and take advantage of it. By relying on the mountains, forests, ravines, rivers, streams, woods and trees, we will have a solid basis for deployment. Meanwhile we should carefully guard the passes and bridges, and know about the advantages of cities, towns and hillocks" [162]. It highlighted the significance of geographical elements, such as the topography, vegetation and water resources, in warfare.

(2) The significance of topography in warfare

In the military books of the Spring and Autumn and Warring States periods, the term *di* sometimes refers to the topography. Due to the underdeveloped technology at that time, the terrain played a pivotal role in winning or losing a war, so all the military writing of this period discussed the relationship between topography and the military. Some works also had chapters devoted to the role of topography in warfare.

The *Sunzi bingfa: Dixing* (the terrain) is a chapter devoted to military geography, with a central emphasis on the influence of topography on the military. In this work, it is argued that "advantageous terrain can be a natural ally in battle. Superior military leadership resides in the ability to access the enemy's situation and create the conditions for victory, to analyse natural hazards and calculate distances", which suggests that topography is one of the factors that a general must take into consideration when taking military action. It also suggests that "to know that the enemy is vulnerable to attack and that your troops are capable of attacking, but not to know that the terrain does not favour you in battle reduces your chances of victory by half" [154]. The use of the terrain in warfare was considered to be the basis for defeating the enemy. Of the 13 chapters in the *Sunzi bingfa*, about seven chapters cover the theory of military topography.

The *Guanzi: Ditu* was an early monograph dedicated to military cartography. It stressed that "leaders of armies must know the maps well. They should be aware of all the conditions of the convoluted paths of deep waters that might inundate chariots and those of famous mountains, huge valleys, big rivers, tablelands, hills, etc. ... And they should be familiar with all the geographical conditions of these places as well. Thus, they can march through these places successfully to attack enemy cities and plan the order of all the military operations according to the advantages provided by favourable geographical conditions [132]." Leaders of armies must use maps, because the maps record the topography of where the battle will take place. In the *Liutao*, it was stressed that different types of soldiers must take advantage of different topographic conditions to win the war: "for the infantry in battle, it is essential to know the enemy's movements and of any changes; for the chariots in battle, it is essential to know the terrain; for the cavalry in battle, it is essential to know the side roads and small bypaths [163]." Hence, ancient scholars had noticed that topography was the main geographical factor affecting victory and defeat in war.

In many wars during the Spring and Autumn and Warring States periods, there were cases where the advantages in the topography were fully utilised to turn the tide of the battle. These cases provided military theorists with rich warfare experiences, and developing theories based on these experiences, they drafted some tactical principles that later became an important part of military topography theory and even military geographical thought.

① Relationship between topography and tactics

In ancient and modern warfare, benefiting from favourable topographic conditions is of prime importance. First of all, this refers to the height of terrain. The *Sunzi bingfa* suggests that "generally speaking, a manoeuvring army prefers high, dry ground to low, wet ground; it prizes the sunny side, and shuns the shady side". It also mentions that "while passing through mountains, stay close to the valleys and pitch camp on high ground facing the sun; if the enemy is on high ground, avoid fighting an uphill battle" [164]. In the *Wuzi bingfa*, it is written that "in warfare, it is preferable to use chariots on high ground rather than on low ground" [165]. The *Sima fa* also offers a similar idea, suggesting that combat and camps should be situated on high ground and avoid swamps or collapsed ground [166].

The military books of that time not only recognised the significance of the terrain in warfare, but also mastered the principles for deploying troops in different geographical areas, such as those involving plains, mountains, hills, rivers and lakes. The *Sunzi bingfa* emphasises the need for caution when marching in dangerous, low-lying and mountainous areas: "Be on guard if the army is marching past precipitous ravines, marshes, reeds and rushes, forested hills and thick undergrowth. These places must be searched thoroughly." While marching through rivers, "move to distance yourself from it (the river); when the advancing enemy is crossing the river, do not meet him in the river, it is to your advantage to wait until he is half-way across and then to strike", so as to gain the upper hand. When engaging the enemy, "do not position your troops near the water to confront your enemy", "take up a position on high ground facing the sun"; "do not take up a position that is downstream from the enemy". In the *Wuzi bingfa*, there is a list of the types of terrain that hinder combat, and the methods for engaging the enemy in these areas: "when they encounter hills, forests, valleys, treacherous mountains, large swamps and other unfavourable terrain, the troops should move through quickly without delay. If the troops are engaging with the enemy in a mountainous area, first of all, beat the drums and shout to disrupt the enemy; then, move the archers forward to the front, and consider the strategy while remaining on guard". In the *Sunbin bingfa*, there is a summary of the effects which different terrains exert upon warfare: "as far as fighting is concerned, the five kinds of terrain can be graded as follows: mountains are better than hills; wooded hills are better than barren hills; barren hills are better than rolling ground; and rolling ground is better than wooded plains ... The following five types of terrain render operations difficult: ravines, rivers, marshes, alkaline land... [167]" In the *Liutao*, strategies for military operations in different terrains are proposed: "They occupy river valleys and dangerous and difficult positions in order to defend against chariots and cavalry. They occupy narrow passes and mountain forests in order to attack a large force with a few soldiers. They occupy low-lying, marshy, deep and dark areas in order to conceal their presence. They clear and open up the ground in order to fight with strength and courage [168]."

② Military classification of terrains

Different types of terrain require different military tactics. In warfare, whether attacking or defending, it is necessary to take advantage of the actual type of terrain to avoid harm. Also, unusual terrain must be utilised to trap the enemy in order to achieve victory. This was especially true during the cold weapon era, and thus explains the classification of terrain according to the interrelationship between topological conditions and military tactics. Ancient strategists did not only classify terrains by their natural features. Instead, they based the classification on the practice of conducting a comprehensive analysis of the terrain, and even of the vegetation, soil, bodies of water and weather conditions, which was distinctly different from that in physical geography.

In the *Sunzi bingfa*, terrains are classified into six categories based on their impact on warfare: "*tong*, that which is accessible; *gua*, that which enmeshes; *zhi*, that which is disadvantageous to both sides; *ai*, that which is narrow and precipitous; *xian*, that which is hazardous; and *yuan*, that which is distant [154]."

Based on a variety of military tactics and combinations of different terrains, the book also divides war zones into nine categories: "dispersive, marginal, contested, open, focused, critical, difficult, beleaguered, and deadly" [169], in which the topographic elements, as well as the combined influence of the geographic location and traffic routes, are taken into consideration.

In addition to the above classification, the book also summarises six kinds of unfavourable terrain for marching and fighting: "steep river gorges, deep shafts, boxed-in recesses, tangled undergrowth, treacherous quagmires or dangerous crevasses." These are difficult terrains that hinder the marching of troops, and hence they should "move away with haste and avoid approaching them if possible" [164].

During the Spring and Autumn and the Warring States periods, not only was there large-scale warfare, but there were also many types of troops involved, including chariots, cavalry, infantry, etc. The chariots and cavalry, in particular, were more restricted by the terrain. The *Liutao* stresses that "in the case of chariots, it is essential to know the terrain" and summarises ten kinds of terrain which impede chariots in battle:

> When you can advance but cannot withdraw, this is fatal terrain for chariots. When you must pass through a restricted area and pursue the enemy over a long distance, this is a terrain that may exhaust the chariots. When the land in front is flat, while the land to the rear is restricted, this is a terrain that may trouble the chariots. When you enter a restricted area from which escape is difficult, this is a terrain where the chariots may be cut off. When the land is collapsing and marshy with black sticky mud, this is a terrain that may cause the chariots to struggle. When it is restricted to the left and flat to the right, and there are sloping fields in front, this is a terrain that is unsuited to chariots. When tall grass runs through the fields and you have to cross deep waters, this is terrain that can defeat the chariots. When the chariots are few in number, the land is flat, and chariots and infantry are not employed in a coordinated way, this is a terrain where the chariots may be defeated. When there are ditches and canals to the rear, deep water to the left, and steep slopes to the right, this is a terrain that may destroy the chariots. When it has been raining incessantly for about ten days, the roads are collapsing, and you can neither advance nor retreat, this is a terrain that may sink the chariots [170].

The classification of these ten types of terrain is based on the characteristics of the chariot, the constraints of the terrain on the chariot, and other factors such as vegetation, water, soil quality, weather conditions, etc. The book also proposes nine unfavourable types of terrain for the cavalry in battle.

In the annexation wars between vassal states, siege battles were frequent. The ancient strategists placed strong emphasis on taking advantage of the unfavourable terrain of the enemy's city when launching a siege battle. The *Sunbin bingfa* provides a comprehensive summary of siege warfare. The book divides cities into two categories according to their topographic characteristics: *xiong* (male), which are difficult to attack, and *pin* (female), which are easy to attack. The topographic features of *xiong* cities are that they "do not have mountains as bulwarks or deep valleys as moats but they have rolling hills all around them...with deep valleys in front and mountains at the back ... built on high land with low land around it, or situated on rolling hills", a description which highlights the height and other features of the terrain. The features of *pin* cities include "no mountains on either flank but only a deep valley behind ... built on poor land ... an army (of the city) which has to rely on stagnant water ... surrounded by marshes but without rolling hills or mountains as a bulwark or deep gullies to form a moat ... situated amidst mountains but with no deep gullies as a moat, nor any rolling hills as a bulwark ... [They] have mountains in front and deep gullies at the back, in other words, high in front and low at the back". [171]

The book not only emphasises the topographic features, but also takes into account the city's soil and water quality.

The military classification of terrain marked the development of thought regarding military geography at that time, since it was a generalisation derived from a large body of experience accumulated from ancient warfare. Such classification had a great influence on the thinking of later generations regarding military geography.

③ The significance of weather, vegetation and celestial phenomena in warfare

Weather conditions must be taken into account during military operations. In the *Sunzi bingfa*, the concept of *tian* refers to "the effects of night and day, of good and bad weather, of winter's cold and summer's heat: in short, the conduct of military operations in accordance with the changes in the natural forces" [172]. When attacking with fire, Sunzi suggests that dry weather is best for using fire attacks. In the *Wuzi bingfa*, eight favourable situations for engaging the enemy are mentioned, among which natural conditions such as the weather are considered.

> First, in strong winds and severe cold, after marching by day and by night, felling trees and crossing rivers despite the hardships faced by the troops. Second, in the midst of scorching summer heat, setting out late, without resting along the way, marching rapidly while suffering from hunger and thirst, and solely focused on reaching a distant location. Third, after being on campaign for a long time, with no food left and the people discontented, rumours spreading uncontrollably, and generals unable to quell the unrest. Fourth, with supplies depleted, little firewood, and continuous rain, with no place left to plunder. Fifth, with limited troops, unable to adapt to the local environment, horses and men falling ill, and with no local reinforcements arriving. Sixth, with troops tired and fearful as the day comes to a close, worn out and hungry, needing to remove their armour and rest. Seventh, with leaders lacking authority, and the army's morale unstable, causing repeated panic among the troops, and with no support from the solitary commander. Eighth, with deployment undecided, and the camp not yet set up, only halfway through traversing the treacherous terrain of mountain and ravine [173].

If the enemy is in any of the aforementioned situations, "it is not necessary to resort to divination or other forms of prediction (no need for precaution) before engaging them in battle". The text discusses weather conditions that can affect the enemy's morale, such as strong winds and severe cold, scorching summer heat, and continuous rain. It also mentions geographical conditions, such as crossing frozen rivers, failing to adapt to the local environment, long distance travel, marching at dusk, and traversing mountains and rivers.

Among the 72 assistants proposed in the *Liutao*, there are "three astrologers who are responsible for observing astronomical phenomena, forecasting the weather, predicting timings, investigating auspicious signs, verifying disasters and abnormalities: thereby assessing whether the signs of Heaven are favourable or unfavourable to military operations" [161]. This highlights the observation of celestial phenomena and weather conditions in warfare. The book also considers "the advantages of high winds and heavy rain" to be an omen of a great victory [174].

Using fire in attacks was common practice in ancient warfare. Many ancient military works discussed the specific ways of employing such a practice. To launch a successful fire attack, it was necessary to consider the weather conditions including the force and direction of the wind, air humidity, etc. As mentioned in *Sunzi bingfa*: "If a fire is set upwind, do not attack from downwind." It concludes that "there is appropriate weather and suitable days for starting a fire. Dry weather is best for using fire attacks and the days when the moon is passing through the constellations of the Winnowing Basket, the Wall, the Wings and the Chariot Platform are the best ones for launching a fire attack because those are generally the days when the winds are strengthening". This recognises the relationship between lunar movement and the wind. It also notes the pattern of wind movement: "If the wind blows persistently during the day, it will probably die down at night [175]."

The *Liutao* also discusses fire attacks. It suggests that when the weather is dry, or in areas with dense grass, the enemy is liable to launch a fire attack. The *Wuzi bingfa* also highlights the influence of the wind on warfare: "Before the battle, it is necessary to observe the direction of the wind. When the wind is favourable (blowing from behind), seize the opportunity and launch an attack. When the wind is unfavourable (blowing towards you), hold the line and wait for the situation to change [176]." There is similar idea in the *Sima fa.*

Due to the necessity for war, ancient strategists were often well-equipped with geographical knowledge. The *Sunzi bingfa* compares the laws governing military operations to those of flowing water: "The laws governing military operations are like those governing the flow of water, which always avoids high points, choosing lower ones instead. To operate the army successfully, we must avoid the enemy's strong points and seek out his weak points. As the water changes its course in accordance with the contours of the terrain, so a warrior changes his tactics in accordance with his enemy's changing situation [177]." This indicates that ancient strategists had already noticed the patterns in the movement of water. Water warfare was common in ancient times, and many military texts have discussed the tactics for such battles, such as the *Wuzi bingfa: Yingbian* (Responding to Changes), *Liutao*: *Baotao: Niaoyun zebing* and so on, which indicates that people at that time clearly had some knowledge of the properties of flowing water.

Vegetation was also recognised as an important factor in warfare. The *Liutao* often mentions the principle of utilising the vegetation in warfare: "...[they] occupy positions in deep and dense grass in order to conceal their retreat and movements. They occupy narrow passes and mountain forests in order to attack a large force with a small one." It suggests that, in warfare, one should take advantage of lush vegetation, mountains and forests to win against a stronger force. The *Liutao: Baotao: Linzhan* specifically discusses methods for fighting in forested areas.

In ancient military texts, there are also discussions of seasonal changes, soil conditions and other natural phenomena. These discussions are scattered throughout different chapters, and although their understanding of natural phenomena remains rudimentary, they do indicate that military strategists during the Spring and Autumn and Warring States periods already had a relatively comprehensive understanding of military geography. In addition, thought regarding military geography developed relatively fast during this period.

3.5.2 Hydraulic Engineering

(1) Overview of hydraulic engineering construction

The use of iron tools greatly facilitated social and economic development during the Spring and Autumn and the Warring States periods, thus enabling the construction of large-scale hydraulic engineering projects. Meanwhile, as the vassal states aspired to dominate China and expand their territory at that time, they were actively developing agricultural production. As the key to agricultural development, hydraulic engineering projects were, therefore, highly valued by the rulers of each state. Moreover, the construction of new ditches and canals could facilitate transportation and military operations. Hence, one of the largest hydraulic engineering projects in Chinese history was motivated by political, economic and military requirements and commenced during the Spring and Autumn and the Warring States periods.

In ancient China, agriculture relied heavily on irrigation to increase crop yields. Initially, this was done manually by carrying water to the fields. As people gained more hydrological knowledge, they began to dig small irrigation channels, which eventually developed into large-scale canal systems for drainage and irrigation. The *Zhushu jinian* ("Bamboo Annals") recorded that in the twenty-ninth year of the Xia dynasty ruler Jie's reign (1516 BC), "people excavate mountains to connect rivers," which was the earliest hydraulic engineering project recorded in China. During the Western Zhou period, a system of drainage and irrigation channels was already in place. In the *Shijing*, it is written that "the water from the pool flows away to the north, flooding the rice fields" [178]. The "pool" was located near Xianyang, and people diverted its water northwards to irrigate the fields. This was a simple water diversion system for irrigation. The *Zhouli* records that the "Dao Ren (official) is in charge of rice cultivation. He builds reservoirs to store the water, constructs embankments to block the water, and digs ditches to drain, divert and channel the water for irrigation" [179]. This describes the small-scale water storage, drainage and irrigation systems of the time.

During the Spring and Autumn and Warring States periods, there was significant progress in both the scale and types of hydraulic engineering projects. Some construction projects that are well-known nowadays were completed at that time. Many canals were excavated, which later formed networks for transportation and irrigation. Some of these infrastructures are still in use today.

① Reservoirs and canal systems

Que Bei, also known as Anfengtang, was built in the late sixth century BC by Sun Shuao of the state of Chu. It is a large-scale water storage and irrigation project located at the south of Shouxian County in Anhui province. As early as the middle of the Spring and Autumn period, this area had become an important agricultural area in the state of Chu. Taking advantage of the local topographical features, Que Bei Reservoir was constructed by building a dam around natural lakes. The "Shuijing: Feishui zhu" records that "the dam has five openings that allow for the intake of rivers flows". [180]

The Que Bei Reservoir is over 100 li long. Before it was constructed, frequent floods and droughts during the rainy and dry seasons caused great harm to agricultural production in the region. After its construction, the region enjoyed continuous good harvests, greatly enhancing the economic strength of the state of Chu. During the Eastern Han dynasty, it was able to irrigate over 10,000 ha of land and continued to function until modern times. It is now a part of the Pishihang Water Conservancy Project.

The Qisi-Yulou Irrigation District is located in the area between Jinzhai County in Anhui province and Gushi County in Henan province. It was built from the early seventh century BC to the late sixth century BC. The *Huainan zi* records that "Sun Shu'ao diverted water from the Qisi River to irrigate the fields of Yulou" [181]. This was the earliest irrigation canal system recorded in China.

The Zhibo Canal is located about 30 li southwest of Taiyuan in Shanxi province, near Jinyang. It was built in the early Warring States period. According to the *Shuijing: Jinshui zhu*, the Zhibo Canal utilised the topographical features of the valley to dam up and store water and used the difference in water levels inside and outside the reservoir to channel water for irrigation [182]. This canal is still in use today. The Zhangshui Twelve Canals, also known as the Ximen Canal, were built in 422 BC. The magistrate of Ye County, Ximen Bao, "diverted the Zhang River to irrigate the county's fields, and thus the state of Wei prospered" [183].

Ye County was located 40 li southwest of present-day Linzhang County. This area was where the Zhang River entered the plain from the mountainous area. Due to the sudden change in terrain, the riverbed was often silted up and, coupled with abundant rainfall in the mountainous area, the Zhang River used to flood frequently. Exploiting the local natural features, including hard soil, an elevated terrain, and a large difference in water levels between the river and the irrigation area, Ximen Bao "dug twelve canals and diverted the river water to irrigate the fields" [184]. After flowing through the Loess Plateau in Shanxi province, the water of the Zhang River contains a large amount of organic matter in its sediment. When the river flows through Ye County, the gradient of the slope decreases, and the flow slows down. Through the irrigation of the Twelve Canals, the original saline-alkali soil was transformed into fertile soil, promoting the local agricultural economy. The Zhangshui Twelve Canals have benefited local people for nearly a thousand years.

The Baiqi Canal, built in the late Warring States period, was an irrigation area created by the diversion of the Yi River (known as the Man River nowadays), a tributary of the Han River. The *Shuijing zhu: Mianshui zhu* records that "when Bai Qi of the Qin planned an attack against the state of Chu, he rebuilt an existing weir which was several hundred li away from the Yan Capital and constructed a large canal to channel the water from the Man River (passing through Changgu Valley in Xishan Mountain) to the Yan Capital. Today's Yundou Bei area had since become a deep reservoir" [185]. Later generations used this canal to irrigate farmland. The Baiqi Canal has now developed into the Changqu irrigation area.

The Que Bei and Qisi-Yulou irrigation areas are located in the undulating hilly areas of southern China, where the topography is low relief and thus favours water storage projects, while the Zhangshui Twelve Canals and the Zhibo Canal in the north are water diversion projects for irrigation. The Baiqi Canal is located in the Huai River basin, which is between north and south China, and according to the "Shuijing: Mianshui zhu", it is an irrigation system that combines canals and ponds which was formed by excavating canals to connect the natural ponds in this area.

During the Warring States period, the state of Qin highly valued the construction of hydraulic engineering projects, among which the Dujiangyan and the Zhengguo Canal were the best-known.

Dujiangyan was built in the 51st year of Qin Zhaowang's reign (256 BC). Its original name is no longer verifiable, and it was called Jianbeng, Jianyan, and Du'an Dayan after the Han and Wei dynasties until it was finally named Dujiangyan in the Song dynasty. It is located on the alluvial fan of the Min River, in Dujiangyan, Sichuan province. It is a comprehensive hydraulic engineering project, and its three most important components are the Yuzui Levee (or Fish Mouth Levee) for water division, the Feishayan (or Flying Sand Weir) for regulating the water flow, and the Baopingkou (Bottle-Neck Channel) for diverting the water for irrigation. The Yuzui Levee divides the Min River into the inner and outer streams. The outer stream is a section of the Min River, while the inner stream is the diversion canal. The Feishayan is an overflow channel used to regulate the inflow of water. The Baopingkou is the water diversion channel which was gouged through Yulei Mountain. The design of Dujiangyan is quite sophisticated and, from a technical point of view, the current Dujiangyan is almost the same as the ancient one. Three stone figures were also placed in the river to monitor changes in the water level: "When there is insufficient water, the level should not be as low as the figure's foot; when there is plenty of water, the level should not exceed the figure's shoulders [186]." Obviously, these three stone figures were the water gauges in ancient times. The construction of Dujiangyan not only helped control droughts and floods on the Chengdu Plain but also facilitated agricultural irrigation and navigation on the Min River. The *Shiji* records that "during his time as the governor of Shu, Li Bing excavated two new channels in Chengdu that were navigable by boat, and he used the extra water diverted from the river to irrigate land, benefiting the local people". The *Huayang guozhi* also records that "in the dry season, water is drawn for irrigation, and in the rainy season, the water gate is shut to control the water flow. Therefore, it is said that 'flood and drought is controlled by people, and there is no famine. The world calls this place the land of abundance'".

The Zhengguo Canal was built in the first year of Qin Shi Huang's reign (246 BC). It starts in Jingyang County in Shaanxi province and flows eastwards into the Luo

River, irrigating the Guanzhong Plain. It is over 300 km long. This was the earliest large-scale hydraulic engineering project built in the Guanzhong area. The Zhengguo Canal fully exploited the topographical advantages of the area with the excavation of a water intake in the downstream section of the Jing River's concave bank. The canal made use of the water flow with the highest velocity and finer sediment on the surface layer flowing from the convex bank to the concave bank, while the water flow containing larger particles of sediment on the bottom layer flowed from the concave bank to the convex bank. This prevented coarse sediment from entering the channel and silting up the canal bed, while ensuring a larger water intake at the canal mouth. At the same time, the fine sediment rich in organic matter also entered the canal used for irrigation. The completion of the Zhengguo Canal promoted the economic development of the Qin dynasty.

② Canal and embankment projects

During the Spring and Autumn and Warring States periods, the construction of reservoirs, dykes, and channels was mostly aimed at flood control and irrigation, while the excavation of canals was mostly for military purposes, which also contributed to local economic development. Canals constructed during this period were mainly located in the north of the middle and lower reaches of the Yangzi River and in the south of the middle and lower reaches of the Yellow River.

The Han Canal, also known as the Han Gou, was excavated by King Fuchai of Wu state in 486 BC with the aim of conquering the Central Plains. It was the earliest canal connecting the Huai River and the Yangzi River. The excavation of the Han Canal took advantage of the many natural rivers and lakes in the area and connected them by means of artificial channels. In addition to the Han Canal, the state of Wu also excavated the Xujie (Zhongjiang) and Xupu (Dongjiang) canals to attack the state of Chu. Chu, in turn, excavated the Lido Canal to attack Wu.

The Hong Canal, also known as the Bian Canal, was excavated by the state of Wei in the tenth year of King Hui's reign (360 BC), with Kaifeng as the centre. It connected the "Ji, Ru, Huai, and Si rivers" [183], as recorded in the *Shiji*: *Hequ shu*. It connected the Yellow River, Huai River and Yangzi River systems, forming a canal network centred around Kaifeng.

The Lingqu Canal was constructed by Emperor Qin Shi Huang from 223 to 214 BC in order to unify the Lingnan region. It connected the Xiang River and Li River, linking the two major river systems of the Yangzi and Zhu rivers, thus opening up water transportation between northern and southern China. Its construction took advantage of the terrain and addressed the problem of seepage in karst caves. It is the world's earliest lock and summit-level canal.

During the Spring and Autumn and Warring States periods, the most famous embankment engineering project was the Yellow River Dyke. As the Yellow River flowed through the Loess Plateau region, it carried a large amount of sand and silt, which was continuously deposited as sediment when the river reached the plains, causing frequent flooding. In the Spring and Autumn period, embankments had been built in the lower and middle reaches of the Yellow River. During the Warring States period, embankments became more widespread, and the construction of embankments along the Yellow River in the territories of Qi, Zhao and Wei began to take shape. Hence, the hydraulic engineering construction projects during the Spring and Autumn and Warring States periods were remarkable at world level.

(2) Hydrologic knowledge

The construction of a hydraulic engineering project requires surveys to be conducted, as well as design and planning in the field; this necessitates a particular type of hydrological knowledge in order to understand the water conditions, the relative position of the water source, the irrigated area, the relative elevation, and the topography, and to be able to tackle a series of potential problems such as drought, flood, erosion and siltation. Many texts in the Spring and Autumn and Warring States periods covered hydrology or hydraulic engineering construction, and monographs on hydrological knowledge, such as the *Yugong: Daoshui* and *Guanzi: Dudi* had appeared. Among them, the *Yugong: Daoshui* is "an ancient water system table based on geographical knowledge, at the very least the earliest hydrological geography in China" [187]. The *Guanzi: Dudi* is a systematic treatise on hydrological knowledge and hydraulic engineering technology.

① Classification of rivers

In a particular watershed, many rivers form a connected network of water systems. Modern hydrology classifies rivers into different levels based on their final outflow destination (ocean, river or lake). Early classification of rivers was based on their origin, flow path and final outflow destination.

> Some bodies of water are big and others are small. Some are nearby and others are far away from us. Waters rising in the mountains and pouring into the sea are referred to as *jingshui*. Those originating from other rivers and pouring either into large rivers or into the sea are referred to as *zhishui*. In the valleys among mountains, some parts have water but other parts do not and this kind of water is referred to as *gushui*. Waters originating from underground and pouring into big rivers or seas are referred to as *chuanshui*. Those originating from underground and becoming stagnant are referred to as *yuanshui* [188].

Although such a classification was not as simple and clear as in modern hydrology, its actual aim was to make rational use of water resources and to build hydraulic engineering facilities for agricultural purposes: "As for these five kinds of waters, they can be guided according to geographical conditions and can also be intercepted or blocked as well. Nevertheless, you cannot prevent them from causing disasters from time to time." In this statement, "guided" refers to the diverting of water for irrigation according to the terrain, while "intercepted or blocked" represents the building of embankments to avoid flooding. It also emphasised that if this approach was adopted over a long period, water disasters will occur. Therefore, the early classification of rivers originated from the practices of production, and its significance was positive as it led to the controlling of the water flow to benefit the people.

② Preliminary understanding of hydrological factors

In hydraulic engineering construction, a series of hydrological factors must be considered, such as the level, velocity, discharge, and quality of the water and its sedimentation conditions.

Changes in water level are the first factor to consider in hydraulic engineering construction. The earliest water gauge in China, the *shiren* (stone figure), appeared as part of the Dujiangyan project, indicating that the ancient people had already mastered the changing patterns in water levels. The water level is subject to the amount of water the river is receiving. Ancient scholars attached great importance to the changing patterns of water levels. The *Zhuangzi* mentions that "the time of the autumnal floods had come, and the hundred streams were all discharging themselves into the river. Its current was greatly swollen" [189]. This recognises the role of rainfall in feeding the flow of the river and the impact of seasonal changes in rainfall on the water level. In the *Mencius*, it was recorded that "there is a spring of water; how it gushes out! It rests neither by day nor by night. It fills up every hole, and then advances, flowing onto the four seas". "It was this which he found in it to praise. But suppose that the water has no spring. In the seventh and eighth month when the rain falls abundantly, the channels in the fields are all filled, but it can be expected that they will be dried up again within a short while" [190]. The two ways of feeding a river had already been noticed: spring water and rainfall.

By mastering the seasonal changes in river water levels, it was possible to conduct flood control and hydraulic engineering construction. In the *Guanzi: Dudi*, it was recommended to "order the official in charge of regulating the rivers and watercourses to inspect the dykes and dams during winter"; "during the three months of the spring season the weather is dry and all the rivers are low. At that time, the frozen rivers begin to thaw, and all the mountains and rivers are short of water." During the time when the water level was low, by "digging out the earth from the watercourses to increase the dykes during spring and winter", the accumulated sediment could be removed and the water-carrying capacity of the watercourse could be increased, thereby reducing water disasters.

In the hydraulic engineering construction of that time, people had a preliminary understanding of the proportional relationship between flow rate and cross section. In the construction of the Dujiangyan project, people had summarised the principle that by maintaining the riverbed at a certain depth, a larger cross section could be created to allow more water to flow. This principle had also been applied in other hydraulic engineering projects of that time.

In the *Guangzi: Shuidi*, the regional differences in water quality are also pointed out. In the *Lüshi Chunqiu: Jichun ji: Jinshu*, water is divided into: "*qing* (short of salinity and minerals)", "*zhong* (too much salinity and minerals)", "*gan* (sweet)", "*xin* (pungent)" and "*ku* (bitter)". However, due to the underdeveloped scientific and technological conditions at that time, people were not able to conduct in-depth research on the differences in water quality.

③ Discovery of the effects of erosion and deposition

Due to the development of hydraulic engineering projects for agricultural purposes, during the Spring and Autumn Period and Warring States periods, people discovered the effect of water flow on landforms during the construction of canal systems. The *Zhouli* recorded that "a good ditch washes itself with water" [191]. Here, the "washing" refers to the erosion caused by the water flow. In the *Dao De Jing*, Laozi argued that "there is nothing in the world softer and weaker than water, and yet for attacking things that are firm and strong there is nothing that can take precedence over it" [192]. Regarding the accumulation of flowing water, the *Zhouli* suggests that "along the great rivers, there must be shores alongside". "Shores" here refers to the alluvial geomorphology formed by the accumulation effect of the flowing water. In the *Zhuangzi*, it is written that "when the stream dries up, the valley is empty; when the mound is levelled, the deep pool (beside it) is filled up" [193]. This recognises the basic mechanism of erosion: the water erodes the objects on its surface, and the sediment sinks to the bottom.

Flowing water can change the terrain, while conversely, the terrain can also constrain the flow of water: "Following its nature, when water comes from high up and heads for lower-lying places, it flows so rapidly that it can sweep away stones on the way. And when it comes from a lower-lying place and heads for a higher one, it will be held up [188]." When this principle is applied to hydraulic engineering construction, "ditches must be dug following the flow of the water, and embankments must be built following the contours of the terrain; a good ditch will wash away the silt with water by itself, while a good embankment will reinforce itself with the sediment carried by the water". Such practices use the erosion, carrying and depositing effects of the flowing water to wash the silt away out of the ditches and reinforce the embankments.

The movement of water is complex. It is influenced by the curvature of the riverbank and the slope of the riverbed. However, it can also change the shape of the riverbanks and riverbed. In the *Guanzi: Dudi*, it is recorded that "because of the nature of water, it will stop flowing and recede when it reaches extremely steep places. If a place is full of water, the water behind will drive the water in front of it and advance meanderingly. It will progress smoothly when it encounters lower-lying places. And it will splash about when a higher place is encountered. If the surface of the ground is undulating, it will cause a land slide. And if the surface is too rough, it will skip over some places. When it skips over some places, it will swirl. When it swirls, there will be whirlpools. Where there are whirlpools, the water will accumulate at some points. Where it accumulates, the mud and sand it carried will be deposited. When the mud and sand is deposited, the watercourse will silt up. Where it is silted up, it will change its original course. When it changes course, it will surge. When it surges, it will flow unrestrained. When it is unrestrained, it will harm people." This statement points out the reflux phenomenon of the water flow at the bend of the riverbed, as well as the impact of the riverbed on the speed of water flow, and it provides a comprehensive analysis of the hydrodynamics and sediment deposition of the river. What is more valuable is that the motivation behind understanding water flow patterns at that time was to be able to formulate specific plans for building river channels in hydraulic engineering construction: "if the water level upstream is raised and the water is led into a pottery channel with about a third of every chi in length sloping downwards, then the water can flow forty-nine li extremely swiftly. And the water can be channelled to distant places by a circuitous route and can also be channelled to higher places as well." The key to a successful canal is the slope. The slope should be "no more than three li in distance; if the base of the canal can be lowered by 49 *cun*, that is a 1/1,000 falling gradient, then the water can flow smoothly" [194].

In the same period, Western scholars' understanding of the effects of erosion and deposition came from their observations of the Nile. The annual flooding of the Nile brought a large amount of sedimentary deposits into the estuary, forming the Nile Delta. Ancient Egypt was located downstream of the Nile. Although the Egyptians knew that the sediment came from upstream, they had little knowledge of the erosive effects of the river, but they had a deeper understanding of the effects of deposition [195].

Herodotus, an ancient Greek scholar, pointed out that the Nile Valley was formed by the accumulation of sediment. He further explained that sedimentation was the reason for the formation of the Nile Delta and extended this theory to other river landforms. Herodotus observed that "the soil in Egypt is different from those in neighbouring Arabia and Libya, or in Syria, rather it is soft black soil, made up of silt and alluvial soil brought by the river from Ethiopia [196]." Therefore, Herodotus provided further evidence for the deposition effect of rivers. Aristotle also suggested that the territory of Egypt was formed from the sediment carried by the Nile [197].

④ Understanding of the water cycle

The water on the Earth's surface evaporates under solar radiation. After rising to high altitude, the vapour is transported by air currents to other areas, where it condenses and precipitates under appropriate conditions to form rainfall, and thus completes the water cycle on the Earth's surface. Although the concept of evaporation and condensation had not yet been clearly formulated during the Spring and Autumn Period and the Warring States periods, the idea of the water cycle had already emerged. In the *Lüshi chunqiu*, it is noted that "clouds move towards the west all the year round, waters flow to the east by day and night. Nonetheless, the source of clouds in the sky will never be exhausted and oceans on Earth will never become full. From another point of view, small can be considered as big, and heavy can be considered as light. That is in accordance with the principle of the round" [198]. This suggests the two aspects of the cycling system (*huandao*, the round): the cloud moving westwards and the water flowing eastwards. It is the gathering of small water which flows into the big ocean and the changing of the heavy water drops into light vapour that ensure "the source of clouds in the sky will never be exhausted and oceans on Earth will never become full", which all together contribute to an everlasting water cycling system.

3.6 The Use of Iron Wares and Mineral Petrology [199]

3.6.1 A Glorious History of Iron Wares

The smelting of iron and the manufacturing of ironware began in Xinjiang at the end of the Shang dynasty and the beginning of the Western Zhou dynasty. In the Central Plains region, iron smelting began in the middle to late Western Zhou dynasty [200]. The prevalence of ironware was the most striking characteristic of the Spring and Autumn and the Warring States periods, indicating that China had gradually transitioned to the Iron Age from the Bronze Age before the Western Zhou dynasty after acquiring full understanding of the properties of iron ore.

Only a few examples of ironware from the Spring and Autumn period has been unearthed, such as iron bars, iron projectiles, swords (Fig. 3.5), daggers, *ding* and so on.

By the Warring States period, the types and quantities of ironware had significantly increased to include shovels, hoes, swords, axes, chisels, iron crucibles, knives, scrapers, hammers, sickles, spears, etc. It is evident that ironware at that time was not only used for agricultural implements but also for weapons, tools for handicrafts and other utensils, suggesting that the use of ironware was already widespread. Iron

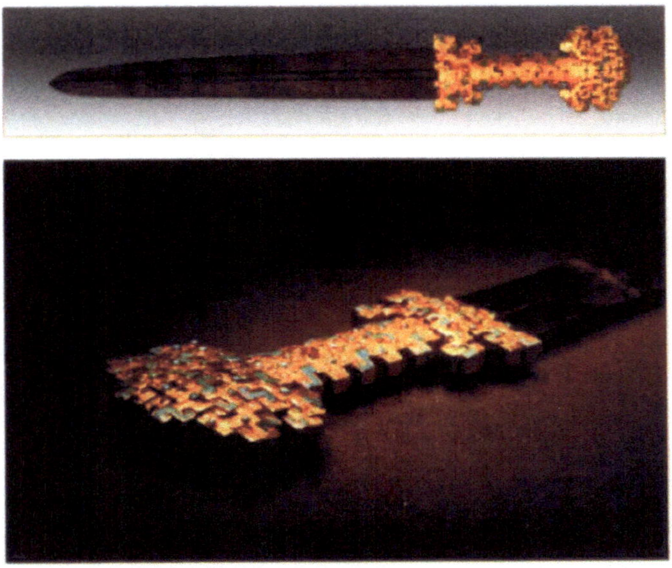

Fig. 3.5 A gold-handled iron sword unearthed from the Spring and Autumn tombs in Baoji. 徐爽 (2018) *Baigonglu: Yanjing bajue huasi xiangqian*百工录: 燕京八绝花丝镶嵌[Hundreds of examples of craftwork: Filigree inlay art—eight unique skills from Yanjing]. Nanjing: Jiangsu Art Press南京: 江苏艺术出版社, p 4

farming implements were more suited to agricultural production, and the implements unearthed have been found in the seven Warring States and Baiyue regions, suggesting the iron processing technology had reached a relatively high level. The state of Chu was not only the earliest region to use iron farming implements, but also the first to cast iron cooking utensils.

Seventy-nine ironware items were excavated from a late Warring States period tomb in Yanxiaodu (now Yi County, Hebei province) in northern China, most of which were production tools and weapons. Among these ironware items, six were pure iron or steel products, and three were wrought iron products that had been softened or untreated.

Carbon steel had already appeared in the late Spring and Autumn period, such as in a steel sword unearthed at a tomb in Changsha, Hunan province, which contained carbon steel with a carbon content of 0.5%.[201] In the late Warring States period, the blooming process was popular, through which sponge iron could be obtained. Then, by increasing the carbon content, high-carbon steel could be produced. Quenching technology had also been mastered by that time.

In a storage pit in Luoyang, Henan province, 126 iron artefacts were unearthed, with a total weight of more than 400 kg, including agricultural tools, handicraft tools, etc., indicating that iron tools had been widely used by the people.

During the Warring States period, the state of Zhao, centred around Handan, had developed iron smelting. There were rich magnetite deposits underground, especially in Ci Mountain in Wu'an, where the magnetite ore was shallow and of high quality, making it easy to mine.

In addition to casting iron tools, iron ore was also used in making pigments, refractory materials, weighty sword handles and other items.

3.6.2 Early Gold Coins and the Extensive Use of Gold Wares

Gold is one of the precious metals that was used relatively early in ancient China. Gold earrings and gold rings from the Xia dynasty have been unearthed [202]. Gold artefacts from the Shang dynasty, such as gold staffs, gold masks, tiger-shaped decorations, fish-shaped ornaments, leaf-shaped decorations and gold ingots were discovered in Sanxingdui, Sichuan [203]. During the Spring and Autumn and Warring States periods, gold was used in various sectors, and an abundance of gold artefacts from this period have been unearthed, such as shellfish, foil, vessels, earrings, etc. The most characteristic object unearthed is gold currency, including gold bars, *yingyuan* (Fig. 3.6) (also known as *yingzijin*; *ying* is the name of Chu's capital; a *yuan* is a unit of weight, and one *yuan* weighs about 12 *zhu*, equivalent to half a *liang*), gold plates, *chenyuan*, *lujin* (gold content 94%-95%), etc. In addition, there were gold spring-shaped objects, gold leaves, belt hooks, spoons, cups, threads, pendants, bead strings, necklaces, chains, crowns, gold animals, and bronze objects with inlaid gold.

Fig. 3.6 The *yingyuan* (gold coin) of the state of Chu (photographed by Liang Qing 梁擎)

This demonstrates the high level of the gold production and processing capacity at the time and reflects people's understanding of the properties of gold.

During this period, the majority of the gold production in the state of Chu was from alluvial sand gold, which was obtained through washing and sifting. It could later be melted into high-purity gold. The mining and refining of gold from primary ores or veins, which entails much higher production costs than alluvial gold, was still rare at this time.

Gold coins have been unearthed in provinces such as Anhui, Jiangsu, Shandong, Henan, Hubei, and Hunan. China had already entered the stage of minting gold coins, and the casting of gold ingots began in the Warring States period.

In addition to iron and gold, there are 83 types of minerals and rocks involved in the cultural relics of this period. Listed in order of frequency these are: copper, jade, pottery, coloured soil, agate, limestone, turquoise, soapstone, earthenware, lead, tin, crystal, granite, sandstone, silver, shale, cinnabar, slate, glass, marble, etc. [204]. The main uses of several of the minerals above are listed below.

Copper ore: mainly used for making bronze utensils. There were numerous types of bronze utensils, and they existed in large quantities. They were widely used in society at that time, including as agricultural tools, handicraft tools, weapons, musical instruments, etc. One ancient tomb yielded 800–1,690 pieces of bronze utensils, illustrating the vast quantity and popularity of bronze. From ancient copper mine sites, it has been discovered that the main types of copper ore mined at that time were native copper, bornite, malachite, chalcopyrite, and covellite.

Jade ore: mainly used for making jade utensils, including decorative objects, ritual vessels, jade chimes, etc. The types of jade included black jade, white jade, yellow jade, grey-white jade, blue-white jade, green jade, mixed jade, etc. During the Spring and Autumn period, many deceased individuals were found with small, delicate jade objects in their mouths. Some of these objects were intricately carved jade pendants and, in some cases, the jade was carved as thin as a sheet of paper, indicating the improvement in jade-carving techniques at that time. The practice of placing jade objects in the mouths of the dead developed during the Warring States period [205]. It was not only the mouth, but jade was also placed in the seven orifices of the body to prevent decay. Poor families who could not afford jade would use stone chips instead. In 1954, 262 Eastern Zhou dynasty tombs were excavated in Luoyang, and nearly 1500 pieces of jade artefacts were unearthed, including many sets of jade masks (made of jade pieces in the shape of eyes, noses, etc.). It can be seen that the custom of using jade as burial objects was prevalent at that time. [63] The idea behind this practice was to endeavour to transform the deceased into a deity or an immortal and preserve their physical form. However, in reality, the body would still decay.

During the Spring and Autumn and the Warring States periods, jade was not only decorative, but it also symbolised the power and privilege of the ruling class. It was also made into ceremonial objects used in religious activities. At that time, not only was the variety of jade used to distinguish between different ranks of authority, but also the size and shape of the jade. The *Zhouli* records that "to worship Heaven with a pale *bi*, Earth with a yellow *cong*, the East with an emerald *gui*, the South with a red *zhang*, the West with a white amber, and the North with a black *huang*".

Concerning the nature of jade, Confucius argued that "it is refined, pleasant and beneficial. Its veining has regular patterns and an orderly arrangement. It is hard and strong and will not be bent. It is sharply angular yet does not cause injury. It will break but will not give away. Its flaws and virtues are both visible. Strike it and its sounds will ring forth clearly and be heard in the distance, and when they cease, there is a sense of sadness" [206]. The *Guanzi: Shuidi* described jade as "magnificent and brilliant" in colour. The understanding of these characteristics of jade was acquired through the long-term practice of jade carving by craftsmen.

As regards the production of jade, in the *Shanhai jing*, over 100 places of jade production are recorded. However, the jade referred to in the book is not only limited to soft jade but also includes various stones, such as talc, marble, crystal, agate, and spinel, which are all termed as jade. Many of the unearthed jade artefacts from this period were also made from these stones, not just from soft jade. Understanding this point is key to understanding why the *Shanhai jing* records so many jade producing areas. The jade produced in Qinghai and Xinjiang mentioned in the book should refer to the source of Kunlun Mountain nephrite, called Kunlun jade and Hetian jade.

During the Spring and Autumn and Warring States periods, Duke Sui received a pearl called Suihouzhu (or the Marquis of Sui's pearl), which is a treasure and is as famous as Heshibi (or Mr He's Jade). Suihouzhu is first mentioned in the *Mozi*, as being a night pearl with a diameter of 23 cm. Yindian Town in Suizhou is rich in fluorite ores which are used in steelmaking. The fluorite ore itself cannot emit light, and it can only fluoresce under ultraviolet irradiation. Geologist Hao Yongwei believes that Suihouzhu may be a round gem-quality diamond, since some spherical olivine (an indicator mineral of kimberlite, and diamond is contained in kimberlite) was discovered in the Dahongshan area of Suizhou in 1972. Later, gem-quality diamonds were indeed found in this area. Ordinary diamonds also have luminous properties, so it is highly possible that the ancients picked up an oversized diamond in Suizhou and offered it to Duke Sui. In the end, the Suihouzhu was buried with Qin Shi Huang.

During the Spring and Autumn and the Warring States periods, limestone ($CaCO_3$) was mainly used as a building material, or for manufacturing lime, stone chimes, whetstones, etc. According to the author's calculations, a total of 161 stone chimes from this period were unearthed. The numbers of chimes found in each tomb varied, with a maximum of 32 and a minimum of 4. Among them, sets of 13 stone chimes were the most common. The raw material for stone chimes was mostly limestone, with a few made of marble, slate or soapstone. There were 32 sets of stone chimes excavated from the tomb of the Marquis Yi of the Zeng state in Suixian County, Hubei province; their main raw material was limestone, and a few were marble.

The density and purity of the material, as well as the direction of extraction in the quarry directly affect the sound and duration of the chimes. The pitch of the chimes is related to their size and thickness, with larger and thinner chimes producing lower pitches and smaller and thicker chimes producing higher pitches.

The areas where the stone chimes were unearthed are Lucheng, Wanrong, Taiyuan, Changzhi in Shanxi, Zhucheng, Tai'an, Yishui in Shandong, Jiangling and Suixian in Hubei, Luoyang and Weichuan in Henan.

Silver was also one of the precious metals used in ancient China. Initially, natural silver was used to make small ornaments, such as the silver earrings and nose rings from the Xia dynasty unearthed in Huoshaogou, Gansu province. During the Spring and Autumn and the Warring States periods, silver coins emerged. The silver coins unearthed in a cellar in Fugou, Henan province (18 coins weighing 3,072.9 g in total and with an average weight of 170.7 g per coin) are the earliest physical evidence of silver coins in China [207]. In addition, there were silver filigree ornaments, coins with a silver outer-layer (lead-silver coins), silver plaques, silver-inlaid bronze wares, silver belt hooks, silver shells, silver dishes, etc.

What types of silver ore were used to extract silver at that time? According to Hou Dejun's research, it is believed that the Chu people were unable to use copper sulfide ores to smelt copper until the Warring States period and, therefore, could not extract silver from silver sulfide ores such as argentite (Ag_2S). They could only obtain silver by melting natural silver. In nature, there is elemental silver (natural silver), which is secondary, while the main silver minerals are argentite, lead-silver or silver-lead ores.

3.6.3 Mineralogical and Petrological Records in Literature

(1) *Wanwu*

The *Wanwu* ("Ten Thousand Things") is a part of the Han dynasty bamboo slips excavated at Fuyang, Anhui province, containing information on medicine and physics. It may have been written in the Warring States or Spring and Autumn periods. It records the use of minerals and rocks as medicinal materials, such as limestone, yellow earth, clay, salt, and vermiculite, making it the earliest example of Chinese literature on mineral medicine.

(2) *Wuzang shanjing*

The "Wuzang shanjing" (Classic of Mountains in Five Directions), also known as the "Shanjing", is a part of the *Shanhai jing*. It was written during the Spring and Autumn and Warring States periods and is one of the earliest works on minerals and rocks in China. The "Shanjing" describes minerals and rocks in great detail, recording as many as 89 kinds of minerals and rocks from over 400 locations. The description includes 11 aspects of the minerals, such as the hardness, colour, lustre, transparency, magnetism, touch (rough or smooth), sound produced when struck, medicinal properties, collective state (soil-like, block-like, egg-like, rice grain-like), symbiotic relationship, and melting (or solubility) [208].

The symbiotic or spatial relationship between minerals in the "Wuzang shanjing" includes: copper-whetstone, iron-wen stone, silver-grindstone, iron-jade-coloured earth, gold-silver, platinum-iron, etc. [209]

In some regions or mountains, there are altitudinal and yin-yang relationships between different minerals, as recorded in the "Zhongshan jing", "the Fuyu mountain has more copper on its yang side and more iron on its yin side". Here, yin and yang refer to the north and south slopes of the mountain. In some cases, yin and yang may refer to differences in elevation, with the peak being yang and the deep valley being yin [210].

In terms of mineral nomenclature, the "Shanjing" classifies minerals into four categories: gold, jade, stone and earth. This was the earliest mineral classification system in the world [211]. In terms of the knowledge of rocks and minerals, the "Shanjing" predated the earliest monograph on this subject in ancient Greece, *De Lapidibus*, by Theophrastus (371-285 BC), and is richer in content. *De Lapidibus* only records 16 types of minerals classified into three categories: gold, stone and earth, without mentioning jade.

(3) *Shangshu: Yugong*

The "Yugong" is a chapter in the *Shangshu* ("Book of Documents") and was written around 500 BC. It records various minerals and rocks, including three colours of copper, *di* (whetstone, made of coarse sandstone), *li* (fine stone, made of fine sandstone), arrowhead stone (such as quartz, granite and quartzite used for making arrowheads), cinnabar, *wusetu* (lit. five-colour earth)*fuqing* (such as limestone, marble and jade for making stone chimes), salt, lead, strange stones (including soapstone, talc, amethyst, scheelite, wollastonite and white dolomite), *yao* (jade), *kun* (referring to beautiful and strange stones), *cuo* (made of fine sandstone or quartzite), *qiu* (also written as *qiong*, referring to beautiful jade), *ling* (referring to hard jade, namely jadeite), *lang gan* (referring to green tourmaline), iron, grindstone, and *lou* (a kind of metallic mineral).

(4) *Yizhoushu: Zhifang jie*

In the *Yizhoushu: Zhifang jie*, various minerals and rocks are recorded, including gold, tin, cinnabar, silver, jade and stone.

(5) *Guanzi*

The minerals and rocks recorded in the *Guanzi* include gold, cinnabar, magnetite, copper, ling stone, lead, tin, red copper, ochre, iron, silver, and so on.

(6) *Shiji*

Although the *Shiji* ("Records of the Grand Historian") was composed in the early Western Han dynasty, some of the minerals and rocks recorded in it may reflect the situation in the Spring and Autumn and Warring States periods, so it is necessary to discuss some of its content.

The *Shiji: Huozhi liezhuan* mentions minerals and rocks such as jade, stone, gold, tin, lead, cinnabar, copper and iron. The *Shiji: Fanju liezhuan* records statements made by Fan Ju: "I heard that the sstate of Zhou has the fine *di'e* [diamond], Song has the precious *jielü* [emerald], Liang has the luminous *xuanli*, and Chu has the *hepu* jade." According to Shen Fuwei, *di'e* is the ancient translation of the Sanskrit word *vadjra*, meaning diamond. *Xuanli* is described by Ban Gu in the *Xidu fu* and Zhang Heng in the *Xijing fu* as a fluorescent stone. *Xuanli* in Sanskrit means milky white opal and these were mainly produced in Pune County near Mumbai, India. According to the *Bieguo dongming ji*, in 116 BC, Emperor Liu Che started to build the Zhaoxian Ge (or Pavilion for Accommodating Immortals) and used *xuanli* and *huoqi* to make the beds. Opals are composed of hydrosilicic acid, a chemical similar to the silicate in *huoqi*. *jielü* was crystal clear and bright in colour. It was produced in the Wadi Sikait-Abu Subera hills region of Roman Egypt and imported into China via Persia. Therefore, the Persian name *jielü* was adopted. Emerald is a variety of beryl, which is beryllium aluminosilicate and is not limited to green [212]. Hepu was the treasure stone discovered by Bian He in the state of Chu, that is, labradorite.

3.6.4 Prospecting Theories

In the *Guanzi: Dishu*, the experience of Chinese people in prospecting for minerals is recorded.

> When cinnabar appears on the surface of a mountain, there must be gold inside. When *cishi* (magnetite) appears on the surface of a mountain, there must be a copper vein inside. When *lingshi* (green stone) appears on the surface of a mountain, there must be lead ore, stannite and cuprite inside the mountain. When reddle appears on the surface of a mountain, there must be iron ore inside. These are all indicators of mines.

There is also another way of expressing it, as follows: if there is vanadium on the surface of a mountain, there is silver beneath it. If there is cinnabar on the surface of a mountain, there is a good mine beneath it. If there is *cishi* on the surface of a mountain, there is cuprite beneath it. These are all indicators of mines.

The relationship between different mineral ores mentioned above can be interpreted in the following three ways.

Firstly, the appearance of a certain mineral outcrop on a vertical ore body or vein may indicate the presence of another mineral deposit below, and such a mineral indicator was called *miao* (seedling) or *yin* (guiding object) in ancient times. The upper and lower parts of some polymetallic deposits may contain different enriched minerals. Hence, the ancient people had some understanding of this vertical zonation of minerals.

Secondly, if a certain mineral appears on a mountain and another mineral appears at the foot of the mountain, and they are each produced in different strata or rocks without their origins being connected in any obvious way, this is a relative spatial relationship.

Thirdly, if a certain primary ore deposit is found on a mountain and another sand ore appears at the foot of the mountain, this is also an upper-lower relationship. This relationship is not necessarily related to the formation of a certain mineral. Therefore, the prospecting theories described in the *Guanzi: Dishu* only involved finding ore deposits through the symbiotic relationship of metallic minerals to some extent, while the rest was through the relative spatial relationship between mineral deposits [213].

This was the earliest systematic prospecting theory in China. It was also one of the important prospecting theories to emerge in the Spring and Autumn and the Warring States periods and laid a theoretical foundation for the development of mining in the future.

3.7 Summary

The Spring and Autumn and the Warring States periods were a time of rapid development for geoscience in Chinese history. The term *dili* also appeared in works during this period, and the emergence of this concept marked a leap forward in geoscientific knowledge.

The literature of this period is quite rich in geoscience content, and the works that have been passed down to the present day include the earliest comprehensive geoscience works in China: *Shanhai jing* and *Yugong*, as well as the earliest travelogue-style geoscience work, *Mutianzi zhuan*. Seminal works on hydrology, soil, geology, maps, military geography, and other aspects of geoscience also emerged, such as the chapters "Diyuan" (types of landform), "Ditu" (map), "Dudi" (survey of the land), and "Shuidi" (water conservancy) in the *Guanzi*, and the chapters "Dixing" (topography) and "Jiudi" (types of battlegrounds) in the *Sunzi bingfa*.

Moreover, during this period, copious ideas and thoughts on geoscience emerged in an unprecedented manner. There were two main trends that characterised geoscientific thinking during this period: firstly, the philosophical methods of speculation and conjecture were directly applied to the explanation of natural phenomena and their laws. Although such interpretation involved mystic features, it was still the first step that humans had taken towards grasping the laws of nature and conquering the natural world. Secondly, some thoughts on geoscience were directly derived from the experiences that people had acquired from their long-term production practices.

The earliest surviving physical maps in China are the *Zhaoyu tu* drawn at the end of the Warring States period and the seven maps unearthed from the Qin tomb at the Fangmatan site. The analysis of these physical maps, as well as the records in the literature of the time, show that during the Spring and Autumn and the Warring States periods people had already mastered a high level of cartographic technology.

Much of the geoscience knowledge during this period came directly from ordinary social life and production practices. For example, the frequent wars between the vassal states during this period promoted the development of military geographical thought, while the construction of hydraulic engineering projects by various vassal states during the Warring States period also promoted progress in hydrological knowledge.

References

1. Li J李久昌 (2004) Dui zhongguo lishidilixue yuanyuan wenti de zairenshi对中国历史地理学渊源问题的再认识 [Rethinking the origins of Chinese historical geography]. In: Shi N史南海 (ed) *Zhongguo dili luncun diyiji*中国历史地理论丛第1辑 [Essays on Chinese historical geography I]. Shaanxi People's Press陕西人民出版社, Xi'an, pp 102–109
2. There are different opinions regarding this. For example, Wang Chengzu believes that the words "*tianwen*" and "*dili*" in the *Xici* carry abstract meanings. See: Wang C王成组 (1988) *Zhongguo dilixue shi*中国地理学史 [A history of Chinese geography]. Beijing Commercial Press商务印书馆, p 4
3. Hutton J (1989) The history of geology. Commercial Press商务印书馆, Beijing, p 23
4. Claval P (2007) A history of geographical thoughts (trans: Shenghua Z郑胜华, Demei L刘德美, Qinghua L刘清华). Bejing University Press北京大学出版社, Beijing, p 7
5. Jiuchang L李久昌 commented in the aforementioned article that "the formation of the concept '*dili*' was a great leap forward in the understanding of historical geography in China. Although its definition was not conceptual, '*dili*' was described in a detailed and practical way. Compared with 'geography' which the ancient Greeks used to refer to the surface of the Earth, the Chinese definition had greater scientific and practical value; it was simply the constraints of the period in time which limited the expression of such a concept to only a basic meaning
6. According to Li E'rong, "*dizhi*" was first used in Wang Bi's 王弼 (226-249 AD) *Zhouyi zhu* 周易注 [Annotations of the *Book of Changes*] during the Empire of Cao-Wei, yet *dizhi* in its modern sense only appeared in the book *Dili quanshu* ("Geography Encyclopedia") published in the 3rd year of the Xianfeng reign (1853), cited from: Li E李鄂荣 (1984) "'Dizhi' yici heshi chuxian yu woguo wenxian" "地质"一词何时出现于我国文献 [An investigation of when the term "dizhi" first appeared in Chinese literature]. *Zhongguo keji shiliao*中国科技史料 [Historical data of Chinese science and technology] (3). But a popular opinion holds that the term was first introduced by Lu Xun鲁迅 in his article *Zhongguo dizhi luelun* 中国地质略论 [A brief discussion of Chinese geology] written in Japanese in 1903
7. Fang H方豪 (2008) *Zhongxi jiaotong shi shangce*中西交通史:上册 [A history of communication between China and the West I]. Shanghai: Shanghai People's Publishing House上海人民出版社, p 33

8. Zhou, Zhenhe 周振鹤 (2009) *Tiguo jingye zhidao*体国经野之道: 中国行政区划沿革 [The approach to governing the country: the evolution of administrative divisions in China]. Shanghai Bookstore Publishing House上海书店出版社, Shanghai, p 88
9. As regards the completion date of the *Shanhai jing*, a traditional view holds that it was written by Yu the Great and Boyi in ancient times, but this has been rejected by the current academic community. Wang Chengzu 王成组believes that the "Mountains" appeared in the late Warring States period, while Gu Jegang believes that the "Mountains" was completed from the late Spring and Autumn to the early Warring States periods and his view has been accepted by most scholars (Gu Jegang 顾颉刚 (1934) "Wuzangshanjing shitan" 五藏山经试探 [A tentative study of the "Wuzang shanjing"], *Shixue luncon* 史学论丛 [Collection of essays on history] p 1). Nor do recent studies differ significantly from Gu's view. There are more debates about the completion dates of the "Seas" and the "Great Wilds", but most scholars argue that these two chapters were written later than the "Mountains", either in the Qin-Han dynasties or even later. Some believe that they predate the "Mountains" (Yang Chao 杨超 *"Shanhai jing" jiqi xiangguan de jige wenti*《山海经》及其相关的几个问题 [The *Shanhai jing* and several related issues], *Shanhai jing xintan* 山海经新探 [A new insight into the *Shanhai jing*], Chengdu: Sichuan Academy of Social Sciences Press 四川省社会科学院出版社, 1986: 6). Recent studies on the history of geoscience have mostly overlooked the geographical value of the "Seas" and the "Great Wilds", but the geoscientific thinking reflected in the writing style and framework is consistent with those of the "Mountains"
10. Hou R侯仁之 (1962) 中国古代地理学简史*Zhongguo Gudai Dilixue Jianshi* [A brief history of geography in ancient China]. Science Press科学出版社, Beijing, p 8
11. Ge Z葛兆光 (2011) *Zhaizi zhongguo: chongjian youguan 'Zhongguo' de lishi lunshu* 宅兹中国: 重建有关"中国"的历史论述 [Living in the "zhongguo" (the centre of the *tianxia*): the reconstruction of the historical statements about "zhongguo"]. Zhonghua Book Company中华书局, Beijing, p 80
12. There is an article "AH Chang and 'Shanhai jing'" 阿长与山海经in Lu Xun's 鲁迅*Zhaohua xishi*朝花夕拾 ("Dawn blossoms plucked at dusk"), which well expresses the attraction of the *Shanhai jing* for youngsters growing up in times past
13. Volume 71 "On *Shanhai jing*" 记山海经, in the *Hui'an ji* 晦庵集 [Collection from the Twilight Studio]; *Zhuzi yulei - juan yibai sanshiba* 朱子语类·卷一百三十八 [A collection of the discourses of Master Zhu, vol 138)
14. Zhang, Z张祝平 (2001) Songren suolun Shanhaijingtu bianzheng宋人所论《山海经图》辩证 [An examination of the statements by the Song people about *the Charts in Shanhai jing*]. *Zhongguo lishi dili luncun*中国历史地理论丛 [Collec Essays Chinese Histor Geogr] 4:66–69
15. According to Tan Qixiang 谭其骧, unlike other parts of the *Shanhai jing*, which are full of mythological accounts, the "Mountains" is valuable in geographical sense, from its form to its content, as it focuses on descriptions of mountains, rivers and natural resources in different areas. See Hua L华林甫 (2009) *Zhongguo lishi dilixue zongshu* 中国历史地理学·综述 [Chinese historical geography: a review]. Shangdong Education Press山东教育出版社, Jinan, pp 440–441
16. Tan Q谭其骧 (1986) "'Wuzang shanjing' de diyu fanwei tiyao" "五藏山经" 的地域范围提要 [A summary of the geographical scope recorded in "Wuzang shanjing"]. In: *Shanhaijing xintan* 山海经新探 [A new insight into the *Shanhai jing*]. Sichuan Academy of Social Sciences Press 四川省社会科学院出版社, Chengdu, pp 13–14
17. Wu F吴枫 (1987) *Jianming zhongguo guji cidian* 简明中国古籍辞典 [A concise dictionary of Chinese literary classics]. Jilin Literature and History Press吉林文史出版社, Changchun, p 48
18. From the "Zhongshan jing" 中山经 [Central mountains] in the *Shanhai jing*
19. Li S李申 (1989) *Zhongguo gudai zhexue he zirankexue*中国古代哲学和自然科学 [Ancient Chinese philosophy and natural sciences]. China Social Sciences Press中国社会科学出版社, Beijing, pp 18–24. Moreover, Nanzhou X徐南洲 (1986) argues in his article "*Shanhaijing* – yibu zhongguo shanggu de keji shishu" [*Shanhai jing*—A historical book on ancient Chinese science and technology] (in China "Shanhai jing" Academic Forum 中国山海经学术讨论

会(ed) *Shanhai jing xintan* 山海经新探 [A new insight into the *Shanhai jing*]. Chengdu: Sichuan Academy of Social Sciences Press 四川省社会科学出版社, pp 241–263) that Kua Fu's pursuit of the sun is the scientific activity of humans exploring the laws of the sun's movement, and Jing Wei's filling up of the sea is a manifestation of human exploration of the navigation of the sea, and it also provides explanations for natural phenomena such as floods in the South and droughts in the North through mythological stories. Ke Y袁珂 (1986) also argues in his article "*Shanhai jing*: 'Gaigu zhi wushu' shitan" 山海经"盖古之巫书"试探 [A tentative study on the *Shanha ijing*: the ancient book of witchcraft] (in *Shanhaijing xintan* 山海经新探 [A new insight into *Shanhai jing*], Sichuan Academy of Social Sciences Press 四川省社会科学院出版社, Chengdu, pp 231–240) that the ancient people "would imbue everything they saw, heard, and thought with a mythological (religious) hue, and we call this a primitive way of deepening thinking"

20. Institute of Geographic Sciences and Natural Resources Research, Chinese Academy of Sciences中国科学院地理研究所 (2006) *Zhongguo gudai dili mingzhu xuandu (diyiji)*中国古代地理名著选读(第一辑) [Selected readings of ancient Chinese geographic classics, vol 1]. Xueyuan Press学苑出版社, Beijing, p 1
21. Zheng Q郑樵, *Tongzhi juan sishi*通志·卷四十 [Comprehensive records, vol 40]
22. Ai N艾南英, *Yugong tuzhu: Xu*禹贡图注·序 [Commentary on the "Yugong" map: preface]
23. Hua L华林甫 (2009) *Zhongguo lishi dilixue zongshu* 中国历史地理学·综述 [Chinese historical geography: a review]. Shandong Education Press山东教育出版社, Jinan, p 429
24. Zhou Z周振鹤 (2009) *Tiguo jingye zhidao* 体国经野之道: 中国行政区划沿革 [The approach to governing the country: the evolution of administrative divisions in China].上海: 上海书店出版社Shanghai Bookstore Publishing House, p 88. Wang Chengzu believes that the *Yugong* was compiled by Confucius, see Chengzu W王成组 (1982) *Zhongguo dilixue shi shangce*中国地理学史上册 [A history of Chinese geography I]. Commercial Press商务印书馆, Beijing, pp 4–6); Jiegang G believes that the *Yugong* "is a work from the early 3rd century BC" (see *Zhongguo gudai dili mingzhu xuandu diyiji* 中国古代地理名著选读第一辑 [Selected readings of ancient Chinese geographic classics, vol 1], p 4. Shi Nianhai believes that the book was written during the Warring States period and could not have been written earlier than 482 BC (see Nianhai S史念海 (1979) Lun *Yugong* de zhuzuo shidai [On the era of the writing of the *Yugong*]. *Shanxi shifan daxue xuebao* 陕西师范大学学报 [J Shaanxi Normal Univ] 1(3); while Shuzhi X辛树帜 (1964) believes that the book was written in the Western Zhou dynasty (see *Yugong xinjie* 禹贡新解 [New interpretation of the *Yugong*], China Agriculture Press 农业出版社). For related arguments, see Hua L华林甫 (2009) *Zhongguo lishi dilixue zongshu* 中国历史地理学·综述 [Chinese historical geography: A review]. Shangdong Education Press山东大学出版社, Jinan
25. Liu S刘盛佳 (1990) *Yugong*: Shijie shang zuizao de quyu renwen dilixue zhuzuo《禹贡》——世界上最早的区域人文地理学著作 [*Yugong*—the earliest regional human geography work in the world]. *Dili xuebao*地理学报 [Acta Geographica Sinica] (4)
26. According to Zhenhe Z: The Jiuzhou system, however, represents a new understanding. Taking major mountains and rivers as boundaries, it divides the known world into nine regions … The division of the regions is based on the geomorphology and climate conditions, with reference to two key parameters of physical regionalisation, the soil and vegetation conditions. Such division presupposes that the world is a whole, which represents the general desire for unification in the Warring States period. Cited from Zhou Z周振鹤 (2009) *Tiguo jingye zhidao* [The approach to governing the country: the evolution of administrative divisions in China]. Shanghai Bookstore Publishing House上海书店出版社, Shanghai, pp 88–91
27. For a comprehensive review of the *Yugong*, please refer to Gu Jiegang's 顾颉刚 introduction to the *Yugong* in the *Zhongguo gudai dili mingzhu xuandu (diyiji)* 中国古代地理名著选读(第一辑) [Selected readings of ancient Chinese geographic classics I]; for recent research on the *Yugong*, please refer to Linfu H 华林甫 (2009) *Zhongguo lishi dilixue zongshu* 中国历史地理学·综述 [Chinese historical geography: A review]. Shangdong Education Press 山东大学出版社, Jinan

28. Peng J彭静中 (1990) *Zhongguo fangzhi jianzhi* 中国方志简史 [A brief history of Chinese local chronicles]. Sichuan University Press四川大学出版社, Chengdu, p 40
29. Gu J顾颉刚 (1951) Mutianzi zhuan jiqi zhuzuo shidai穆天子传及其著作时代 ["Mutianzi zhuan" and its time of writing]. *Wenshizhe*文史哲 [J Literature History Philos] (2). Shi W 史为乐 (1992) argues in his article "Mutianzi xizheng shitan"穆天子西征试探 [Discussion on the westward journey of Mutianzi] (published in the *Zhongguoshi yanjiu* 中国史研究 [J Chin Histor Stud (3)]) that the *Mutianzi zhuan* was written during the 20th year of the reign of King Xiang of Wei (299 BC) during the Warring States period
30. Wang Y王庸 (1960) *Zhongguo dilixue shi* 中国地理学史 [A history of Chinese geography]. Commercial Press商务印书馆, Beijing, p 9
31. Chen Z陈正祥 (1983) *Zhongguo wenhua dili* 中国文化地理 [Chinese cultural geography]. SDX Joint Publishing Company北京: 三联书店, Beijing, p 203
32. Wang C王成组 (1988) *Zhongguo dilixue shi*中国地理学史 [A history of Chinese geography]. Commercial Press商务印书馆, Beijing, p 9
33. Shi W史为乐 (1992) Mutianzi xizheng shitan穆天子西征试探 [Discussion on the *westward journey of King Mu*]. *Zhongguoshi yanjiu* 中国史研究 [J Chin Histor Stud] (3)
34. The French scholar Émmanuel-Édouard Chavannes annotated the *Wang xiwangmuguo jicheng* ("Journey to the Kingdom of Xiwangmu") and believed that it was Duke Mu of Qin, instead of Duke Mu of Zhou, who visited the Xiwangmu
35. Wang S王守春 (2000) *Mutianzi zhuan* diyu fanwei shixi "穆天子传" 地域范围试析 [Analysis of the geographical scope recorded in the *Mutianzi zhuan*] *Zhongguo lishidili luncong* 中国历史地理论丛 [Collections of essays on Chinese historical geography] (01)
36. According to Fang Hao's 方豪 *Zhongxi jiaotong shi* ("History of Communication between China and the West"), after the end of the Qing dynasty, an increasing number of researchers were studying the *Mutianzi zhuan*. During the period of the Republic of China, Ding Qian 丁谦 wrote the *Mutianzi zhuan dili kaozheng* ("Geographical research on the *Mutianzi zhuan*"), and Gu Shi 顾实 wrote the *Mutianzi zhuan xizheng jindi kao* ("Research on the Present Location Recorded in the Biography of Mutianzi's Westward Journey"); they both agreed that the "Kunlun zhiqiu (Kunlun Mountains)" are along the southern border of the Kingdom of Khotan. As for the "Kingdom of Xiwangmu", Ding believed it was the Kingdom of Chaldea, while Gu thought it was Persia. German sinologist Friedrich Hirth argued that the journey of King Mu of Zhou would not have crossed the Great Wall, and that it was impossible to determine the ethnic groups encountered during his westward journey. When he returned eastwards, he could certainly have brought back some legends about the mysterious regions of Central Asia, but this would not have influenced the culture of the Zhou dynasty
37. In the book *Zhaizi zhongguo: Chongjian youguan "Zhongguo" de lishi lunshu* 宅兹中国:重建有关 "中国" 的历史论述, Ge Zhaoguang 葛兆光quotes Xu Youren's 许有壬 introduction to the *Annan zhi lue* 安南志略·序: "A scholar should study everything that the universe and heavens comprise, but in some uninhabited places or places where communication is obstructed, very limited information can be known of, not to mention in even poorer and remote settlements and mountainous areas. Yuanming read the *Zhouwang zhuan* ("Biography of the King of Zhou") and *Shanhe tu* ("Map of Mountains and Rivers") for pleasure, which shows his vision and noble ideas.
38. *Zhouli: Diguan situ: Cao Ren*周礼·地官司徒·草人 [Rites of Zhou: The Earth officials Si Tu: Cao Ren]
39. Xia W夏纬瑛 (1979) *"Zhouli" shuzhong youguan nongye tiaowen de jieshi*《周礼》书中有关农业条文的解释 [Interpretation of the provisions on agriculture in the *Zhouli*]. China Agriculture Press北京农业出版社, Beijing, p 39
40. A typical example is the *Shijing*: *Youfeng: Qiyue*诗经·豳风·七月 [Book of poetry: Odes of the Bin State: July]
41. *Shanhai jing: Haiwai xijing* 山海经·海外西经 [Classic of mountains and seas: classic of regions beyond the seas: The West]
42. *Huainanzi: Qisu xun*淮南子·齐俗训 [Huainanzi: respecting the customs of different places]
43. *Yugong* 禹贡 [Tribute of Yu]

44. *Menzi, Tengwen gong xia*孟子·滕文公下 [Mencius: Duke Tengwen II]
45. *Guanzi: Chengma*管子·乘马 [Guanzi: military taxes]
46. *Guanzi: Dudi*管子·度地 [Guanzi: appraising terrains]
47. *Zhouli: Zhifangshi*周礼·职方氏 [Rites of Zhou: the map officers]
48. Jin S靳生禾 (1987) *Zhongguo lishi dili wenxian gailun*中国历史地理文献概论 [Introduction to Chinese historical and geographical documents]. Shanxi People's Publishing House山西人民出版社, Taiyuan, p 20
49. *Shanhai jing: Haiwai nanjing*山海经·海外南经 [Classic of mountains and seas: classic of regions beyond the seas: The South]
50. *Shanhai jing: Haiwai beijing*山海经·海外北经 [Classic of mountains and seas: classic of regions beyond the seas: The North]
51. *Shanhai jing: Haiwai xijing*山海经·海外西经 [Classic of mountains and seas: classic of regions beyond the seas: The West]
52. Yuan K袁珂 (1985) *Shanhaijing jiaoyi*山海经校译 [Collation and translation of *Shanhai jing*]. Shanghai Ancient Books Press上海古籍出版社, Shanghai, p 197
53. All the men and women in the state are registered by the government, and those who are newborn will be noted, and those who die are removed (*Shangjun shu: Jingnei*商君书·境内 [Book of Lord Shang: within the realm])
54. *Zhouli: tianguan zhongzai*周礼·天官冢宰 [Rites of Zhou: The Heaven officials]
55. *Lüshi Chunqiu: Shenshi*吕氏春秋·审时 [Master Lü's Spring and Autumn annals: Assessing situations before military action]
56. Cang X仓孝和 (1988) *Ziran kexueshi jianbian*自然科学史简编 [A brief history of natural science]. Beijing Publishing Group北京出版社, Beijing, p 205
57. Pang P庞朴 (1984) "Yinyang wuxing tanyuan" 阴阳五行探源 [On searching for the origins of yin-yang and wuxing]. *Zhongguo shehui kexue*中国社会科学 [Soc Scie China] (3)
58. *Shijing: Gong Liu*诗经·公刘 [Book of poetry: Lord Liu]
59. *Shuowen jiezi*说文解字 [Discussing writing and explaining characters] (an early 2nd century Chinese dictionary from the Han dynasty)
60. *Zhouyi: xici shang*周易·系辞上 [Book of changes: The relationship of the hexagrams I]
61. *Zhouyi: xiangua*周易·咸卦 [Book of changes: the symbol of influence]
62. Tang X唐锡仁 (1981) Zhongguo gudai yinyang xueshuo dui tianqi xianxiang de jieshi中国古代阴阳学说对天气现象的解释 [An explanation of the meteorological references in the theories of yin-yang and wuxing]. *Zhongguo zhexueshi yanjiu* 中国哲学史研究 [Stud History Chin Philos] (2)
63. *Dao De Jing, Chapter 42* 道德经·四十二章
64. *Shangshu: Hongfan*尚书·洪范 [Book of documents: Grand laws of governance]
65. *Guoyu: Zhengyu*国语·郑语 [Discourses of the states: discourses of Zheng]
66. *Chunqiu fanlu: Wangdao tongsan*春秋繁露·王道通三 [Luxuriant dew of the Spring and Autumn annals: a King's understanding of the three elements]
67. *Liji: Yueling*礼记·月令 [Book of rites: monthly policies]
68. *Zuozhuan: Zhaogong si nian*左传·昭公四年 [The commentary of Zuo: the 4th year of Duke Zhao]
69. *Zuozhuan: Xianggong ershiba nian*左传·襄公二十八年 [The commentary of Zuo: the 28th year of Duke Xiang]
70. *Huainanzi: Tianwen xun* 淮南子·天文训 [Huainanzi: patterns of heavens]
71. Wan G万国鼎 (1955) Zhongguo gudai dui turang zhonglei jiqi fenbude renshi中国古代对土壤种类及其分布的知识 [Ancient Chinese knowledge of the type of soil and its distribution]. *Nanjing nongxueyuan xuebao*南京农学院学 [J Nanjing Agric Coll] (1)
72. *Guoyu: Zhouyu shang*国语·周语上 [Discourses of the states: Discourses of Zhou I]
73. *Chunqiu zuozhuan zhengyi: Zhaogong ershisi nian*春秋左传正义·昭公二十四年 [Canonical commentaries on the Spring and Autumn annals: the 24th year of Duke Zhao]
74. *Dao De Jing, Chapter 42* 道德经四十二章
75. *Xunzi: Lilun*荀子·礼论 [Xunzi: Discourse on ritual principles]

76. *Huangdi neijing: Suwen: Tianyuanji dalun* 黄帝内经·素问·天元纪大论 [The inner canon of the Yellow Emperor: essential questions: comprehensive discourse on the arrangements of the principles of heaven]
77. *Shanhai jing: Haiwai dongjing*山海经·海外东经 [Classic of mountains and seas: classic of regions beyond the seas: The East]
78. *Mengzi: Wanzhang zhangju shang*孟子·万章章句上 [Mencius: words from Wan Zhang I]
79. Tong S童书业 (1962) *Zhongguo gudai dili kaozheng lunwen ji*中国古代地理考证论文集 [A collection of essays on ancient Chinese geography]. Zhonghua Book Company中华书局, Beijing, p 7
80. *Lüshi chunqiu: Youshi lan*吕氏春秋·有始览 [Master Lü's Spring and Autumn annals: views on the beginning of the world]
81. *Jinshu: Tianwen*晋书: 天文 [Book of the Jin: astronomy]
82. *Dadai liji: Zengzi: Tianyuan*大戴礼记·曾子·天圆 [Records of ritual matters by Dai the Elder: Zengzi: celestial heavens]
83. *Zhoubi suanjing: juanxia*周髀算经·卷下 [Arithmetical classic of the gnomon and the circular paths of heaven II]
84. *Zhuangzi: Tianxia pian*庄子·天下篇 [Zhuangzi: Land under heaven]
85. Division of Chinese Philosophy at the Chinese Philosophy Research Centre, China Academy of Science & Teaching and the Research Office of Chinese Philosophy History at the School of Philosophy, Beijing University. 中国科学院哲学研究所中国哲学史组, 北京大学哲学系中国哲学史教研室 (1962) *Zhongguo zhexueshi ziliao jianbian (xianqin)* 中国哲学史资料简编 (先秦) [Compilation of historical texts of Chinese philosophy (pre-Qin periods)]. Zhonghua Book Company中华书局, Beijing, p 303
86. *Shiji: Fengshan shu*史记·封禅书 [Records of the grand historian: book of sacrificial ceremonies]: "The King of Wei and King of Xuan from the Qi state, and the King of Zhao from the Yan state all dispatched sea explorers to seek immortals in Penglai, Fangzhang and Yingzhou."
87. *Shiji*: *Mengzi xunqin liezhuan*史记·孟子荀卿列传 [Records of the grand historian: biographies of Mengzi and Xunzi]
88. Guo, Y郭永芳, Song Z宋正海 (1984) Dajiuzhou shuo——zhongguo gudai yizhong feizhengtongde haiyang kaifangxing diqiuguan大九州说——中国古代一种非正统的海洋开放型地球观 [The theory of the great nine states: a non-traditional open-marine perspective of the earth in ancient China]. *Daziran tansuo*大自然探索 [Discovery of nature] (2)
89. Liu S刘盛佳 (1990) *Dilixue sixiang shi*地理学思想史 [The history of geography ideas]. Central China Normal University Press华中师范大学出版社, Wuhan, p 53
90. *Zhouli: Diguan situ*周礼·地官司徒 [The Earth officials Si Tu]. According to Jia Gongyan's 贾公彦 (1979) annotation: "The place where the sun is in the south is called *jinnan* (close to the south), and where the sun is in the north is called *jinbei* (close to the north)." See *Shisan jing zhushu* 十三经注疏 [Commentary and annotations on the *Thirteen Classics of Chinese Literature*]. Zhonghua Book Company中华书局, Beijing, p 704
91. *Zhouli: Kaogong ji: juan sanshijiu*周礼·考工记·卷三十九 [Rites of Zhou: records of diverse crafts, vol 39]
92. Geoscience History Group, Institute of the History of Natural Sciences, Chinese Academy of Sciences中国科学院自然科学史研究所地学史组 (1984) *Zhongguo gudai dilixue shi*中国古代地理学史 [History of ancient Chinese geography]. Science Press北京科学出版社, Beijing, p 190
93. Aristotle's *Meteorologica* (1959). London edition, pp 179–183
94. Li X李星星, Zhang X张旭, Luo Y罗勇 (1989) *Kuozhang yu jiaowang—Quyu lishi wenhua jianlun*扩张与交往——区域历史文化简论 [Expansion and communication: a brief discussion on regional history and culture]. Chengdu Bashu Press成都巴蜀书社, Chengdu, p 15
95. *The geography of Strabo*, Book 2, p 315
96. In the *Yugong*, the *jiuzhou* (nine regions) are divided into: Ji, Chong, Qing, Xu, Yang, Jing, Yu, Liang and Yong. However, in the *Lüshi chunqiu: Youshilan,* the Liang is replaced by You; and in the *Zhouli: Zhifang*, the Xu and Liang are replaced by You and Bing respectively

97. *Zhouli: Xiaguan: Zhifang*周礼·夏官·职方 [Rites of zhou: Xia officials: the map officers]
98. Tang X唐锡仁 (1988) Lun xianqing shiqi de rendi guan"论先秦时期的人地观 [The perspective of the human-Earth relationship in pre-Qin periods]. *Ziran kexue shi yanjiu* 自然科学史研究 [Stud History Nat Sci] (4):311–317
99. *Shijing: Xiaoya: Shiyue zhijiao*诗经·小雅·十月之交 [Book of poetry: Minor odes: at the turn of the tenth month].
100. *Shisan jing zhushu*十三经注疏 [Commentary and annotations on the *Thirteen Classics of Chinese Literature*] (1979). Zhonghua Book Company中华书局, Beijing, p 446
101. *Zuozhuan: Xuangong shiwu nian* 左传·宣公十五年 [The commentary of Zuo: the 15th year of Duke Xuan]
102. *Mozi: Shangtong zhong* 墨子·尚同中 [Mozi: Exaltation of the virtuous II]
103. Herodotus. *The Histories*. pp 391–392, 447
104. *Liji: Wangzhi*礼记·王制 [Book of Rites: imperial regulations]
105. *Guanzi: Shuidi*管子·水地 [Guanzi: water and earth]
106. *Dadai liji: Yi benming di bashiyi*大戴礼·易本命第八十一 [Records of ritual matters by Dai the Elder, vol 81]
107. *Zhouli: Diguan dasitu* 周礼·地官大司徒 [Rites of Zhou: the Earth officials Si Tu: Grand Si Tu]
108. Plato (1970) The laws. Penguin Books Ltd. 704 (D)–705 (A)
109. *Xunzi: Tianlun*荀子·天论 [Xunzi: discourse on nature]
110. Zhou, C周春堤 (1980) "Mozi de dili sixiang" 墨子的地理思想 [Mozi's geographical thought]. *Dili yanjiu baogao*地理研究报告 [Report on geographical research], p 5
111. *Guanzi: Neiye*管子·内业 [Guanzi: Inner enterprise]
112. *Mengzi: Gaozi shang*孟子·告子上 [Mencius: Gao Zi I]
113. *Guanzi: Qingzhong jia*管子·轻重甲 [Guanzi: The art of balancing]
114. *Guanzi: Jincang*管子·禁藏 [Guanzi: prohibitions and storage]
115. *Xunzi: Wangzhi*荀子·王制 [Xunzi: regulations of a king]
116. *Encyclopedia of China: environmental science*, p 502
117. *Lüshi chunqiu: Shangnong*吕氏春秋·上农 [Master Lü's Spring and Autumn annals: the importance of agriculture]
118. *Mengzi: Lianghuiwang zhangju*孟子·梁惠王章句 [Mencius: discourses of King Liang Hui]
119. *Guanzi: Baguan*管子·八观 [Guanzi: examining a state in eight aspects]
120. *Guanzi: Dishu*管子·地数 [Guanzi: benefits of geographical conditions]
121. *Zhouli: Diguan*周礼·地官 [Rites of Zhou: the Earth officials]
122. *Guanzi: Fajing*管子·法禁 [Guanzi: things forbidden by law]
123. *Guanzi: Lizheng*管子·立政 [Guanzi: establishing the correct policies]
124. *Liji: Quli*礼记·曲礼 [Book of rites: summary of the rules of propriety]
125. *Guoyu: Zhouyu san*国语·周语三 [Discourses of the states: Discourses of Zhou, vol 3]
126. Yuan Q袁清林 (1985) Xianqin huanjin baohu de ruogan wenti先秦环境保护的若干问题 [Several matters relating to environmental protection in the pre-Qin period]. *Zhongguo keji shiliao*中国科技史料 [Historical data of Chinese science and technology] (1)
127. *Zuozhuan: Zhaogong ershi nian*左传·昭公二十年 [The Commentary of Zuo: the 20th year of Duke Zhao]
128. *Zhoulii, Diguan situ*周礼·地官司徒 [Rites of Zhou: the Earth officials Si Tu]
129. Geoscience History Group, Institute of History of Natural Sciences, Chinese Academy of Sciences中国科学院自然科学史研究所地学史组 (1984) *Zhongguo gudai dilixue shi*中国古代地理学史 [A history of ancient Chinese geography]. Science Press北京科学出版社, Beijing, p 276
130. *Hanfeizi: Wudu*韩非子·五蠹 [Hanfeizi: five kinds of harmful insects]
131. *Lunyu: Xiangdang*论语·乡党 [Analects of confucius: home town]
132. *Guanzi: Ditu*管子·地图 [Guanzi: maps]
133. *Zhanguo ce*: *Zhaoce*战国策·赵策 [The strategies of the Warring States: strategies of Zhao]
134. *Zhanguo ce*: *Dongzhou ce*战国策·东周策 [The strategies of the Warring States, strategies of the Eastern Zhou]

135. Yuan B毕沅 (Qing dynasty) *Shanhaijing xinjiaozheng*山海经新校正 [New corrections of the *Shanhai jing*]
136. Shen Y杨慎 (Ming dynasty) *Shanhaijing buzhu* 山海经补注 [Supplementary notes to the *Shanhai jing*]
137. Peng J彭静中 (1990) *Zhongguo fangzhi jianzhi* 中国方志简史 [A brief history of Chinese local chronicles]. Sichuan University Press四川大学出版社, Chengdu, p 39
138. *Zhouli: Zhifang*周礼·职方 [Rites of Zhou: the map officers]
139. *Zhouli: Xiaguan sima*周礼·夏官司马 [Rites of Zhou: the Xia officials Si Ma]
140. According to Weiying X夏纬瑛 (1979), the Tu Xun was an official who told the king about the gods and spirits related to the land, and superstitious elements were a feature of his accounts. See: "*Zhouli*" *shuzhong youguan nongye tiaowen de jieshi* "周礼" 中有关农业条文的解释 [Interpretation of the provisions on agriculture in the *Zhouli*]. China Agriculture Press中国农业出版社, Beijing
141. *Guanzi: juan'er (qifa)*管子·卷二 (七法) [Guanzi, vol 2 (seven main principles)]
142. *Guanzi: juanshi (ditu)*管子·卷十 (地图) [Guanzi, vol 10 (on geographical conditions)]
143. "Maps are different from books. Any literate person can transcribe a book, whilst reproducing a map requires the necessary materials, tools and skills. Before printing was invented or progressed to the point where maps could be printed, maps were much more difficult to reproduce and preserve than books. Once the original was lost, the map could never be recovered. The main reason for the scarcity of ancient maps, however, is that maps have been used as symbols of political power in China from the very beginning, and thus the reproduction and preservation of maps became the privilege of the rulers." See Jianxiong G葛剑雄 (2007) *Ren zai shikong zhijian* [In between time and space]. Zhonghua Book Company中华书局, Beijing, p 42.
144. *Zhouli: Chunguan: Zhong Ren*周礼·春官·冢人 [Rites of Zhou: The Chun official: Public Tomb Officers].
145. *Zhouli: Chunguan: Mu Dafu*周礼·春官·墓大夫 [Rites of Zhou, The Chun officials: Private Tomb Officers]
146. Zhang S张守中, Zheng M郑名桢, Liu L刘来成 (1979) Hebeisheng pingshanxian zhanguoshiqi zhongshanguo jizang fajue jianbao河北省平山县战国时期中山国墓葬发掘简报 [A brief report on the excavation of the Zhongshan state tombs from the Warring States period in Pingshan County, Hebei Province] *Wenwu*文物 [Relics] (1):1–31
147. Cao W曹婉如 (1989) Youguan tianshui fangmatan qinmu chutu ditu de jige wenti有关天水放马滩秦墓出土地图的几个问题 [Several questions about the maps excavated from the Qin tombs in Fangmatan, Tianshui]. *Wenwu*文物 [Relics] (12)
148. He S何双全 (1989) Tianshui fangmatan qinmu chutu ditu chutan天水放马滩秦墓出土地图初探 [A preliminary study of the maps excavated from the Qin tombs in Fangmatan, Tianshui]. *Wenwu*文物 [Relics] (2). Zhang Xiugui has a different opinion regarding the drawing time of this map. He believes that the map was drawn around 300 BC, before the 8th year of the reign of King Zhaoxiang of the Qin dynasty. See: Zhang X张修桂 (1991) Tianshui Fangmatan ditu de huizhi niandai [The drawing date of the Tianshui Fangmatan maps] *Fudan xuebao*复旦学报 [Fudan J] (1)
149. He S何双全 (1989) Tianshui fangmatan qinmu chutu ditu chutan天水放马滩秦墓出土地图初探 [A preliminary study of the maps excavated from the Qin tombs in Fangmatan, Tianshui]. *Wenwu* 文物 [Relics] (2)
150. Cao W曹婉如 (1989) Youguan tianshui fangmatan qinmu chutu ditu de jige wenti有关天水放马滩秦墓出土地图的几个问题 [Several questions about the maps excavacated from the Qin tombs in Fangmatan, Tianshui]. *Wenwu*文物 [Relics] (12)
151. According to the *Shanhai jing: Dahuang bei jing* 山海经·大荒北经 [The classic of the Great Wilderness: The North]: "Chi You 蚩尤started wars with the Yellow Emperor, so the Yellow Emperor ordered Ying Long 应龙 to meet him face-to-face in the wilderness of Jizhou. Ying Long accumulated a large amount of water, and Chi You invited Fengbo (Marquis of Wind) and Yushi (Master of Rain) to create massive storms. The Yellow Emperor then summoned a heavenly maiden named Mei to stop the wind and rain. As a result, Ying Long was able to kill Chi You"

152. Yu Q余起棻 (1988) *Junshi kexue gaiyao*军事科学概要 [Summary of military science]. The People's Liberation Army Press北京解放军出版社, Beijing, pp 75–76
153. *Taipin yulan: disice*太平御览·第四册 [Imperial reader of the Taiping era, vol 4], p 3413
154. *Sunzi bingfa: Dixing*孙子兵法·地形 [Sunzi's art of war: Terrains]
155. *Sunzi bingfa: Shiji*孙子兵法·始计 [Sunzi's art of war: the start of assessments]
156. *Sunzi bingfa: Xing*孙子兵法·形 [Sunzi's art of war: foundations]
157. *Sunbin bingfa: Bazhen*孙膑兵法·八阵 [Sunbin's art of war: eight formations]
158. *Wuzi bingfa: Lunjiang*吴子兵法·论将 [Wuzi's art of war: Military generals]
159. *Sima fa: Renben*司马法·仁本 [Regulations issued by Si Ma: motivated by kindness]
160. *Sima fa: Dinjue*司马法·定爵 [Regulations issued by Si Ma: granting titles of nobility]
161. *Liutao: Longtao: Wangyi*六韬·龙韬·王翼 [The six strategies: the dragon strategy: generals and commanders]
162. *Liutao: Hutao: Juedao*六韬·虎韬·绝道 [The six strategies: the tiger strategy: the blocking of routes]
163. *Liutao: Quantao*六韬·犬韬 [The six strategies: the canine strategy]
164. *Sunzi bingfa: Xingjun*孙子兵法·行军 [Sunzi's art of war: deploying the troops]
165. *Wuzi bingfa: Yingbian* 吴子兵法·应变 [Wuzi's art of war: responding to emergencies]
166. *Sima fa: Yongzhong*司马法·用众 [Regulations issued by Si Ma: orderliness]
167. *Sunbin bingfa: Dibao*孙膑兵法·地葆 [Sun Bin's art of war: significance of the ground]
168. *Liutao: Longtao: Qibing*六韬·龙韬·奇兵 [The six strategies: The dragon strategy: Novel tactics]
169. *Sunzi bingfa: Jiudi*孙子兵法·九地 [Sunzi's art of war: Nine regions]
170. *Liutao: Quantao: Zhanche*六韬·犬韬·战车 [The six strategies: The canine strategy: Battle chariots].
171. Yinqueshan hanmu zhujian zhengli xiaozu银雀山汉墓竹简整理小组 (1975) [Preservation group for the bamboo slips from the Han tombs in Yinque Mountain]. *Sunbing bingfa: xiabian* 孙膑兵法:下编 [Sunbing bingfa II]. Cultural Relics Press北京文物出版社, Beijing
172. *Sunzi bingfa: Jipian*孙子兵法·计篇 [Sunzi's art of war: the start of assessments]
173. *Wuzi bingfa: Liaodi*吴子兵法·料敌 [Wuzi's art of war: analysing the enemy]
174. *Liutao: Longtao: Bingzheng*六韬·龙韬·兵征 [The six strategies: the dragon strategy: Army's indications]
175. *Sunzi bingfa: huogong*孙子兵法·火攻篇 [Sunzi's art of war: fire attacks]
176. *Wuzi bingfa: zhibing* 吴子兵法·治兵[Wuzi's art of war: managing soldiers]
177. *Sunzi bingfa: Xushi* 孙子兵法·虚实篇 [Sunzi's art of war: weaknesses and strengths]
178. *Shijng: Xiaoya: Baihua* 诗经·小雅·白华 [Book of poetry: minor odes: white flower]
179. *Zhouli: Diguan: Dao Ren*周礼·地官·稻人 [Rites of Zhou: the Earth officials Si Tu: Dao Ren]
180. *Shuijing: Feishui zhu*水经·肥水注 [Water classic: commentary on the Fei River]
181. *Huainanzi: Renjian*淮南子·人间训 [Huainanzi: in the world of Man].
182. *Shuijing: Jinshui zhu*水经·晋水注 [Water classic: Commentary on the Jin River]
183. *Shiji: Hequ shu*史记·河渠书 [Records of the grand historian: book of rivers and canals]
184. *Shiji: Huaji liezhuan*史记·滑稽列传 [Records of the Grand Historian: collected biographies of the eloquent]
185. *Shuijing: Mianshui zhu*水经·沔水注 [Water classic: commentary on the Mian River]
186. *Huayang Guozhi: Shuzhi:juansan*华阳国志· 蜀志, 卷三 [Chronicles of Huayang: records of Shu, vol 3].
187. Research Institute of Geographic Sciences and Natural Resources, Chinese Academy of Sciences中国科学院地理研究所 (2005) *Zhongguo gudai dili mingzhu xuandu (diyiji)* 中国古代地理名著选读 (第一辑) [Selected readings of ancient Chinese geographic classics, vol 1]. Xueyuan Press北京学苑出版社, Beijing, p 36
188. *Guanzi: Dudi*管子·度地 [Guanzi: on examining the geographical conditions]
189. *Zhuangzi: Qiushui*庄子·秋水 [Zhuangzi: the floods in autumn]
190. *Mengzi* [Mencius]*: Li Lou*孟子·离娄
191. *Zhouli: Kaogong ji*周礼·考工记 [Rites of Zhou: records of diverse crafts]
192. *Dao De Jing*道德经

193. *Zhuangzi: Waipian: Quqie*庄子·外篇·胠箧 [Zhuangzi: additional chapters: prising open containers]
194. Editing Team of the *History of water conservancy in China*《中国水利史稿》编写组from Wuhan Water Resources and Electric Power College武汉水利电力学院 and the Water Resources and Electric Science Research Institute水利水电科学研究院 (1979) *Zhongguo shuili shigao shangce*中国水利史稿: 上册 [History of water conservancy in China I]. China Water & Power Press北京水利电力出版, Beijing, p 104
195. The process of erosion was discussed by Theophrastus. In: Thomson JO (ed) History of ancient geography, p 153
196. Herodotus (1987). The histories. The University of Chicago Press, Chicago, p 136 (Book 2)
197. Aristotle's *Meteorologica*. London edition, 1959:111
198. *Lüshi Chunqiu:Jichun ji: Yuandao*吕氏春秋·季春纪·圜道 [Master Lü's Spring and Autumn annals: records on the third month of the spring season: laws of heaven]
199. Wenheng Y杨文衡provided the supplementary material in this section
200. Tang J唐际根 (1993) *Zhongguo yantieshu de qiyuan wenti* 中国冶铁术的起源问题[The origin of iron metallurgy in China]. *Kaogu*考古 [Archaeology] (6)
201. *Wenwu*文物 [Relics] (1978) 10:44
202. *Wenwu*文物 [Relics] (1984) 10:88
203. Chen D陈德安, Wei X魏学峰, Li W李伟纲 (1998) *Sanxingdui——changjiang shangyou wenming zhongxin tansuo*三星堆——长江上游文明中心探索 [Sanxingdui—Study on the ancient centre of civilisation in the Upper Reaches of the Yangzi River]. Sichuan People's Publishing House四川人民出版社, Chengdu
204. The author has collated the statistics derived from information in the *Kaogu* and *Wenwu* journals in tabular form, but due to the length of this chapter, the listed tables cannot be published, and only the statistical conclusions are presented. The data are up to 1990.
205. Guo郭宝钧 (1956) Luoyang xijiao handai juzhu yizhi洛阳西郊汉代居住遗址 [Residential site of the Han dynasty in the western suburbs of Luoyang]. *Kaogu tongxun*考古通讯 [Archaeological communication] (1)
206. *Xunzi: Faxing*荀子·法行 [Xunzi: following the example of conduct]
207. *Wenwu* 文物 [Relics] (1980) 10:61
208. The modern names of these ancient rocks and minerals explained in parentheses are quoted from: Chen, Guoshen 陈国生, Yang X杨晓霞 (1997) "Wuzangshanjing Zhong kuangwumingcheng kaoshi ji qita dili fenbu. "五藏山经"中矿物名称考释及其他地理分布研究 [Study on the names of minerals and records of geographical distributions in the "Wuzangshanjing]. 自然科学史研究 [Stud History Nat Sci] (4):368–383
209. Zhang Y张贻侠 (1962) 'Shanhai jing'—Shijie shang zui gulao de kuangchan dizhi wenxian"《山海经》——世界上最古老的矿产地质文献 ["Shanhai jing"—The world's oldest literature on minerals and geology]. *Guangming ribao*光明日报 [Guangming daily], 28 August
210. Xia X夏湘蓉, Li Z李仲均, Wang G王根元 (1986) *Zhongguo gudai kuangye kaifa shi*中国古代矿业开发史 [History of mining development in ancient China]. Geology Press北京地质出版社, Beijing, p 319.
211. Wang Z王子贤, Wang H王恒礼 (1985) *Jianming dizhixue shi*简明地质学史 [A brief history of geology]. Henan Science and Technology Press河南科技出版社, Zhengzhou, p 5
212. 沈福伟Shen F (1983). 结绿和埃及宝石贸易*Jielü he aiji baoshi maoyi* ["Jielü" and the Egyptian gem trade]. In: Dongrun Z et al. 朱东润等 (eds) *Zhongguo wenhuashi luncong disiji*中华文史论丛: 第4辑 [J Chin Literature History 4] (Shanghai Ancient Books Press上海古籍出版社, Shanghai)
213. Xia X夏湘蓉, Li Z李仲均, Wang G王根元 (1980) *Zhongguo gudai kuangye kaifashi*中国古代矿业开发史 [The history of mining development in ancient China]. Geology Press 北京地质出版社, Beijing, p 319
214. Jizhou冀州, Yanzhou兖州, Zhou Q青州, Xuzhou徐州, Yangzhou扬州, Jingzhou荆州, Liangzhou梁州, Yongzhou雍州, Yuzhou豫州. Li C李长傅 (2007) *Lichangbo wenji*李长傅文集 [Lichangbo collected works]. Henan University Press河南大学出版社, Kaifeng, p 269

215. Sohu (2020) Popular science: the number ones of the history of Chinese maps. https://www.sohu.com/a/395898615_772793. Accessed 18 April 2023.
216. Xu S徐爽 (2018) *Baigonglu: Yanjing bajue huasi xiangqian*百工录: 燕京八绝花丝镶嵌[Hundreds of examples of craftwork: filigree inlay art—eight unique skills from Yanjing]. Jiangsu Art Press南京: 江苏艺术出版社, Nanjing, p 4

Chapter 4
Geoscience in the Qin and Han Dynasties

4.1 The Impact of the Social Environment on Geoscience

The Qin and Han dynasties established the earliest unified feudal empire in China. The social environment during this period had the following ten major impacts on geoscience.

First, the unified political situation played an essential role in promoting national integration, interaction and exchange, enabling the rapid accumulation of geoscience knowledge across all nation states.

Second, the policies and measures implemented by the First Qin Emperor (Qin Shi Huang), such as "the standardisation of the wheelbase, the uniformity of the script, the unification of weights and measures, and the stipulation on highway width", provided a unified system of rules and institutions for Chinese culture and science, which was far-reaching, and also laid the foundation for the development of road traffic and transportation geography.

Third, the First Qin Emperor's administrative division plan of "dividing the world into 36 prefectures and installing magistrates, commandants and supervisory censors in the prefectures" established the principle of administrative division in China, which has existed for more than 2,000 years, and provided the basis for China's administrative geography. This was an idea and governance measure that was adapted to the feudal autocracy and, according to Zhou Qingchen, an official in the Qin dynasty, ensured that "by re-dividing the territories of feudal princes into prefectures and counties, everyone will enjoy peace and happiness. There will be no war, and the division principle will be passed on to all generations" [1].

Fourth, the policy of attaching importance to agriculture in the Qin and Han dynasties, which was both a continuation of the past and the beginning of the future, contributed to the prosperity of the country in the Qin and Han dynasties, and it also promoted the accumulation of scientific knowledge in irrigation, water conservancy and agricultural meteorology.

W. Yang and Q. Yang, *A History of Ancient Chinese Geoscience*, Understanding China,
https://doi.org/10.1007/978-981-97-4457-2_4

Fifth, in the Qin and Han dynasties, the emperors' pursuit of immortality and the elixir of life, a negative thought and a foolish practice, led people towards a maritime civilisation, broadened their geographical vision and promoted the accumulation of knowledge of maritime geography.

Sixth, the policies of "burn books and bury the literati in pits" and of "proscribing all non-Confucian schools of thought and declaring Confucianism the state ideology" in the Qin and Han dynasties had the historical effect of consolidating feudal unification and spreading feudal culture, but at the same time, they seriously suppressed the liberal progression of ideas, imprisoned people's minds for a lengthy period and, to a certain extent, hindered the development of science and technology.

Seventh, sea and land transportation during the Qin and Han dynasties were relatively developed, which promoted economic, political, scientific and cultural exchanges between China and foreign countries. The discovery of the statuette of a human head carved from a shell during the late Western Zhou dynasty, at the Zhouyuan site, suggests that the Central Plains of China had engaged with the Western Regions (including Central Asia) before the eighth century BC [2]. Seven hundred years later, during the Western Han dynasty, Zhang Qian, the diplomat famed for his exploits in Inner Asia, was sent to the Western Regions, promoting exchanges between China and the Western Regions, and providing the Chinese people with additional geographical knowledge.

Eighth, the theology of divination of the Han dynasty shed some light on geoscience. It is impossible to avoid considering the geoscientific knowledge behind the theology of divination when discussing this (theology) in the Han dynasty.

Ninth, geoscience was boosted by the mining industry, including mining for salt, iron, copper, gold, silver and cinnabar, and the craft industry, such as pottery and lacquerware, as well as by medicine and alchemy in the Qin and Han dynasties.

Tenth, the papermaking techniques of the Qin and Han dynasties, one of the four great inventions of ancient China, made significant contributions to the development and dissemination of civilisation in China and around the world. Cai Lun of the Eastern Han dynasty was a distinguished practitioner of this technology.

4.2 A Wider Geographical Vision: Trade Routes in China and Beyond

4.2.1 Zhang Qian's Journey to the Western Regions

Zhang Qian (175?–114 BC) was an outstanding diplomat and explorer in early China. His diplomatic activities and the experience from his voyages of discovery pioneered the Silk Road, built a bridge for political and cultural exchanges between China and the Western Regions, promoted amicable relations among countries, broadened the geographical vision of the Chinese people, and made outstanding contributions to the development of Chinese geography.

Zhang Qian's expedition to the Western Territories did not form part of his personal ambitions, but was in the political interests of the Western Han dynasty at that time. To remove the military threat of the Huns, an ancient people in China, Emperor Wu of the Han dynasty wanted to dispatch envoys to seek military alliances with the countries in the Western Regions, and especially with the Dayuezhi (modern Uzbekistan and Tajikistan) who bore a grudge against the Huns, in order to fight jointly against the Huns. For this reason, Emperor Wu of the Han dynasty recruited talented people to go on this mission. Zhang Qian "responded to the call as an official and was sent on a mission to Yuezhi" [3]. With its obvious political motivation, this mission was certain to encounter opposition from the Huns and he was faced with various difficulties and dangers.

Zhang Qian, born in Chenggu County, Hanzhong Prefecture (in the east of present-day Chenggu County, Shaanxi province), was an official during the Jianyuan reign (140–135 BC). In 139 BC, the 2nd year of the Jianyuan reign [4], leading more than 100 people, including Tang Yifu, his interpreter, he set out from Longxi (present-day Taonan in Gansu province), and was captured and detained for ten years when passing through the territory of the Huns. He was forced to marry a Hun woman and have a son with her. By so doing, the Huns intended to undermine Zhang Qian's intention to go to Dayuezhi and to tie him down, so that he would end up living amongst the Huns. However, Zhang Qian's loyalty to the Han emperor never wavered and his mission was constantly at the forefront of his mind. He kept his guile hidden so that the Huns' dropped their guard and relaxed their supervision over him. This plan worked. Finally, one day, he seized the opportunity to escape with a few followers and "went west for tens of days to the country of Dawan (present-day Fergana in Central Asia)".

The king of Dawan had desired contact with the rich and powerful Western Han dynasty for a long time. Learning of Zhang Qian's arrival, he was very glad and asked him, "Where are you going?" Zhang Qian replied, "I'm going to Dayuezhi. Please send someone with me, your Majesty." The king of Dawan thus dispatched attendants to escort Zhang Qian to Kangju (a place between the east of Amu Darya and the west of Balkash Lake in Central Asia today), and then Kangju people accompanied Zhang Qian to Dayuezhi.

After arriving in Dayuezhi, Zhang Qian found that things had changed. The old king had been killed by the Huns, the prince had been crowned king and had conquered Daxia. The new king believed that the land of Dayuezhi was fertile, and that people lived and worked in peace and contentment. Besides, since Dayuezhi was far away from the Han dynasty, there was no need to retaliate against the Huns. Zhang Qian could find no way out, so he left Dayuezhi to return to Daxia.

In the first year of the Yuanshuo reign (128 BC), Zhang Qian passed by the northern foot of the Kunlun Mountains and through the residential area of the Qiang people in the Qaidam Basin of Qinghai, aiming to return home. As the Huns controlled the Qiang people, Zhang Qian was recaptured by them. After years of detention, at a time when the Huns were experiencing civil strife, Zhang Qian, his wife and Tang Yifu fled from the Huns and returned to the Han dynasty. To commend Zhang Qian's achievements, Emperor Wu of the Han dynasty conferred on Zhang Qian the title of Taizhong Daifu, a senior personal counsellor to the emperor, and appointed Tang Yifu to the position of an imperial emissary.

Although Zhang Qian did not achieve the political objectives of Emperor Wu of the Han dynasty, he had gained an abundance of geographical knowledge about the western territories through his travels. He documented this knowledge and reported it to Emperor Wu of the Han dynasty. Later, Sima Qian wrote the *Dawan liezhuan* ("Dawan Commentary") based on this information, and this has been preserved to the present day.

As Zhang Qian had once travelled to Daxia as an envoy, he knew more about the Huns, so he served as a guide when he accompanied General Weiqing on a major military raid against the Huns in the sixth year of the Yuanshuo reign (123 BC). Zhang Qian "knew where the crops and water were, so the army was spared from hunger and thirst" [5]. Emperor Wu granted Zhang Qian the title of Marquis of Bowang, and three years later he was honoured with the title of General.

In the first year of the Yuanshou reign (122 BC), Zhang Qian reported to Emperor Wu:

> When I was in Daxia, I saw the bamboo poles and cloth produced in the Shu Prefecture. I asked where they sourced these. The Daxia people said that the merchants went to Shendu (today's India) to buy these products. Shendu is some kilometres distant from the southeast of Daxia. Just as in Daxia, the inhabitants' custom is to find flat land and settle down. But their terrain is low lying, and the climate is hot and humid. Their people ride elephants in battle. Their land is close to the Ganges River. Now, Shendu is to the southeast of Daxia with some goods originating from the Shu Prefecture, which indicates that the country of Shendu is not far away from the Shu. Now, when we send envoys to Daxia, we must pass through the homeland of the Qiang people who hate us. If we go a little further north, we will be captured by the Huns. However, the route in the Shu Prefecture is unimpeded, without enemy interference [3].

Zhang Qian boldly speculated about the geographical direction and distance: that is, Shendu was not far from Shu. If setting out from Shu to Shendu, and then continuing to Daxia, the mission would not be hindered by aggressors and the route would be straightforward. Emperor Wu of the Han dynasty was glad to hear that and immediately asked Zhang Qian to implement this plan. Thus, Zhang Qian sent envoys from Shu Prefecture and Qianwei Prefecture along four different pathways, heading towards the southwest via Ran, Mang, Qian, Qiong, Bo and other places. After merely one or two thousand kilometres, all the teams came to a standstill because of the unintelligible languages spoken by the ethnic minorities, who denied the Han envoys permission to cross their land. The team heading south stopped at Kunming, and the northbound group halted at Zuodu County.

Although Zhang Qian did not get through the route from the southwest to India during his expedition, he heard that, thousands of miles away from the west of Kunming, there was a kingdom called Dianyue (now around Tengchong, Yunnan province), where the residents rode elephants and that Sichuan businessmen often traded there. It is thus clear that the route from southwest China to India had existed for a long time. This geographical knowledge laid the foundation for further expansion of the southwest region. At the same time, it proved that the route envisaged by Zhang

Qian was correct. However, due to the constraints of the social conditions, Zhang's vision failed to be realised.

During the Yuanshou reign, Zhang Qian led a third expedition to the western territories on the order of Emperor Wu. After arriving in Wusun, Zhang Qian dispatched dozens of assistant envoys to Dawan, Kangju, Dayuezhi, Daxia, Anshi, Shendu, Khotan, Mi and other countries. After returning to China, Zhang Qian was promoted to Daxing, as one of the heads of the nine central government departments in charge of the relations between the dynasty and vassal states. A year later, this distinguished explorer and imperial diplomat passed away. Subsequently, China's road to the western territories was completely opened up and a continuous stream of envoys and businessmen travelled back and forward. This route was called the "Silk Road" by later generations (see Fig. 4.1). In order to commemorate Zhang Qian's achievements in his expedition to the Western Regions, "later envoys called Zhang Qian Marquis of Bowang".

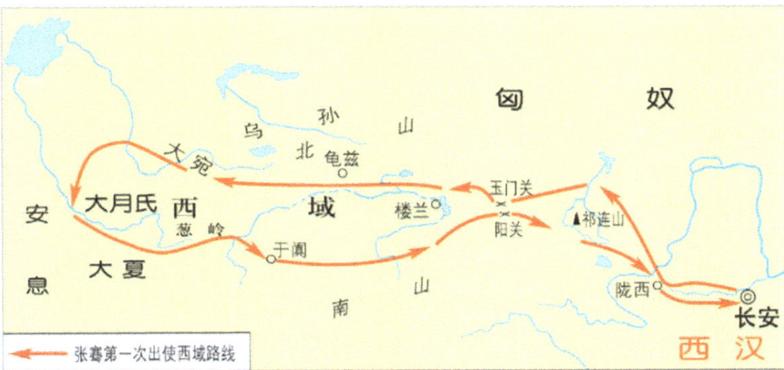

Fig. 4.1 Map of Zhang Qian's expedition to the Western Regions (cited from the Ministry of Education 教育部 (2018). *Zhongguo lishi* 中国历史 [Chinese history]. Beijing: People's Education Press 人民教育出版社, p 68)
Chang'an 长安—in present-day Xi'an, Shanxi.
Longxi 陇西—in present-day Lintao, Gansu.
Qiuci 龟兹—in present-day Kuqa, Xinjiang.
Loulan 楼兰—in the present-day west of the Lop Nor.
Dawan 大宛—in present-day Fergana Valley.
Anxi 安息—on the present-day Iranian plateau and in Mesopotamia.
Dayuezhi 大月氏—in the present-day Amu Darya area.
Congling 葱岭—in present-day Pamirs.
Yutian 于阗—in present-day Hetian, Xinjiang.
Wusun 乌孙—in the present-day Ili River and Lake Issyk-Kul area.
Yumen Pass 玉门关—in present-day Dunhuang, Gansu.
Yang Pass 阳关—in present-day Dunhuang, Gansu.
Daxia 大夏—between the present-day Hindu Kush Mountains and Amu Darya.

Zhang Qian's expedition to the Western Regions not only enriched the Chinese people's geographical knowledge and broadened their geographical vision, but also directly fostered the material and cultural exchanges between China and the West. Chinese exquisite handicrafts, especially silk, lacquerware, jade and bronze ware, were exported to the West, while products indigenous to the Western Regions, such as alfalfa, grapes, walnuts, pomegranates, sesame seeds, broad beans, cucumbers, garlic, carrots, and various woollen goods, furs, horses, camels, lions and ostriches were progressively imported to China. Western music, dance, painting, sculpture and acrobatics were also introduced to China, which had a positive impact on ancient Chinese culture and art.

4.2.2 The Travellers: Ban Chao, Ban Yong and Gan Ying

Ban Chao (32–102 AD), Ban Yong and Gan Ying were famous travellers and diplomats in the Eastern Han dynasty after Zhang Qian. Ban Chao, the younger brother of the historian Ban Gu, was born in Pingling County, Fufeng Prefecture (now the northwest of Xianyang, Shaanxi province). Ambitious and gifted with a silver tongue since childhood, he loved working and reading. When his family was trapped in poverty, Ban Chao "rescripted documents for officials and earned some money to support his mother". He admired Zhang Qian and said: "If a man does not have ambition and a strategy, he should still follow the example of Fu Jiezi, in the reign of Emperor Zhao of the Han dynasty, and of Zhang Qian, in the reign of Emperor Wu, in making a contribution in the Western Territories and obtaining the title of Marquis. How could any man be satisfied by only engaging in clerical work for a long time?".

In the 16th year of the Yongping period (73 AD), accompanied by 41-year-old Ban Chao, Dou Gu led troops to fight against the Huns. They attacked Yiwu (today's Hami City, Xinjiang province) in the Puliuhai (today's Balikun Lake) and won the battle. Dou Gu appreciated Ban Chao's talents and sent him to the Western Territories. In Shanshan, Khotan, Shule and other countries, Ban Chao skilfully employed stratagems to integrate these places into the territory of the Han dynasty. In the 8th year of the Jianchu reign (83 AD), Ban Chao was appointed as adjutant with military powers. In the 3rd year of the Yongyuan period (91 AD), Ban Chao was promoted to Du Hu, the position of the highest administrative commanding officer in the border area, and was stationed in Kucha. In the 6th year of the Yongyuan reign (94 AD), Ban Chao vanquished Yanqi, so the path south of Tianshan Mountain was unimpeded, and more than 50 nations in the Western Territories were assimilated into the domains of the Han dynasty. "The nations of Tiaozhi and Anshi, as well as the coastal ones 40 thousand li away, all came to pay tribute through interpreters." In the 7th year of the Yongyuan reign (95 AD), Ban Chao was granted the title of Marquis of Dingyuan, and was given an income from the taxes of a thousand families. In the 9th year of the Yongyuan period (97 AD), Ban Chao "sent Gan Ying to the westernmost sea and then he returned. These were places that had not been visited by any of his predecessors and they were not recorded in the "Mountains" section of the *Shanhai jing* in detail. The envoy had a thorough understanding of the customs and geographical environment of the places, and he brought back local treasure. In this way, nations like Mengqi and Doule from far away came to seek patronage and sent envoys to pay tribute" [6].

In the 12th year of the Yongyuan reign (100 AD), Ban Chao was elderly and wanted to return to his homeland, so he wrote to the emperor, hoping to "cross through the Yumenguan Pass, the main route from the Western Regions back to the Han dynasty, while he was still alive". In the second year, his sister, Ban Zhao, also wrote to the emperor to intercede for her brother, saying that Ban Chao "is almost 70 years old now, aged and ill, his hair is not black, his hands are not responsive, and his ears and eyes are not sharp. Only with a stick can he stagger around". "We have not seen each other for 30 years. We siblings are separated, and now fear that we may not recognise each other" [7]. The emperor was very moved after reading the siblings' letters and ordered Ban Chao to return to the capital. In August of the fourteenth year of the Yongyuan period (102 AD), Ban Chao returned to Luoyang and was honoured as the Shesheng commandant. Later, in September, he died at the age of 71.

Ban Yong had lived in the Western Regions with his father, Ban Chao, when he was a child. When he grew up, he inherited his father's ambition and returned to the Western Regions. In the first year of the Yongchu period (107 AD), Ban Yong worked as the military minister. In the second year of the Yanguang period (123 AD), he was appointed as Senior Clerk of the Western Regions, and "leading 500 prisoners who had just been released to the west, he was stationed in Liuzhong city. Ban Yong then defeated and pacified the Jushi Kingdom. From the Jianwu period to the Yanguang period, the Western Territories cut off contact with the Central Plains three times and resumed exchanges three times. In the 2nd year of Emperor Shun's Yongjian reign (127 AD), Ban Yong conquered the Yanqi Kingdom. At that time, seventeen nations, such as Kucha, Shule, Yushi, and Shache, came to surrender [6]." Ban Yong spent almost his entire life in the Western Regions, trekking from the South Route to the North Route. He recorded his personal experience in the book *Xiyu ji* ("Record of the Western Regions"), which recorded details of the routes, locations, climate, topography, products and customs of the nations in the Western Territories. Later, when Song Ye of the Southern dynasty wrote the *Book of the Later Han: Treatise on the Western Regions*, he collated the main content of the *Xiyu zhuan* ("Accounts of the Western Regions") which has been passed down to this day. This is the significant contribution made by Ban Yong to ancient Chinese geography.

The life story of Gan Ying is unknown due to the lack of any biography in historical books. There are only three passages recorded in the *Book of the Later Han: Treatise on the Western Regions*. First, a quotation in the reference above meantions Gan Ying was mentioned. Second, "In the ninth year of the reign of Emperor Yongyuan, Ban Chao sent Gan Ying to the Daquin Empire (Rome), arriving in the Tiaozhi Kingdom. Facing the sea, Gan Ying wanted to sail across it, but the boatman on the western border of Parthia said to Gan Ying: 'The sea is vast. It takes three months for people to cross it with a favourable wind, and two years if the wind is slow. Therefore, people who set sail take food for three years. On the way, people are often homesick, and many people die.' Gan Ying gave up after hearing that." Third, "after that, Gan Ying arrived in the Tiaozhi Kingdom and Parthia, went to the West Sea and looked into the Daquin Empire (Rome). He had travelled everywhere within about 40,000 li from the Yumenguan Pass and the Yangguan Pass." According to the research of scholars, Gan Ying possibly started his expedition from Kucha, passed through Shule and Shache, climbed over the Congling ridges, travelled through Puli and Wulei to Dayuezhi, and proceeded westwards to Mulu, Hedu, Aman and Ctesiphon, finally reaching Yuluo in the Tiaozhi Kingdom. When returning home, Gan Ying travelled eastwards from the Tiaozhi Kingdom, passed through Alexandria, Kophen, the Xuandu Mountains, Wuxie and Pishan, and returned to Qiuci to report the progress of the mission [8]. Although Gan Ying did not complete the task of visiting the Daquin Empire (Rome) at that time, he had promoted the exploration of the Western Regions initiated by Zhang Qian, Ban Chao and others, and made a long, arduous journey as far as the Tiaozhi Kingdom in West Asia. He was the first Chinese explorer to reach the Persian Gulf, the geographical value of which is self-evident.

4.2.3 South Sea Routes

After the First Qin Emperor (or Qin Shi Huang) unified China, great importance was attached to maritime navigation. He himself set sail on the sea on several occasions. In particular, the First Qin Emperor sent thousands of boys and girls, led by Xu Fu, to seek immortality. Legend has it that Xu Fu voyaged to Japan, which is possible. Therefore, Xu Fu was seen as the pioneer of maritime navigation in ancient China and the pathfinder of maritime routes.

In the Han dynasty, Emperor Wu paid great attention to the subject of navigation. In order to control the sea routes, he dispatched Yan Zhu, Zhu Meichen and others to establish a navy. From then on, extending from the Bohai Sea in the north to the coast of present-day Vietnam in the south, all the sea routes to the east and south of China became unimpeded, the maritime channels with Southeast Asia and countries along the Indian Ocean were established, and the first China-India ocean route in ancient China was opened up. This route is recorded in the *Book of the Han: Treatise of Geography*, including the purpose of the mission, the country or regional names throughout the expedition, and the time required for each voyage. In the book it is said:

> From the county borders of Xuwen, Hepu or Rinan (now near Hue, Vietnam), the ship sailed for five months to Duyuan (now near the Basai River in the northwest of the island of Sumatra); then it voyaged for four months to the Yilumo Kingdom (recorded as Homeritai in Greek text, possibly the Ruhmi/Rahma known to the Arabs, now a direct connection from the estuary of the Salween River in southern Myanmar); and it continued to navigate for more than twenty days to the Chenli Kingdom (possibly Clysma or Arsinoe, now near the city of Bagan, Myanmar). After journeying for more than ten days, they arrived at the Fugandoulu Kingdom (possibly Damietta, now near Taigong, Myanmar). It took about two months to travel by ship from the Fugandoulu to the Huangzhi Kingdom (possibly Kanchipuram, now Conyephron in southeastern India), which shared similar customs with

the Zhuya Prefecture. The Huangzhi Kingdom, with a large population, boasted a vast territory and various treasures. Since the reign of Emperor Wu, they had all paid tribute to the court. Sailing from the Huangzhi Kingdom to China for eight months, the ship arrived at Pizong (possibly Pulau Pisang, now a sugar cane island at the southwestern tip of the Malay Peninsula) and travelled for two months to the borders of Rinan, Xiang and Lin prefectures. To the south of the Huangzhi Kingdom, lay the Yichengbu Kingdom (possibly today's Siri Lanka), and the Han dynasty interpreter returned from there.

In the 9th year of the Yanxi (166 AD) in the Eastern Han dynasty, "Emperor Marcus Aurelius Antoninus of the Daquin sent envoys from the border of Rinan to offer ivory, rhino horn, and tortoiseshell" [6]. It can be seen that during the Qin and Han dynasties, the two major empires of the East and the West, the Eastern Han dynasty and Rome, already had maritime traffic and shipping routes.

4.3 Cartographic Documentation

The unified political and social environment of the Qin and Han dynasties created favourable conditions for the development of cartography. Before the discovery of the maps at the Fangmatan and Mawangdui sites, people's understanding of the maps of the Qin and Han dynasties was often based on Pei Xiu, a cartographer in the Jin dynasty. Pei said: "Today, there are neither ancient maps nor books collected by Xiao He in the palace, only miscellaneous maps such as the *Yudi* and *Kuodi* of the Han dynasty. These maps do not set a unified scale, do not verify the orientation, and the records of famous mountains and rivers are not complete. Although these maps provide an approximate schema, they are not precise and cannot be relied on" [9].

This view has been circulating for more than 1,700 years. When the Mawangdui map and the Fangmatan map were unearthed in 1973 and 1986 respectively, people were amazed. Thus, although Pei Xiu's views on the Qin and Han maps cannot not be completely refuted, they should at least be revised appropriately.

4.3.1 Maps Referenced in Literature

The Qin dynasty had various types of maps: the unearthed Fangmatan map and documentary records provide evidence for this. According to the *Book of the Han: Biography of Xiao He*, Liu Bang entered Xianyang and "only Xiao He was allowed into the imperial palace to collect the legal provisions from the prime minister and assistant prime minister of the Qin dynasty". This action helped Liu Bang a lot. Later, "the reason why Liu Bang knew about the dangers of the world, population distribution, the strengths and weaknesses of places, and the hardship of the people was because Xiao He had obtained the complete collection of the literature about the Qin". The *Qin ditu* ("Map of Qin") is mentioned twice in the *Book of the Han: Treatise of Geography*. Since the map was marked with the names of the two countries situated far apart from each other, Langya Prefecture and Dai Prefecture, this type of map may have covered the whole territory of the Qin dynasty. In the *Records of the Grand Historian: Biography of Dawan*, a kind of "ancient annotated map" is also mentioned. Based on this ancient map, Emperor Wu of the Han dynasty gave the mountain range at the source of the Yellow River the name of Kunlun Mountains. This kind of map may have been a map of the Qin dynasty, or it may have been the legendary *Shanhai jing tu* ("Map of Mountains and Seas"), which extended over a wide area, including the mountain range in the south of Xinjiang province today.

In the Han dynasty, the national map was the *Yu ditu* ("Map of China"). From this map, it can be seen that "there are 106 counties and states in *tianxia* (under heaven)" [10]. In addition, the documents that mention the *Yudi tu* include the *Records of the Grand Historian: The Family of Three Kings*, *Book of the Han: Biography of the Huainan King*, *Book of the Han: Biography of Liu Fei*, *Book of the Later Han: Annals of Empress Mingde*, *Book of the Later Han: Biography of Deng Yu*, etc. Yu Xi of the Jin dynasty stated in the *Zhilin* that the "*Yu ditu*, drawn by a cartographer in the Han dynasty, is not from ancient times" [11].

Dating from the Han dynasty, the *Junzhen tu* ("Battle Array Schemas") [12] and the *Zhujun tu* ("Garrison Maps") were unearthed at the Mawangdui site. Taking the *Zhujun tu* as an example, in addition to mountains, rivers, roads and residential areas, this type of map mainly represented the deployment of the local army, defence zone boundaries, fortresses, etc. The *Junzhen tu* contains *topographic maps* that supported military activities, such as the one from Mawangdui. This is a more accurate topographic map drawn after taking measurements. In addition, as recorded in the *Book of the Han: Biography of Li Ling*, Li Guang once led a cavalry of 800 troops, penetrated more than 1,000 km into the Huns' territory, and passed beyond the frontier fortress of Juyan to explore the terrain. In the 2nd year of the Tianhan reign (99 BC), Li Ling led an infantry of 5,000 soldiers out of Juyan, "walked north for thirty days, and arrived at Junji Mountain where they stopped and set up camp. Along the way, he had ordered people to sketch the mountains and rivers they passed and he sent his cavalryman, Chen Bule, back to Chang'an". This is a relatively detailed map that might have been based on actual surveying.

There were schematic diagrams which were relatively simple and temporary. For example, Zhang Qianqiu, Zhang Tang's grandson, "and Huo Yu, the son of Huo Guang, were both Zhonglang generals. They led the troops to attack Wuhuan with the Duliao General Fan Mingyou. Once, Huo Guang asked Zhang Qianqiu about the military strategies, as well as the geographical position of mountains and rivers. Zhang answered him casually, drawing a map on the ground without any omissions" [13].

In the 8th year of the Jianwu period (32 AD), Ma Yuan "used rice to simulate the terrain of the valley, analysed the military situation for the emperor, and pointed out which mountain road the army should enter and exit from. The analysis of the twists and turns was very clear. The emperor said that 'the enemy is already within sight' [10]." According to the *Xie Chengshu*, Zang Min was "the Zhonglang General of the Huns. When he returned to the capital, Yuan Feng, the supreme government official, asked him about the land, customs and tribal groups of the kingdoms in the Western

Regions. Zang Min answered all the questions in detail, such as that "the Western Regions were originally composed of 36 kingdoms, which were later divided into 55, and then were split up into more than a hundred. He dictated information about the size, the length of the journey, the number of different nations, the climate, mountains and rivers, grass and trees, birds and beasts and all the kinds of things not available in China, and then sketched out the terrain" [14]. Here, Ma Yuan was the inventor of a simple geographical model achieved by stacking rice, which was the earliest form of the sand table used by later generations of the army. What Zhang Qianqiu and Zang Min sketched, either on the ground or on something else, was a schematic diagram.

In the *Book of the Han: Record of Art and Culture*, many military books are accompanied by maps, such as:

Military Science of Sun Tzu, 82 chapters, Volume 9.

Emperor of Wei, 21 chapters, Volume 10.

In total there are 284 chapters in 14 military books, which included 47 volumes of maps.

There were also the local *juntu* (prefectural maps) [15] and maps of remote areas in the Han dynasty. For example, in the first month of the lunar year in the 4th year of the Jianzhao reign (35 BC), Emperor Liu Shi of the Han dynasty "offered sacrifices to the ancestors in his ancestral temple before executing the Xiongnu ruler Zhizhi Chanyu, the king of the Huns, and declaring a general amnesty. The prefecture officials celebrated with a banquet and Emperor Liu Shi displayed an annotated map recording the victory over Zhizhi Chanyu to the noblemen" [16]. This map may have indicated the remote location where Zhizhi Chanyu ruled.

Among the local maps, there was also a kind of *dixing tu* (topographic map), which may have been similar to the military topographic maps whilst serving a different purpose. For example, the *Dixing tu* drawn by Zhang Heng in the Eastern Han dynasty may be a landscape painting map of mountains, rivers, cities and towns in local areas [17]. In addition, another example is that of Li Xun, in the Eastern Han dynasty, who "went to Youzhou on a mission. After arriving in Youzhou, he announced that the emperor had granted favours to his subjects, he appeased the Beidi, and he sketched the mountains, fields and villages in the areas he passed along the way onto a map, amounting to a total of more than 100 volumes. All the books were reported to the throne, and Li Xun was praised by Emperor Zhang of the Han dynasty" [18].

In the Han dynasty, an official post "Yushi Zhongcheng" was created to manage the atlas and confidential documentations [19]. According to Ban Gu's *Dongdu fu*, "emperors received maps of the world." Therefore, the Han dynasty may have had a system in which the prefectures and states submitted maps to the throne [20].

4.3.2 Map Relics

According to current archaeological data, the map relics of the Qin and Han dynasties mainly comprise two items: the first is the set of maps unearthed at the Mawangdui site and the second is the map of a marketplace on the brick reliefs unearthed from the tomb of the Eastern Han dynasty in Sichuan province.

4.3.2.1 Maps Unearthed at Mawangdui

In 1973, three silk maps from the Western Han dynasty, dating back more than 2,100 years, were unearthed in the No. 3 Han Tomb at the Mawangdui site, Changsha, Hunan province. The level of cartography was quite high at that time. Based on actual measurements, there were azimuthal coordinates, scales and unified legends. The focus of each of the three maps was different: they include a *Topographic Map, a Garrison Map* and a *City Map*. The geographical scope included in the topographic map is mainly the southern part of the vassal state of Changsha at that time, that is, the present-day Xiao River and the Nanling Mountains. This is the main area of the map, with the highest level of precision in the drawing. The part which stretches as far south as the South China Sea was the prefecture of the Nanyue King, the feudal prince of the Western Han dynasty. The scale of this part became smaller moving away from the centre, the accuracy dropped significantly, and the coastline was inaccurate. Any curve drawn was almost symbolic, not to mention the scale (see Fig. 4.2).

The *Garrison Map* (see Fig. 4.3) is the earliest extant coloured silk map in the world. In addition to the mountains, rivers, roads and residential areas, the map focuses on the deployment of the nine garrisons' troops, the boundaries of defence areas, and watchtowers. Different geographical elements are distinguished by colours on the map. For example, mountains are represented by black mountain glyphs; rivers and lakes are drawn in cyan; red boxes on a black background indicate the stations and military engineering buildings of the troops; red dotted lines stand for the army's action routes; red triangles represent castles; black circles denote settlements; the colour red marks the boundaries of garrison areas, etc.

In a word, the three maps unearthed at Mawangdui not only show the high-level and precision of mapmaking in the Western Han dynasty, but also illustrate that not every map in the Western Han dynasty is accurate, and that even the same map may contain different levels of accuracy. Therefore, Pei Xiu's argument can still be considered reasonable, just not comprehensive enough.

(1) Map of the marketplace on the brick reliefs unearthed from the tomb of the Eastern Han dynasty in Sichuan province

This is a local city map. The marketplace is a part of the city, and the map is marked with directions and has no scale. The marketplace painted does not actually refer to that of a particular city, but rather reflects the general picture of a town in the prefectures and counties of the Han dynasty. It is a landscape map, used as artistic decoration, not for pragmatic use [21].

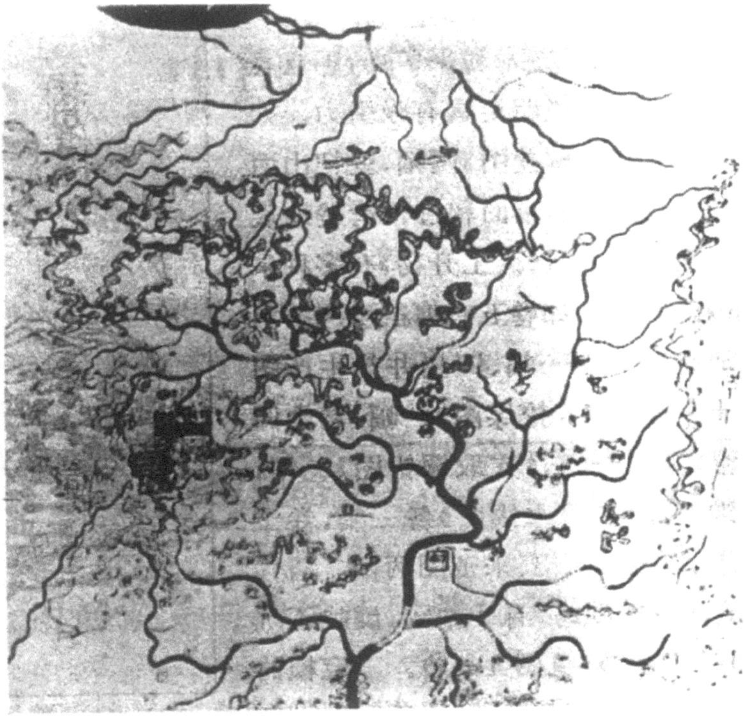

Fig. 4.2 The *Topographic Map* of Mawangdui

4.4 Geographical Literature

4.4.1 Geoscience in Dictionaries

The dictionaries of the Qin and Han dynasties, such as *Erya*, *Shuowen Jiezi* and *Shiming,* contained both social science and natural science knowledge.

(1) *Erya*

There are 19 chapters in *Erya*, which was the first dictionary to exist in ancient China. This book may date back to antediluvian times. Over a long period of time, many authors added to it and it was finally completed during the Han dynasty [22]. The chapters, "Shidi", "Shiqiu", "Shishan", and "Shishui" explained geographical names and organised the different geographical terms from the preceding dynasties, entailing rich geographical knowledge and the development of geography at that time. For example, in the book, the description of the different forms of sedimentary geomorphology caused by flowing water is very concise, saying that "a habitable place in the water is called *zhou*, a smaller one is *zhu*, an even smaller one is *zhi*, and a still smaller is *di*".

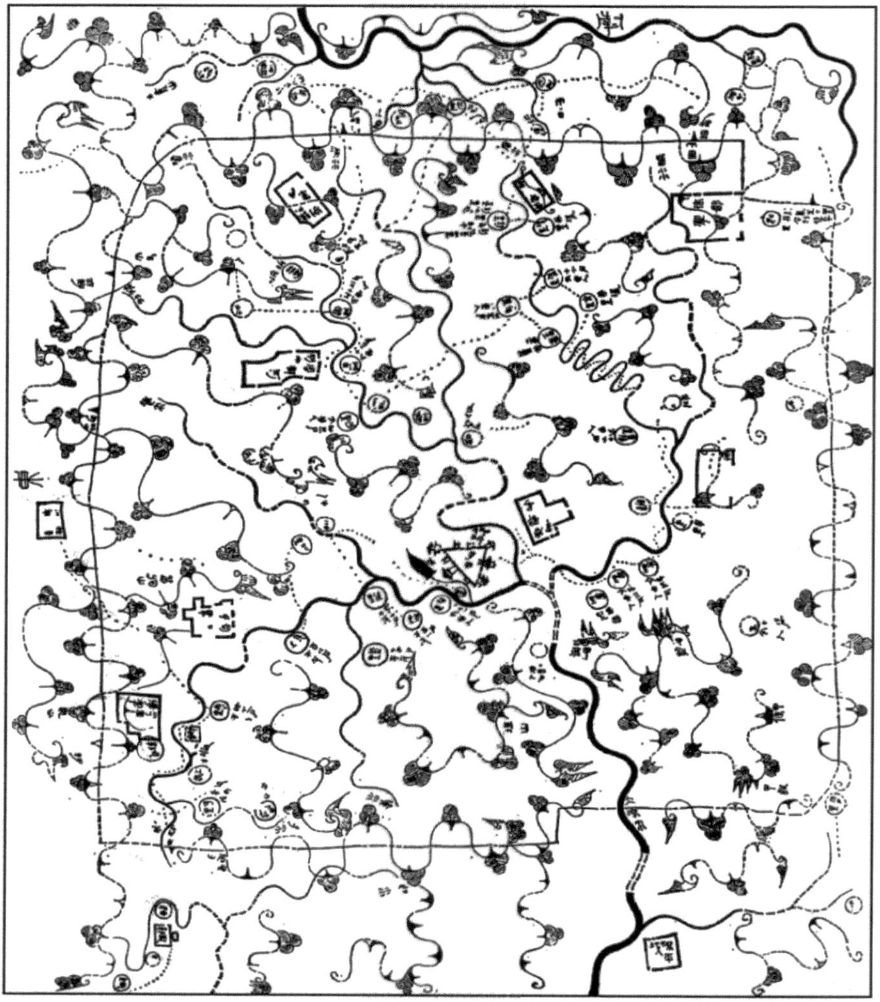

Fig. 4.3 The restored *Garrison Map* on silk fabric excavated from Mawangdui Han tomb number 3

(2) *Shuowen jiezi*

The *Shuowen jiezi,* compiled by Xu Shen in the Eastern Han dynasty, was the first dictionary in China to analyse word forms, explains word meanings and identifies word sounds [23]. The book contains 9,353 single characters and 1,163 repeated characters. Xu Shen interpreted 133,441 words and divided them into 540 sections [24]. The book is rich in geography and, according to the author, there are 575 geographical terms, including 169 names of ancient places, 146 rivers, 17 names of mountain, three swamps, one puddle, one pool and four mounds. There are nine more names of watercourses recorded in the *Shuowen jiezi* than in the "Shuijing" which appeared later. In addition, there are 111 terms for landforms, including 84 terms for hills, 25 terms for flowing water, three terms for plains, one term for a desert, 50 meteorological terms and 15 terms for soil. The explanation of the terms in the book is remarkable and is a reflection of the author's geographical knowledge. For example, there are 41 terms of mountainous landforms together with their explanations in the "Mountain" section, which is subdivided into mountainous areas, mountains near rivers, and islands in the sea. Peaks, valleys, caves and hillocks in mountainous areas are also introduced. There are also different terms depending to the condition of the vegetation on the mountain, the size of the stones, and whether the mountain is damaged or not, which is very detailed and comprehensive. A few items inherit the definitions used in the *Erya*, such as cén ("岑", high hill), qiáo ("峤", high and pointed mountain), *mì* ("密", hidden or open spaces in the mountain), *hù* ("岵", forested mountain), and *xué* ("嶨", rocky mountain).

(3) *Shiming*

The *Shiming* ("Explaining Terms"), compiled by Liu Xi at the end of the Han dynasty, was the first dictionary marking and comparing the phonetic notation of certain words in China, with 27 chapters in total. Known as an encyclopaedic dictionary of nouns, it was incomplete at the time of discovery. The purpose of this classic was to explore the origins of various terms. After the definition of each word, there is a further explanation of the underlying reason for the name. The geographic content of this book is mainly reflected in the sections which include the "Interpretation of Heaven", "Interpretation of Earth", "Interpretation of Mountains", "Interpretation of Water", "Interpretation of Hills" and "Interpretation of Provinces".

The "Interpretation of Heaven" is about astronomical phenomena, weather, "heavenly stems and earthly branches". For example, "Snow is a natural phenomenon. When water vapour in the sky condenses with cold air, the snow falls slowly."

The "Interpretation of Earth" refers to the land, soil, and plain landforms, with the five colours of soil being the most prominent features.

The "Interpretation of Mountains" is about twenty kinds of mountain landforms. Among these, "the foot of the mountain which is called *lù* ('麓') and the body of the mountain which is called *shí* ('石')" are two new concepts.

The "Interpretation of Water" introduces the names of water bodies, hydrological features and flowing water geomorphology, including 19 names for water bodies, two hydrological characteristics and seven flowing water features.

The "Interpretation of Hills" explains the hill landforms. There are 24 kinds of hill landforms, some explanations of which are quite vivid. For example, "the lower central part is called *wanqiu* in the shape of a bowl. The higher part is called *heqiu*. *He* means loads on a carrying pole, hence the shape of the hill looks as if one side of a person's shoulder is lower than the other due to the pole he is carrying on one side. The concept *muqiu* refers to a hill which covers one *mu* [666.67 square metres] of the land." The "Interpretation of Provinces" elaborates on the origin of the names of twelve ancient states, 13 vassal states and 18 prefectures, as well as their geographical locations, their city names and sizes, and their grassroots civil organisations.

4.4.2 Sima Qian and Geoscience

Sima Qian (145–86 BC) was a distinguished Chinese historian, and his book, the *Shiji* ("Records of the Grand Historian"), is not only a historical work and a literary work, but also a masterpiece with a rich geographic content. Chapters such as *Biography of the Profiteers*, *Book of the River and Drainage*, *Biography of the Huns*, *Biography of Dawan*, and *Biography of the Southwest Barbarians* are all valuable geographical documents. How was Sima Qian able to write these geographical chapters? That was due to his family and personal experience.

Sima Qian "was born in Longmen and led a farming and herding life in the southern foothills of Longmen Mountain. At the age of ten, he could already read ancient works". In order to fulfil his father's last wishes, Sima Qian gathered written material while also conducting social surveys and field trips to collect the various anecdotes and historical facts scattered among the folk.

In 126 BC, when he was 20 years old, he left home and travelled around famous mountains and rivers to investigate nature and society. The first time he set out from Chang'an, "he travelled south to the Jianghuai region, ascended Kuaiji Mountain, explored the cave of Yu the Great, visited Jiuyi Mountain, and rafted between the Yuan and Xiang rivers. In the north, he crossed the Wen and Si rivers, studied in the capitals of Qi and Lu, observed the style of Confucius' education, and learned and practised archery rituals in Zou Mountain. When he travelled to Fan, Xue and Pengcheng, he encountered danger but finally managed to return to Chang'an after passing through the states of Liang and Chu" [25].

In 116–112 BC, he travelled westwards with Emperor Wu of the Han dynasty to western Shaanxi and eastern Gansu; eastwards across the Yellow River to Shanxi and Henan; to the southwest to Sichuan and then crossing the Dadu River he reached the north of the Jinsha River. In 110 BC, he also travelled east to Mount Tai, northeast to Liaoxi (east of present-day Lulong County, Hebei province), west to Jiuyuan (present-day Wuyuan, Inner Mongolia), and then back to Chang'an via Yan'an in northern Shaanxi [26].

These experiences as a traveller enabled him to understand the geographical conditions, natural scenery, social economy, culture, customs and other conditions of various places, and this rich accumulation of first-hand material allowed him to later write the geographic chapters in *Records of the Grand Historian.*

Records of the Grand Historian: Biography of Profiteers was the genesis of economic geography in ancient China, outlining the economic geography of China two thousand years ago. Its main economic geographical points are as follows.

First, the economy has laws, just as "water flows downwards, day and night…If we consider the situation of surpluses or shortages of commodities, we will understand the price equilibrium. If the prices are too high, they will be lowered. When prices are extremely low, they will rise."

Secondly, economic development can "make the country prosperous and make families rich". The social environment will also improve: "when a gentleman is rich, he is willing to be benevolent; when the common people are rich, they can regulate their own labour"; "when people are rich, benevolence and righteousness will naturally follow". However, he also pointed out that "those who become rich through farming are of the upper class, through business are of a lower class, and through thieving are of the lowest class. Without the noble character of a hermit, there would be shame in claiming benevolence and righteousness, simply because of having been poor for a long time".

Thirdly, the unified feudal political situation of the Han dynasty promoted economic development. "With the rise of the Han dynasty and the unification of the world, customs checkpoints were opened and the ban on exploiting mountains and rivers was lifted. Therefore, rich merchants were able to travel throughout the world and transactions were unimpeded."

Fourthly, there were regional differences in economic development across the country. The reasons for the differences included natural conditions and human factors. The natural resources in China were different from east to west, north to south. For example, "to the west of the Taihang Mountains is rich in wood, bamboo, millet, wild hemp, yak tail hair and jade, and to the east of the Taihang Mountains is rich in fish, salt, lacquer, silk and song and dance. *Nanmu* (a precious type of arbor wood in mainly subtropical areas), catalpa, ginger, Osmanthus, gold, tin, lead, cinnabar, rhinoceros, hawksbill sea turtles, beads, ivory and animal skins are produced at the south of the Yangzi River. The area north of Longmen and Jieshi Mountain is rich in horses, cattle, sheep, fur felt, animal tendons and horns; thousands of miles of mountains are dotted with copper and iron mines, like a chessboard filled with chess pieces. This is general information about the distribution of products in various regions". Consequently, because of the varied natural conditions and the resources in the different places, the people of each place developed their own economy with local features, such as the Guanzhong Plain, where the capital city (Chang'an) was located, which had different regional characteristics from the neighbouring areas:

> The Guanzhong area stretches from the east of two counties, Qian and Yong, to the Yellow River and Mount Huashan, the fertile fields covering a vast area. The Han dynasty chose Xianyang as its capital. Zhuling, near Chang'an, was a focal point for people and things all over the world. The place was very small and the population was large, so the local people became more and more skilful and engaged in business. To the south of Guanzhong, lie the Ba Prefecture and Shu Prefecture, which are also fertile areas, rich in gardenias, ginger, cinnabar and instruments made from stone, copper, iron, bamboo and wood. However, the Bashu area is encircled by a thousand-mile plank road leading to Guanzhong. Only the Baoxie route could control its entrance, connect with the four roads, and barter surplus goods for shortages. Tianshui, Longxi, Beidi and Shangjun had the same customs as Guanzhong. In the west, Qiangzhing had geographical advantages and, in the north, there was the livestock of Rongdi, whose animal husbandry was ranked first in the world. However, the terrain there was dangerous and only Chang'an, the capital, offered safe passage. Therefore, the whole Guanzhong area occupied one third of the world, and its population only accounted for three tenths of the world's population. However, the wealth there represented six tenths of that of the world.

For the whole country, the differences were even more distinctive:

> Chu and Yue were sparsely populated areas. People there ate rice as their staple food and fish as a usual dish. They used the "slash-and-burn" farming method and filled the fields with water to extirpate weeds. They were self-sufficient in melons, fruits, conches and clams without buying from other places. With a favourable terrain, food was abundant, and there was no famine. People just remained alive without any significant effort, so they lived in poverty and had no savings. Therefore, there were neither starving people nor rich families south of the Jianghuai regions. The area north of the Yi and Si rivers was suitable for planting grain, mulberry and hemp and raising livestock. It had a small land area and the population was large. The people had suffered many floods and droughts. Since people there were fond of storing goods, the Qin, Xia, Liang and Lu regions were diligent in their agricultural practices and valued their labour force. This was also the case in the "three *he*" regions (Hedong, Henei and Henan), as well as in Wan and Chen. The residents in the Qi and Zhao areas were smart and nimble, and they speculated for profit; residents in the Yan and Dai areas could farm produce, raise livestock and farm silkworms.

Fifthly, the rise and distribution of cities in the Han dynasty was discussed, focusing on their geographical location and the development of transportation and economic systems. For example, an important economic capital Tao, known as "the centre of *tianxia*", was located along the Ji River, which interconnected with the Yellow River and its large and small tributaries, forming a network of land and waterways. It was surrounded by a wide and fertile plain, which was rich in grain; the nearby Qi-Lu area, with mulberry and hemp growing everywhere, produced an abundance of silk fabric. Tao was in the centre of this affluent area, making it the most prosperous city at that time. In addition, the *Records of the Grand Historian: Biography of Profiteers* also recorded Handan along the Zhang River, the capital of Yan between the Bohai Sea and Jieshi Mountain, Linzi between the Bohai Sea and Mount Tai, Suiyang between the Liang and Song states, and a dozen larger cities such as Luoyang, Zou, Lu, Yingdu, Chen, Wu and Panyu.

The text above outlines how Sima Qian's analysis of the characteristics of a local economy cannot be divorced from consideration of the natural conditions and, more importantly, of human labour. In places with good natural conditions, economic development is not necessarily prosperous, but, conversely, in places with poor natural conditions, as long as the people are good at exploiting these, the economy will also have the potential for significant improvement. The idea of developing the economy in harmony with local conditions still has relevance today.

Although the *Records of the Grand Historian: Biography of Profiteers* was not yet a pure work of economic geography, it attached great importance to the geographical elements in its accounts of the economic issues; it addressed the characteristics of economic phenomena in its depictions; and it approached a breadth of economic geography which was commensurate with the scope of the field under discussion. The descriptions of the regional economies, the accounts of agriculture, forestry, animal husbandry and minerals and the analyses of cities pioneered regional economic geography and sectoral economic geography [27].

The *Records of the Grand Historian: Biography of Dawan* was compiled by Sima Qian based on the report that Zhang Qian had written to the emperor after going to the Western Territories as an envoy. The first sentence in the text is that "the signs of Dawan were seen by Zhang Qian", which affirmed the value of Zhang

Qian's expedition to the West. Then it describes Zhang Qian's expedition to the West. Since Zhang Qian arrived in Dawan first, the biography concentrates on Dawan, and the distance to other countries or cities in the Western Regions was based on the mileage from Dawan. The location of Dawan was "to the west of the Han, thousands of li away from Han". The records of the economy, population, military strength, boundaries, products, hydrology, etc. of Dawan and other western countries are the earliest documentation of the Fergana Basin and the geographical conditions of the western regions in China, and they are also the earliest passages about China's extraterritorial geography. It was recorded that "in the west of Khotan, the water flows westwards into the West Sea (i.e. the Aral Sea); in the east, the water flows into Yanze Lake (i.e. Lop Nor)", obviously referring to the water system on both sides of the watershed of the Pamirs Plateau, which is correct. However, it was incorrect to say that the Yellow River originated in the Salt Sea. This error affected future generations for 800–900 years [28].

In addition, the biographies of the Huns and of the people of South Vietnam, Korea, etc. in the *Records of the Grand Historian* recorded the relationship between the Han dynasty and the minorities on the borders or in the neighbouring countries. The geographical situation of these regions was the earliest chapter in the frontier and extraterritorial geography of China.

4.4.3 Ban Gu and His Historical Writing, **Hanshu**

Ban Gu (32–92 AD) and Sima Qian share several similarities: first, they are both famous historians, literary scholars and geographers of the Han dynasty, though a time difference of 177 years separates them. Secondly, they both inherited their fathers' legacy. Ban Gu's father, Ban Biao, was also a famous historian, and he was "extremely talented and good at research, so he chose the field of history". Sima Qian wrote the *Records of the Grand Historian* only before the reign of Emperor Wu of the Han dynasty, but not after that. Ban Biao then "followed up the historical references and observed anecdotes around him, composing dozens of biographies." Ban Biao died before his writing plan was completed, so Ban Gu "studied his father's work with great care in order to fulfil his father's mission". While Ban Gu was concentrating on his book, someone complained to Emperor Ming of the Han dynasty that Ban Gu had "distorted national history". Without pursuing the truth, Emperor Ming of the Han dynasty suddenly confiscated Ban Gu's manuscripts and put him in prison. Later, he was rescued from prison by his younger brother, Ban Chao, and regained the trust of Emperor Ming, so he stayed in the capital as a magistrate of Lantai and continued to write books. Before Ban Gu finished the *Book of the Han*, he was implicated in the defeat of Dou Xian's army, and the Luoyang magistrate, Zhong Ke, seized the opportunity to retaliate, causing Ban Gu to be executed in Luoyang prison at the age of 61 [29]. Later generations often used the terms "Ban and Ma" or "Qian and Gu" or "Shi and Han" to praise the achievements of Ban Gu and the *Book of the Han.*

The chapters of the *Book of the Han* with geographical content are: Treatise on Geography, Treatise on *Wuxing* (the Five Elements/Agents), Treatise on Rivers and Canals, Treatise on Food and Commodities, Account of Profiteers, Account of the Huns, Account of the Southwestern Yi, Account of Southern Yue and Min-Yue, Account of the Koreans, Account of the Western Territories etc., among which Treatise on Geography is the most prominent.

The *Book of the Han: Treatise on Geography* ("*Hanzhi*" for short), originally written by Ban Gu, was the first Chinese work with geography in its title and it provides the earliest accounts of the borderland administration during the preceding dynasties. Its style set the standard for Chinese works on historical geography for two thousand years [30].

There are three strands to the content of the *Hanzhi.*

First, it describes the changes in the boundaries from the time of the Yellow Emperor to the beginning of the Han dynasty and provides transcriptions of the full texts of the *Yugong* and the *Zhouli: Zhifang.*

Second, it focuses on the geography of the Han dynasty. Due to the frequent transformation of the administrative regions, the setting of the prefectures and counties often changed. Therefore, the *Hanzhi* is specific to a certain period: the second year of Emperor Ping of the Han dynasty (2 AD). The history of 103 prefectures (states) and 1,587 counties (*dao, yi, houguo*) under their jurisdiction are recorded, as well as their populations, mountains, rivers, natural resources, water conservancy facilities, important settlements, customs and historical sites. This approach to writing has two advantages: first, it uses the administrative regions as the framework and presents a clear perspective; second, there are important geographical facts attached to the administrative districts. It is easy to find out which products are available in each administrative region. The disadvantage is that the classification of mountain and river systems is disrupted by the administrative regions, and the physical geography of the country lacks integrity. The geographic content is insufficient and overly simple, neglecting the transport infrastructure. Nevertheless, it still contains many precious geographical data. For example, it has the earliest records of the oil producing areas, of the officials in charge of the salt, iron, copper, gold, wood, mining, ponds, lakes, clothing, oranges, Yunmeng Lake, *junshu* (tribute collecting), etc., as well as of the statistics of the national population. Using these figures, the national population distribution map, the mineral distribution map and the industrial distribution map could be drawn in the second year of Emperor Ping of the Han dynasty.

Thirdly, it contains a transcription of Liu Xiang's *Yufen* ("Division of the Domain") and Zhu Gan's *Fengsu* ("Customs and Traditions"). The *Yufen* is about the correspondence between place and constellation, for example, "the land of the Qin Empire corresponds to the location between the Dojing Constellation and the Yugui Constellation." This kind of division has no geographical significance, but rather reflects the ancient concept of the "unity of heaven, earth and man". The *Fengsu* focuses on the economy, products, customs and history, and discusses the local characteristics of each region, as well as some foreign geography and sea routes. The content relating to the customs, products and weapons of Hainan Island is the earliest surviving documentary material relating to this area.

The five volumes of *Wuxing zhi* ("Treatise on Five Elements/Agents") bring together the *Spring and Autumn Annals* written by Dong Zhongshu, Liu Xiang, Liu Xin and others: derived from the doctrine of yin-yang and discussing various explanations for woe and fortune, it is a historical source for the intellectual history of the Western Han dynasty. The solar eclipses, lunar eclipses, astral variations and various natural disasters recorded in the book are valuable sources for studying the history of science, especially historical natural science.

The *Shihuo zhi* ("Treatise on Food and Commodities") was a seminal work by Ban Gu, covering the legendary ancient social and economic conditions, the collapse of the square-fields system, the economic conditions in the Spring and Autumn and the Warring States periods, the Shang Yang Reform, the economic depression in the early Han dynasty, the economic conditions and regional differences in the Western Han dynasty, the destruction of the social economy at the end of the Wang Mang's rule, the currency system, immigration, land reclamation, and the management of the commercial market. It is an important document in the study of the economic geography of the Western Han dynasty. From that time onwards, almost every dynasty had its version of the *Shihuo zhi* up until the appearance of the *Qin shigao* ("Manuscript of the History of the Qing Dynasty").

The *Xiyu zhuan* ("Account of the Western Regions") was first created by Ban Gu also, compensating for the shortcomings of the *Records of the Grand Historian: Biography of Dawan.* It documents 51 countries in the Western Regions, and specifically describes the capital city of each country, the mileage from Chang'an, the number of households, the number of troops, local products, customs, and material and cultural exchanges with the Central Plains. In terms of physical geography, it mentions Congling Mountain (the Pamirs), saying that, to the east of Congling, "there are mountains in the north and south, rivers in the centre, and these rivers have two sources, one in the Congling Mountains and the other in Khotan. Khotan is at the foot of the mountain to the south, and its river, which flows north, joins with the Congling River, and flows into Puchanghai (Lop Nur) in the east. Puchanghai, a salt marsh, is more than 300 li from the Yumenguan Pass, covering an area of 300 li, and its range does not increase or decrease with the seasons." The book also mentions the desert and says that the Shanshan Kingdom was "near the Bailongdui and lacked water and grass"; the land was "sandy with few fields, so the citizens had to farm in the nearby countries. The Kingdom was abundant in jade, reeds, Tamarix, and Pennisetum". These records are accurate. However, a similar mistake was made as in the *Records of the Grand Historian* in saying that the water in Puchanghai goes underground and flows southwards into the Yellow River.

4.4.4 Scholarship in Water Conservancy and Canals

During the Qin and Han dynasties, because the rulers attached great importance to the farmland irrigation systems, water conservancy and canal water transportation, a number of water conservancy projects appeared and led to the spectacular scenes of the harnessing of the Yellow River and the controlling of water disasters. Although the Qin dynasty did not exist for long, during the 15-year rulership of the Qin dynasty, giant construction projects including the Great Wall, highways, water courses and irrigation canals—the Qin Canal—were completed one after another. In addition to the Lingqu Canal, the Jiangnan Canal was also excavated during the Qin dynasty. Before the First Emperor of Qin unified the six countries, the state of Qin had constructed the Dujiangyan irrigation system, Zhengguo Canal and other large-scale water conservancy projects. During the Han dynasty, there had been many irrigation projects, including Shanhe Weir, Liufu Canal, Bai Canal, Chengguo Canal, Longshou Canal, Liumen Dam, Hongxi Reservoir, and Jian Lake, etc. The harnessing of the Yellow River was a major water conservancy project in the Han dynasty, with Wang Jing as the water expert involved in this project.

The *Records of the Grand Historian: Book of Rivers and Canals*, originally written by Sima Qian, systematically describes the historical facts of water management in previous generations and the contemporary flood control, navigation, and management of the Yellow River, and the construction of irrigation and water conservancy systems throughout China. The book not only lists the water conservancy projects, but also highlights the geographical characteristics of these projects. For example, in the book it is written that "from Hanzhong Prefecture to Shu Prefecture, there are many mountains and steep slopes along the old road, which is rugged and remote. Today, if we cut through the Baoxie path, with fewer hillside tracks, the journey can be four hundred li shorter than it would be if we took the old road".

The *Book of the Han: Treatise on Rivers and Canals* is a continuation of the *Records of the Grand Historian: Book of Rivers and Canals*. Therefore, at the beginning, about one third of the content is taken from the "Book of Rivers and Canals", with only a few words added, deleted or modified. The later chapters, following on from the 6th year of the Yuanding period (111 BC), form the new content of *Treatise on Rivers and Canals*. They describe the origin and construction process of the Liufu Canal and Bai Canal by diverting the Jing River; the bold ideas of the Qi people and their suggestion of drawing the river into the desserts of Inner Mongolia and letting it flow eastwards into the sea; and the controversies over the history and governance policy of the breaching and diversion of the Yellow River. When Wang Mang was in power, Wang Heng quoted the *Zhoupu* ("Zhou Manual") in his discussion of river management which said that "the river was diverted in the 5th year of King Ding (602 BC)". This is the earliest record of a Yellow River diversion [31]. These records have high historical value, enlightening future generations on flood control.

During the Qin and Han dynasties, there were two treatises on water systems: first, the *Shuijing* of the Qin dynasty and, second, the *Shuijing* in the late Han dynasty.

The *Shuijing* of the Qin dynasty is known by a name given by people in modern times [32]. It is attached to the end of the *Shanhai jing: Hainei dongjing* ("Classic of Eastern Regions Within the Seas"). The full text is nearly 500 words in length, and describes the origin, flow direction and destination of 26 waterways. The basic geographical facts in this book are in accord with the *Hanzhi* and *Shuijing* of the Han dynasty. The distribution range of the waterways is generally consistent with the territory of the Qin empire. Zhou Zhenhe believed that it was written during the Qin dynasty. Some of the waterways it contains, such as the Jing River, Wen River, Bai River (not the Wenshui and Baishui in the Han dynasty *Shuijing*), are not even included in the Han dynasty *Shuijing*. According to the written order in the book, the names of the 26 waterways were: Dajiang, Beijiang, Nanjiang, Zhejiang, Lujiang, Huaishui, Xiangshui, Hanshui, Mengshui, Yingshui, Rushui, Jingshui, Weishui, Baishui, Yuanshui, Ganshui, Sishui, Yushui, Yishui, Huangshui, Luoshui, Fenshui, Jishui, Liaoshui, Hutuoshui and Zhangshui.

Compared with the chapter "Yugong", which only records nine main rivers, the *Shuijing* of the Qin dynasty had made great progress.

The *Shuijing* written at the end of the Han dynasty is a rather controversial work. Its date of writing and author are still uncertain. As Li Daoyuan did not indicate the author of the *Shuijing* when he annotated it, the problems increased subsequently. The *Tang liudian: Gongbu: Shuibu yuanwailang zhu* ("Six Statutes of the Tang Dynasty: Ministry of Works: Note of Water Ministry Officials") first recorded that the author of the *Shuijing* was Sang Qin. The *Book of the Sui: Bibliographical treatise* documents two versions of the *Shuijing* at the same time. One is "Guo Pu's annotation of the three-volume *Shuijing*", and the other is "Li Shanchang's annotation of the forty-volume *Shuijing*". It was obviously incorrect to have recorded "Guo's composition" (rather than "Guo's annotation") in the *Old Book of the Tang: Bibliographical treatise*; similarly, "Guo's annotation" should not have been changed to "Guo's composition" in the *Book of Sui: Bibliographical treatise*. In the *New Book of the Tang: Art and Culture Annals*, it is said that the three volumes of Sang Qin's *Shuijing* were once composed by Guo Pu who adapted two statements from the *Six Statutes of the Tang Dynasty* and the *Old Book of the Tang: Bibliographical treatise*, yet the *Tongzhi: Yiwen lue* by Zheng Qiao mentions that the three-volume *Shuijing* was "annotated by Guo Pu", believing that Sang Qin was of Han nationality [33]. Likewise, Chao Gongwu stated in the *Junzhai dushu zhi* that "the four volumes of the *Shuijing* were written by Sang Qin (32–7 BC) during the time of Emperor Cheng of the Han dynasty", but he did not provide the basis for this conclusion. According to research by the Qing scholars, most people believed that the *Shuijing* was written in the early Wei dynasty of the Three Kingdoms and the author was unknown. However, in the waterway sections of the *Hanzhi*, Sang Qin was quoted seven times, and Li Daoyuan also quoted Sang Qin in his notes. Therefore, there were grounds for the Tang people to designate Sang Qin as the author of the *Shuijing*. We believe that it is appropriate to place this work at the end of the Han dynasty, which does not fundamentally contradict the preceding arguments.

According to the *Tang liudian*, the *Shuijing* in the late Han dynasty recorded a total of "one hundred and thirty-seven" waterways, one chapter for each waterway, and attached 60 waterways from the "Location of Waters and Mountains in the Yugong". In the description, the watercourses provide an outline for the description of the origin, flow direction, flow area, destination, and distribution of each main river and the flow of the branches of the watercourse, etc., in order to fully capture an overview of the river systems throughout the country. The original book has been lost. Fortunately, Li Daoyuan annotated it and saved it in the *Shuijing zhu*, which has been handed down to this day. However, there are only 123 articles, 14 fewer than those documented in the *Tang liudian*. The recorded waterways vary in complexity with some errors. Li Daoyuan corrected more than 60 points when he made notes.

Compared with the *Shuijing* in the Qin dynasty, the *Shuijing* in the late Han dynasty had evolved: in terms of the number of waterways, there are more than five times as many in the Han *Shuijing* as in the Qin dynasty's version; in terms of the number of words, there are about 20 times more in the former than in the latter.

4.5 The Emergence of Local Chronicles

The combination of Chinese history and geography resulted in local chronicles, which also emerged during the development of historiography. During the Qin and Han dynasties, due to the establishment and development of a centralised system, especially the prefecture and county level regulations, the central government urgently needed to understand the geographical and political and economic conditions of all parts of the country. During the reign of Emperor Wu of the Han dynasty, the prefectures and states were ordered to report information about the local properties, tributes and roads, together with submitting the *jiche* (record books of chariots) to the court historians. When Emperor Cheng of the Western Han dynasty was on the throne, Zhang Yu, the prime minister in charge of properties, ordered his subordinate Zhu Gan to record the local customs, and expanded the scope of the collection of local materials [34].

4.5.1 National Chronicles

During the Eastern Han dynasty, the national chronicles included Ying Shao's *Shisanzhou ji* ("*Records of Thirteen Prefectures*") and *Dili fengsu ji* ("Records of Geographical Customs").

The original version of the *Dili fengsu ji* has been lost and, nowadays, there exists an edited collection of a national chronicle: *Han tang dili shuchao* ("Geography Book of the Han and Tang Dynasties") by Wang Mo. The book contains the records of Dunhuang, Jiuquan, Yulin, Rinan and other prefectures. Although this book generally covers all parts of the country, it actually focuses on the places of You, Ji, Qing and Qi. Because Ying Shao once worked as Mount Tai Magistrate, and later assisted the powerful warlord Yuan Shao for a long time, he was most familiar with the surroundings of Mount Tai and the east side of the Taihang Mountains. Records of the Luo River state that "it flows northeast to Qiancheng and into the sea. When the stream is large, it flows into the big river and the sea. When the water flow is tiny, it becomes a mere trickle". The record from Qiancheng County mentions that "there is a big river fifty li away from the northwest of Qiancheng County, and Luowo is in the north of Qiancheng County". The record regarding Fu River shows that "the Fu River flows from Fuyang County to the sea. Every day the tide rises and falls, until there is no water in the channel". These are not only records of the past, but they are also features demonstrated in the present day. It can be seen that not only did Ying Shao write the *Records of Geographical Customs* with reference to literature, but it is possible that he also obtained the material from field investigation [35].

The *Shisanzhou ji* can only be found in several articles cited in the *Shuijing: Sishui zhu.* Yao Zhenzong's *Book of the Later Han: Record of Art and Culture* states that "the *Shisanzhou zhi* is similar to the *Book of the Han: Treatise of Geography.*"

4.5.2 Local Chronicles

The official term *(di)fangzhi* (local chronicle or gazetteer) has been found in the *Book of the Later Han: Treatise on the Western Regions*, and there are a number of *fangzhi* dating from the Han dynasty. According to the *Research on Ancient Chinese Local Records*, about 50 kinds of Han dynasty *fangzhi* have been identified so far, but most of them are lost. Introductions to several representative works follow.

The *Sanfu huangtu* ("Imperial Maps of Sanfu") first appears as a volume in the *Book of the Sui: Annals of Classics and Records*, then as three volumes in the *Junzhai dushu zhi by* Chao Gongwu, and as two volumes in the *Zhizhai shulu jieti* by Chen Zhensun. The present-day six volumes of the *Sanfu huangtu* are not the original version from the Eastern Han dynasty, but rather they were written by people after the Three Kingdoms. The original illustrations have been lost. According to quotations in other works, the *Sanfu huangtu* is a local chronicle recording the Daoist temples, mausoleums, *mingtang* [ritual halls for emperors to worship the ancestors] and *piyong* [the ditch encircling the exterior of the mingtang], and *jiaozhi* [places where the gods of heaven and earth were worshipped] of the three prefectural districts near the capital city Chang'an, namely Jingzhao, Fengyi and Fufeng.

The original version of Mr Xin's *Sanqin ji* has been lost and an edited book has been handed down. "The mountains, rivers, capitals and palaces in this book are all geographical features in the Qin and Han dynasties without taking any account of the Wei and Jin dynasties. Thus, this book must have been written by the Han people [36]." The desert topography recorded in this book constitutes one of the earliest documents in China, and states that "the Shajiao Mountain, located to the west of the Yellow River, is more precipitous than the stone mountain. Its sands are as coarse as grains of rice in size. To the south of the mountain, there is a spring called the sand well, which was not filled with sand until recent days. When someone climbs the mountain, he is bound to step into a sandpit. At that moment, the sand will make a sound like a 'drum and horn' and even the feet can feel the vibration of the sand grains" [37].

The *Bajun tujing* ("Brochure of Ba Prefecture") is the work of people during the Eastern Han dynasty. According to Dan Wang's writing, "the *Bajun tujing* involves a territory of four thousand li from north to south, five thousand li from east to west, further extending into the surrounding area by over ten thousand li. The Ba Prefecture has fourteen counties with prime ministers and five officials overseeing salt and iron. There are four hundred and sixty-four thousand, seven hundred and eighty households, and eight hundred and seventy-five thousand, five hundred and thirty-five people. Due to the large area of the Ba Prefecture, the remote county is 1,200–1,500 li away from the prefecture, and the township is 300 to 1,000 li away from the county [38]". It is evident that this book records the boundary, subordinate counties, salt and iron officials, and household registration of the Ba Prefecture in the Han dynasty.

4.6 Geoscientific Information in *Weishu*

The term *chenwei* ("divination"), together with *weishu* ("augury books") refers to prophecy books that reflected a mystical version of Confucianism during the Han dynasty.

Weishu was prominent in the ideology and culture of the Han dynasty and had an extensive influence from the imperial court to civil society, as well as in intellectual and bureaucratic circles. It flourished from the late Western Han dynasty to the early Eastern Han dynasty. In substance, *Weishu* discusses yin-yang and *wuxing* ("five agents"), the interaction and unity between heaven and man, astronomy and the calendar, geography, customs, history, accounting, etc. Revised by the authorities and taught in official schools for a long time, it catered to both high and low classes in society. The practice of divination often conflicted with imperial power, so it was repeatedly banned from the Three Kingdoms period onwards. Very few books dating from after the Sui dynasty have survived. Fortunately, various collated versions exist from the beginning of the Ming dynasty, for example, the *Gu weishu* by Sun Jue of the Ming dynasty, 55 divination topics in the *Collectanea of the Han Studies Lodge* by Huang Shi, 40 augury segments in the *Yuhan shanfang ji yishu* by Ma Guohan, and *Qiwei* compiled by Zhao Zaihan. Of these, the *Weishu jicheng* ("Wei Shu Integration"), compiled by the Japanese scholars Kōzan Yasui and Shohachi Nakamura, is the most complete version [39]. It was compiled on the basis of the editions from the Ming and Qing dynasties and was published from 1959 to 1964.

There are mainly five aspects to the geoscientific knowledge in *weishu*.

1. Knowledge about the earth's sphericity and state of motion

The theory of the celestial sphere in *weishu* slightly predates Zhang Heng's theory, but is not as explicit as Zhang Heng's. The *Spring and Autumn Annals: Yuanming bao* says that "the heavens are like an egg. The sky is large and the earth is small with water on the surface. The heavens and the earth stand on qi, and float on water". Zhang Heng argued in the *Huntianyi zhu* ("Note to Armillary Sphere") that "the celestial sphere is in the shape of an egg, and the earth is like the egg yolk, lying in the egg. The sky is big and the ground is small ... The heavens wrap the earth like the eggshell enveloping the yolk. The heavens and the earth stand on qi, and float on water". In comparison, Zhang Heng's words are much clearer.

As for the state of motion of Earth, the *Hetu* ("River Map") suggests that "the earth is constantly moving, just like people sitting on a big boat with the windows closed overlooking the movement of the boat" [40]. "There are four directions to the earth's movement. In winter, the earth rises up, thirty thousand li from the north to the west; in summer, it goes down, thirty thousand li from the south to the east; in spring and autumn, it moves towards the centre. The earth is always moving but people are unaware of this, just as when a person takes a boat and does not feel the boat travelling [41]." The so-called "earth movement" and "four directions" refer to the fact that Earth is constantly moving through the four seasons, giving a scientific explanation for the causes of the solar terms. The *weishu* of the Han dynasty is remarkable in that it reached such a high level of understanding of the Earth's movement.

2. Knowledge about weather and climate

The *Spring and Autumn Annals: Yuanming bao* says that "the yin and yang converge to form clouds, harmonise to form rain, congeal to form snow, condense to form frost, and disperse to form dew". The *Spring and Autumn Annals: Ganjing fu* says that "the yin qi specialises in condensing water into hail". The *Spring and Autumn Annals: Shuo tici* records that "warm air causes the rain, on encountering the yin qi, it will coagulate into hail; cold air leads to snow, and when it meets up with the yang qi, it will turn into graupel" [42]. This explanation suggests that the formation of clouds, rain, snow, frost, dew, hail and graupel results from the changes in yin and yang, which was the theoretical conclusion drawn by the ancient people based on their life experience.

As for knowledge of weather forecasting, in the *Spring and Autumn Annals: Han hanzi* is is written that "the ants hiding in the cave can predict the rain first; before the clouds completely cover the sunlight, the fish have already swum to the surface to breathe with their mouths open. The bird that lives in the nest knows the wind first; when trees shake, the birds fly". This was a way of predicting changes in the weather from animal activity, which is the most primitive method of forecasting the weather.

3. Knowledge about soil

The *Xiaojing: Yuanshen qi* ("Classic of Filial Piety") records that "the yellow and white soil is suitable for growing millet, the fertile soil is suitable for growing wheat, the deep red soil is suitable for growing beans, and the wet soil is suitable for growing rice". This is based on the experience of agricultural production, knowing which soil is suitable for which crops.

The *Hetu: Kuodi xiang* talks about the geographical distribution of soil. The soil types vary on the different regions (states). "In the southeast of Shenzhou, it is called *chen* ("晨", morning) soil, in the south of Qiongzhou it is called *shen* ("深", deep) soil, in the southwest of Rongzhou it is called *tao* ("滔", ill) soil, in the west of Yanzhou it is called *bing* (flat) soil, in central depending Jizhou it is called *bai* ("白", pure/rammed) soil, in the northwest of Zhuzhou it is called *fei* ("肥", fertile) soil, in the north of Yuanzhou it is called *cheng* ("成") soil, in the northeast of Xianzhou it is called *yin* ("隐", shaded) soil, and in the east of Yangzhou it is called *xin* ("信", under the sun) soil."

4. Impact of water, soil and environment on human character, language and voice

The *Hetu* says that "there are different themes across *jiuzhou* (nine regions of ancient China), and the springs are different in hardness and softness. In Qingzhou and Xuzhou, where the springs are salty and sour, the pitch of people's voice is usually Jue (mi) and Yu (la), the notes of the ancient Chinese five-tone scale, and they keep their mouth wide, speaking gently and at a slow pace. In Jingzhou and Yangzhou, where the springs are sour and bitter, the pitch of people's voices tends to be Jue (mi) and Hui (sol), with a light air and a clipped delivery. In Liangzhou, where the springs are bitter and spicy, people often stress the Shang (re) and Hui (sol), speaking strongly and loudly, and their delivery is obstructed. In Yanzhou and Yuzhou, where the springs are sweet and bitter, it is customary for the pitch of people's voices to be Gong (do) and Hui (sol), calm and steady, and their delivery is clear. In Yongzhou and Jizhou, where the springs are spicy and salty, the pitch of people's voices tends to be Shang (re) and Yu (la), and their speaking style is swift and intense [43]". The author believed that due to the different natures of the five tones, and of the land and springs in *jiuzhou*, there were differences in people's temperament, voice and language.

5. Knowledge about iron ore

The *Hetu: Kuodi xiang* highlights that "Yingzhou is rich in *kunwu* stones which are smelted into iron to forge swords. The stones are as bright as crystal and that is the iron ore" (cited from Volume 32 of the *Gu weishu*). The *Longyu River Map* says that "Liuzhou island is in the West Sea, covering 3000 li, with many mountains and rivers. The *kunwu* stones here can be smelted into iron to forge swords which are as bright as the stars and so sharp that they can slice jade like cutting mud" (cited from Volume 34 of the *Gu weishu*). These two quotations refer to iron ores, with different locations but fundamentally the same nature. This is the earliest account of iron ore in China.

4.7 Meteorological Knowledge and Geological Instruments

4.7.1 Meteorological Knowledge

The knowledge of meteorology in the Qin and Han dynasties had made remarkable progress compared with the previous dynasties. This is not only reflected in the records of various meteorological phenomena, but also in the discussion of diversified meteorological causes. Moreover, primitive meteorological instruments and weather forecasting had come into being. The following describes the understanding of physical atmospheric phenomena in the Qin and Han dynasties.

(1) Wind

What is wind? According to Wang Chong, "wind is air [44]." This understanding was correct and advanced at that time.

There are various kinds of wind. The *Lüshi chunqiu: You shilan* records eight names for eight directions of wind. In the Han dynasty, there also existed the concept of the Eight Winds based on the wind direction and the eight solar terms. For example, the *Yiwei* ("Interpretation of the Book of Changes") says that "the wind of the eight solar terms is called the Eight Winds. The wind which blows at the start of spring is called *tiaofeng*; at the spring equinox *mingshu feng*; at the start of summer *Qingming feng*; at the summer solstice *jingfeng*; at the start of autumn *liangfeng*; at the autumn equinox *change feng*; at the start of winter *buzhou feng*; at the winter solstice *guangmo feng*" [45]. The *Erya* says that "the east wind is the *gufeng*; the south *kaifeng*; the west *taifeng*; and the north *liangfeng*". The *Erya* also talks about wind movement: "a storm dies from top to bottom and builds from bottom to top. The movement named *piao* means a whirling wind, *bao* means a wind at sunrise, *yi* means the sky is dark and windy, and *mai* refers to a wind accompanied by rain [45]."

In the Han dynasty, people benefited from the mountain-valley wind. Zheng Hong of the Han dynasty "led a very frugal life. He always cut trees in person and used food for trade. Whenever he crossed the Xijin, he felt the divine wind helping him, because the boat could move without paddling. Thus, the villagers made full use of the wind to sail [46]". In the *Kuaiji ji*, Kong Lingfu documented the story of Zheng Hong returning an arrow lost by an immortal. He wished for the south wind in the morning and the north wind in the evening, so that people could be assisted by a tail wind when carrying firewood by boat [47]. This legend deifies Zheng Hong, but the description is clearer than in the *Shuijing zhu*, that is, the wind direction is the opposite in the morning and evening, which is a typical feature of mountain-valley winds.

(2) Cloud and rain

Clouds and rain are closely related, and the ancients often described them together.

The *Huangdi neijing* ("Yellow Emperor's Inner Canon"): *Suwen* says that "the air rises from the ground to form clouds, and the air drops from heaven as rain. Rain can fall because of absorption by the earth. Clouds can rise because of evaporation by the sun in heaven".

The *Shiming* ("Explanation of Names") says "the rain comes from the clouds".

Wang Chong stated that "the air from the mountain turns into clouds, and before rising into the sky, it may drop as rain [48]". "The clouds and rain come from the hills. If it falls, a cloud will become rain [49]." "The rain comes out of the mountain, carried by clouds. If the clouds scatter, the raindrops fall. The cloud and the rain are the same [50]." "When it is going to rain, clouds come out of the mountain first, and then turn into rain [51]." The above statements by Wang Chong are correct. However, he only spoke of the small cycle of water and gas over the mainland, not about the large cycle of water and gas from the ocean to the mainland, which is mentioned in the *Lüshi chunqiu: Yuandao*. On the subject of the relationship between clouds and rain, Wang Chong's arguments were the best at that time.

The duration of precipitation was recorded in the Han dynasty. For example, according to the *Book of the Han: Treatise on Five Agents*, "In the summer of the third year of Emperor Cheng (30 BC), there was a heavy rainfall in Sanfu for more than thirty days."

With regard to the shape and name of clouds, the *Records of the Grand Historian: Tianguan shu* mentions five kinds of clouds with different shapes. The *Huainanzi: Dixingxun* makes a reference to blue clouds, red clouds, white clouds and black clouds, naming the cloud based on its colour. In addition, the section also mentions the relationship between clouds and precipitation patterns. In the *Yifeihou*, Jing Fang vividly describes the shape of clouds, namely, cumulus, cirrus, stratiform and direct spreading clouds [52].

(3) Fog, dew, frost, snow, graupel and hail

The *Dadaili ji* says that "if the yin qi (negative energy) prevails over yang qi (positive energy), the air condenses as frost". "The heaven and earth gather the yin qi. When the yin qi is warm, it turns into rain. When it is cold, it turns into snow [53]."

The *Shiming* says that "graupel is like the stars. The competition between the rain and snow is like the scattering of stars".

The *Erya* says that "the harmony between rain and snow causes graupel".

The *Spring and Autumn Annals: Shuotici* says that "the strong yin qi coagulates into snow. If the yang qi comes, it will be dispersed as graupel [54]".

The *Baihu tong* says: "Dew is the beginning of frost, and it will turn into frost when the air is cold". The relation between dew and frost here is accurate. In this regard, Liu Xiang further elaborates in the *Wujing tongyi* that dew and frost do not fall from the sky, but are formed by the water vapour on the ground when the temperature drops.

As for the cause of snow, Dong Zhongshu believed that "in cold weather, the rain congealed into snow, which was so light that it could be blown by the wind [55]".

Wang Chong of the Eastern Han dynasty summarised that "the clouds and mist are a sign of rain. In summer, they turn into dew, and in winter, they form frost. With warm air, they become rain, and with cold air, they condense into snow. Condensation in the form of rain and dew rises from the ground, instead of falling from the sky

[50].” He clarified the essence of fog, dew, rain, snow and frost and their relationship with each other: essentially, that water vapour condensing on the ground formed fog, dew, rain, snow and frost depending on differences in temperature.

Dong Zhongshu composed an article about rain and hail. In July of the first year of the Yuanguang (134 BC), when rain and hail fell in the capital, Bao Chang asked Dong Zhongshu: “What are hailstones? What makes them?” Dong Zhongshu answered, “Hail is a kind of graupel. On encountering the yin qi, rain will condense into hail [56].” Graupel is formed in the same way as hail, but the diameter of graupel particles is smaller and the diameter of hail is larger.

(4) Temperature

As for temperature, the *Huangdi neijing: Suwen* reflected on the idea that temperature decreases with the increase in height, indicating thinking regarding atmospheric stratification, which was more than 1,700 years earlier than in the western world.

As for the highest and lowest temperatures in a year, the *Yi: Jilan tu* (a type of *weishu*) says that “the highest temperature falls within the thirty days after the summer solstice, the lowest temperature is within the thirty days after the winter solstice [57]”.

Wang Chong pointed out that the reason for the four seasons is related to the relative position of the sun and the earth. The *Lunheng: Shuori* says that “in summer, the sun is located in the Dongjing constellation and, in winter, the sun moves to the position of the Qianniu constellation. It is a very long distance to Qianniu, so the day becomes short, while Dongjing is much closer, so the day becomes longer”. From this, he further concluded that “the spring is warm, the summer is hot, the autumn is cool, and the winter is cold. The four seasons are a natural phenomenon [58]”. Wang Chong’s understanding was the result of long-term astronomical observations in China.

(5) Humidity

Humidity is mentioned several times in the *Huainanzi*. For example, “When water vapour rises from mountains, the cornerstone of the pillar is wet.” “By hanging feathers and dry carbon, which have different levels of hygroscopicity, on each end of the balance scales, the humidity of the air can be inferred from the tilt angle of the balance.” “A high level of humidity means that rain is likely [59].” Wang Chong also pointed out that there were regional differences in humidity, and he said that “the north of the Yangzi River is dry, which is suitable for wasps and scorpions; the south is humid, which makes a good habitat for snakes [60]”. Wang Chong further indicated the rule of periodic changes between dry and wet weather, droughts and downpours, saying that “sunny days and rainy days are like day and night. Rain will stop naturally and a drought will be alleviated by rain spontaneously. When the phenomenon of yang peaks, the phenomenon of yin will emerge, and vice versa [61]”.

(6) Weather forecasts

Books on weather forecasting dating from the Han dynasty mainly include the *Huainanzi*, *Yifei hou*, *Zhoudi tuji*, *Lunheng*, and Cui Shi's *Nongjia yan* ("Farmer's Proverbs"). There were two main ways of predicting the weather: one was by observing the activities of animals, and the other was by watching cloud shapes.

The first way to forecast the weather was by observing animal activity. The *Huainanzi: Xiuwu xun* says "if it is about to be windy, although the grass and trees are still static, the birds will have flown away; if it is going to rain, before the clouds completely cover the sunlight, the fish will already have swum to the surface to breathe".

The *Huainanzi: Miaocheng xun* says that "the magpie living in a nest can discern the wind first; the marmots living in an underground burrow can sense the depth of the water. The *zhenniao* (possibly a hawk-like bird with poisonous feathers) can predict sunny days, and the *yinxie* (the female of the *zhenniao*) can foresee rainy days". Gao Yinzhu said that "magpies can predict the amount of wind in a year. If there is to be a great deal of wind, their nest will be built in a lowdown position; the marmots will avoid digging a burrow where the water will reach." "If the sky is going to clear up, the bird called the *huimu* will tweet, and if there is going to be rain, the *yinxie* will sing."

The *Lunheng: Biandong* says that "ants will migrate and earthworms will emerge on a potentially rainy day. Insects in nests will migrate on a windy day, and animals in dens will move on a rainy day. The wind and rain can be perceived by creatures".

The second way of forecasting the weather is based on cloud shapes. Jingfang's *Fengjiao yaojue* says that "the rain can be predicted by the size of a cloud. If the black cloud is like a piece of cloth, it will rain for a whole day; two pieces of cloth are two days' rain, and so on [52]".

Jingfang's *Yizhuan* records that "the blue, white, red and black clouds are in the east, west, north and south. When they appear, it is going to rain. If there are black clouds in Hanchuan as large as mats, this indicates that it will rain in five days [62]".

The *Zhoudi tuji* says that "on Taibai Mountain, Mei County (now Mei County, Shaanxi province), if there is going to be a storm, clouds in the shape of waterfalls will appear from the mountain [63]".

4.7.2 Geoscientific Instruments

In the Han dynasty, several geoscientific instruments emerged. The invention of geological instruments undoubtedly promoted the development of China's geosciences. Among them, the seismograph invented by Zhang Heng is the most famous. In addition, there were some meteorological instruments.

(1) Seismograph

In the first year of the Yangjia period in the Eastern Han dynasty (132 AD), Zhang Heng invented and created the first seismograph in human history. It was "made of refined copper, with a diameter of eight feet and a raised lid, resembling a wine bottle [64]". The inner structure was mainly comprised of a column in the middle (equivalent to the heavy hammer of the modern seismograph, working as a pendulum) and the eight channels around it (eight groups of lever machines facing in eight directions connected with the instrument body around the pendulum). Outside the seismograph, there were eight dragons with small copper balls in their mouths facing in eight directions. On the ground, there were eight corresponding toads raising their heads and opening their mouths. When a strong earthquake occurred in a certain direction, the seismic wave made the column deflect. Then a lever was triggered, enabling the dragon to open its mouth, and the copper ball fell. The frog caught the copper ball and made a clattering sound. When the observers heard the sound, they knew the time and direction of the earthquake. After the seismograph was made, it was placed in Luoyang, the capital at that time.

On 1st March, in the 3rd year of the Yonghe period (138 AD), the dragon on the west opened its mouth and the copper ball dropped. However, people in Luoyang did not feel the earthquake, so they doubted the validity of the instrument. But a few days later, a messenger came from Longxi and reported to the government that a big earthquake had occurred in Longxi (today's southeast of Gansu province), which dispelled people's doubts. Luoyang was about 700 km away from the epicentre. The seismometer can detect earthquakes that people cannot feel, which shows that it is quite accurate. The success in predicting the Longxi earthquake by seismograph heralded a new era in which human beings could measure and report earthquakes with scientific instruments. From then on, the feudal government "made the historiographer record the direction of the earthquakes". Compared with foreign countries, Zhang Heng's seismometer was invented more than 1,500 years earlier than the first modern one in the West [65].

(2) Wind measuring instrument

The wind measuring instrument was invented very early in China. The oracle bone inscriptions excavated from the Yin Ruins already contained the word "伣" (*qian*) which refers to a cloth, feather or grass hanging on a pole for measuring the wind direction [66]. This wind vane indicates the wind direction, and the wind speed can be estimated according to the angle of the wind vane. During the Han dynasty, modelled on a simple wind indicator, the "綄" (*huan*) was created, which is recorded in the *Huainanzi: Qisu xun*. Both Xu Zhu and Gao Youzhu pointed out that the weight of the *huan* amounted to five *liang* [a unit of weight, 1 *liang* = 50 g]. Moreover, *qian* were still in use in the Han dynasty. The *Huainanzi: Qisu xun* provides the earliest recorded evidence of a *qian*.

Apart from the *qian* and *huan*, the *tongfenghuang* (copper phoenix) and *xiangfengwu* (wind-indicating bird) were also used to measure wind direction. According to the *Sanfu huangtu*, "in the first year of the reign of Emperor Wu (104

BC), he built the Jianzhang Palace and the 25 *zhang* high Zuofeng Watchtower". The *Sanfu jiushi* ("Old Stories") says that "there is a round watchtower at the north of the Jianzhang Palace's gate which is 25 *zhang* high with a copper phoenix on the top" [67]. The *Guanzhong ji* records that "the round watchtower of the Jianzhang Palace is adjacent to the north road, and there is a golden phoenix on it, so it is called the Phoenix Watchtower". The *Hanwu gushi* ("Stories of Hanwudi") mentions that "there is a five-*chi tongfenghuang*, decorated with gold, standing on the Jianzhang Palace. When the wind blows, the pivot under the bird rotates with the wind, making the bird revolve as if it is going to fly" [68].

In the Eastern Han dynasty, a more sophisticated wind detector, the *xiangfeng tongwu*, appeared. It was supposedly invented by Zhang Heng, a well-known scientist. Placed on the Nanlingtai of Chang'an Palace, it moved on encountering the wind [69]. Guo Yuansheng of the Jin dynasty also said in his *Shuzheng ji* that "there is a *xiangfeng tongwu* in the Nanlingtai of Chang'an Palace; if the wind blows, the copper bird will revolve with the wind" [70].

Mural paintings in an Eastern Han dynasty tomb discovered during archaeological excavations displayed the *xiangfengwu* and anemometer banners. For example, in the tomb of the Eastern Han dynasty excavated in Lujia Village, Anping County, Hebei province in 1971, there was a mural depicting a clock tower on which stood a *xiangfengwu* and an anemometer banner [71]. This is the earliest extant depiction of the *xiangfengwu* and anemometer banner in China (see Fig. 4.4).

(3) Humidity measuring instrument

In the Western Han dynasty, the instrument for measuring humidity was a balance-type hygrometer. The *Huainanzi: Shuoshan xun* says that "the dry and wet air is known by hanging feathers and charcoal". Because charcoal is highly hygroscopic and dries out easily, its weight changes with the humidity in the air. "If dry, the charcoal is light; if wet, the charcoal is heavy." Moreover, feathers or a clod of earth have poor hygroscopicity, and their weight is little affected by changes in humidity. Hang the feathers or soil and dry charcoal on the two ends of a balance for a while. When the humidity changes, the charcoal will change in weight with the humidity, while the weight of the feather or soil is almost unchanged. In this way, the humidity can be estimated from the tilt angle of the balance. This is explained in detail in the *Collection of Historical Records.*

The measurement of precipitation also began in the Han dynasty. The *Book of the Later Han: Etiquette Annals* says that "the prefectures and states should report the precipitation at the beginning of spring, the beginning of summer, and the beginning of autumn". This shows that the central government required local governments to report rainfall on time.

Fig. 4.4 A Xiangfengwu and anemometer banner in the mural paintings in the Eastern Han dynasty tomb. Jiangxi Provincial Museum 江西省博物馆 (2018) *Jingshi dafaxian: Nanchang handai haihunhouguo kaogu chengguozhan* 惊世大发现: 南昌汉代海昏侯国考古成果展 [Discoveries that astounded the world: Exhibition on the archaeological achievements in Haihunhou state during the Han dynasty in present-day Nanchang]. Nanchang: Jiangxi Art Press 江西美术出版社, p 77

4.8 Views of the Shape of the Earth

In the Qin and Han dynasties, there were roughly three views of the shape of the Earth: one was the view that the Earth was square; the second was the concept of a flat Earth; the third was the theory of a spherical Earth.

4.8.1 The Square Earth View

The concept of a square Earth in ancient China originated very early. Tomb 45 in Xishuipo, Puyang, Henan province, from over 6,000 years ago, reflects the philosophy of "a square earth with a round sky above". Zhou Bijia in the Western Han dynasty posited that "the earth is as square as a chessboard" (cited from *Book of the Jin: Astronomical Treatise*). The *Zhoubi suanjing* records that "the sky is square and the land is round". Wang Chong believed that the sky and the earth were flat. The *Lunheng: Shuori* said that Heaven and Earth were parallel, which was obviously incorrect. From that time, the view of a square Earth became mainstream, and it was still widely held until the end of the Qing dynasty.

4.8.2 The Flat Earth Theory

The theory of a flat Earth was not widespread in ancient China. The belief that the ground looks like an upside-down plate and that heaven and earth are both in the shape of round arches, parallel to each other, appeared in the *Zhoubi suanjing* of the Western Han dynasty. The *Book of the Jin: Treatise on Astronomy* quotes *Zhoubi*'s words and further elaborates that "the sky and the earth are curved surfaces, rather than planes, which are high in the middle and lower in all four directions. The North Pole is the highest part of the sky, which has a downward incline on all sides." Although the flat Earth theory was not correct, it was a step forward compared with the square Earth theory.

4.8.3 The Spherical Earth Theory

The astronomical and mathematical work of the Western Han dynasty, *Zhoubi suanjing*, quoted Chen Zi as describing that "when the East is in the middle of the day, the West is in the middle of the night. When the West is in the middle of the day, the East is in the midnight hours". "When the sun rotates to the extreme north, the north is in the middle of the day and it is midnight in the south; when the sun rotates to the extreme south, the south is in the middle of the day, and it is midnight in the north." This is only possible if the Earth is spherical.

According to the Hun Tian (Celestial Sphere) theory, the dome of the sky is in the shape of an eggshell, and the earth is like an egg yolk in the sky, which clearly indicates that the Earth is shaped like a sphere. The representative work of Hun Tian theory, Zhang Heng's *Note to Armillary Sphere* in the Eastern Han dynasty proposed that "the celestial sphere is as round as an egg, and the ground is like the egg yolk, lying in the egg. The sky is big and the ground is small, and there is water on the surface of the sky. The heavens wrap the earth like the eggshell enveloping the yolk" [72].

In the stone reliefs of the late Eastern Han dynasty, one person stands on a hemisphere, which is carved with mountains and rivers, representing the shape of the Earth [73]. This sculpture reflects the thought regarding the integration of heaven, earth and people at the end of the Eastern Han dynasty, and also reveals the idea of a round Earth at that time, which makes it the earliest cultural relic relating to the theory of a spherical Earth (see Fig. 4.5).

Fig. 4.5 The Han stone portrait on Jinshan Mountain, Ganyu, Jiangsu. Wang, Kai 王恺 (1985) *Kaogu* 考古 [Archaeology] (9):793–798

4.9 Thought, Policies and Regulations on Environmental Protection

The environmental protection thought, policies and regulations of the Qin and Han dynasties were developed based on those of the pre-Qin dynasty. The following is a description of two aspects.

4.9.1 Thought on Environmental Protection

The environmental protection thinking of the Qin dynasty is recorded in detail in the *Lüshi chunqiu: Twelve Annals*, which sets out the different environmental protection measures to be implemented according to the different times and months of the year. Here are some examples.

In the first month of spring, people should be forbidden from "logging, hunting birds, killing baby insects and stealing birds' eggs". This section also talks about the relationship between people and the environment, and the relationship between people's longevity and food and clothing.

In mid spring, people should not "drain the rivers and ponds, deforest the land, and burn down the forest." This section also talks about the relationship between people and the environment and the relationship between animals and the environment.

[…]

The *Lüshi chunqiu: You shilan* also discusses the relationship between animals and the environment: "If you overturn the nest and destroy the eggs, the phoenix will not fly back; if you cut open the beast's belly and eat its foetus, the kylin will not come back. If you drain the river for fishing, the fish and dragons will not swim back."

The *Lüshi chunqiu* further posits the idea of "following the laws of heaven and earth", which means that human activities should adapt to heaven and the earth and should "have no impact". The meaning of "no action" here is different from that of Laozi. Here means that people should follow the laws of nature and not violate the nature of things, according to Wang Sunnian's annotations. Therefore, the *Lüshi chunqiu: Twelve Annals* had compiled a unified and inclusive system for natural changes and social activities. At the same time, it emphasised that human activities must follow the laws of nature and make rational use of natural resources. The idea of environmental protection is exactly based on the idea of "following the laws of heaven and earth".

In the *Huainanzi: Zhushu xun*, Liu An of the Western Han dynasty, reiterated the "previous emperors' methods" of protecting the environment, this is, people should not hunt the whole herd, or kill the immature; should not drain the water to catch fish, or burn the forest to hunt; hunting and exploration should be in harmony with the seasons". As a result of this environmental protection, there would be such a biological flourishing that "the vegetation will spout like steam, the beasts will return like a flowing spring, and the birds will fly back like smoke".

To obtain natural resources, the *Huainanzi: Shuoshan xun* advocates optimising the natural environment, in that "those who want to attract fish should first supply the water, and those who want to incur birds should first plant trees; fish gather when water accumulates, and birds gather when trees flourish". This thought is more advanced than the previous emperors' methods, changing passive protection into active protection, giving full reign to people's subjectivity, correctly handling the relationship between production and environmental protection, and attaching a high economic value to environmental protection.

The *Huainanzi: Benjing xun* also puts forward the view that troubled times will worsen the environment, saying that in times of chaos without a unified authority and legal system, the natural resources cannot be well protected, leading to overexploitation, deforestation, over-hunting and loss of habitat. All creatures are impacted by calamities. Therefore, the quality of the environment is closely related to the social order.

Conversely, the destruction of natural resources and the deterioration of the environment will also lead to social unrest. This is also explained in the *Huainanzi: Benjing xun* which says that, in troubled times, all kinds of resources are lost, chiefly the following five elements in particular. First, timber loss is caused by the massive construction of palaces, terraces and plank roads with gorgeous decoration and delicate engravings. Second, water loss is the result of the excavation of lakes and rivers, and the construction of artificial fountains and waterfalls for a flashy lifestyle. Third, soil loss can be attributed to the rampant construction of city walls, tall pavilions, long fences, and smooth roads. Fourth, a scarcity of metals can be ascribed to people's unsatisfied desire for ornaments. They open mines, smelt metals, and cast exquisite utensils and ornaments. Fifth, fire related loss stems from food production and the smelting of metal through the burning of combustible material. The obsessive use of fire also pollutes the air and damages property on the ground as a result. At the end, the author emphasised that "these five shortages of resources are enough to destroy the world" which was the first time in history that importance had been attached to the environment and that the quality of the environment had been linked with the survival of the world. It is true that normal production and life need to be guaranteed. What the author opposed was a minority of people plundering natural resources without restraint to satisfy their selfish desires, causing environmental deterioration. Therefore, the author then recommended the same view of moderation as in the *Book of Changes*.

The *Book of the Han: Biography of Gongyu* records Gong Yu's views on environmental protection: he believed that mining "for hundreds of *zhang* would undermine the energy of the land leading to climate deterioration, and that the excessive cutting down of trees might cause floods and droughts". Here, he posited that mining would cause the climate to deteriorate, which was an inference that could be drawn from the traditional yin and yang theory, but this was not borne out in reality. However, he was quite right to propose the theory that if people cut down forests without restraint, they will cause floods and droughts, which is still the case today.

In the Han dynasty, in addition to environmental protection thinking being discussed in the literature, it is also reflected in the cultural and artistic relics, as some examples suggest below.

The bronze ox lamp unearthed at the north gate of Guihua Garden, Changsha, in 1949, is a cultural relic of the Western Han dynasty. The ox's ears are pendulous and its legs are short. Two horns rise from its back, merge into one pipe, and then extend downwards to form an inverted bowl-shaped exhaust pipe. There is a hole in the centre of its back, on which lies a round panel topped with a cone, and a groove is set at the edge of the panel to fix a wind screen to prevent the light from being blown out. The stomach is hollow so that liquid can be poured in. When the lamp is lit, the smoke reaches the abdomen through the smoke tube (i.e. the ox horn), serving the dual purpose of suppressing smoke and removing dust [74]. This bronze ox lamp is tangible evidence of environmental protection in the Western Han dynasty.

The Changxin Palace Lamp, unearthed in 1968 in the Western Han Tomb in Mancheng, Hebei, is also an environmentally-friendly household product. The lamp, which is in the shape of a maid holding a lamp with both hands, can be taken apart. The lamp panel can be rotated, and the lampshade can be opened and closed. Therefore, the brightness and the angle of the direction of illumination can be adjusted at will. The head of the copper maid can also be disassembled. Her body is hollow and her right arm is connected to the flue, channelling smoke into the body to keep the room clean [75].

In addition, there is a swan goose lamp which was excavated from the Western Han dynasty tomb in Shuo County, Shanxi province, and phoenix lanterns excavated from the Han dynasty tomb in Hepu, Guangxi, cattle lanterns excavated from the Han dynasty tomb in Ganquan County, Hanjiang district, Jiangsu province [76], a swan goose lamp excavated from Shenmu, Shaanxi province, a swan goose lamp excavated from the Han dynasty tomb in Wuxingzhuang, Xiangfen County, Shanxi province [77], a copper princess lamp excavated from the Liujia Dashan outside the east gate of Changsha, and a copper lamp of the Western Han dynasty excavated from the Yangjia Dashan, Changsha [78]. All these lamps share the same basic structure and are all indoor environmental protection products. Since the number of copper lamps of this type excavated is unparalleled in any previous dynasty, this is further proof that people in the Han dynasty had a strong sense of environmental protection.

4.9.2 Policies and Regulations

In 1975, the bamboo slips recording Qin dynasty laws were unearthed at Yunmeng, from the Shuihudi Tomb in Hubei province. These include nearly 20 titles such as *Tianlü, Jiuyuan* and *Canglü*. The *Tianlü* ("Laws of Farming") stipulates ten rules governing environmental protection and the preservation of natural resources. First, in February, it is forbidden to cut down trees in the mountains. Second, a waterway must not be blocked. Third, it is prohibited to burn grass as fertiliser except in summer. Fourth, it is prohibited to gather newly sprouted plants. Fifth, it is forbidden to capture young wild animals or to gather eggs and young birds. Sixth, poisoning fish and turtles is prohibited. Seventh, traps and nets should not be set up to catch birds and animals. This ban is lifted only in July. Eighth, people who live near pens for cattle and horses and other imperial gardens shall not take hunting dogs to hunt when there are young animals. Ninth, when the hunting dogs enter the imperial gardens, if they have not chased or injured the wild animals in the garden, these shall not be killed through choice. Tenth, if a hunting dog is killed in a specially guarded area, the complete body of the dog should be handed over to the government. If a hunting dog is killed in other imperial gardens, the dog meat can be eaten and only the dog skin should be handed over.

In terms of protecting natural resources, the *Tianlü* emphasises that we should harvest at the proper time. This is the earliest legal document on the protection of natural resources discovered in China so far [79].

The legal texts of the Han dynasty included *lü*, *ling*, *ke* and *bi*. *Lü* have been lost, but a few of them have been quoted in historical accounts. For example, the *Fengsu tong* contains the regulation about protecting young animals. *Ling* refers to the imperial edicts, and some of those on environmental protection have been preserved in the archives, such as the following.

In the 6th year of Emperor Wu of the Han dynasty (105 BC), an edict was issued in the first month: "no cutting of vegetation". And another imperial edict in the second month described the experience of encountering a group of cranes without capturing them [80].

In the 3rd year of the Yuankang reign, Emperor Xuan of the Han dynasty (63 BC) issued an imperial edict in June, stipulating that "in spring and summer, it is prohibited to destroy any nests, collect eggs or shoot birds" [81].

In the 3rd year of Emperor Yuan of the Han dynasty (46 BC), the Emperor issued an edict to warn all officials "not to violate the four-season ban" [82].

In the 2nd year of Emperor Zhang of the Eastern Han dynasty (85 AD), an imperial edict stated that "in spring, all plants sprout. It is appropriate to fully utilise the yang qi to nurture all things" [83].

In February of the third year of the Yuanhe reign (808 AD), the Sikong (imperial censor) stated: "It is spring now. We should not harm any creatures. The horse and carriage should bypass the plants as far as possible. As the poem expressed in the poem, 'don't let cattle or sheep trample on the reeds.' In the *Liji*, it is said that a faithful man should cut down vegetation at the right time, and conform to the rules of heaven. It is exactly what I am hoping for [83]."

In the Han dynasty, strict laws were formulated against illegal logging and poaching in restricted areas. According to the *Sanfu jiushi*, "the thief who cuts down cypress trees in tombs will be executed in the crowded town centre" [84]. The *Han gongchen biao* ("List of Han Meritorious Officials") also documents that "in the 4th year of Yuanding (113 BC), Marquis Zhang Shi was imprisoned for stealing deer".

In the Han dynasty, some places were short of water. In order to make fair and reasonable use of water resources, laws were passed. Magistrate Zhao Xinchen of Nanyang Prefecture "formulated water resources allocation rules and carved them on river stones to prevent disputes" [85]. This was a local administrative regulation to ensure the rational allocation of water resources.

4.10 Knowledge in the Field of Mineralogy and Petrology

Evidence of the knowledge of minerals and rocks in the Qin and Han dynasties mainly encompasses three aspects: first, the knowledge of minerals reflected in archaeological relics; secondly, the mineral knowledge recorded in literature; third, the understanding and utilisation of rocks in the Qin and Han dynasties.

4.10.1 Mineralogy in Archaeological Artefacts and Documentation

According to the statistical analyses in the journals of *Archaeology*, *Cultural Relics* and other archaeological publications, there is evidence of four types of minerals and rocks in the cultural relics of the Qin and Han dynasties.

(1) Metallic minerals

There was use of 26 kinds of metallic minerals, including gold, silver, copper (chalcopyrite, malachite, bornite, chalcocite, blue vitriol, azurite, *shilü* [verdigris] and *zengqing* [a kind of azurite]), iron (magnetite, hematite, limonite, pyrite, lodestone, *tuhong* [red earth], ochre and black ironstone), tin, lead (cerussite), mercury (cinnabar), and zinc (lead zinc ore), among which copper and iron were the most common, with 8 kinds of copper and 8 kinds of iron. However, the copper and iron, as well as other metallic minerals, were not all used in metal smelting at that time, but rather were used as traditional Chinese medicine or paint pigments. For example, azurite, *shilü*, *zengqing*, *tuhong*, and ochre etc. were all used as pigments, while hematite, cinnabar, cerussite, etc. were used for both metal smelting and pigments. Some uses also applied to construction, for example, Qin Shi Huang used magnets to build the North Gate of Xianyang. The gold coinage in the Qin and Han dynasties was more advanced than that of the Spring and Autumn and the Warring States periods, such as the horse hoof-shaped gold and the kylin toe-shaped gold (Fig. 4.6).

(2) Non-metallic minerals

There was use of 28 kinds of non-metallic minerals, including graphite, talc, white gypsum mud, *wuhuatu*, green gypsum mud, graphite clay, mica, fluorite, white quartz, red halloysite, stalactite, sulfur, coal, black amber, kaolin, *shihuang* [a kind of pyrite], bauxite, feldspar, halite, gypsum, realgar, saltpetre, red ochre, limonitum, *yushi* [arsenic ore], orpiment, silverstone, quartz, etc. These non-metallic minerals had extensive applications: some were used as traditional Chinese medicine, some as

Fig. 4.6 Horse hoof-shaped gold and kylin toe-shaped gold (photographed by Liang Qing 梁擎)

pigments, some as filling for tombs, some as sources of energy, some for porcelain and pottery and some as craft decorations in craft work.

(3) Gems

There was use of 14 kinds of gemstones, including jade, agate, amber, white marble, ruby, emerald, crystal, lapis lazuli, turquoise, red coral, goldstone, amethyst and pearl, which were mainly used as craft decoration, and a few were used for traditional Chinese medicine or in architecture, sculpture, and as seals and (coloured) glass.

(4) Rocks

There was use of 13 kinds of rocks, including limestone, sandstone, marble, micaceous rock, slate, granite, Qiyang stone (a kind of argillaceous slate), mica schist, mud shale, shale, conglomerate, ophicalcite and sandy shale. Some of these were used as building materials, some were used as stone tools, such as millstones, stone inkstones, mortars, pestles, stone sharpeners, stone tables, whetstones, and some were used in handicrafts. In addition to burning to make lime, limestone was also used as a building material, and for carving portraits and sculptures.

Archaeological material shows that/reveals that there was use of 81 kinds of minerals and rocks during the Qin and Han dynasties, including 68 kinds of minerals (including gemstones) and 13 kinds of rocks. The most common minerals and rocks were copper, iron, gold, silver, lead, cinnabar, jade, agate, talc, crystal, sandstone, amber, turquoise, limestone, white marble, marble, salt, and realgar: making a total of 18 kinds.

According to excavated murals, there was use of 12 kinds of mineral pigments in the Qin and Han dynasties, including earth red, cinnabar, vermilion, ochre, mineral yellow, azurite, verdigris, graphite, cerussite, gypsum, aluminium powder and zinc white.

4.10.2 Mineralogical References in Literature

(1) Silk book *prescriptions for fifty-two diseases*

The silk book *Prescriptions for Fifty-two Diseases,* one of the silk books excavated from the No 3 Han dynasty tomb at Mawangdui, Changsha, is the oldest medical prescription found in China. The font used is seal script, with some use of curve script. It would have been written at the end of the third century BC, recording 21 kinds of mineral drugs, including saltpetre, feldspar, talc, humus soil, stove ash, stove loess, mud deposited in wells, long immersed mud, frozen soil, salt, halite, *yushi* [arsenic ore], cinnabar, realgar, mercury, iron, ferrosic oxide, gold, *yanjishui* [yellow earth water], etc.

(2) *Huainan wanbi shu*

The *Huainan wanbi shu* was mainly written by Liu An, the King of Huainan in the Western Han dynasty (179–122 BC). Today, only a compilation of the fragments by Sun Fengyi in the Qing dynasty remains, with details of about nine minerals including basic copper carbonate, mica, azurite cinnabar, magnetite, gold, salt, realgar and iron [86].

(3) *Zixu fu* by Sima Xiangru

Sima Xiangru's *Zixu fu* ("Master Nil") in the Western Han dynasty records 20 kinds of minerals and rocks, including *dan* (cinnabar), *qing* (hollow azurite, azuritum, basic copper carbonate, etc.), *zhe* (terra-cotta, red ochre, hematite), chalk (limestone, white clay, kaolin), orpiment (AS_2S_3), *baifu* (white quartz), tin (SnO_2), jasper, gold (Au), silver (Ag), *chiyu* (akadama, a kind of nephrite), rhodonite, *lin* (a kind of jade), *min* (marble, white marble), *kunwu* (jade-like stone: the River Map says that "Yingzhou abounds in *kunwu* stones which can be smelted into iron to forge swords. The stones are as bright as crystal"), *jian* (jade-like stone), *le* (jade-like stone), *xuanli* (the *Hanshu yinyi* has a record of "*xuanli*, black whetstone", that is, of millstone or whetstone), *ruanshi* (a soft stone similar to jade, such as talc, agalmatolite and Qingtian stone), and *wufu* (jade-like stone).

(4) *Shen Nong bencao jing*

The *Shen Nong bencao jing* was composed at the end of the Western Han dynasty and the beginning of the Eastern Han dynasty. The book is anonymous and was compiled by later scholars who founded the ancient literature of Shen Nong. There are 39 kinds of minerals recorded with their medicinal properties explained in great detail. The characteristics of the minerals are embodied in their naming and classification, such as colour, magnetism, fluidity, tenacity, taste, etc. The minerals are divided into upper, middle and lower grades according to their medicinal value with little relevance to mineralogy.

(5) *Book of the Han: Treatise on geography*

The *Book of the Han: Treatise on Geography* was written by Ban Gu of the Eastern Han dynasty, recording 12 kinds of minerals and rocks, with details about Lantian jade, iron, salt, gypsum, copper, gold, silver, lead, tin, oil, natural gas, aragonite, etc. Compared with the *Shen Nong bencao jing*, the list of minerals and rocks recorded in this book was inadequate.

(6) *Jiran wanwu lu*

According to the thirteenth volume of *Yuhan shanfang ji yishu*, the *Jiran wanwulu* ("Catalogue of Everything in Nature") was written by the people of the Eastern Han dynasty who created the ancient literary tradition. There are a total of 17 kinds of minerals in the text, as well as one kind included in Volume 987 of the *Taiping yulan*. They are described in detail as follows.

① The nephrite is from Lantian (today's Lantian County, Shaanxi province).
② The lead powder, also known as lead monoxide, is oxidised from galena and other lead containing minerals and can be used as yellow pigment.
③ The *shiliuhuang* (sulfur) is distributed around Hanzhong.
④ The *shidan* [copper sulfate ($CuSO_4$)], comes from Longxi (today's north of Linzhao, Gansu province) and Qiangdao (today's north of Zhouqu County, Gansu province).
⑤ The *chishizhi* (kaolinite, mainly composed of Al_4 [Si_4O_{10}] $(OH)_8$), can be found in Hedong (today's northwest of Xixia County, Shanxi province), and red kaolin is the best quality.
⑥ The gypsum rubrum, also known as anhydrite ($CaSO_4$), can be found in Hedong, and white gypsum is the best.
⑦ The stalactite, also known as calcium carbonate ($CaCO_3$), comes from Wudu (now the west of Chengxian County, Gansu province), and the yellow and white types are the best.
⑧ The limonitum, also known as limonite ($Fe_2O_3{\cdot}nH_2O$), is distributed in Hedong.
⑨ The saltpetre (KNO_3), which can be found in soil, on rock surfaces or in caves in dry areas, is abundant in Long County (now Zhangjiachuan Hui Autonomous County in Gansu province).
⑩ The superior talc ($Mg_3[Si_4O_{10}](OH)_2$) should be pale and smooth.
⑪ The alunite ($KAl(SO_4)_2(OH)_6$) comes from Wudu.
⑫ The hollow azurite (malachite) and azuritum (azurite) abound in the Ba Prefecture (on today's north bank area of the Jialing River, Chongqing).
⑬ The *baiqing*, also known as weathered copper sulfate ($CuSO_4$), comes from Xingan, and the green form is the best. It can also be found in the Ba Prefecture.
⑭ The *luqing* is native to Hongnong (now the north of Lingbao, Henan province) and Yuzhang (Nanchang, Jiangxi province).
⑮ The ochre (Fe_2O_3) comes from Qi Prefecture (now Zibo, Shandong province), and is red in colour. And some ochres are from Shu Prefecture (now Chengdu, Sichuan province).

⑯ The green chalk, also known as graphite (C), comes from Sanfu (now the central Shaanxi province).

⑰ The *longgu*, or dragon's bone, the common name for the skeleton fossils and tooth fossils of vertebrates in China, comes from Hedong.

⑱ The *yushi*, also known as arsenopyrite (FeAsS), comes from Hanzhong, and the white form is preferred (This entry is not in the *Yuhan shanfang ji yishu* compilation, but can be found in Volume 987 of the *Taiping yulan*. The crystals are columnar, and the aggregates are often granular or in dense lumps, tin white in colour, and with a metallic lustre. Its hardness is around 5.5–6, the specific gravity is 5.9–6.2, and it is brittle.) [87].

(7) *Hanjian*

The *Hanjian* ("Bamboo Slips of the Han Dynasty") refer to the medical slips discovered in the early Eastern Han tombs in Hantanpo, Wuwei city, Gansu province, in 1973. There are 17 kinds of mineral medicine listed on the bamboo slips, namely stalactite, magnetite, feldspar, halite, cinnabar, black stone, alunite, dragon's bone, gypsum, realgar, azuritum, saltpetre, talc, red ochre, kaolinite, limonitum, and earthstone. Most of these can be found in the *Shen Nong bencao jing* and were used as mineral medicine at that time.

(8) *Shuowen jiezi*

The *Shuowen jiezi* by Xu Shen in the Eastern Han dynasty covers an extensive range of knowledge of minerals and rocks. For example, characters with the radical "玉" (pinyin: yu, jade) include 21 types of jade; characters with the radical "丹" (pinyin: dan, red) include 2 types of minerals; characters with the radical "厂" (pinyin: chang, shelter on cliff) cover 8 names of rocks; characters with the radical "石" (pinyin: shi involve 2 names of minerals and 15 names of rocks; characters with the radical "金" (pinyin: jin, gold) represent 16 names of ores; the characters with the radical "土" (pinyin: tu, soil) relate to 15 names of soil, adding up to 79 minerals (including soil) and rocks in total. Many of these written in ancient script are now obsolescent. But there are still some in common use today, such as gold, silver, copper, iron, lead, tin, soil, earth, chime stone, inkstone, cinnabar, jade, etc.

(9) *Zhouyi cantong qi*

The Daoist classic *Zhouyi cantong qi* ("Kinship of the Three According to the Book of Changes"), written by Wei Boyang of the Eastern Han dynasty, records 17 kinds of alchemical mineral medicines, such as lead, gold, cinnabar, realgar, mica, hollow azurite, sulfur, halite, saltpetre, orpiment, mercury, lead powder, alum, magnetite, pearl-like stone, *yushi* [arsenic ore] and sal ammoniac.

(10) *Chu jinkuang tulu*

The one-volume *Chu jinkuang tulu* ("Catalogue of the Gold Ore"), written by Hu Gangzi at the end of the Eastern Han dynasty, specifically described the smelting of gold and silver ores. The original work has been lost, but the key points are included in Volume 9 of the *Jiudan jingjue* ("Book of the Nine Elixirs"). Hu Gangzi first talked about the geological distribution of gold and silver ores, pointing out that gold deposits were found either in water or in mountains, that is, alluvial gold and lode gold.

On the subject of alluvial gold, the book says that "alluvial gold is distributed in the water, and its shape is like bran, chess pieces, jujube beans and millet, etc.; if the water flows from north to south, the gold is on the east bank, three to seven *cun* into the sandstone; if the water flows from east to west, the gold is on the south bank, five to nine inches into the sandstone". These were high grade deposits of gold with a fine texture.

As for lode gold, the book says that "the gold buried in the mountain is round in shape. If the mountain is east–west, the gold seam is located in the north of the mountain, if the mountain is north–south, the gold seam is in the west of the mountain, nine to ninety feet below ground, mixed with sand and stone, reddish yellow in colour, delicate and slippery, and so resilient that it is resistant to folding; it will turn white and silver when melted. And it is used to make magical pellets with other medicines, by putting them into the Bafeng Furnace together".

These words indicate that gold veins were mined as early as the Han dynasty and that the distribution of gold was well understood. In addition, Hu Gangzi also pointed out that attention should be paid to the identification of pseudo gold deposits while discussing the search for gold veins. In the book, it is written that "an iron mine is similar to a gold one to some extent, but it is not as useful as a gold mine. It is also located in the water or exposed on the mountain, but not in the place where the yin and yang gather. Moreover, iron ore is angular in shape and yellow and green in colour, which distinguishes it from gold". The iron ore mentioned above was the *jinya* stone later described by Ge Hong and Tao Hongjing, which essentially was a kind of copper sulfide iron ore ($CuFeS_2$) or pyrite (FeS_2).

The book also refers to the smelting of gold ores. Hu Gangzi suggested various technological processes depending on the type of gold ore. In the smelting process, other minerals were also added into the montmorillonite pot in addition to salt, such as alunite (ferric sulfate), sulfur, azuritum, alum, cinnabar, realgar (As_2S_2), sodium sulfate, sal ammoniac, alum, mercury, tin, lithargite (PbO), amethyst, orpiment (As_2S_3), lead, arsenic oxide.

Hu Gangzi also talked about the method for extracting silver from silver ores. For example, in the process of cupellation, alunite, sal ammoniac, lead and other minerals were mentioned [88].

4.10.3 Applications of Petrology

In 4.10.1 "Mineralogy in Archaeological Artefacts and Documentation", the rocks used in the Qin and Han dynasties are listed. Based on this list, we will further describe the utilisation of rocks.

(1) Limestone

Limestone was mainly used for burning into lime or making stone chimes (chimes unearthed at the Yangjiawan Han Tombs in Xianyang), inkstones (a set of blue stone inkstones unearthed at the Eastern Han Tombs in Chenzhou, Hunan, a blue stone inkstone unearthed at the Eastern Han Tomb in Jiayuguan, and two blue stone inkstones unearthed at the Eastern Han Tombs in Hebei, Cang County), stone portraits (stone relief tomb of the Han dynasty in Yinan, Shandong), stone statues (two nude figure statues from the Han dynasty, one male and one female, in Shijiazhuang), building material (the limestone pillars and steps unearthed at the architectural site of the Qin Shi Huang mausoleum, the stone coffin and stone field model unearthed at the Eastern Han Tomb site in Pi County, Sichuan province, and the tomb built with more than 1,000 cubic metres of limestone in the Eastern Han tombs in Mi County, Henan province), household implements (manger, mortar, pestle, stone mill, stone table, and stone cabinet), etc.

(2) Sandstone

Sandstone was mainly used for making stoneware, foundation stones and sculptures. For example, the Han dynasty stone statue unearthed in Dujiangyan, the inkstone unearthed in the Fenghuangshan Han tombs in Jiangling, Hubei Province, the ploughshare mould excavated from the iron workshop site of the Han dynasty in Nanyang, Henan, the stone carvings from the Eastern Han Tombs Gate in Pi County, Sichuan province, the coffin chamber of the Nanyue King of the Western Han dynasty, five millstones unearthed at the Western Han Tomb site in Xianyang, the stone mill unearthed at the Western Han Tomb site in Yizheng, Jiangsu, the whetstone unearthed at the Han tombs in Yancheng, Jiangsu, and the stone coffin of the Eastern Han dynasty made of more than 4,000 pieces of sandstone in Dingxian County, Hebei province.

(3) Shale

Shale was mainly used for making inkstones and sharpening stones.

(4) Conglomerate

Conglomerate was mainly used as building material. For example, part of the stone relief tomb in Yinan, Shandong, was built with gravel.

(5) Marble

Marble also had many uses, such as for making the stone tablets placed in the mouths of the dead, and stone plugs, stone pigs, stone pillows, stone sheep, stone tables, stone nose plugs, stone eyeshades, stone ear plugs, stone anus plugs.

(6) Ophicalcite

Ophicalcite, which was only used in traditional Chinese medicine according to historical records, was found in the medicine storage tank unearthed at the Eastern Han Tomb in Chengdu, Sichuan.

(7) Granite

Granite was mainly used in the art of sculpture, as a building material and for stoneware, such as the granite mortar unearthed at the Eastern Han Tomb in Taizhou, Jiangsu province, 11 human and animal granite statues discovered around Xi'an, 16 large stone sculptures in front of Huo Qubing's tomb in Xingping, Shaanxi, also known as the giant round sculptures, and the sarcophagus unearthed at the stone coffin tomb of the Qin and Han dynasties in Yongji, Jilin.

(8) Sandy shale

Sandy shale was only used for whetstones in recorded history, such as the examples unearthed at the Eastern Han Tomb in Yushu, Jilin province.

(9) Alunite

Alunite has only been unearthed in one place so far, that is, at the Eastern Han Tomb in Tongguan, Shaanxi:14 inkstones, 2 stone mortars and 1 flatiron-shaped boulder.

(10) Slate

Slate was mainly used for making stoneware, such as inkstones.

(11) Micaceous rock

Micaceous rock was used as an inkstone, for example, the two pieces of inkstone for grinding eyebrow powder unearthed at the Han Tomb in Yancheng, Jiangsu.

(12) Qiyang stone

Twenty-one stone items unearthed at the Western Han Tomb at Weijiadui, Changsha, were carved from Qiyang stone, including rings, seals, handguards, hooks, pendants, sheaths, mirrors and liquor cups.

References

1. *Shiji: Qinshihuang benji*史记·秦始皇本纪 [Records of the grand historian: the basic annals of the first Qin emperor]
2. *Wenwu*文物 [Relics] (1986) 1:49
3. *Shiji: Dawan liezhuan*史记·大宛列传 [Records of the grand historian: collected biographies of Dawan]
4. The specific time of Zhang Qian's 张骞 mission is not clearly recorded in historical references. It is inferred here that the 2nd year of the Jianyuan reign (126 BC) was his departure time, because according to the *Tongjian* 通鉴 [Comprehensive mirror in aid of governance], Zhang Qian returned to the imperial court in the 3rd year of the Yuanshuo reign (126 BC), and 13 years earlier than that would have been the 2nd year of the Jianyuan reign
5. *Shiji: Weijiangjun piaqji liezhuan*史记·卫将军骠骑列传 [Records of the grand historian: biographies of general Wei]
6. *Hou hanshu: Xiyu zhuan*后汉书·西域传 [Book of the later Han: treatise on the Western regions]
7. *Hou hanshu: Banchao zhuan*后汉书·班超传 [Book of the later Han: biography of Ban Chao]

8. Mo R莫任南 (1982) Ganying chushi daqin luxian ji qi gongxian甘英出使大秦路线及其贡献 [Gan Ying's mission and contribution to the Daquin empire] *Shijie lishi*世界历史 [World history] 2
9. *Jinshu: Peixiu zhuan*晋书·裴秀传 [Book of the Jin: biography of Pei Xiu]
10. *Hou hanshu: Mayuan zhuan*后汉书·马援传 [Book of the later Han: biography of Ma Yuan]
11. *Shiji: Huainan hengshan liezhuan*史记·淮南衡山列传 [Records of the grand historian: treatise on Heng Mountain in Huainan], cited from the *Shiji suoyin*史记索隐 [Searching for the obscure in the *Records of the Grand Historian*]
12. *Hanshu: Jiangdu yiwang zhuan*汉书·江都易王传 [Book of the Han: biography of Liu Fei as Lord Yi of Jiangdu]
13. *Hanshu: Zhangtang zhuan*汉书·张汤传 [Book of the Han: biography of Zhang Tang]
14. *Hou Hanshu: Zanghong zhuan: zhu'er*后汉书·臧洪传注二 [Book of the later Han: biography of Zang Hong, annotation II]
15. *Hanshu: Kuangheng zhuan*汉书·匡衡传 [Book of the Han: biography of Kuang Heng]
16. *Hanshu: Yuandi ji*汉书·元帝纪 [Book of the Han: annals of Emperor Yuan of the Han dynasty]
17. Zhang Y张彦远 (Tang dynasty) *Lidai minghua ji: juansan* 历代名画记卷三 [Records of famous paintings from past dynasties, vol 3]
18. *Hou hanshu: Lixun zhuan*后汉书·李恂传 [Book of the later Han: biography of Li Xun]
19. *Hanshu: Baiguan gongqing biao*汉书·百官公卿表 [Book of the Han: memorial of a hundred officials and secretaries]
20. Chen Z陈正祥 (1979) *Zhongguo dituxue shi* 中国地图学史 [A history of Chinese cartography]. Commercial Press Hong Kong Branch商务印书馆香港分馆, Hong Kong: , p 7.
21. Cao W曹婉如 (1985) Donghan chengshi jubu ditu de yanjiu东汉城市局部地图的研究 [Research on local maps of cities in the Eastern Han dynasty]. *Ziran kexueshi yanjiu*自然科学史研究 [Research on the history of natural science] 2
22. Liu Y刘叶秋 (1983) *Zhongguo zidian shilue*中国字典史略 [A brief history of Chinese dictionaries]. Zhonghua Book Company 中华书局, Beijing, p 28
23. Lu Z陆宗达 (1984) *Shuowen jiezi tonglun*说文解字通论 [General discussion on *An explication of written characters*]. Beijing Publishing House 北京出版社, Beijing, p 5
24. *Shuowen jiezi juanshiwu: Xu*说文解字卷十五: 叙 [An explication of written characters vol 15: narration]
25. *Shiji: Taishigong zixu*史记·太史公自序 [Records of the grand historian: Sima Qian's preface]
26. Zhai Z翟忠义 (1962) *Zhongguo gudai dili xuejia ji lüxingjia*中国古代地理学家及旅行家 [Chinese ancient geographers and travellers]. Shandong People's Publishing House山东人民出版社, Jinan, pp 12–14
27. Yang Z杨正泰 (1987) *Zhongguo lishi dili yaoji jieshao* 中国历史地理要籍介绍 [Introduction to important historical and geographical books of China]. Sichuan People's Publishing House 四川人民出版社, Chengdu, p 43
28. The misunderstanding lasted for 894 years, from 93 BC when the *Shiji* was written, to AD 801 when Du Youli 杜佑力 persuasively argued that the Yellow River did not rise in the Kunlun Mountains
29. *Hou hanshu: Banbiao zhuan*后汉书·班彪传 [Book of later Han: biography of Ban Biao]
30. Hou R侯仁之 (1962) *Zhongguo gudai dilixue jianshi*中国古代地理学简史 [A brief history of ancient Chinese geography]. Science Press北京科学出版社, Beijing, p 14
31. Wang C王成组 (1982) *Zhongguo dilixue shi*中国地理学史 [A history of Chinese geography]. Commercial Press商务印书馆, Beijing, p 37
32. Zhou Z周振鹤 (1986) Beihushi le de qindai *shuijing*被忽视了的秦代"水经" [The neglected *Shuijing* of the Qin dynasty]. *Ziran kexue shi yanjiu* 自然科学史研究 [Research on the history of natural science] 1
33. Wang, Yinglin 王应麟 pointed out in his *Kunxue jiwen* 困学纪闻 that Sang Qin 桑钦, whose name appears in the *Hanshu: rulin zhuan* 汉书·儒林传 [Book of the Han: Biographies of scholars], was born at the end of the Western Han dynasty and the beginning of the Eastern Han dynasty

34. *Suishu: Jingji zhi: Dili jilei xu*隋书·经籍志·地理记类叙 [Book of the Sui: bibliographical treatise: geographical records and narratives]
35. Shi N史念海, Cao E曹尔琴 (1986) *Fangzhi chuyi*方志刍议 [Preliminary discussion on local chronicles]. Zhejiang People's Publishing House浙江人民出版社, Hangzhou, p 88
36. Mo W王谟, *Hangtang dili shuchao* 汉唐地理书钞 [Han and Tang geographical notes]
37. Cited from the *Taiping yulan: juan wushi: Dibu shiwu: Shajiao shan*太平御览卷五十地部十五沙角山 [Imperial reader of the Taiping era, vol 50 "The Earth", Chapter 15 "Shajiao Mountain"]
38. Cited from the *Huayang guozhi: Bazhi* 华阳国志巴志 [Records of Huayang state: treatise on Ba prefecture]
39. Liu Z刘泽华 (1993) Handai *weishu* zhong shen, ziran, ren yitihua de zhengzhi guannian汉代"纬书"中神、自然、人一体化的政治观念 [The political concept of the integration of God, Nature and Man in the Augury Book of the Han dynasty]. *Wen shi zhe*文史哲 [Art, history and philosophy] 1:35
40. *Baikong liutie: juanyi*白孔六帖卷一[Six scrolls of jottings from Bai and Kong, vol 1], cited from *Tang leihan: juanshi'er*唐类函卷十二 [Collected works from the Tang dynasty, vol 12]
41. Annotations to *Wenxuan: Lizhishi zhu*文选·励志诗注 [Selected works: annotations of inspirational poems]
42. *Guwei shu: juanqi & shiyi*古微书·卷七、十一[Ancient augury book, vols.7 & 11]. In: JueS 孙穀 (ed) (1939) *Congshu jicheng chubian* 丛书集成初编: 第0690–0693册 [A preliminary compilation of the integrated series]: vols 0690–0693. Beijing: Commercial Press 商务印书馆
43. *Chuxue ji: juanba*初学记·卷八 [Writings for elementary instruction, vol 8]
44. Chong W王充 *Lunheng*: *Ganxu*论衡·感虚 [Balanced discussions: superstitions]
45. Jian X徐坚 et al (1985) (Tang dynasty) *Chuxue ji: juanyi* 初学记·卷一 [Writings for elementary instruction, vol 1]. Zhonghua Book Company中华书局, Beijing
46. Daoyuan L郦道元 *Shuijing: Jianjiangshui zhu* 水经·渐江水注 [Water classic: commentary on the Jian River]
47. Junfu L凌骏甫 (Ming dynasty) *Xiangyan lu*湘烟录
48. Chong W王充 *Lunheng*: *Ganxu* 论衡·艺增 [Balanced discussions: exaggerations]
49. *Lunheng*: *Ganxu*论衡·感虚 [Balanced discussions: superstitions]
50. *Lunheng*: *Shuori*论衡·说日 [Balanced discussions: the Sun]
51. *Lunheng*: *Shungu*论衡·顺鼓 [Balanced discussions: floods]
52. *Taiping yulan: juanba*太平御览·卷八 [Imperial reader of the Taiping era, vol 8]
53. *Chuxue ji: juan'er*初学记·卷二 [Writings for elementary instruction, vol 2]
54. *Yiwen leiju: juan'er*艺文类聚·卷二 [Classified collection based on the classics and other literature, vol 2]
55. Hong G葛洪*Xijing zaji: juan'er* 西京杂记·卷二 [Miscellaneous records of Xijing, vol 2]
56. *Guwenyuan: juan shiyi*古文苑·卷十一 [Anthology of poems, vol 11]
57. *Gu weishu: juan shiwu*古微书·卷十五 [Ancient augury book, Vol 15]
58. *Lunheng: Hanwen*论衡·寒温 [Balanced discussions: cold and warm]
59. See "Shuolinxun"说林训 [Lessons from nature] and "Taizuxun"泰族训 [Laws of the past and present] in *Huainanzi* 淮南子 [Master of Huainan]
60. *Lunheng*: *Yandu*论衡·言毒 [Balanced discussions: poisonous substances]
61. See "Shungu"顺鼓 [Floods] and "Mingyu"明雩 [Rituals and ceremonies of worships for rain] in *Lunheng*: *Shungu*论衡·[balanced discussions]
62. *Han-wei yishu chao*汉魏遗书钞 [Collection of texts from the Han-Wei periods]
63. *Han-tang dili shuchao*汉唐地理书钞 [Geographic texts from the Han-Tang periods]
64. *Hou hanshu*: *Zhangheng zhuan*后汉书·张衡传 [Book of the later Han: biography of Zhang Heng]
65. Needham J李约瑟 (1976) *Zhongguo kexue jishushi (zhongyiben)*中国科学技术史 [History of science and technology in China (trans)], Vol 5, Book 2. Science Press科学出版社, Beijing, p 361. See also: Yang W杨文衡 (1998) *Zhonghua wenhua tongzhi: Kexue jishu dian: Dixue zhi*中华文化通志·科学技术典·地学志 [General annals of Chinese culture: science and technology dictionary: Geography]. Shanghai People's Publishing House上海人民出版社, Shanghai, p 359

66. Liu Z刘昭民 (1970) Zhongguo gudai qixiang yiqi he qixiang guance hongju de faming中国古代气象仪器和气象观测工具的发明 [Invention of ancient Chinese meteorological instruments and observation tools]. *Mingri shijie (Taiwan)*明日世界(台湾) [Tomorrow's world (Taiwan)] 2
67. Chen Z陈直 (1985) *Sanfu huangtu jiaozheng*三辅黄图校证 [Verification of the *imperial plan of the three capital regions*]. Shaanxi People's Publishing House陕西人民出版社, Xi'an, pp 40–43
68. For the above two quotations, see *Shuijing: Weishui zhu* 水经渭水注 [Water classic: commentary on the Wei river]
69. *Gujin tushu jicheng: Lixiang huibian: Qianxiang dian*古今图书集成·历象汇编·乾象典 [Collection of ancient and present-day books: Collected texts on meteorology and astronomy: astronomical observations]
70. *Taiping Yulan: juanjiu*太平御览·卷九 [Imperial reader of the Taiping era, vol 9]
71. *Guangming ribao*光明日报 [Guangming daily] (1972), 22 June
72. *Kaiyuan Zhanjing: juanyi*开元占经·卷一 [Classic of astrology from the Kaiyuan era, vol 1]
73. *Kaogu*考古 [Archaeology] (1985) 9
74. Zhixi G高至喜 (1959) Niu Deng牛灯 [Ox Lamp]. *Wenwu* 文物 [Relics] 7
75. *Kaogu*考古 [Archaeology] (1972) 1:8
76. *Wenwu*文物 [Relics] (1987) 6:1
77. *Kaogu*考古[Archaeology] (1989) 11:981
78. *Wenwu*文物 [Relics] (1979) 7:92
79. Luo G罗桂环, Wang Y王耀先, Yang C杨朝飞 (1995) *Zhongguo huanjing baohu shigao* 中国环境保护史稿 [Historical manuscript of China's environmental protection]. China Environmental Science Press中国环境科学出版社, Beijing, p 51
80. *Hanshu: Wudi ji*汉书·武帝纪 [Book of the Han: annals of Emperor Wu]
81. *Hanshu: Xuandi ji*汉书·宣帝纪 [Book of the Han: annals of Emperor Xuan]
82. *Hanshu: Yuandi ji*汉书元帝纪 [Book of the Han: annals of Emperor Yuan]
83. *Hou hanshu: Zhangdi ji*后汉书·章帝纪 [Book of the Later Han: annals of Emperor Zhang]
84. *Taiping yulan: juan jiubai wushisi*太平御览卷·九百五十四 [Imperial reader of the Taiping era, vol 954]
85. *Hanshu: Zhaoxinchen zhuan*汉书·召信臣传 [Book of the Han: biography of Zhao Xinchen]
86. Included in Book 0694 of the *Congshu jicheng chubian* 丛书集成初编 [A preliminary compilation of the integrated series]
87. See Zhongjun's L李仲均 (1990) *Jiran Wanwu Lu* kuangwuyao shuzheng《计然万物录》矿物药疏证 [Demonstration of mineral medicine in the *Catalogue of All Things on Earth*], published in Hebei dizhi xueyuan xuebao河北地质学院学报 13(3); also published in the *Li Zhongjun wenji*李仲均文集 [Collected works of Li Zhongjun] (1999) Xi'an Map Publishing House西安地图出版社, pp 99–105
88. Kuanghua Z赵匡华 (1984) Hu Gangzi jiqi dui zhongguo gudai huaxue de zhuoyue gongxian 狐刚子及其对中国古代化学的卓越贡献 [Hu Gangzi and his outstanding contribution to ancient Chinese chemistry]. *Ziran kexue shi yanjiu*自然科学史研究 [Research on the history of natural science] 3:224–235

Chapter 5
Geoscience in the Wei, Jin, Southern and Northern Dynasties

5.1 The Impact of the Social Environment on Geoscience

The period of the Wei, Jin, Southern and Northern dynasties saw frequent wars and numerous regime changes. During this lengthy stage in history, except for a brief period of unification in the Western Jin dynasty, China was divided between the north and the south for a long time. Under such social and historical conditions, China's geosciences did not remain stagnant, nor did they experience a "dark age" as was the case in Europe at the same time, but rather they developed continuously and achieved eminent results in some fields. We believe the developments in the study of geoscience during this period were greatly promoted by the factors outlined below [1].

5.1.1 Han Chinese Culture as the Foundation

Despite the turbulent environment of fragmentation and division during the periods of the Wei, Jin, Southern and Northern dynasties, ancient Chinese cultural traditions were well preserved and passed down rather than being destroyed. The basic cultural and ideological traditions of the Han nation were not abandoned, not during political changes within the Han nation nor during the regimes of invading ethnic minorities, some of which even absorbed and developed the cultural and ideological traditions of the Han nation.

Such was the historical environment in which geoscience during these periods achieved fruitful results based on past studies. The most important achievement of geoscience, the *Shuijing zhu* ("Commentary on the Water Classic"), which drew on 477 texts from the body of existing literature and 358 examples of inscriptions on stone tablets, is a masterpiece based on the collection, summary and synthesis of past data. Similar national geographic works include the *Jifu jing* ("Classic of the Capital

W. Yang and Q. Yang, *A History of Ancient Chinese Geoscience*, Understanding China,
https://doi.org/10.1007/978-981-97-4457-2_5

and the Outer Territories") by Yuzhi in the Jin dynasty and *Shisanzhou ji* ("Records of the Thirteen Prefectures") by Kan Yin in the Northern Wei dynasty. The rich mineralogical knowledge contained in *Baopuzi* ("He Who Holds to Simplicity") by Ge Hong in the Eastern Jin dynasty was basically inherited from his teachers and ancestors [2]. In terms of climatology, it was the Northern Wei dynasty, which practised the Han system, that first included the 72 pentads in the calendar to guide agricultural activities. Also in land transportation, although measures of this period deviated from the highly integrated pattern of the Qin and Han dynasties, it inherited the governance system of the Qin and Han dynasties.

5.1.2 Han Chinese Migration to the South

For a long time, the middle and lower reaches of the Yellow River had been the political, economic and cultural centre of the country. However, from the separation of the Three Kingdoms at the end of the Eastern Han dynasty, the northern central plains gradually lost their special status as the sole centre of the country. For instance, in the Three Kingdoms period, the Sun Wu regime was concentrated in Suzhou and Hangzhou, while the Liu Shu regime was based in the Sichuan Basin and Hanzhong Basin. During the Uprising of the Five Barbarians at the end of the Western Jin dynasty, the northern ethnic groups flooded into the south on a large scale. By contrast, the Eastern Jin dynasty, the political and cultural symbol of the Han nation, established its capital in the south of the Yangzi River, and began to rapidly develop the vast areas south of the Yangzi River.

Professor Chen Qiaoyi, a well-known Chinese geographer, referred to the massive geographical changes experienced by the Chinese nation from the beginning of the fourth century to the end of the sixth century as the Great Geographical Exchange, and compared it to the Great Geological Discovery in the West [3].

This large-scale geographical migration of human history involved a wide range of areas and a large population. For almost three centuries during the Southern and Northern dynasties, about 900,000 northerners migrated to the south, accounting for nearly one sixth of the total population of the Southern dynasties [4].

The Great Geographical Exchange significantly altered the views of ancient geoscience in China. For example, a large group of nomads living in the northern grassland forsook their abundant natural resources, crossed the Great Wall to enter the Central Plains, settled down in this foreign land, and engaged in agricultural activities. On the other hand, the vast majority of the Han people who had originally settled in the Central Plains gave up the dry and flat wheat-producing areas where they had lived for generations and moved to the low-lying and wet rice growing areas in the south of the Yangzi River.

The Great Geographical Exchange also triggered tremendous transformation of ancient Chinese geoscience. Restricted by subjective and objective conditions, Chinese geoscientists before the Qin and Han dynasties had been limited both in the scope and depth of the relevant knowledge. They had often resorted to myths and hypotheses when describing and explaining geographical phenomena. This was true even of the most important works of geoscience during the Spring and Autumn and the Warring States periods, such as the *Yugong* and *Shanhai jing*, whose maps, even the most exquisite Western Han maps excavated from the Han dynasty Mawangdui tombs "define neither the plotting scales nor the precise directions".

Moreover, the Great Geographical Exchange provided good opportunities for geoscientists to broaden their horizons and enhance their perceptual knowledge of the geosciences. Since they actively participated in fieldwork to obtain first-hand information, a new trend in geological research had developed which valued on-site investigation. A large number of vivid and detailed diaries and descriptions exemplify their immersion in the practice of geology. The diligent field survey master Li Daoyuan of the Northern Wei dynasty travelled to Henan, Shandong, Shanxi, Hebei, Anhui, Jiangsu, Inner Mongolia and other places. He conducted careful observations and in-depth studies wherever he went, and finally completed the most famous work on river hydrological in the history of geoscience in China, the *Shuijing zhu*. Fa Xian, an eminent monk in the Eastern Jin dynasty, conducted meticulous fieldwork during his unprecedentedly difficult journey of thousands of miles. Ultimately, he completed one of the oldest travelogues in the world—the *Foguo ji* ("Records of the Kingdom of Buddha").

It was during the era of the Great Geographical Exchange that literary works in various forms, such as monographs, travelogues and fieldnotes, were published one after another, in an unparalleled number compared to their predecessors. For example, six versions of the *Jinzhou ji* ("Journal of Jin Zhou") were written by authors who included Fan Wang, Sheng Hongzhi, Yu Zhongyong, Guo Zhongchan, Liu Chengzhi and others. Many of the geological records described the southern regions, and the most popular ones are those named-*yiwu zhi* ("record of rarities") in the Central Plains, such as the *Jiaozhou yiwu zhi* by Yang Fu, *Linhai shuitu yiwu zhi* by Shen Ying, *Funan yiwu zhi* by Zhu Ying, etc. The development of regional geographic records is the inevitable result of the Great Geographical Exchange. The abundant and detailed descriptions of various karst landforms widely distributed in the south and of the special plum rain and typhoon weather in the south are the most representative examples of the boom in geoscience.

5.1.3 A Wide Diversification of Thought and Philosophy

Confucianism was the philosophy fundamental to national unification during the Han dynasty. In the Wei and Jin dynasties, different philosophies emerged and coexisted with each other, such as Buddhism, Confucianism and Daoism. The competition and debates between academic schools and religions did not restrict or stifle the progress of the rudimentary geoscience of the day. Instead, religiously motivated practical activities such as expeditions, alchemy and pharmacy promoted the objective development of geoscience.

In the Wei, Jin, Southern and Northern dynasties, thanks to the stimulation of these competing schools of thought, all "Six Schools of Heaven" came into being. Moreover, during the Southern Liang period, there was lively debate on the theory of celestial bodies, leading to some new understandings of fundamental geographical issues such as the shape of the Earth and its place in the universe.

China's native Daoism reached its zenith during the Wei, Jin, Southern and Northern dynasties. Many Daoists were hermits aiming to become immortal by cultivating their virtues and producing elixirs. During this period, Daoists, especially the alchemists, paid special attention to the various minerals used in elixirs, and thus conducted more in-depth observation and research on the properties and distribution of such minerals. Therefore, Daoists greatly inspired the development of mineralogical knowledge during this period. Two Daoist masters, Ge Hong and Tao Hongjing, were both keen on alchemy, and made substantial contributions to both alchemy and mineralogy.

In addition, as an important raw material of elixirs, stalactite minerals were much sought after by Daoists. They paid special attention to the origin of stalactite minerals—the karst caves. The enthusiasm of the Daoists prompted the rapid development of understanding of the karst cave landform in this period. During the Wei, Jin, Southern and Northern dynasties, not only were a variety of karst landforms described in detail, but the origin of stalactites was also identified correctly for the first time.

After its introduction into China during the Eastern Han dynasty, Buddhism flourished during the Wei and Jin dynasties. To search for the original meaning of Buddhism and worship in the sacred places, many Buddhists braved hardships and travelled across mountains and rivers to Tianzhu (modern day India), the birthplace of Buddhism. The upsurge in Buddhists' westward pursuit of Dharma greatly promoted the deepening of geographical understanding in Central Asia, South Asia and other regions. Many geographical works were written by Buddhists, of which the *Foguo ji* is the most famous.

5.2 Advances in the Shape of the Earth Cosmology

In the Wei, Jin, Southern and Northern dynasties, new theories about the structure of the cosmos, such as the Xintian theory, the Qiongtian theory and the Antian theory, were formulated in addition to the three unique theories formed in the Han dynasty, the Huntian theory, the Gaitian theory and the Xuanye theory, with the Huntian theory in the dominant position. On the other hand, heated academic debate on the shape of the Earth highlighted people's strong interest in geographical issues such as the Earth's form and its position in the universe.

5.2.1 Huntian, Gaitian and Shuenye Hypotheses

(1) The Huntian hypothesis

This hypothesis was formed in the Han dynasty and developed rapidly during the Wei, Jin, Southern and Northern dynasties, gradually becoming the dominant cosmic theory.

Lu Ji (187–219 AD) in the Wu state of the Three Kingdoms was the leading scholar of the Huntian hypothesis. In the *Huntianyi shuo* ("Theory of the Armillary Sphere"), he abandoned Zhang Heng's view of the heavens and the earth having the same diameter, and explicitly put forward the theory that the earth was spherical:

> The Earth is smaller than Heaven and is covered by it, with half above the ground and the other half beneath it, like the egg white around the yolk [5].

Wang Fan of the state of Wu (228–266 AD) wrote the *Huntianxiang shuo* ("Theory of the Huntian") to illustrate the superiority of the Huntian theory over the Gaitian and the Xuanye theories:

> Former scholars believed that the structure of Heaven and Earth was like that of an egg enclosed in its shell, rotating without poles in a chaotic pattern, thus the name Huntian (or the Chaotic Heaven).
>
> All celestial bodies are like balls, and the Earth cuts Heaven in half, with Yang Cheng (present-day Dengfeng, Henan) in the centre [5].

We can tell that Wang Fan had inherited Zhang Heng's flat-Earth theory and had introduced Yang Cheng as the centre of the earth in his Huntian theory.

Liu Zhi, who lived from Cao Wei to the Western Jin dynasty, said in his *Treatise on Heaven*:

> The armillary sphere covers the Earth with the sky, and the Earth is held up by the air...
>
> Now the Sun has a diameter of thousands of miles, yet the Earth can cover it, so the shaded area should be more than half of the whole area [5].

Liu Zhi boldly proposed that the area between heaven and the earth was filled with air and as a result of this support the earth was suspended in heaven (see Fig. 5.1 [6]). However, he still adopted the fallacy in the Huntian theory that the earth was too large.

In order to demonstrate the celestial phenomena and to honour the correct nature of the Huntian theory, the Huntian scholars also designed related instruments. The scholar Ge Heng in the Eastern Wu advanced the theory of Huntian: "The Wu national, Ge Heng, style name Sizhen, having a clear understanding of Heaven in mind, made an armillary sphere, with the Earth at the centre and Heaven revolving around it in order to calculate the degrees on a dial [7]." This armillary sphere clearly points out that "the Earth is in the centre" (Fig. 5.2 [8]).

The celestial model designed by Luji resembles the shape of an egg [5]. This model described the cosmos as an ellipse, measuring slightly longer from east to west than from north to south [9].

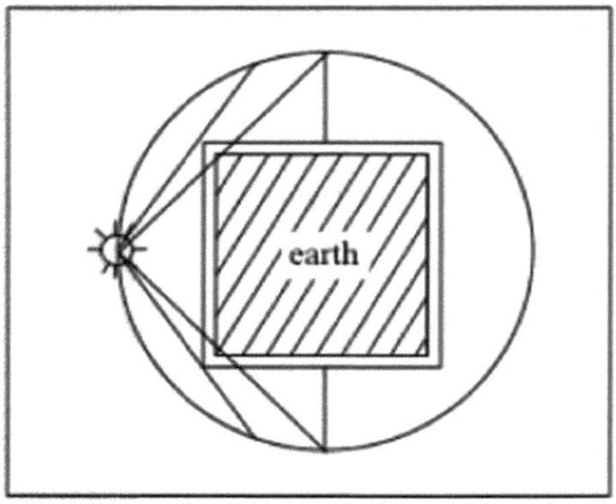

Fig. 5.1 The size of the earth indicated by Liu Zhi's Huntian theory

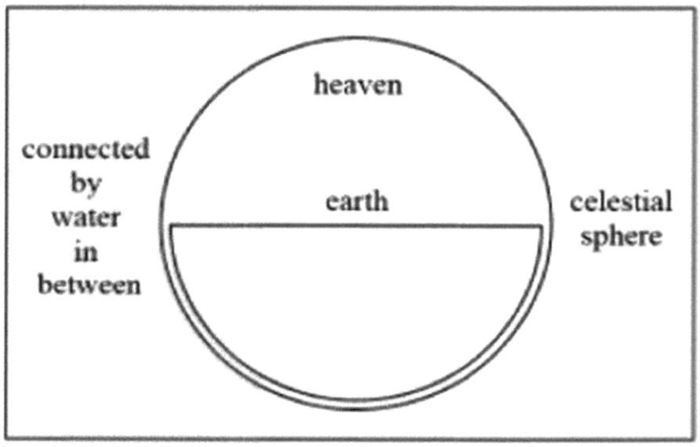

Fig. 5.2 The sketch map of the Huntian theory conceived by Ge Heng

In the 6th Year (323 AD) of the Guangchu reign of the Former Zhao regime of the Eastern Jin dynasty, the assistant officer of Nanyang County, the historian Kong Ting, built a bronze armillary sphere. The *Book of the Sui: Astronomical Treatise* detailed the structure of the armillary sphere as being made of two loops and a sighting tube (*heng*) [10]. This is the earliest detailed description of the armillary sphere, which stood in front of the Hualin Chongyun Temple until the Liang dynasty [11].

In the 13th year of the Yuanjia reign by Liu Song (436 AD), an imperial astronomer, Qian Lezhi, melted copper to mould the Huntian armillary sphere:

> The diameter is six *chi* (31.68cm) and 8.25 *fen* (31.68mm), and the perimeter measures 12 *chi* and 6.25 *fen*, the Earth is inside the Heaven... [12]

The astronomer Hulan made an iron model for the Imperial Astronomic Bureau [13]. This is the only armillary sphere made of the transition metals recorded in China's history.

(2) The Gaitian hypothesis

The Gaitian hypothesis originated in the Yin and Zhou dynasties and was further developed into the second Gaitian theory (Fig. 5.3) in the Wei, Jin and Southern

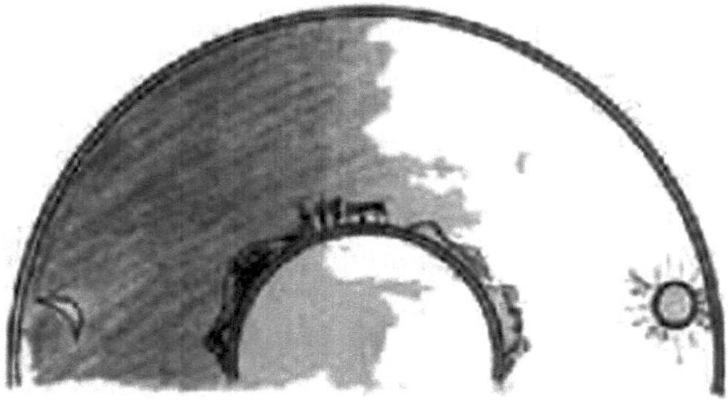

Fig. 5.3 Illustration of the second Gaitian theory (cited from [14])

and Northern dynasties. *The Book of the Jin: Astronomical Treatise* recorded the following relevant information:

> The sky is like a covering lid, while the Earth is like a reversed basin. They are both higher in the middle and lower around the margins. The centre of Heaven and Earth is directly beneath the North Pole and, therefore, is the highest point on the Earth. As a result, the waters flow away from the centre in the four directions and the three lights (the sun, the moon, and the stars) hide and appear regularly, making the day and the night.

The second Gaitian theory differs from the first in that it claimed that the earth was an arch rather than a flat plain. The Gaitian theory was so popular during that period that folk songs were composed about it. An example is the *Chi Le Songs* composed by Hu Lyu Jin, a singer of the Xianbei ethnic group. during the Southern and Northern dynasties:

> Under Yinshan Mountain lies the Chile River. The heavens are like an arch that covers the surrounding fields. Greyish dark are the heavens, and the fields are a vast expanse. Under the tall grass are cattle and sheep that the wind reveals [15].

(3) The Xuanye hypothesis

According to legend, the Xuanye theory originated in the Yin era and was recorded by the scholar Qie Meng in the Han dynasty (second century AD). The theory also developed during that time as well. Yang Quan of the Three Kingdoms further delineated the infinite universe philosophy of the Xuanye theory in his work *Wuli lun* ("On the Laws of Nature"):

> The earth has form, yet the heavens are formless. The earth is like the ash at the bottom and the heavens are like the smoke on the top.

The text written under the pseudonym Liezi in the Wei and Jin dynasties included the famous story of the Qi people worrying about the heavens falling down in the chapter entitled "Tianrui" of *Liezi*. Marvellous arguments were made about the Xuanye theory, together with the following claims regarding the earth.

> The Qi people asked: how damaged is the earth? The wise man answered: the earth is nothing but piles of soil and rocks, which fill the void in all directions. Therefore, there is no need to worry about its collapse if you walk on it every day.

5.2.2 *Xintian, Antian and Qiongtian Models*

Three new models of the earth's form came into being during the Wei, Jin, Southern and Northern dynasties. Despite all conforming to the Gaitian theory which held the earth to be a massive disc, these new models all advanced understanding of the shape of the earth and observation of the cosmos building on the Gaitian theory.

(1) Xintian model by Wu Yao Xin

Yao Xin, the Chamberlain of Ceremonials of Wu of the Three Kingdoms, composed the one-volume *Xintian lun*, proposing the Xintian model and giving his explanation of the theory. The book was lost for a long time, and the *Book of the Jin: Astronomical Treatise* recorded his theory as follows:

> Humans are spiritual beings, so their form most closely resembles that of heaven. Humans have a large jaw over the chest, yet their neck cannot cover the back. Comparing the human body with the cosmos, it is known that heaven is lower in the south, and higher in the north. Also, just as heaven rotates towards the south and Polaris in the north is lower when winter comes, the sun moves away from us and the northern Big Dipper moves closer to us, bringing the chilling coldness of the northern heavens. With the arrival of summer, the Pole stars rise, the heavens rotate towards the north and the northern Big Dipper moves further away, whilst the sun gets closer to us, thus the southern heathland ushers in the steaming hot weather. When the Pole stars are higher, the sun's passage beneath the earth is shallow, so the night is short and the heavens are further away from the earth, meaning that the days are longer. When the Pole stars are lower, the sun's journey is deep beneath the earth, so the nights become longer and the heavens are closer to the earth, signifying shorter days.

In general, the Xintian model by Yao Xin still regarded the Earth as a block under heaven, rather than as a sphere.

(2) Qiongtian model by brothers Yu Bing and Yu Song

In the Eastern Jin dynasty, the Qiongtian model was created to provide a reasonable explanation for the fact that the structure of the heavens and the earth does not collapse. According to Chen Meidong's study, brothers Yu Bing and Yu Song gave their respective explanations of the Qiongtian model [16].

Volume Two of the *Taiping yulan* ("Imperial Reader of the Taiping Era") describes Yu Bing's interpretation of this model:

> The shape of Heaven is like a bamboo hat that covers the surface of the Earth. Heaven floats above the *yuanqi* (vigorous energy). Similar to a big, inverted bowl covering the water, Heaven does not submerge because it is filled with air. The sun rotates at both ends of Heaven, setting in the west (south [17]) and rising again in the east, and the sun is not buried under the Earth.

Yu Rong's interpretation of the Qiongtian model is documented in the *Books of Jin: Astronomical Treatise*:

> The arch of Heaven is like an egg, with the shell covering the boundaries that connect with the four oceans. Heaven floats above the *yuanqi* (vigorous energy), and is similar to a big, inverted bowl covering the water; Heaven does not submerge because it is filled with air. The sun rotates at both ends of Heaven, setting in the west and rising again in the east, and the sun is not buried under the Earth. Heaven has a zenith, just as the bamboo hat has a tip.

The interpretations of the Qiongtian model proposed by the Yu brothers still fell within the Gaitian framework, but they made further developments based on it. Fundamentally, they accepted Zou Yan's Jiuzhou theory, believing that the earth was embraced by the ocean in four directions and the oceans were connected to the sky. Moreover, they proposed that the space between the earth and heaven air was filled with air which supported heaven so that it did not collapse, avoiding the need for legendary sky-supporting pillars like the Buzhou Mountain.

(3) Antian model by Yu Xi

During the Xiankang reign of the Eastern Jin dynasty (335-342AD), "Yu Xi from Kuaiji composed the *Antian lun* ("Book of Heaven") to analyse the Huntian and Gaitian theories" [18]. All six chapters of the *Antian lun* have been lost. However, this model is described in the *Book of Jin: Astronomy*:

> The height of Heaven is infinite and the depth of the Earth is immeasurable. The eternal shape of Heaven is above while the still shape of the earth lies beneath. They cover each other; therefore, they are either both squares or both circles. It is not possible for one to be square in shape and the other to be circular. The stars that are scattered in Heaven move automatically like the tides of the seas and rivers. All things in the world move and stop following a pattern.

The Antian model is a theory of the cosmos built on the Xuanye theory. Many people believed that the sun, the moon and the stars might fall out of heaven, if, as in the Xuanye theory, they merely floated on the air and heaven did not have a hard shell. According to the *Liezi*: "Some people in the Qi state are worried that Heaven might collapse, depriving them of their lives and houses, so they cannot sleep or eat well."

5.2.3 The Polemic Between Huntian and Gaitian

In the Han dynasty, the more advanced Huntian theory replaced the more intuitive and elementary Gaitian theory developed before the Qin dynasty. At that time, and until the Wei, Jin, Southern and Northern dynasties, the Gaitian theory was re-vitalised with new understandings and perceptions of the celestial spheres. The three cosmos theories, the Xintian, Qiongtian and Antian models, were further improved, leading to the completion of the "Six theories on Heaven" [14].

The publication of these novel cosmos theories sparked heated debate among people about the understanding of the earth and thus led to the first Great Debate between the Huntian theory and the Gaitian theory.

According to the *Book of the Sui*: *Astronomical Treatise,* in the 6th year of the Putong reign of the Southern dynasty (525 AD), Emperor Xiao Yan (464–549) organised a gathering of Confucian scholars in Changchun Hall to observe the celestial bodies and to write texts to promote the Gaitian theory, making it the official theory. During the Putong reign (520–527), Tongtai Temple was built to demonstrate the beliefs of Emperor Wu in supporting the *Jin Gang Jing* ("Diamond Sutra"). It is probably the most detailed and largest architectural model illustrating the Gaitian theory in ancient China [19].

During the controversy between the two theories, Cui Ling'en, the consultant for the emperors in the Directorate of Education, tried to combine the two theories into a single new one:

> In the past, there was divergence between scholars who supported either the Huntian or the Gaitian theories. They believed the two theories were not compatible with each other. I believe that the two theories can be integrated with one another [20].

Xin Dufang, during the Northern Qi dynasty, also claimed in the introduction to his mathematical work *Sishu zhoubi zong* that the Gaitian and Huntian theories were "fundamentally the same":

> The Huntian theory looks downwards from the top while the Gaitian theory looks up from the Earth. According to the methods in the *Zhoubi suanjing*, the top-down view and the bottom-up view are fundamentally the same despite their distinct perspectives [21].

5.2.4 Worlds of Mount Sumeru in Buddhist Cosmology

Until the late Western Jin dynasty, the Indian cosmos theory had been introduced into China with the spread of Buddhism. The Mount Sumeru theory was one of the main views regarding the Heaven-Earth structure in ancient India.

According to the Dirgagamas co-translated by Buddhayaśas and Zhu Fo Nian:

> Floating on the sea are eight thousand worlds around Mount Sumeru and the Four Continents, with the Great Kumgang Mountain and the Second Kumgang Mountain outside the Four Continents. In between Mount Sumeru and the Four Continents there are interconnecting mountains and waters, and Mount Sumeru lies at the centre of the Four Continents. The sun, the moon and the stars rotate around Mount Sumeru from east to west at the central summit [22].

In comparison with the Gaitian theory, the Sumeru theory had a more detailed description of the earth and was more imaginative. Although the theory did not specify the overall form of the earth, its belief was that the Four Continents were much lower than Mount Sumeru and that the earth was rooted in the water, which is similar to the claims in the Huntian theory [23]. Figure 5.4 is a diagrammatic representation of the Mount Sumeru theory [24]. In the figure, "西利天宫" means

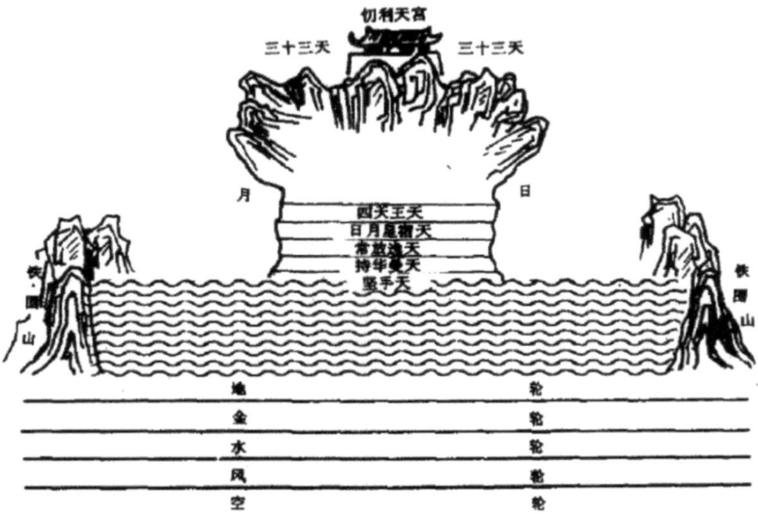

Fig. 5.4 The Mount Sumeru theoretical model [25]

"Trayastrimsha Heaven", "三十三天" means "33 days", "四天文天" means "Heaven of the Four Kings", "日月星宿天" means "Heaven of the Sun, Moon, and Stars", "常放逸天" means "Paranirmita Vasavartin", "持华曼天" means "Heaven of Holding Flower Garlands", "坚手天" means "Kumbhira", and "铁围山" means "Cakravāda-parvata".

The pious Buddhist Emperor Wu of the Liang dynasty proposed the Kamgang Mountain theory based on the Mount Sumeru theory:

> On the periphery of the Four Oceans are Kamgang Mountain, also called Iron Circle Mountain and, to its north, Mount Black. The sun and the moon rotate around the mountains and complete a full rotation within a day and a night. The sun and the moon will be visible when they rotate to the southern side and will be hidden on the northern side [5].

We can see that his theory of Kamgang Mountain simply replaced Mount Sumeru with Kamgang Mountain, with no other modifications. Figures 5.5 and 5.6 are diagrammatic representations of the Mount Sumeru theory and Kamgang Mountain theory drawn by Chen Meidong [26].

5.3 Advances in Regional Geography in China

During the Wei, Jin, Northern and Southern dynasties, great progress was made in understanding domestic regional geography, mainly in two respects: firstly, the flourishing of *fangzhi* (local chronicle or gazetteer), characterised by comprehensive records of natural and human conditions in a certain region of the country, greatly

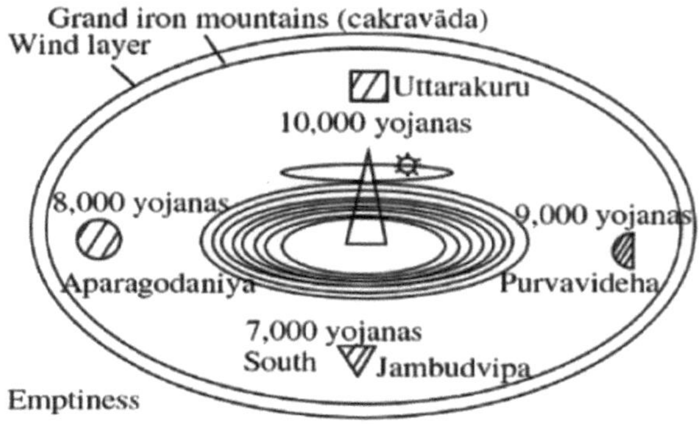

Fig. 5.5 Diagrammatic representation of Mount Sumeru drawn by Chen Meidong

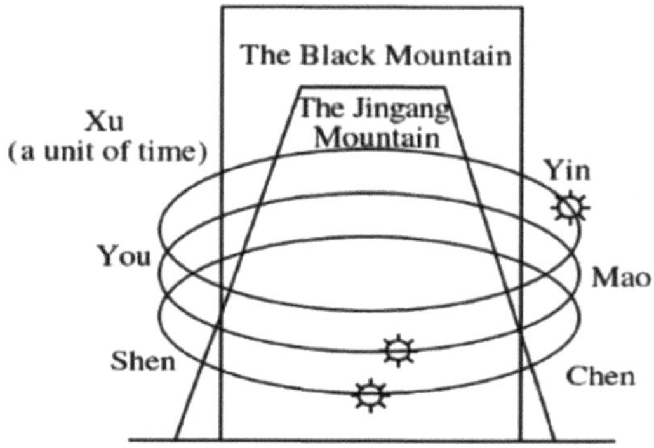

Fig. 5.6 Diagrammatic representation of Kamgang Mountain drawn by Chen Meidong

enhanced the understanding of domestic regional geography; secondly, the understanding of domestic physical geographical features advanced further, especially in terrestrial hydrography, karst geomorphology and biogeography.

5.3.1 An Upsurge in Works of Regional and Historical Geography

During these periods, a number of regional historical and geographic texts emerged [27], making the records of the knowledge of physical and human geography and the regions more detailed and accurate than those of previous generations. Overall, the published literature can be classified according to the following four categories.

(1) Treatise on regional geography and national geography

The *Yugong*, written during the Warring States period, pioneered the approach of describing the geography of each region from a historical perspective, which can be regarded as a prototype of historical geography writing. The large number of maps and administrative records in the Qin and Han dynasties laid the intellectual groundwork for the compilation of treatises on national geography. In the Eastern Han dynasty, the *Hanshu: Dili zhi* ("Book of the Han: Administrative Geography") by Ban Gu was the first treatise on historical geography with a structure based on regional features, including mountains, rivers and produce. This set the standard for the future writing of *dili zhi* (geographic records) in a biographical style. During the Wei, Jin and Southern and Northern dynasties, the genre of *dili zhi*, at both regional and national levels, offered richer content in a relatively complete format.

Four treatises on administrative geography existed during this era.

The *Xu Hanshu: Junguo zhi* ("Sequel to the Book of the Han: Geography of Provinces and States") was compiled by Sima Biao of the Jin dynasty (240–306 AD). The *Xu Hanshu* was lost long ago, but the section entitled "Geography of Provinces and States" has survived thanks to its also being incorporated into Fan Ye's *Hou Hanshu* ("Book of the Later Han"). This geographical treatise omits the details available in the *Hanshu: Dili zhi*, but rather focuses on the new prefectures and states or on those removed in the Eastern Han dynasty compared with the previous dynasties, with a clearer classification system than in the *Hanshu: Dili zhi*.

The *Songshu: Zhoujun zhi* ("Book of the Song: Geography of States and Prefectures") was compiled by Shen Yue of the Liang dynasty (441–513 AD). This book, based on the *Taikang dili zhi* [28], not only makes up for the lack of historical geography during the Wei and Jin dynasties, but also provides precise and detailed descriptions. Apart from the geographic evolution and household statistics, the book also documents the distribution of overseas states and counties and the mileage by water and land to the capital city.

The *Nanqishu: Zhoujun zhi* ("Book of the Southern Qi: Geography of States and Prefectures") was compiled by Xiao Zixian of the Liang dynasty (448–537 AD). This book was adapted by Jiang Yan's *Qishi: Zhoujun zhi* ("History of Qi: Geography of States and Prefectures") [29] and quite accurately reflected the administrative division and its changes (excluding the household population) during the Southern Qi period.

Three volumes of the *Weishu: Dixing zhi* ("Book of the Wei: Topography") were compiled by Wei Shou of the Northern Qi dynasty (506–572 AD). The treatise lists the number of states, prefectures, and household and population size. There are also notes on the history of the administrative units, cities and moats, produce, climate, plants, mountains, rivers, water conservancy projects, landforms, shrines and tombs, etc. It is one of the most detailed works of historical geography written in a biographical style [30].

During the era of the Wei, Jin and Southern and Northern dynasties, works on the general geography of the entire country included: the *Dili ji* ("Geography") and *Taikang sannian diji* ("Geographical Records for the Third Year of the Taikang Era of Jin") by Zhang Yan of the Wei dynasty, *Didao ji* by Wang Yin of the Jin dynasty, *Wulu dili zhi* ("Geography of the Wu Kingdom") by Zhang Bo, *Jiuzhou yaoji* ("Essentials of the Nine Regions") by Le Zi, *Jifu jing* (The Classic of the Capital and the Outer Territories) by Zhi Yu, *Yudi zhi* ("A Collection of Geographical Essays") by Gu Yewang of the Liang dynasty in southern China, *Shisan zhouzhi* ("Records of Thirteen States") by Kan Yan of the Northern Wei dynasty, and so on [31]. Among them, the most complete text was the *Jifu jing*. As the *Suishu: Jingji zhi* ("Book of the Sui: Scriptures and Records") says:

> During the Jin dynasty, Zhi Yu compiled the *Jifu jing* based on the *Yugong* and *Zhouguan* (i.e. *Zhouli*, "Rites of the Zhou"). The series of documents describing the states, provinces, prefectures, borders, events, countries, mountains and hills, bodies of water, pavilions, roads, fields, ethnic groups, customs, ancestors ... all 170 volumes, have already been lost.

It can be seen that the *Jifu jing* was more than a book on geography— it included social and human elements in contrast with other geographic volumes such as the *Hanshu: Dili zhi*.

(2) Books on geographical topics and urban constructions

During the period of the Wei, Jin and Southern and Northern dynasties, the private practice of collecting local histories became popular, and geographical records of local environments in various forms emerged one after another. For example, just the *Suishi* ("Book of the Sui") alone references a list of more than 100 and contains 1,400 volumes. It generally covers places that were rich in geographical information and the biographies of key people.

Geographical texts during the Wei and Jin dynasties were called *diji* (地记), or *dizhi* (地志). Lu Cheng from the Southern Qi dynasty collected 160 local annals from which he compiled the 149 volumes of the *Dili shu* ("Book of Geography") and its one-volume catalogue. The Liang dynasty scholar Ren Fang added 84 works to the *Dili shu* and completed the *Diji* ("Record of Geography") composed of 252 volumes. Table 5.1 shows part of a bibliography of geographic literature in this period.

Many works in Table 5.1 have been lost, except the *Huayang guozhi* ("Records of the Countries to the South of Mount Hua"/ "Huayang Chronicle") by Chang Qu and *Jiujiang zhi* by He Yan. Some of the lost articles can be found in the *Hantang dili shuchao* ("Copies of Geography Books from the Han and Tang Dynasties") by Wang Mo and in the *Jinxiang diji* by Chen Yunrong. From these copies and records, we can perhaps catch a glimpse of the lost books: they would have mainly described the territory, place names, products, mountains and rivers, relics, and myths and legends, and, of these, the produce would have been described in particular detail.

The geographic texts from this period roughly fall into the following categories:

① Landscape records. *Records of Mountains and Rivers of Yidu* by Yuan Shansong during the Eastern Jin dynasty describes the natural landscape, famous scenic spots, local products and customs of Yidu (now Yichang, Hubei province). Of the places described, the best-known place is the site of the geothermal hot springs in the east of Henshan County (now Duzhenwan, Changyang County, Hunan province). *The Names of Wuxing Mountains* by Zhang Xuanzhi during the Jin dynasty documents the origins of the names of mountains, rivers and lakes in present-day Zhejiang province, including Sanshan Mountain, Jinshan Mountain, Zhushan Mountain, the Ying Stream and Wucheng Lake, as well as their production, stories and scenery. The book is a significant source of information regarding the geography and toponymy of the region.

Table 5.1 Geographic literature of the Wei, Jin, and Southern and Northern dynasties listed in Zhang Guojian's *Examination of Ancient Chinese Fangzhi*

Types	Geographic literature
Geographic literature designated *ji*	During the three kingdoms period: In the kingdom of Wei: *Qinji* and *Yiyang ji* by Ruan ji, *Guiyangji* by Yang Yuanfeng In the kingdom of Wu: *Loudi ji* by Gu Qizhi, *Wuxian ji* by Gu Wei In the kingdom of Shu: *Sanba ji* by Qiao Zhou, *Huayang ji* by Monk Ren Xian During the Western and Eastern Jin dynasties: *Luoyang ji* by Lu Ji, *Wudi ji* by Zhang Bo, *Jizhou ji* by Pei Xiu, *Guanzhong ji* by Pan Yue, *Guangzhou ji* by Gu We, *Wujun ji* by Gu Yi, *Xiangzhong ji* by Luo Han, *Yidu Shanchuan ji* ("Record of Mountains and Rivers") by Yuan Shansong During the Southern Song dynasty: *Kuaiji ji* by Kong Lingfu, *Poyang ji* by Liu Chengzhi, *Yuzhang ji* by Lei Jizong, *Nankang ji* by Deng Deming, *Jinzhou ji* by Sheng Hongzhi, *Jinzhou ji* by Guo Zhongchan, *Danyang ji* by Shan Qianzhi During the Southern Qi dynasty: *Wu diji* by Lu Daozhan, *Wuling ji* and *Yuanling ji* by Huang Min During the Southern Liang dynasty: *Diji* by Ren Fang, *Wuling ji* by Bao Jian, *Wuling ji* by Wu Anpin During the Southern Chen dynasty: *Jian'an diji* by Gu Yewang
Geography literature designated *zhi*	In the kingdom of Shu: *Yizhou zhi* by Qiao Zhou, *Wujunguo zhi* by Wu Weizhao During the Western and Eastern Jin dynasties: *Huangyang guozhi* by Chang Qu, *Jiujiang zhi* by He Yan, *Shuzhi* by Chang Kuan, *Yongning dizhi* and *Kuaijijun shicheng dizhi* by unknown authors During the Southern and Northern Song dynasties: *The Sixth Year of Yuanjia dizhi*, *Wu dizhi* by Dong Lan, *Nanyue zhi* by Shen Huaiyuan During the Southern Qi dynasty: *Dizhi* by unknown authors, *Haidai zhi* by Cui Weizu During the Southern Liang dynasty: *Jinan dizhi* by Xiao Yi During the Southern Chen dynasty: *Yudi zhi* by Gu Yewang

② Records of the terroir and customs of a region. These were most popular during the Han and Tang dynasties. The books mostly contain a concise description of local geography and customs.

③ Records of "alien objects" or *yiwu zhi* in the south of the Yangzi River, such as unknown grasses, trees, animals and minerals. The region was unfamiliar to the Chinese who usually lived in the north of the Yangzi River or in the Central Plains. Records of "alien objects" were initially created to inform the northern literati about the region after the mass migration, and these books mainly focused on the resources and economy of the south. The Wei, Jin and Southern and Northern dynasties was the time when the genre of *yiwu zhi* prospered, including *Funan yiwu zhi* by Zhu Ying, *Nanzhou yiwu zhi* by Wanzhen, *Linhai shuitu yiwu zhi* by Shenying from the Wu Kingdom, *Yiwu zhi* by Qiao Zhou from the Shu Kingdom, *Yiwu zhi* by Xu Xian, *Yiwu zhi* and *Liangzhou yiwu zhi* by Song Ying and *Nanfang yiwu zhi* by Chen Qichang during the Jin. However, many have been lost [32]. In Wu Shenying's *Linhai shuitu yiwu zhi* ("Alien Objects of the Coastal Land and Water"), the geography of Taiwan was first mentioned:

Yizhou (present-day Taiwan) is in the area of the sea to the southeast (the seat of governance is near the coast of present-day Zhejiang), two thousand li away from the prefecture. Frost and snow never occur on the land and the flora remains alive all year long. Crisscrossed by mountain streams, it has fertile soil suited to five types of grain and abundant resources of fish. The men all shave off their hair and pierce their ears, but the women do not [33].

④ Records of particular sites or *zhuanzhi*. Yang Xuanzhi of the Northern Wei dynasty wrote the five volumes of *Luoyang qielan ji* ("Record of Buddhist Temples in Luoyang") (547 AD) that provide the best-known record of temples. The rise and fall of Buddhist temples in Luoyang, the capital of the Northern Wei dynasty, is well documented in the book. *Duyi bu* ("Records of Cities") is dedicated to chronicling capital cities, including the walls and moats, palaces, streets, temples and gardens in the metropolises.

Most of the *duyi bu* (records of particular cities) from the Wei and Jin dynasties have been lost, but one which has survived—*Sanfu huangtu* ("Plan of the Three Capital Regions")—was anthologised by Bi Yuan and Sun Xingyan during the Qing dynasty. From their anthology we can catch a glimpse of the geography of the ancient Xianyang capital, the ancient Chang'an capital, and the palaces, city gates, gardens, schools, suburbs, warehouses, barns, bridges and other constructions and folklore of the Qin and Han dynasties. It is an indispensable historical reference for the study of the Chang'an capital and of the Han people.

In addition, the aforementioned *Luoyang qielan ji* comprises five volumes which address the interior, the east, the south, the west and the north of the city respectively, with each volume starting with a specific gate into the city. Its clear structure and accurate depictions make it an excellent *duyi bu* on Luoyang.

(3) Tujing

During the Han and Jin dynasties, the drawing of maps was relatively developed. Some of the maps at that time were supplemented by written descriptions. Some scholars referred to these as "text with maps" (from the *Yu ditu*, "Territorial Maps") while others called them *tujing* (from the *Guanglingjun tujing*, "Map Record of Guangling Prefecture" by Wang Yi). As a genre, *tujing* were separate from the ancient maps and formed their own system.

A *tujing* is a scholarly work that contains illustrations of a location consisting primarily of maps or of pictures combined with text. It is derived from the *diji* (record of a place) but with much more detailed information. The earliest known *tujing* is the *Bajun tujing* of the Eastern Han dynasty. Examples from the Wei, Jin and Southern and Northern dynasties in the *Suishu: Jingji zhi* include the *Youzhou tujing*, *Jizhou tujing* and *Qizhou tujing*.

We can only have a basic understanding of this period from the *Bajun tujing* as referenced in a citation in the *Huayang guozhi: Bazhi:*

> The territory is 4,000 li north to south, 5,000 li east to west, and the circumference is more than 10,000 li. There are fourteen counties, five salt and iron officials, and each place has a prime minister assigned to it. The number of households is 464,780 and the people registered

amount to 1,875,535. The distance from faraway counties to the prefecture is about 1,200 to 1,500 li and from a village to the city it can be three to four hundred li or a thousand li.

This quote shows that the map was rich in geographic content, including the concepts of institutions, territory, counties, officials, households and distance.

(4) Comprehensive geographical records

From the time of the Han and Wei dynasties, geographic literature was further enriched, with the perspective changing from narrow to comprehensive, integrating the description of places with those of people. During the Wei, Jin and Southern and Northern dynasties, the representative text of the comprehensive geography genre was the *Huayang guozhi* (Fig. 5.7) by Chang Qu of the Eastern Jin dynasty.

Chang Qu, styled as Daojiang, a native of Jiang Yuan in the Shu Prefecture (present-day Chongzhou in Sichuan province), was a former soldier who lived between the late third century to the mid-fourth century AD approximately. The *Huayang guozhi* was completed around 348–354. It is named "Huayang" because the region recorded in this work relates to the area of Liangzhou recorded in the *Yugong*: "Liangzhou is located from the north of Hua Mountain to the shores of the Hei River."

Starting with *Bazhi* ("巴志"), or the Chronicle of Ba Area, this book describes the geography and history of the southwest region centred on Bashu (the area of and around the Sichuan Basin) and covers more than a thousand years, from its

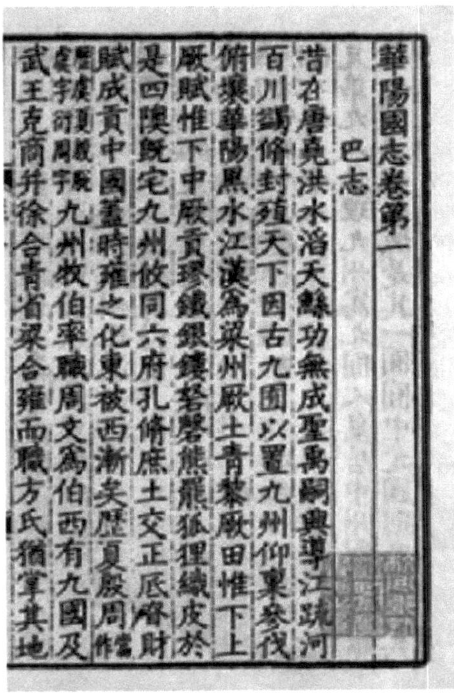
華陽國志卷第一
巴志
昔在唐堯洪水滔天鯀功無成聖禹嗣興導江疏河
百川蠲脩封殖天下因古九囿以置九州仰禀參伐
俯壤華陽黑水江漢為梁州厥土青黎厥田惟下上
厥賦惟下中厥貢璆鐵銀鏤砮磬熊羆狐狸織皮於
是四隩既宅九州攸同六府孔脩庶土交正底慎財
賦成貢中國蓋時雍之化東被西漸矣歷夏殷周
九州牧伯率職周文為伯西有九國及
武王克商并徐合青省梁合雍而職方氏猶掌其地

Fig. 5.7 A facsimile of a page from the *Huayang guozhi*

very beginnings to the end of the 3rd year of the Yonghe reign (347) [34]. The twelve volumes of the book can be divided into three main parts: Volumes One to Four are structured by regions and introduce the historical geography of each region; Volumes Five to Nine chronicle the historical events; Volumes 10 to 11 document the distinguished men and women of the region. Furthermore, Volume 12 incorporates the "Preface" and the "Catalogue of Daughters of Officials in Three States". Thus, Chang Qu combined chronicles, *dili zhi* and biographies in the *Huayang guozhi*, creating a new genre and becoming the most seminal work of the early comprehensive chronicles.

The *Huayang guozhi* has four main features in terms of geography [35].

First, it records the history of the 33 prefectures and 180 counties in the region, the changes in border defences, administrative divisions, demography, climate, topography, material resources, agricultural and water resources, transportation, industry and commerce, urban architecture, monuments and scenic spots, customs and habits, etc. It covers all aspects of historical, physical and human geography, with much more detailed content than in the official dynastic history.

Secondly, unlike geographical records written in biographical style, the prefectures and counties included were not limited to a specific period of time, but rather extended to many that once existed but had since been merged, thus better reflecting the evolution of administrative divisions.

Thirdly, much of the description of geographical evolution concentrated on the Han and Jin dynasties, which is a weak link in works on administrative geography. Since there was no historical geography literature during the Three Kingdoms and the late-published *Jinshu: Dili zhi* ("Book of the Jin: Geography") only provides a general review of the administrative area of the early Jin dynasty, consequently, what is recorded in the *Huayang guozhi* is exceptionally valuable.

Fourth, it attaches great importance to the ethnic minorities in southwest China and reports the history and status quo of more than 30 minority groups or tribes in this region, including their names, distribution, customs, relations with the Han and their legends and myths. The *Houhanshu: Nanyi xiyi liezhuan* ("Book of the Later Han: Foreign Regions in the South and the North") draws heavily on the *Huayang guozhi.*

The *Huayang guozhi* also records mining and smelting, materials production, and agriculture in various places. Three contributions make it stand out: firstly, it notes the use of fire wells to cook salt in the Wenjing River area, Linqiong, Sichuan province, which is the earliest record of the use of natural gas in the world; secondly, it documents the process of digging salt wells by Li Bing at Guangdu (the present-day Renshou and Shuangliu areas of Sichuan) during the Warring States period, which is the earliest record of well-drilling in the world; thirdly, it is the first record of the Dujiangyan irrigation system and provides great detail.

To sum up, the flourishing of writings on historical geography during the Wei, Jin, and Southern and Northern dynasties greatly contributed to people's understanding of domestic (and neighbouring) regional geography.

5.3.2 *Synthetic Studies of Terrestrial Hydrology*

This section draws on Chen Qiaoyi' study of the *Shuiijng zhu* [36].

As early as the pre-Qin dynasty, China already had treatises on hydrology and geography, the *Yugong: Daoshui (*guiding water) and *Guanzi: Dudi* (terrain), but before the Wei and Jin dynasties, these accounts of land hydrology lacked detail. In the *Yugong*, more than 40 rivers and lakes were mentioned, but only nine were introduced with details; in the *Mountains*, although more than 350 rivers and lakes were named, the descriptions were mostly vague. *Shiji: Hequ shu* and *Hanshu: Dili zhi* reviewed some rivers and water conditions more rigorously, but the descriptions remained superficial. The period of the Wei, Jin and Southern and Northern dynasties witnessed a breakthrough in the study of Chinese terrestrial hydrography, which was primarily revealed thanks to the discovery of two monographs.

(1) *Shuijing*: the first monograph that systematically reviewed the water system in China

It is argued that the work *Shuijing* was completed first, written by an author from the Three Kingdoms era. In fact, there remains controversy over who wrote it and at what time [37]. The *Book of the Sui: Scriptures and Classics* records that the three volumes of the *Shuijing* were annotated by "Guo Pu", but the author is not mentioned. Volume 69 of the *Siku quanshu zongmu* ("General Catalogue of the Complete Library of the Four Treasuries") points out:

> In the section "Pei River", Guanghan is already called Guangwei, which definitely does not date from the Han period; Jinning is still called Weining in the section "Zhong River", which predates the Jin dynasty. Judging from the text, it may date from the Three Kingdoms period.

During the Qing dynasty, both Quan Zuwang (1705–1755) in his *Wujiao chaoben: Tici* ("Proofread Manuscript of *Shuijing zhu:* Foreword") and Yang Shoujing in the *Shuijing zhushu* ("Detailed Commentary on the Water Classic") further pointed out that the *Shuijing* was written by authors from the Wei Kingdom.

As the first systematic review of the water system in China, the *Shuijing* accurately records the details of 137 major rivers, including their sources, directions, destinations and watercourse features, and maps out the spatial relationship between the major and minor waterways. For example, in accounting for the Qi River:

> The Qi River flows out of the west of Dahao Mountain in Longji County. It passes through the south of Neihuang County and is known as the white ditch. It flows through the northeast of the county and joins the Huan River. The course of the river also runs northeast to the north of Guantao County, on to the west of Qingyuan County, on to the east of Guangzong County as the Qing River, and on to the west of Dongwucheng County. It then curves north to the east of Guangchuan County and then east to the south of Xiu County. It continues northeast to the west of Guang County, on to the west of Pi County, on to west of Fuyang County, on to the north of Huyi, on to the south of Xiangyi, and finally flows into the sea after Piaoyuyi in the northeast [38].

This is a very clear description of where the Qi River begins its entry into the sea and where it converges with its tributaries. The account of rivers in the *Shuijing* is not only rich in content, but also supports the approach of "tracing the history of a place through the records documenting rivers", marking a new development in terrestrial hydrography in ancient China.

Guo Pu of the Eastern Jin dynasty [39] and Li Daoyuan of the Northern Wei dynasty both annotated *Shuijing* following its publication. After the Tang dynasty, Guo Pu's version was lost, while Li Daoyuan's *Shuijing zhu* was disseminated. The modern version contains only 123 articles.

(2) *Shuijing zhu*: an encyclopaedia of terrestrial hydrology in ancient China

Li Daoyuan (Northern Wei dynasty?—27AD) had been extremely interested in geographical books and natural landscapes since his childhood. As an adult, Li led many expeditions when he was appointed as Emperor Xiaowen's attendant and to a position as a local official. His footprints spread across the south of the Great Wall and the north of the Huai River area [40]. He read extensively and collected detailed information about people's deeds, stone inscriptions, local maps and scriptures, etc. [41]. In the reigns of Yanchang and Zhengguang (515–524 AD) [42], Li completed the *Shuijing zhu*, a geographical masterpiece communicating unprecedentedly diverse knowledge of terrestrial hydrology. It comprised 40 volumes (of which five volumes were lost in the Song dynasty) and over 300,000 words in total (Fig. 5.8). Based on the then territory of the Western Han dynasty, the *Shuijing zhu* reviews areas both at home and abroad, incorporating findings in the fields of physical and human geography.

① Accounts of river hydrology

Before the Wei and Jin dynasties, the records of hydrology mostly focused on key rivers and their valleys, the descriptions of which remained rather simple. The *Shuijing zhu*, however, recorded the river hydrology of the whole country for the first time and included 1,252 rivers [43]. After Li Daoyuan, the 10-volume *Shan shuijing* ("Abbreviated Water Classic") by Li Jifu and the three-volume *Bu shuijing* ("Supplementary correction of the Water Classic") by Jin Caigui were unfortunately lost; other studies of national rivers and watercourses that are available today include the *Jin shuiijng* ("Modern Water Classic") by Huang Zongxi in the early Qing dynasty and the 28 volumes of the *Shuidao tigang* ("Outline of Watercourses") by Qi Zhaonan during the Qianlong reign. The latter is extensive but only shows marked development in bodies of water in the south of the Yangzi River area [44].

The coverage of rivers in the *Shuijing zhu* starts from Anzhou (today's Longhua in Hebei) in the north, to Nanjun (today's central Vietnam) in the south, to the sea in the east and to the west of India; the river basins mentioned include the Luan River, the Hai River, the Yellow River, the rivers of the Shandong Peninsula, the Huai River, the Yangzi River, the Pearl River, the Tarim River, the Yuan River and up to the Red River basins in China, and the Indus and Ganges basins.

Details of each river include not only its source and destination, but also its mainstem, tributaries, valley widths, riverbed depths, seasonal changes in water volume and level, sand volume, glacial stage, as well as the river travelling through volcanoes, waterfalls, rapids, lakes, etc.

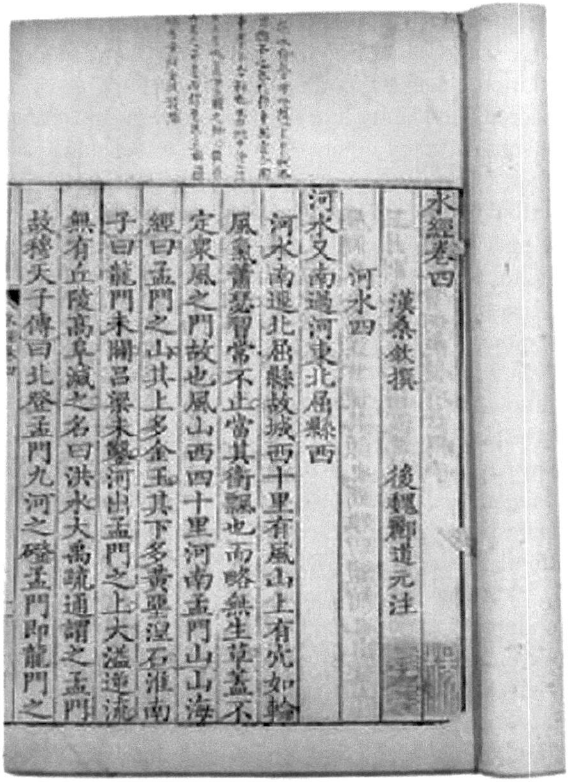
水經卷四
漢桑欽撰　後魏酈道元注
河水四
河水又南過河東北屈縣西
河水南逕北屈縣故城西十里有風山上有穴如輪
風氣蕭瑟習常不止當其衝飄也而略無生草蓋不
定眾風之門故也風山西四十里河南孟門山山海
經曰孟門之山其上多金玉其下多黃堊涅石淮南
子曰龍門未闢呂梁未鑿河出孟門之上大溢逆流
無有丘陵高阜滅之名曰洪水大禹疏通謂之孟門
故穆天子傳曰北登孟門九河之隥孟門即龍門之

Fig. 5.8 A facsimile of a page entitled *Shuijing juansi* ("水经卷四", the fourth volume of Water Classic) from the *Shuijing zhu*

Volume One "River Water" sets the standards for classifying and naming rivers at all levels according to their stem and tributary relationships, length and size, and whether they flow alone into the sea or merge into a large river:

> Water can be measured by size or distance; the water out of the mountains and into the sea is termed Jingshui; [45] into the large river and the sea it is termed Zhishui; out of the ditch, flowing into the large river and the sea it is termed Chuanshui.

These river classification and naming standards appeared in the earliest Chinese hydrographic work *Guanzi: Dudi* during the pre-Qin period, but Li Daoyuan only draws on three types of water bodies among the five mentioned in the *Guanzi*, and they become the style and norms for recording water systems in the *Shuijing zhu.*

The descriptions of the river sources in *Shuijing zhu* are vivid and extensive, and they also closely capture the physical and geographical characteristics of the origin of each river. For example, Volume 17 states that "the Wei River runs from the Heishui Gorge to the Cen Gorge, where eleven rivers from the north and south flow into it. Connecting the hot springs in Pingxiang County, its north has thermal water. Its northeast branch passes through the south of Pingxiang County".

The *Shuijing zhu* sets out the details the courses of some rivers, such as in Volume 33 "Jiangshui": "The (Min) River starts from Tianpengque and flows eastwards through Wenguan, passing through the north of Didao County… from the Baima Ridge over 20 li to Longhe, and 80 li to Canling County, and 60 li south to Shijing, and more than 60 li northwards; after a hundred or so *bu*, westwards more than one hundred and twenty li to the former Wenshan County, after over 200 steps, towards the southwest for about 100 li where, on reaching Shiban, the river becomes slightly larger." The description of the Min River from Tianpengque is given section by section along its course. It not only includes the locations the river passes through, but also records the length and width of the river, which have proved to be valuable data in the study of ancient physical geography.

Seasonal changes in the volume of water are also discussed in the *Shuijing zhu*. For example, Volume 23 "Xiashui" (Xia River) mentions that, as a tributary of the Han River, Xiashui is "dry in winter and lively in summer", and Volume Four "Heshui" (Yellow River) records that the Yellow River's tributary, Jiao River, is "a dry one as the water stops in winter and flows in summer". Many accounts of the rivers' seasonal changes are available, such as in Volume 21 "Weishui" (Wei River) where it is mentioned that "the east of Ru River joins the Zhangmo Spring, the water level increases at the north mound, and when the river is full spate during the spring and summer, it flows to the Ru River in the south". Volume Five "Heshui" records when the river runs dry and when it is fast-flowing and the specific water level of the flooding: "In the east is the Bailu Abyss that measures 300 *bu* from north to south, over a thousand *bu* east to west, and more than three *zhang* deep. Its water is clear in winter and turbid in summer, accumulated but not flowing. Floods in summer can reach a depth of five *zhang*."

Furthermore, the *Shuijing* records the periods of ice cover of certain northern rivers. Volume Five "Heshui" states that "the imperial government sets up ice chambers at Sifu (referring to Shouyang Mountain, in the northwest of present-day Yanshi County, Henan province), and the chamber contains ice wells". Quoting the *Zuozhuan* ("Zuo's Commentary"): "When the sun's orbit tilts between the Emptiness and the Roof, it is time to store ice. In December, ice can often be collected at the mouths of rivers, in the corners of gorges and in north-facing shadows." Volume One "Heshui" describes the ice thickness of the Mengjin River section: "according to the *Shuzheng ji*, the Mengjin and Hejin rivers remain constantly turbid, narrower than the Yangzi River, but broader than the Huai and Ji rivers. In the coldest period, the ice can be up to several *zhang* thick. When the ice begins to melt, no carriage dares to cross until people see foxes crossing over."

The *Shuijing zhu* is equally rich in its accounts of river gorges. There are nearly 300 gorges noted in the book, such as the Mengmen, Longmen and Sanmen gorges of the Yellow River, Yique Gorge of the Luo River, the Three Gorges of the Yangzi River, Gaoyao Gorge of the Pearl River (Volume 37 "Yinshui"), and Kongleng Gorge of the Xiang River (Volume 37 "Xiangshui"). Some important gorges have more detailed descriptions, including in Volume Four "Heshui" where it is said that "the river flows out of Mengmen Mountain higher up, an excessive flow rate causes a counterflow, and there are no hills or high mounds … the Mengmen gorge is at the upper mouth of the Longmen, a truly dangerous place in the river".

② Accounts of lake hydrology

Geographical information on lakes dating from before the Northern Wei dynasty is relatively sparse. The *Shuijing zhu* was the first to provide a comprehensive record of more than 500 lakes [46]. Its coverage starts at the present-day Liao River basin in the east, continues to the present-day Pearl River basin in the south, west to the Xinjiang's internal flow area, north to Inner Mongolia and other places, and even as far as Tianzhu (present-day India), Linyi (present-day southern Vietnam) and other areas outside the border.

Various types of lakes are documented in the *Shuijing zhu*, including endorheic lakes (Volume One "Hehui", Pu Changhai Lake), exhorheic lakes (Volume 37 "Yeyushui", Yeyuze Lake), anthropogenic lakes (Volume 40 "Jianjiang shui", Chang Lake), seasonal lakes (Volume Five "Heshui", Machangrong Lake) and lagoons (Volume 36 "Wenshui", Lurongpu Lake) and so on.

The book pays special attention to the size, colour, location and other hydrological characteristics of the lakes. As mentioned in Volume 40 "Jianjiang shui": "The stream in the northeast of Zhejiang meets Chang Lake which is five li wide and 130 li from east to west. Along the lake there are 69 water gates which can be opened to irrigate the thousands of fields beneath, and the lake stretches to the Yangzi River in the north." Zuo Lake "turns red, shimmering like cinnabar. It is fed by the Puyang River and it has an outlet which flows into the Zhe River". Volume 29 "Mianshui" records that "Taihu Lake has a surface area of 500 square metres". The process of paludification in certain lakes is also noted. Volume 22 "Zhenshui zhu" states that "The (Putian) Lake is in the west of Zhongmou County (east of present-day Zhongmou County, Henan province); it borders the Great Wall in the west and ends at Guandu in the east, and its northern end is connected to the Qushui River. The distance from east to west measures about forty li, and from north to south it is about 20 li. There are 24 pools, each with a name, and the waterways connect with one another". Before the Han dynasty, it had been a large lake in the Central Plains but, during Li Daoyuan's time, it had become fragmented into 24 pools through paludification.

The *Shuijing zhu* also attaches great importance to the economic significance of the lakes. For example, Volume 11 "Koushui zhu" describes the extensive use of Yangchengdian Lake saying that "it extends to the east through Yangcheng County where it is reduced to pools and swamps. When the water rises, these wetlands teem with bulrushes and bamboos, as well as water chestnuts and lotus roots. Many children and teenagers are attracted to harvest them by boat, singing in praise of the genial spring".

③ Accounts of waterfalls

Before the Northern Wei dynasty, details of waterfalls in geographical literature were rather limited, but the *Shuijing zhu* fills the gap admirably [47], covering a wide range of river basins including the Yellow River, the Huai River, the Yangzi River and the Pearl River. Diverse names are used to refer to the waterfalls in the book. Occasionally the modern generic term "waterfall" is used, e.g. Volume Five "Heshui" mentions that "on the (Shicheng) mountain top, springs gush and waterfalls roar";

the description in Volume 13 says that "the Luo River leaves the mountain in the south, rushing down from the ten *zhang* high cliffs to the abyss with a loud noise. This section is called Luomahong, the torrents of which are similar to the Mengmen water". Other terms for a waterfall often follow the description of their essential features, e.g. *xuanshui* (hanging water), *xuanliu* (hanging flow), *xuanhong* (hanging flood), *xuanjian* (hanging ravine), feiquan (flying spring), feibo (flying wave), etc.

The height was mentioned in the case of nearly half of the waterfalls in the *Shuijing zhu*: "its height reaches more than 200 *zhang*, looking like a hanging cloud" (Volume 40 "Jianshui"); "seven or eight *chi* away from the flat ground there hangs the spring water" (Volume 27 "Mianshui").

The geomorphic phenomenon of the "pothole" that often co-exists with waterfalls abounds in *Shuijing zhu*. Referred to as *tanzhu* (islet) and *jing* (well), it is a round hole formed by water impact and eddies, as described in Volume 11 "Koushui" and Volume 39 "Lujiang shui".

Shuijing zhu describes the phenomenon of multiple waterfalls occurring along the same fall-making band in several rivers. Volume 20 "Yangshui" states that "the Xihan River flows southwest…the Gaowanggu River joins it on the righthand side, further west there is the Xi Stream, and still further west is the Huanghuagu River; they all originate in Bei Mountain and flow southwards into the Xihan River with blowing sprays". This describes the three tributaries of the Xihan River along a north–south axis, and several waterwalls appearing along the east–west fall-maker line.

The accurate account of the locations and heights of the waterfalls meant that the *Shuijing zhu* provided extremely valuable information enabling future studies to calculate the speed of river erosion from that baseline. China's celebrated historical geographer Shi Nianhai compared the Mengmen Waterfalls (now Hukou Waterfalls) in the *Shuijing zhu* with its location as marked in the *Yuanhe junxian tuzhi* ("Illustrated Gazetteer of the Yuanhe Prefectures and Counties") during the Tang dynasty, and he reached the conclusion that from the 3rd year of the Xiaochang reign of the Northern Wei to the 8th year of the Yuanhe reign (527–813 AD) in the Tang dynasty, the waterfall had retreated by 51 m per year on average [48].

④ Accounts of hot springs

The *Shuijing zhu* recorded over 30 hot spring locations in Shandong, Liaodong, Shaanxi, Yunnan, Guizhou, Huaiyang, Fujian and Guangdong, all of which are mainly found in the Taihang Mountains and the Shaanxi-Gansu regions [49].

The water temperature of hot springs in the *Shuijing zhu* is described in detail and they can be roughly divided into: 1) springs with a cooler water temperature, which are mostly described as "warm during a cold winter" (Volume Four "Heshui", Shichengshan Spring) or "warm in winter and cold (cool) in summer" (Volume 37 "Yishui", Lan Spring in the Wuzhong Mountain; Volume 39 "Leishui", Zhenggong Spring; Volume 40 "Jianjiang shui", Houji Spring); 2) springs with a lower water temperature, often described as "a consistent temperature in winter and summer" (Volume 31 "Zhishui", Zishantang Valley) and "warm in summer and hot in winter" (Volume 37 "Yishui"); 3) hot springs with a relatively high water temperature are described as yan (hot), zhuo (scalding) and tang (boiling soup); and 4) springs with a high temperature and excessively hot springs are recorded by making comparisons with the temperature for cooking food, such as "its heat can be used to cook chicken" (the hot spring in Xinyang County mentioned in Volume 31 "Yunshui").

The records of hot springs in the *Shuijing zhu* cover a wide range of fields [50]. In terms of industrial and agricultural production, hot springs were mainly used for boiling salt and irrigation. For example, Volume 39 "Leishui" mentions that "the hot springs are located along the Bian County borders, northwest of Chen County (now part of Hunan); located along their banks are several thousand *mu* of fields that rely on the springs for irrigation. These are often planted in December and harvested the following March. Water cools after passing here, hence there are no rice seedlings further on. Wherever the warm water manages to flow, rice is harvested three times a year". The use of hot springs shortened the growth period of crops and increased crop yields. In healthcare, water from the hot springs was widely applied to treat skin diseases and rheumatism and was effective in keeping people healthy through drinking and bathing in the water. For example, Volume 11 "Koushui" suggests that "the hot spring comes from the Xuan Valley in the northwest where the spring water at its source is as hot as boiling soup and can be used for bathing to heal many diseases". A description in "Yishui" explains that "the Yishui (Yi River) flows further to the east and joins the three hot springs; hot springs appear on the south and north banks of the Daxi Stream. Misty air rises above the springs and bathing in the spring can heal wounds, pruritus and hundreds of other diseases". Volume 29 clearly points out that the hot springs in Li Mountain can "burn and wash any wounds". In addition, hot springs proved to be popular in tourism and religious activities.

The *Shuijing* was the product and symbol of hydraulic development in ancient China. According to Li Daoyuan's comment in the *Preface to the Shuijing zhu*, the book described rivers from their source, with the tributaries, river forks and so on down to their estuary mouth, thus establishing the concept of a "water system" that had a profound influence on Chinese hydrography.

5.3.3 Rapid Developments in Geomorphology

During the period of the Three Kingdoms, Wei, Jin and Southern and Northern dynasties, the Yangzi River basin, the Min River basin and the Pearl River basin were developed one after another, which greatly expanded people's horizons. The complex and diverse landscapes in these areas attracted many travellers and scholars, which led to great advances in people's understanding of landforms. In addition, the activities and travelogues of monks enriched the knowledge about western deserts and high mountain glaciers.

(1) Rapid Development of the understanding of karst topography

Before the Han dynasty, the accounts of karst topography were mainly based on northern China [51]. During the Three Kingdoms, Wei, Jin and Southern and Northern dynasties, one of the most important periods in the development of knowledge relating to karst landforms in ancient China, the understanding of karst landforms advanced significantly, and a large number of accounts of karst caves emerged (Table 5.2). These records covered almost all areas of karst landscapes in China; the accounts were rich in content and have left many wonderful descriptions of stalactites, caves, surface karst features and other aspects in particular. In addition, the causes and characteristics of karst landscapes were also being recognised.

The main reasons for the rapid understanding of karst topography in this period were as follows. (1) The geographical horizon had expanded from the north to the south. The south was the most widely distributed area of karst landforms in China. (2) There were active tourists and adventurers. The peculiar karst phenomenon attracted scholars and adventurers to move south. (3) There were also medical needs. Stalactites—limestone deposits in limestone caves—were considered medicine [52]. The consumption of stalactites as drugs naturally led to the exploration of limestone caves.

(2) Progress in the understanding of fluvial geomorphology

Before the Wei and Jin dynasties, there were many records about the influence of fluids on landforms. In the period of the Wei, Jin and Southern and Northern dynasties, the body of records of fluvial geomorphology was expanding and understanding had also made significant progress.

Regarding the erosion of rivers, Li Daoyuan in the Northern Wei dynasty had already recorded the erosion effect of river water on river valleys in Volume 4 of the *Shuijing zhu*: "As an ancient saying goes: water is not a chisel, but it can carve the stone. Now I believe it!".

As regards the topography of river deposits, Sheng Hongzhi, a scholar of the South Liu Song dynasty, recorded the deposition of the Yangzi River in the form of sand bars: "There are ninety-nine islets in Zhijiang County (in Hubei province) … and suddenly a new islet emerged [53]."

There was already a record of floods causing landslides as early as in the *Book of Songs: Xiaoya.* In the Northern Wei dynasty, Volume 34 "Jiangshui" of the *Shuijing zhu* described the "New Avalanche Beach" formed by a landslide in the Wu Gorge: "The mountain collapsed in the 12th year of the Yongyuan reign of the Han dynasty and once again in the 2nd year of the Taiyuan reign of the Jin dynasty. When it collapsed, the water flowed back more than a hundred miles and raised high waves. Now, some of the stones on the beach are round, and some are square. Many like this were rolled down from the collapsed cliff, making the river even more turbulent, so it is called Xin Beng Tan (literally 'newly collapsed beach')."

(3) Progress in the understanding of structural landforms

Before the Wei and Jin dynasties, there were accounts of geological structures and landforms formed by crustal movements. In the period of the Wei, Jin and Southern and Northern dynasties, there were some new understandings of structural landform.

In terms of volcanic landforms, the term "volcano" already appeared in Guo Pu's *Shanhai jing tuzan* ("Eulogistic Picture of the Classic of Mountains and Seas")

Table 5.2 Records of karst topography during the three kingdoms, Wei, Jin and North and South dynasties

[Dynasty] author	Book title	Historical sources on the characteristics of underground karst geomorphic phenomena	Characteristics	Source
[Wu] Gu Qiqi	*Records of Loudi*	There are three caves in Dongting Mountain, and there are clear springs flowing out, except a cave in the northwest … has big bats like birds … the cave has soda straw and stalactites, which are accessible in cold weather but not in spring or summer	A record of the hydrology, deposits, creatures, climate and other phenomena in the cave	*Beitang shuchao* ("Notes of the Book of the North Hall") Vol 158
[Wei] Wu Pu	*Wupu Herbal*	Stalactite … born in the shady valley place under the cliff, resembling yellow-white milk, connected with each other in mid-air	The book correctly identified the causes and characteristics of stalactites	*Zhongyi guji zhengli congshu* ("Series of Ancient Books on Chinese Medicine"), 1987
[Jin] Zhang Bo	Wu's Records: Geographic Journals	There are cave mountains in Shi'an and Shi Yang, and there are caves in the mountains where stone stalactites are born		*Taiping yulan* ("Imperial Reader of the Taiping Era") Vol 987
[Jin] Zheng Jizhi	*Records of Dongyang*	In Longqiu Mountain, there are nine beautiful rocks with white surfaces. If you look at them from a distance, they all resemble lotus flowers. A hermit, Long Qiuchang, dwells among the rocks, hence the mountain's name. There are window-like caves in the peaks, with stone beds inside	Vivid descriptions of geomorphological phenomena in peaks and caves	*Hou hanshu* ("Book of the Later Han") Vol 32, annotated by Liu Zhao
		(In Jinhua) there is a stone bed on the west cliff of the north mountain, watered by running water, and at its side there is a stone field like a rice paddy cloud		*Taiping yulan* ("Imperial Reader of the Taiping Era") Vol 54

(continued)

Table 5.2 (continued)

[Dynasty] author	Book title	Historical sources on the characteristics of underground karst geomorphic phenomena	Characteristics	Source
[Jin] Yuan Shansong	*Records of Yidu Mountain*	Fifty li from the northwest of the salt water, there is a steep mountain, named Nanliu City. Climbing the mountain from the west side, I reached a cave. After walking a hundred li, a rock can be seen … it is also said that there is a stream in Cushan County (in the west of present-day Changyang County): water flows into a stone cave, which is the source of the stream	Descriptions of the connection between surface water and caves and underground rivers	*Yuding yuanjian han* ("Royal Decree of the Fountainhead") Vol 26
[Jin] Gu Kaizhi		Lou River has a stone cave south of Ma'an Mountain, more than ten feet high, allowing a dozen people to enter it		*Beitang shuchao* ("Notes of the Book of the North Hall") Vol 158
[Jin] Ge Hong	*Baopuzi neipian: Xianyao* ("He Who Holds to Simplicity: Inner Chapter: Immortal Medicine")	When "stone honey" grows in the cave of Shaoshi Mountain, there will be a deep valley in the mountain which people cannot cross over. If one tosses a stone into the valley, no sound will be heard for a long while. There are stone pillars rising to ten *zhang* outside, with stone above the pillars. The length of the diameter can measure ten *zhang*. The stalactite grows from the stone. Over a long time, a drop constantly falls there. It is like in a house with a leak after a shower of rain. However, the dense stalactites keep growing, even though the ceiling does not overflow	Puddles and stone stalactites ("stone honey") in cavernous terrain	*Taiping yulan* ("Imperial Reader of the Taiping Era") Vol 985
[Liang] Xiao Zikai	*Notes in Jian'an*	Tianji Mountain is 20 li south of Jiangle County. At the bottom of the mountain there is Baohua Cave, that is, where Chisongzi (the legendary rain master during the first Yan Emperor period) collected medicine. In the cave there are stone swallows, stone bats, stone rooms, stone pillars, and stone mortar, stone wells … above the streams there are springs drip onto them, and for those who reach the top of the mountain, it is like ascending into the blue sky	Multiple geomorphological phenomena in one cavern	*Taiping yulan* ("Imperial Reader of the Taiping Era") Vol 47

(continued)

Table 5.2 (continued)

[Dynasty] author	Book title	Historical sources on the characteristics of underground karst geomorphic phenomena	Characteristics	Source
[Southern] Wang Shaozhi	*Notes in Shixing*	There is a stone chamber in Zhongsu County (in modern day Hedongbao, northwest of Qingyuan County, Guangdong province) … inside there are hanging stones whose sound is as loud as a chime	Hanging stones are stalactites	*Congshu jicheng chubian* ("A Preliminary Compilation of the Integrated Series")
		There is a view of the gorge in Zhongsu County, with a chain of mountains and steep cliffs	Steep slopes of the limestone river valley	
		Dianyang Gorge is in the downstream section of the Liang River and Xian River. It is more than 20 li long, with high mountains and running rivers		
[Southern] Wang Shaozhi	*Notes in Shenjing*	Hejiayan is thirty li north of Xingyang County. There is a cave nearby, which has a very dark and narrow entrance. It used to be a place where people harvested stalactites		*Taiping yulan* ("Imperial Reader of the Taiping Era") Vol 54
Deng Deming (Jin)	*Notes in Nankang*	Tongtian Cave is located in the southwest of the county. There are multi-coloured stones on its walls, stone beds on its floor and stone pellets around		*Beitang shuchao* ("Notes of the Book of the North Hall") Vol 158
Sheng Hongzhi (Southern)	*Notes in Jinzhou*	Qin Mountain is in Fengsheng County (the capital is in the northeast of present-day Fuchuan County, Guangxi province) … Six or seven li into the mountain cave, there are stone pillars, stone drums, and stone pellets	Shi Danwan (stone pellets) are travertine deposits, now called stone beads	
		In Yangshan County (present-day southeast of Qinglian Town, north of the Lian River) Guangdong province), a stream of water flows out of the cliff cave. The water overflows ten times and dries up ten times, just like the tides following a timing	Multiple tidal springs in the cave	

(continued)

Table 5.2 (continued)

[Dynasty] author	Book title	Historical sources on the characteristics of underground karst geomorphic phenomena	Characteristics	Source
		In Henshan County, Yidu, there is a mountain which has a wind cave with an opening several *chi* wide, called the wind well, where the wind comes out in summer and enters in winter	The airflow conditions inside and outside the cave	
Li Daoyuan (Northern Wei)	*Commentary on the Water Classic*	The source of the River Yun is in Dahong Mountain, in the southeast of Caiyang County (present-day Caiyang, southeast of Zaoyang, Hubei province). At the foot of the mountain, there is a stone gate, with stacked cliffs on both sides which are extremely precipitous, hundreds of *chi* high. After entering the stone gate, there is a stalactite cave above which the bare cliff is as steep as a wall, a place off the beaten track. There are many stalactites formed like paste in the cave, and become narrower from the top of the cave downwards, looking like snow-white ice cones with dripping water. The cave is extremely deep, and no one has ever penetrated to the end (Vol 31 "River Yun")	Vivid and detailed descriptions of both surface and subsurface karst phenomena	*Siku quanshu* ("Complete Library of the Four Treasuries")
		The water originates in the Sacred River valley in the southwest of the county and flows southeast through the eastern end of the Dafangling Ridge. There is a stone cave at the foot of the mountain, the entrance to which faces northeast and is four or five *chi* high and wide, becoming higher after the entrance to the cave, stretching straight into the depth, with water in the cave (Vol 12 "Sacred River")		
		The Yi River flows eastwards along the northern foot of Kong Mountain. There are stalactite caves, which produce very beautiful stalactites … About a li into the cave, there is a shallow underground pool which people can wade through…there is another large hole at the top of the cave, which is wide open to the sky (Vol 11 "Yi River")		

during the Jin dynasty [54]. Wei Shou, a historian of the Northern Qi dynasty, gave an account of volcanic eruption in Yueban Kingdom (in the northwest of Lake Barkash) in the *Book of the Wei: Western Regions* [55]: "Yueban Kingdom, in the northwest of Wusun, 5,465 km from the Dai … the Kingdom has a volcano on the southern boundary, the stones beside the mountain are all scorched and melted, flowing dozens of miles before solidifying. This is sulfur, and people take it as medicine [56]." Wang Jiayin believes that it is a volcano in Ili: "It was erupted during the Northern Wei dynasty [57]."

Before the Qin and Han dynasties, there were many records about earthquake topography, but they were mostly expressed as "cracked valley" or "cracks on the ground". In the Wei, Jin, Southern and Northern dynasties period, the terrain formed by earthquakes was described in more detail in books such as *Songshu* and *Weishu*. In the *Weishu: Zhengling zhi*, an earthquake in the first year of the Yanchang reign (512 AD) made "mountains fall, the earth rift, and springs gush" in Hengzhou (governed by present-day Luoyang, Henan province).

In the Liang dynasty, Jiang Yan (444–505 AD) wrote the *Fugue of the Mountains on the River*, describing how the mountains were formed of multiple layers of hard rocks [58].

(4) Progress in the understanding of desert landforms

A desert was termed *liusha* (flowing sand) in ancient China. The *Yugong,* written during the pre-Qin period, recorded the desert in the northwest of China. After the Jin dynasty, scholars knew more about desert topography and the phenomenon of booming sands and introduced the idea of a wider distribution of deserts.

① Desert terrain and the phenomenon of booming sands

The *Taikang diji* mentions that "in the Juyan County, the quicksand is shaped like the fifth and sixth day of the moon" [59], suggesting that that the desert terrain is mostly crescent-shaped sand dunes. In the Northern Wei dynasty, Li Daoyuan, in Volume 40 "Mountains and Waters in Yugong" of the *Shuijing zhu,* also proposed that "the shape of the so-called quicksand in the *Shangshu* is like the moon on the fifth day of the month".

The *Stories of the Three Regions of Qin* (by Mr Xin) outlines the phenomenon of the booming sands of the desert in the present-day Hexi Corridor Region of Gansu province: "There is a sand dune called *shajiao* in the Hexi area, even more steep and dangerous than stone mountains. The sands are thick and yellow, like dry grains. At the south of the dunes, there is a spring with a long history and clear water, and people call it a 'sand well'. Stepping on the sand, one can hear the booming and burping of sands, vibrating under the feet like the sound of horns and drums [60]." Liu Jingshu in the Liu dynasty also described booming sands as he explained the formation of the dunes: "There is a sand mountain in the west of Liangzhou (contemporary Wuwei, Gansu province) … the wind blew the sands into a dune from which the sound of drums and horns can be heard from time to time [61]." The exact location of the mountain was noted by Li Daoyuan of the Northern Wei dynasty in Volume 40 of the *Shuijing zhu*: "Mingsha Mountain is five kilometres south of Dunhuang city."

② A wide range of deserts

According to the *Faxian zhuan* ("Biography of Fa Xian") which appeared in the Eastern Jin dynasty, the desert of Khotan (present-day Hetian County, Xinjiang) was as follows: "There is no resident in the south west, and it is difficult to walk on the sand." Liang Huijiao (497–554 AD) commented in his work *Gaoseng zhuan* ("Biography of the Monks") that "he (Monk Fa Yong) travelled through the desert and reached Gaochang County (in the southeast of present-day Turpan city, Xinjiang)". The desert through which the two monks Fa Xian and Fa Yong travelled westward was the Taklamakan Desert in the south of Xinjiang province.

Both the *Shuijing* of the Three Kingdoms era and Volume 40 "Mountains and Waters in Yugong" of *Shuijing zhu* mention the "sandy land" in the northeast of Zhangye Juyan County (located in the southeast of Ejina Banner), which is the Ejina Desert of Inner Mongolia.

In the Northern Wei dynasty, Li Daoyuan noted the present-day Erdos Desert in Volume 3 "Heshui" of the *Shuijing zhu.*

(5) Progress in the understanding of glacial landforms

Before the Han dynasty, glacial mountains were termed *lingshan.* In the period of the Wei, Jin and Southern and Northern dynasties, the fact that many travellers trekked through the glacial mountains during the westward journey to study Buddhism meant that they knew and wrote more about glacier topography.

In the Northern Wei dynasty, Yang Xiaozhi's *Luoyang qielanji* recorded the snow mountains in Bohe (located in the Wakhan Valley east of the border between present-day Tajikistan and Afghanistan): "...there are big snow-capped mountains, which melt in the morning and freeze in the evening, looking like jade peaks [62]."

Crossing the Congling Mountains (today the Pamir Plateau and the Karakorum Mountains), the Eastern Jin Monk Fa Xian, on his trip to the West Regions, kept a travelogue [63] of the glacial landscape, documenting both the gravels and rock outcrops formed by the weathering effect of ice cracking in the high mountains, and the steep gorges formed by the ice and snow melt water with their extremely strong erosion force.

(6) Progress in the understanding of coastal landforms

Coral terrain is the most characteristic biological coastal landform. Kang Tai of the Wu during the Three Kingdoms described coral reefs in the present-day South China Sea in the *Funan ji* [64]. He also commented in the *Waiguo zazhuan* that "in the South Sea during the Qin dynasty, there were coral shoals and parts of these were often fished up by people with metal nets" [65].

In the Wei and Jin dynasties, accounts of the estuary delta were abundant, including in Volume 10 "Qingzhang shui" of the *Shuijing zhu* by Li Daoyuan: "The Qingzhang River torrents its way to the east of the sea." Volume 5 "Heshui" described in great detail the estuary delta of the Luo River, one of the downstream tributaries of the Yellow River.

5.3.4 Advances in Biogeography

During the period of the Wei, Jin and Southern and Northern dynasties, biogeography made tremendous progress, with many biogeographic treatises and relevant literature arriving on the scene.

(1) On regional phytogeography: *Nanfang caomu zhuang* by Ji Han

In the first year of Yong'an of the Western Jin dynasty (304 AD), Ji Han (263–306) compiled a regional phytogeographical work *Nanfang caomu zhuang* ("Records of Plants and Trees in the Southern Region") based on his observation and literature review. The book was included in the *Siku quanshu* ("Complete Library of the Four Treasuries").

As the earliest book of regional phytogeography in the world, *Nanfang caomu zhuang* is the first to include more than 80 kinds of plants in places such as the Lingnan area, Hepu County, the South Sea and northern Vietnam. The geographical distribution of the plants is mainly as follows: first, in multiple counties; second, in a certain place; and third, in a certain environment, etc.

In the book, while recording the plants in Lingnan, it also attaches great importance to the differences in the distribution of plants in different parts of Lingnan, such as plants that are unique to a certain place. The book also notes the variations in a certain plant in different areas. The book also proposes for the first time that the Nanling Mountains are a boundary line for the distribution of plants in China. The Nanling Mountains serve as the dividing line between the south and central subtropical zones. The south and the north sides of the mountains are different not only in climate but also in biological distribution.

(2) On biogeographic knowledge: *Shuijing zhu* by Li Daoyuan

① Phytogeography in the *Shuijing zhu* [66]

There are at least 140 species of plants in the *Shuijing zhu*, which is very rich in its diversity, from conifers to broadleaf plants, from Chinese peach and litchi to overseas Sal and Bo tree, etc. The book describes the latitudinal zonality of vegetation distribution in China and neighbouring countries. In Volume 36 "Wenshui", it describes the tropical rainforest evergreen broad-leaved forest in Linyi country (present-day southern Vietnam), while Volume Two records temperate forests in Jincheng County (in Gansu province).

The book also introduces the longitude zonality of vegetation distribution, for example: Volume 22 records the swampy land in the eastern part of China's humid zone; Volume Two records the grassland and desert in the dry area of Loulan in western China (today's Lop Nor): "The area is sandy and saline with few paddies, and food is supplied by the neighbouring country, which produces jade, reeds, tamarisk, paulownia, white grass and so on. Its capital is on the eastern border, in front of the White Dragon Pile, and lacks water and grass."

The book further describes the vertical zonality of vegetation distribution in China, for example, Volume 4 records that on top of Fushan Mountain, "there is little grass and few trees for five or six li away from the mountain top". In addition, Volume 9 "Qishui" records the change in the bamboo in the Qishui watershed from "produced millions of arrows by bamboo" in the Later Han dynasty to "nothing left" in the Northern Wei dynasty.

② Zoogeography in *Shuijing zhu* [67]

There are more than 100 species of animals in the *Shuijing zhu*, including vertebrates, arthropods and molluscs. The book records the regional distribution of animals. For instance, in Volume 33 "Jiangshui", it is written that "there are many apes in the gorge, but they are not seen on the northern bank. Similarly, someone moved apes to the northern mountain, but their sounds cannot be heard any more. Probably apes cannot live in the north". The book also notes the seasonality of animal activities. In Volume 37 "Yeyu River", it is written that "Diaoniao Mountain is eighty li northwest of Yeyu County. It is a place where thousands of birds flock together, whining and chirping. They come every July and August and stay for 16 or 17 days, six flocks each year".

(3) On Bambusoideae of the family Poaceae: *Zhupu* by Dai Kaizhi

Published in the late fifth century, the book was based on field survey and previous studies on bamboo.

With a length of about 3000 words, the *Zhupu* ("Book of Bamboo") begins with a general discussion on the classification, location, morphological characteristics, habitat and geographical distribution of bamboos, followed by descriptions of each kind of bamboo. This is the first monograph in ancient China to comprehensively study the subfamily Bambusoideae of the family Poaceae. The contributions of this book to phytogeography are mainly threefold:

① The book clarified the natural taxonomic position of bamboo. Before the Jin dynasty, bamboo was regarded as a grass, but Dai Kaizhi began to notice the difference between bamboo and grasses: "Among the plants, there is something called bamboo. It is neither hard nor soft, neither grass nor wood", "Grass, wood and bamboo are different plants, just like fish, birds and breasts are different animals" (cited from *Siku quanshu*).
② The book elucidated that there are differences in the growing environment of various types of bamboos. The author discovered that some bamboos grow on the banks of streams and rivers, while others grow on hills, foothills or slopes, and on the hillocks. Therefore, it has been concluded that "some thrive near sand and water, while others live on rocks".
③ The author set out the arguments for there being geographical distribution of bamboo and the reasons for this for the first time: bamboos are sensitive to temperature. Bamboos "cannot survive the cold", and therefore they thrive in the Nanling Mountains where the weather is warm. The book also elaborates on the distribution of various bamboos, including Dendrocalamus brandishii (Suma

bamboo). The author roughly takes the Qinling Mountains and the Huai River as the borders of bamboo distribution in China [68]

(4) Biogeographical knowledge in other literature

Ge Hong's monumental work, the *Baopuzi: neipian* clearly points out the difference between pine growing in dry uplands and in wet water: "Guan Song that grows in uplands has short roots and a sweet odour; while those growing in water have slim and yellow leaves, long roots and a bad smell."

The *Shudu fu* ("Ode to the Capital of the Shu") by Zuo Si described vertical differences in that coniferous plants are distributed at a higher height than broad-leaved forests on the same mountain.

In Volume 10 of the *Qimin yaoshu* ("Essential Techniques for the Welfare of the People"), it was recorded that betel nuts can only grow in tropical areas: "(betel nuts) will not thrive in cold weather, and thus cannot grow in the north but only in Hainan Island." There is a similar description in the *Shuijing zhu.*

5.4 Advances in Borderland and Extraterritorial Geography

Borderland and extra-territorial geography sprang up in the pre-Qin period and made enormous strides in the period of the Wei, Jin and Southern and Northern dynasties. China's communication with the outside world was not interrupted by the long period of division, but on the contrary, due to the ongoing commerce and trade and the spread of Buddhism, the Sino-foreign exchanges in some areas became even more active, thus promoting the development of the borderlands and of geography in these areas. During this period, the scope of extraterritorial geography did not change much when compared with that of the Western Han dynasty, but people's understanding of certain of its aspects was greatly improved, with the geographical conditions of quite a few countries and regions recorded in great detail for the first time.

5.4.1 East Asia, Southeast Asia and South Asia

(1) Geography of the Japanese Islands

In the *Shiji* and *Hanshu*, accounts of the territory to the east of the Chinese mainland are only included in the section "Collected Biographies of Korea". From the reign of Emperor Wu of the Western Han dynasty, more than 30 small countries in the Japanese archipelago had dealings with the empire [69]. During the Wei and Jin dynasties, Japan had frequent contact with China [70], even during the war years. In the course of the long-term exchanges between the two countries, the empire gradually developed a sound grasp of the geography of the Japanese islands. The section "Woren zhuan"

(Biography of the Japanese) appeared in the *Sanguo zhi: Weizhi: Wuwan xianbei dongyi zhuan: Woren zhuan* ("History of the Three Kingdoms: Biography of the Wei: Wuwan, Xianbei and Eastern Ethnic Groups: Biography of the Japanese") compiled by Chen Shou of the Jin dynasty (233–297 AD). This was the first document in China's official history to describe the geographical and other conditions of the Japanese islands. It not only recorded the location, climate, minerals, plants and customs of Japan, but also the directions, mileage, number of households, flora and mountains of the major countries to the southeast of Japan in the third century.

It is commonly believed that the *Sanguo zhi* by Chen Shou was largely based upon the *Weilue* ("Outline of Wei") by Yu Huan during the Taikang reign (280–289 AD) in the Jin dynasty [71]. The fact that the latter has been lost makes the former even more precious: first, there is no other account of the Japanese islands in the third century; second, all dynastic records of Japan that followed were based on this piece of writing either adding or subtracting some of the content, including the *Hou hanshu: wozhuan* ("The Book of the Later Han: Commentaries on Japan").

(2) The geographical writing of the Funan kingdom

Since the Han and Wei dynasties, China had developed a greater geographical understanding of the Funan Kingdom through its engagement with Southeast Asian countries [72]. Funan, also known as Banan, with territory situated in the area of present-day Cambodia and southern Vietnam, was annihilated by the northern vassal state Chenla in the middle of the seventh century. The *Yiwu zhi* by Yang Fu during the Han dynasty had already recorded the name "Funan" [73]. The start of diplomatic relations between Funan and China dates from around the Three Kingdoms. During the Huanglong reign of Eastern Wu (229–231 AD), Zhu Ying and Kang Tai stayed in Funan for many years when they visited Southeast Asia [74]. After returning to China, Kang Tai completed the earliest geographical work with the name of "Funan" in the title—*Funan ji* ("Records of Funan"). In the *Sanguo zhi: Wushu*, it can also be read that "in December of the 6th year (243 AD), King Fan Zhan dispatched an envoy to present a musical performance and local specialties to the imperial government".

The *Sanguo zhi* was the earliest official chronicle to record Funan. By the Jin dynasty, the two countries were in frequent contact with each other. The section "Record of Funan" is included in the *Jinshu: Siyi zhuan* ("Book of the Jin: Four Foreign Countries"). It begins with Funan's geographical features: "Located in a large sea gulf (referring to the Gulf of Siam), Funan is three thousand li away from Linyi. The land is three thousand li wide and contains cities, capitals and palaces. The people have darker skin...and make a living from agriculture." Similar statements also appear in the *Book of the Southern Qi: Records of Southeast Foreign Countries: Funan*. Later, in the *Book of the Liang: Records of Foreign Countries: Funan*, there is the earliest systematic description of Funan:

> Funan is located in the south of the Rilan Prefecture (its capital is in the centre of today's Vietnam) … The city is 500 li from the sea and is crossed by a large river (i.e. the Mekong River), ten li wide, which flows northwest and joins the sea in the east. The country is more than 3,000 li wide with mostly low-lying land. Its climate and customs are similar to those of Linyi. It produces gold, silver, copper, tin, eaglewood, ivory, jadeite and five-colour parrots.

> Three thousand li away from its southern boundary lies Dunxun (in present-day lower Tenasserim, Myanmar) which is a thousand li wide and ten li away from the sea. Its eastern border is shared with Jiaozhou (its capital is around present-day Beining in Vietnam). Its western border runs along Tianzhu (present-day India) and Anxi (Parthia, at the southwestern Caspian Sea) …
>
> Outside Dunxun and among the sea islands, there also lies Pisai 8,000 li away from Funan…
>
> The eastern border of Funan is formed by the Dazhang Sea (the present-day South China Sea) where the large islet of Zhubo (the old name for the island of Java in Indonesia) was located in the middle. To its east lies Wuma Island (an unidentified place), and it was possible to reach the Daziran islands (the East Indies) which lay over 1,000 li further east.

It can be seen that by the Southern and Northern dynasties the official historical geography of the area of the present-day Cambodia and southern Vietnam region was largely available and accurate.

(3) The geographical description of Shiziguo ("the Lion Country")

In the Eastern Jin dynasty, Shiziguo (present-day Sri Lanka) began to develop diplomatic relations with China. According to Volume 54 of the *Book of the Liang*: "the first trip of an ambassador from Shiziguo to China to pay tribute took 10 years." Exchanges between the two countries continued thereafter. The geographical conditions of Shiziguo became clearer as more monks journeyed to the west during the Jin dynasty. The *Book of the Song* first mentioned Shiziguo without providing any geographical details. From the seventh to the eighth year of the Yixi reign (411–412 AD), a monk called Fa Xian stayed on the island for two years on his way back from obtaining scriptures from Tianzhu (India). In his travelogue *Foguo ji*, the geography of Shiziguo was introduced for the first time:

> The country is on a large island, measuring 50 *youyan* (a distance measuring unit used in ancient India) from east to west and 30 *youyan* from north to south. Hundreds of small islets surround it, each one at a distance of 10, 20 or 200 li. They all belong to the big island and produce jewellery and pearls.
>
> …The country has no native residents…because of trade and the business people…it has become a big country. It is a suitable place to live as there is no difference between winter and summer and the flora is always luxuriant. Seeds can be sown at any time depending on people's availability.

Apart from the description of the width and length of Shiziguo which is incorrect in the above quotation (it should be longer from north to south and shorter from east to west), other records remain largely realistic. Afterwards, the *Book of the Liang: Record of Shiziguo* reported the geography of this country for the first time and its text was almost entirely taken from the *Foguo ji*.

5.4.2 Detailed Examinations of the Geography of the Western Regions

Since the introduction of Buddhism to China in the early Eastern Han dynasty, a large number of western monks had come to the East to translate the scriptures and teachings. Meanwhile, many local adherents were not satisfied with the materials brought by foreign monks and undertook pilgrimages to the West. Therefore, the frequent westward journeys in the late Jin and early Song dynasties greatly advanced understanding of the geography of the Western Regions and Central Asia.

According to Fang Hao's statistics [75], nearly 150 followers travelled west to seek the Dharma from the time of the Western Jin to the Southern and Northern dynasties. Zhu Shixing (?-282 AD) was the earliest believer who made such a journey, but he only reached Khotan (modern Hetian in Xinjian). In 260 AD, the 5th year of the Ganlu reign of the Wei dynasty, he set out from Chang'an, Yongzhou (present-day Chang'an, Shaanxi), crossed the western deserts and arrived in Khotan after a journey of ten thousand li [76]. In 518 AD, in the later Wei, the Empress Dowager Hu ordered the Monk Hui Sheng to depart from the Chongli Temple in Luoyang to seek Buddhist sutras in the western regions. He was accompanied by a non-believer, Song Yun, a native of Dunhuang, both of whom were received by the King of India. This trip was recorded in the *Jiaji* by Song Yun and in the *Xingji* by Monk Hui Sheng.

According to Tang Yongtong [77], works on the history and geography of the western territories during this period included: *Fotudiao zhuan* by Fo Tu Diao, *Xiyu zhi* ("Record of the Western Territories"), *Xiyu tu* ("Maps of the Western Territories") and *Sihai baichuan shuiyuan ji* ("Record of Sources of Rivers and Seas") by Dao An (314–385 AD), *Waiguo shi* ("Matters of Foreign Countries") by Zhi Seng Zai, the five-volume *Waiguo zhuan* ("Records of Foreign Countries") by Tan Jing, *Funan ji* by Zhu Zhi from the Liu Song period, *Liguo zhuan* by Fa Sheng, *Foguo ji* by Zhu Fa Wei, *Bieji* by Fa Xian of the Qi, and *Shijieji* ("Records of the World") by Seng You of the Qi (445–518 AD). The best-known wayfarer monk was Fa Xian.

(1) Fa Xian and *Foguo ji*

In 399 AD, Fa Xian and other monks set out from Chang'an (present-day Xi'an), passed through Long (present-day Gansu Long County), Zhangye and Dunhuang, crossed the Taklamakan Desert and Khotan over the Congling Ridge (the present-day Pamir Mountains) in the west, and reached Bafu, the capital of Magadha (present-day Patna of Bihar state in northern India) via Jiecha (Taxkorgan), Fulousha (Purusapura, present-day Peshawar, Pakistan), Dravidia (present-day Punjab) and other places. In the year 408, Faxian went to the Tamauri Empire (present-day Tamlaq, southwest of Calcutta, India). A year later, he boarded a merchant ship and sailed south to Shiziguo, changed ships to head north via Yabati (present-day Sumatra), sailed around the South China Sea and the East China Sea, and finally, in July 412, he landed at Laoshan in Qingzhou (present-day Laoshan, Qingdao) [78]. Fa Xian's westward pilgrimage lasted for 14 years, travelling through about 30 countries and regions: thus he became one of the first few Chinese explorers to cross the high mountains of the western territorial borders and to reach India, and the first Chinese traveller to journey to India by land and return home by sea with a written record of the visit.

In 416, Fa Xian completed his travelogue *Foguo ji* (Fig. 5.9). This was the first complete travelogue on Central Asia, India and Southeast Asia in ancient China, and the earliest record of sea and land routes between China and India. Containing nearly 14,000 words and with the visits presented in chronological order, the book documents 33 countries, covering as far as northwest China, Central Asia, South Asia, the Indian Ocean, Southeast Asia and the southeast coast of China. This book has become a classic and is significant for the reasons outlined below [79].

First of all, it is a realistic account of an investigation conducted through fieldwork. The study strongly emphasised the distance between various places and applied a variety of methods: in the western territories the unit of measurement adopted was "li", to record the journey in Central Asia this was "day"; whilst in South Asia it was *youyan* (Indian unit for one day's mileage), and in some places the distance was measured by *bu*, *chi* or visual estimation. As most of the ancient countries in the western territories no longer exist, the *Foguo ji* has become a key text in the understanding of the historical geography of the western territories and South Asia.

Fig. 5.9 The cover of a modern edition of the *Foguo ji* (annotated by Zhang, Xun 章巽)

Secondly, it is also an outstanding nautical log as Fa Xian followed a sea route for the whole of his return journey. The *Foguo ji* is the earliest systematic record of the trade winds and the sailing ships of the Southern Ocean, as it provides an in-depth description of the full details of his experience of voyaging by sea.

Once completed, it not only became a classic work of Buddhism, but was also applauded by geoscience scholars. It was cited in nearly 30 places in the *Shuijing zhu* by Li Daoyuan of the Northern Wei dynasty. In the late Qing dynasty, Ding Qian published his monograph *Foguo ji dili kaozheng* ("A Geographical Study of Foguoji"), the first Chinese study of *Foguo ji.*

(2) The concept of Xiyu

The term Xiyu ("西域", Western Regions) mainly appeared in the literature of the Han, Wei, Jin and Southern and Northern dynasties. It broadly refers to the vast area west of the Yumen Pass and Yangguan Pass, which is now called Central Asia. In its narrow sense, Xiyu indicates the Tarim Basin and its surrounding areas [80]. Since Zhang Qian had conducted his first mission to Xiyu in the Eastern Han dynasty and blazed the trail for others, understanding of the geographical partition of Xiyu had greatly improved.

Following the *Hanshu: Xiyu zhuan* in which the southern and northern roads to Xiyu and various countries along the roads were introduced, the *Weilue: Xirongzhuan* and *Xiyu tuji* ("Illustrative Record of the Western Regions") by Pei Ju focused on three roads to describe the geography of Xiyu, outlining the structural understanding of a region according to its main traffic routes.

In the 3rd year of the Taiyan reign (437 AD), Emperor Taiwu of the Northern Wei dynasty (personal name Tuoba Tao) dispatched Dong Wan and others to Xiyu. In their accounts of their experiences in Xiyu, the idea of marking the different zones in Xiyu was proposed for the first time, and the potential extent of the four geographical regions of Xiyu was described briefly in the *Weishu: Xiyu zhuan* [81]: the first region was the area south of the present-day Tianshan Mountain in Xinjiang; the second region referred to the southern end of the present-day Caspian Sea; the third region was the area along the southern and northern shores of the middle and upper reaches of the Amu Darya; and the fourth region may have referred to the northern part of the present-day Caspian Sea [82].

(3) A deepening understanding of the geography of the western regions

Since the Wei and Jin dynasties, the increasing number of westward journeys to seek the Dharma had promoted people's understanding of the geography of the Western Regions [83].

Among the official histories recorded during the Wei, Jin, and Southern and Northern dynasties, the *Hou Hanshu: Xiyu zhuan* and *Wei Lue: Xirong zhuan* added accounts of the mountains and rivers around the Mediterranean Sea. For example, the *Hou Hanshu: Xiyu zhuan* mentions that "the King of Yanqi lives in Nanhe city"; "the country is surrounded by big mountains and connected to Qiuci, with dangerous roads before the entrance to the city. Sea water curves along the mountains surrounding the city for more than 30 li". The *Jinshu: Xirong zhuan* contains similar accounts: "big mountains on all sides" is a description of the Yanqi Basin (now part of Xinjiang), and the inlet of sea water may refer to the water system with Bosten Lake at the centre [84].

The *Weishu: Xiyuzhuan* provides extra information regarding several mountains and bodies of water west of the Congling Ridge [85]. For example, "The sea voyage from Persia to the Mexican Gulf across the Atlantic Ocean is a distance of 10,000 li … its sea resembles the Bohai Sea but lies opposite to the Bohai Sea in the east. (The country) is located between two seas, with a flat land and a small population."

There were few records of the deserts in the Western Regions before the Han dynasty, but the situation improved after the Eastern Jin. Two monks, Fa Xian of the Eastern Jin and Fa Yong of the Liu Song, both passed through the Taklamakan Desert on their way to the Western Regions. *Foguo ji* and *Gaoseng zhuan* ("Biographies of Eminent Monks") described these more accurately (see 5.2 for more details). Volume 15 of the *Chu Sanzang jiji* ("Triptitaka") by Seng You mentions that "out of the Yangguang in the west is a desert stretching over 2,000 li, without water, grass or nomads. It is freezing cold in the winter and scorching hot in the summer".

Since Zhang Qian's mission to the Western Regions, the Congling (i.e., the present-day Pamir Plateau) had become a vital junction for any exchanges between the Central Plains and Xiyu, but very few details about this area were conserved in the Han dynasty. The *Foguo ji* was the first to describe the vegetation and glacial landforms of the Congling Ridge region (see 5.2 for details). Song Yun of the Northern Wei dynasty also vividly noted the high, rugged and dangerous mountains and the frigid climate of the Pamir region in his *Xingji* [86].

5.4.3 Increased Insights into the Geography of the Southern Borders and Extraterritoriality

Before the Han dynasty, people knew very little about the geography of the extraterritorial countries along the southern borders. From the Wei and Jin dynasties, the boom in trade and commerce and the journeys westwards to seek Buddhism greatly enriched the knowledge of those southern regions at home and abroad. According to the *Nanshi: Yimo* ("History of the South: Ethnic Minorities"), Lü Dai under Emperor Sun Quan of the Wu dynasty dispatched Zhu Ying and Tai Kang on a diplomatic mission to Southeast Asia between 226-331AD. The countries and regions they passed through included Linyi (the present-day south of central Vietnam), Funan (present-day Cambodia), "Southwestern Large Island" (present-day Malaysia), India, the legendary empire of Daquin and over a hundred other countries.

(1) Geography of the South China Sea

In ancient times, the South Sea was called Zhanghai (the "Rising Sea") or Dazhanghai (the "Great Rising Sea"). By the time of the Three Kingdoms, Kang Tai began to record the geography of the South Sea in the *Funan zhuan*, such as the topography of the coral reefs: "above the Zhangai [sea] appears a coral reef; on the bottom there are rocks where the coral is born." The reefs are formed by coral polyps and the "bottom" refers to the volcanic cone or rocks in the sea [53].

Around the same time, Wan Zhen, in the *Nanzhou yiwu zhi,* described his voyage from the Malay Peninsula to China: "Along the northeast line there are huge rocks and one can find magnetic stones in the shallow water [87]." The reference to "magnetic stones" indicates the hidden reefs beneath the water which, if ships encountered them, would cause accidents to happen, just like magnetic stones attracting the ships. This suggests that people had some knowledge of the form and cause of coral reefs and sandbars in the South China Sea.

Kang Tai, in his book, also described the animals and plants of certain islands in the South China Sea: "East of Funan is Zhanghai, there are islets there, home to the parrots with five different colours, and the white one among the five colours which looks like a hen [88]." "East of Funan, in Zhanghai, there is an island called Dahuozhou; on the island, if the trees receive the spring rains their bark turns black, and if they catch fire their bark turns white [89]."

(2) Southeast Asian countries in geographical books

Zhu Ying and Kang Tai completed two geographical works about the South Sea region. These were the most ancient materials seen in history that officially recorded China's mission to Southeast Asia.

The one-volume *Funan yiwu zhi was* written by Zhu Ying. The *Suishu: Jingji zhi* and *Tangshu: Wenyi zhi* have included its catalogue. Although it has been lost, the book provided many references for later literature to draw on, including *Liangshu: Zhuyi liezhuan* and *Hainan zhuguo zhuan* [90].

The *Wushi waiguo zhuan* ("Accounts of Foreign Countries during Wu Dynasty"), also known as *Funan ji* and *Funan tusu,* was lost and its content was only partially conserved in books such as the *Shuijing zhu, Yiwen leiju* ("Classified Collection Based on the Classics and Other Literature") and *Taiping yulan* ("Imperial Reader of the Taiping Era"). These existing materials document the location, distance, produce, population, customs, climate, trade, religion and crafts of more than 30 countries and regions.

Wan Zhen, the prefecture chief of Danyang in the Wu dynasty, was the author of the *Nanzhou yiwu zhi.* Although this book was named after exotic objects, most of its accounts are about the local customs of Southeast Asian countries, and many countries mentioned in the book have not been seen elsewhere before. For example, "Linyang (now the area from the western part of Thailand to Myanmar and the northern part of the Malay Peninsula) is over 7,000 li to the west of Funan; the land is flat with over 100,000 households; men and women are kind and everyone practises Buddhism [91]."

During the Liu Song dynasty, Zhu Zhi, who had visited Funan personally, also wrote a book called *Funan ji,* yet it was lost and information from the book can only be partially retrieved from references. Some southeast Asian countries other than

Funan were mentioned, such as: "The country Piqian (unknown, some think it is at the estuary of the Pahang River in Malaya) is 8,000 li away from Funan, in the middle of the sea [92]."

(3) Geography of the Southeast Asian countries

The *Songshu* ("Book of the Song") by Shen Yao (441–513 AD) and *Nanqi shu* ("Book of Nanqi") by Xiao Zixian (448–537 AD) gave an account of Southeast Asian countries in the sections *Manyi liezhuan* ("Biographies of Barbarians"). Yao Silian (557–637 AD), the Later Tang author of *Liangshu: Zhuyi liezhuan* ("Book of the Liang: A History of the Southern Nations"), for the first time, drafted a chapter focusing on the topic "Hainan zhuguozhuan", which systematically reviewed the knowledge of the historical geography of the Southeast Asian region.

The countries detailed in the *Liangshu* include Linyi (in present-day central Vietnam), Funan (present-day Cambodia), Panpan (in present-day Thailand, along the coast of the Gulf of Vanlun), Dandan (in the present-day Malay Peninsula), Kadaram (in present-day Kedah on the Malay Peninsula), Langkasuka (in present-day Thailand, in the southern part of Pattani), Poliguo (in present-day Bali, east of Java), Central Tianzhu (in the Indus area), Shiziguo (in present-day Sri Lanka), and a number of geographical locations of nations unknown today. Many geographical accounts of these nations are particularly interesting, for example:

> The Linyi nation was once Xianglin County at the southern border of the Rinan Prefecture during the Han dynasty. General Fu Bo with his army explored here and established an administrative body. The land may be up to six hundred li wide, and the city is one hundred and twenty li away from the sea and over 400 li away from the southern border; its north borders the Jiude Prefecture. Its southern boundary has a waterside path which is 200 li long... the nation has gold mountains, with red rocks that can produce gold…
>
> Poliguo is the island in the southeast sea of Guangzhou. Travelling to Guangzhou takes two months. Walking from the eastern coast to the west takes ten days, and from the south to the north 20 days … it is as hot as China in midsummer. Fields can yield two crops a year and the flora flourishes all year long.
>
> It is a day's journey to Guangzhou in February. The border of the country is a journey of 50 days between the borders at the east and west of the country and 20 days between north and south. There are one hundred and thirty-six hamlets. The soil is as hot as in the summer in China. The grain ripens twice within a year, and the grass and trees always thrive..."

5.5 Geography of Historical Administrations and Toponymy

5.5.1 Early Examples of Administrative Geography

As a crucial part of traditional Chinese geography, the development of its administrative geography, which had emerged in the Han dynasty, reached maturity during the Wei, Jin and the Southern and Northern dynasties. Administrative geography as a field of study focusing on the changes in boundary and regional administration was also taking shape.

(1) Geography of historical administrations

Authors during the Wei, Jin and the Southern and Northern dynasties built on the pioneering work of the *Shiji: Hequ zhi* and *Hanshu: Dili zhi* and reviewed the geographical evolution of earlier dynasties in a similar style. The *Songshu: Zhoujun zhi* ("Book of the Song: Record of States and Prefectures") is the most typical example. It also provides one of the most detailed records of administrative geography, with much better granularity on historical changes than the *Jinshu:Dili zhi* that was written after it.

(2) Administrative geography in classics

During the Han and Jin dynasties, scholarly examination of ancient mountains and rivers and place names appearing in the classics was one distinct approach to the study of administrative geography. During the Southern and Northern Jin dynasties, many advances were achieved in this field, but unfortunately few of the studies have survived to the present day.

The *Collected Explanations of Zuozhuan* by Du Yu (222–284 AD) in the Western Jin dynasty entailed the enriching of the geographical interpretations of the Spring and Autumn states. In his other work *Chunqiu shili* ("Commentary on the Spring and Autumn Annals"), he dedicated three volumes to the "Names of the Lands". The names are divided into three categories: names of vassal states, of four barbarians, and of mountains and rivers. All the names listed can be explained in terms of locations and directions. For example, in "Duke Xuan two years, Shoushan", "Shoushan" is annotated as "Shouyang Mountain at the southeast of Puban County to the east of the river".

At the beginning of the Western Jin dynasty, the geographer Jing Xiangfan also completed a monograph on toponymy: *Chunqiu diming* ("Location names during the Spring and Autumn") [93]. The original text was lost long ago, but from the surviving entries it can be seen that: (1) the place names are mostly annotated with the then location and distance. For example, "Shimen, the land of Qi, is 60 li southwest of the former Lu County in Jibei today [94]" and (2) attention was given to the origin of the place names. For example, "Fifteen li southwest of Puyang city is '沮丘' (pinyin: jǔ qiū), '沮' (pinyin: jǔ) and '楚' (pinyin: chǔ) were homophones in the Six Kingdoms period, so it is wrong to consider that it is '楚丘' (pinyin: chǔ qiū) [95]".

The Jin dynasty master of scripture Guo Pu was especially adept in the exegesis of ancient documents, and his exegesis of the scriptures contains a wealth of historical and geographical content. His 18-volume work *Shanhaijing zhuan* ("Biography of the Mountain and Sea Scriptures") is the earliest work on the exegesis of the "Mountain and Sea Scriptures" currently available. In this work, Guo Pu examined the contemporary location of dozens of mountains in the *Shanhai jing*, such as in the commentary "Shanhaijing: Zhongshan jing: Taimuro's Mountain", which reads, "That is, Zhongyue Songshan Mountain is also high, now located in the west of Yangcheng County."

Guo Pu provides many geographical references in his annotation of the classics. The 18-volume *Shanhaijing zhuan* by Guo Pu is the earliest annotated commentary of the *Shanhai jing*. In it, Guo Pu identified the locations of scores of mountains. For example, he explained that "this is Songshan Mountain located in the west of present-day Yangcheng County". Guo Pu was also diligent in his exegesis of the *Erya* and wrote the *Eryazhu* and *Erya tuzan*. For example, in the section "Explain the Places: Jiuzhou", he mapped out the geographical extent of each of the *jiuzhou* (nine regions of ancient China): "from the river flowing from east to west is Jizhou".

(3) Administrative geography as an aspect of physical geography

During the period of the Wei, Jin and Southern and Northern dynasties, the thriving nature of physical geography greatly promoted the development of administrative geography and made it an integral part of geographical studies.

Since the Wei and Jin dynasties, the genre of *fangzhi* (local chronicle or gazetteer) had been a core feature of ancient Chinese geographical literature. Most *fangzhi* chronicled the administration of borderlands. The term *yange* (administration) first appeared in the initial article "Sanfu yange" in the *Sanfu huangtu* ("Plan of the Three Capital Regions") as a *fangzhi* text [96]. Originating in the Western Jin dynasty, geographical records beyond the local level, known as *zongzhi* (general chronicles), also tended to include administrative changes. Zhi Yu in his *Jifu jing* pointed out that "the former Zhounan is present-day Luoyang". Both the *fangzhi* and *zongzhi* of later dynasties inherited the style of the Wei and Jin and included accounts of the administrative status of certain places.

Although the *Shuijing zhu* mainly focused on the water systems, details of administrative geography abound [97]. First, it recorded the setup, development and naming principles of prefectures; second, it recorded the administrative history of many states that could no longer be found elsewhere; and third, most of the 2,500 county names were traced back to the pre-Qin period, for example:

> Xiang County belonged to the Song; in the 23rd year of Qin Shi Huang (224 BC) it was named Sishui Prefecture; in the 4th year of Emperor Gao of Han (203 BC) this changed to Pei Prefecture; in the 6th year of Emperor Wu of Han (117 BC) it became a state called Xiangcheng until Wang Mang made changes and called it Wufu Prefecture and Wufuting County.

This brief account explained the history of the administrative geography of Xiang County over the previous few hundred years.

Other representative works in this field of study include the *Chunqiu shili: Tudi ming* ("The Rules and Formulae of the Spring and Autumn Annals: Names of the Land") by Du Yu, *Yugong diyu tu* by Pei Xiu and *Chunqiu tudi ming* by Jing Xiangfan. A few important ones appeared later, such as the twenty volumes of *Yongchushanchuan gujijin* by Liu Chengzhi in the Southern Song dynasty and *Gujin zhoujun ji* ("The Past and Current Records of States and Prefectures") by Tao Hongjing in the Southern Liang dynasty. However, they were lost a long time ago, and the surviving texts mainly focus on the name changes of the places.

5.5.2 The Formation of Toponymy

Accounts of location names can date back to the Shang dynasty or even earlier. Tremendous progress was made during the Wei, Jin and Southern and Northern periods, the main features of which included: varied explanations for the origin of place names; systematic reviews of the naming rules; valuable accounts of the administrative impact, spelling and pronunciation; toponymical dictionaries and specialised studies of toponymy.

(1) Varied explanations for place names

The rich variety of explanations for place names went beyond those listed in the *Yuanhe junxian tuzhi* ("Illustrated Gazetteer of the Yuanhe Prefectures and Counties") in the Tang dynasty. Some places were named after the mountains, bodies of water, terrains, flora and fauna, minerals, soil, features of hydrology, climate and meteorology, colours and numbers, etc. Other place names might be based mainly on officials, characters, temples, tombs, historical sites, myths and legends, specialties, reign titles, ethnic minorities, immigrants and auspicious rites, as well as on images, lexical meanings, metonymy and compounds.

(2) Summary of the naming rules

① Mountains

It was common to include mountains in place names as can be found in the names of prefectures, counties, cities, rivers, lakes, valleys, palaces, villages, etc. and especially in the names of countries. The *Xuhanshu: Junguo zhi* by Simabiao states that "Tiancheng is named after Tianshi Mountain". The *Shisan zhouzhi* ("Record of Thirteen States') by Kan Yin writes that "Houshi County in Henan adopts the mountain's name".

② Bodies of water

Ever since *Shiming* by Liu Xi from the Eastern Han had proposed the naming method with water, this was commonly used in the Jin dynasty.

③ Similar names for neighbouring places

Places that were interconnected with one another tended to adopt similar names, e.g. in the *Yu dizhi* by Gu Yewang "using the temple name for the place".

④ Plants

Many plant names were adopted for locations, such as "Liaoshao (peony) County" in the *Shisan zhouzhi* by Kan Yin.

⑤ Reign titles

Yang Xuan in Northern Wei explicitly pointed out in Volumes Two and Three of the *Luoyang qielan ji* that “Zhengshi Temple obtained its name from the reign of Zhengshi”.

⑥ Alias

One reason for the existence of various names is that some places were referred to in more than one way.

⑦ Principles for the naming of administrations

In his work *Shuijing zhu,* Li Daoyuan summarised the principles for the naming of administrative levels since ancient times: “Regarding state names: Chen, Lu, Qi and Wu alike; for former cities, Changsha and Danyang alike; for mountains and hills, names like Taishan (Mount Tai) and Shanyang; for rivers, names such as Hexi (lit. River West), Hedong (lit. River East); for specialities, Jincheng city has *jin* (gold), the air in Jiuquan smells like *jiu* (alcohol), Yuzhang produces camphor trees, Yanmen is where *yan* (wild geese) breed…”.

(3) Evolution of place names and pronunciation

In the period of the Wei, Jin and Southern and Northern dynasties, geographers paid more attention to the study of the evolution of geographical names and word sounds. However, these accounts did not deal with the natural landscape or socio-economic changes. Therefore, the study of the evolution of place names was still part of administrative geography.

5.6 Advances in Cartography

The development of cartography was hampered to a certain extent by the division of the country and occasional wars during the Wei, Jin and Southern and Northern dynasties. However, there were still commendable advances in cartography. During the Three Kingdoms period, Liu Hui wrote the *Haidao suanjing* (“Sea Island Mathematical Manual”) and undertook a mathematical exposition of the Chinese *chong cha* solution procedures. As a glorious era for Chinese cartography, the Western Jin dynasty witnessed the most outstanding cartographer Pei Xiu. He proposed the *Zhitu liuti* (“Six Mapmaking Principles of Cartography) which had a significant impact on the development of ancient Chinese cartography.

5.6.1 Map Compilation and Geographic Modelling

It was difficult to conduct large-scale territorial surveys under national partition, but regional and thematic surveying and mapping had never stopped, especially the compilation of maps for feudal division, the equalisation of fields, and warfare, etc., which repeatedly appeared in historical literature. During the Three Kingdoms period when battles and revolutions arose endlessly, people tended to rely on collecting old maps or surveying regional conditions. Before attacking Yizhou, Zhang Song (?-212 AD) once drafted a topographic map for Liu Bei (161–223 AD): "a map outlines the mountains, rivers and settlements" [98]. The sister of the prime minister Zhao Da "weaved a map of the states, including patterns of mountains, rivers and seas, cities and formations, on a piece of silk and presented it to the King of Wu" [99]. Once the Western Jin had been unified, Pei Xiu led a large-scale cartographic work project and produced the historical atlas *Yugong diyu tu,* and he applied the 1:1,800,000 scale to shrink the *Map of Tianxia* to create the *Fangzhang Map* [100].

Although the number of maps discovered dating from this period is not large, there were some outstanding achievements in the production of historical maps, geographical modelling and Daoist maps.

(1) *Chunqiu Menghui Tu* ("a Chart of Interstate Meetings During the Spring and Autumn Period") and *Yugong Diyu Tu* ("Yugong Regional Map")

The *Chunqiu menghui tu* by Du Yu (222–284 AD) during the Jin dynasty was a historical map that reflected the capitals and meeting venues of the allied states. Around the 4th to the 7th year of the Jin dynasty (268–271 AD), Pei Xiu, with the help of his supporter Jing Xiangfan, completed the earliest written historical atlas in China, the 18-part *Yugong diyu tu.* In its prelude, cited from *Jinshu: Peixiu zhuan,* it is written:

> I hereby examine and verify the mountains and rivers, hills and pools, the former nine states and the updated 16 states in the Jin dynasty, the prefectures and counties, borders and corners, as well as the old names of the interstate meeting venues and the land and water routes, all compiled in the 18-part atlas.

It pioneered the writing of regional and administrative history and introduced the comparison of obsolete and new place names, which had a great influence on later styles of literature. However, this atlas was seldom cited afterwards and may have been lost long ago.

(2) Geographic modelling: *Wooden Fangzhang Map*

During the Qin and Han dynasties, simple geographic models were already used in combat, as observed in the tomb of Qin Shi Huang which was "filled with mercury to represent hundreds of rivers and seas [101]". In the Southern dynasty, a more realistic geographic model was created by Xie Zhuang (421–466 AD):

> …a board the size of a square *zhang* is made; on it, mountains and rivers in various locations are marked with their particular systems and organisation. Each one on its own is a unique state but combined together it forms a map of *tianxia* [102].

This exquisite but practical geographic model could be taken apart or recombined as a whole, making it easy to maintain.

(3) The Daoist map: *Wuyue zhenxing tu* ("Diagram of the Forms of the Five Mountains")

Around the time of the Han and Wei dynasties, a very important diagram appeared in Daoism—the *Wuyue zhengxing tu*. This Daoist diagram, inheriting the norms of traditional cartography and with some innovation, was an alternative school of Chinese cartography during this period.

The *Han Wudi neizhuan* ("Intimate Life of Emperor Wu of Han") describes how Emperor Wu of Han met the Queen Mother of the West and received the *Wuyue zhengxing tu*.

According to the research conducted by Cao Wanru and Zheng Xihuang, the diagram was created before the Eastern Jin dynasty as a plan of the Five Mountains [103]. Jiang Sheng has argued that the massifs, peaks, valleys, caves, water sources, climbing routes, specialties and other aspects of Mount Tai are all expressed in the form of map symbols or annotations and, to varying degrees, they are largely geographically realistic expressions, thus forming a relatively accurate topographic map using a method close to approximate contour" [104].

5.6.2 Cartographer Pei Xiu and the Six Observable Principles of Map-Making

Before the Jin dynasty, the craft of map-making had reached a high level and there were abundant accounts of cartography, yet no record of cartographic theory has been found. Pei Xiu, who was a talented official concurrently serving as a Shangshuling (chief of the secretarial staff) and prefect, realised that having "only the Han maps at his disposal, they lack *fenlü* (proportional scale) and *zhunwang* (relative location), nor do they mark the names of the mountains or rivers, leaving only unreliable rough shapes deficient in scientific precision" [105]. He then collected information, conducted research, and completed the production of the 18-part *Yugong diyu tu*, *Fangzhang Map* and other maps. Pei Xiu explained the Six Observable Principles of Map-Making:

> There are six principles of cartography: *fenlü* (proportional scale) to distinguish range and size; *zhunwang* (orientation) to determine the positional relationship; *daoli* (road distance) to measure the distance between elements; *gaoxia* (measurement of elevation), *fangxie* (diagonal distance) and *yuzhi* (curvature) which are subject to various safe or difficult local conditions… Only by referring to these six principles can the actual distance be determined from the scale, the exact relationship between elements be determined by the orientation, and the specific interconnecting routes by the road mileage... Therefore, although there are mountains and seas, the great differences in borderlands and foreign nations, the complexity of terrain and the twists and turns of roads can be determined according to these principles [106].

Before Pei Xiu, although China had accumulated rich practical experience in cartography, the lack of theoretical generalisation and guidance had been a problem.

Pei Xiu's Six Principles of Cartography had always been an essential part of ancient Chinese mapping [107].

5.6.3 *Mathematician Liu Hui and His Work,* the Sea Island Mathematical Manual

In the Western Han dynasty, the *chong cha* (double differences) solution emerged, that is, a method of measuring height, length and depth based on the proportional relationship of the corresponding sides of similar right-angled triangles. In the fourth year of King Chenliu of the Wei dynasty (263 AD), Liu Hui, an eminent mathematician, completed his magnum opus, the 10-volume *Jiuzhang suanshu zhu* ("Annotations of Nine Chapters of Arithmetic"), the tenth volume of which contained a discussion of the application of *chong cha* in surveying by Liu Hui himself. Around the late Southern and Northern dynasties, this volume became a standalone publication, and because its first problem concerns the determination of distances involving a remote "sea island", it is called *Haidao suanjing* ("Sea Island Mathematical Manual").

Liu Hui commented in the preface: "Where it is necessary to look at the very high, measure the absolute depth and know the distance, the *chong cha* solution must be applied and serves as the ratio for triangles." There are nine problems in the extant version, in which three basic methods are necessary for measuring the height and distance—these are the *chong cha*, the *lian suo* (lit. "connecting the rope on the surveyor's pole") and the *lei ju* solutions [108]. The *Haidao suanjing* became the mathematical basis for ancient surveying. At a time when the matrix, ruler, metre and rope were the main measuring tools, the *chong cha* method broadened the scope of indirect measurement and promoted the development of ancient surveying [109].

The Western Jin dynasty was a glorious era in the development of ancient Chinese cartography. The main reasons for the success of the maps and Pei Xiu's cartographical achievements include: (1) Chinese map surveying and mapping technology had a long history of development, and a good foundation had been established in the Qin and Han dynasties; (2) after the destruction of the Three Kingdoms, the redivision of prefectures and counties and the changes of geographical names across the nation motivated the creation of new maps and new surveying techniques; and (3) Pei Xiu had fought with the army and thus valued the importance of maps. The fact that he was a senior local official and an intellectual gave him the opportunity to collect and study maps.

5.7 Advances in Geology and Mineral-Petrology

5.7.1 *Innovative Findings in Geology*

During the Wei and Jin dynasties, the development of geological knowledge was mainly manifested in two aspects: the formation of the idea of water-to-land evolution formed and the deepening of the understanding of fossils.

(1) Transformations between oceans and continents

① Observations regarding the rise and fall of the Earth's crust

By the time of the Wei and Jin dynasties, the topography formed by the lifting movement of the Earth's crust had increased, and observations regarding the lakes formed by the subsidence of the Earth's crust were more detailed. During the Wu period of the Three Kingdoms, Zhang Bo wrote in the *Records of Wudi*: "Danghu Lake to the east of Pinghu (present-day Zhejiang) was more than 40 li wide. The lake collapsed in the Han dynasty and dried up; the remains of the lake floor were faintly visible [110]."

Gan Bao's *Soushen ji* ("In Search of the Supernatural") in the Jin dynasty mentions the formation of another collapsed lake: "Quanxian County (governed from the south of present-day Jiaxing, Zhejiang) was known as Changshui County during the Qin dynasty. During the time of Qin Shi Huang (who reigned 246–210 BC), it was reduced to a lake."

② Hypothesis on the transformations between oceans and continents

In the early years of the Western Jin dynasty, Du Qian (222–284 AD) proposed the idea of sea and land changes:

> As the saying goes: high banks make valleys and deep valleys make hills. There are two monuments carved in memory of the Du family's contributions, with one standing on the top of Xian Mountain (alias Xianshou Mountain, present-day Xiangfan, Hubei) and the other buried in a deep valley, who knows whether they will swap location in the future or not? [111]

Ge Hong of the Eastern Jin first proposed the idea that "alterations between the East Sea and the land have occurred three times now". In the Tang dynasty this thought was widely acknowledged as a geological concept describing how the Earth's activity leads to transformation where the land meets the sea [112].

In addition, Liu Yiqing (403–444 AD) of the Southern Song dynasty compiled the *Shishuo xinyu* ("New Account of Tales of the World") [113] which also contains notes of topographic changes in the first and second volumes:

> Guo Jingchun crossed the river and settled in Jiyang County. Their family tomb was less than a hundred steps from the river. Some thought it was too close, but Jingchun replied: "It will become land." Now the sand has risen and the area dozens of li away from the tomb has become fields."

(2) Better understanding of fossils

As early as the pre-Qin era, there were records of fossils. By the time of the Wei and Jin dynasties, the understanding of fossils had deepened in terms not only of the variety of fossils, but also of their characteristics, origins, causes, etc.

① *Spirifer*

During the Wei and Jin dynasties, there were many accounts of Spirifer, that is, the Spiriferidae family of brachiopods.

Luo Han in the Jin dynasty mentioned *shiyan* (spirifers) in the *Xiangzhong ji* ("Record of Hunan"): "There were *shiyan* in Lingling County (governed by the district that is modern Lingling, Yongzhou, Hunan province). With the thunder and wind, pieces are flying off, attracting local people to collect them as medication [114]." There is a similar account in Yu Zhongyong's *Xiangzhong ji* during the Liu Song dynasty [115].

In the Northern Wei dynasty, Li Daoyuan noted in the *Shuijing zhu: Xiangshui* that "its mountain (Shiyan Mountain, in present-day Qiyang, Hunan) has clams that turned to stones and they look like swallows. Large and small, the stone pair resembles a mother and son. In times of storm, the fossils will fly like real swallows".

② Fish

Works such as the *Junguo zhi* ("Records of the States and Prefectures") by Sima Biao of the Jin dynasty [116], *Jingzhou zhi* by Sheng Hongzhi of the Southern dynasty [117], *Nanyue zhi* ("Records of Southern Yue") by Shen Huaiyuan of the Southern dynasty and *Shuijing zhu: Lianshui* by Li Daoyuan of the Northern Wei dynasty all documented the area best-known for producing ichthyolites in ancient China—Shiyu Mountain near Xiangshui (modern Xiangxiang County in Hunan province). Among these works, the *Nanyue zhi* by Sheng Huaiyuan not only meticulously described the location, buried beds and preservation conditions of the fossil fish, but also proposed an identification method—using fire to burn the fossils [118].

③ Longgu

The term *longgu* (dragon bone) appeared in the *Shanhai jing: Zhongshan jing* referring to the fossilised skeletons of mammals such as elephants, rhinos, three-toed horses, etc. In the *Wupu bencao* ("Wu Pu's Materia Medica") from the Wei period, the colour of these fossils was described as "bluish white".

Accounts from this period suggested that mammal fossils were mainly discovered in the following areas: Shi'an (near present-day Hankou, Hubei), Wucheng County (east of present-day Zhongjiang County, Sichuan), Jin (present-day Shanxi), Taishan, Yizhou (governed from present-day Chengdu, Sichuan), Liangzhou (the east of present-day Hanzhong, Shaanxi), and Bazhong County (governed from the east of present-day Mianyang, Sichuan) [119].

④ Amber

Zhang Hua in his work *Bowu zhi* ("A Treatise on Manifold Topics") during the Jin dynasty pointed out that amber is petrified tree resin: "Pine resin enters the ground", and "a lifelike bee in the amber...the bee might have been trapped in the pine resin as it fell to the ground." The *Xuanzhong ji* by Guo Pu further explains that its formation took a long time: "Maple grease dried up within the ground, and thousands of autumns afterwards became amber [120]." The *Bencao jizhu* ("An Annotated Collection of the Pharmaceutical Canon") by Tao Hongjing in the Liang dynasty noted that "amber smells like and is the colour of pine when heated, and the red examples are better quality [121]".

During the Wei and Jin dynasties, amber still mainly came from extraterritorial areas, but some sources of amber were also found in present-day Yunnan province, such as in Yongchang (governed by present-day Baoshan, Yunnan province) [122] and Ningzhou (present-day Xiangyun County, Yunnan province) [123].

5.7.2 In-Depth Investigations in Mineral-Petrology

During the Wei, Jin, and Southern and Northern dynasties, although China's mining production was experiencing a period of slow development [124], the widespread use of mineral medicine and the wealth of stones as medicine in alchemy greatly promoted the development of mineral and petrological knowledge.

(1) Mineral-petrology literature

① Works on herbalism: *Wupu bencao* and *Bencao jizhu*

Although the field of Materia Medica developed rapidly, many of the herbal works cannot be found today, with only fragments cited in texts by later generations. These texts not only inherited the tradition of the *Shennong Materia Medica*, but also attached great importance to detailing the mineralogical characteristics and the origin of mineral medicines, rendering herbal studies one of the most important written resources for understanding mineralogy in ancient China.

Wu Pu, a student of Hua Tuo during the Three Kingdoms period and a native of Guangling (present-day Jiangdu, Jiangsu), completed the *Wupu bencao*. The original book was lost very early on but was re-compiled by Shang Zhijun [125] and others to include 33 kinds of mineral medicines, such as red sand, stalactite, gravel, saltpetre, mirabilite, chalcanthite, azurite, white quartz, purple quartz, five-colour gedanite, realgar, sulfur, mercury, magnetite, actinolite, griotte, feldspar, white gravel, halite, brine salt and cretaceous rocks. The *Wupu bencao* surpassed the *Shennong bencao jing* in many aspects. For example, in the description of the mineralogical properties of white quartz, the *Wupu bencao* not only recorded its cause, shape, colour, origin, etc., but also classified quartz minerals according to colour. These improvements were significant as they allowed the doctors, pharmacists and members of the public to identify various materials, and they have been widely adopted as a writing style by later generations of medical herbalists. Furthermore, they provide precious historical evidence for determining the composition and current names of these mineral medicines.

From the first year of Jianwu of Qi to the second year of the Yongyuan reign (494–500 AD), Tao Hongjing (456–536 AD) completed one of the most important works of Materia Medica after the *Shennong baicao jing*, namely *Bencao jizhu*. The book has long been dispersed, but it can still be seen from the citations in later literature that it encompassed an extremely rich knowledge of mineralogy. It had added 44 kinds of mineral medicines to the lists used by famous doctors since the Wei and Jin, bringing the total number of mineral medicines to 94, almost double the number in the *Shennong baicao jing*. These mineral drugs had three characteristics: first, they were diverse, and they had been widely extended to every major category in the contemporary inorganic mineral classification system; second, their composition was mainly mineral, supplemented by rocks; third, they included gold and silver mineral drugs. This book offered the earliest in-depth and comprehensive description of the properties and origins of mineral medicines of all the herbal studies, and it is the most important document for understanding mineralogy and petrology before the Tang dynasty [126].

② Pharmaceutical monograph: *Leigong paozhi lun*

In addition to the works of Materia Medica, there were also many accounts of mineralogical properties in the medical literature, among which the *Leigong paozhi lun* ("Master Lei's Discourse on Drug Processing") by Lei Xiao of the Liu Song dynasty was the most prominent. The earliest pharmaceutical monograph in China, it introduced more than 40 varieties of mineral medicines such as cinnabar, mica, stalactite, white alum, verdigris, *shizhong huang*. [literally yellow liquid inside a stone], *luanzhong huang* [literally yellow liquid inside a pebble] and *heiji shi* [literally black chicken stone, probably obsidian]. Before reviewing the preparation method of each medicine, the author would first describe its characteristics, especially the method for identifying its authenticity, and the advantages and disadvantages. The identification method of mineral drugs described in the book was mainly based on external characteristics, that is, mineral drugs were distinguished according to differences in colour, light transmission, lustre, shape, etc. The most important contribution of this book was the innovative use of the "streak method" to identify mineral drugs.

③ Texts on alchemy: *Baopuzi: Xianyao*

Alchemy entered its golden age during the Wei, Jin, Southern and Northern dynasties. Because a Fangshi (alchemist) dedicated himself to seeking external substances to make himself stronger, taking mineral rich herbs was prevalent. This prompted Fangshi to pay more attention to the various minerals that could purify elixirs and to conduct more in-depth observation and research on the properties and distribution of these minerals. Therefore, most of the alchemical writings of this period entailed a relatively rich knowledge of mineralogy and petrology.

The most important works of alchemy are the sections of "Jindan" (golden elixir), "Xianyao" (immortal medicine) and "Huangbai" (gold and silver) in the *Bappuzi* by

Ge Hong of the Jin dynasty. These three sections document over ten kind of mineral herbs and their features, especially Volume 11, "Xianyao", which vividly describes the characteristics, origin, production and causes of a variety of *shizhi* (stalactites): "it takes a while to form merely one drip, like the drips from a leaking roof after the rain, which may only be heard from time to time." It also accurately depicts the physical features and production areas of realgar (known as arsenic disulfide today): "Realgar can be found in Wudu Mountain, pure and inclusion-free. Its redness resembles the colour of a cockscomb, and only the bright varieties can be used for elixir, not those in pure yellow without any reddish lustre [113]."

④ Geographical text: *Shuijing zhu*

Many of the geographical texts of this period mentioned the geographical distribution of mineral rocks, as well as their properties and uses, among which the *Shuiing zhu* by Li Daoyuan of the Northern Wei was the most detailed. The minerals described in the book include coal, oil and gas in fuel minerals; gold, silver, copper, iron, tin and mercury in metallic minerals; realgar, orpiment, sulfur, salt, graphite, mica, quartz, amber and jade in non-metallic minerals. Volume Six describes the four main states of salt—sea salt, rock salt, lake salt and well salt, whilst also pointing out the location, topography, and the total area and salt mining methods of the pond salt in Xiezhou, Shanxi province. Volume 33 further introduces the regularly conjoined shape of halite—literally "umbrella salt" (a type of well salt so-called because its shape looks like an umbrella).

(2) Improved understanding of mineralogy and petrology

① Growing number of mineral species

The term *toushi* (chalcopyrite) mostly refers to golden or yellow ores such as copper pyrites ($CuFeS_2$) and pyrite (FeS_2) [127], as recorded in both Zhong Hui's *Treatise on Chu Yao* in the Three Kingdoms and in the *Jade Chapter* by Guye Wang of the Chen period of the Southern dynasties: "*toushi* looks like gold". In the *Guangzhi* by Guo Yigong of the Jin dynasty, the method of separating *toushi* from gold is mentioned: "rinsing the *toushi* with water makes it possible to see if it has mixed with gold [128]." In the *Jinchu suishi ji* ("Seasons and Customs of the Chu Kingdom") by Liang Zonglin of the Southern dynasties, *toushi* is referred to as pinchbeck alloy.

Volume Four of Chang Xuan's *Huayang guozhi* from the Eastern Jin dynasty notes the name and origin of cupronickel: "Tanglang County (today's Huize, Qiaojia and Dongchuan in Yunnan) is named after the mountain, and here silver, lead, argentan and various herbal medicines are produced.

Liu Daohui of the Southern dynasty had already mentioned *jingang shi* (diamond) in his *Notes on the Residence of Jin*: in the third year of Xianning of the Western Jin dynasty (277 AD), "Dunhuang sent gemstones…they can cut jade [129]." Ge Hong of the Eastern Jin dynasty described the properties of diamond in the *Baopuzi*: "The Funan Kingdom (in present-day Cambodia and southern Vietnam) produced a gemstone which can be carved like jade, similar to purple quartz. It is formed on stone in water one hundred *zhang* deep, like a stone stalactite. If a metal hammer is used to retrieve it, the hammer will be damaged [130]."

② Method of mineral identification

During the Wei and Jin dynasties, given the growing understanding of mineral rocks, some began to categorise them using novel identification methods beyond analysing their colours or shapes in comparison with those with similar appearances or compositions.

During the Eastern Jin dynasty, the stripe method was used to identify the characteristics of minerals. By the time of the Liu and Song of the Southern dynasty, Volume One of the *Leigong paozhi lun* had recorded the five kinds of talc minerals that had been verified by the stripe method: the white, green, black, gold and blue steatites [131]. Experts had realised that the colour of mineral streaks and of the powdered mineral was more stable than those of a chunk.

The *Wupu bencao* of the Three Kingdoms records that by burning it, sulfur can be identified by judging its flame colour when it is being burned in a fire [132]. Similarly, Ge Hong in his *Baopuzi* also pointed out that burning gold in fire will not make it melt, and burying gold in the earth will not make it rot. The chemically stable composition of gold can be used to distinguish it from similar metals.

(3) Progress in prospecting

① Symbiotic relationship of minerals

Building on the understanding of the symbiotic relationship between minerals discovered in the pre-Qin dynasty, more linkages between minerals were identified during this period.

Guo Pu in his *Liuzhe zan* posited that iron and earthy hematite could be symbiotic [133], just as Ge Hong discovered that gold and mercury could be in his *Baopuzi*. In the *Bencao jizhu*, Liang Hongjing also documented azurite and malachite.

In the Southern dynasties, prospectors considered quartz veins to be a sign of gold and called those brown in colour with black heads "gold companions" according to the *Shixing ji* by Wang Shaozhi of the Southern dynasties, quoted in Volume 15 of the *Guangdong xinyu* ("A New Account of Guangdong") by Qu Dajun of the Ming dynasty. Lu Benshan's study confirms that an obvious sign for locating gold ore in today's Qinling mountains is a brown-coloured quartz vein. The *Shixing ji* also features an early record of the symbiotic relationship between mercury and quartz.

② Indicator plants of mineral deposits

During the Wei and Jin dynasties, it was discovered that certain plants could be enriched by certain minerals. Zhang Hua's *Bowu zhi* ("Records of Diverse Matters"): *Dili,* written in the Jin dynasty says: "If mugwort is burned after three years of growth, its bodily fluids contain lead and tin." In the *Leigong paozhi lun*, mercury is mentioned "whoever uses it, do not take it from the grass", indicating that a certain grass is known to be rich in mercury.

The *Bowuzhi* from the Jin dynasty clearly pointed out that certain plants can be used as indicators of mineral deposits: "a lot of knotweed on the ground indicates that hematite is around" [134].

The *Dijing tu* ("Map of the Earth Mirror"), written in the Southern Liang dynasty (502–557 AD), systematically reviewed the rich empirical knowledge of plant prospecting. This book was lost in the Tang dynasty but the parts collated in the Qing dynasty posit that [135]: (1) some "green onions" may grow in soil close to natural silver or silver-bearing minerals; (2) mutant varieties of plants grow in areas rich in certain minerals: "a green stem looking red means there is lead underneath", "a green stem with tainted yellow means there is copper under it"; and (3) the discovery of plant mutations in a certain season can be used as a positive sign for prospecting. Although these accounts do not necessarily correspond to reality, and some may be based on insufficient assumptions, they reveal the start of the development of geobotanical prospecting or biogeochemical prospecting, which provided new techniques allowing people to locate underground mineral deposits [136].

5.8 Advances in Meteorology and Climatology

5.8.1 *Innovative Findings in Meteorology*

The art of weather forecasting dates back to a very early time in history, and it flourished during the period of the Wei, Jin and Southern and Northern dynasties. It made great advances in theory and practice, which included a lot of weather and climate forecasting.

(1) Zhanhou and related works

In this period, there were many more well-known Zhanhou (people who predicted the future based on geographic changes such as in the weather) (see Table 5.3), many of whom were particularly good at observing qi and predicting the weather [137].

During the period of the Wei, Jin and Southern and Northern dynasties, some Zhanhou works had already been published, such as the *Wind Divination of Duke Hou*, *A Comprehensive Collection of Wind Divination* and *Important Techniques of Wind Divination,* all of which are recorded in the *Book of the Sui: Treatises of Scriptures.*

During this period, *zhanhou shu*, the art of interpreting omens, was more widely employed in military action because of the frequent wars, and there were also a number of books. For example, in the *Book of the Sui: Treatises of Scriptures*, there were 13 volumes of the "Fengjiao zabinghou" (wind divination), 12 volumes of the "Fengjiao bingfa" (wind divination as a military strategy), and 2 volumes of the "Conjecture of the Military Strategy in the 12th Year of the Yuanjia reign".

Table 5.3 Some of the documented Zhanhou in the Wei, Jin, Southern and Northern dynasties

Dynasty	Name	Original text	Source
Wu dynasty, one of the Three Kingdoms	Wu Fan	Wu Fan, courtesy name Wenze, is good at interpreting omens	Volume 2 of the *Taiping yulan*, citing from the *Wulu*
Shu dynasty, one of the Three Kingdoms	Zhang Yu	Zhang Yu, a military officer of the Shu Kingdom, had the extraordinary talents of a Zhanhou	*Record of The Three Kingdoms: Shu Kingdom: Zhou Qun Biography*
Wei dynasty, one of the Three Kingdoms	Guan Lu	Ni Qinghe asked Lu for the date of the rains because Qinghe County had been suffering from drought for a long time. Lu said it would rain that night, but no one believed him as it was a hot, sunny day. However, at twilight it rained heavily with thunder and clouds, as Lu had predicted	*Record of The Three Kingdoms: Wei Kingdom: Guan Lu Biography*, annotated by Pei Songzhi of the Southern dynasty
The Northern Wei dynasty	Wang Zao	When the emperor was on the way back to the capital, it had not rained for a long time. The emperor asked Wang Zao: "When will it rain?" Wang answered: "It will rain heavily between 3–5 pm" However, there were still no clouds between 1–3 pm, and the emperor was starting to doubt him. Zao said: "Please wait for a little longer." Later, the clouds gathered, and the rain poured down as Zao had predicted	*Book of the Wei: Wang Zao Biography*
The Northern Wei dynasty	Zhang Yuan	An expert in reading omens, and someone who knows the inner and outer star divisions. He once wrote the *Guanxiangfu* ("Ode to Meteorological observation")	*Book of the Wei: Zhang Yuan Biography*
The Northern Qi dynasty	Zhang Zixin	Xi Yongluo and Zhang Zixin were sitting opposite each other when magpies sang in the tree, fought and fell down from the tree. Zixin said, "Not good, there will be wind at dusk from the southwest." … After Zixin left, the wind came as he had predicted	*History of the North: Biography of Zhang Zixin*
The Northern Qi dynasty	Wu Zunshi	Wu Zunshi … an expert in divination … Wen Xiang recommended him as office secretary to the great general. During a tour to East Mountain, it was cloudy, and people feared that the rain would ruin the hunting … Li Yexing said: "According to the *Yijing*, a mountain (*gen*) sits above the ground (*kun*), so when clouds appear around mountains, there will be rain," but Zun Shi argued: "The ground (*kun*) can control water, so there will be no rain." … In a moment, the clouds dispersed, and Zun Shi was rewarded	*History of the North: Biography of Wu Zunshi*

(2) Progress in weather forecasting

In the Wei, Jin, Southern and Northern dynasties period, the art of Zhanhou continued to develop, building on the Han dynasty, and new theories and methods of meteorological forecasting emerged.

The method of predicting the humidity of the air from hanging charcoal, which originated in the Han dynasty, further developed into the main method of predicting the weather in this period. In the Western Jin dynasty, Zhang Hua's *Ganyinglei congzhi* says: "Wind can be predicted with ashes, rain can be predicted by hanging charcoal. Grind elm into ashes indoors, and if it is going to rain, the ashes will spread. When you put the same weight of soil and coal on scales, if it is rainy, the coal will be heavier (because it absorbs water); if it is sunny, the coal will be lighter (because the water in the coal evaporates when there are low levels of humidity)" [138].

In the Northern Wei dynasty, Jia Siwei clearly presented the weather characteristics before a frost in the *Qimin yaoshu: Trees Planting (Number Thirty-Two)*: "If the sky is clear after rain, and the wind is freezingly cold, it is certain that there will be a frost at night time." This means that before the arrival of a cold wave, there are clouds and rain with dry, cold air approaching, and the weather becomes cold with a cold wind blowing from the north; later at night, a lot of heat is released from the ground, and there is a frost. The book also describes using fumigation for frost prevention: "…light a fire and afterwards when the smoke emerges, the ground will be free from frost." This simple measure is still widely used in rural areas of China.

(3) The widespread use of a wind detecting machine

Zhang Heng of the Eastern Han dynasty created the wind detecting machine *xiangfengqi* in order to observe wind direction. In the Wei, Jin and Southern and Northern dynasties, the wind detecting machine was widely used [139] except in the Eastern Jin dynasty when the use of the machine was suspended [140]. Many poems and works of fiction even incorporated "*xiangfeng*" (measuring wind) in their title, such as the *Xiangfeng fu* written by Zhang Hua, Fu Xuan, Sun Chu, and Pan Yue, etc. [141].

During the Eastern Jin and Southern and Northern dynasties, five *liang* of chicken feathers [a *liang* is a unit of weight in Chinese and five *liang* is equal to 250 g] were used to make the instrument, that is, "five *liang*" as in its name [142].

5.8.2 In-Depth Investigations in Climatology

(1) Atmospheric extinction

The phenomenon of atmospheric extinction had already been noticed by Jiang Yan in the later Qin dynasty. According to the *Book of the Sui: Astronomy*:

> Jiang Ji says: "The sun is the essence of pure Yang. The light shines outside and dazzles people's eyes. Therefore, people regard the sun as small. When it first appears, the earth has a

shimmering veil of energy which shields it from the sun's rays and it - the large, red sun-does not dazzle the human eye. When there is no energy, the sun is white and not very substantial. The earth is not as strong as the sky, therefore, during a day, the sun is the colour red in the morning and evening, but white in the middle of the day. If the energy from the earth rises, merges with and equals the sky, so it also becomes red, despite the time of day. The sun and fire are similar; the flames of a fire are red and yellow, and the sun is red. However, when the (daytime) sun is red, it is like fire without flames. If the light dims out of order unexpectedly, it is different.

The shorter the wavelength of radiation, the stronger the effect of extinction. The atmospheric absorption of blue light at dawn is stronger than that of red light at noon and, therefore, the sun appears to be red near the horizon or white at noon [143].

(2) Plum rain

After the Eastern Jin dynasty and the Southern and Northern dynasties, the economy of the Yangzi River basin developed rapidly, and more and more literature about the plum rains was published. The *Fengtu ji* ("Record of Wind and Land") by Zhou Di of the Western Jin dynasty and the *Jinchu suishiji* ("Seasons and Customs of the Chu Kingdom") are early documents that use the term Plum Rain. The book *Liang Yuandi compilation* clarified the reason for the naming of the Plum Rain: "it rains after the plums ripen, hence the name Plum Rain."

(3) Trade winds and cold waves

Zhou Chu of the Western Jin dynasty in his work *Fengtu ji*, provided an earlier comprehensive account of the trade wind: In the month of June in the south, a strong wind blows from the southeast for a long time; then the wind stops in June. It is commonly known as the *huangque changfeng* [literally the siskin long wind] [143]. This described the timing and characteristics of the southeast wind blowing to the Chinese mainland. A siskin is a kind of migratory bird, which spends the summer in the northeast and, in autumn, is found in the southeast of Zhejiang and Fujian. The time of its migration and the timing of the southeast wind are quite close.

The *Jinchu suiyue ji* first recorded the time when the cold wave moved south: "At the Chong Yang festival, there are often sparse winds and cold rain." The book also recorded the cold wave in each solar term, or the time it moved south: "the three symbols of the Minor Cold: plum blossom, camellia, narcissus; the three symbols of the Major Cold: *ruixiang* (*Daphne odora*), orchid, mountain; the three symbols of the Start of Spring: *yingchun* (*Jasminum nudiflorum*), cherry, *wangchun* (Flos Magnoliae); the three symbols of Rain Water: camellia, apricot, plum blossom; the three symbols of the Waking of Insects: peach blossom, *ditang* (*Kerria japonica*), rose." These are the names of the flowers which represented each cold wave.

(4) Typhoons

Shen Huaiyuan first used the term *jufeng* (typhoon) in the Southern Liu and Song dynasties and gave a comprehensive account of typhoons, outlining the timing and characteristics of typhoons, as well as the abnormal phenomena and serious destructiveness before their arrival: "during the reign of Xi'an, there were many typhoon days."

(5) *Shujiu hantian* [the coldest days of the year]

The *Dongjiu ge* ("Song of Counting the Nine Solar Terms from Winter to Spring"), which was popular in ancient China to describe the process of periodic change in winter, already existed in the Southern dynasty. Liang Zonglin's *Jinchu suiyue ji* says: "After eighty-one days, counting from the winter solstice, winter comes to an end." Unfortunately, no specific content was recorded.

References

1. Zhao, R赵荣 (1994) Weijin nanbeichao shiqi de zhongguo dilixue yanjiu魏晋南北朝时期的中国地理学研究 [China's geography research during the Wei, Jin, Southern and Northern dynasties]. *Ziran kexueshi yanjiu*自然科学史研究 [Studies on the history of natural science] 13(1):65–75
2. Zhao, K赵匡华, Cai J蔡景峰 (1992) *Zhongguo gudai kexuejia zhuanji: Ge Hong shang*中国古代科学家传记·葛洪:上[The ancient Chinese scientists: biography of Ge Hong I]. Science Press北京科学出版社, Beijing, p 197
3. Chen, Q陈桥驿 (1992) Lidaoyuan he shuijing zhu yiji zai dixueshi shang de diwei郦道元和"水经注"以及在地学史上的地位 [Li Daoyuan and his significant work *Shuijing zhu* in the history of geoscience]. In: *Lixue xinlun: Taiyuan*郦学新论 [New insights into the study of Lidaoyuan]. Shanxi People's Publishing House山西人民出版社, Taiyuan, pp 50–52
4. Ma, Z马正林 (1987) *Zhongguo lishi dili jianlun*中国历史地理简论 [A brief introduction to the history and geography of China]. Shanxi People's Publishing House陕西人民出版社, Xi'an
5. Xida QT瞿昙悉达 (Tang dynasty). *Kaiyuaun zhanjing: juanyi*开元占经·卷一[Classic of astrology from the Kaiyuan era, vol 1]
6. Chen, M陈美东 (2008) *Zhongguo tianwenxue daxi: Zhongguo gudai tianwenxue sixiang*中国天文学大系·中国古代天文学思想 [Chinese astronomy series: Ancient Chinese thought on astronomy]. Science and Technology Press of China中国科学技术出版社, Beijing, p 139
7. Chen, S陈寿 (Jin dynasty), *Sanguozhi: Wushu: Zhaoda zhuan*三国志·吴书·赵达传 [A history of the three kingdoms: book of Wu: biography of Zhaoda], annotated and quoted from *Jinyangqiu*晋阳秋
8. Pan N潘鼐 (2005) *Zhongguo gu tianwen yiqi shi (caitu ben)*中国古天文仪器史 (彩图本) [History of ancient astronomical instruments in China (coloured edition)]. Shanxi Education Press山西教育出版社, Taiyuan, p 5
9. Chen, M陈美东 (2003) *Zhongguo kexuejishu shi: Tianwenxue juan*中国科学技术史· 天文学卷 [History of science and technology in China: astronomy]. Science Press北京科学出版社, Beijing, p 232
10. Chen M陈美东 (2003) *Zhongguo kexuejishu shi: Tianwenxue juan*中国科学技术史·天文学卷 [History of science and technology in China: astronomy]. Science Press北京科学出版社, Beijing, pp 248–249
11. Zheng W魏徵 (Tang dynasty), *Suishu: Tianwenzhi shang*隋书·天文志上 [Book of the Sui: astronomy I]
12. Yue S沈约 (Liang dynasty), *Songshu: Tianwenzhi yi*宋书·天文志一[Book of the song: astronomy, vol 1]
13. Wei Z魏徵 (Tang dynasty). *Suishu: Tianwenzhi shang*隋书·天文志上 [Book of the Sui: astronomy I]
14. Zheng W郑文光, Xi Z席泽宗 (1975) *Zhongguo lishi shang de yuzhou lilun*中国历史上的宇宙理论 [Theories of the universe in Chinese history]. People's Publishing House北京人民出版社, Beijing, pp 58–87

15. Editorial Department of People's Literature Publishing House人民文学出版社编辑部 (1959) *Hanwei liuchao minge xuan*汉魏六朝民歌选 [Selected folk songs of the Wei, Jin and Southern and Northern dynasties]. People's Literature Publishing House人民文学出版社, Beijing, p 55
16. Chen M陈美东 (2008) *Zhongguo tianwenxue daxi: Zhongguo gudai tianwenxue sixiang*中国天文学大系·中国古代天文学思想 [Chinese astronomy series: ancient Chinese thought on astronomy]. Science and Technology Press of China中国科学技术出版社, Beijing, pp 99–100
17. Corrected by Kejun Y (Qing dynasty)严可均 *Quan sanguo liuchao wen: Quan sanguo wen: juan bashi'er* 全三国六朝文·全三国文·卷八十二 [Complete collection of prose writings from the earliest antiquity to the pre-dynastic Tang: the three kingdoms texts, vol 82]
18. Xuanling P房玄龄 (Tang dynasty). *Jinshu: Yuxi zhuan*晋书·虞喜传 [Book of the Jin: biography of Yu Xi]
19. Chen M陈美东 (2008) *Zhongguo tianwenxue daxi: Zhongguo gudai tianwenxue sixiang*中国天文学大系·中国古代天文学思想 [Chinese astronomy series: ancient Chinese thought on astronomy]. Science and Technology Press of China中国科学技术出版社, Beijing, p 109
20. Silian Y姚思廉 (Tang dynasty) (1974) *Liangshu: Rulin zhuan: Cuilingen zhuan*梁书·儒林传·崔灵恩传 [Book of the Liang: biography of the Rulin scholars: biography of Cui Ling'en]. People's Publishing House北京人民出版社, Beijing, p 677
21. Yanshou L李延寿 (Tang dyansty), *Beishi: Yishu zhuan: shang*北史·艺术传上 [History of the Northern dynasties: treatise on arts I]
22. Shi D道世 (Tang dynasty), *Fayuan zhulin: juansi*法苑珠林·卷四 [Forest of gems in the Garden of the Dharma, vol 4]
23. Chen M陈美东 (2003) *Zhongguo kexuejishu shi: Tianwenxue juan*中国科学技术史·天文学卷 [History of science and technology in China: astronomy]. Science Press北京科学出版社, Beijing, pp 257–259
24. Wang HL王海林 (2005) Sanqian daqianshijie guanyu fodu yuzhouguan de duihua三千大千世界关于佛都宇宙观的对话 [Dialogue on the cosmos between a great chiliocosm and the Buddhism Centre]. In: Nai P潘鼐(ed) *Zhongguo gutianwen yiqishi*中国古天文仪器史(彩图本) [History of ancient astronomical instruments in China (coloured edition)]. Shanxi Education Press山西教育出版社, Taiyuan, p 248
25. Wang H王海林 (1992) *Sanqian daqian shijie: Guanyu fojiao yuzhoulun de duihua* 三千大千世界: 关于佛教宇宙论的对话 [A great chiliocosm: dialogues on Buddhist cosmology]. Today's China Press北京: 今日中国出版社, Beijing, p 28
26. Chen M陈美东 (2003) *Zhongguo kexuejishu shi Tianwenxue juan*中国科学技术史·天文学卷 [History of science and technology in China: astronomy]. Science Press北京科学出版社, Beijing, pp 257–259
27. Huang Y黄燕生 (1988) Zhongguo lidai difangzhi gaishu中国历代地方志概述 [A summary of Chinese regional geography by dynasty]. In: Xinxia L来新夏 (ed) *Zhongguo difangzhi zonglan* (1949–1987)中国地方志综览 [An overview of Chinese regional geography (1949–1987)]. Huangshan Press合肥: 黄山出版社, Hefei
28. Yue S沈约 (Liang dynasty) (1974) *Songshu: Zhoujunzhi*宋书·州郡志 [Book of the song: geography of states and prefectures]. Chinese Publishing House华文出版社, Beijing, Preface
29. Zixian X萧子显 (Liang dyansty) (1972) *Nanqi shu: Zhoujun zhi*南齐书·州郡志 [Book of the Southern Qi: geography of states and prefectures]. Chinese Publishing House华文出版社, Beijing, Preface
30. Yang W杨文衡 (1994) *Shijie dili xueshi*世界地理学史 [A history of world geography]. Jilin Education Press吉林教育出版社, Changchun, pp 120, 230–233
31. Zhang G张国淦 (1962) *Zhongguo gufang zhikao*中国古方志考 [Local chronicles in ancient China]. Chinese Publishing House华文出版社, Beijing
32. Wang J王晶波 (2000) *Hantang jian yiyi yiwuzhi kaoshu*汉唐间已佚 "异物志" 考述 [A study of the lost texts: "records of alien matters" during the Han and Tang dynasties]. Beijing daxue xuebao北京大学学报 [Journal of Beijing University], a special issue for visiting scholars and trainee teachers

33. Xian L et al李贤等 (Tang dynasty), *Houhanshu: Dongyi liezhuan: Woren zhuan*后汉书·东夷列传·倭人传 [Book of the Later Han: records of the eastern foreigners: biography of Japanese pirates]
34. Qu C常璩 (Eastern Jin dynasty) (1984) *Huangyang guozhi juan ershi: Xuzhi*华阳国志·卷二十· 序志 [Chronicles of Huayang, vol 20: Preface]. Bashu Publishing House, Chengdu
35. Qingquan C陈清泉, Su S苏双碧, Li G李桂海 (1988) *Zhongguo shixuejia pingzhuan shang ce* 中国史学家评传: 上册 [A commentary on the Chinese historians I]. Zhongzhou Publishing House of Ancient Books郑州古籍出版社, Zhengzhou, pp 133–134
36. Qiaoyi C陈桥驿 (1985) *Shuijing zhu yanjiu*水经注研究 [A study of the *Shuijing zhu*]. Tianjin Ancient Books Publishing House天津古籍出版社
37. *Suishu: Jingji zhi*隋书·经籍志 [Book of the Sui: Bibliographical treatise] has the earliest record: "Three volumes of the 'Shuijing' 水经are annotated by Guo Pu郭璞". But it does not mention who the author is
38. Li D郦道元 (Northern Wei dynasty), *Shuiijng zhu: Qishui*水经注·淇水 [Commentary on the water classic: Qi river] (Quotation from the *Shuiijng zhu* in this section all come from the *Siku quanshu*)
39. *Suishu: Jingji zhi*隋书·经籍志 [Book of the Sui: bibliographical treatise] records that "the three volumes of the 'Shuijing'水经are annotated by Guo Pu郭璞". Du You杜佑 (Tang dynasty) said in his *Tongdian: Zhoujun zhi* 通典·州郡志 [Encyclopedia: Statutes of provinces and prefectures]: "The 'Shuijing' annotated by Guo Pu has three volumes … the notes lack detail and are incoherent"
40. Yang W杨文衡 (1992) *Zhongguo gudai kexuejia zhuanji: Li Daoyuan (shang ce)*中国古代科学家传记·郦道元: 上册 [Biographies of ancient Chinese scientists: Li Daoyuan I]. Science Publishing House北京科学出版社, Beijing, pp 249–290
41. Chen Q陈桥驿 (1990) *Zhongguo dabaike quanshu: dili juan: Li Daoyuan*中国大百科全书·地理卷·郦道元 [Encyclopedia of China: geography: Li Daoyuan]. Encyclopedia of China Publishing House中国大百科全书出版社, Beijing, p 284
42. He C贺昌群 (1955) Yingyin Shuijing zhu shu de shuoming影印"水经注疏"的说明 [The notes of the facsimile of the *Commentaries and Explanations to the "Shuijing zhushu"*] In: Yang S杨守敬, Xiong H熊会贞 (eds) Shuijing zhushu (yingyin ben)水经注疏影印本 [Commentaries and explanations to the *Shuijing zhushu* (facsimile edition)]. Science Press北京科学出版社, Beijing
43. Chen Q陈桥驿 (1985) *Shuijing zhu yanjiu*水经注研究 [Research on *commentary on the water classic*]. Tianjin Ancient Books Publishing House天津古籍出版社, Tianjin, pp 29–42
44. Tang X唐锡仁, Yang W杨文衡 (2000) *Zhongguo kexue jishu shi: Dixue juan*中国科学技术史·地学卷 [History of science and technology in China: Geoscience]. Science Press北京科学出版社, Beijing, p 446
45. Yiqing Z赵一清 (Qing dynasty) wrote the *Shuijing zhushi juan yi*水经注释·卷一[Commentary on the *water classic*, vol 1]; and Bingxun S沈炳巽 (Qing dynasty) wrote the Shuizhu jishi ding'e: juanyi水注集释订讹·卷一[Correction of the annotations of the *Commentary on the Water Classic*, Vol 1]: "Tuo" 佗 should be "ta"他 (he) (from the *Siku quanshu* 四库全书本)
46. Chen Q陈桥驿 (1985) *Shuijing zhu yanjiu*水经注研究 [Research on *commentary on the water classic*]. Tianjin Ancient Books Publishing House天津古籍出版社, Tianjin, pp 65–77, 36–37
47. Chen Q陈桥驿 (1985) *Shuijing zhu yanjiu*水经注研究 [Research on *commentary on the water classic*]. Tianjin Ancient Books Publishing House天津古籍出版社, Tianjin, pp 65–77, 49–64
48. Shi N史念海 (1982) *Heshanji: er ji*河山集: 二集 [Collected works of rivers and mountains II]. SDX Joint Publishing Company 北京三联书店, Beijing
49. Chen Q陈桥驿 (1985) *Shuijing zhu yanjiu*水经注研究 [Research on *commentary on the water classic*]. Tianjin Ancient Books Publishing House天津古籍出版社, Tianjin, pp 78–87
50. Gong S龚胜生 (1996) Zhongguo songdai yiqian kuangquan de dilifenbu jiqi kaifa liyong中国宋代以前矿泉的地理分布及其开发利用 [The geographical distribution of mineral springs and their utilisation before the Song dynasty]. *Ziran kexueshi yanjiu*自然科学史研究 [Studies on the history of natural science] 15(4):346–351

51. Li Z李仲均 (1973) Woguo guji zhong jizai yanrong dongxue shilüe我国古籍中记载岩溶洞穴史略 [Records of karst cave in ancient Chinese literature]. *Gujizhun dongwu yu renlei xuebao*古脊椎动物与人类学报 [Journal of ancient vertebrate and humans] 11(2):201–205
52. *Shennong Bencao jing*神农本草经 [The Herbal classic of Shennong]
53. *Taiping yulan: juan liushijiu*太平御览·卷六十九 [Imperial reader of the Taiping era, vol 69]
54. Mo W王谟 (Qing dynasty) *Hangtang dili shuchao* 汉唐地理书钞 [Han and Tang geographical notes]
55. Shilong T陶世龙in the *Huoshan bian* 火山辩 [Debate on volcanoes] posits that the "volcanoes" in the Yuepan Kingdom are formed by spontaneous combustion of the coal layer (http://www.newsmth.net/bbsanc.php?path)
56. Shou W魏收 (Northern Qi dynasty) *Weishu: juan yibai ling er*魏书·卷一百零二 [Book of Wei, vol 102]. Chinese Publishing House华文出版社, Beijing, p 2268
57. Wang J王嘉荫 (1963) *Zhongguo dizhi shiliao*中国地质史料 [Historical data of Chinese geology]. Science Press北京科学出版社, Beijing, p 54
58. Yan J江淹 (Liang dynasty) Jiangwentong wenji: juanyi江文通文集·卷一 [Collection of Jiang Wentong, vol 1]. In: *Siku quanshu*四库全书 [Complete library of the four treasuries]
59. Gongzhan D杜公瞻 (Sui dynasty) *Bianzhu juanyi*: Tiandi bu编珠·卷一·天地部 [Composition of pearls, vol 1 heaven and earth]. In: *Siku quanshu*四库全书 [Complete library of the four treasuries]
60. Zongyi T陶宗仪 (Yuan dynasty) *Shuofu: juan liushiyi*说郛·卷六十一 [Persuasion of the suburbs, vol 61]. In: *Siku quanshu*四库全书 [Complete library of the four treasuries]
61. Jingshu L刘敬叔 (Liu Song dynasty) *Yiyuan: juanyi*异苑·卷一[Alien places, vol 1]. In: *Siku quanshu*四库全书 [Complete library of the four treasuries]
62. Yang X杨衒之 (Later Wei dynasty). *Luoyang qielan ji*: *juanwu*洛阳伽蓝记卷五 [Record of Buddhist temples in Luoyang, vol 5]. In: *Siku quanshu*四库全书 [Complete library of the four treasuries]
63. Zhang X章巽 (1985) *Faxian zhuan jiaozhu*法显传校注 [Annotated biography of Fa Xian]. Shanghai Ancient Books Publishing House上海古籍出版社, Shanghai
64. *Taiping yulan: juan liushijiu*太平御览·卷六十九 [Imperial reader of the Taiping era, vol 69]
65. *Chuxue ji: juanliu*初学记·卷六 [Writings for elementary instruction, vol 6]
66. Chen Q陈桥驿 (1985) Shuijing zhu jizai de zhiwudili"水经注"记载的植物地理 [Phytogeography in the *commentary on the water classic*]. In: *Shuijing zhu yanjiu* 水经注研究 [Research on the *commentary on the water classic*]. Tianjin Ancient Books Publishing House天津古籍出版社, Tianjin, pp 111–123
67. Chen Q陈桥驿 (1985) Shuijing zhu jizai de zhiwudili"水经注"记载的动物地理 [Zoogeography in the *commentary on the water classic*]. In: *Shuijing zhu yanjiu*水经注研究 [Research on the *commentary on the water classic*]. Tianjin Ancient Books Publishing House天津古籍出版社, Tianjin, pp 124–131
68. Gou C苟萃华 (1991) Daikaizhi zhupu tanxi戴凯之 "竹谱" 探析 [Analysis of "Book of the Bamboo" by Dai Kaizhi]. *Ziran Kexueshi Yanjiu*自然科学史研究. [Studies on the history of natural science] 10(4):342–348
69. According to the *Hou hanshu: Dongyi liezhuan* ["Book of the Later Han: Biographies of eastern foreigners"], Japan dispatched an envoy to China for the first time in the 2nd year of Emperor Guangwu's reign (57 AD): "in the 2nd year of Jianwu Zhongyuan (57), the Japanese Kingdom offered tributes to the Emperor and called themselves "Dafu". They were from the southern border of Japan. Emperor Guangwu awarded them a seal and ribbon in return." Afterwards, Japan sent emissaries to China four times: in the first year of Yongchu's reign of the Han dynasty (107), the 3rd year of Jingchu's reign of the Wei dynasty (239), the 4th year of Zhengshi's reign of the Wei dynasty (243), and the 2nd year of Taishi's reign of the Western Jin dynasty (266). On the fourth time, Japan sent a large delegation of 20 people. In the first year of Zhengshi's reign of the Wei (240), a Chinese emissary escorted the Japanese envoys back to their home for the first time. More details can be found in *Rizhong liangqian nian* 日中两千年 [Two thousand years of China-Japan relations], by Shintaro Nakamura 中村新太郎 (1980). Changchun: Jilin People's Press吉林人民出版社, p 13

70. According to *Sanguozhi: Weizhi: Dongyi liezhuan* 三国志·魏志·东夷传 [History of the Three Kingdoms: Record of the Wei dynasty: Biography of eastern foreigners], the Japanese queen sent emissaries to pay tribute to the Wei dynasty in the 2nd year of the reign of the Ming Emperor (238 AD). Liu Xia, the chief of Daifang Prefecture County, dispatched officials to help them reach the capital city. The *Jinshu: Wudi benji* 晋书·武帝本纪 [Book of the Jin: Imperial biography of Emperor Wu] records that "in the 2nd year of the Taishi reign (266), Japanese envoys paid tribute to the Chinese emperor on the 16th day of November (Chinese lunar calendar)". The *Suishu: Dongyi zhuan* 隋书·东夷传 [Book of the Sui: Biography of eastern foreigners] also mentions that there was constant communication between China and Japan from the Wei to the Qi and Liang dynasties
71. According to He Yuanjing 何远景 (1994) "'Weizhi: Woren zhuan' qian siduan chuzi 'Dongguan hanji' kao" "魏志·倭人传"前四段出自"东观汉记"考 [The first four paragraphs of the "Records of Wei: Japanese pirates" come from the "Biography of the Later Han"] *Neimenggu shifan daxue xuebao (sheke ban* 内蒙古师范大学学报(社科版)[Journal of Inner Mongolia Normal University (social science edition)] (3), the first four paragraphs of the *Sanguo zhi: Weizhi: Wuwan xianbei dongyi zhuan di sanshi: Woren zhuan* 三国志魏志乌丸鲜卑东夷传第三十: 倭人传 [Records of the Three Kingdoms: Records of Wei: Biographies of the Wuwan, Xianbei and eastern foreign nations No 30: the Japanese] come from the *Dongguan hanji: Dili zhi*东观汉记·地理志 [Biography of the Later Han: Geography]
72. Lu J陆峻岭, Zhou S周绍泉 (1986) *Zhongguo guji zhong youguan jianpuzhai ziliao huibian*中国古籍中有关柬埔寨资料汇编 [A compilation of Cambodian materials in ancient Chinese texts]. Chinese Publishing House华文出版社, Beijing, pp 1–21 and 32–51
73. *Taiping yulan: juan qibaijiushi: Siyibu: Jinlin guo*太平御览卷七百九十·四夷部·金邻国 [Imperial reader of the Taiping era, vol 790, four Barbarians: the Jinlin Kingdom]
74. Wen J汶江 (1989) *Gudai zhongguo yu yafei diqu de haishang jiaotong* 古代中国与亚非地区的海上交通 [Maritime transportation between China and Asian-African regions in ancient times]. Sichuan Academy of Social Science Press四川社会科学院出版社, Chengdu, p 49
75. Fang H方豪 (1987) *Zhongxi jiaotong shi*中西交通史 [A history of communication in China and western countries]. Yuelu Press 岳麓书社, Changsha, pp 211–212
76. Xuan D道宣 (Tang dynasty) *Shijia fangzhi* 释迦方志 [Chronicle of Buddha Sakyamuni]. Zhonghua Book Company中华书局, Beijing, pp 95–99
77. Tang Y汤用彤. *Han wei liangjin nanbeichao fojiao shi*汉魏两晋南北朝佛教史 [A History of Buddhism in the Han, Wei, Western and Eastern Jin dynasties]. Zhonghua Book Company 中华书局, Beijing, pp 418–424
78. Adachi K足立喜六 (1937) Faxian zhuan kaozheng 法显传考证 [Collected papers on examinations of the *Biography of Fa Xian*]. Jianmin H何健民, Xiaoliu Z张小柳 (trans). National Institute for Compilation and Translation & Commercial Press上海国立编译馆、商务印书馆, Shanghai, Preface
79. Jin H靳禾生 (1980) Fa Xian jiqi foguoji de jige wenti法显及其 "佛国记"的几个问题 [Several questions regarding Fa Xian and his work *A Travel to the Buddhist Countries*]. *Shanxi daxue xuebao*山西大学学报 [Journal of Shanxi University] 1
80. Yu T余太山 (1995) *Lianghan weijin nanbeichao yu xiyu guanxishi yanjiu*两汉魏晋南北朝与西域关系史研究 [Historical research on the relationship between the Han, Wei, Jin, Southern and Northern dynasties and the Western Regions]. China Social Sciences Press中国社会北京科学出版社, Beijing, Introduction
81. See Yu T余太山 (2001) Bufen zhengshi *Xiyu zhuan* suojian xiyu shanshui 部分正史《西域传》所见西域山水 [Landscape descriptions of the West regions in the *account of the western regions* in some official histories]. Shilin史林 [Historical review] 3:55
82. Zhang X章巽 (1986) Gudai zhongyang yaxiya yidai de diyu qufen古代中央亚细亚一带的地域区分 [Geographic division of Central Asia in ancient times]. In: *Zhangxun wenji*章巽文集 [Collected works of Zhang Xun]. Ocean Press海洋出版社, Beijing, pp 212–218
83. Niu Z钮仲勋 (1996) Woguo gudai dui zhongya de dili kaocha he renshi我国古代对中亚的地理考察和认识 [Geographical exploration and understanding of Central Asia by the

ancient Chinese]. *Dili xueshi yanjiu*地理学史研究 [Research on the history of geography]. Geological Publishing House地质出版社, Beijing, pp 14–18

84. Yu T余太山 (2001) Bufen zhengshi Xiyuzhuan suojian xiyu shanshui部分正史《西域传》所见西域山水 [Landscape descriptions of the Western Regions in the *account of the western regions* in some official histories]. *Shilin*史林 [Historical review] 3:54
85. Yu T余太山 (2001) Bufen zhengshi Xiyuzhuan suojian xiyu shanshui部分正史《西域传》所见西域山水 [Landscape descriptions of the western regions in the *account of the western regions* in some official histories]. *Shilin* 史林 [Historical review] 3:55–56
86. See Xuanzhi Y杨炫之 (Northern Wei dynasty) *Luoyang qielan ji juan wu*洛阳伽蓝记卷五 [A record of Buddhist monasteries in Luoyang, vol 5]
87. *Taiping yulan: juan jiubai bashiba*太平御览·卷九百八十八 [Imperial reader of the Taiping era, vol 988]
88. Ouyang X et al.欧阳询等 (Tang dynasty). *Yiwen leiju juan jiushiyi*艺文类聚·卷九十一 [Classified anthology of literary works, vol 91]
89. *Taiping yulan: juan qibai bashiliu*太平御览·卷七百八十六 [Imperial reader of the Taiping era, vol 786]
90. Xiang D向达 (1957) Hantang jian xiyu ji hainan zhuguo gu dilishu xulu汉唐间西域及海南诸国古地理书叙录 [Description of ancient geographical texts on the countries in the West Regions and South Sea during the Han and Tang dynasties]. In: *Tangdai Chang'an yu xiyu wenming*唐代长安与西域文明 [Civilisations of Chang'an and the West Regions during the Tang dynasty]. SDX Joint Publishing Company北京三联书店, Beijing, p 569
91. *Taiping yulan: juan qibai bashiqi*太平御览·卷七百八十七 [Imperial reader of the Taiping era, vol 787]
92. *Taiping yulan: juan qibai bashiab*太平御览·卷七百八十八 [Imperial reader of the Taiping era, vol 788]
93. Liu S刘盛佳 (1987) Jindai jiechu de ditu xuejia: Jing Xiangfan晋代杰出的地图学家——京相璠 [An outstanding cartographer in the Jin dynasty: Jing Xiangfan]. *Ziran kexueshi yanjiu* 自然科学史研究 [Studies on the history of natural science] 6(1):60–63
94. Daoyuan L郦道元 (Northern Wei dynasty) *Shuijing zhu: Jishui*水经注: 济水 [Commentary on the water classic: Ji River]
95. Daoyuan L郦道元 (Northern Wei dynasty) *Shuijing zhu: Jishui*水经注 [Commentary on the water classic: Hu River]
96. See *Sanfu huangtu jiaozhu*三辅黄图校注 [Corrections and annotations of the *Imperial Plan of the Three Capitals Region*] by He Qinggu 何清谷 (1995). Xi'an: Sanqin Publishing House 三秦出版社
97. Chen Q陈桥驿 (1985) Shuijing zhu jizai de xingzhengquhua"水经注" 记载的行政区划 [Administrative divisions in the *commentary on the water classic*], *Shuijing zhu yanjiu* 水经注研究 [Research on the *commentary on the water classic*]. Tianjing Ancient Books Publishing House天津古籍出版社, Tianjin, pp 149–162
98. *Sanguo zhi: Shuzhi: Xianzhu zhuan di'er*三国志·蜀志·先主传·第二 [Records of the Three Kingdoms: Shu Kingdom: Biography of Liu Bei II], cited from *Wushu* 吴书 [Book of Wu]
99. Jia W王嘉 (Jin dynasty) *Shiyi ji: juanba*拾遗记·卷八 [Record of gleanings, vol 8]
100. Shinan Y虞世南 (Tang dynasty) *Beitang shuchao jiushiliu* 北堂书钞 [Excerpts of books from the Northern Hall, vol 69], cited from *Jin zhugong zan*晋诸公赞 [Commentary on the nobles of the Jin]
101. Qian S司马迁 (Han dynasty) (1974) *Shiji: Qinshihuang benji*史记·秦始皇本纪 [Records of the grand historian: imperial biography of Qin Shi Huang]. Zhonghua Book Company中华书局, Beijing
102. Yue S沈约 (Liang of the Southern dynasty) (1974) *Songshu: Xiezhuang zhuan*宋书·谢庄传 [Book of the song—biography of Xie Zhuang]. Zhonghua Book Company中华书局, Beijing
103. Cao W曹婉如, Zheng X郑锡煌 (1987) Shilun daojiao de wuyue zhenxing tu试论道教的五岳真形图 [Daoist plan of the five Great Mountains]. *Ziran kexueshi yanjiu*自然科学史研究 [Studies on the history of natural science] 6(1):52–57

104. Jiang S姜生 (2008) Dongyue zhenxingtu de dituxue yanjiu东岳真形图的地图学研究 [A cartographic study on the plan of Mount Tai]. *Lishi yanjiu*历史研究 [Historical study] 6:34–51
105. Pang X et al房玄龄等 (Tang dynasty), *Jinshu: Peixiu zhuan* 晋书·裴秀传 [Book of the Jin: Biography of Pei Xiu]. Zhonghua Book Company中华书局, Beijing
106. There are serious omissions of parts of the Six Observable Principles of Map-Making in the *Jinshu: Peixiu zhuan*, while in the *Yiwen leiju* ("Classified Collection Based on the Classics and Other Literature") and *Chuxue ji* ("Writings for Elementary Instruction"), the complete text is recorded
107. Wei H胡渭 (Qing dynasty) in his *Yugong zhuizhi: Yugongtu houshi* 禹贡锥指·禹贡图后识 An analysis of the *Tribute of Yu*], applauded Pei Xun's unparalleled achievements in cartographic techniques since the Qin and Han dynasties
108. Solutions can be found in Qian B钱宝琮 (1981) *Zhongguo shuxue shi*中国数学史 [A history of mathematics in China]. Science Press北京科学出版社, Beijing, pp 72–75
109. Wang S王树连 (1997) Weijin nanbeichao cehuishi kao魏晋南北朝测绘史考 [History ofsurveying during the Wei, Jin, Southern and Northern dynasties]. *Celiang gongcheng*测量工程 [Surveying engineering] 6(3):42–43
110. Ying Z张英 (Qing dynasty) (1932) *Yuanjian leihan: dibu: hu'er*渊鉴类涵·地部·湖二 [Categorised collection in the Yuanjian Study: Earth: Lake]. Shanghai saoye shanfang上海埽叶山房
111. Xuanling P et al房玄龄等 (Tang dyansty) *Jinshui sanshijiu juan*晋书·三十九卷 [Book of the Jin, vol 39]. Zhonghua Book Company中华书局, Beijing
112. Li Z李仲均 (1982) Woguo gudai guanyu "hailubianqian" dizhi sixiang ziliao kaobian我国古代关于"海陆变迁"地质思想资料考辨 [Investigation into the geological theories about "changes of sea and land" in ancient China]. *Kexueshi jikan: dishiji* 科学史集刊: 第10辑 [Collected papers on the history of science, vol 10]. Geology Press地质出版社, Beijing, pp 19–20
113. *Siku quanshu*四库全书 [Complete library of the four treasuries]
114. *Taiping yulan: juan sishijiu*太平御览·卷四十九 [Imperial reader of the Taiping era, vol 49]. Spirifers from Quanling County are recorded in Volume 49
115. *Taiping yulan: juanjiu*太平御览·卷九 [Imperial reader of the Taiping era, vol 9]
116. *Taiping yulan: juan liushiwu*太平御览·卷六十五 [Imperial reader of the Taiping era, vol 65]
117. *Taiping yulan: juan jiubai sanshiliu*太平御览·卷九百三十六 [Imperial reader of the Taiping era, vol 936]
118. Zongyi T陶宗仪 (Yuan dynasty) (1988) *Shuofu: juan liushiyi*说郛·卷六十一[Persuasion of the suburbs, Vol 61], cited from *Shuofu sanzhong* 说郛三种 [Three records of the *Shuofu*]. Shanghai Ancient Books Publishing House上海古籍出版社, Shanghai, p 2833
119. Tang X唐锡仁, Yang W杨文衡 (2000) *Zhongguo kexue jishu shi: dixue juan*中国科学技术史·地学卷 [History of science and technology in China: geoscience]. Science Press北京科学出版社, Beijing, p 263
120. *Taiping yulan: juan babai lingba*太平御览·卷八百零八 [Imperial reader of the Taiping era, vol 808]
121. Cited from Xiaozhong's C曹孝忠 (Song dynasty) *Zhenghe xinxiu jingshi zhenglei beiyong bencao* 政和新修经史证类备用本草[Newly edited *zhenglei bencao* in the Zhenghe reign], originally written by Tang Shenwei 唐慎微 (Song dynasty). People's Medical Publishing House人民卫生出版社, Beijing
122. Hongjing T陶弘景 (Liang dynasty), *Mingyi bielu*名医别录 [Anecdotes of famous doctors]. Zhijun S尚志钧 (ed). People's Medical Publishing House人民医学出版社, Beijing, p 70
123. Yue S沈约 (Liang dynasty), *Songshu: Wudiji*宋书·武帝纪 [Book of the Song: imperial biography of Emperor Wu]
124. Xia X et al夏湘蓉等 (1986) *Zhongguo gudai kuangye kaifashi*中国古代矿业开发史 [History of mining industries in ancient China]. Geological Publishing House地质出版社, Beijing
125. Pu W吴普 (Wei dynasty) (1987) *Wupu bencao*吴普本草 [Wu Pu's Materia Medica]. Zhijun S尚志钧(ed). People's Medical Publishing House人民医学出版社, Beijing

126. Ai S艾素珍 (1994) Lun bencaojizhu zhong kuangwuxue zhishi ji zai zhongguo kuangwuxueshi shang de diwei论"本草集"注中矿物学知识及在中国矿物学史上的地位 [Mineralogical knowledge in "Bencao jizhu" and its status in the history of Chinese mineralogy]. *Ziran kexueshi yanjiu*自然科学史研究 [Studies on the history of natural science] 13(3):273–283
127. Zhao K赵匡华 (1987) Zhongguo lidai huangtong kaoshi中国历代黄铜考释 [Intepretation of brass across Chinese dynasties]. *Ziran kexueshi yanjiu*自然科学史研究 [Studies on the history of natural science] 6(4):232–331
128. *Taiping yulan juan: babai yishisan* 太平御览·卷八百一十三 [Imperial reader of the Taiping era, vol 813]
129. *Taiping yulan juan: qibai bashiliu* 太平御览·卷七百八十六 [Imperial reader of the Taiping era, vol 786]
130. Xiao Z张斅 (Southern Song dynasty) (1986) 雷公炮炙论*Leigong paozhi lun* [Lei Gong's treatise on the preparation and broiling of materia medica]. Wang Xingfa 王兴法 (ed). Shanghai College of Chinese Medicines, Shanghai, p 6
131. *Taiping yulan: juan jiubai bashiqi: Liuhuang*太平御览·卷九百八十七: 硫磺 [Imperial reader of the Taiping era, vol 987: Sulfur]
132. Pu Z张溥 (Qing dyansty) *Han wei liuchao baisanjia ji juan wushiqi*汉魏六朝百三家集卷五十七 [Collected works from 103 scholars of the Wei, Jin, Southern and Northern dynasties, vol 57]
133. The original version was lost, but relevant text is found in Volume 2 "Yu yuliang" 禹余粮of the *Tujing yangyi bencao*图经衍义本草, cited from the *Congshu jicheng chubian*丛书集成初编 [A preliminary compilation of the integrated series]
134. Although the six volumes of the *Dijing tu* can be found in the *Book of the Sui: Scriptures and Classics,* they are not recorded in the *Tangshu: Yiwen zhi.* This quote comes from the *Hantang dili shuchao* ("Geographical Notes in the Han and Tang Dynasties") by Wang Mo 王谟 (Qing dynasty), republished by Zhonghua Book Company中华书局, Beijing, pp 53–54
135. Li E李鄂荣 (1987) Dijingtu de dizhixue sixiang 地镜图的地质学思想 [Geological thinking in the "Dijingtu"]. *Zhongguo dizhi*中国地质 [Geology in China] 4
136. Liu Z刘昭民 (1994) Zhongguo songdai zhiqian de zhanhoujia中国宋代之前的占候家 [The offical Zhanhou before the Song dynasty in China]. *Zhongguo keji shilliao*中国科技史料 [Historical data of Chinese science and technology] 15(2):14–15
137. Hua Z张华 (Western Jin dynasty) *Shuofu: juan ershisi*说郛 卷二十四 [Persuasion of the suburbs, vol 24]. In: *Ganyinglei congzhi*感应类从志. The Commercial Press商务印书馆, Beijing
138. The *Jinshu: Yufu zhi* ("Book of the Jin: Carriages and Costumes") recorded that wind instruments were used in the Grand Imperial Processions. Zhang Hua mentioned in his *Xiangfeng fu that* there was a wind detecting machine in the west part of the city. Guo Yuansheng also said in the *Zhengshu ji*: "On the observatory to the south of Chang'an Palace, there was a *xiangfeng tongwu.* Some said it would move with the wind." Ge Hong of the Jin dynasty described in the *Xijing zaji* that "according to Jin rituals, a bird sculpture is placed on the pole at the front of the carriage (like a sail that indicates the wind direction). Xiao Yan, the son of Duke of Changsha, presented the *Xiangfeng fu* to the emperor during his trip to Hualin Garden
139. In the *Book of the Jin* 晋书, it is recorded that "when the 'deposed Emperor' of the Eastern Jin dynasty ascended the throne, there were flocks of wild pheasants crowding into the Xiangfeng, a wind detecting machine, which was later abolished by Huan Wen"
140. Kejun Y严可均 (Qing dynasty) *Quan shanggu sandai qinhan sanguo liuchao wen — quan jin wen*全上古三代秦汉三国六朝文·全晋文 [A complete collection of prose writings from the earliest antiquity to the pre-dynastic Tang: the Jin texts]. Zhonghua Book Company 中华书局, Beijing
141. Ding F丁福保 (1959) *Quanhan sanguo jin nanbeichao shi: quan qi shi*全汉三国晋南北朝诗·全齐诗 [A complete collection of ancient poems from the Han, Three Kingdoms, Southern and Northern dynasties: the Qi poems, vol 4]. Zhonghua Book Company 中华书局, Beijing

142. Chen J陈久金 (2008) *Zhongguo tianwenxueshi daxi: Zhongguo gudai tianwen xuejia*中国天文学史大系·中国古代天文学家 [Series of Chinese astronomical history: Ancient Chinese astronomers]. Science and Technology Press of China中国科学技术出版社, Beijing, pp 160–164
143. *Taiping yulan: juanjiu* 太平御览·卷九 [Imperial reader of the Taiping era, vol 9].

Chapter 6
Geoscience in the Sui, Tang and Five Dynasties

6.1 The Impact of the Social Environment on Geoscience

The impact of the social environment on geoscience in the Sui, Tang and Five Dynasties is mainly evident in the following six areas.

First, the unified political situation of the Sui and Tang dynasties contributed to the construction of large-scale water conservancy projects, such as the Grand Canal. It also contributed to the development of water and land transportation, especially international water and land transport systems. For example, several land routes leading to Central Asia and India were built at that time and, by sea, one could go to Japan, Korea, Southeast Asia, India and East Africa. The situation was also conducive to exploration for the source of the Yellow River, which was a major contribution made by people in the Tang dynasty to ancient Chinese geography. Leading figures included Hou Junji, Li Daozong, Liu Yuanding and others.

Second, the well-developed transport systems in the Sui and Tang dynasties provided convenience to travellers and boosted the development and success of the field of geography. Many adventurers left significant marks on the geography of the Sui and Tang dynasties, such as Xuan Zang, Yi Jing, Wang Xuanze, Du Huan, Li Ao, Zhang Kuangye, Gao Juhui and others through their travel expeditions.

Third, the needs of the ruling class exerted considerable influence on people's understanding of geography. For example, Jia Dan's maps, local chronicles culminating in *tujing*, theories of administrative geography by Li Jifu and Du You, etc.

Fourth, the military influenced the development of geography, which can mainly be seen in maps, geomorphology, meteorology and water conservancy. Some examples are the *Hebei xianyao tu* ("River North Strategic Feature") by Li Jifu, military frontier maps by Li Deyu and the *Jiuzhou shexian tu* ("Maps of Battlefields across Jiuzhou") by Wei Yuanzhong, etc. As regards meteorology, examples are the *Wangqi Jing* by Shao E and *Xiangyu shu* by Huang Zifa. The most commendable example

W. Yang and Q. Yang, *A History of Ancient Chinese Geoscience*, Understanding China,
https://doi.org/10.1007/978-981-97-4457-2_6

in the field of water conservancy is the excavation of the Grand Canal in the Sui dynasty.

Fifth, the development of the economy contributed to geology, such as in the handicraft industry and mining in particular. In addition, water conservancy projects and seafaring promoted the development of hydrology, which was reflected in the marine tidal theories and other works by Dou Shumeng, Feng Yan, Lu Zhao, Qiu Guangting.

Sixth, geoscience in the Sui and Tang dynasties also benefited from the boom in religion, especially from Daoism and Buddhism in the field of mineralogy and geography. For example, Daoism's contributions to ancient Chinese mineralogy include the *Jinshibu wujiu shujue* by an unknown author, *Shiyao erya* by Mei Biao, *Yudong dashen dansha zhenyaojue* ("The Secret Recipe of Alchemy by Yudong God") by Zhang Guo, *Yaopu* ("Pharmacopoeia") by Hou Ningji, *Longhu huandanjue* ("Alchemical Recipes") by Jinlingzi, *Danfang jianyuan* ("The Original Ingredients of Alchemy") by Du Gutao, *Taiqing shibi ji* ("Records of Taiqing Stone Wall") by Su Yuanming and adapted by Chu Ze, and *Baozang lun* ("A Treatise on Stored Treasures") by Xuanyuan Shu of the Five Dynasties. Contributions by Buddhists to ancient Chinese extraterritorial geography include the *Datang xiyu ji* ("Pilgrimage to the West in the Tang dynasty"), *Zhongtian zhuguo xingj*i ("A Trip to Middle Indian Countries"), *Datang xiyu qiufa gaoseng zhuan* ("Biographies of Eminent Monks who Sought for the Dharma in the Western Regions in the Tang dynasty"), and *Nanhai jigui neifa zhuan* ("An Account of the Inner Law Sent Home from the South Sea").

6.2 Surveying and Mapmaking

6.2.1 Achievements in Land Surveying

According to the *Suishu: Tianwen zhi*, Liu Zhuo (554–610) of the Sui dynasty strongly advocated that the incorrect statement that "the difference in the sun's shadow is one inch for every thousand li" should be proved by measurement. He suggested that "a water worker and an arithmetician should be invited, and then chose a flat area of land in the north of the Yellow River and measured hundreds of li which lay at between longitude 114.2 degrees and 114.5 degrees east, with the help of a timer and a rope to measure the length of a sundial's shadow at different times. Then he calculated the difference in arriving at the length of the meridian." In 607, Emperor Yangdi of the Sui ordered various regions to measure the shadow, but because Liu Zhuo died soon afterwards, his proposed approach was put on hold and never realised.

More than a hundred years later, Liu Zhuo's idea was acted on by Monk Yi Xing (683–727), originally known as Zhang Sui, who was a famous astronomer in the Tang dynasty. He led and organised a large-scale astronomical geodetic measurement project with Nangong Yue in 724 and achieved extremely valuable results in astronomy and metrology.

The purpose of astronomical geodesy in the Tang dynasty was to complete Yi Xing's task of correcting the calendar system, not for ground mapping. The measurement involved many steps, and the work directly related to surveying mainly involved two: one was the measurement of the length of shadows on the vernal and autumn equinoxes and the summer and winter solstices; the other was the measurement of the height of the "north celestial pole".

While Yi Xing designed and led the project, Nangong Yue implemented the measurement work. There were 13 test sites established throughout the empire for this project: (1) Linyi (at about latitude 18°N, in the middle of present-day Vietnam); (2) Anan protectorate (in present-day Vietnam); (3) Wuling County, Longzhou (present-day Changde, Hunan province); (4) Xiangzhou (present-day Xiangyang, Hubei province); (5) Wujin, Shangcai County, Caizhou (present-day Shangcai, Henan province); (6) Fugou, Xuzhou (present-day Fugou County, Henan province); (7) Junyi Taiyuetai, Bianzhou (present-day northwest of Kaifeng, Henan province); (8) Baima, Hua Zhou (present-day Huazhou County, Henan); (9) Taiyuan Prefecture (present-day Taiyuan, Shanxi); (10) Hengye Army, Yuzhou (present-day Yuzhou, Hebei); (11) Tiele (at about latitude 51°N, in present-day Mongolia); (12) Yangcheng (present-day Gaocheng Town, Dengfeng, Henan); (13) Luoyang. These vast areas range from latitude 18°N to latitude 51°N. The surveying team led by Nangong Yue measured four of these locations, and this was carried out according to Liu Zhuo's plan. The results were of great significance.

According to the *Xin tangshu: tianwen zhi* ("New Book of the Tang: Records of Astronomy"), Nangong Yue started from Baima in Hua County, through Junyi and Fugou, then to Shangcai County; he measured the shadow lengths in winter, summer, spring and autumn with the eight-foot rod, and measured the height of Polaris in various places with the complex matrix, and then measured the distance between each of these four locations. It was found that the distance from Hua County to Shangcai County was 526 li and 270 *bu*, but the difference in the sun's shadow on the summer solstice was more than two inches, which completely disproved using the different shadow lengths of the sun observed at the same time in two places to determine the ground distance between them. More importantly, Yixing made a new discovery when analysing and comparing a large number of the observations made by Nangong Yue and others, namely, the ratio of shadow difference to the north–south distance was not a constant at all. However, the ratio of the difference between latitudes to the distance between the two places was a definite number. They reached an important conclusion: the length of one degree of latitude equalled the length of a one-degree meridian line.

During the Sui and Tang dynasties, the concept of the Spherical Earth was not accepted by Chinese geographers, so they were not aware that the height of the "north celestial pole" was geographical latitude. However, the height of Polaris had already been used to mark the geographical location of certain places, reflecting the latitude of such places. In this sense, it can be said that Yixing and others had already carried out large-scale latitude measurement work.

The length of one degree of meridian measured in the Tang dynasty was slightly larger than the current 111.2 km. However, the flat terrain and the absence of rivers from Shangcai to Junyi made the measurements more accurate, and the difference compared with the modern measurement is only 11.3 km [1].

In the Tang dynasty, level measuring technology was fairly advanced. Li Quan's *Tabaiyin jing* ("Scripture of Venus and Yin"), written in 759, introduced the level tools and measurement methods in detail.

> The cuboid is two *chi* and four *cun* long. There are three rectangular cuts perpendicular to the length of the cuboid, one at each of its ends and one in the middle. Each rectangular cut is one *cun* and eight *fen* long, one *cun* wide, and one *cun* and three *fen* deep. The distance between the rectangular cuts is one *chi* and four *cun* (precisely one *chi* and five *cun* upon verification). There is a groove filled with water running along the length of the cuboid that connects the rectangular cuts. The groove is three *fen* wide and one *cun* and three *fen* deep. A rectangular wood block is placed in the middle of each cut, floating on the water. On top of each wood block there is a sawtooth ridge eight *fen* high, one *cun* and seven *fen* wide, and one *fen* thick. Underneath the [centre of the] cuboid is a pedestal supporting the mechanism.
>
> The *zhaoban* (target board) is shaped like a square hand fan. It is four *chi* tall, two *chi* for the black half at the bottom and two *chi* for the upper white half. It is three *chi* thick and the handle, which it can be held by, is one *chi* long. The measuring pole is two *zhang* tall, carved with divisions of 200 *cun* and 2,000 *fen*. [2]

The structure and usage of this kind of level are also similarly recorded in Du You's *Tongdian* and Xu Dong's *Huqian jing* of the Northern Song dynasty. In the *Wujing zongyao* by Zeng Gongliang of the Northern Song dynasty, not only are the measuring tools and measuring methods described in detail, but detailed drawings are also attached [3]. It is evident that the latter three records were taken from the *Taibai yinjing*, given this description: a target was used by the holder and the observer to obtain the measurement. The holder moved the target up and down according to the observer's signal, until the three floating pieces of wood were lined up with the black and white demarcation lines. The holder then wrote down the corresponding dimensions on the level (degree pole). When measuring the height difference between two points, the level was moved between the two points, on each of the two points respectively, on a stand designed to hold the level ruler. If the two points to be measured were far from each other or if there was a big difference in height, the instrument needed to be set up several times to measure segments, which is called compound level measurement in modern surveying. The measurement principle and method used at that time was consistent with modern-day approaches [1].

6.2.2 The Drawing of Official and Private Maps

During the Sui, Tang and Five Dynasties, the ruling class attached great importance to maps. After conquering Danyang, Yang Guang, the King of Jin "ordered (Pei) Ju and Gao Jiong to collect maps of Chen" [4]. In the Tang dynasty (621), "Emperor Taizong entered the palatial area (Luoyang, the eastern capital) and ordered Fang Xuanling to collect maps of the Sui" [5]. The Tang court clearly stipulated: "The maps were entrusted to the prefectures to be updated once every three years, with the household registry maps being submitted to the relevant province [6]. Every time a foreign guest came to the capital, Honglu officials would enquire about the landscape and customs of his country and record the information on maps." After 780, the maps were updated every five years. "If a state or county had changed the names of its mountains and rivers, the update was not limited to five years [7]." Tang Taizong also ordered Lü Cai to create the *Fangyu tu* and *Jiaofeiqi zhanzhen tu* ("Cavalry Battle Array Map") [8], as well as to collect maps of the vassal states and neighbouring countries. Xu Jingzong compiled the texts and maps written and drawn by the emissaries who went to Kang and Tukhara into the 60 volumes of the *Xiyu tuzhi* ("Maps of and Notes about the Western Regions") and presented them to the emperor. After Jia Yanzhong arrived in Liaoning, he also drew the *Liaodong shanchuan dishi tu*. Wang Yanwei presented the *Zhan'e tu* and the *Gongjun tu* ("Garrison Map") to the emperor [9]. Yuan Zhen drafted the road map of Huiji Khanate. Yan Liben sketched the *Xiyu zhuguo fengsu tu* ("Map of the Customs in the States of the Western Regions") [10].

From the *New Book of the Tang*: *Treatise on Literature*, we learn that the governments of the Tang dynasty commissioned the sketching of 13 volumes of the *Chang'an sinian shidao tu* ("Map of Ten Regions Compiled in the Fourth Year of Chang'an") and 10 volumes of the *Kaiyuan sannian shidao tu* ("Map of Ten Regions Compiled in the Third Year of Kaiyuan"). Li Jifu drew 10 volumes of the *Shidao tu* in the Yuanhe reign. These maps were developed based on the maps submitted by various prefectures. In February of the 3rd year of Changxing (932), Li Zanhua, the Military Commissioner of the Huaihua Army, sent in the *Map of Khitan*. In April of the 3rd year of the Changxing, the Secretariat announced that "the three capitals and the prefectures of the provinces are permitted to redefine the order of the map". The *Shidao tu* could follow the old approach or adopt a new one [11]. In February of the 4th year of the Changxing reign (933), the Puzhou Prefecture recommended the revision of the *River Levee Map*, "the names of the places along the river are clearly identifiable". It can be seen that from the Tang to the Five Dynasties, this type of *shidao tu* had to be modified in accordance with any change in the political situation. These maps have all been lost.

Maps drawn by Li Jifu were some of those collected in the *Yuanhe junxian tuzhi* ("Maps and Notes of Prefectures and Counties in the Yuanhe Reign"), the *Shidao tu* and the *Hebei xianyao tu* which contained all the military sites and fortifications in the north of the Yellow River and which Emperor Xianzong had hanging on the wall of his bathing hall for reference [12]. Li Deyu, the son of Li Jifu, also attached great importance to military maps. In the Choubian Tower that he built, there was a map of the mountains and rivers on the road to the southern barbarians on the left wall and a map of the mountains and rivers on the road to Tubo on the right wall. Whenever there was a military exercise, the generals were called to "discuss in front of the maps so that they would know everything about the situation of the enemies" [13]. Prior to Li Deyu, during the period of Emperor Zhongzong, the commander of the entire army, Wei Yuanzhong (637–707), had already drawn the *Jiuzhou shexian tu* containing the victories and defeats in earlier wars [14].

Due to the need for water transport, maps of water transport routes were produced in the Tang dynasty. In the 8th year of the Yuanhe reign (813), Wang Bo presented the *Pipagou Transport Map* to the Emperor. In the 2nd year of the Changqing reign (822), Wang Ban presented another *New Yin River Mouth Map* [15]. In the 3rd year of the Changxing reign (932), Zhao Dejun presented the *New South-eastern Rivers Map* [16].

During the Yuanhe period, the five-colour *Dizhi tu* drawn by Li Gai, a native of Guangling, was a distinctive map. According to Lü Wen, who had seen this map, there were mountains, rivers, topography, products, cities and towns, historical sites, borders, roads, etc. in the map, distinguished by different colours [17]. Unfortunately, the map has been lost.

During the Sui and Tang dynasties, there was another special school of cartography whose maps were interwoven with astronomy, Taoism and calendrics. For example, Li Ban, a Taoist priest in the Sui dynasty, compiled the *Fangzhi tu* (or *Fangyu tu*). Li Ban, the father of Li Chunfeng, was a Gaotang County commandant in the Sui dynasty and, feeling frustrated because of his low rank, he abandoned his official position and became a Taoist priest. He had a literary talent and called himself Huang Guanzi. He interpreted some of the theories of Laozi, compiled the *Fangzhi tu* and the 10 volumes of the *Wenji* ("Collected Works"), which were both popular at that time" [18]. Li Chunfeng revised *Fangzhi tu* carefully; his annotations of prefectures and counties were the ones most used afterwards. They were divided up by mountains and rivers. Li Gai, the grandson of Li Ban, compiled the *Dizhi tu* (as mentioned above). Lü Cai compiled the *Fangyu tu*. "Lü Cai, was a native of Qingping, Bozhou. He studied hard at a young age and was well versed in the theories of yin-yang and medicine ... Emperor Taizong ordered him to compile the *Fangyu tu*. These maps are like local chronicles. Although they depict mountains and rivers, they are different from standard maps but rather are maps "related to Taoism and calendrics" [19].

The Tang dynasty saw the emergence of a cartographic method that resembled the Mercator projection and which was specifically applied to astronomical star charts. For example, a star chart from the Tang dynasty found in the Dunhuang Caves uses two projections to mark all the stars in the sky. One shows the stars near the North Pole on a chart with poles at the centre, and the other shows the stars near the equator on a "horizontal chart" concentrated along the equator. Such a "horizontal chart" is essentially an approximate Mercator projection. The characteristic of this projection method is that all longitude lines run from the north to the south, and all lines of latitudes run east to west, and these are both sets of parallel lines at right angles to each other, with equal intervals between longitude lines and increasing intervals between latitude lines as latitude increases. Although the Dunhuang star chart does not indicate latitude and longitude, the positions of stars near the equator are consistent with those on the astronomical star chart drawn by Su Song from the Northern Song dynasty using cylindrical projection [20]. Regrettably, this method of cartography was only used in star charts and was not extended to land maps.

6.2.3 Maps in Tujing

In the Sui and Tang dynasties, local chronicles feature *tujing*, or graphical records, as a genre. These graphics include maps, architectural drawings, pictures of local products, etc. Therefore, many maps can be found in *tujing*. For example, the *Zhuzhoujun tujing* ("Graphical Records of Prefectures and Counties") by Lang Weizhi in the Sui dynasty, *Quyu tuzhi* ("Graphical Records of Different Regions") by Yu Shiji, *Xiyu tuji* ("Graphics and Records of the Western Regions") by Pei Ju, *Dongdu tuji* ("Graphical Records of the Eastern Capital") by Yu Wenkai, *Yuanhe junxian tuzhi* ("Graphical Records of Counties in the Yuanhei Reign") by Li Jifu in the Tang dynasty, *Xiyu tuzhi* ("Graphical Records of the Western Region") by Xu Jingzong and other *tujing* from across the entire country all included maps. Unfortunately, these maps have all been lost. In these *tujing*, there were administrative maps, city planning maps, and maps of mountain and rivers. For example, the *Quyu tuzhi* "has a map at the beginning. It is two *chi* long. If the chapter is about mountains and rivers, there is a related picture at the beginning of the volume. If the chapter is about states and prefectures, there are maps of administrative units" [21]. In order to compose the *Quyu tuzhi*, the imperial court ordered all the prefectures to compile "maps of customs and local products and to submit them to the Shang Shu (Department of State Affairs)" [22].

The *Xiyu tuji* encompassed pictures of the customs, local products and clothing of the countries in the Western Regions. Its preface said that "the king and common people wear different clothes to show their status" [23]. There may have been maps of the routes to the Western Regions and maps of the mountains and rivers of these western states. The *Xiyu tuji*'s map covered an area of 20,000 li, including a vast area of the eastern coast of the Mediterranean Sea and to the south of the Aral Sea.

The 20-volume *Dongdu tuji* ("Maps of and Notes about the Eastern Capital") recorded the planning and description of the construction of the Eastern Capital, epitomising urban planning thought before the Sui dynasty.

The 60-volume *Xiyu tuzhi* was an illustrated local chronicle about the countries and nations in the Western Regions. According to the *New Book of the Tang: Treatise on Literature*, "Emperor Gao Zong sent envoys to the Kang Kingdom and Tukhara to explore their customs and products and to draw pictures to report back". Then Xu Jingzong was asked to compile them into the *Xiyu tuzhi*.

All the maps contained in the above books have been lost and there is no means of establishing their exact details.

6.2.4 Well-Known Cartographers

The most famous cartographer of the Tang dynasty was Jia Dan (730–805), courtesy name Dunshi. He was a native of Nanpi, Cangzhou (now Nanpi, Hebei). Over a period of 47 years, Jia Dan served in different positions including 13 years as a minister, and his political achievements were outstanding. As a young man he was fascinated by dialects, and he was intrigued to geography for thirty years. Considering the development needs of the country, he used every opportunity to study the geography and draw maps within the context of political and military activities. He would talk to anyone who come from or went to faraway places and question them about geographic information, based on which he drew maps of perilous sites and pictures of different customs. "Because the Tibetan regime had occupied Longyou for years and the Tang dynasty had stopped conquering outlying places by then, it was not possible to return to the towns and counties of olden times. [24]" He had long served in important official positions in local and central government and witnessed the decline of the country and the inability to tackle various incidents at the frontier. He could only express his deep concern. He said: "I have served in many positions, but there is still a great deal I did not do. I carried mountains and rivers in my heart." With an ambition to rejuvenate the Tang dynasty, he actively produced geographical works and drew maps that could help to recover its lost territories and re-establish effective rule in the frontier areas, as evidenced in the Preface to the *Weiguogong zhenyuan shidao lu* by Quan Dejiu [25]. His main geographical works and maps are listed below.

In the 14th year of the Zhenyuan reign (798), a scroll map of the *Guanzhong longyou shannan jiuzhou* ("Topographical Records of Guanzhong, Longyou, Shannan and Nine Regions") and the six volumes of the annotated text of this map, and four volumes of the *Tubo huanghe lu* ("Records of Tibet and the Yellow River Reaches") [26].

In the 17th year of the Zhenyuan reign (801), the scroll *Hainei huayi tu* ("Map of Chinese and non-Chinese Territories in the World"), and the 40-volume *Gujin junguo xiandao siyi shu* ("Discussion of the Four Kinds of Barbarians in China, Past and Present) [27].

In the 18th year of the Zhenyuan reign (802), four volumes of the *Zhenyuan shidao lu* ("Topographical Records of Ten Prefectures in the Zhenyuan Reign of the Tang Dynasty"). In this work, the distances between counties, provinces and the two capitals were clearly recorded along with the changes to counties or provinces [28].

From the 17th to the 21st year of the Zhenyuan reign (801–805), 10 volumes of the *Ditu* and *Shiyizhou ditu*, which cannot be found today; the 10-volume *Huanghua sida ji* ("Records of the Imperial Glory Extending in Four Directions"), which have also been lost but were recorded in the *New Book of the Tang*: *Treatise on Literature* and in the *History of Song: Treatise on Literature* [29].

Jia Dan inherited Pei Xiu's Six Mapmaking Principles and believed that the "six principles are new ideas and models for making maps". His work *Guanzhong longyou shannan jiuzhou tu* (lost) adopted Pei Xiu's model and showed the mountains, rivers, passes, roads, bridges, military and urban setups of some places in Longyou and Guanzhong. Jia Dan commented that it was difficult to use symbols to represent the geographical content of the map, such as the areas of the administrative regions, the number of households, the mountains and rivers, so he provided explanatory annotations and compiled these in books, namely the *Gunazhong longyou shannan jiuzhou bielu* and *Tubo huanghe lu.*

Jia Dan also instructed draughtsmen to draw a scroll of *Hainei huayi tu* which was "3 *zhang* wide and 3.3 *zhang* long. One inch equals a hundred li". According to the *Old Book of the Tang: Biography of Jia Dan Jia,* "Jia condensed the world into a map. Although the universe is vast, opening the scroll does not occupy the full court. The map is inclusive ... with the ancient states drawn in black and the present-day ones in red. Marking them with different colours makes it straightforward to read the map" (*Jiutangshu: Jiadan zhuan*). In order to draw this map, Jia Dan spent more than thirty years reading literature, surveying and interviewing. Although it has been lost, its recorded features are as follows.

First, this was the first large-scale map in China's history, and it covered not only the administrative regions of the Tang dynasty, but also some neighbouring countries. The scale was equivalent to 1:1.8 million. The map covered 30,000 li from east to west and more than 30,000 li from north to south, making it a giant map of nearly 100 square metres, which was much larger than Pei Xiu's *Fangzhang tu.* On the map were recorded the course of changes and developments in borderland areas, administrative divisions, names and locations of the past, as well as prefectures and counties, mountains and rivers, transportation routes, and so on. It reflected the geography, topography and history of the world at that time. It was more complex than the ancient maps in both style and content, demonstrating Jia Dan's expertise and superb cartography.

Second, Jia Dan was the first to use the cartographic method of "differentiating places with black and red". This method was followed by later historical cartographers, such as Li Zhaoluo in the Qing dynasty and Yang Shoujing in more modern times [30].

Another cartographer in the Tang dynasty worth mentioning is Yuan Zhen (779–831), courtesy name Weizhi, a native of Henei (modern Qinyang, Henan province). He lost his father at an early age and was raised and educated by his mother. He was a famous poet in the Tang dynasty, known as one of the "Yuan Bai" along with the poet Bai Juyi. He served in more than 14 official positions in different places. In his early years, he was a man of rectitude who yielded neither to the nobles nor to the wealthy and was, therefore, repeatedly relegated. When he was middle-aged, he had a close relationship with eunuchs, which was criticised by some but defended by others in later years. Throughout his remaining years, he remained motivated and still secured political achievements despite being relegated.

During the Yongzhen period, Yuan discussed northwest border issues with Emperor Xianzong. In Volume 24 of the *Yuanshi changqing ji* ("Collected works of Mr Yuan of the Changqing Period"), two articles written by Yuan Zhen, namely, "Jin xibeibian tujingzhuang" and "Jin xibeibian tuzhuang" explained the reason behind and the process of drawing several maps for the emperor to observe. Yuan Zhen finished drawing quickly and, in only a few days, he accurately completed the *Jingxi jingbei zhouzhen fengshudaolu tu* ("Maps of Beacon-fire Frontier Routes in the Jingxi and Jingbei Areas"). It is evident that Yuan Zhen was a skilled draughtsman and a superb cartographer in the Tang dynasty.

6.3 Illustrated Treatises and Atlases as Chronicles

Tuzhi and *tujing* as local chronicles had already emerged in the Eastern dynasty. From the Southern and Northern dynasties, many more *tujing* appeared and their popularity peaked in the Sui, Tang and Northern Song dynasties.

6.3.1 National Chronicles

There was a limited number of *fangzhi* in the Sui dynasty, but "because every prefecture was ordered to collect its customs, local products, and maps and present the compilation to the central government, there were the 151-volume *Zhujun wuchan tusu ji* ("Produce and Customs of All Prefectures"), the 129-volume *Quyu tuzhi* ("Maps and Treatises of All Domains"), and the 100-volume *Sui zhuzhou tujing ji* ("Collections of Map Guides to All Regions in the Sui"), and many other records" [22].

In the 6th year of the Daye reign (610), Emperor Yang ordered Yu Shiji to compile the *Quyu tuzhi*, with a map at the beginning of each volume [31]. This was a large-scale collection of national maps, but unfortunately it was lost. The *Sui zhuzhou tujingji* compiled by Lang Weizhi was a collection of *tujing* based on the local chronicles submitted by each state and prefecture in the form of maps with comments. This book has also been lost.

The Tang dynasty attached great importance to the compilation of *tujing*. Its central government had a special official in charge of this and stipulated that every three years (once altered to five years), every county and province should review their *tujing* and send the update to the military department of the Imperial Secretariat. "When there are no additions or removal of provinces or counties, the *tujing* will be updated every five years and presented to the central government together with the documents relating to households [32]." The central government would then compile a national chronicle based on the local *tujing* presented to them.

During the Zhenguan reign, Li Tai, Head of the Wei and son of Emperor Taizong, invited Xiao Deyan and other scholars to compile the 550 volumes of the *Kuo dizhi* ("Comprehensive Gazetteer") based on the districts determined by the divisions in the *Book of the Thirteenth Year of the Zhenguan Reign*. This comprehensively described the evolution of 10 provinces, 360 prefectures (including 41 governorates) and 1,557 counties, as well as their mountainous terrain, rivers, customs and local products, relics and famous people. It was a territory chronicle known as the *Zhenguan dizhi* in the Tang and Song dynasties [33]. The original version was lost in the Southern Song dynasty, but we can still read Sun Xingyan's edited version. He commented in the preface: "The book … contains the content of many geographical books across six dynasties and is thus more valuable than the *Yuanhe junxian tuzhi*" (an important general history of the early Tang dynasty without pictures or maps, which is different from the Tujing genre). Since the *Kuo dizhi* cited many geographical references over six dynasties, it was often used by later scholars to clarify ancient place names.

In the middle of the Tang dynasty, Kong Shurui (729–800) "was proficient in geography and revised the *Dili zhi*, which was praised by people at that time for being detailed and well researched [34]."

Jia Dan compiled the 40 volumes of the *Gujin junguo xiandao siyi shu*, the explanatory notes for the *Hainei huayi tu*. Later, Jia Dan summarised it in the 4 volumes of the *Zhenyuan shidao lu* but they were lost. In the 1970s, a remnant of the *Zhenyuan shidao lu* was found in the Dunhuang Caves, making it the earliest surviving copy of the national chronicles [35].

Li Jifu (758–814), a renowned geographer in the Tang dynasty, was a native of Zanhuang County, Zhao Prefecture (now Zanhuang, Hebei province). He was a talented and knowledgeable scholar who served in positions such as governor, Emperor's advisor, etc. He wrote the 30 volumes of the *Liudai lue* ("A Concise Report of the Six Ages"), ten volumes of the *Shidao tu* ("Ten-District Map"), three volumes of the *Gujin diming* ("Place Names of the Past and the Present"), ten volumes of the *Shan shuijing* ("Brief Water Classic"), and ten volumes of the *Yuanhe guoji pu* ("Book of National Accounts from the Yuanhe Reign"). All his works have been lost, except *Yuanhe junxian tuzhi* which has survived to this day. The original book consisted of 42 volumes, with only 34 volumes surviving today. "Yuanhe" was the era name of Emperor Xianzong's reign (806–820). The geographical facts described in the *Yuanhe junxian tuzhi* are also limited to the eight years of the Yuanhe, including the rank, number of households and villages, history, mountains and rivers, the tributes, salt and iron, fields, military facilities, troops and horses, and ancient relics of each province and prefecture belonging to the ten divisions of the country at that time. Each volume is accompanied by a map at the beginning. By the Northern Song dynasty, the pictures had faded, so the name of the book was changed to the *Yuanhe junxian zhi* ("Treatise on all districts from the Yuanhe reign-period"). When Zhang Jizhong first copied this book in the 3rd year of the Chunhe reign in the Southern Song dynasty (1176), the book was already incomplete, but still maintained the structure of the 42 volumes. After the Song dynasty, its index had been lost, and Volumes 19, 20, 23–26 were missing. Only 34 volumes have been passed down to this day.

Li Jifu's main objective in writing the *Yuanhe junxian tuzhi* was to focus on practicality and to oppose dwelling on the past and neglecting the present. Like Jia Dan, he advocated regaining lost territories. At the time of Emperor Xianzong, more than 50 prefectures in the north and south of the Yellow River were occupied by vassals, and the western part of Sichuan had fallen into the hands of the Tubo Kingdom. However, ten sections (each dealing with one province) still remained in this book, suggesting that his intention was to restore the territorial integrity of the country.

In describing the history of the administrative regions, the *Yuanhe junxian tuzhi* traced events back to the Zhou, Qin, and Han dynasties. Its record of the administrative history of the Eastern Jin and Northern and Southern dynasties is particularly valuable. Many of the important villages, towns, passes, harbours, tombs and other locations recorded in the literature before the Tang dynasty were discovered only by referring to this book.

Moreover, the *Yuanhe junxian tuzhi* also recorded the natural geographical conditions, the economic geography and the population geography of the time. This book has consistently been considered to be the best one of all the general chronicles. The *Siku quanshu zongmu tiyao* says that it is the oldest and "its style is also the best. Although some content has been lost over time, no other book can surpass it". It had a great influence on the compilation of later *fangzhi*.

Based on the *Chorography of Dunhuang Grottoes* by Wang Zhongluo, the following remnants of geographical literature can be identified.

(1) *Tang Tianbaochunian dizhi canjuan*

The structure consists of a list of the names of the sections and the number of provinces, followed by the names of the provinces and prefectures, then the distance from the western capital Chang'an, the distance from the eastern capital Luoyang, the tributes, the amount of capital for public expenditure, then the name and rank of the counties, the number of towns and their amount of capital for public expenditure. It is a national chronicle which can compensate for the shortcomings of the *Book of Tang*: *Dili Zhi*. There are 65 kinds of tributes listed, grouped into six categories: (1) daily necessities, such as candles, chess pieces, mats, arrow feathers, fans, mirrors, ironware, etc.; (2) medicines, such as Chinese goji berries, cypress roots, *Cistanche*, musk, *Asarum*, Poria mushrooms, Chinese sage, ginseng, fossil fragments, liquorice, Dendrobium orchids, *Pueraria*, stalactites, gypsum, mica, etc.; (3) silk, linen, and woollen textiles, such as saddle felt, fine linen, green cloth, white cloth, silk, etc.; (4) animal skins, such as wild horse skins, bear skins, etc.; (5) minerals, such as gold dust, silver, salt, andesite, red copper, white quartz, etc.; (6) agricultural produce, such as beans, Chinese alpine rush mats, nettle, hemp, etc. From these tributes, we can see the produce and its distribution in the country at that time, as well as the large amount of goods that the ruling class took from the people.

(2) The remnant "Jannan dao" of the *Zhenyuan shidao lu*

There were records of twelve provinces and prefectures located in present-day Sichuan and Yunnan provinces. The tributes included musk, yak tails, angelica, Shu horses, *Notopterygium* roots, yaks, jaggery, cohosh, rhubarb, etc.

(3) The remnant of the *Zhudao Shanhediming yaolue* ("Synopsis of Place Names of Mountains and Rivers of all Areas")

It first introduces the history of the establishment of administrations, followed by the events, the name of counties, the name of the bodies of water, mountains, customs, produce and tributes. Intended for Emperor Xuanzong to browse at night, the text is simple and concise. The references cited are not the same as in the original literature but were edited to suggest only the main idea. The book was completed in July of the 8th year of the Xiantong reign (867).

6.3.2 Extraterritorial Chronicles

During the rule of Emperor Yang of the Sui dynasty, the Eastern Turkic Khaganate declined due to civil turmoil and the Western Turkic Khaganate surrendered. With the ending of the Tuyuhun Kingdom, the obstacles preventing the Sui dynasty from resuming close relations with the Western Regions were removed. During the Daye reign, Emperor Yang of the Sui "created the position of Officer of the Western Garrison to satisfy a need" [36]. Pei Ju served in this position: "at that time, most of the western vassals came to Zhangye and made deals with the Chinese. The emperor ordered Ju to take charge of it" [4]. During his tenure, Pei Ju explored the mountains, customs and politics of the countries in the Western Regions and found that, since the Eastern Han dynasty, the situation of the countries in the western region was not well understood. "Although we have known about the number of households since the Dawan, we do not know anything about the landscapes, family names, customs, costumes and produce of the countries in the Western Regions … I recorded the different appearances and behaviours from the king to the common people according to the costumes and manners of their own countries, drew them out and compiled the *Xiyu tuji*". "There are three volumes, recording 44 countries. I also drew a map of the area which stretches from the far west to the south of the North Sea, which is nearly 20,000 li. I think this is an area which merchants have always visited, so we can learn about the situations in different countries from here. Still there are some places about which it is difficult to learn everything quickly, because of their remoteness, but I cannot imagine things out of nothing, so there is something missing … There are now more than a thousand households codified in the household records. They enjoy the benefits of the Western Sea, which produces many precious and exotic goods. Those who live in the mountains do not live in a named country, and as for the small tribes, most of them are not recorded either [4]."

From the above preface written by Pei Ju, we learn the purpose and content of the *Xiyu tuji:* first, to record the costumes, customs and produce of the countries in the Western Regions. Second, to describe the three roads leading to the countries in the Western Regions: the northern route in Tianshan; the central route, or the northern road in Tianshan's southern network; and the southern route in Tianshan. Third, to draw the maps of the mountains and rivers of the Western Regions, which covered an area of 40,000 square li, including areas to the south of the Aral Sea and along the east coast of the Mediterranean Sea.

In the 3rd year of the Xianqing reign of the Tang dynasty (658), Xu Jingzong completed the *Xiyu tuzhi,* a 60-volume book on the profiles of the countries in the Western Regions. But the book cannot be traced nowadays.

Apart from the aforementioned ones, there are also chapters on extraterritorial geography in official historical records, such as the *Suishu: Waiguo liezhua*n ("History of the Sui Dynasty: Foreign Biographies"), *Jiu tangshu: Tujuedeng liezhuan* ("Old Book of the Tang: Biographies of Turks), *Xin tangshu: Tujuedeng liezhuan*

("New Book of the Tang: Biographies of Turks"), *Waiguo liezhuan* ("Foreign Biographies") and *Siyi fulu* ("Appendix of Four Barbarians") in the old and new *Wudai shi* ("History of Five Dynasties").

The 16-volume *Tongdian: Bianfang dian* ("Comprehensive statutes: Frontier Defence") by Du You of the Tang dynasty records the geography of neighbouring countries; this is an important work on the extra-territorial geography of the Tang dynasty.

6.3.3 Local Chronicles

According to Zhang Guogan's *Zhongguo gufang zhikao* ("Study of China's Ancient Chronicles"), there were only seven kinds and 40 volumes of local chronicles in the Sui dynasty, such as the *Shanggujun tujing* ("Graphical Records of Shanggu County"), *Jiangdu tujing* ("Graphical Records of Jiangdu"), *Jingsh ilu* ("Records of the Captial"), *Dongdu tuji* ("Illustrated Records of the Eastern Capital"), *Luoyang gujin ji* ("Records of Luoyang's Past and Present"), *Yizhou ji* ("Records of Yizhou"), and the *Yongzhou tujing* ("Graphical Records of Yongzhou").

The local chronicle as a genre of the Tang dynasty had names which included *zhuan, lu, ji, tujing, zhi, shu,* and *shiji*. From the *Taiping yulan* ("Imperial Reader of the Taiping Era") and the *Taiping huanyu ji* ("Universal Geography of the Taiping Era [976–983]"), it is known that more than 50 provinces in the Tang dynasty had *tujing*, covering almost the whole country. Although these *tujing* have been lost for a long time, the two remnants – *Shazhou tujing* ("Graphical Records of Shazhou") and *Xizhou tujing* ("Graphical Records of Xizhou") – found in Dunhuang show that they recorded not only the administrative agencies and regions, but also the rivers, lakes, dykes and weirs, post roads, ancient cities, schools, and proverbs in the region. If the content of the *tujing* of remote areas was so complete, then the *tujing* of the interior may have been even more detailed.

The *tujing* found at Dunhuang, apart from the two mentioned above, also include the *Shazhou dudufu tujing* ("Graphical Records from Shazhou Governor's House"), *Shazhou dijing* ("Topography and Environment of Shazhou"), *Shazhou dizhi* ("Shazhou Chronicle"), *Guazhou yixi canzhi* ("The Remains of Guazhou Yixi Chronicle"), etc. The *Shouchangxian dijing* ("Topography and Environment of Shouchang County") written in the 9th year of the Tianfu reign of the Five Dynasties (944) is complete, containing the distance to provinces, official institutions, households, towns, history, temples, towns, fortresses, mountains and rivers, bodies of water, drains and streams, pavilions, and cities. It still followed the format of the Tang dynasty.

According to Wang Zhongluo's *Chorography of Dunhuang Grottoe*s, it is recognised that the content of the *Shashou dudufu tujing* remnants is very special.

First, the desert landscape recorded is very precise: "to the east of the water is Mingshaliu Mountain. The mountain is flowing, the mountain peak is not constant, sometimes there is a deep valley, sometimes the valley turns into high cliffs".

Second, the recorded hydrological characteristics are clear. "To the west of the water are rock mountains with no grass and trees. And 80 li northeast of the water, people have built a large dyke called Maquankou. Its weir is 150 *bu* from north to south, 20 *bu* wide and two *zhang* high, with a total of five gates to distribute water to irrigate the fields. Because there is little rain, the valleys are often covered with snow, after the Start of Summer (the seventh solar term), the snow melts, and the meltwater runs into the river."

Third, the local salt ponds are documented. "The righthand side of the province is vast. Most of it is desert and most of the remainder consists of salt ponds. There are three salt ponds. The East Salt Pond, 50 li to the east of the province, is two hundred *bu* wide and three li long. The saltwater evaporates forming crystals, people get them out of water and dry them. The taste is lighter than Hedong Salt, but the shape is similar. The West Salt Pond is usually called Shaquan Salt. It lies 117 li to the north of the province ... People get salt out of the water: the big pieces are the size of horse teeth. The salt tastes delicious and looks like snow. Though many people take salt from it, the supply of salt is inexhaustible."

Fourth, there are records of the names, locations and changes of the post stations.

Fifth, there are records of the state schools, county schools, medicines and the imperial divine temples.

Sixth, the four gods are recorded, namely, the God of Earth, the God of Wind, the God of Rain and the God of Heaven.

Seventh, the book documents temples, mounds, halls, the ancient cities, ancient Great Wall, ancient frontier cities, the pool used by Zhang Zhi to wash away the ink after his calligraphy practice, auspicious omens and ballads. The book was completed after the 4th year of the Kaiyuan reign (716).

There were no fewer than six types of Yunnan local chronicles in the Tang dynasty, among which Fan Chuo's *Manshu* ("Book of the Southern Barbarians") is the best [37]. With the exception of the *Manshu*, all the others have been lost. When drafting *Manshu*, Fan Chuo referred to his own personal experience and investigations as well as to the *Yunann ji* and *Yunnan xingji* ("Travelogue of Yunnan") as references. In great detail, he noted information about the transport routes, important mountains, rivers, lakes, cities and towns in Yunnan, the economic life of various ethnic groups, production techniques, customs and produce, the history of the Nanzhao Kingdom, military organisations, and the geography of the neighbouring countries. This book is the earliest and most detailed local chronicle of Yunnan preserved to this day and is of great value.

Moreover, there were also some local records in the Tang and Five Dynasties, such as the three volumes of the *Guilin fengtu ji* ("The Customs of Guilin") by Mo Xiufu, two volumes of which have been lost, three volumes of the *Beihu lu* ("Records of Northern Regions") by Duan Gonglu, three volumes of the *Lingbiao luyi* ("Customs and Produce of Linnan Region") by Liu Xun, and one volume of the *Xianlu ji* ("Stories of Being Captured") by Hu Qiao.

6.4 History of Administrative Geography

Aside from the inheritance of the historical geography typical in the *Shiji* and *Hanshu*, the administrative geography of the Sui, Tang and Five Dynasties benefited from the appearance of new works of administrative geography and the course of the changes in local administrations was also stressed in *fangzhi* as a genre.

6.4.1 Administrative Geography in Official Histography

The "Shizhi" in the *Suishu* ("Book of the Sui"), originally named *Wudai shizhi* ("Historical Treatise on the Five Dynasties"), describes historical events beyond the Five Dynasties and is rich in authentic material and accurate in its exposition. The three volumes of the *Dili zhi* record the number of households and the population of the whole country in the 5th year of the Daye (609). The evolution of the establishment, the mountains and lakes, types of landforms, minerals and mining areas, and places of interest are recorded. The history of the establishment of the counties dates back to the Later Wei dynasty and to the Daye period. For example: "Jingzhao Jun (prefecture): in the 3rd year of the Kaihuang reign, Yongzhou was established. The place is 18 li and 115 *bu* from east to west, 15 li and 175 *bu* from north to south. In the east, there are three gates –Tonghua, Chunming and Yanxing gates; in the south are Qixia, Mingde and Anhua gates; in the west are Yanping, Jinguang and Kaiyuan gates; in the north is the Guanghua Gate … In the 3rd year of the Daye reign, the administrative level of Yongzhou was changed from a *zhou* to a *jun*, and Official Yin was appointed to govern 22 counties and 38,499 households. In the 3rd year of the Kaihuang reign, Daxing was established: the Later Zhou dynasty established Wannian County which was later renamed Daxing by Gaozu, the future Emperor Wen of the Sui dyansty. The Changle Palace was located in Daxing. There was also the location of the Ducheng County of the Later Wei, and of Xibacheng County and Shanbei County of the Western Wei, which were all abolished in the Later Zhou dynasty."

The four volumes of the *Old Book of the Tang: Dili zhi* compiled by Liu Xu and others basically do not record natural geography. In the 28th year of the Kaiyuan (740), the number of prefectures came to 328, and there were 1,573 counties, 8,412,871 households and 48,143,609 people. Compared with the number of households in the 5th year of the Daye of the Sui dynasty, although the number of households had decreased, the population had increased by more than two million.

The seven-volume *New Book of the Tang: Dili zhi* compiled by Song Qi and Ouyang Xiu records the number of households in the Tianbao period, the establishment of *zhou* (states) and *jun* (prefectures) in the Tianyou period, and the boundaries of each prefecture which were based on the 15 *dao* (administrative unit) of the Kaiyuan period. In addition to the names and the history of the administrative regions, the work also records some natural geography. The mineral production areas and hydraulic engineering facilities are especially detailed. For example, "Henan section: covering the ancient areas of Henan, Yan, Qing, Xu, Han, Hongnong, Yingchuan, Runan, Chenliu, Pei, Taishan, Jiyin, Jinan, Donglai, Qi, Shanyang, Donghai, Langxie, Beihai, Qiansheng, Dongjun … there is a local government, and 29 prefectures and 196 counties. Its famous mountains include Sanyao, Shaoshi, Dizhu, Meng, Yi, Songgao, Taiyue. The rivers include Yi, Luo, Ru, Ying, Yi, Si, Huai and Ji. The eastern capital was established in the Sui dynasty and was abandoned in the 4th year of the Wude reign. In the 6th year of the Zhenguan reign, it was called Luoyang Palace; in the 2nd year of the Xianqing reign, it was called the eastern capital; in the first year of the Guangzhai reign, it was called Shendu; in the first year of the Shenlong reign, it was called the eastern capital once again; in the first year of the Tianbao, it was called Dongjing; in the 2nd year of the Shangyuan reign, it was abandoned; and in the first year of the Suzong reign, it became the eastern capital once more."

The *Old History of the Five Dynasties: Records of Prefectures and Counties* compiled by Xue Juzheng and others was based on the *Kaiyuan shidao tu*, which "only contains the reformation of the Five Dynasties, while the old system which still exists in the Tang dynasty has been omitted". It focuses entirely on the evolution of the establishment and contains no natural geography. Generally, the evolution of state and county settings only relates to the Five Dynasties and does not go back to the previous periods.

Ouyang Xiu's single volume of the *New History of Five Dynasties: Zhifang kao* listed the 288 *zhou* that appeared in the Five Dynasties in a table, and below it described the status of the Liang, Tang, Jin, Han and Zhou dynasties. For example, if a certain dynasty had a particular state, then it was noted "yes"; if a certain dynasty had established a capital in this state, then it was noted "capital"; if a certain dynasty had this state, but the name was different, then it was noted "had xx state"; if a certain dynasty had established this state, then it was noted "xx was established"; if there was no such state in a certain dynasty, it was left blank. This table created by Ouyang Xiu is useful for verifying the names of states in the Five Dynasties. The table is followed by a detailed description of the evolution of the establishment of each state.

6.4.2 Administrative Geography in Tongdian

Although Du You's *Tongdian* is a book on general political history, the *Zhoujun dian* ("Statutes of Provinces and Prefectures") and *Bianfang dian* ("Statutes of Frontier Defence") in it are special sections on administrative geography in ancient China [38]. It breaks the rule that the geography in an official history should only record the current dynasty or backtrack a little, but instead it traces the history of an administrative unit as it develops, generally from the Spring and Autumn period. This format was invented by Du You. It was later followed by many others and acted as a template for the series of *Shitong* ("Ten Comprehensives").

The evolution of the political regions recorded in the 14-volume *Zhoujun dian* dates from the ancient times of the Yellow Emperor and ends at the end of the Tianbao reign of the Tang dynasty. The first volume describes the evolution from ancient times, from the Tangyu Age to the Sui dynasty. The second to the thirteenth volumes describe the evolution before the Tianbao reign of the Tang dynasty, based on the concept of *jiuzhou* in the *Yugong*, and they record the states and counties in the past. The fourteenth volume describes the evolution of the 71 states and 294 counties of the Tang dynasty.

6.4.3 Administrative Geography in the Chronicles

Most of the *fangzhi* of the Sui and Tang dynasties have been lost, but from the few which have survived, there is no doubt that the evolution of the establishment of the counties was recorded in *fangzhi*. For example, in the *Yuanhe junxian tuzhi*, the evolution of political regions is often traced back to the Zhou, Qin, and Han dynasties. The remnant of the *Shazhou zhi* contains the evolution of Shouchang County. The Five Dynasties' *Shouchang xiandi jing* and Sun Xingyan's *Kuo dizhi* also contain evolutionary geography. Jia Dan's *Hainei huayi tu* was the first map of evolutionary geography, with the names of ancient places noted in black and the names of present-day places marked in red. Later studies were deeply influenced by it.

6.5 Hydraulic Engineering and Hydrology

6.5.1 Water Conservancy Projects

The Sui and Tang dynasties were the period when there were great advances in water projects. The most famous one is the Grand Canal of the Sui dynasty, which included the Yongji Canal, the Tongji Canal, the Shanyang Ditch and the Jiangnan Canal, with a total length of more than 5,000 li, flowing through today's Hebei, Henan, Anhui, Jiangsu and Zhejiang, and connecting the Yangzi, Huai, Yellow, Hai and Qiantang rivers. It stretched to the Hebei Plain in the north, the Guanzhong Basin in the west, and the Taihu Plain and the Qiantang River in the south, linking northern China, southern China and the Guanzhong area where the capital was located, and forming a national canal network with Luoyang as the centre. This water project was unprecedented in history.

The Grand Canal of the Sui dynasty brought great benefits to the Tang dynasty. The Tang made full use of the Grand Canal to form a well-connected water transportation network throughout the entire country [39]. After the An Shi Rebellion in particular, the economic centre of the Tang dynasty gradually shifted to the south of the Yangzi River, making the Grand Canal the traffic lifeline of the Tang dynasty and underpinning the survival of the country.

In order to ensure smooth passage through the Grand Canal and improve its efficiency, the Tang Imperial dedicated a great deal of effort to the maintenance and expansion of the Canal and dredged it frequently. At the same time, it organised the Guangtong Canal and expanded the water transportation network in Guanzhong; it also expanded the Bian Canal so that the canals near Taishan were also included in the transportation network of the Bian Canal; it improved the Shanyang Dou and the Jiangnan Canal and expanded the Yongji Canal to transport goods from Jiangnan to the north, which further consolidated national unification.

The well-developed agricultural water projects of the Tang dynasty contributed to the success of agriculture. According to the statistics in the *China's Water Project History (Mid Volume)*', 253 farmland water projects were constructed during the Tang dynasty, including 33 covering an irrigated area of more than 1,000 hectares and 15 covering 500–1,000 hectares. In terms of geographical distribution, the centre shifted from the Yellow River in the north to the Yangzi River in the south, reflecting a shift in the economic hub. The excavation of the Grand Canal and the construction of a large number of agricultural water projects in the Sui and Tang dynasties reflected the rich knowledge of hydrology and water engineering techniques that were at the forefront of the world at that time.

6.5.2 The Development of Tidology

During the Sui and Tang dynasties, sea navigation made advances. Due to the frequent occurrence of marine disasters, people were forced to study the ocean tides, so ocean tidology developed rapidly in this period and resulted in remarkable achievements.

During the years of the Baoying and the Dali in the Tang dynasty (762–779), there was a monograph on the study of ocean tides – the *Haitao zhi* (Records of Ocean Tides) by Dou Shumeng, also called *Haiqiao zhi*. Dou Shumeng was a folk scientist whose life story is unknown, apart from the fact that he was a native of Zhendong [40]. The monograph *Haitao zhi* is the first comparative study on tides in China still in existence [41]. The book is divided into six chapters, discussing the causes of ocean tides and the pattern of ocean tidal movement, calculating the number of tidal cycles over a long period of time, creating a scientific and unique charting method for the projection of high and low tides, providing a vivid description of the corresponding changes of tides and the moon in a lunar month, and pointing out the phenomenon of differences in tidal movement because of changes in the distance between the earth and the moon.

Dou Shumeng's contributions to ocean tidology are evident in four aspects.

First, building on earlier work, he further clarified the relationship between the tides and the moon's movement. The relationship between the two is extremely close. Dou Shumeng observed the relationship between the tides and the movement of the moon quite carefully.

Second, his calculation of the tidal cycle was very accurate. Dou Shumeng worked out that the tidal cycle was 12 h 25 min 14.02 s, which is very close to the modern calculation of the regular diurnal tide which is generally delayed by 50 min per day. Dou Shumeng also pointed out that there were three cycles of tidal movement, two tidal cycles a day; two spring tides and two neap tides each lunar month; and two equinoctial tides in a tropical year.

Third, he elaborated on the equinoctial tide. Dou Shumeng said that when the new moon appeared after "Jianglou" or "Shouxing" [names of two of the twelve signs of the zodiac in ancient Chinese astronomy], three days and 18 days afterwards, the moon would be located in "Dalou" or "Dahuo" and then there would be a spring tide. The equinoctial tide was not in "Jianglou" or "Shouxing" on the first or fifteenth day of the lunar month, but it occurred three days later when the moon reached "Daliang" or "Dahuo". This means that the theoretical projection of the equinoctial tide at that time had been corrected by the tidal data obtained from actual measurements.

Fourth, Dou Shumeng first invented the method of using charts to predict tidal times. This involves coordinates on vertical and horizontal axes: the horizontal axis at the top lists the phases of the moon in a lunar month; the vertical axis at the side lists the twelve hours of a solar day. The coordinates of the points indicating the phase of the moon and the moon passing through the upper and lower transits are marked on the graph, and the points are linked to form a line to calculate the tide times in a lunar month. If the moon phase of the day is known, the time of the spring tide on that day can be found on this chart. This is Dou Shumeng's tide chart. This chart predates the earliest one in Europe by five centuries; that was the High Water at London Bridge tide table in the thirteenth century. In this regard, Joseph Needham fairly pointed out: "The systematic compilation of tide charts was obviously earlier in China than in the West. As far as we have seen, it dates back to the ninth century at least [42]."

Another famous tidologist of the Tang dynasty was Feng Yan, a native of Tiao (present-day Jing County, Hebei province), who lived in the eighth century. He became the Minister of Personnel and Imperial Minister. He wrote the ten volumes of the *Fengshi wenjian ji* ("Master Feng's Records of Hearsay and Personal Experience") and an article "Shuochao" (On Tides). The *Fengshi wenjian ji* was not completed and was included in Volume 440 of the *Quantangwen* ("A Compilation of Tang literature"). Since he "lived in Huaihai when he was young and watched the tide every day", he had his own views regarding the causes of the tides and the changing patterns of the tides in a lunar month. And he was the first to point out gravity by explaining the relationship between the moon and the tides.

Lu Zhao, a native of Yichun, Yuanzhou (now Yichun, Jiangxi province), was a scholar in the late Tang dynasty. In the 2nd year of the Huichang (842), he became a Gongshi after passing the county exam, and came first in the highest imperial exam in the following year. He served as the governor of the She, Xuan, Chi and Ji Prefectures. As the author of the *Wenbiao ji*. he explained his views on tides in one of the articles "Haichaofu bingxu" in this collection. Some of his views were correct, but some were wrong. Following Ge Hong, he argued explicitly that the tides were related to the sun, which was progress. However, he did not carry out actual observation and exaggerated the role of the sun in the development of tides. He also raised 14 questions about the tides and provided answers, reflecting the state of tidal research in the Tang dynasty and contributing to the study of tides in later times.

Li Jifu, a geographer of the Tang dynasty, recorded the phenomenon of late tides in the *Yuanhe junxian tuzhi*. He wrote: "The river tide appears every day and night. It is often at its lowest on the tenth and twenty-fifth day of the month, and highest on the third and eighteenth day of the month. Every year, on the 18th of August, men and women from all over will gather to watch people conquering the tide—the activity of *nongtao* [43]." Li Jifu not only knew that there were two high tides and two low tides in one lunar month, but also knew that the high tide was not exactly on the first and fifteenth day of the lunar month, and the low tide was not exactly on the seventh and twenty-second day of the lunar month, but two to three days later. The discovery of the late tide phenomenon promoted the development of local tide charts.

Qiu Guangting, a tide expert in the Five Dynasties, was a native of Wucheng (present-day Wuxing, Zhejiang province), and was a senior scholar at the Imperial College and author of the book *Jianming shu*. His article "Haichao lun" (on Sea Tides) was included in the *Haichao jishuo* ("Collected Writings on Sea Tides"). His view was that the formation of sea tides was not due to the expansion and contraction of sea water, but instead was due to the movement of the land. He believed that sea water did not move, but rather that the land moved up and down so that tides were formed. This view was very valuable. However, it was inconsistent with the facts because the role of the solid tide of the earth's crust was overstated and it was understood as being the single factor that explained the formation of ocean tides.

6.5.3 The Development of Land Surface Hydrology

During the Sui and Tang dynasties, there were no massive floods of the Yellow River, which was historically rare and a special point in the history of continental hydrology. There were no historical books like the *Hequ shu* ("Book of Rivers and Canals") or *Gouxu zhi* ("Records of Canals") in the Sui, Tang and Five Dynasties, and there was no work like the *Shuijing zhu* recording the river systems in the country. Although 72 rivers were recorded in the *New Book of the Tang: Dili zhi*, more than 550 in the *Yuanhe junxian tuzhi* and 218 in *the Wudai shi: dilizhi* ("History of the Five Dynasties: Treatise on Geography") most of them only recorded the names of rivers without any description of the water system.

During the Tang dynasty, people already knew that rivers had seasonal changes in water volume, with periods of normal water levels, floods and droughts. Following the Southern and Northern dynasties, fish motifs sculpted on stone inscriptions were used as a water gauge to measure the levels of the rivers. For example, the inscriptions recording the water level of the Yangzi River in the dry period, in the second year of the Guangde reign of the Tang dynasty (764 AD), still exist today [44].

In the Tang dynasty, the hydrological characteristics of the hillside rivers that relied on being replenished by the snow meltwater were accurately described as "the water level decreases in the morning and rises at night". An example such as the *Shazhou dudufu tujing* written after the fourth year of the Kaiyuan reign (716) says: "In spring, there is little rain and much snow. In summer, the snow melts and the water runs into the rivers, the water level decreases in the morning and rises at night." Down the slopes of the snowy mountains, the rivers are recharged by the meltwater; because of the great difference in temperature between daytime and night time, the rate of the snow melt also varies between day and night. At night, the temperature is low, the snow melts more slowly, so the water level in the morning decreases; during the day, the temperature is high, the snow melts faster, so the water level rises in the evening. This hydrological feature had never been highlighted before the Tang dynasty.

People in the Tang dynasty concentrated a great deal of attention on the quality of the water. Li Weigong measured the weight of water from rivers and springs all over the country to compare the quality of the water. He lived in the capital (now Xi'an) but did not drink its water. He made tea with water from Huishan spring in Changzhou. A monk once said to him: "(You) draw water from thousands of li away, is it worthwhile? … The well behind the kitchen of Haotianguan in Beijing is said to be connected to the Huishan spring." Li Wei Gong then weighed water from different places, and only the Huishan water was equal in weight to the Haotian water. Since then, he stopped using Huishan water [45].

Lu Yu (733–804) divided the water into three levels when talking about water quality: "mountain water is the best, river water is in the middle, well water is the lowest. The water from rivers refers to those which are far away from people and the well water refers to the ones used by many people." He also said that poor water quality would make people sick. He said: "The water in the mountains is so turbulent that it cannot be drunk. People who drink it for a long time will develop neck disease (goitre) [46]." Sun Simiao also believed that mountain water would lead to the development of a goitre [47].

Some water can cure diseases, as documented in Volume Ten of the *Youyang zazu* ("Miscellaneous Morsels from Youyang") by Duan Chengshi: "Leiping Mountain in Huayang (now Yang County, Shaanxi province) has the Tiangong Spring. If you drink its water, you can cure bowel disease."

During the Wei and Jin dynasties, the Chinese had invented medicinal wells to produce artificial mineral water, which basically involved adding minerals beneficial to human health to the water to improve water quality. According to the *Wanghuai lu* ("Record of Longings Forgotten") by Shen Kuo of the Northern Song dynasty, Li Wenrao of the Tang dynasty had a medicinal well into which he put cinnabar, sulfur, gold, pearls and jade.

6.5.4 Seeking the Source of the Yellow River

During the Tang dynasty, new progress was made regarding exploration of the source of the Yellow River, correcting the misconceptions of previous generations.

In the 9th year of the Zhenguan reign (635), Hou Junji and Li Daozong went on an expedition in the region of the source of the river: "travelling 2,000 li in the wilderness, there was frost in summer. Soldiers ate wild vegetables and ice, the horses were fed with snow. After a month, they reached the Xingxiu Sea and finally reached Bai Lake (also known as Zhaling Lake). They looked at Jishi Mountain and saw the source of the Yellow River [48]." Thus we learn that people in the early years of the Tang dynasty had already explored as far as the Xingxiu Sea. Later the route between the Tang and the Tubo Kingdom (modern Tibet) also passed through Heyuan, such as for the marriage of Songtsen Gampo and Princess Wencheng in the 15th year of the Zhenguan reign (641) [49]. After 181 years, Liu Yuanding travelled to Heyuan in the second year of the Changqing reign (822) and gained a better understanding of the geographical situation there. "In the upper region of the river, 2,000 li southwest of Hongjiliang, the water channel is narrow. In spring, it can be forded, and in autumn and summer, boats can pass. There are three mountains 300 li to the south, which are elevated in the centre. The region is called Zishan and is next to the Datongyang Kingdom. Mount Menmoli in Tubo is found there, 5,000 li east of Chang'an, where the Yellow River originates. The water is clear and flows slowly and gradually merges with the water of various rivers. The water becomes red in colour, and the farther it flows, the more other water converges with it, and the water becomes muddy. The Hehuang area is where the Xirong nation resides [50]." This is according to Liu Yuanding in his report *Shi Tubo jingjian jilue* ("A Brief Travelogue to Tubo Kingdom"), describing his visit to the source of the Yellow River. Unfortunately, the book has been lost. Liu Yuanding's experience as noted in the *Old Book of the Tang: Record of Tubo* is slightly different from what is mentioned in the *New Book of the Tang: Record of Tubo.*

Du You was the first person to repudiate the earlier theory on the source of the Yellow River. At the end of his book *Tongdian: Zhoujun dian,* completed in 801, he firmly opposed the idea that there was a connection between the Puchang Sea (today's Lop Nor) and Jishi and he posited that the source of the river was in Xizhi (today's Hequ region in southeast Qinghai). Although he was still vague about the real situation, as the first to correct the erroneous understanding, this was remarkable and epoch-making.

Jia Dan was also very interested in the situation of the upstream section of the Yellow River. In the 14th year of the Zhengyuan reign (798), he completed the first work named after the Yellow River in China's history: the ten volumes of the *Tubo huanghe lu.* The text is illustrated with pictures, documenting the "source of all mountains and waters [24]" in the territory of Tubo. Unfortunately, the book has been lost.

6.6 Meteorological and Climatological Observations and Records

6.6.1 *Meteorology and Climatology Literature*

In the Sui, Tang and Five Dynasties, besides the historical accounts that included meteorological and climate information, such as the *Tianwen zhi*, *Sitian kao*, *Lüli zhi* and *Wuxing zhi*, similar statements could also be found in texts on agriculture, astronomy and geography and in different styles of writing such as in notes and *fangzhi*. The topics included various types of precipitation, temperature anomalies, gales, droughts, weather forecasts, etc. What was special and unprecedented in the Tang dynasty was the appearance of monographs on weather and climate: Shao E's *Wangqi jing* ("Book on Observing Climates") and Huang Zifa's *Xiangyu shu* ("Book on Observing Rain").

The *Wangqi jing* was a military book originally and it discusses the relationship between clouds and battlefield strategies. It states that "when two armies are fighting against each other, there must be clouds, as well as sun haloes, which are recorded on each side. Above the winning side, the clouds are thick and vice versa." It also talks about the differences between clouds in different geographic places, some are different in colour and some are different in shape. It should be noted that the shapes and colours of clouds in different places were not fixed, but rather they changed at random. The *Wangqi jing* is unrealistic in regarding the shape and colour of clouds in different places as a fixed form, but what is valuable is that it identified the shapes of various clouds, which was knowledge accumulated through observation.

The *Xiangyu shu* raised five important points regarding weather forecasting.

1. To forecast the time of rain and the scale of rain from the shape, colour, location and speed of the clouds. For example, "if the cloud is the shape of a fish, it may rain on the same day, if it swims fast; if it swims slowly, there may be no rain or it may rain later". "If there are clouds like sheep and pigs in all directions, there will be rain immediately."
2. To predict the changes in the solar term. For example, if the sky is white in the morning and red in the evening, this is a sign that a gale is approaching.
3. To observe the rainbow. For example, "if there is a broken rainbow in the evening, it will rain from midnight to midday the following day".
4. To observe the fog. For example, "if there is fog in the south when the sun rises, there will be a shower of rain from 7 to 9 am".
5. To forecast the weather by observing creatures. For example, "every evening, set a stalk of grass alight, if it burns (away) completely, it will be clear the next day and vice versa. If there is water in hollows in trees, it is raining. When the rain falling in a well produces bubbles, the downpour will become heavier. If the fish leap out of the water, it will be cloudy and rainy. If there are water drops on the wall, it will rain heavily. If the stone is dripping with liquid, it will rain for several days [51]."

Only 10 items from the *Xiangyu shu* were collected in the *Shuofu*. In contrast, more than 20 items were gathered together in the *Xiangyu shu* and Fang Hui wrote a preface to it in the eighth year of the Dade (1304) of the Yuan dynasty. From the preface by Fang Hui, it is known that "there are ten chapters in the *Xiangyu shu*. 30 items are about predicting the changes of solar terms; 50 are about observing clouds; 31 about inspecting the sun and the moon and the stars; four about observing the wind; seven about analysing sounds; 12 about calculating the time; 14 about observing different things; seven about waiting for the rain to stop and the sky to clear; three about praying for rain; nine about praying for the sky to clear, making a total of 169 items. They are all accurate". Thus, the *Xiangyu shu* was very rich in content, but unfortunately it was lost early on. The *Shuofu* and the version of the *Xiangyu shu* published in the eighth year of the Dade are both excerpts, accounting for only 6% and 13% of the original text.

6.6.2 Meteorological Records

Li Zhao of the Tang dynasty mentioned several different trade winds in his *Tang guoshi bu* ("Supplement to the history of the Tang dynasty"), saying that "boats from Jianghuai travel upstream and wait for the northeast wind, which is called the trade wind. In July and August there is the Shangxin (referring to the early autumn wind and rain, marking the beginning of the autumn trade wind), in March there is the Niaoxin (cuckoos frequently appear in the rush ploughing season, when there is often spring rain), in May there is the Maxin (the wind and rain during the wheat harvest)".

Li Chunfeng in the early Tang dynasty talked about the wind speed and the distance and signs of wind in the *Yisi zhan: Houfeng fa*. He said that "when a wind's speed is slow at first and fast later, it is blowing from far away and vice versa. If the wind blows and the leaves are swaying, it will have an effect up to ten li away; creaking branches, 100 li; shaking branches, 200 li; falling leaves, 300 li; breaking small branches, 400 li; breaking large branches, 500 li; breaking branches with sawdust, 1,000 li; uprooting large trees, 5,000 li. When the wind is very strong, blowing for three days and three nights, the whole country will be affected. For two days and two nights, half of the country will be affected. For one day and one night, there will be an impact up to 10,000 li away". In this book, Li Chunfeng talked about eight levels of wind, and if a gentle breeze and no wind are included, there are ten levels. This is similar to the wind levels set by the Englishman Francis Beaufort in 1804, but Li Chunfeng was more than 1,100 years ahead of Beaufort.

Liu Xun of the Tang dynasty observed and described tropical storms more carefully than Shen Huaiyuan of the Southern and Northern dynasties; he began with the precursors of tropical storms and then he described the disasters caused by tropical storms [52].

The New Book of the Tang: Record of Wuxing mentioned various forms of precipitation: in addition to rainfall, there was "cloudy fog that freezes trees for several days", "fog that remains for days", "rain which freezes trees", "snow-like fog", "rain and snow", "constant rain", "long rain", "continuous heavy rain", etc.

6.6.3 Meteorological Instruments

The meteorological instruments of the Tang dynasty included *xiangfengjing* [flags that could indicate the direction of the wind], *yubao* [poles with chicken feathers], *muwu* [wooden anemometer masts], *fengxiangji* (anemometers), and *zhanfengduo* (wind observing instruments).

According to the *Kaiyuan tianbao yishi* ("Bequeathed matters from the Kaiyuan (713–741) and Tianbao (742–756) Periods"), "in each of the five palaces, there are long poles in the hall with five-colour flags and bells hanging at the top of the poles. If the bells ring, the attendants can look at the direction of the flags to learn the direction of the wind." "In the palace of Prince Qi, pieces of jade were hung in the bamboo grove, and when the sound of the pieces striking each other was heard, one could tell that there was wind. The device was called a *zhanfengduo*."

Wang Rui wrote in the *Zhi guzi* that "on boats, people carved wood into *wu* on which flags were hung, waiting for the wind from any direction; this weather vane was known as a *wuliang* pole. The army used goose feathers to gauge the wind strength; this device was known as a *xiangfengwu* (wind-indicating bird)".

In his work *Yisi zhan: Houfengfa*, Li Chunfeng introduced the contemporary wind measuring equipment and described it in detail. He said: "Five-*zhang*-long poles ($16\frac{2}{3}$ metres) are set on the plains with chicken feathers on top of them to monitor the wind. Disks can also be used at the top of the poles along with a three-foot *wu* with a flower in its mouth which can spin when it senses there is wind". Chunfeng also said: "The feathers must be chicken feathers since chickens know the time. Each feather weighs eight taels to imitate the eight winds. The pole is five *zhang* long, in order to follow the norms of pentatonicism. It is said that *wu* is the essence of the sun, so it is the first to feel the wind in its nest. The pole does not have to be too long, but it should not be concealed. When the wind starts, it moves and can then be seen. On a permanent residence, it is appropriate to use *wu*. For military settings, it is appropriate to use feathers. The chicken feathers will be broken in the middle and the splintered point fastened to a thin rope. It should come to three or four feet in length and then the ropes of feathers will be hung on poles. There are different animal shapes on the top of the poles. The meteorological instruments of the Tang dynasty included *xiangfengjing* (flags that can indicate the direction of the wind), *yubao* (poles with chicken feathers), *muwu* (wooden anemometer masts), anemometers, and *zhanfengduo* (wind observing equipment)."

Fang Xinru of the Song dynasty recorded in the *Nanhai baiyong* that an anemometer was installed on the Huaisheng Pagoda in Guangzhou during the Tang dynasty.

6.6.4 Phenological Records

In the Tang dynasty, Bai Juyi clearly explained the relationship between topography and climate in his poem "Peach Blossoms at the Temple of the Great Forest" and in its preface. The preface says, "Because the mountain is high, the seasons change slowly. Even when the season is already early summer, the weather is still like February. Pear and peach trees are just flowering, the grass in the mountain has not yet grown. Everything is different from the villages on the low-lying lands. It feels like entering another world. So I wrote a poem: 'In late spring, the flowers all withered far and wide, but the peach blossoms are in full bloom on this mountainside. I'm often sorry that spring has departed without leaving any trace; not knowing it came to adorn this place [53].'" This means that the temperature is lower in the mountains than in the plains at the same time of year, and this can be shown via the delays in the solar terms. His observation was accurate.

As regards phenology in the Sui, Tang and Five Dynasties, the phenology in the *Suishu: Lvli zhi* is exactly the same as that in *Weishu*, with only a few words changed. The phenology in the *Old Book of the Tang: Calendric Record* is almost the same as that in the *Yizhoushu* ("Superfluous [chapters of the] Book of the Zhou"). The solar terms in the *Old Book of the Tang: Calendric Record* are earlier than the ones in the *Suishu: Lüli zhi*. The reason for this may have been due to a change in climate or related to the different source material on which the authors relied.

In addition, there was also knowledge of phenology in the poems of the Tang dynasty. For example, Bai Juyi's "Grass on the Ancient Plain – Farewell to a Friend" says: "Wild grasses spread o'er the ancient plain; with spring and autumn they come and go. Wildfire burns but can't destroy them; They spring up again when vernal breezes blow [54]." These four verses point out two important patterns in phenology: firstly, there is an annual cycle of grass growing and withering; secondly, the cycle of grass growing and withering follows the seasons, and when the spring breeze blows, the grass grows [55].

6.7 Travellers' Contributions to Chinese and World Geography

The travellers of the Sui, Tang and Five Dynasties were either diplomatic envoys of the state who were appointed to travel; religious representatives who travelled in order to obtain scriptures in the west; or officials. Although they travelled in different ways, they had one common characteristic, that is, they focused on geographic investigation during their travels and captured their insights in various written works, thus making contributions to the geography of both China and the world. The following are some representatives and their deeds in chronological order.

6.7.1 Chang Jun

In October of the 3rd year of the Daiye reign (607), Emperor Yang of the Sui dynasty sent Chang Jun, head of Garrison Reclamation, and Wang Junzheng, head of the Yu Ministry, on a diplomatic mission to the Chitu Kingdom (the southern part of the Malay Peninsula today). They set off by boat from Nanhai Prefecture (present-day Guangzhou) and it was smooth sailing for twenty days. They passed through the Xisha Islands in China, some islands to the south of Vietnam, and then entered Siam Bay (now known as the Gulf of Thailand) and sailed along the coast until they reached the Chitu Kingdom. It took more than a month to reach the capital. The two countries exchanged gifts and the host welcomed the guests with a banquet. In the spring of the 6th year of the Daye reign (610), Chang Jun returned to Chang'an together with Na Xiejia, the ambassador of the Chitu Kingdom. After returning, he wrote about his experiences in the two volumes of the *Chituguo ji* ("Record of the Red Earth Kingdom"). The book has been lost, but some parts of it have been preserved in the *Suishu: Chituguo zhuan*. These mainly relate to the size, location, architecture in the capital, customs, clothing, systems, religion, climate and the produce of the country. This book provided Chinese people a better understanding of the geographical environment and social conditions of the Chitu Kingdom.

6.7.2 Xuan Zang

The most famous religious traveller of the Tang dynasty was Xuan Zang (602–664) [56]. His given name was Yi and he was a native of Goushi, Luozhou (present-day Goushi Town, Yanshi, Henan province). Born into a Confucian family, he studied the *Xiaojing* ("Classic of Filial Piety") and was a filial son. He adhered to the spirit of Confucianism, "learned the classics and respected the ancestors and sages. He only read the classics and acted as the sages did [57]." Xuan Zang was the youngest of four brothers. He had gone to the Jingtu Temple with his second brother Chen Su to listen to sermons since he was a child and became a Buddhist monk at the age of 13. Afterwards, he went to Chang'an with his second brother, and then he went to Chengdu via Ziwu Valley and Hanchuan. He received his ordination in Chengdu at the age of 21, before starting to travel with merchants, sailing down the Three Gorges, and reaching the Tianhuang Temple in Jingzhou. He then travelled north to Xiangzhou (present-day Anyang, Henan province) and Zhaozhou (present-day Zhao County, Hebei province), and arrived in Chang'an, where he stayed at the Dajue Temple and became well known. He was familiar with the Buddhist scriptures and studied the doctrines. He found that there were many disputes and theoretical inconsistencies within Buddhism, which affected the development of Buddhism. In order to source authoritative Buddhist classics, he was determined to go to Tianzhu (India) to obtain scared texts.

In the first year of the Zhenguan reign (627), Xuan Zang set out from Chang'an and travelled to Liangzhou via Qinzhou and Lanzhou. At that time, the Tang dynasty was newly founded; therefore, people were forbidden from leaving the country. Li Daliang, the governor of Liangzhou, forced Xuan Zang to return to the capital. Hui Wei, a Buddhist leader, supported Xuan Zang's goal and sent two of his disciples to escort him. They did not dare to be seen in public and had to travel at night. After Hui Wei's disciples returned, Xuan Zang followed the merchants to cross the national border, passing through Guazhou, the Yumen Pass, Iwu (present-day Hami), Yanqi, Gaochang (present-day Turpan), westwards along the southern foot of the Tianshan Mountains, and crossing Mount Ling (today's Jengish Chokusu). He continued westwards along the southern shore of the Daqing Pond (present-day Issyk-Kul) to the Huoliximijia Kingdom (present-day Karakalpak), and then turned southeastwards heading out of Tiemen (present-day Badak Mountain). Later, he passed through the snowy mountains (present-day Hindu Kush) and the Black Mountains, and finally reached North India. In India, he visited the Ganges and the Indus as well as the southeast coast of India. Everywhere he went, he met with famous scholars and discussed the classics of Buddhism and Brahman. He favoured Mahayana Buddhism but did not reject the study of Hinayana Buddhism. He also actively studied some of the Indian sciences, such as logic and grammar.

After achieving satisfactory results in India, Xuan Zang returned to Chang'an in 645. When he had set out, the Tang court was refusing him permission to leave the country, but its attitude changed dramatically on his return. Emperor Taizong "was overjoyed" and asked him to "come and meet me at once". He also sent the Duke of Liang, Fang Xuanling, to welcome him. On the day of his return to Chang'an, "the people who heard of him rushed to see him, and they trampled on each other in order to take a look at him" [58].

In the 20th year of the Zhenguan reign, Xuan Zang finished the 12-volume *Datang xiyu ji* ("Great Tang Records on the Western Region") within one year and presented it to Emperor Taizong. This is an outstanding geographical work from the Tang dynasty. The geographic achievements of the *Datang xiyu ji* are mainly found in four aspects.

(1) New geographical knowledge

Since the Han dynasty, the high mountainous region west of the Kunlun Mountains had been named Cong Ling. Volume 12 of the book pointed out that Bomiluo (Pamirs) was a part of Cong Ling. This was the first time that the Pamirs had been mentioned in an ancient Chinese geographical work [59].

In Volume One, the rising of Mount Alunao (in present-day Afghanistan) was mentioned: "The mountain rises hundreds of feet every year." This figure may have been exaggerated, but it was the first time that specific figures had been used to describe the rising of the Earth's crust, which is of great significance in the history of geology.

(2) The detailed description of the geography and history of Central Asia and India surpasses all previous work. For example, the overview of India in Volume Two,

which includes the territory, population, cities, clothing, food, writing system, education, Buddhism, family names, martial arts, penal code, rituals, illnesses and deaths, taxation and produce, can be regarded as a geographical chronicle of India. For example, Xuan Zang translated the name "induo" into Chinese as "Indu", based on the pronunciation of "Indu" in a certain region. From then on, India became the name of South Asia. The *Datang xiyu ji* contains many major events in the ancient and medieval history of India. Discussion of King Kanishka is a popular topic among historians of many countries around the world. The book refers to him four or five times, which provides valuable information.

The book covers more than a hundred countries. The length of the sections varies, but there is a fixed pattern: size of territory, size of capital cities, geographical situation, agriculture, commerce, customs, literature, language, writing systems, currency, kings, religion, etc. Xuan Zang was able to depict a large number of facts in extremely simple language, which was not only precise but also vivid. Therefore, it can be said that the *Datang xiyu ji* is a unique book which is unparalleled by any other book.

(3) The description of a particular region contains both natural and human geography, which makes it an essential document for modern day study of the historical geography of the Central Asian and Indian region. For example, the book describes Jālandhara as, "the country is more than a thousand li from east to west and more than 800 li from north to south. The capital city of the country is 12 or 13 li in circumference. It is suited to the cultivation of grain crops and japonica rice as well as fruits and plants. There are more than fifty *shichidō garans*, more than 2,000 monks and disciples, and two schools. There are three *deva-kula* and more than 500 believers of other religions. The king of this country believed in another religion but, after meeting Rohan, he converted to Buddhism and ruled the country according to the principles of Buddhism".

(4) The first and last two volumes are on the geography of the Chinese frontier, while the remaining volumes are all on the geography of other places, making it one of the extant monographs on the geography of the Chinese frontier and regions outside China. The book has received much attention from scholars all over the world and has been translated into French, English and Japanese since the latter half of the nineteenth century.

6.7.3 Wang Xuance

Wang Xuance, a famous diplomat and traveller in the Zhenguan reign of the Tang dynasty, made three trips to India: in the 17th year of the Zhenguan reign (643), the 21st year of the Zhenguan reign (647), and the 2nd year of the Xianqing reign (657). According to the *Old Book of the Tang: Xirong liezhuan*, "At first, Wang Xuanze was sent to Tianzhu, and the King of Tianzhu sent an ambassador to pay tribute. However, when King Harsha Sīlāditya died, there was unrest across the whole country and a usurper seized the throne. He sent an army to attack Xuance.

Xuance was defeated and captured. He was robbed of the tributes paid by the other countries. After that, Xuance walked to Tubo and returned with 1,200 soldiers and he was finally victorious after a three-day battle. Subsequently, Xuance was promoted to the position of Chaosan Dafu." It is evident that Wang Xuance was a bold and tactful diplomat, general and traveller. He wrote the *Zhongtianzhu guo xingji* ("Travel Notes of Central India") [60]. The book provided a detailed and realistic description of the geography, landscape, mountains and rivers, scenic spots, religion, culture, politics, economy, and social customs of the five countries in Tianzhu at that time. The book was lost after the Song dynasty. There were ten volumes of text and three volumes of pictures, totalling 13 volumes [61]. Some of the contents are recorded in the *Fayuan zhulin* ("A Forest of Pearls from the Dharma Garden"). For example, Volume 24 quotes information regarding a famous hot spring in Nepa from the *Xiguo xingzhuan*, which is more detailed than in Volume Seven of the *Datang xiyu ji*. It can be seen that the geographic value of the *Zhongtianzhuguo xingji* was no less than that of the *Datang xiyuji*. The *Old Book of the Tang: Biographies of Xirong: Nepal* also recorded, "During the Zhengguan period, Li Yibiao went to Tianzhu and crossed Nepal. Narendradeva was very happy and went to visit the hot spring with Yibiao. The spring is around 20 *bu* in circumference, its water is always boiling and the water level is consistent. If things are thrown into the spring, steam rises. If kettles are hung above it, the food will soon be cooked. After Wang Xuance was captured by Tianzhu, cavalry from Nepal and Tubo significantly helped in defeating Tianzhu."

6.7.4 Du Huan

The first Chinese traveller to visit Egypt and leave behind works was Du Huan of the Tang dynasty. He was a descendant of Du You (735–812). Du You recorded in Volume 191 of the *Tongdian* that "Du Huan accompanied Gao Xianzhi, the Zhenxi Commissioner, on his western expedition and arrived at the West Sea in the tenth year of the Tianbao reign (751). At the beginning of the Baoying reign (762), he returned from Guangzhou with merchant ships and wrote the *Jingxing ji* (Record of Travels)". This was the only text recording information about Du Huan.

In 751, Gao Xianzhi, a general of the Tang dynasty, started a war with Chaj, which was defeated and begged for help from the Arab Empire. Gao Xianzhi led his army to the city of Aulie Ata, where he "fought with the Arab Empire for five days. There were traitors who attacked his army together with the Arab Empire. Xianzhi lost the war and only several thousand soldiers survived. Li Siye, the Youweiwei General, advised Xianzhi to escape by night and sacrificed himself to save Xianzhi" [62]. This is a brief account of the Battle of Talas.

Du Huan was captured in the 10th year of the Tianbao when he fought alongside Gao Xianzhi in the city of Talas (present-day Jambul, Kazakhstan) against the Arabs. After that, he travelled westwards with the Arab army and travelled all around Arabia as a prisoner for nearly ten years. Later, he got the opportunity to travel to some African countries and returned home by merchant ship at the beginning of the Baoying. After his return, he wrote the *Jingxing ji.* The book was lost, and only around 1,500 words were recorded in the *Tongdian.*

It was the earliest work in China to record the teachings of Islam and the spreading of manufacturing techniques throughout the Arab Empire by Chinese artisans. Volume 193 of the *Tongdian* quotes from the book, outlining the westward transmission of the manufacturing techniques of the Tang dynasty. The *Jingxing ji* also recorded the location, distance, ethnicity, number of troops, mountains, topography, climate, trading markets, history, produce and customs of several countries in Asia and Africa. It has received high praise from scholars all over the world. The relevant parts have been translated into English, French, Japanese and other languages.

6.7.5 Yi Jing

Yi Jing (635–713), whose family name was Zhang, was a native of Qizhou (now Jinan, Shandong province) in the Tang dynasty. His ancestral hometown was Fangyang (today's Zhuozhou, Hebei province). In 641, at the age of seven, Yi Jing became a monk and was ordained in 655. For the next five years, he studied the rituals. After the 5th year of the Huanqing reign (660), he began his study tours, which improved his understanding of Buddhism.

In 670, Yi Jing and other monks had the desire to go to India to seek the Dharma. At the beginning of 671, Yi Jing travelled south from Qi Zhou, passing through Puzhou and Caozhou until he reached Yangzhou. In autumn, he went to Guangzhou with Feng Xiaoquan. In November, he boarded a Persian merchant's cargo ship with Shan Xing and started his journey to India. In fewer than twenty days, he arrived in Sri Vijaya (now a giant port in Indonesia). He stayed there for six months to study Sanskrit to prepare for his journey to India. In February of the fourth year of the Xianheng, he arrived in East India. He stayed there for one year to pursue further study of Sanskrit and then went to Central India with a Chinese monk. They travelled to various Buddhist sites and studied in Nalanda for ten years. He left Nalanda in 685 and returned to the east by sea. He stayed in the South China Sea region for nearly ten years and arrived in Luoyang in the month of May in 695. From then on, he translated Buddhist scriptures in both Luoyang and Chang'an until his death [63].

In 691, when Yi Jing was travelling east from India to Sri Vijaya, he completed two works: the *Datang xiyu qiufa gaoseng zhuan* ("The Great Tang Biographies of Eminent Monks who Sought the Dharma in the Western Regions") and *Nanhai jigui neifa zhuan* ("A Record of Buddhist Practices Sent Home from the Southern Sea").

The *Datang xiyu qiufa gaoseng zhuan* is a biography of 57 monks (including Yi Jing and monks from Korea, Vietnam, Afghanistan and Uzbekistan) who travelled to the South China Sea and India to seek the Dharma during the 40 years from the 15th year of the Zhenguan to the 2nd year of the Tianshou reign. Attached to the book is the *Chonggui Nanhai zhuan*, in which the four Chinese monks who accompanied Yi Jing to Sri Vijaya in 689 were recorded, and the book was sent back to Chang'an after its completion, together with the *Nanhai jigui neifa zhuan* and other translated Buddhist works.

The *Datang xiyu qiufa gaoseng zhuan* is an important work of literature for the study of the history of Sino-Indian relations, transportation systems and religious figures. It recorded the newly opened roads to India through present-day Tibet and Nepal in the early Tang dynasty and also recorded the transportation routes from the South China Sea to India in detail. According to Yi Jing's record, there were many routes across the South China Sea. One could set off from Guangzhou, Jiaohai or Zhanbo, and via Sri Vijaya, Java, Langjashu, or Nārikira and arrived in East India; or one could go to South India first, then travel on to Simhala, finally arriving in East India or West India. It is evident that sea traffic was common at that time and it covered a large area.

This book recorded that most of the Chinese monks who went to India to seek the Dharma travelled by sea. This fact indicates that the connection between China and India by sea did exist before Yi Jing, even if the road through Xinjiang and Central Asia was the main route. However, thereafter, most people travelled by sea.

The book also mentioned a road from present-day Yunnan to India. This route existed before Emperor Wu of Han's rule. However, because it was a difficult road, not many people travelled along it, and so few people wrote about it.

The *Nanhai jigui neifa zhuan* recorded the history, geography, economy and culture related to India and the South China Sea at that time. Although it was mainly about Buddhism and provided reliable materials for studying the religious history of these regions in ancient times, it also provided information for the study of the social, economic, cultural, medical and healthcare history of the region.

The two aforementioned works were valued by scholars all over the world and were translated into French, English and Japanese at the end of the nineteenth century.

6.7.6 Li Ao

Li Ao (772–844) was a native of Longxi. Born into an aristocratic family, he studied Confucianism seriously from an early age and was a man of elegance. In the 14th year of the Zhengyuan reign (798), he became a Jinshi and was granted the title of Jiaoshulang (in charge of writing and editing official books). At the beginning of the Yuanhe reign, he was transferred to the position of Guozi Boshi (professor at the Imperial College) and worked as a historiographer. He was a relative and student of Han Yu, who recommended him to become an official in Lingnan. In the first month of 809, Li Ao left Luoyang for Guangzhou to serve in his post. Starting from Luoyang, he sailed through the Luo River into the Yellow River, turned into the Bian Canal, followed the Shanyang Canal, passed through Yangzhou, and then passed Suzhou and Hangzhou along the Jiangnan Canal. Then he travelled along the Qiantang River, crossed the Xinjiang River and the Boyang Lake into the Ganjiang River, crossed Dayuling Mountain, followed the Zhen River and the Bei River, and finally reached Guangzhou. The whole journey took 124 days. He recorded this journey in a diary and named it *Lainan lu* ("Records of the Journey to the South") [64]; this was 846 words long and was of significant geographical value. In particular, the mileage he recorded along the way provides valuable information for the study of transportation in the Tang dynasty.

6.7.7 Zhang Kuangye

In 938, Khotan King Li Shengtian sent an envoy, Ma Jirong, to pay tributes of red salt, Radix, yak tails and jade products. The Later Jin dynasty appointed Zhang Kuangye as the Secretary of Honglu, Gao Juhui as Panguan (magistrate), and Shengtian as the King of Khotan. In December of that year, Kuangye and others travelled for two years from Lingzhou (present-day south of Lingwu, Ningxia) to Khotan, and they returned in the winter of 942. They passed through Liangzhou (now Wuwei, Gansu), Ganzhou (now Zhangye, Gansu), Suzhou (now Jiuquan, Gansu), Guazhou (now in the east of Anxi, Gansu), and Shazhou (now in the west of Dunhuang, Gansu). Gao Juzhen wrote the *Xingji*, recording the geography and mountains and rivers they saw during this mission. The book has been lost, but part of it has been preserved in the *New History of the Five Dynasties: Four Barbarians Appendix*. The text is a valuable historical reference for the study of the history and geography of the region, as it describes the crossing of deserts, the produce of Khotan, rivers and jade fishing in rivers. However, he was wrong in treating the tributaries of the upper Tarim River as the source of that river.

6.8 Environmental Protection and Human Geography

6.8.1 Philosophies and Policies

The environmental protection ideas and measures of the Sui, Tang and Five Dynasties were rich in content and can be summarised under three headings: (1) protection of natural resources by the legal system; (2) afforestation; and (3) environmental protection in cities.

(1) Protection of natural resources by the legal system

The idea of environmental protection in the Tang dynasty was widely emphasised by society. Government departments included the idea in various legal regulations and scholars incorporated the idea in their poems, as exemplified below.

The *Datang liudian: Shangshu gongbu* ("The Six Statutes of the Tang dynasty") recorded a decree for the protection of biological resources. The *Tang dazhaoling ji: Jinyi liechi* ("Collected Grand Edicts and Decrees of the Tang Dynasty") was more explicit: it stipulated that people should protect the environment and prevent crops from being harmed.

According to the *Old Book of the Tang: Xuanzong benji*, during the Kaiyuan period, "the world was prohibited from catching carp", "woodcutting was prohibited on Mount Li", and "woodcutting was prohibited on Mount Tai and within 10 li of the mountain".

In the *New Book of the Tang: Dili Zhi*, there were records of environmental protection measures in different places. In Yunyang, Jingzhao Prefecture, "woodcutting is prohibited". In Guancheng County, Zhengzhou, "there is Pushe Pond, which was renamed Guangren Pond in the 6th year of the Tianbao reign, and where fishing is forbidden". In Kaifeng County, Bianzhou, "in Fuyuan Pond, which was renamed Benpeng Pond in the 6th year of the Tianbao reign, fishing is banned". In Qushan County, Haizhou, "there is Yongan Dyke 20 li to the east, which is joined to the mountain in the north and surrounds the city for 10 li to protect the city from the tide".

In the *Yuanhe junxian zhi*, the protective measures for tombs in various places were recorded: near the tomb of Duke Huan of Qi in Linzi County, Qingzhou, "woodcutting is forbidden within 20 *bu*". "Woodcutting is banned within 15 *bu* of Yan Ying's tomb". And herding was forbidden near Emperor Yap's tomb. At the tomb of Bo Yi in Hedong County, Hezhong province, "woodcutting is forbidden". At the tomb of Duke Wen of Jin in Jiangzhou County, "woodcutting is forbidden within 50 *bu*". Also, at the tomb of Zhuge Liang in Xi County, Xinyuan Fu, "woodcutting is forbidden". At the Xiuling tomb of Emperor Wu, Xiao Yan, in Danyang County, Runzhou, "woodcutting is prohibited within a hundred *bu*".

The poem *Shizhu* by Li Qunyu of the Tang dynasty [65] reflected the damage to the environment caused by the ceramic industry, mining, metallurgy and noise. First, the forest was destroyed. Second, the land was destroyed. Third, the air was polluted and there was serious noise pollution.

(2) Afforestation

Emperor Li Yuan stipulated that "the fields should be planted with elm, mulberry, jujube and any other suitable trees" [66]. There were more specific regulations in the *Tanglv shuyi* ("Law Code of Tang and Commentary"): "every *mu* of a field should be planted with more than 50 mulberry trees or more than 10 elm trees or jujube trees. Those who do not obey will be sentenced according to the law." And those who broke the law would be flogged 40 times as punishment.

In the Tang dynasty, planting trees at the imperial tombs was an important mission. In the first and second months of the year, and in the seventh and eighth months, when the weather was suitable, people were ordered to plant trees. After the planting, the officials concerned had to report the number of trees [67].

In the Tang dynasty, afforestation of the roads was also considered important. From Tang poetry, we know that apart from planting acacia and willow, fruit trees were also planted on both sides of the roads at that time. For example, Cen Shen's "green locust-trees line broad roads", Wu Yuanheng's "green locust-trees line long post roads", and Du Mu's "tens of thousands of trees can be seen in Hannan". In

740, Tang Xuanzong ordered the planting of fruit trees on the roads between the two capitals [68].

During the Sui dynasty, willow trees were planted on both sides of the Grand Canal to stabilise the embankments and to improve the environment. According to the *Yangdi kaihe* ji ("A Record of Digging the Canal by Emperor Yang"), these willow trees "have scattered roots, which can strengthen the river bank; they can also shade the people on the boats; thirdly, the sheep that lead the boats can eat their leaves". The *Daye zaji* (Miscellaneous Accounts of the Daye period) recorded that along the canal "more than 2000 li were shaded by trees". Du Mu's poem *Sui diliu* ("Willow at the Dyke of the Sui Dynasty") also showed that the willows planted in the Sui dynasty were still well protected in the Tang dynasty [69].

(3) Environmental protection in cities

During the Tang dynasty, the population of Chang'an reached more than one million, and there was a vast amount of daily refuse. In order to dispose of the waste, there were people who made a living by cleaning up the refuse, and thus made a fortune. According to the *Chaoye zaqian* ("Draft Notes from the Court and the Country"), "Luo Hui, a rich man in Chang'an, made his living by disposing of excrement". In the *Taiping guangji* ("Extensive Records Assembled in the Taiping Era"), it is said that "Pei Mingli, a native of Hedong, was good at managing his business. He collected discarded things, accumulated and sold them, through which he made a sizable fortune".

The Tang dynasty managed the urban environment through legislation. Emperor Xuanzong ordered relevant departments to manage the environmental conditions of Chang'an and Luoyang: "The two capital cities are the places where imperial residences exist; the streets must be renovated. There will be punishment for pouring out sewage or littering [70]." The building of sewers in Chang'an was common in the Tang dynasty. In addition to the drains on both sides of the main roads, there were also drains along the sides of the street between two neighbourhoods. The drains were fitted with perforated iron panels and stone frames, which were early examples of environmental protection facilities made of iron anywhere around the world, indicating the importance the Tang dynasty attached to urban environmental protection.

City landscaping is an important part of urban environmental protection, and the Tang dynasty advocated planting trees in cities many times in the form of imperial decrees. In 766, Emperor Daizong ordered "the planting of trees in the six streets of the city" [71] and banned cutting down or destroying the trees [72].

6.8.2 Human Geography

(1) Industrial geography

Some of the industrial geography of the Tang dynasty, such as the types of minerals, their geographical distribution and production, was recorded in the *New Book of the Tang: Dili zhi*, *New Book of the Tang: Shihuo zhi* and *Yuanhe junxian tuzhi*. For example, from the *New Book of the Tang: Dili zhi*, we learn that the largest amount of iron was produced in 32 places in Huainan and Jiangnan. The next most prolific were the Hedong and Hebei regions, with 26 places. The third was Jiannan, with 18 places. The fourth was Henan, with 8 places. The fifth was Guannei, with 6 places. The sixth was Lingnan, with 5 places. In addition, there were deposits of salt, gold, silver, copper, tin, lead and cinnabar. According to the *Statistics of Minerals of the Tang dynasty* by Yang Yuan, the production areas of the nine major minerals at that time were:

① Gold. Lingnan Dao had the most gold producing areas, a total of 32; Jiannan Dao, second, had 28; Jiangnan Dao, third, had 15.
② Silver. Lingnan Dao had the most silver producing areas, a total of 51; Jiangnan Dao, second, had 28; Henan Dao, had 6; the others had fewer than 3.
③ Copper. Most places of copper production were in Jiangnan Dao, a total of 39; Hedong Dao, second, had 10; Jiannan Dao, had 8; the others had fewer than 5.
④ Iron. Jiangnan Dao had 29 iron ore producing areas; Jiannan Dao had 26; Hedong Dao had 17; Shannan Dao had 12; Hebei Dao had 10; Henan Dao had 9; Guannei Dao had 7; Lingnan Dao had 6; the rest had fewer than 4.
⑤ Tin. Tin mining was not a prosperous activity in the Tang dynasty. There was only a total of 19 sites, 3 in present-day Jiangxi; 2 each in Shaanxi, Henan, Zhejiang, Hunan, Guangdong, Guangxi and Sichuan; and 1 each in Shanxi and Shandong.
⑥ Lead. The mining of lead was not a prosperous activity either in the Tang dynasty, with a total of 13 sites. It was the least developed of the metal mining industries and all of the sites were south of the Yangzi River, in locations such as Jurong in Jiangsu; Xuancheng, Jixi and Gui Chi in Anhui; Dayu and Shangrao in Jiangxi; Yizhang and Guiyang in Hunan; Guangning, Kaiping and Yangchun in Guangdong; Xingye and Tengxian in Guangxi. Each place had one site.
⑦ Salt. Most of the salt production was in Jiannan Dao, with a total of 52 sites; Shannan Dao was the second most productive area with 20 sites; there were 18 in Jiangnan Dao, 15 in Henan Dao, 13 in Guannei in Longyou Dao, 10 in Lingnan Dao, 2 in Hebei Dao and 2 each in Hedong Dao and Huainan Dao.
⑧ Cinnabar. The largest number of locations producing cinnabar was in present-day Hunan amounting to a total of 11; there were 4 in Sichuan; 3 in Guangxi; 1 each in Shaoxing in Zhejiang, in Lianzhou in Guangdong and, and in Guizhou; 2 in Lueyang, Shaanxi; 1 in Minxian, Gansu. The total number was 24.
⑨ Mercury. Mercury mining in the Tang dynasty was also not a prosperous activity. There were only 10 production areas, which were all in the south. Six of them

were in present-day Hunan; and there was 1 each in Guangdong, Guangxi, Sichuan and Guizhou.

In Volume 7 of Fanchuo's *Yunnan zhi*, the deposits of salt, gold, silver, tin, azurite, amber and realgar in Yunnan were recorded, which supplemented the information in the three books above.

(2) Agricultural geography

In the early Tang dynasty, great importance was attached to the military border garrisons. For example, Lou Shide farmed in the Hetao area and in areas between the Yellow River and Huangshui, and he achieved remarkable success. Guo Yuanzhen farmed in Hexi and "taking advantage of the environment, the rice harvest was abundant. The silk industry and grazing were also prosperous [73]."

Agriculture in the Jiangnan area during the Tang dynasty was no longer backward and Jiangnan had developed into an important agricultural area. For example, Yuezhou was a rich town in the late Tang dynasty: a place where "its production can feed half of the country [74]". As for the Wu, Wuxing and Danyang Prefectures, "half of the country's resources were in the area [75]". Changzhou was "a large county which furnished the army with food and paid a huge amount of tax" [76].

In the middle of the Tang dynasty, Li Ao's record of land use was a step forward compared with the *Liji: Wangzhi* ("Record of Rites"). It says, "A field measuring a square li is 540 *mu*, a field measuring 10 li is 54,000 *mu*, a field measuring 100 li is 540 million *mu*, and a field measuring 1,000 li is 50.4 billion *mu*. Within a square li, 10 *mu* is for the house, and the rest is for vegetables and livestock. For a region of 100 li, the city, roads and cemeteries only take up 36 li and farmland covers 190.045 million *mu*, the rest of the estate grows mulberry [77]." The "billion" here equates to the modern "100,000", which is different from the concept of "billion" today. In the era described in the *Zhouli*, farmland accounted for 67% of an area measuring 100 square li, while in the Tang dynasty, farmland accounted for 64%, which was a slight decrease. However, within an area measuring a square li, 2% was used to build a city, a palace and lanes.

Fanchuo's *Yunnan zhi* provides a summary of the agricultural geography of Yunnan by recording the agricultural production practices, farming tools, the regional distribution of crops and horses, etc.

(3) Urban planning

Chang'an, the capital city of the Sui and Tang dynasties, was a newly-built capital city whose planning and design were extremely formal. There were seven aspects to the city planning.

① The city plan was square, with three gates on each side, and each gate usually had three doorways, except for the Mingde Gate, which had five doorways. The palace was in the middle.

② Palaces were located to the north of the centre of the city with the main palace facing south.

③ South of the palaces was the imperial city: the buildings for civil and military officials, temples, the altar, official handicraft workshops, military camps, etc.
④ The route from the Chengtian Gate through the Zhuque Gate to the Mingde Gate was the city's central axis and the city was symmetrical on each side.
⑤ The north and south gates were directly opposite each other, as were the east and west gates, and in the middle were the main streets of the city. The neighbourhoods divided by the streets were also symmetrical from east to west. The east–west and north–south streets were perpendicular to each other, forming an extremely neat chessboard grid, and within each grid square was a neighbourhood or a market.
⑥ The business district of the city was orientated towards the south, deviating from the traditional planning convention of locating the "city behind palaces".
⑦ The markets were still not open to outsiders and commercial activities in the city were time limited.

(4) Military geography

Du You's *Tongdian* contained 15 volumes of the *Bingdian* ("Military Statute"), in which the experience gained from all the wars won and lost before the Tang dynasty and the principles of the art of war were summarised. One of the volumes, "Victory or Defeat by Topography", was on military geography that explained the close relationship between the features of the land and the military.

Li Quan's *Taibai yinjing* [78] showed a deeper understanding of the relationship between the terrain and the military. It clearly stated that the danger presented by the terrain was different for different people and that the winning or losing of wars was mainly determined by people rather than geographical advantage. He viewed the army and the terrain as two forces that acted upon each other: "the army becomes invincible because of the terrain and the terrain becomes impregnable because of the army".

The *New Book of the Tang – Biography of Li Deyu* recorded Li Deyu's frontier defence: "He built the Choubian Building and drew maps of his enemies on the walls of the buildings so that he could know everything about the other countries."

Li Jifu of the Tang dynasty once stressed that geographic study should serve the military so that the pros and cons of attack or defence could be analysed. In the *Yuanhe junxian tuzhi,* he marked the location of forts, garrisons with warhorses, passes, ports, etc.

6.9 Geology and Mineralogy

In the Sui, Tang and Five Dynasties, apart from the knowledge inherited from former scholars, new things also emerged in the fields of geology and mineralogy. There were monographs on minerals. During this period, the mining, smelting and ceramics industries boomed, alchemy flourished, and the use of minerals in Chinese medicine and paint pigments increased, all of which contributed to the accumulation of mineral knowledge and laid the foundation for the formation of ancient Chinese mineralogy. The following discussion is divided into seven parts.

6.9.1 Yan Zhenqing and Others' Theories on Sea Level Changes

The term "the vast seas changed into fields" is an ancient term that conveys the geological idea of sea and land changes in China. It originated very early, according to the literature; the earliest record was Xu Yue's *Shushu jiyi* ("Memoir on the Methods of Numbering") in the Han dynasty. In 771, Yan Zhenqing first proved this geological phenomenon with fossils in his prose text *Fuzhou nanchengxian magushan xiantan ji* ("The Story of Magu's Fairy Altar"), which elevated the hypothesis put forward by Ge Hong in the Jin dynasty through the mouths of the spirits to a scientific level. He stated that "there are still snail and mussel shells in stones, which may have formed in the sea" [79]. This not only explained why seashells appear in the rocks on high mountains, but also recognised that sea and land changes had occurred there, thus making the saying more scientific.

6.9.2 Geology and Mineralogy in Handicraft Production and Archaeological Artefacts

(1) Handicraft production

The mining, smelting, ceramics, construction and other handicraft industries in the Sui, Tang and Five Dynasties conveyed the knowledge and utilisation of minerals and rocks at that time. In the above section on "Industrial Geography", there was a description of the deposits and geographical distribution of nine kinds of minerals in the *New Book of the Tang: Dili zhi*. The *Yuanhe junxian tuzhi* recorded 35 kinds of minerals and rocks, including gold, silver, copper, iron, tin, lead, cinnabar, mercury, azurite, copper blue, petroleum, mineral fuels, salt, gypsum, realgar, orpiment, talc, jade, mica, white quartz, amethyst, kaolinite, stalactite, tonalite, fossil fragments, fish fossils, spirifer, marble, fungid coral, saltpetre, etc. In addition, the ceramics industry required clay, porcelain clay, quartz sand, mica and other minerals. In the construction industry, granite, marble, limestone and sandstone were often used.

(2) Archaeological artefacts

From the data in the information on the minerals and rocks in the journals of *Archaeology* and *Cultural Relics* before 1990, the author found that a total of 24 kinds of minerals had been utilised in the Sui, Tang and Five Dynasties, including gold, silver, copper, iron, tin, lead, jade, etc. (Fig. 6.1); 9 kinds of rocks had been used, including

Fig. 6.1 Mineral medicine as archaeological artefacts of the Tang dynasty (photographed by Liang Qing 梁擎)

slate, sandstone, basalt, limestone, gravel, granite, shale, red sandstone, marble, etc. The details of the utilisation are as follows.

Gold: gold coins, gliding, mirrors, gold teeth, gold utensils (cups, plates), golden accessories, medicine, alchemy, etc.

Silver: sycees, silver "cake" coins, mirrors, silver coffins, silver utensils and jewellery (cups, plates, lids, necklaces), silver plaques, silver jugs (Fig. 6.2), etc.

Bronze: statues, bronze coffins, mirrors, daggers, spoons, chopsticks, bowls, incense burners, brassware, etc.

Iron: rulers, armour, figurines, locks, keys, agricultural tools, handicraft tools, weapons, etc.

Tin: tinware, gilt tin cups, etc.

Lead: coins, alchemical medicines.

Jade: jade objects such as bracelets, pendants, slips, etc.

Fig. 6.2 Silver jug from the Tang dynasty (photographed by Liang Qing 梁擎)

White marble: epigraphs, architectural features, statues, medicine mills.

Quartz: plates and bowls.

China stone, feldspar: porcelain.

Gypsum: epigraphs, alchemy.

Talc: stone pillows, alchemy.

Cinnabar, stalactite, amethyst, white quartz, amber, lithargite: Chinese medicine, alchemy.

Shale: inkstone.

Marble: architectural features, stone boxes.

Limestone: architectural features, statues, stone carvings (lions, warrior-servants, tomb doors, outer coffins, epigraphs, stone men), chimes.

Red sandstone: architectural features, grinding stones, sculptures.

Granite: architectural features.

Sandstone: statues, tomb records.

The chromogenic agents in Tang-sancai wares (Tang ceramics with a three-coloured glaze) were various metal oxides, such as verdigris, antimony yellow, ferric oxide yellow, brown iron, copper blue, cobalt blue, manganese violet, etc. Of these, the yellow glaze was obtained by using iron oxide and the green glaze was copper sulfate.

The minerals used as pigments were ochre, copper blue, realgar, azurite, vermilion, etc.

Knowledge about rocks in the Sui, Tang and Five Dynasties was recorded, for example the *Kuo dizhi* ("Comprehensive Gazetteer") described the chime stone of the Sishui area. The stone chimes produced in Jiuzhen county, Lingnan Dao (present-day Thanh Hoa, Vietnam) were better than the stone resonators produced in Lingling, Hunan, as recorded in the *New Book of the Tang: Dili zhi* and *Yuanhe junxian tuzhi*.

In the preface to a poem by Bai Juyi is entitled *Huayuan qin: ciyuegong feiqirenye* ("Huayuan Chime: It's Not Others Who Criticise the Musician"), it is written: "During the Tianbao period, the Sibin chime was replaced by the Huayuan chime. People who knew the chime said that the Sibin chime's tone was not as harmonious as the Huayuan's. Thereafter, chimes were all made of Huayuan stone [80]."

Yuan Zhen, a poet of the Tang dynasty, also wrote a poem about Huayuan chime stones, the preface to which stated: "According to Li Zhuan, during the Tianbao period, the Sibin chime was replaced by the Huayuan chime." The poem goes, "Sibin stones were made into chimes, and few people listened to this kind of ancient music … Huayuan stones are soft and are easy to carve [81]."

The Huayuan mentioned in the poems is the present-day Yaozhou district of Tongchuan, Shaanxi. From the above, we can see that the stones used for making chimes in the Tang dynasty did not come from a single place, but came from Anhui, Hunan, and Shaanxi, as well as from Qing Hua in Vietnam.

During the Tang dynasty, there were many kinds of stones used to make inkstones. The four famous types of inkstones (*yan*) are Duan inkstone (*Duanyan*), She inkstone (*Sheyan*), Tao inkstone (*Sheyan*) and Lu inkstone (*Luyan*).

Duan inkstone was so named because it was found in Duanzhou (now Zhaoqing, Guangdong). According to *Shiyin yantan,* "Duanxi stones were first used in the Wude era of the Tang." In the early Tang dynasty, Duan inkstones were mainly for practical use, focusing on the quality of the stone rather than on any carving. Li He's *Yangsheng qinghua zishi yange* ("A poem for Yang's Duan Inkstone made of Purple Stone with Green Patterns") read: "The Duanzhou stoneworkers are highly dexterous as craftsmen. They sharpen their stone sculpting knives and quarry the stones on the summit of the mountain, just like stepping on the sky and carving purple clouds. The inkstones are evenly shaped. Filled with water, they look like lips and the patterns on the inkstone appear from time to time." Liu Yuxi's *Tang xiucai zeng Duanzhou zishiyan: yishi dazhi* ("A Poem to thank Scholar Tang for his Purple Inkstone") read, "the inkstone from Duanzhou is very precious; you give it to me because you know I am indifferent to fame and wealth." The purple rocks that the poets referred to are the rocks found at the bottom of the Duanxi River, which are "grey and pale when dry, but green and purple when moistened" [82]. This is a kind of silty sericite schist [83]. "It cannot be damaged and is not delicate like jade. To add water and be ground on, it will be oily and the ink will become thick [84]."

She inkstone was named after She County in Anhui province and had its origin in the Kaiyuan reign of the Tang dynasty. According to the *Shezhou yanpu* by Tang Jie of the Song dynasty, a hunter named Ye chased a beast to Wuyuan city in the middle of the Kaiyuan reign and found a beautiful stone. He picked up a piece and made it into an inkstone, which was very agreeable to use. In 1976, a She inkstone was unearthed from a Tang tomb in Hefei; the stone was 20 cm long, 11 cm wide at the top, 15 cm wide at the bottom, and 35 cm high. The stone was smooth, greenish-blue in colour and the lines were even and smooth. The tomb dated from the 5th year of the Kaicheng (840) [85]. She inkstone is green and crystal, with a fine texture, as firm as jade, and makes no sound when ink is being ground. It consists of black slate and phylloid sand, and its main characteristics are that it is hard and moist [86]. Its average Mohs amounts to 4.

Tao inkstone was so named because it was found in ancient Taozhou (present-day Lintan County, Gansu Gannan Autonomous Prefecture). The inkstone already existed in the Tang dynasty and it became very famous by the Song dynasty. Also, in the Song dynasty, in Volume Three of Gao Sisun's *Yanjian*, there is an article "Taoshi Yan" which says, "The stone comes from Lintao. The green stones in Tao River are soft in nature and cannot be ground durably." The nature of Tao stone is clearly explained. Poems by Zhang Wenqian, Zhao Wujiu and Yang Xinxiang show that Tao inkstone was popular because of its green colour. Tao inkstones were made of different kinds of stones that were delicate and crystalline. Their colours were bright and the patterns were curvy, like waves and curling clouds. Because Tao stones came from the deep bed of the Lintao River, extraction was difficult, so there were very few Tao inkstones in the world [87]. Tao stone is a kind of hydromica argillaceous slate [86].

Lu inkstone was named after Lu County (present-day Yanzhou, Shandong province), where it was found. The name Lu Inkstone was not confined to the inkstones which came from Lu County, but also referred to the geographical area of the ancient Lu state and, later, to the province of Shandong. Shandong produced many famous inkstones. The best of them was Hongsi inkstone from Qingzhou (now the Weifang and Zibo area in Shandong). Records show that these were of various colours and textures. Some of them resembled clouds and mountains, while others were like flowers, birds and animals. The stone was often interspersed with stone glints, ice patterns, spirals, stripes and spots which constituted their unique decorations. As a rule, the inkstones were sedimentary rocks and were coloured stones. The appearance of the four famous inkstones of the Tang dynasty was indicative of the new achievements of the Tang dynasty in the use of rocks.

Building bridges from stone was one of the principal features of the Chinese culture of stone architecture. There were many stone bridges built in the Sui, Tang and Five Dynasties, such as: the Zhaozhou Bridge in Hebei, designed and built by Li Chun in the Sui dynasty with limestone; the arch bridge in the Sui dynasty in Linying, Henan province, which was built with red sandstone.

The stone carvings in tombs of the Sui, Tang and Five Dynasties include watch towers, sculpted corners, steles with and without inscriptions, ornamental columns, stone-carved *tianlu* and ostriches, stone men, stone horses, stone lions and so on. The underground parts of tombs or objects in the tombs include burial chambers, entrances, stone coffins, sarcophaguses, epigraphs, portraits, stone tools, figurines, various animals, tables, chairs, land purchase coupons, ornaments, seals, stone mirrors, chess boards, etc. In short, there were many objects and architectural features made of stones in this period, which showed that the use of rocks was very broad.

6.9.3 The Properties of Minerals and Rocks

In the Sui, Tang and Five Dynasties, there was a certain degree of understanding of the properties of minerals. For example, the *Xinxiu bencao* ("Newly Revised Canon of Materia Medica"), compiled by Su Jing (who later changed his name to Su Gong) and others in the Tang dynasty, described the colour of streaks as light cyan [88]. The description of cinnabar cleavage was: "If broken, it is like mica." The description of the hardness of talc was "extremely soft and smooth". The description of the hardness of talc in the *Jinshibu wujiu shujue* ("Book of Alchemical Metal and Stones") was similar.

As regards the specific gravity of minerals, the *Longhu huandan jue* ("Dragon and Tiger Formulae for Returning Mercury to Cinnabar") by Jinlingzi in the middle of the Tang dynasty recorded the specific gravity of eight minerals: the values and words used differed slightly from those in *Sunzi suanjing* ("Sun Zi's Mathematical Manual"). "1 square *cun* of gold is 1 *jin*; 1 square *cun* of silver is 14 *liang*; lead and tin are 9.5 *liang*, iron is 6 *liang*, jade is 9 *liang*, *baishi* [lit. white stone] is 3 *liang*, and earth is 2 *liang* [89]." Jinlingzi's book omitted the specific gravity of copper and increased the specific gravity of earth and tin. The word "stone" was changed to "white stone". The specific gravity of jade was reduced to only nine *liang*. Finally, it also clearly pointed out that the different specific gravity of various minerals was due to their own material differences. The weights recorded in *Longhu huandan jue* are credible and close to today's data, which is of important scientific value. For example, the specific gravity of gold in this book is about 13.54, which is close to today's 19.3; the specific gravity of silver is about 11.85, which is close to today's 10.5. Tin is around 8.04 in the book, also close to today's 7.31.

The magnetic properties of magnetite were recorded in Chen Zangqi's *Bencao shiyi* ("Supplement to Materia Medica"): "a magnetic stone is like the mother of iron. It can attract iron as a mother summons her son [90]." The *Jinshibu wujiu shujue* also recorded that "magnetites are from Cizhou, those that can attract six or seven needles are of the lowest quality." In the *Xinxiu bencao* it is written that "good ones can attract ten needles, even iron knives can be moved."

Concerning the properties of bluestone, the *Longhu huandan jue* recorded that "in Qianshan County, Xinzhou (present-day Shangrao, Jiangxi province) there is a 'bitter' spring. Boiling the water from this spring can produce blue vitriol; cooking the blue vitriol can produce copper, while boiling it with iron for a while can also produce copper." It is clear that a small-scale industry of copper refining already existed in the the Tang dynasty [91]. This method was further developed in the Five Dynasties according to the *Baozang lun* ("Treasure Store Treatise"): "Using 'bitter water' containing copper sulfate to soak iron will produce red coal, and to continued boiling will turn the coal dark and hard." This suggests that in the mid-Tang era the new technology of wet-process copper making already existed.

As for the shape and structure of mineral crystals, the description of amethyst crystals by Xun Liu of the Tang in the *Lingbiao luyi* ("The Records of the Unusual from Lingnan") stated: "It has five edges and its two heads are like arrowheads." It is inaccurate to talk about five edges; instead, it should be six-sided with six edges. The *Xinxiu bencao* described the monomeric form of cinnabar: "There are dozens of levels of cinnabar. The best ones are the bright ones. The big ones are like eggs, the small ones are like jujube chestnuts. Cinnabar looks like the hibiscus flower and mica when broken." The monolithic form of calcite was described in the *Xinxiu bencao* as "the large ones are a square *chi* in size" [92].

Liu Zongyuan discussed the differences in the properties of stalactites depending on their different origins and occurrences in the *Yu Cui Lianzhou lun chizhongru shu* ("Discussion about Stalactites with Cui Lianzhou").

There are eight types of mineral aggregates described in the *Xinxiu bencao*: (1) granular, such as cinnabar; (2) amorphous, such as red halloysite; (3) powdery, such as powdered cinnabar; (4) stratiform, such as mica; (5) lumpy, such as talc; (6) bead-like, such as pyrolusite; (7) lamellar, such as calcite; (8) tetragonal, such as pyrolusite.

Regarding the symbiotic relationship between minerals, the *Jinshibu wujiushujue* recorded the symbiotic relationship between pyrite (iron sulfide) and copper ore, stating: "Wherever there is copper ore, there is pyrite." The symbiotic relationship between *lujian* (halides), Taiyin xuanjing stone (cryptocrystalline quartz) and salt was apparent in that: "the three are found together in the salt ponds of Xie County in Hedong." The *Xinxiu bencao* mentioned the symbiotic relationship between copper ore and various blue minerals like azurite and malachite, saying: "Where copper is found, various blue minerals can also be found. But chrysocolla is rare … sometimes there are voids inside [93]." Chrysocolla is a type of blue copper ore which is hollow: it is shaped like a stalactite or kidney, or is grape-like in shape and is often found together with malachite (a green copper ore). Chen zangqi wrote in the *Bencao shiyi* that gold coexisted with *fenzishi* (limonite). Limonite, whose composition includes ferrous oxide and copper oxide, is a black-brown mineral. This symbiotic relationship can be used to locate mineral deposits. As Chen Zangqi said, "When searching for gold deposits, people usually dig to a depth of over 10 *chi* until the limonite layer is reached. Limonite is black on one side. Once a limonite deposit is discovered, gold deposits can often be found beneath it [94]."

With regards to mineral identification, the *Longhu huandan jue* recorded that "if chalcanthite is burned, it will turn white". Yan Shigu used colours to identify the five metals: "Yellow is gold, white is silver, red is copper, green is lead, and black is iron [95]." In the *Biebao jing* ("Book of Special Treasures"), it is written that jade fluorescence was used to identify jade: "If there is jade in the stone, it will reflect red light under a lamp [94]."

6.9.4 Indicator Plants for Minerals

Volume 16 of the *Youyang zazu* ("Miscellaneous Morsels") by Duan Chengshi of the Tang dynasty, and *Yunxi youyi* ("Subjects Discussed with Friends of Yunxi") by Fan Shu both recorded the method of using indicator plants to find minerals. Duan Chengshi stated that "when there are shallots on the mountain, there will be silver underneath; when there are alliums, there will be gold underneath; when there is ginger above ground, there will be copper and tin underneath." According to modern geological and botanical research, indicator plants do exist in nature, for example, the indicator plant for the copper mines in western Sichuan is amaranth; the indicator plant for sandstone-type copper deposits in Hunan is sorrel; the indicator plant for copper ores in the downstream area of the Yangzi River and western Zhejiang is *Elsholtzia haichowensis*. The same mineral has different indicator plants in different areas. Moreover, the same plant can also indicate different minerals. For example, in addition to copper, amaranth can also absorb zinc, lead, silver and other metals; *Thlaspi arvense* seed and plantain are indicator plants for zinc; fir, pine and spruce are indicator plants for gold; and *Buddleja davidii* is an indicator plant for mercury. It is evident that the various indicator plants proposed by the people of ancient times were based on experience.

6.9.5 Works on Mineralogy and Petrology

There were two kinds of mineralogy works in the Sui, Tang and Five Dynasties: first, the knowledge of minerals in works on herbs and plants, or *bencao*; second, the knowledge of minerals in alchemical works.

(1) *Bencao*

① *Xinxiu bencao*

During 650–655, Li, Yu Zhining and others were ordered to compile 7 volumes of the *Yinggong tang bencao* ("Tang Materia Medica"). In 659, Su Jing, Changsun Wuji and 22 other authors were ordered to revise it, and they renamed the book *Xinxiu bencao* (or *Tang bencao*); it consisted of 20 volumes, plus 1 volume of indices, 25 volumes of pictures of medicines, 7 volumes of Tujing, making 53 volumes in total. There were 11 parts, containing a total of 844 kinds of medicines, including 109 kinds of inorganic medicines. These two books were lost after the Song dynasty, parts of the books only were preserved in the *Qianjin yifang* ("Supplemental Prescriptions Worth a Thousand in Gold") by Sun Simiao, *Chongxiu zhenghe zhenglei bencao* ("Zhenghe Materia Medica") and *Bencao gangmu*. At the end of the Qing dynasty, Fu Yunlong found 10 volumes of a facsimile of the *Tang bencao* in Japan in 731 and the third volume of the *Zhenghe bencao* copied by the Japanese Kojima Moku, making a total of 11 volumes which were published in the *Zhuan xilu Series*. In 1959, Shanghai Scientific & Technical Publishers published a facsimile of the *Xinxiu bencao*. Thereafter, the book began to be known around the world. This book was the first pharmacopoeia issued by the state in China and was also the earliest national pharmacopoeia in the world [96].

According to the articles from the *Xinxiu bencao* cited in the *Chongxiu zhenghe zhenglei bencao* and *Bencao gangmu*, the following is known.

Firstly, the recorded minerals and rocks included cinnabar, stalactite, alunite, saltpetre, mirabilite, talcum, azurite, etc., totalling 66 kinds. Some of them were synthetic compounds, such as silver paste, tin powder, etc.; some of them had different names, but were actually the same thing.

Secondly, the origin, properties, kinds of different rocks and minerals (for example, there were five kinds of alunite: green, white, yellow, black and red) were recorded. The quality, identification methods, synthetic methods, pharmaceutical properties, and the occurrence of minerals and rocks in different places were also recorded. These descriptions were closely related to mineral petrology and are valuable historical materials from ancient China.

② *Bencao shiyi*

In 713–741, Chen Zangqi wrote 10 volumes of the *Bencao shiyi*. The original book has been lost, but there are some quotations in the *Chongxiu zhenghe zhenglei bencao* and *Bencao gangmu*. From these quotations, we know the following.

First, there were 24 kinds of minerals and rocks recorded, such as coral, agate, lead, copper green, magnets, gold, etc. Some of them are synthetic compounds, such as glass, glaze, etc.

Second, it recorded the origin, occurrence, properties, morphology, pharmaceutical properties, applications, identification methods, and the different qualities of various minerals and rocks from different deposits.

③ *Haiyao Bencao*

Li Xun (c. 855–930), a native of the former Shu in the Five Dynasties, lived in Zizhou (present-day Santai County, Sichuan) [97]. His ancestors were from Persia (present-day Iran). He wrote the six volumes of *Haiyao bencao* ("Overseas Pharmacopoeia"), but the original book has been lost. There are some citations in the *Chongxiu zhenghe zhenglei bencao* and *Bencao gangmu*. From the citations, we know the following.

Firstly, there were 11 kinds of minerals and rocks recorded, such as jade, sulfur, silver, black stone, coral, etc.

Secondly, he was very knowledgeable about drugs from overseas, which may have been related to his ancestors being Persian or may have been related to the increase in traffic between China and foreign countries during the Tang dynasty, with many foreign drugs being imported into China. For example, "green salt, according to the *Gujin lu*, was born from a stone in the Persia Empire, and the stone tasted salty. It can be made into atacamite and its colour will remain true for a long time"; "alum: the potassium alum that comes from Persia and Daquin is white and clear. It has vertical patterns. It is also more effective in alchemy than Hexin or Shimen varieties. Recently, it can regularly be seen in Wenzhou. Persia also produces lowigite, which has vertical golden patterns inside."

(2) Alchemical works

① *Taiqing shibi ji*

Su Yuanming (Qingxiazi), a Taoist priest from Luofu Mountain in the Sui dynasty, wrote three volumes of the *Taiqing shibi ji* ("Records of Taiqing Stone Wall"). Some were edited by Chu Ze. Cao Yuanyu, Yuan Hanqing, Zhang Zigao and Yoshida Koubon believed it was written by Su Yuanming in the late third century and compiled by Chu Ze in the early sixth century. Zhang Jueren and Zhu Sheng believed it was written in the Song dynasty. It described the art of Chinese alchemy and anonymity was not adopted [98]. It was included in the *Daozang* ("Daoist Canon"), in books Nos 582–583. In the various alchemical formulas, the minerals mentioned included realgar, stalactite, orpiment, cinnabar, mica, white quartz, etc.: 47 kinds altogether, but there were no descriptions of their characteristics.

② *Jinshibu wujiu shujue*

This book was written in 664 and the author remains unknown. It was included in book No 589 of the *Daozang*, with more than 2,000 words recording 45 kinds of minerals used for alchemical purposes. The origin of the minerals recorded includes 47 present-day counties and cities in 17 provinces and autonomous regions in China. There were also four other countries or regions. The content recorded 17 categories of data relating to minerals, such as name, origin, form, colour, transparency, lustre, quality, percussive sound, degree of dryness, fracture, identification method, magnetism, taste, odour, symbiotic relationship, usage, etc. The content was very informative and completely focused on mineralogy. It is a true example of mineralogical literature of significant scientific value. The author outlined his objective in writing the book before the main text: "if you want to learn Taoism and alchemy, you must first know the rocks. You should learn their shapes and characteristics." This was a major contribution by Chinese Taoism to ancient Chinese mineralogy.

The minerals recorded included cinnabar, realgar, jade, white quartz, mica, *tianmingsha* (borax), actinolite, azurite, sulfur, etc. A few examples will be given below to illustrate the scientific value of the records.

Cinnabar: from Chenzhou and Jinzhou. A cinnabar piece which is as large as a peach or a date, as bright and transparent as a pomegranate is a good one. The second best are those the size of horse teeth. The lowest quality ones are purple and cannot be used.

Realgar: from Wudu. The colour resembles a cockle. The ones which are delicate and red are the best and the ones from Persia which are dark red are the least valuable.

White quartz: from Shouyang and Zezhou. There are many kinds of white quartz, but those which are bright without any stain on the surface and are the same colour as water are the best, no matter what their thickness is.

Mica: from Langya, Pengcheng, Qingzhou, Qizhou, Lu and other states. There are six kinds of mica, and they can be clearly distinguished in the sunlight. The yellowish white ones with greenish patterns are called *yunying*, the greenish yellow and white ones are called *yunye*, the pure white ones are called *yunjing*, and the greenish white ones with blackish patterns are called *yunmu*. The best ones are the shiny ones.

Tianmingsha (also known as borax): from Persia. It can be used to solder hardware. There are many counterfeit drugs. The authentic ones taste sour and salty, not bitter; their colour is bluish white; they do not reach boiling point when boiled; the juice is like water and as sticky as glue. If there is purple gas and lacquer when burned, it is also genuine.

③ *Shiyao erya*

The *Shiyao erya* ("Encyclopedia of Mineral Medicines") is a two-volume work, written in 806, imitating the form of "Erya" and recording many obscure drug names. The author, Mei Biao, was a native of Jiangyuan (present-day Chongzhou, Sichuan). "He had been interested in Taoism since he was young, and he was an expert in alchemy. He read numerous books on alchemy and learned them by heart [99]." In the *Shiyao erya: Preface*, he explained the aim of writing the book: "At that time) cryptic names were used for all the drugs, and there were various versions of these cryptic names. Without knowing the name, people cannot understand the prescriptions. To solve this problem, the book records the different names, special formulas, and unique techniques of various panaceas. It emulates the form of "Erya" and comprises six chapters which have been compiled into two volumes. For those with alchemical expertise, the book lists the information necessary. The formulas that can be easily prepared have been put at the beginning of the book. Those that require special techniques or are difficult to make have been listed at the end with explanations of their complexity."

In the book, there are 68 kinds of minerals and chemical compounds in total, and 347 examples of cryptic names and aliases. Among these, the greatest number of cryptic names for one mineral or compound was 30, the minimum was 1, and the average was about 5. For example, the cryptic names of lead included: *xuanhuang hua*, *qinghua*, *qianfei*, *feiliu*, *huodan*, *liangfei*, etc. As can be seen, the primary purpose of the book was to eliminate the mysterious names that alchemists deliberately added and to contribute to the popularisation of alchemical drug names.

④ *Yudong dashen dansha zhenyaojue* ("The Secret Recipe of Alchemy by Yudong God")

The author was Zhang Guo of the Tang dynasty, a native of Gusheshan. He lived in seclusion on Zhongtiao Mountain in Hengzhou and styled himself as Tongxuan Xiansheng. He travelled between the Fen and Jin areas, and it was said that he had the secret of longevity and was hundreds of years old [100]. This work was included in book No 587 of the *Daozang*, mainly focusing on mineral identification. For example, the Method of Identifying Alchemical Ingredients described the occurrence and quality of the cinnabar deposits, which was important in mineralogy. It is written that "cinnabar … the best quality is from stone caves in Chenzhou and Jinzhou, and there are several colours. Average quality cinnabars come from Jiaozhou and Guizhou. There are also several types. The lowest grade comes from Hengzhou and Shaozhou, and there are also several kinds. The top-grade *guangmingsha* [a kind of cinnabar] comes from Chenzhou and Jinshan (mountain) stones, and one can find twelve pieces at a time. Their colours are like blossoming red lotus flowers, and they shine brightly. Sometimes there may only be nine at a time … Place the big ones on red silk at night, and the bright ones are of the top quality. The ones as big as horse teeth or which shine brightly are top-quality examples. The ones as white as mica are average-quality *maya sha* [a kind of cinnabar which looks like horse teeth]". "*Shidan* (copper sulfate) pieces comes from Song Mountain and Zhongjiao Mountain in Puzhou. It is green. To identify *shidan*, it can be applied on iron and copper and burnt with fire, and the colour will turn red."

⑤ *Youyang zazu: yuge: Caoyao yihao*

Duan Chengshi, the author of the *Youyang zazu* ("Miscellaneous Morsels"), was born in Zouping, Linzi, Shandong province; he was born around 803 and died in 863. He served in different positions. When he was an official in Jiangnan Dao, he was "good in governance". In his later years, he wrote 20 volumes of the *Youyang zazu* and10 volumes of the *Xuji*.

In the *Youyang zazu: Yuge: Caoyao yihao*, Duan Chengshi imitated the style of the *Shiyao erya* and recorded the cryptic names of ten kinds of minerals or compounds (in parentheses), including *danshan hun* (realgar), *qingyaonu* (azurite), *beidi xuanzhu* (saltpetre), etc.

⑥ *Longhu huandan jue*

The two volumes of the *Longhu huandan jue* were written by Jinlingzi, whose real name and biography are unknown. This work is included in book No 590 of the

Daozang. According to Chen Guofu, the book was completed between 686 to 741 or 758 to 760 [101]. The book recorded the deposits, occurrence and quality of alchemical mineral ingredients. At the beginning of the book, the author pointed out that identifying the quality was a fundamental skill for alchemists: "Those who are good at Daoism and aspire to longevity must first identify the quality of minerals". The following is a sequential narrative of how he identified the mineral medicines.

Identification of mercury:

First are the various names for mercury. "According to the *Xianjing*, the names for mercury include Heshang Chanu ('fair lady on the river'), Changshengzi ('immortal man'), Gong (mercury), Taiyang Liuzhu ('droplets of the sun'), Shengjiao ('magic gum'), Lingyangzi, Xuanminglong ('mysterious dragon'), Xuanshui ('mysterious river'), Baihunao ('white tiger's brain') and Jinyinxi ('gold and silver mat')."

Second is the origin. "There are several kinds of mercury from different places … the best ones come from stone caves in Chenzhou and Jinzhou and exist in several colours. The average ones come from Jiaozhou and Guizhou. The worst ones come from Heng and Shao."

Third is the occurrence. "The top-grade *guangmingsha* (natural cinnabar) come from Chenzhou and Jinshan stones, twelve or nine pieces at a time. The second best has seven or five at a time. Each aggregate contains a large bead which can weigh more than ten liang. The small ones weigh eight or nine liang, and there are also six or seven liang ones around the big ones. There are also low-quality crystals around the centre."

Fourth is the quality. "Place the big ones on red silk at night, the bright ones are also of the highest quality. The ones as big as horse teeth and which are bright are the top-quality *baimayasha* (a kind of cinnabar which looks like horse teeth). The ones like bamboo shoots which are red purple in colour are top-quality *zilingsha* (a kind of cinnabar which is red and purple in colour). If it looks like the edge of a stone flake and reflects green light, it is subpar *zilingsha.* If there are three bright beads weighing one *jin*, they are of average quality. Bright pieces are of low quality. The low-quality *zilingsha* also come from Shao and Heng and are obtained when the rocks are struck."

Fifth is the relationship between the quality and the medicinal effect. "The best ones will lengthen one's lifespan. The low-quality ones should not be used."

Sixth is the relationship between the quality and the amount of refined mercury. "A one *jin guangmingsha* contains only two *liang* of shiqi (sulfur dioxide), leading to the extraction of 14 *liang* of mercury; a one *jin baimayasha* contains four *liang* of shiqi, leading to the extraction of 12 *liang* of mercury; *zilingsha* contains 6 *liang* of shiqi, from which 10 *liang* of mercury can be extracted; *xisha* [a kind of cinnabar which comes from a brook] contains 7.5 *liang* of shiqi, producing 8.5 *liang* of mercury; other kinds contain 8.5 *liang* of shiqi, producing 7.5 *liang* of mercury."

In addition, there is also the identification of lead, copper salt, azurite, etc., each of which follows roughly the same format as the identification of mercury but with less detail. Therefore, no repetition is needed.

⑦ *Danfang jingyuan*

The book was finished between the Kaibao and Tianyuan periods and has been lost. The segments which have survived were included in the *Qiangong jiageng zhibao jicheng* ("Collection of the Locations of Lead and Mercury") according to book No 5 of the *Daozang*. *Danfang jingyuan* chiefly focused on the appearance, properties, and pre-treatment of drugs used in alchemy, thus it can be considered to have been a drugs manual. Some of its contents are found in the *Chongxiu zhenghe zhenglei bencao*. For example, mica: "mica powder can produce mercury, pyrargyrite and can also be eaten"; jade: "jade powder can nourish cinnabar"; stalactite: "stalactite can be used as an ice box"; alum stone: "purple alum stones can produce mercury"; mirabilite: "mirabilite can tame realgar", and so on.

From the quotations above, we can grasp a basic overview of the *Danfang jiangyuan* ("The Original Ingredients of Alchemy"). It contained 31 kinds of minerals and rocks, as well as their origins, properties, medicinal properties, qualities and uses, which made it an important document of ancient mineralogy and petrology.

⑧ *Yaopu*

Hou Ningji of the later Tang dynasty of the Five Dynasties became a Jinshi in 928. He wrote the *Yaopu* ("Pharmacopoeia") which was included in the third book of the *Tangdai congshu* ("Tang Dynasty Book Series"). He recorded 16 minerals and 15 cryptic names of minerals or compounds (in parentheses) in the style of the *Shiyao erya*. They were: sulfur, no cryptic name; Fengwei Tuantou ("flavoured clump", fructus amomi); Taiqing Zunzhe (mirabilite); Shi Zhongning (talc); Jinshan Lishi (native copper); Qing Fen ("light powder", mercury); Wulingzhi (trogopterus dung) and so on.

⑨ *Danfang jianyuan*

The *Danfang* [chamber] *jianyuan* ("Recompilation of The Original Ingredients of Alchemy"), also known as *Danfang* [prescription] *jianyuan*, was included in book No 596 of the *Daozang*. The author was Dugu Tao in the Southern Tang dynasty of the Five Dynasties, whose life story is unknown. This work shared its origins with the *Danfang jingyuan* of the Tang dynasty, but this book had more in-depth content and was better organised. It is particularly remarkable that it had perfected the classification of alchemical drugs, which were divided into 24 chapters. This was a great advance in the study of alchemical drugs. The book gave a brief introduction to the source, origin, properties, functions, and usage of each drug in a clear and organised manner, and sometimes the artificial manufacturing process was also included. Consequently, it was a highly scientific manual of minerals and medicines [102]. It recorded 16 kinds of minerals, including raw silver, Wuchang copper, realgar, etc. It is an important mineralogical text from the Five Dynasties.

⑩ *Baozang lun*

The author of the *Baozang lun* was Xuanyuan Shu of the early Five Dynasties. Some believe it was written during the Kaihuang period of the Sui dynasty. Tang Rihua also wrote the *Baozang lun*. The original article has been lost, but there are some citations in the *Chongxiu zhenghe zhenglei bencao* and *Bencao gangmu*. From these citations, we learn the following.

First, there were 11 kinds of minerals and rocks recorded, such as cinnabar, jade, saltpetre, etc.

Second, the subspecies of minerals were recorded, for example, regarding iron: "there are five kinds of iron, Jing iron is from Danyang; its colour is purple and it is strong. Shangrao iron ranks second. Bin iron is from Persia, and it is so strong that it can cut gold and jade. The iron from Taiyuan and Shu Mountain is blunt. Gang iron is produced from the mountain rocks in southwest Zhanghai: it looks like purple quartz and cannot be damaged by water or fire, and it can also cut jade." Lead: "There are several kinds of lead. Persian lead is firm and white, ranked first in the world. Caojie lead comes from Qianwei, and it is the essence of silver. Lead which comes from silver deposits contains five colours. The lead from Leping in Shangrao is similar in quality to that from Persia and Caojie. Fuban lead cannot be used. Japanese lead can form an alloy in order to attain the gold."

Third, it pointed out the genuine ones. Raw silver was one such example: "there are 17 pieces of silver, only 5 of them are true, the rest are false. In silver pits, between the stone gaps, there is raw silver bursting out, the natives call it 'old man's beard'."

Fourth, it specified the medicinal properties and usage, such as in the case of realgar: "if the herb is cooked and refined into a liquid, its colour does not change. Its juice can be consumed and can also be used to change copper and silver into gold."

6.9.6 Accounts of Fossils

In the Sui, Tang and Five Dynasties, there was knowledge of fossils scattered throughout the literature. For example, in the *Xinxiu bencao* and *Yuanhe junxian tuzhi*, it was said that Qiyang in Yongzhou (now Qiyang, Hunan) produced *shiyan* (literally stone swallow): "They are small and shaped like clams, and they are as hard as heavy stones. It is said that *shiyan* come out of stone caves and fall down with the rain due to thunderstorms, which is incorrect [88]." "Shiyan Mountain, which produces *shiyan* that can be used in medicine, lies 110 li north of Qiyang County [103]." The *shiyan* here is the fossil of an ocean-dwelling brachiopod. Because the shell is particularly wide on both sides and the shape is like a flying swallow spreading its wings, it is called a "stone swallow".

With reference to the fossilised mammal bones, or *longgu* (literally dragon bones), in Volume 12 of the *Yuanhe junxian tuzhi*, it is stated that "in Puban, two li east of Hedong County, *longgu* are found". This means that the fossil fragments were found in the southwest of Yongji, Shanxi province. In Li Zhao's *Tangguo shibu* ("Supplement to the Tang History") from the Tang dynasty, it was recorded that "in the caves of the Bin and Jin mountains, there are many *longgu* which have five colours. People use them in medicine". Bin refers to present-day Bin County in

Shaanxi, and Jin refers to Shanxi. *Longgu* are found in both these locations. *Longgu* refer to the fossils of mammal bones and teeth from the late Cenozoic period.

As regards *shiyu* (literally stone fish), in Volume 29 of the *Yuanhe junxian tuzhi*, it is said that "the stones on Shiyu Mountain are the colour of mica. The shape of a fish appears when a layer is chipped away, and the fishy smell of fish paste is emitted when it is burned with fire. The mountain is 15 li to the west in Xiangxiang County". In Volume Ten of the *Youyang zazu*, "there is Shiyu Mountain in Xiangxiang County, Hengyang. The stones of the mountain are black. If there is orpiment, a fish will appear when a layer is excavated. The fishes are several *cun* long and will smell fishy if burned." *Shiyu* here refers to ancient fish fossils.

As for fossil molluscs, the "snail and mussel shells" Yan Zhenqing mentioned in the *Fuzhou nanchengxian magushan xiantan ji* is a reference to the fossil of a mollusc.

Concerning plant fossils, Lu Guimeng's *Eryi shixu* ("Preface to a Poem for Fossils") records that "Dongyang (now Dongyang, Zhejiang) has many famous mountains, and Jinhua is the largest. The ancient pines on the mountain are often turned into stones". Du Guangting's *Yiluji* records: "There is a dead pine tree in Yongkang County, Wuzhou Prefecture (now Yongkang, Zhejiang province), which was broken and accidentally fell into the water and turned into stone. When the unchanged part was put into the water, it also turned into stone. The branch, trunk and bark are the same as pinewood, but stronger [104]." It is likely that the author had mistaken petrified wood for dead pine. Petrified wood, also known as fossilised wood, is one of the common plant fossils. The secondary xylem cells are all replicated with silica at a molecular level and at the same speed, so that the petrified wood not only preserves the annual rings, but also the fine structure of the plant, resulting in it looking the same as a dead tree which was easily misunderstood by ancient people.

With reference to amber fossils, they were formed during tens of millions of years of geological processes after the resin of gymnosperms had flowed into the ground. Volume Seven of Fanchuo's *Yunnan zhi* from the Tang dynasty recorded the amber produced in Yongchang (now Baoshan city) in Yunnan, saying: "Amber is dug up in Amber Mountain that is located at a distance of an 18-day journey to the west of the frontier of Yongchang city, far from the pine forest. Large pieces weigh more than 20 *jin*. In 794, Meng Yimou of Nanzhao presented a piece of amber, weighing 26 *jin*. It was thought to be rare at that time." The *Xinxiu bencao*, on the other hand, argued that at different time periods, different kinds of amber were formed. For example, pine resin entered the ground and turned into poria after 1,000 years, then turned into amber after 1000 years, and then *jiangzhu* (lit. river pearl, black amber) after another 1,000 years [105]. Han Baosheng's *Shu bencao* recorded that maple resin could also turn into amber after 1000 years and that most tree resins would be transformed after flowing into the ground, but they were not as durable as maple and pine resin [106].

6.9.7 Accounts of Fountains and Springs

There are some records in the literature of the Sui, Tang and Five Dynasties regarding the geographical distribution and causes of hot springs. For example, the *Wudai shizhi: Dili zhi* records three hot springs in Zhouzhi and in Lintong, Shaanxi; and in Sanmenxia, Henan. The *New Book of the Tang: Dili zhi* records that there were five hot springs altogether in Lintong and in Meixian County, Shaanxi; in Lingwu, Ningxia; in Linru, Henan; and in Mangbuzhuang, Liugu, 450 li northwest of Lhasa in Tibet. It is said that the hot springs in Linru "can cook rice" and that the hot springs in Lhasa, Tibet, "are two *zhang* high, with vapour like clouds of smoke, and are able to cook rice" and that the water temperature was quite high. The *Yuanhe junxian tuzhi* records nine hot springs in Henan, Shanxi, Hubei, Hunan and Sichuan. "They can be used for boiling rice and treating all kinds of diseases." Some of them were "green in colour"; some people used hot spring water to "irrigate the fields, and the harvest is multiplied".

Chen Zangqi of the Tang dynasty inherited Zhang Hua's theory of the genesis of hot springs, believing that "there is sulfur underneath: even if the water is hot, there is still a smell of sulfur" [107]. Xuan Zang put forward a new theory about the cause of hot springs in Volume Nine of the *Datang xiyu ji* saying that the springs "flowed through 500 small hot hells, which were heating the water". Although this theory was influenced by the Buddhist concept of hell, it is basically consistent with the modern geological view that underground magma heats up the groundwater. Due to the limitations of the time, Xuan Zang could not use the word "magma", only the word "hell", but the scientific concept expressed was an advanced one.

6.9.8 Views of the Shape of the Earth

In Li Chunfeng's Huntian Theory of the Tang dynasty, he proposed that the earth was spherical. In Li Hua's *Wujiang song* ("Wish for the Emperor's Longevity"), it is written that "the earth is spherical and the sky is high" [108].

Some people have pointed out that several dozen gold coins from the era of the Eastern Roman Empire have been excavated in China, six of which belonged to the period of 577 to 713. There was a design of the Earth on the gold coins excavated in the tombs [109]. This indicates that the western concept of the Earth had been introduced into China as early as the Sui and Tang dynasties, and it had some influence on the people who encountered these gold coins. The introduction of the Jiuzhi Calendar and Dayan Calendar during the Kaiyuan reign, when the concept of a virtual radius of the Earth was mentioned in discussions on the calculation of lunar eclipses, is an indication of the influence of the ancient Indian Spherical Earth theory. These circumstances show that the foreign idea of a Spherical Earth had spread in China during the Sui and Tang dynasties and a few Chinese scholars had mentioned it.

References

1. Feng L冯立升 (1995) *Zhongguo gudai celiangxue shi*中国古代测量学史 [A history of ancient Chinese metrology]. IMU Press呼和浩特: 内蒙古大学出版社, Hohhot, pp 117–121
2. “Taibai yinjing juansi ‘shui gongju pian di sanshiqi’” 太白阴经·卷四, “水攻具篇” 第三十七 [The Taibai Yin Meridian, Vol 4, Water Instruments 37]. In: *Shousange congshu di’qishiqi ce*守山阁丛书·第七十七册 [Collectanea of the Shoushan studio, vol 77]
3. *Wujing zongyao qianji: juan shiyi:‘Shuigongpian’*武经总要前集·卷十一, “水攻篇” [The complete list of essentials in the military classics, first section, vol 11, Water attack]
4. *Suishu: Peiju zhuan*隋书·裴矩传 [Book of the Sui: biography of Pei Ju]
5. *Jiu tangshu: Taizong benji*旧唐书·太宗本纪 [Old book of the Tang: biography of Emperor Taizong]
6. *Tang liudian: Zhifang langzhong*唐六典·职方郎中 [The six statutes of the Tang dynasty: the map officers]
7. *Tang huiyao juan wushijiu: Zhifang yuanwailang*唐会要·卷五十九 职方员外郎 [Institutional history of the Tang dynasty, vol 59, The Assistant Map Officers]
8. *Jiu tangshu: Lücai zhuan*旧唐书·吕才传 [Old book of the Tang: biography of Lücai]
9. *Xin tangshu: Wang yanwei zhuan*新唐书·王彦威传 [New book of the Tang: biography of Wang Yanwei]
10. Lu L卢良志 (1984) *Zhongguo dituxue shi*中国地图学史 [History of Chinese cartography]. Surveying and Mapping Press北京测绘出版社, Beijing, p 59
11. *Jiu wudai shi: Mingzongji*旧五代史·明宗纪 [Old history of the five dynasties: biography of Emperor Mingzong]
12. *Jiu tangshu: Li jifu zhuan*旧唐书·李吉甫传 [Old book of the Tang: biography of Li Jifu]
13. *Xin tangshu: Li deyu zhuan*新唐书·李德裕传 [New book of the Tang: biography of Li Deyu]
14. *Jiu tangshu: Wei yuanzhong zhuan*旧唐书·魏元忠传 [Old book of the Tang: biography of Wei Yuanzhong]
15. *Cefu yuangui juan sibai jiushiqi: Hequ*册府元龟·卷四百九十七 河渠 [Models from the archives, vol 497 The waterways]
16. *Jiu wudai shi: Tang mingzong ji*旧五代史· (唐) 明宗纪 [Old history of the five dynasties: biography of Emperor Mingzong (of the Tang)]
17. *Lüheshu wenji juansan: Dizhitui xu* 吕和叔文集·卷三, 地志图序 [Collected works of Lü Heshu, vol 3, Preface to the “Maps of Regional Geography”]
18. *Jiu tangshu: Lichunfeng zhuan*旧唐书·李淳风传 [Old book of the Tang: biography of Li Chunfeng]
19. Wang Y王庸 (1938) *Zhongguo dilixue shi*中国地理学史 [A history of Chinese geography]. Commercial Press上海商务印书馆, Shanghai
20. Lu L卢良志 (1984) *Zhongguo dituxue shi*中国地图学史 [A history of Chinese cartography]. Surveying and Mapping Press北京测绘出版社, pp 75–76, Beijing
21. *Daye shiyi ji*大业拾遗记 [Anecdotes in Emperor Yang of Sui’s journey to Jiangdu]
22. *Suishu: Jingji zhi* 隋书·经籍志 [Book of the Sui: bibliographical treatise]
23. *Suishu: Jingji zhi* 隋书· 裴矩传 [[Book of the Sui: biography of Pei Ju]
24. *Jiu tangshu: Jiadan zhuan* 旧唐书·贾耽传 [Old book of the Tang: biography of Jia Dan]
25. *Wenyuan yinghua: juanqibai sanshiqi*文苑英华·卷七百三十七第3841页 [Finest blossoms in the garden of literature, vol 737: 3841]
26. *Cefu yuangui wubai liushi & liubai wushisi*册府元龟·卷五百六十, 六百五十四 [Models from the archives, Vols 560 and 654]
27. *Jiu tangshu: Dezong ji xia*旧唐书·德宗纪下 [Old book of the Tang: annals of Emperor Dezong II]
28. *Quantang wen juan sibai jiushisan: Quandeyu xuwen*全唐文·卷四百九十三·权得舆的序文 [A complete collection of the Tang literature, Vol 493, Preface by Quan Deyu]

29. Tang X唐锡仁, Yang W杨文衡 (2000) *Zhongguo kexue jishu shi dilijuan*中国科学技术史·地学卷 [History of Chinese science and technology: geology]. Science Publishing北京科学出版社, Beijing, p 446
30. Bai S白寿彝, Shi N史念海 (eds) (1997) *Zhongguo tongshi: diliu juan dishi ce*中国通史: 第六卷第10册 [A comprehensive history of China, Vol 6, Book 10]. Shanghai People's Publishing House上海人民出版社, Shanghai, p 1886
31. *Taiping yulan·juan liubai ling'er: Sui daye shiyi*太平御览·卷六百零二·隋大业拾遗 [Imperial reader of the Taiping era, vol 602, Gleanings from the records of the Great Sui]
32. *Xin tangshu: Baiguan zhi*新唐书·百官志 [New book of the Tang: treatise on official posts]
33. He C贺次君 (1980) *Kuodizhi jijiao: Qianyan*括地志辑校·前言 [Preface to the annotated Comprehensive Gazetteer]. Zhonghua Book Company中华书局, Beijing
34. *Jiutang shu: Kong shurui zhuan*旧唐书·孔述睿传 [Old book of the Tang: biography of Kong Shurui]
35. Chen Z陈正祥 (1979) *Zhongguo ditxue shi*中国地图学史 [A history of Chinese cartography]. The Commercial Press (HK) Ltd.商务印书馆香港分馆, Hong Kong, p 19
36. *Beishi: Xiyu zhuan*北史·西域传 [History of the Northern dynasties: treatise on the Western Regions]
37. *Manshu*蛮书 [Book of the barbarians] is also known as *Yunnan zhi*, *Yunnan ji*, *Nanyi zhi*, *Nanman shuji*, etc.
38. Wang C王成组 (1982) *Zhongguo dilixue shi*中国地理学史 [A history of Chinese geography]. The Commercial Press 商务印书馆, Beijing, p 53
39. *Jiutang shu: Cuirong zhuan*旧唐书·崔融传 [Old book of the Tang: biography of Cui Rong]
40. *Quan tangwen· juan sibai sihi*全唐文·卷四百四十 [A complete collection of Tang literature, vol 440]
41. Xu Y徐瑜 (1978) Tangdai chaoxi xuejia dou shumeng jiqi haitaozhi唐代潮汐学家窦叔蒙及其 "海涛志" [Tidal scientist Dou Shumeng and his *Records of Ocean Tides*]. *Lishi yanjiu* 历史研究 [Historical study] 6
42. Needham J李约瑟 (1975). *Zhongguo kexue jishu shi (zhongyiben) di'si juan*中国科学技术史(中译本)第四卷 [History of science and technology (trans), vol 4]. Science Publishing House科学出版社, Beijing, p 786
43. *Yuanhe junxian tuzhi 26: Jiangnan dao hangzhou qiantang xian*元和郡县图志·卷二十六, 江南道杭州钱塘县 [The Yuanhe maps and records of prefectures and counties, vol 26, Qiantang County, Hangzhou, Jiangnan Dao]
44. Changjiang guihua bangongshi长江规划办公室 [Yangzi River Planning Office] (1974) *Changjiang shangyou yiyuduan lishi kushui diaocha*长江上游宜渝段历史枯水调查 [Investigation into the historical dry seasons in the Yiyu section of the upper reaches of the Yangzi River]. *Wenwu*文物 [Relics] 8
45. *Wang Dang* 王谠 (Song dynasty) *Tang yulin* 唐语林 [Forest of conversations on the Tang]
46. *Chajing: juanxia*茶经·卷下 [The classic of tea II]
47. *Qianjin yaofang*千金要方 [Essential prescriptions worth a thousand pieces of gold]
48. *Xin tangshu: Tuguhun zhuan*新唐书·吐谷浑传 [New book of the Tang: biography of Tuguhun]
49. *Jiu tangshu: Tubo zhuan*旧唐书·吐蕃传 [Old book of the Tang: treatise on Tubo]
50. *Xin tangshu: Tubo zhuan*: *xia*新唐书·吐蕃传 (下) [New book of the Tang: treatise on Tubo II
51. *Xiangyu shu, dade banian kan*相雨书·大德八年刊 [Book of weather forecasting, issued in the 8th year of the Dade reign]
52. *Lingbiao luyi, juan shang*岭表录异·卷上 [Records of the unusual from Lingnan I]
53. Gu X顾学颉, Zhou R周汝昌 (1982) *Bai juyi shixuan*白居易诗选 [Selected poems by Bai Juyi]. Beijing People's Publishing House人民文学出版社, Beijing, p 237
54. Gu X顾学颉, Zhou R周汝昌 (1982) *Bai juyi shixuan*白居易诗选 [Selected poems by Bai Juyi]. Beijing People's Publishing House人民文学出版社, Beijing, p 2
55. Zhu K竺可桢, Wan M宛敏渭 (1984) *Wuhouxue*物候学 [Phenology]. Science Publishing科学出版社, Beijing, pp 17–18.

56. There are different opinions regarding Xuan Zang's dates of birth and death, according to Ji, Xianlin季羡林 (1985) *Datang xiyu ji jiaozhu: Qianyan*大唐西域记校注: 前言 [Annotations of the "Records from the regions west of the Great Tang": Preface] Beijing: Zhonghua Book Company中华书局
57. Li H慧立, Yan C彦悰 (1983) *Daciensi sanzang fashi zhuan* 大慈恩寺三藏法师传 [Biography of Tripitaka Dharma Master from Daci'en Temple]. Zhonghua Book Company中华书局, Beijing, p 5
58. *Daciensi sanzang fashi zhuan juan wu*大慈恩寺三藏法师传 [Biography of the Tripitaka Dharma Master from Daci'en Temple]
59. Li L郦隶彬 (1977) *Datang xiyuji: Qianyan*大唐西域记·前言 [Records from the regions west of the Great Tang: preface]. Shanghai People's Publishing House上海人民出版社, Shanghai
60. *Zhongtianzhu guo xingji* 中天竺国行记 [Journey to the middle Indian countries] is also known as *Wang xuance xingzhuan* 王玄策行传 [Travels of Wang Xuance], *Xiyu xingzhuan* 西域行传 [Journey to the Western Regions], *Xiguo xingzhuan* 西国行传 [Journey to the western countries]
61. Sun X孙修身 (1994) Tangchao jiechu waijiao huodongjia wang xuance shiji yanjiu唐朝杰出外交活动家王玄策史迹研究 [Research on Wang Xuance: A prominent traveller and diplomat of the Tang dynasty]. *Dunhuang yanjiu*敦煌研究 [Dunhuang research] 3
62. *Zizhi tongjian juan erbai yishiliu, Tianbao shizai tiao*资质通鉴·卷二百一十六, "天宝十载"条 [Comprehensive mirror in aid of governance, Vol 216, Ten years of the Tianbao reign]
63. Wang B王邦维 (1988) *Datang xiyu 1iufa gaoseng zhuan: Qianyan*大唐西域求法高僧传·前言 [Biographies of eminent monks seeking the Dharma in the Western Regions in the Great Tang: Preface]. Zhonghua Book Company中华书局, Beijing
64. *Liwengong ji: juan shiba*李文公集·卷十八 [Collection of Li Wengong's works, Vol 18]. Also published in Book 7 of the *Quan tangwen* 全唐文 [A complete collection of Tang literature] (1983). Beijing: Zhonghua Book Company中华书局, pp 6442–6443
65. *Quantang shi juan wubai liushi jiu* 全唐诗·卷五百六十九 [A complete collection of Tang poems, Vol 569]
66. *Jiu tangshu: Shihuo zhi*旧唐书·食货志 [Old book of the Tang: treatise on food and commodities]
67. *Tang huiyao: juan ershiyi*唐会要·卷二十一 [The comprehensive statutes of the Tang dynasty, vol 21]
68. *Tang huiyao: juan liushiba*唐会要·卷六十八 [The comprehensive statutes of the Tang dynasty, vol 68]
69. *Fanchuan wenji: juansan* 樊川文集·卷三 [Literary collection of Fanchuan, vol 3]
70. *Tang huiyao: Jiexiang* 唐会要·街巷 [The comprehensive statutes of the Tang dynasty: streets and alleys]
71. *Cefu yuangu:juan shisi*册府元龟·卷十四 [Models from the archives, vol 14]
72. *Tang huiyao: Daolu*唐会要·道路 [The comprehensive statutes of the Tang dynasty: districts]
73. *Xin tangshu: Guo yuanzhen zhuan*新唐书·郭元振传 [New book of the Tang: biography of Guo Yuanzhen]
74. Mu D杜牧, "Shou Lina zhedong guanchashi jian yushidafu zhi"授李纳浙东观察使兼御史大夫制 [The appointment of Li Na to the post of supervisor and imperial counsellor in Zhedong]. In: *Quantang wen juan qibai sishiba*全唐文·卷七百四十八 [A complete collection of Tang literature, vol 748]
75. Mu D杜牧, "Cuigong xingzhuang"崔公行状 [Cui Gong's deeds]. *Wenyuan yinghua juan jiubai qishiqi*文苑英华·卷九百七十七 [Finest blossoms in the garden of literature, vol 977]
76. Fu L梁甫, "Dugugong xingzhuang" 独孤公行状 [Dugu Gong's deeds] in *Quan tangwen juan wubai ershi'er*全唐文·卷五百二十二 [A complete collection of the Tang literature, vol 522]
77. Ao L李翱, "Pingfu shu"平赋书 [Equal farmland tax] *Quan tangwen juan liubai sanshiba*全唐文·卷六百三十八 [A complete collection of the Tang literature, vol 638]
78. See *Shoushange congshu* 守山阁丛书 [Collectanea of the Shoushan Studio] and the military section in the S*iku quanshu*四库全书 [Complete library of the four treasuries]
79. *Yanlugong wenji: juanshisan*颜鲁公文集·卷十三 [Collected works of Yanlu Gong, vol 13]

80. *Baixiangshan shiji: juansan*白香山诗集·卷三 [Collection of Baixiangshan poems, vol 3]
81. *Yuanshi changqing ji: juan ershisi* 元氏长庆集·卷二十四 [Collected works of Yuan Zhen during the Changqing reign, vol 24]
82. *Duanxi yanpu* 端溪砚谱 [Catalogue of Duanxi inkstones]
83. Wang F王福泉 (1985) *Baoshi tonglun*宝石通论 [A review of gemstones]. Science Publishing 科学出版社, Beijing, p 215
84. Gao Z高兆 (Qing dynasty), *Duanxi yanshikao*端溪砚石考 [Research on Duanxi inkstone]. In: *Tanji congshu* 檀几丛书 (1992) [Tanji series]. Shanghai: Shanghai Ancient Books Publishing House 上海古籍出版社
85. Zhang B张秉伦, Wu X吴孝铣, Gao Y高有德 (1990) *Anhui kexue jishu shigao*安徽科学技术史稿 [Historical manuscript of Anhui science and technology]. Anhui Science and Technology Publishing House安徽科学技术出版社, Hefei
86. Wang F王福泉 (1985) *Baoshi tonglun*宝石通论 [A review of gemstones]. Science Publishing 科学出版社, Beijing, p 216
87. Qi J齐儆 (1991) *Zhongguode wenfang sibao*中国的文房四宝 [Four treasures of a study in China]. Commercial Press商务印书馆, Beijing, p 60
88. *Bencao gangmu juanshi*本草纲目·卷十 [Compendium of materia medica, vol 10]
89. *Daozang*: *Dongshenbu zhongshulei di wubai jiushi ce*道藏洞神部众术类·第590册 [Daoist canon: Cave god: Techniques and methods, Book 590]
90. *Chongxiu zhenghe zhenglei bencao: juansi*重修政和证类本草·卷四 [Revised Zhenghe Materia Medica, vol 4]
91. Guo Z郭正谊 (1981) "Shuifa liantong shiliao suyuan"水法炼铜史料溯源 [History of water-based copper refining]. In: *Zhongguo keji shiliao*中国科技史料 [Historical data of Chinese science and technology] 4
92. *Bencao gangmu: juanjiu*本草纲目·卷九 [Compendium of materia medica, vol 9]
93. *Bencao gangmu: juanshi*本草纲目·卷十 [Compendium of materia medica, vol 10]
94. *Bencao gangmu: juanba*本草纲目·卷八 [Compendium of materia medica, vol 8]
95. Yan S颜师古 *Hanshu: shihuozhi zhu*汉书·食货志注 [Annotation of the *book of the Han: Food and commodities*]
96. Si Y姒元翼 (1984) *Zhongguo yixue shi*中国医学史 [A history of Chinese medicine]. People's Medical Publishing House人民卫生出版社, Beijing, p 38
97. Li Z李仲均 (1999) "Woguo bencaoxuezhong jizaide yaoyong kuangwu shilue" 我国本草学中记载的药用矿物史略 [A brief history of medicinal minerals recorded in Chinese Materia Medica]. In: *Li zhongjun wenji*李仲均文集 [Collected works of Li Zhongjun]. Xi'an Map Press西安地图出版社, Xi'an
98. Zhu Y朱越利 (1996) *Daozang fenlei jieti* 道藏分类解题 [The classification of the Daoist canon]. Huaxia Publishing House华夏出版社, p 244, Beijing
99. *Shiyao erya: Xu* 石药尔雅·序 [Preface to the *Encyclopedia of Mineral Medicines*]. Quoted from *Daozang di wubai bashiba ce* 道藏·第五百八十八册 [Daoist cannon, Book 588]
100. *Jiutangshu: zhangguo zhuan*旧唐书·张果传 [Old book of the Tang: biography of Zhang Guo]
101. Chen G陈国符 (1979) "Daozangjing zhong waidan huangbaishu cailiao de zhengli" 道藏经中外丹黄白术材料的整理 [Materials for alchemy in the Daoist canon]. *Huaxue tongbao*化学通报 [Chemistry bulletin] 6
102. Zhao K赵匡华, Zhou J周嘉华 (1998) *Zhongguo kexue jishu shi: Huaxue juan*中国科学技术史·化学卷 [History of science and technology: chemistry]. Science Publishing科学出版社, Beijing, p 292
103. *Yuanhe junxian tuzhi· juan ershi jiu*元和郡县图志·卷二十九 [The Yuanhe maps and records of prefectures and counties, Vol 29]
104. Recorded in *Longwei mishu disihan* 龙威秘书·第四函 [Secret books of the dragon force, Slipcase 4]
105. *Wei jiangzhou ji: juanba*韦江州集·卷八 [Collected works of Wei Jiangzhou, vol 8]
106. *Bencao gangmu juan sanshiqi*本草纲目·卷三十七 [Compendium of materia medica, vol 37]
107. *Bencao gangmu: juanwu*本草纲目·卷五 [Compendium of materia medica, vol 5]

108. *Wenyuan yinghua· juan qibai qishisi*文苑英华·卷七百七十四 [Finest blossoms in the garden of literature, vol 774]
109. Wang Q汪前进 (1995) "Cong chutu dongluoma jinbishang de diqiu tu'an tantao xifang diyuan gainian suitang shiqi zai zhongguo de chuanbo he yingxiang从出土东罗马金币上的地球图案探讨西方地圆概念隋唐时期在中国的传播和影响 [Exploring the influence of the Western Spherical Earth theory in China during the Sui and Tang dynasties from the earth image on excavated Eastern Roman gold coins]. In: *Keshi xinzhuan* 科史薪传 [History of science through generations]. Shenyang: Liaoning Education Press辽宁教育出版社

Chapter 7
Geoscience in the Song, Liao, Jin and Yuan Dynasties

7.1 The Impact of the Social Environment on Geoscience

During the Song, Liao, Jin and Yuan dynasties, China's ancient Earth science reached its peak with a series of glorious achievements. This great progress was closely related to the social background, which favoured the development of Earth science in the following eight respects.

(1) The maritime industry was well developed. The rise of coastal port cities contributed to the accumulation of knowledge relating to extraterritorial geography and ocean tides.

From the mid-tenth century to the early twelfth century, the Northern Song, Liao and Western Xia were in tripartite confrontation. On the land border, the three regimes established Uighur rulers centred on Gaochang (East of Turpan), Kucha (Kucha) and Khotan (Hotian), Kara-Khanid Khanate across the Pamir Mountains, the Tibetan Empire which was in chaos on the Tibetan plateau, and the Dali Kingdom dominated by the Bai and Yi ethnic groups in the Yunnan region. The Northern Song dynasty was at a disadvantage and was forced to provide tribute, in exchange for "peace". Under such circumstances, unable to connect with Central Asia, West Asia, Europe and Africa by land, the Song government had no choice but to connect with the outside world by sea instead of along the historical Silk Road. This resulted in a more developed maritime industry in the Northern and Southern Song dynasties—the compass was used for navigation, coastal port cities emerged, and foreign trade flourished. During the Yuan dynasty in particular, the political situation of the great unification ensured the further development of sea transportation. Such a social environment was conducive to better understanding of extraterritorial geography and the study of ocean tides. As a result, several books on extraterritorial geography and marine tides emerged.

W. Yang and Q. Yang, *A History of Ancient Chinese Geoscience*, Understanding China,
https://doi.org/10.1007/978-981-97-4457-2_7

(2) The hedonistic ruling class in the Northern Song dynasty built large-scale construction projects, and searched for strange stones and foreign objects, which led to more knowledge about minerals and rocks.

The ruling class, led by Emperor Huizong of Song, was politically dark and corrupt and lived in luxury. In order to collect more treasures, from the first year of the Chongning reign (1102), the Zaozuo ju (Bureau of Construction) was founded in Suzhou, Hangzhou and other places, specialising in the production of various exotic and precious crafts for Emperor Huizong and his favourites to enjoy. Huizong had a passion for marble, and Cai Jing and his friends catered to his hobby. During the Zhenghe reign (1111–1118), the Yingfeng Ju (Bureau of Provision) was established in Suzhou to search for exotic flowers, trees and stones for the emperor. Most of the goods were transported to Kaifeng by waterways, on ships called the Flowers and Stones Fleet. In his more than twenty-year reign, major projects such as Bright Hall, Hall of Preserving Harmony, Extended Fortune Palace and Boundless Longevity Mountain (later renamed "Mount Genyue") were constructed. To find materials for the imperial park of Genyue, huge quantities of rarities were collected from all corners of the empire for more than twenty years, coupled with the most skilful craftsmanship. The search for exotic stones and treasures helped people in the Song dynasty quickly learn a lot about minerals and rocks, and stone and inkstone catalogues, or *shipu*, proliferated, such as the *Yuyanggong Stone Catalogue*, *Xuanhe Stone Catalogue*, *Yunlin Stone Catalogue*, *Record of Taihu Stone*, *Inkstone Catalogue*, *On Inkstones*, *Shezhou Inkstone Catalogue*, *Duanxi Inkstone Catalogue*, and so on.

(3) The political, economic and military needs of the ruling class contributed to the advancement of local chronicles and cartography, and a number of representative gazetteers and maps of various types appeared.

Administrative geography also progressed further. For instance, in the first year of Emperor Huizong's reign (1107), "the court founded the Bureau of Gazetteers of the Nine Regions and ordered the prefectures and counties to compile their *tujing* [map guides] [1]", and then submit them to the Bureau. To strengthen his reign, the ruler had to have a complete grasp of the mountains and rivers, customs and folklore, water and land transportation, and the tribute of goods and services throughout the country. Therefore, there were repeated orders and requests to compile and edit *tujing* in various places, which promoted the widespread development of map guides and local chronicles.

(4) The unified political situation of the Yuan dynasty provided ideal conditions for exploration to locate the source of the Yellow River, and the earliest and most detailed monograph on the source of the river, the *Heyuan ji* ("The Record of the Source of the Yellow River") appeared.
(5) Due to the requirements of agricultural production, military affairs, navigation and other domains, significant advances in meteorological climatology occurred. For the first time, there were diaries recording phenology day by day.

(6) Hydrology made remarkable progress because of the needs of agriculture, flood control and water conservancy projects.
(7) Thanks to the development of ceramics, mining, alchemy and traditional Chinese medicine, mineral knowledge was amassed and several works on mineral records were compiled.
(8) During the Song and Yuan dynasties, especially during the Song dynasty, thought and speech were relatively free; the culture of lecturing was prevalent, with academies in various regions and different schools of thought exchanging ideas and competing with each other. The rulers of the Song dynasty did not explicitly support one school of thought or suppress another. Multiple ideas coexisted and contended with each other, creating a favourable social environment for academics to prosper.

7.2 Chronicles: A Thriving Genre of Historical Accounts

The Song and Yuan dynasties were a turning point in the history of Chinese *fangzhi* (local chronicles/gazetteers). The period was not only notable for the variety and quantity of works, but also for the emergence of many excellent local gazetteer theories and styles, which laid the foundation for the flourishing of local chronicles in the Ming and Qing dynasties.

During the Han and Tang dynasties, which lasted 1,100 years, fewer than 400 *fangzhi* were compiled, about 100 of which were completed during the Sui and Tang dynasties. In the Song dynasty, 1,016 *fangzhi* emerged within 320 years [2], which shows the rapid increase in their number.

The Song people were not confined to traditional ideas, but rather they innovated and expressed their own opinions. Most of the *fangzhi* of the Han and Tang dynasties were devoted to a certain aspect, and there were few all-encompassing "encyclopaedias". In the early Northern Song dynasty, after the *Taiping huanyu ji* ("Universal Geography of the Taiping Era [976–983]"), a geographical treatise, *fangzhi* gradually became encyclopaedic, focusing more on history and human geography and less on physical geography.

Although there were no theories relating to *fangzhi* during the Song dynasty, many prefaces and postscripts to *fangzhi* began to theoretically clarify or discuss the nature, function, principles and methods of compilation. According to Zheng Xingyi and Ma Guangzu, the function of *fangzhi* was to aid governance; *fangzhi* should fall into the category of history and could not be defined as being simple geography books. While Zhao Yanruo thought the principle underlying compiling the local record should "be used in governance" and "emphasise the present whilst skimming over the past".

7.2.1 General Chronicles

The Northern Song unified the empire, ending the upheaval of the Five Dynasties and Ten Kingdoms period. To consolidate power and strengthen their control over the people, the Song rulers needed to compile a general national record. Of the 40 general chronicles (or *zongzhi*) of the Song dynasty, 35 have been lost and 5 are extant. They are described as follows.

(1) *Taiping huanyu ji*

The *Taiping huanyu ji* is a book comprising 200 volumes, of which 113–119 are missing. The authorship is attributed to Yue Shi (930–1007), courtesy name Zizheng. As a scholar, his official position gave him access to the royal collection and the local records of various states. It took eight years from 979 to 987 to complete this book. Instead of following the *Yuanhe junxian zhi* and *Huayang guozhi*, when describing circuits (or *dao*), prefectures and counties, the book innovatively added content relating to human geography and economic geography, such as customs, surnames, people, art and local products, thus combining history and geography which was considered to be a milestone in the history of Chinese *fangzhi*. At the date of the writing of the book, the administrative divisions of the circuits in the Song dynasty had not yet been fixed, so the book was still based on the ten circuits of the Tang dynasty, with prefectures (military and supervision) as the category and counties as sub-categories. Each prefecture began with the historical records of the mountains and places since ancient times, the meanings of place names, migration, governance, especially the details of separation and changes of names during the late Tang dynasty and the Five Dynasties period. The sub-categories were the counties of the abolished prefectures, the boundaries, the number of households, customs, names, people, and local products. Each county started with the number of townships, their history, and the governance, and then moved onto mountains, scenic spots, passes, specialties, roads, temples, and so on.

In addition to the combination of history and geography, there were three other breakthroughs in this book. Firstly, in terms of demographic geography, the book initiated the separation of the statistics for *zhuhu* (landowning households) and *kehu* (households without land or peasant migrants). When it came to economic geography, it created a column of both tribute and non-tribute "local produce", which was richer than the "tribute" in the *Yuanhe junxian zhi* and could better reflect the economic outlook of a region. The records of the date of establishment, scope, production scale, production technology, management and taxation of the Jianchang (a supervisory unit) revealed the regional distribution of handicraft production, production conditions and relations in the early Northern Song dynasty. For example, a silver mine had been discovered in Jian'an County, Jianzhou (Modern Jian'ou, Fujian). In the 8th year of the Kaibao reign (975), a field was established to mine copper and silver there. Later, in the 3rd year of the Taiping xinguo period (978), the Longbei supervisory unit was set up to supervise seven fields nearby. It was also the first book in China to cite the literary sources which made it easier for the readers to refer to these and increased their historical value. Overall, given these four innovative aspects, the book is considered to be a groundbreaking masterpiece. Of course, this book also has obvious shortcomings, mainly the heroic historical view of feudal historians, the Han nationalism, and the miscellaneous content. Nevertheless, such defects do not outweigh the merits.

(2) *Yuanfeng jiuyu zhi*

The *Yuanfeng jiuyu zhi* ("Yuanfeng Treatise of the Nine Regions") is a ten-volume geographical treatise written by the Song dynasty scholars Wang Cun, Zeng Zhao and Li Dechu [Yuanfeng is an era name corresponding to the years 1078–1085 during Emperor Shenzong of Song's reign].

The *Yuanfeng jiuyu zhi* originated from the *Shidao tu* of the Tang dynasty. In the 6th year of the Dazhong xiangfu reign (1013), Wang Zeng and Li Zong'e compiled the *Jiuyu tu* ("Map of Nine Regions"), which was then used to determine official salaries, servanthood and penalties. In the 8th year of the Xining reign (1075), due to changes in the boundaries of political districts, the *Jiuyu tu* was no longer relevant, so Zeng Zhao and Zhao Yanruo were ordered to revise it, on the advice of the magistrate Liu Shidan. Later Zhao Yanruo was dismissed, and Wang Cun and Li Dechu participated in the revision, with Wang Cun as the main editor. The book was completed in the 3rd year of the Yuanfeng reign (1080) and renamed *Jiuyu zhi*. Since this book was the reworking of a map, the text only served as notes and descriptions of the map, so it was relatively brief. It was revised several times later, and the system described was actually that of the 8th year of the Yuanfeng reign (1085), while official publication took place in the first year of the Yuanyou reign (1086) after the first lunar month [3].

First, the *Yuanfeng jiuyu zhi* listed the four capitals, which were the eastern capital—Kaifeng, the western capital—Henan, the northern capital—Daming, and the southern capital—Yingtian; then it explained the 23 *lu* (political divisions, equivalent to provinces), and finally the abolished counties and prefectures, the prefectures outside the civilised world, and the semiautonomous "haltered and bridled" prefectures (Jimi fuzhou). The outline begins with the *lu*, the primary administrative division, under which the figures for the number of provinces, prefectures, troops, and counties are listed. It then lists the secondary level of administrative division starting with the provinces and prefectures, while it also provides information about the subordinate districts, cities, administrative changes, fields, household registrations, tributes of local products, the number of counties and their history of development. Furthermore, the third level of administrative divisions is also defined, headed by the counties (*jian*, industrial prefectures containing mines and salterns) and followed by streets, townships, villages, mountains, rivers, lakes or marshes, and springs. The book traced the abolition, creation and redistribution of counties and prefectures over time. As regards the uncivilised areas, only the *lu* each one belonged to was recorded. The book also described the military commissioner and corresponding jurisdiction of the Jimi fuzhou and counties (tribe, town).

This *zongzhi*, which focused on geographical territories, was as all-encompassing as the *Taiping huanyu ji*. Topics regarding historical sites, people, customs and other items that were less relevant to the contemporary people were not included.

(3) *Yudi guangji*

The *Yudi guangji* was compiled by the Southern Song period scholar Ouyang Min, who was probably a descendant of the great historian Ouyang Xiu. It was written during the reign of Zhenghe (1111–1118). The first three chapters of the book describe the regions of China through history and the fourth chapter lists the names of prefectures and counties in the Song. The rest of the book is a detailed description of the local administration of the Song period, including "prefectures outside civilisation" (*huawai zhou*), such as the sixteen prefectures (*shiliu zhou*) in the northeast that were ceded to the Khitan Empire (907–1125). The book focused on the historical changes in administrations, while geographical records of households, folklore or local produce were not discussed. It drew on the existing historical literature and local chronicles, providing a detailed account of the geography from the Yuanfeng reign to the Zhenghe reign (1078–1085). As regards the source of the Yellow River, people used to believe that it was a convergence of multiple rivers flowing from the Kunlun Mountains; the author was opposed to this idea and instead agreed with Liu Yuanding who considered the Purple Mountain or Mount Menmoli as the location of the source of the Yellow River.

(4) *Yudi jisheng*

The *Yudi jisheng* ("Record of the Landscape of the Imperial Domain"), comprising 200 volumes, is a geographical chronicle compiled by Wang Xiangzhi, courtesy name Yifu. When he was young, he accompanied Emperor Guangzong of the Song dynasty, Zhao Dun, on travels to Jiang, Huai (Jianghuai, a geographical area referring to the plain between the Yangzi and Huai rivers, in the modern provinces of Anhui and Jiangsu), Jing (modern Hubei province) and Min (modern Fujian province), and learned from his elder brother about the southwest; therefore, he accumulated a wealth of geographical knowledge. Later, he worked as an official in Jiangning (modern Nangjing) and other places and completed this book in 1227, the 3rd year of Baoqing reign [4]. The book opens with the then capital Lin'an (modern Hangzhou, Zhejiang) and describes all 166 prefectures (first-class, *fu*; second-class, *zhou*; military, *jun*; and industrial, *jian*). For the prefectures, Wang provided twelve volumes dealing with the administrative history, the districts within each prefecture, the customs and habits, landscape and territory, tourist spots of interest, eminent officials, eminent personalities, eminent monks, tombstones with inscriptions, monasteries, and so on. The book described in detail the local scenery and the beauty of the mountains or rivers for scholars to refer to. It was also of high historical value because of its accurate treatment of the primary sources used by Wang.

(5) *Fangyu shenglan*

The 70-volume *Fangyu shenglan* ("Overview of the Unified Territories of the Great Yuan)") was completed in the third year of the Jiaxi reign (1239), by Zhu Mu who hailed from Jianyang (modern Jianyang, Fujian) and was orphaned at a young age. Zhu Mu, together with his brother Zhu Kui, was a disciple of Zhu Xi (1130–1200), and he was appointed gentleman of meritorious achievements (*digonglang*) and later became director (*shanzhang*) of the Hanjiang Academy in the military prefecture of Xinghua. The book was divided into sections relating to the seventeen provinces or *lu* of the Southern Song and provided information regarding the names of all the prefectures through history, their administrative boundaries, the field tax revenue and the number of registered households, the passes and strategic places, customs and habits, mountains and rivers, important temples, monasteries and official buildings. Zhu made use of countless sources, including prose and poems. The book's purpose was also to serve as a source of information for the literati and scholars, so the *Siku quanshu zongmu tiyao* says that "instead of being a geography book, it is more like a genre of reference book".

(6) *Dayuan dayitong zhi*

The Yuan emperor Qubilai (posthumously Yuan Shizu, 1260–1294) had an imperial geography compiled of the Yuan Empire. In 1291, the first version consisting of a total of 755 volumes was finished, compiled under the guidance of Jamāl al-Dīn (pinyin: Zhama luding) and Yu Yinglong and many other scholars. This first version was called *Dayitong zhi and* collected in the Mi Fu, a secret place for hiding books and other archives in the Forbidden City [5]. This was the first national gazetteer to use the name *Yitong zhi* ("yitong" means grand unity). In 1303, Bei Lan Xi and Yue Xuan completed the final edition which incorporated the information in other geographical works such as the *Yunnan tuzhi*, *Gansu tuzhi* and *Liaoyang tuzhi*, resulting in a version 1,300 volumes long. The reference material for all the provinces south of the Yellow River was mostly cited from the *Yudi jisheng* and from the previous gazetteers of the Song and Yuan dynasties, while for the provinces north of the river, most information was cited from the *Yuanhe junxian tuzhi*, *Taiping huanyu ji* and some old records of the Jin and Yuan dynasties. The structure of the book was designed around 11 provinces, subdivided by *lu* or *fu,* and further subdivided by counties. In most cases, each volume recorded one prefecture, and occasionally a single prefecture was recorded in two to three volumes. The book gave a detailed account of ten aspects: history, townships, distances to strategic places, mountains and rivers, local products, customs and habits, historic sites, official buildings, people, and the immortals, and also included maps and graphs.

The book is considered one of the largest geographical descriptions of the Chinese Empire. Due to ethnic conflicts, it endured for just a short time, since the Ming dynasty refused to retain it and private individuals were unable to preserve it, and so it was soon lost. It can be assumed that some of its contents were collated into later geographical literature, such as the *Yongle dadian* ("Yongle Encyclopedia"), *Daming tongyizhi* ("Gazetteer of the Unity of the Great Ming"), *Manzhou yuanliu kao* ("Research on Manchu Origins") and *Rehe zhi* ("Local Gazetteer of Rehe"). Until the Qing dynasty, only fragments of 15 volumes had been preserved unfortunately. Based on the *Yuanshi: Dili zhi* and other sources of literature, Zhao Wanli reconstructed the Yuan imperial geography between 1944 and 1965. His work *Yuan tongyi zhi* was published in 1966 (Beijing: Zhonghua Book Company, 1966), and is 10 volumes and about 45,000 Chinese characters in length [6].

(7) *Jiuyu zhi*

The 80 chapters of the *Jiuyu zhi* were written by Zhu Siben, a famous monk, who was born in Linchuan, Jiangxi. At a very young age, he loved to read geography books and knew a lot about the distribution of Chinese mountains and rivers and the nine prefectural divisions. He admired Sima Qian's travels nationwide. Later, he was granted just such an opportunity, for the emperor ordered him to visit famous mountains, rivers and seas all over the country to make sacrifices, so his long-cherished wish to travel around China was realised [7]. During his travels, he collected geographical information from many places, examined the geographical environment in the field and gathered rich geographical knowledge, which made him a famous geographer and cartographer in the Yuan dynasty.

The *Jiuyu zhi* by Zhu Siben was engraved and published in the first year of the Dade reign (1297). Since it was not reprinted after the Ming dynasty, the original engraving has long been incomplete and lost, making it a rare and precious book. As the official *Yitong zhi* of the Yuan government had appeared earlier, this private work did not play much of a role in history.

7.2.2 Illustrated Treatises and Atlases

The development of *tujing* (illustrated treatises and atlases) reached its peak during the Northern Song dynasty. In the 8th year of the Kaibao reign (975), Song Huai was ordered to compile the *Kaibao zhudao tujing* [8]. In the 3rd year of the Dazhong xiangfu reign (1010), Li Zong'e and Wang Zeng led the compilation of the *Xiangfu zhouxian tujing*, which was 1566 volumes long. In the second year of the Kangding reign (1041), Zhao Xun presented the five-chapter-long *Kangding jumi tujing*, which "studies the geographical advantages and difficulties" [9]. In the first year of the Daguan reign (1107), "the court set up the Tuzhi Bureau and commanded the prefectures and counties to compile their own *tujing*" [10]. The establishment of an imperial office for the compilation of *tujing* was a huge boost to the development of local gazetteers. The *Daguan mingzhou tujing* by Li Maocheng was finished during this period.

In the Song dynasty, a *tujing* was no longer in the format of a single picture accompanied by a textual description. Instead, there were more written records and fewer maps. Some map guides even consisted only of texts. In this sense, it could be said that map guides had transformed into gazetteers. For example, in the case of the *Yanzhou tujing* written by Dong Fen, in the three volumes still in existence out of a total of eight volumes, except for the nine maps on the frontispiece, there are no maps in Volume Two or Three. Thus, the maps accounted for very little of the entire book. The content of the text in the book was mostly unrelated to maps. And the mountains and roads drawn on the maps were not covered, or only partially covered, in the textual account, which meant the treatises were no longer a textual description of the map. When the *Yanzhou tujing* was reprinted during the Shaoxing reign (1131–1162), the nine maps originally attached at the beginning were simply removed, and the book was renamed *Xin'an zhi*. In this sense, starting from the time of the Sui and Tang dynasties, *tujing* had fully developed into local chronicles by the Southern Song dynasty. Nevertheless, in the later local chronicles and more general chronicles, several maps were usually included at the beginning to show the regional boundaries, mountains, rivers, and the distribution of towns and settlements within the jurisdiction. This was partly for the convenience of readers, and partly because the tradition of maps being included in the map guides had been passed down and inherited.

7.2.3 Local Chronicles

According to Liu Weiyi's survey, there were 976 local chronicles in the Song dynasty, 947 lost and 29 extant. Of these, 14 are from Zhejiang, 9 from Jiangsu, 2 from Fujian and Shaanxi, and 1 each from Hunan and Anhui respectively [2]. The Yuan dynasty had 13 local chronicles [11] but only 11 have survived [12].

(1) Local chronicles of the Song dynasty

The best among the Song dynasty's local chronicles were well appreciated by scholars from the Qing dynasty onwards. For example, the *(Chunxi) Sanshan zhi* ("Records of the Three Mountains") is praised for the detailed issues that are not touched on in the official and semi-official histories of the region. The *(Shaoxi) Yunjian zhi* is concise, and the *(Jiatai) Kuaiji zhi* is well organised. The *(Jingding) Jiankang zhi* used maps, tables, *zhi* (chonicles) and biographies to record facts and is emblematic of local gazetteers. What is more, the *Ganshui zhi* by Chang Tang was a pioneer township gazetteer; the *Lin'an zhi*, a capital gazetteer and prefectural annal, which was later revised three times, is considered a masterpiece in its field [13]. Below are more examples:

① *Chang'an zhi*

The *Chang'an zhi* ("Records of Chang'an") is a gazetteer-style description of the old capital Chang'an (modern Xi'an, Shaanxi), its palaces, markets and wards, as well as of the districts surrounding Chang'an. It was written during the Northern Song period (960–1126) by Song Minqiu (1019–1079) and is 20 volumes long. The first volume contains a summary and nine other items: astronomy, local products, local tributes, customs and habits, geography and neighbouring regions, district administration, households and miscellaneous regulations. The second volume covers the province of Yongzhou, the capital, the metropolitan magistrate and local administration offices. From Volume 3 to 6, it describes the palaces and other buildings. Volumes 7, 8, 9 and 10 discuss the imperial ("forbidden") city and the capital cities during the Tang period. Volume 11 provides information concerning various districts.

② *Wujun zhi*

The *Wujun zhi* ("Gazetteer of the Commandery of Wu") was compiled by Fan Chengda (1126–1193), writing in the Southern Song period (1127–1279). The 50-volume book was finished around 1192 and later revised and then printed in the second year of the Shaoding reign (1229). One of the earliest and most complete gazetteers in the Southern Song dynasty, it is praised for its rich content, the logic of the division into categories and its detailed style of description.

③ *Xin'an zhi*

The *Xin'an zhi* was compiled by Luo Yuan (1135–1184). It starts with a preface and a list of the contents. The ten-volume book gives a detailed description of prefectures, local products, the six counties (Shexian, Xiuning, Qimen, Wuyuan, Jixi, Yixian), eminent scholars, graduates of the metropolitan examination, people of righteous conduct, the immortals, governors and miscellaneous records. The main feature is its focus on the economy and the humanities.

(2) Local chronicles from the Yuan dynasty

Local chronicles in the Yuan dynasty basically followed the pattern of those in the Song dynasty, most were from the south and only a minority from the north.

① *Chang'an zhitu*

The *Chang'an zhitu* ("Illustrated record of Chang'an") was written by Li Haowen. The three-volume book was finished in 1344 and presented a richly illustrated account of history. There were fifteen maps in Volume One, twelve are still in existence, three are missing and there were no accompanying illustrations. Volume Two contains five maps and twelve illustrated accounts. In Volume Three, there are two maps and eight treatises. Some of the treatises are about 100 Chinese characters long while others are over 10,000 characters. For example, one volume called "Jingqu tushuo", which is more than 11,000 characters long, gives a detailed description of the history of canal excavation, irrigation areas, canal length, canal supporting facilities and so on, and is viewed as a special treatise on water conservancy engineering.

② *Zhenjiang zhi*

The *Zhenjiang zhi* was compiled by Tuo Yin and Yu Xilu and completed in the 3rd year of the Zhishun reign (1332). It classifies households into four types: aborigines, expatriates, the impoverished and monks. In terms of schools, there are four types: Mongolian language, Confucianism, yin-yang and medicine, reflecting the characteristics of the Mongol rule [14].

③ *Leibian chang'an zhi*

The Yuan period (1279–1368) scholar Luo Tianxiang wrote the ten chapters of *Leibian chang'an zhi*, which described the old capital Chang'an (modern Xi'an, Shaanxi), its palaces, mountains and rivers, tombs, and so on. The original version of this book no longer exists; the Beijing Library has five copies of the remnants of the Ming copy, with one or two volumes missing. In addition to historical information, the book also contained poems and writings from the Jin and Yuan dynasties.

7.2.4 The Methodology of the Local Chronicle

Research on *fangzhi* (local chronicles) during the Song and Yuan dynasties was conducted from three perspectives.

(1) The origins of *fangzhi*

There were four mainstream views on the origins of the *fangzhi*. First, that it was a type of ancient geographical literature, which was supported by Zhu Changwen [15]: "Previous Confucian scholars valued the study of *fangzhi*, so Zhu Gan recorded the local customs of various regions; Gu Yewang created geographical records of the empire; and Jia Dan wrote *Shidao lu* ("Record of the Ten Circuits"). These works constituted history." Wang Xiangzhi also commented: "There are decent works discussing geography nowadays, prefectures and counties have *zhi*, the nine regions have *zhi*, the world has *ji*, and the empire also has *ji* [16]".

Second, *fangzhi* should be perceived as a historical reference, as Zheng Xingyi argued: "A prefecture which has its *zhi* is equivalent to a country which has its history: with *zhi* local customs and traditions can be traced, history can be reflected on and lessons can be learned for the future [17]." This view was embraced and further developed by the Qing dynasty *fangzhi* specialist Zhang Xuecheng.

Third, that the *fangzhi* could be traced back to political books. Sima Guang took the *Zhouguan* ("Officers of Zhou") as a precedent and believed that because the book outlined the local, regional and national duties of officials at different levels, later scholars continued to inform the imperial rulers of the pros and cons of the geographical features of a certain place. This idea was also supported by Ma Guangzu who clarified that *fangzhi* existed to support the aims of local governance.

Finally, *fangzhi* was considered to have multiple origins combining the three views above. Li Zong'e stated: "The Xia dynasty had a territory structured by *wufu*, the official Zhifang in the Zhou dynasty managed maps of the world and the uses of the land. Later *dizhi* have been created by historians, and prefectural records have been drawing on local practices [18]."

(2) The function of *fangzhi*

There were mainly two views regarding the functions of *fangzhi*. First, to preserve history. Liang Kejia [19] and Zhang Jian [20] explained from the perspective of historiography that it was essential to collect information and examine historical facts in a timely manner, which actually emphasised the role of the preservation of history in the books. Second, to facilitate governance. Yue Shi [21] discussed the political and military value of the books. Ding Darong advocated that the books should reflect the living conditions of the people, to serve as a basis for promoting what was beneficial and abolishing what was harmful. Dong Fen believed that the compilation of *fangzhi* would help collect local information for the imperial court, and would inspire officials, the younger generation and the nobles, thus benefiting social morality [22].

(3) The approach to the compilation of *fangzhi*

Song and Yuan scholars explained how to compile *fangzhi* as follows.

① Choose a chief author or editor with these three qualities: educated, knowledgeable, and an expert in documentation.
② Highlight the procedures. Zhou Yinghe proposed in the *Jingding jiankang zhi* that the procedures for writing a gazetteer should include determining the form and main content, assigning tasks, searching widely for information and verifying the materials in detail and carefully. These four steps were consistent with the practice of compilation, the contribution of Song scholars to the production of gazetteers.
③ Correctly deal with the relationship between the ancient and the modern. Lü Nangong advocated taking account of reality, instead of adhering to the past. He argued that a gazetteer should combine the past and the present, but with particular focus on the latter. This point of view is still valuable.
④ Value historical facts and prioritise the truth. Only record information supported by evidence and omit anything which is unsubstantiated.
⑤ Adopt a standard format and style. The way of writing *fangzhi* was established in the Song dynasty, and its main feature was the structure of the text and the approach to classification [23].

7.3 Geography of Historical Administrations

Following the Han and Tang dynasties, evolutionary geography developed significantly during the Song and Yuan dynasties. The style of writing can be divided into three major categories.

7.3.1 Administrative Geography in Historical Records

(1) The *History of the Song: Record of Geography,* compiled by Tuo Tuo and other scholars, is a total of six volumes in length. It only describes the establishment and divisions of circuits, prefectures and counties based on the Yuanfeng period.
(2) The *History of the Liao: Record of Geography*, written by Tuo Tuo and comprising five volumes, provides information about the establishment of prefectures and counties. It describes the establishment of the Liao dynasty without tracing the previous dynasty.
(3) The *History of the Jin*: *Record of Geography*, written by Tuo Tuo and comprising three volumes, gives an account of the establishment of circuits, prefectures and counties. It describes the establishment of the Jin dynasty without tracing the previous dynasty.

(4) The *History of the Yuan: Record of Geography* was compiled by Song Lian, Wang Yi and others, and this six-volume book gives an account of the establishment of circuits, prefectures and counties, which it traces back to the Tang dynasty.

7.3.2 Administrative Geography in Local Chronicles

Both the general and local chronicles reflected the knowledge of administrative geography, such as the *Taiping huanyu ji*, *Yuanfeng jiuyu zhi*, *Chang'an zhi* and *Yanzhou tujing* among the general chronicles, and *Chang'an zhi* by Song Mnqiu, *Yanzhou tujing* and *Qiandao lin'an zhi* by Chen Gongliang, *Xin'an zhi* by Luo Yuan, *Zhiyuan jiahe zhi* by Xu Shuo and other local chronicles.

7.3.3 Studies of Administrative Geography

(1) The volumes "Dili lue" (administrative geography) and "Duyi lue" (capitals and cities) of the *Tongzhi* compiled by Zheng Qiao are considered to be important references for administrative geography.
(2) The 14-volume *Tongjian dili tongshi* by Wang Yinglin was completed in the 17th year of the Zhiyuan (1280). Just from its outline, it can be concluded that this book reflected the long course of changes in administrative geography over more than a thousand years. As it summarised the successes and failures in the governance of certain regions, the book provided good lessons for the rulers [24]. The author also wrote the 100-volume long masterpiece *Tongjian dili kao* on administrative geography, but unfortunately it was lost.
(3) The *Wenxian tongkao* ("Comprehensive Investigations Based on Literary and Documentary Sources") is an administrative history written by Ma Duanlin. The structure of the book is based on 24 "investigative" (*kao*) categories and comprises a total of 348 volumes. The 9-volume section "Yudi kao" (Imperial geography) and the 25-volume "Siyi kao" (Four barbarians) are mostly closely related to administrative geography.

7.4 Surveying and Mapmaking

During the Song and Yuan dynasties, surveying and cartography were more developed than in the Tang dynasty, and it was the peak period for the development of ancient Chinese cartography. Among the surviving documents, there are a variety of maps from the Song and Yuan dynasties, including national maps, foreign maps, border maps; maps of various prefectures, states and counties; maps of mountains and rivers, water conservancy, river management, transportation, land, postal stations, cities and capitals; and historical maps. There are maps made of different materials, such as on silk, paper and stone, and wooden geographic models and globes also exist. The following is a description of four aspects.

7.4.1 Surveying Techniques and Cartographic Theories

The improvements in measurement technology in the Song and Yuan dynasties had a lot to do with the development of mathematics at that time. For example, Shen Kuo's *Huiyuanshu* ("Method to Meet the Round") was of great significance, whether for measuring terrain or in mapping. Qin Jiushao made considerable achievements in indirect measurement methods. His book *Shushu jiuzhang* ("The Mathematical Treatise in Nine Sections") was seen as the most comprehensive work to discuss various measurement methods after the *Haidao suanjing*. Yang Hui's research on the *chong cha* (double difference) method significantly advanced the level of measurement technology. In terms of measuring tools, he adopted a device with a peephole, which had not been documented before. During the Yuan dynasty, surveying and measuring technology continued to develop. The most outstanding aspect was astronomical geodesy on an unprecedented scale. In addition, important achievements were also made in water surveying and mapping. Mathematicians such as Li Ye and Zhu Shijie introduced Tianyuan Shu [a Chinese system of algebra for polynomial equations created in the thirteenth century] into indirect measurement calculations, which provided advanced methods for solving complex measurement problems [25].

According to Zeng Gongliang's *Wujing zongyao*, Li Jie's *Yingzao fashi*, Shen Kuo's *Mengxi bitan* and Zeng Sanyi's *Yinhua lu*, the Song dynasty had already used or invented levelling instruments, the floating fish compass, and other measuring instruments.

In 1072, Shen Kuo conducted a topographic survey to measure the difference in elevation from Kaifeng to Huaikou in Sizhou and his survey was more than 600 years ahead of the Russian topographic survey of the Don River, which began in 1696 [26].

During the Yuan dynasty, Guo Shoujing (1231–1316) also made important contributions to surveying. In the 12th year of the Zhiyuan era (1275), he was ordered to inspect the waterways in the lower reaches of the Yellow River and to conduct extensive topographic surveys [27]. The concept of elevation was of great importance in surveying and he also introduced the concept of altitude for the first time. Thanks to this, the comparison of elevations in different places could be conducted [28].

During the Song and Yuan dynasties, two leading researchers, Shen Kuo and Zhu Siben, developed theories of cartography founded on Pei Xiu's *Six Principles for Map-making*.

In Shen Kuo's famous paper on cartography, it is remarked that although li numbers (distance) were shown on previous maps, distances were normally measured by *bu* along the roads. The road might be straight or have bends, so after finishing mapping, the li number marked on the map could not be consistent with the real distance. Shen Kuo argued that distance should be recalculated according to the four directions of a straight-line distance map, viewed like a bird flying directly to its destination and unaffected by geographic elevations. Hence, he proposed the seven principles of mapping, namely *fenlü* (scale), *zhunwang* (regulated view), *yarong* (mutual inclusions), *bangyan* (collateral or indirect verification), *gaoxia* (lowering of heights), *fangxie* (diagonal distance) and *yuzhi* (straightening of curves) [29]. Shen Kuo's opinion was critical and correct, but he did not further develop Pei Xiu's theory, as pointed out by Hu Daojing and Cao Wanru [30].

Zhu Siben (1273–1333) spent ten years working on the *Yudi tu* ("Map of the Empire"). It was drawn using the method of *Jilihua fang* [31] [drawing grids based on a scale, said to have been invented by Pei Xiu, 224–271] and was the result of an amalgamation of single maps of individual regions which did not correspond to latitudes and longitudes.

7.4.2 Maps Produced on Silk and Paper

Maps drawn on silk and paper during the Song and Yuan dynasties can be classified into six major categories as follows.

(1) General maps

The rulers of the Song dynasty attached great importance to map-making, from the emperors down to the officials at all levels. They used maps extensively as well as organising a workforce dedicated to the drawing of new maps. During the Song dynasty, the census booklet was called the *Runnian tu* ("Leap Year Map") because it was a requirement to compile a register of households every leap year. In the 2nd year of the Taiping Xinguo reign (977), several prefectures presented a Leap Year Map to the court. At the beginning of the Xiangfu reign, Emperor Zhenzong appointed Wang Zeng to compile the three-volume *Jiuyu tu* [32].

In the 9th year of the Xining reign (1076), Shen Kuo was ordered to compile the *Tianxia zhouxian tu*, also known as *Shouling tu*, which it took 12 years of effort to complete. The scale of this map was 1:90,000 and it incorporated a 24-direction compass [33]. However, it has been lost.

In the Song dynasty, the number of administrative division circuits (*lu*) was changed frequently, sometimes there were as many as twenty-six *lu*, sometimes only fifteen *lu*. Since the period when there were 18 *lu* lasted longest, after the Tianxi reign, there appeared a new type of influential national map, the *shibalu tu* (maps of eighteen circuits). The first to draw this type of map was Yan Shu, who drafted the *Shibalu zhoujun tu* ("The Military Maps of Eighteen Circuits and Prefectures") in the first year of Emperor Renzong's reign (1023) and submitted it to the imperial court [33].

In the Song dynasty, maps of the empire also existed as a type of skilled craft. In July of the 3rd year of the Qiandao reign (1167), Emperor Xiaozong of the Song asked the craftsmen to carve a map on a large gold lacquer screen behind the Dragon's Seat, which was called *Yudian pingfeng huayi tu* ("Map of China and Barbarian Countries on the Palace Screen") [35].

During the Yuan dynasty, the cartographer Zhu Siben compiled the two famous chapters of the *Yuditu*. He drew upon the experience in mapping since the Tang and Song dynasties, the available reference books and maps, and his own travel practices. It was likely that he had also absorbed Western mapping techniques. This map was very influential and was an important template for a large number of later maps. Zhu Siben loved travelling and gained a wealth of geographical knowledge from his travels. He found many deficiencies in the maps drawn by his predecessors, which focused his mind on drawing to correct them. From 1311 to 1320, he looked up many historical documents, first drew small maps of each place, and then combined them into a full map. It took him 10 years to produce a map that exceeded the accuracy of his predecessors. However, he took a cautious approach to the remote areas he had not visited, unfamiliar areas, and areas where information was insufficient or not very reliable, so as not to mislead future generations [7].

It is known from the *Guangyu tu* ("Enlarged Terrestrial Atlas") that the shape of Africa drawn in Zhu Siben's *Yudi tu* ("Geographical Map of the Empire") was triangular; this was the first time the Chinese had made a map of Africa, pre-dating the European and African maps [36].

During the Yuan dynasty, there were two other maps of imperial territory, namely the *Shengjiao guangbei tu* sketched by Li Zemin around 1330 and the *Hunyi Jiangli tu* by the Tiantai monk Qing Jun, which has been lost.

(2) Regional maps

During the Song dynasty, local officials valued regional maps. There were few independent maps of local administrations, mountains, rivers and drains and water conservancy, and most of these were attached to *tujing* and *fangzhi*. For example, at the beginning of the extant three-volume *Yanzhou tujing*, there are maps of the Jiande Prefecture and some counties.

There were 19 maps in the *Jingding jiankang zhi* and most of these maps were drawn to scale, but they pointed in various directions, not always to the north. The topographical features were basically accurate, and the contents were detailed, so they were viewed as relatively good regional maps [37].

In the *Chicheng zhi* finished in the 16th year of the Jiading reign (1223), there were thirteen maps and nine of them still existed.

In the *Siming zhi* completed in the 3rd year of the Baoqing reign (1227), there were nine maps.

In the *Lintingzhi* completed in the 3rd year of Kaiyuan reign (1259), there were nine maps.

Of the maps completed in the 4th year of the Xianchun reign (1268), the *Piling zhi* had seven maps and the *Lin'an zhi* had thirteen.

In the 3rd year of the Zhizheng (1343), the *Jinling xinzhi* was compiled, accompanied by 21 maps.

With regard to regions, there were the *Qidan ditu* ("Map of Khitan"), the *Qidan jiangyu tu* ("Map of the territory of the Khitan"), the *Hexi longyou tu*, and the *Xiyu tu*. There were also maps of imperial borders including the *Daliaoguo duijing tu*, the *Xijie duijing tu*, the *Hedong dijie tu* and so on [38].

(3) Maps of exterior regions

During the Song and Yuan dynasties, maritime transportation was very prosperous, and there were corresponding maps of foreign lands. In the month of September of the 8th year of the Dazhong Xiangfu reign (1015), a minister advised the emperor to record the clothing of people in the tributary countries, to draw its shapes and record its traditions.

In the first month of the 3rd year of the Taiping xingguo reign (978), Li Fu from Guangzhou submitted the *Haiwai zhuyu tu* ("Map of the Various Overseas Areas") and the *Lingbiao huamu tu* to the imperial court.

In the 6th year of the Xianping reign (1003), Ling Ce from Guangzhou presented the *Haiwai zhusai dili tu.*

(4) Specialised maps

Specialised maps dating from the Song and Yuan dynasties include the *Yulin tu* ("Fish Scale Map") which was used to confirm the land-use of households, three chapters of maps in the *Daohe xingsheng shu* by Li Chui as a guide to hydraulic engineering, the *Sijieshou erguhe tu* by Han Zhi in the fifth year of the Jiayou reign (1060), the *Zhizhi yanbian junbeitang zhudidao tiaoshi tu* made in the seventh year of the Xining reign (1074), and the map *Heyuan tu* ("Map of River Sources") appended to the *Chuogeng lu* ("A Literati's Pondering in the Breaks of His Ploughing").

(5) City maps

The earliest extant city maps from the Song dynasty are the 14 maps in Song Minqiu's *Henan zhi.* Most of them were historical maps, and only one was a map of the Song dynasty. The *Henan zhi* was lost in the Qing dynasty, and Xu Song reconstructed it based on the *Yongle dadian.*

On the 5th day of the first month of the 3rd year of the Yuanfeng reign (1080), Lü Dafeng commanded Liu Jingyang and Lü Dalin to examine and chart the *Chang'an tu* [39]. In the same year, this map was carved on a stone tablet. The stone tablet was broken at a later date and the remnants were found in the Qing dynasty. The fragments are now conserved by the National Library (formerly Beijing Library), the Chinese History Museum, and the Institute of Natural Science History. The scale, orientation and symbols can still be seen on the fragments [40]. In particular, the street east of the Zhuque Gate, the Imperial City, the Daming Palace, the Inner Court, and the Forbidden Garden are clear.

Cheng Dachang's (1123–1195) *Yonglu* ("Records of Yong") contained 32 maps, forming a historical map collection of Han and Tang capital palaces. The number of maps varied from volume to volume, with some containing as many as 11 maps and some as few as a single map. The maps contained barely any symbols and only truncated lines, and sometimes places were even marked with written words and without any delineating lines, so they could hardly be called maps. Nevertheless, they had immense historical value and form one of the earliest surviving historical map collections of Chang'an city (modern Xi'an, Shaanxi) in China.

(6) Historical maps

The Northern Song dynasty's *Lidai dili zhizhang tu* ("Handy geographical maps throughout the ages") is the earliest surviving historical atlas. The authorship of this work is still under debate. Texts dating from the Song era indicate that the book might be attributed to Shui Anli or Su Shi, or another unknown person [41].

Forty-four maps of the empire covering the entirety of Chinese history were accompanied by illustrations. These maps had a rich and precise content. Unluckily, the original Song edition disappeared and the present-day copy is an engraving from the Ming dynasty.

The Song dynasty engraving of the *Liujing tu* ("Map of Six Scriptures") by Yang Jia is now conserved in the National Library. It consists of two maps: one is the *Shiwu guofeng dili tu* ("Geographical map of the fifteen [states] in the Guofeng [a section of the Book of Songs]"), and the other is the *Wenwu fengkou tu*. Published around the 25th year of the Shaoxing reign (1155), these two maps are the earliest surviving printed maps in China, and the earliest printed maps in the world [42].

The *Yugong shuoduan* ("Analysis of Explanations on the Tribute of Yu") is a commentary on the *Yugong*, a chapter of the Confucian Classic *Shangshu*, written by Fu Yin (1148–1215). Maps of rivers in western China are included in it. Compared with the *Liujing tu*, its printing technology had improved. For example, the lines representing the water system are smooth and flow in a natural direction; the place names are arranged in appropriate and orderly positions; the outline of the map is incised in straight lines which are uniform in width and fit for purpose.

The *Chang'an zhitu* ("Illustrated record of Chang'an") is a history of Chang'an (modern Xi'an, Shaanxi) written during the Yuan period (1279–1368) by Li Haowen. There were 22 original maps, 19 of which are extant. The size of the maps varies, and the scale and orientation are not consistent. They present a richly illustrated account of the history of the ancient capital of China, with maps of the administrative divisions, cities, tombs, palaces, forbidden gardens, places of interest, and the irrigation and water conservancy channels in Chang'an. This is the earliest example of a map of water conservancy and irrigation channels, and the quality of the mapping is better than average [43].

7.4.3 Maps Produced in Rock Carvings

Most of the existing stone-carved maps date from the Song dynasty. Stone maps are maps carved on stone monuments or cliffs, and they range from general national maps to local maps and city maps.

(1) *Jiuyu shouling tu*

The *Jiuyu shouling tu* ("An Administrative Map of Nine Regions") was carved on a stone tablet in the 3rd year of the Xuanhe era of the Northern Song dynasty (1121), and it was discovered by Sichuan cultural relics workers in 1964 and is now preserved in the Sichuan Provincial Museum. The tablet is 130 cm long and 100 cm wide, and the centre of each side is engraved with the four directions: east, west, south and north. This indicates the directions of the map, showing north at the top, south at the bottom, west to the left and east to the right. The scale is not marked on the map (about 1:1.9 million based on modern research). The map covers the administrative regions of the country, with a total of more than 1400 administrative place names, which is the highest number among the Song dynasty maps. The lowest rank of administrative division is the county, and this is the earliest administrative map of China with counties as units. The outline of the map and the relative positions of the prefectures and counties are fairly accurate, especially the outlines of Shandong Peninsula, Leizhou Peninsula and Hainan Island, which are close to the present-day map. Furthermore, it is the most accurate drawing of the coastline among the extant Song dynasty maps. The waterways in the map, except for the bend of the Yellow River in the present-day Inner Mongolia Autonomous Region and Ningxia Hui Autonomous Region, and the upstream part above the bend, are accurate and generally close to the modern map. There is almost no major difference in the map of the Jialing River compared with today's one in particular. There is a uniformity in the use of symbols in the map, such as special symbols for water systems, mountains, and forests, and the size of the text is used to distinguish the level of administrative division [44].

(2) *Yuji tu*

The *Yuji tu* ("Map of Traces of Yu in China") and *Huayi tu* ("Map of China and Barbarian Countries"), which are preserved in the Xi'an Stele Forest, were both carved in 1136, but the *Huayi tu* was carved half a year earlier: the former was carved in April and the latter in October. The two maps are carved on the same stone tablet, one on the front and the other on the back. Each one is about 77 cm in length and width.

The *Yuji tu* (Fig. 7.1) is drawn by the traditional method of "Jilihua fang" (calculating distance and drawing squares), with 70 squares horizontally and 73 squares vertically, totalling 5,110 squares, each of which is equivalent to 100 li of land (about 1:1.5 million), although no extant earlier map shows the grid. The layout of the rivers and coastal situation in the map is relatively accurate, especially the shape of the main channel of the Yellow River, more accurate than in the *Jiuyu shouling tu* and *Huayi tu*. Compared with the *Jiuyu shuling tu*, there are fewer administrative place names (about 380), but more names of rivers, lakes and mountains. The weak points of this map are: the Liaodong Peninsula is not included; the western part of the Shandong Peninsula is not drawn wide enough; there is an extra-large bay in the south of Hangzhou Bay; Hainan Island is too wide between the east and west; some of the large rivers in the south have not been carved, such as the Dongjiang and Hanjiang in Guangdong; the Nujiang in Yunnan; the Xiushui, Xinshui and Fushui in Jiangxi, etc.

There are three stone carvings dating from the Southern Song dynasty: Xi'an Stele Forest, carved in 1136; one in Zhenjiang, engraved in 1142; and one in Jixian, Shanxi, with no date.

(3) *Huayi tu*

The *Huayi tu* (Fig. 7.2) is based on Jia Dan's *Hainei huayi tu* and is marked with the directions of upper north, lower south, left west and right east, and is a map of Asia with China as its main focus. The location of mountains, lakes, and prefectures in the Chinese part is fundamentally precise. There are about 500 place names, 13 rivers, 10 mountains, and 4 lakes on the map. The four neighbouring countries are

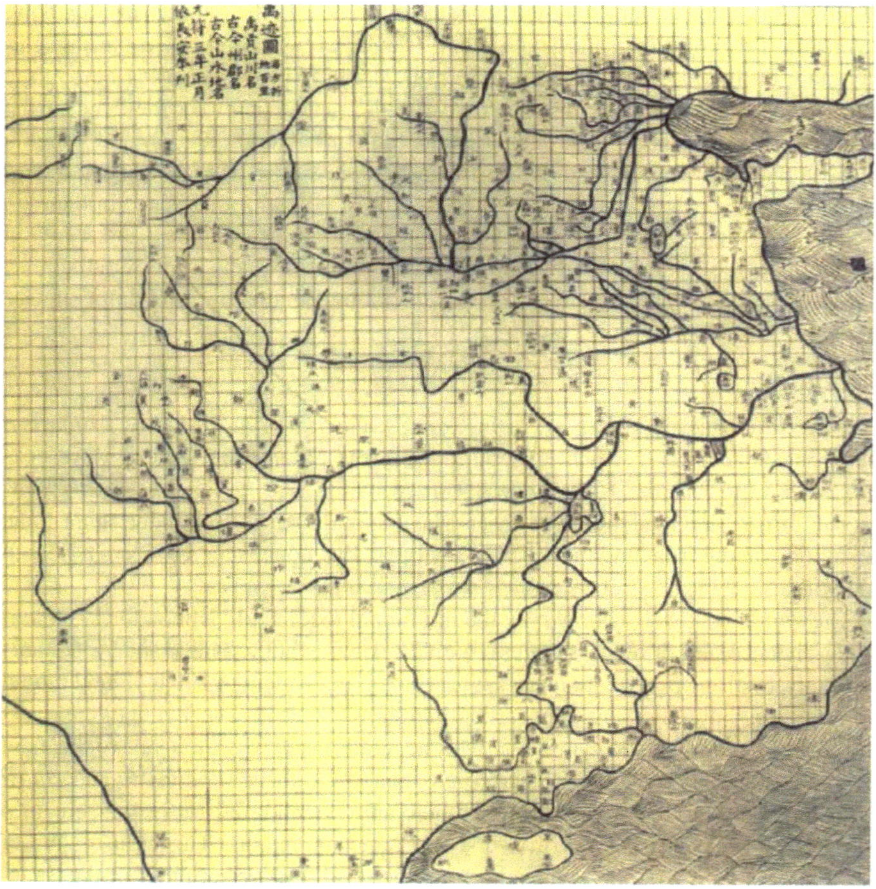

Fig. 7.1 A partial ink-drawn *Map of Traces of Yu*. The editorial board of the *Water Culture Series* 水文化丛书编委会 (2018) *Shuili guibao* 水利瑰宝 [The jewel of water conservancy]. Nanjing: Hohai University Press 南京: 河海大学出版社, p 177

noted with Chinese characters. The outline of the map is relatively rough and not as accurate as that of the *Jiuyu shouling tu* and *Yuji tu*. The most obvious mistake is that the Shandong Peninsula, Liaodong Peninsula, and Leizhou Peninsula are not drawn as land protruding into the sea, while the coast west of the Leizhou Peninsula extends westwards too much; the outline of Hainan Island is distorted; the shape of the main stream of the Yellow River is incorrect above the Hetao area and in the section near Mount Hua; and there are also errors in the drawing of the two ends of the Great Wall. Nevertheless, a strong point is that the symbols representing the Great Wall and the mountains are better [45].

(4) *Dili tu*

The *Dili tu* ("Geographical Map") carved in stone by Wang Zhiyuan in the 7th year of the Chunyou in the Southern Song dynasty (1247) is now kept in the Suzhou Museum. The base map was drawn by Huang Shang between 1189 and 1190. The

Fig. 7.2 A partial ink-drawn *Huayi tu* 华夷图 [Map of China and barbarian countries]. Liang, Erping 梁二平 (2018) *Shuizai shijiede zhongyang: gudai zhongguo de tianxiaguan* 谁在世界的中央 古代中国的天下观 [Who is in the centre of the world: The view of *tianxia* in ancient China]. Shanghai: Shanghai Jiao Tong University Press 上海交通大学出版社, p 156

map is 221 cm high and 106 cm wide. Wang Zhiyuan obtained this map in Sichuan and brought it back to Suzhou to be engraved in stone. It is more accurate than the *Huayi tu*, though no square grid is drawn on the map. There are more than 120 mountain ranges, 60 rivers and 410 administrative names. All the names of rivers are encircled with ovals and the place names are framed with rectangles. Many places are also marked with forest symbols. The names of the capitals of each circuit and a few military centres are carved in relief engraved in *yangwen*, while general place names are engraved in *yinwen* to show the difference in rank (*yangwen* characters are cut in relief and *yinwen* characters are incised in intaglio). Because of the pictographic symbols used for the mountains, the exaggerated proportions and the boxed place names, the map looks crowded and not clear enough.

(5) *Pingjiang tu*

The *Pingjiang tu*, crafted and engraved in stone in the second year of the Shaoding of the Southern Song dynasty (1229), served as a map of the city of Pingjiang Prefecture (present-day Suzhou, Jiangsu). The map, 197 cm long and 136 cm wide, is now in the Museum of Engraved Stone Tablets in Suzhou and is the most complete city planning map extant in China. The map is quite informative, with 10 types of water bodies and 4 types of landforms, as well as vegetation, city walls, streets, houses, bridges, places, and military camps. From the map, it can be concluded that the city of Pingjiang Prefecture was a water city, and the city consisted of crisscrossing river streets and the corresponding land streets, called the double chessboard network. It relied on more than 300 bridges connected into one coherent system, creating the unique appearance of a water town in the south of the Yangzi River. The city was rectangular in shape, about 9 li long from north to south and 7 li wide from east to west, with a circumference of 32 li. There was an inner city in the centre of the city, and the buildings in the city did not adopt a symmetrical approach in terms of layout, but important buildings, like offices lay along a central axis, while the rest of the buildings were scattered on both sides of the central axis. In order to show the prestige and nobleness of the emperor, the palace was located in the best position in the city. The city's artificially dug rivers not only provided the city with water, fire protection and drainage, but also formed a waterway transportation network between the city and its suburbs. The scale of the map is not very precise, especially in the suburbs [46, 47].

(6) *Jingjiang fucheng tu*

The *Jingjiang fucheng tu* ("Map of Jingjiang"), the largest city map (Fig. 7.3) ever seen in China, is carved on a rock cliff behind the Sanmian Pavilion at the southern foot of Yingwu Mountain, north of Guilin. Jingjiangfu city is the modern city of Guilin. The map is 34 m high and 3 m wide. It was carved in the 6th year of the Xianchun (1270). Most of the map is still legible, except for a few parts that have fallen off. The city plan fully reflected the military demands. In the map, only the main roads were indicated, and no names were noted, while the city trenches, military camps, official offices, bridges, and ports were quite detailed. The entire city of Jingjiangfu was composed of six parts. It made full use of the natural rivers, and when there

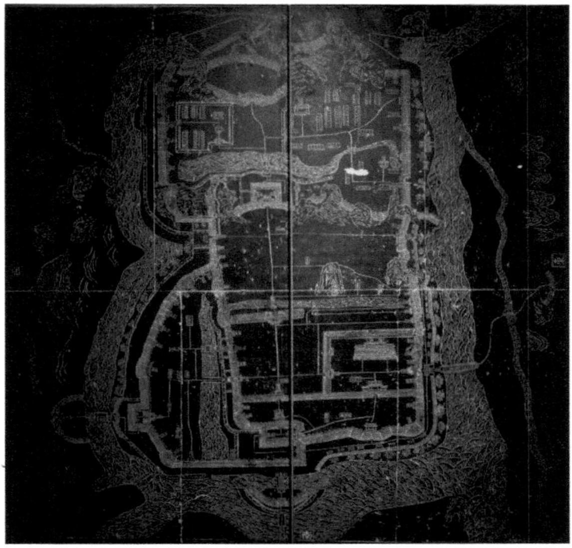

Fig. 7.3 The *Jingjiang fucheng tu* 静江府城图 [Map of Jingjiangfu] (photographed by Liang Qing 梁擎)

were no natural rivers, new moats were excavated. Thus, the most striking feature of the map was the moat on the east, south and west sides, which acted as the first military line of defence for the entire city by using water bodies as protective facilities. Roads and bridges were also designed for military purposes. Most of the bridges were temporary or semi-temporary, guarded by soldiers and easily dismantled. People from outside the city had to cross a bridge to enter the city. This map had a strong military orientation and was actually seen as a city defence map. The map was marked with the orientation, which showed that north was at the top, but not marked with scale, so it was not as accurate as the *Pingjiang tu*.

7.4.4 Geographic Models

The production of geographic models in China dates back to the Qin dynasty, more than 2,000 years ago. In the Eastern Han dynasty, *Jumi tu* by Ma Yuan was a temporary military geography model made by piling up rice grains [48]. Xie Zhuang of the Southern dynasties created the first map on wooden plates, called *Mu fangzhang tu*: "Mountains, rivers and lands, each type is marked by symbols. They are different in each state and prefecture, but combined they constitute the entire empire [49]."

Shen Kuo, a famous scientist in the Northern Song dynasty, tried making three kinds of geographic models with different materials. When he was commanded to inspect the frontier, he made a geographic model of the frontier geography. At first, he made a geographic model with wooden plates like Xie Zhuang, but it was time-consuming. Then he decided to make one on wooden cases with battered wood chips. Due to the cold weather, the paste and wood chips cracked easily, so he used molten wax. The wax geography model was light and easy to carry, but it could not be preserved for a long time. Therefore, he returned to the office and then transferred the wax geography model to a wooden board, turning it into a woodcut model. The emperor was very satisfied with the wood carved geography model and summoned his ministers to see it. He ordered all the border prefectures to submit wood carved maps like Shen Kuo's [50]. In Europe, geographical models made of wax and pulp appeared in the seventeenth century in Switzerland, more than 600 years later than Shen Kuo's [51].

Huang Shang (1044–1130), who lived from the end of the Northern Song dynasty to the Southern Song dynasty, "made the *Yudi tu* and carved it on wooden plates" [34]. Instead of Shen Kuo's approach, Huang adopted Xie Zhuang's methods to make wooden maps [52]. Later Zhu Xi visited Huang Shang to see his wooden map and intended following Xie Zhuang's method as well. In short, the production of geographical models reached its heyday in the Song dynasty, which was not only earlier than in Western Europe, but was also on a much larger scale.

In the 4th year of the Yuan dynasty (1267), Jamal al-Din made the first globe in China. He made a round ball out of wood and then painted various geographical elements on it. The "Tianwen zhi: Treatise on the Astrology" of the *Yuanshi* ("The History of the Yuan Dynasty") said: "It is a round wooden ball, seventy per cent is water painted green; thirty per cent is land painted white. Rivers, lakes and seas run over it. Small squares are marked to calculate the vastness of the territory and the distance." The small squares on the earth may be a grid of latitude and longitude lines, and the proportion of water and land distribution is basically the same as that measured in modern times. The emergence of this globe was of great historical significance for Chinese geography and cartography, both in terms of the shape of the earth and the proportion of the global distribution of land and water areas. However, because the Chinese feudal ruling class stifled people's thought and locked the globe away in secret places, the general public could not see it, let alone understand its value and significance. More than 500 years later, at the end of the eighteenth century, the geographer Li Zhaoluo made two more globes, one in copper and one in wood. The exact content and sizes are not recorded [53].

7.5 Achievements in Hydrology

The Song and Yuan dynasties were a period of vigorous development in ancient Chinese hydrology. This was mainly reflected in two aspects: river and lake hydrology and ocean tidology.

7.5.1 Hydrography of Rivers and Lakes

(1) Observation of river flooding

During the Song and Yuan dynasties, it was recognised that there were differences between the normal water period, the flood period and the dry period based on the changes in water volume. Before the Yuan You reign (1086–1094) of the Northern Song dynasty, *shuize* (stone steles used to measure water levels) had been built on the stone cliffs of Dujiangyan: "with each *chi* higher a mark (*ze*) will be made on the stone until there are ten levels carved. When the water level reaches the sixth *ze*, there will be sufficient volume of water for use; if the water level exceeds the sixth *ze*, the water will be discharged from the Shilang Weir and returned to the river" [54].

In the Southern Song dynasty, *shuize* steles appeared in the Taihu Lake area. The Wujiang Shuize Steles, seven *chi* long, were erected near the Chuihong Pavilion. The left *shuize* stele was used to observe and record special water level changes. There were seven horizontal levels, the lowest one was level one and the highest one was the seventh level [55]. The records had shown that there was a quantitative relationship between water level changes and agricultural land on different terrains, suggesting that the purpose of establishing steles was to understand the long-term pattern of water level changes. The right *shuize* stele was used to observe and record the water level changes in each month of the year. At the sixth level, there were carved the characters that "the water is here in the 5th year of the Shaoxi reign of the Great Song (1194)", and at the seventh level it was marked "the 23rd year of the Zhiyuan reign of the Great Yuan (1286)". The names of the twelve months of the year are engraved on the *shuize* tablet, and each month is divided into three phases: early (the first ten days), mid (the middle ten days), and late (the last ten days). By combining the left and right steles, it was possible to anticipate the local short-term (within a year) and long-term water level changes. The design was very scientific and was an innovation in the history of ancient Chinese hydrographic measurements.

Stone fish on the natural stone ridge were used to record the lowest water level of the year. According to the *Fuzhou shiyu wenzi suojian lu* compiled by Yao Jiyuan and Qian Baotang, "There are 22 kinds in the Northern Song dynasty and 64 kinds in the Southern Song dynasty." People in the Song dynasty had already concluded that the stone fish appeared periodically "once every three, five or ten years" [56]. It was believed that when the water receded and the stone fish appeared, the next year would see bumper crops as the saying predicted "out of the water appear the rock fish, out of the field come rich crops". People of different dynasties carved onto the stone ridge the time of the appearance of the stone fish and the distance from the stone fish to the low water line.

In the Southern Song dynasty, Fan Chengda wrote a detailed account of the hydrology of the Three Gorges of the Yangzi River in his book *Wuchuan lu* ("Record of a Boat Trip to Wu"). The book provides a faithful description of the floodplain cross-section. This is formed because in a straight section of river, when the flood rushes in, the downstream water withdraws first and the upstream water withdraws later, and the rate of reduction in the water level and flow velocity is large in the middle and small on both banks, resulting in a water level in the middle lower than that on both banks, thus forming a concave-shaped centre. The higher the water level is, and the more concave water surface there is, the greater the horizontal ratio drop. Fan Chengda's record of floodplain cross sections is the earliest in China and the world [57].

With better knowledge of river floods, people realised that the flooding of rivers and the weather were closely connected; therefore, in the Song dynasty people established a set of floods named after the weather. For example, the *Songshi: Hequ zhi* mentioned that "if the water rises by one *cun* after the Beginning of Spring [a solar term, one of the twenty-four periods in the traditional Chinese lunisolar calendar], it will reach one *chi* in summer and autumn. Since the prediction is fairly accurate, after the Beginning of Spring, the water of the Yellow River is called *xinshui* (literally creditable water). In February and March when the peach trees begin to put forth their blossoms, the meltwater joins the river and forms fierce waves after the rain, and the water of the Yellow River is called *taohua shui* (literally peach blossom water)'. At the end of spring when the *wuqing* [*Brassica rapa* L.] flowers appear, it is called *caihua shui* (literally vegetable-flower water)". This text shows that the Song people had understood the general pattern of the flooding of the Yellow River in the course of a year. Based on this, they started to make flood forecasts.

(2) Groundwater exploration

The way people in the Song dynasty explored the underground water was to observe whether the ground was covered with snow. The temperature of the groundwater is more stable than that of the ground, warm in winter and cool in summer. Therefore, if there are places where snow does not accumulate in winter, there may be groundwater (springs) underneath.

(3) Hydrological relationship between lakes and rivers

People in the Song dynasty had clearly realised that lakes played an important role in regulating the water volume of rivers. During floods, a lake stores flood water, thereby alleviating flood disasters; during dry periods, it replenishes the water of rivers, facilitating navigation and irrigation. Based on this recognition, knowledgeable people asked the authorities to formulate policies to protect lakes. In the *Jianhu shuo* ("On Jianhu Lake"), Wang Shipeng elaborated on the important role of Jianhu Lake saying: "Jianhu used to irrigate more than 9,000 ha of fields. Later, an increasing number of people reclaimed the land. Previously, there were no floods or droughts in the vicinity thanks to Jianhu. After more than 700 ha of the lake were turned into arable land, each time it rained or the sun shone for a few days, the nearby districts and counties would suffer from flooding or droughts. Now that the lake could not accommodate the rising water from the lakes and rivers nearby, the risk of floods increased significantly [58]."

(4) Study of ancient river channels

During the Song and Yuan dynasties, some people turned to fieldwork and documentary comparisons to determine ancient river courses. In the *Lainan lu* ("A Record of a

Journey South") by Li Ao, it is written that "(the Hangou) flowed from the Huai River to Gaoyou, and then backwards to the Yangzi River." As *Mencius* states: "Dredge the Ru River and Han River, remove the blockages from the Huai River and Si River, and let them flow into the Yangzi River", it could be concluded that the discharge of the Huai River and the Si River used to flow into the Yangzi River. Shen Kuo "had sent an envoy to conduct a survey: the old channel was just as described in *Mencius*. But the Huai River was already deep, and its flow could no longer reach Gaoyou" [59].

Fan Chengda recommended analysis of ancient river courses based on the accumulated water conditions. He had once found that the accumulated water in Ruizhou (modern Xun County, Henan) seemed like a river and deduced that this might be "the remaining water of the Yellow River", i.e. the former course of the Yellow River [60].

(5) Official field research on the origin of the Yellow River

During the Yuan dynasty, the first official field trip took place in 1280. The trip was led by Du Shi who arrived in Hezhou and found that 60 li east of Hezhou had a station for Ning River, 60 li southwest of the station is Shamaguan Mountain. He climbed over the mountain and trudged westward. After four months the team arrived at the origin of the river and reported its location to the emperor [61]. Thirty-five years later, in the 2nd year of the Yangyou (1315), Pan Angxiao, a Hanlin Scholar, wrote *Heyuan zhi*, the first monograph on the origin of rivers, based on the oral narrative of Du Shi's elder brother, Kuo Kuo. At the same time, Zhu Siben obtained a book in Sanskrit that also contained an account of the river's origin and translated it into Chinese. It was very valuable, but unfortunately was lost, and only part of it was preserved in the appendix "Dili zhi" to the *History of the Yuan*.

According to Pan's *Heyuan zhi*, the source of the Yellow River was the Xingxiu Sea (called "cuocha" in Tibetan). Moreover, Zhu Siben's translated work also indicated that the source of the Yellow River was more than 100 li southwest of the Xingxiu Sea, which some researchers nowadays think is the current Kala canal [62].

In addition to the above two books, Liang Yin from the Yuan dynasty also wrote a *Heyuan ji* [63], similar to Pan's *Heyuan zhi*, but less detailed. As well as the source of the Yellow River, the three books also documented the geography of the river source area and refuted the error of the "multiple origins of the Yellow River".

(6) Studies on water quality

Water quality was mentioned in the writings of the Song and Yuan dynasties. For example, You Pi said in *Wanliu xibian jiuhua:* "In Xushe Mountain (now the southwest of Wuxi), the water from the wells is salty and bitter, and people who drink stream water in summer often suffer from abdominal diseases [64]." Li Jiben in the late Yuan dynasty argued that water quality varied from place to place because of the thickness of the soil (or strata), which was reasonable. On the one hand, because the spring water running through the stratum dissolves the elements in the stratum;

on the other hand, because it also affects the changes in the water temperature [65]. Therefore, the water quality differs.

(7) Knowledge of fluvial erosion and sedimentation

Shen Kuo (1031–1095) explained the formation of Yandang Mountain (now in Wenzhou, Zhejiang) by water erosion and sedimentation. He then extended the principle of water erosion to explain the formation of all gullies in rocks with different characteristics, especially the gullies of the Loess Plateau [66]. In West Asia, Avicenna (980–1037), an Arab of almost the same period as Shen Kuo, also proposed water erosion as an explanation [67]. James Hutton (1726–1797), the father of modern geology, did not advance the theory of water erosion until 1788, more than 600 years later than Shen Kuo [68]. If we only refer to water erosion, Laozi made this point in the Spring and Autumn period (771–476 BC), much earlier than in the West.

Through examination of the fossils on the cliffs, Shen Kuo explained the formation of the North China Plain with the theory of fluvial sedimentation [54]. Ouyang Xiu already recognised that the siltation of rivers started downstream but did not state the reason for this. During the Yuan You period, Fan Bailu analysed that "after the river flows to the plain, the flow speed slows down and the sediment contained in the water is immediately deposited" [54].

7.5.2 *Tidology*

(1) The calculation of tide times and tide tables

In the Northern Song dynasty, Zhang Junfang refined Dou Shumeng's tide table in his book *Chaoshuo* ("Notes on Tides"). He made two improvements: firstly, the horizontal coordinate was changed from the phase of the moon to the number of degrees of the moon to the ecliptic plane; the second was that the vertical coordinate used "a quarter" as the smallest unit of measurement instead of "hour". These two changes made his new chart much more sophisticated.

Yan Su, a scientist of the Northern Song dynasty (c. 961–1040), followed the method of Dou and Zhang and improved the prediction of theoretical tide times. He calculated the time the tide was postponed day by day and found that, in a 30-day month, the time of the delay was 3.72 quarters and, in a 29-day month, the figure was 3.735 quarters. Yan's work *Haichao* lun was once carved on a tablet in Kuaiji, Zhejiang province. Yao Kuan saw it and copied the text into his book *Xixi congyu*. Yao Kuan also drew *Haichao tu* when he worked in Ming Zhou (today Ningbo), but it has been lost.

In 1056, Lü Changming also compiled the *Zhejiang chaohou tu*, a table of tide times of practical value. The Yuan dynasty's Xuan Bojiong happened to find Lü's tide chart in a local gazetteer. He viewed it as a treasure and wrote the *Chaohou tushuo* ("Illustrated Explanation of the Timing of Tides"), and then engraved the chart and his work together on a stone tablet and put it behind the wall of the Zhejiang Pavilion by the Qiantang River for passers-by to read.

Shen Kuo also made an important contribution with regard to the tidal cycle. He pointed out that the time of the high tide was different depending on the distance of the harbour from the sea [69].

(2) The causes of tides

Zhang Junfang's theory of the cause of tides, influenced by Ge Hong and Lu Zhao, asserted that: "The moon travels more quickly than the sun. When the moon aligns with the sun, they interact with each other. At that time, yin is at its peak, and yang begins to grow. As a result, the interaction of yin energy and the sea water leads to a large tide. Fifteen days later, after the moon has gradually moved to the east, the sun and moon will be on the opposite side of the earth. The yin energy dissipates and the sea water also recedes, forming another huge tide [70]." He believed tides were the result of the combined effect of the moon and the sun, especially the moon.

Yu Jing (1000–1064) wrote the *Haichao tuxu* based on his long-term observations of the tides in the East and South China seas. He disagreed with Lu Zhao's view and stressed the moon's influence and the synchronisation of the tides and the moon [71].

(3) The cause of tidal bores

Yan Su found that the tidal bore at the estuary of the Qiantang River was related to the existence of a north–south stretch of sand there [72]. It has been proved that, because of the sand bar at the mouth, this makes the propagation speed of the tidal trough much lower than that of the tidal peak, so that when the tide surges forward, the vertical position between the peak and the trough becomes closer and steeper; on reaching the north bank of the estuary near Dajian Mountain, the tidal peak catches up with the valley current, and the tidal wave before the slope tends to be steep, like a stack of water wall, rushing forwards. Yan Su's theory was a major contribution to ancient Chinese tidal theory, but he denied the important role of the trumpet-shaped estuary on the formation of high tide, which was inaccurate.

Xuan Zhao maintained the belief that the higher-than-normal high tides gathered because of the estuarine terrain. He explained that near the trumpet-shaped river mouth in Zhejiang province, there were two mountains: Kan Mountain in the south and Zhe Mountain in the north. They were located at the Haimen junction (literally Gateway of Sea) where the river joins the sea. Lower than Haimen was a sand bar stretching over 300 li from east to west, like a threshold holding back the seawater, causing the spectacular tidal bore [73].

In conclusion, the Tang and Song dynasties were highly developed in ancient Chinese tidology, leading the world at that time. Joseph Lee made a fair assessment that "in the middle of the eleventh century, that is, before the Renaissance, they (meaning Chinese scholars) were always much more advanced than the Europeans in tidal theory". "Until modern times, China in general had more knowledge of and interest in tidal phenomena than Europe [74]."

7.6 Advances in Meteorology and Climatology

7.6.1 Precipitation and Methods of Measurement

Some scholars in the Song dynasty made a clear distinction between the various types of precipitation and elaborated on their causes. According to Cheng Yi (1033–1107), "The dew is pure and descendant. The ancestors believed that 'dew becomes frost', which is indeed true. However, dew is different from frost: while dew can nourish life, frost damages it. Snow and frost are also different: frost can harm things, but snow does not. Rain and dew are also not alike: rain is murky, but dew is clear. Dew and fog are distinct in nature: dew is bright, but fog is dingy [75]."

Zhu Xi (1130–1200) stated, "Steam condenses into rain. When a rice steamer is covered with a lid, the steam becomes dense and may crystallise into drops of water. If not covered with a lid, its steam disperses." "Why is there no frost and dew in the mountains? I say, the upper air is clear, the wind is strong, although there is a slight fog, it is blown away. Snow is just rain condensed by the cold, so it is in cold high places that snow is first formed [76].

People in the Song dynasty were very interested in the shape of ice, snow and hail. In the *Chunming tuichao lu* ("Spring Morning Records after Withdrawing from the Court"), it is written that "in the winter frost, the roof tiles are in the shape of various flowers, and I drew them on paper." Shen Kuo in Volume 21 of the *Mengxi bitan* ("Dream Pool Essays") describes that "in He Prefecture, there once fell hail, some large like chicken eggs, some small like lotus seeds, each one resembling a human head with ears, eyes, mouth and nose, as if they have been carved".

Regarding the measurement of precipitation, the mathematician Qin Jiushao of the Southern Song dynasty recorded the fact that all the states and counties in the country used the Tianchi basin to measure rainfall in his *Shushu jiuzhang* ("Mathematical Treatise in Nine Sections"). According to Qin's record, there were many kinds of rain measuring instruments in the Song dynasty, and there was no standardised one. Since "the shape of the instrument is different, the amount of rain received is also different, and it is impossible to calculate the amount directly" [77]. Therefore, through calculation, the amount of rainfall collected in the rain gauge was converted into the amount of rainfall on the flat earth.

There were also two methods for measuring snow. To know the depth of the accumulated snow on a flat surface, it was only necessary to calculate the thickness of the snow on an inclined surface [specifically, to put a board at an angle against the wall, and by knowing the resulting dimensions of the triangle and the thickness of the snow on the board, the depth of the snow layer could be determined]. This was the first method. The second was to use a [truncated cone-shaped] bamboo basket to estimate the snow layer on the ground through calculation.

These instruments for measuring rain and snow were still very inconvenient to use, not sufficiently accurate, and needed to be improved. But nonetheless, these were the earliest instruments for measuring precipitation recorded in Chinese literature, a meaningful innovation, and one commonly used. They existed about 400 years earlier than the rain gauge created by Benedetto Castelli in Italy in 1639.

7.6.2 Phenological Observations and Theories

Before the Song dynasty, records of the weather in China were generally produced monthly or per pentad (in the ancient Chinese calendar, roughly 5 days), and there was no daily record of the weather. As far as the current document is concerned, the *Gengzi: Xinchou riji* written by Lü Zuqian (1137–1181) in the Southern Song dynasty was the earliest record of daily observations of the physical weather in China. It recorded the weather in Jinhua, Zhejiang province, for one year and seven months, including the flowering time of twenty flowers and the first time insects were heard chirping in the autumn.

One of the first theoretical expositions of phenology was made by Shen Kuo. He discussed in detail the relationship between phenology and geographical altitude, latitude, and plant species, and said: "Some flowers bloom in March in the flatlands, but in the deep mountains, they bloom in April. This is common sense and is due to the difference between the high and low terrain. Some bamboo shoots can be harvested in February, some in March or April and some even in May, which are called late bamboo. Different kinds of rice ripen at different times, perhaps in July, or in August and September, or in October. Those that are harvested in October are called late rice. These differences are in the nature of things [78]."

7.6.3 Factors Influencing Local Temperatures

During the Song and Yuan dynasties, people had a comparatively clear understanding of the factors affecting temperature in different places. In terms of latitude, the difference in average annual temperature from south to north was obvious.

Zhu Xi and Fan Chengda were aware of the influence of topography on temperature. Zhu Xi mentioned: "There are many high mountains in northwest Fujian and the terrain is quite low in southeast Fujian. The high mountains block the invasion of the cold wave, so in Fujian it is warm in winter. There are mountains in the south of Jiangsu and Zhejiang. With uplands in these two prefectures and lowlands in the north, cold waves sweep in easily [79]." This explains the cold weather in winter.

After personally visiting Emei Mountain in Sichuan, Fan Chengda recorded in detail the fact that the temperature on Emei Mountain decreases as the height increases. He stated, "At first, I wore summer clothes, but it got colder as I climbed higher; at the top of the mountain, I started to put on a scarf and boots, but I was still so cold that I lit the stove [80]."

7.6.4 Wind and Anemoscopes

(1) Tornadoes

Shen Kuo's account of China's tornadoes is a valuable source in the history of Chinese meteorology. In the past, people thought that tornadoes were unique to the Americas, but Shen Kuo's notes affirmed that there had been tornadoes in China as early as the eleventh century, which helps explain the regional distribution of tornadoes in the ancient world. Shen Kuo said: "In the ninth year of Xining (1076), in Wucheng County of Enzhou Prefecture (now Wucheng, Shandong province), there was a whirlwind from the southeast, penetrating into the sky like a ram's horn and even the huge trees were uprooted. Soon, the whirlwind approached the county, and people were all engulfed by the clouds. Many people were killed or injured, and the county was entirely in ruins [81]."

(2) Trade winds

During the Song and Yuan dynasties, people wrote a lot about the trade winds. Su Shi wrote in his poem about *bo feng* (the strong monsoon wind at the end of the rainy season and the beginning of summer) [82]. Chen Yanxiao in *Gengxi shihua* also stated: "During summer, a wind blows from the southeast for several days. People named it *bo feng* and believed that the wind was endowed with power by the gods to bring people back." Most accounts refer to the southeast monsoon, one of the trade winds. In the late Yuan dynasty, Lou Yuanli in his book *Tianjia wuxing* ("Husbandman's Five Agents"), a book on the theory of the *wuxing* (Five Agents), also gave a relatively detailed description of several kinds of trade winds, such as "before or after the Lantern Festival, there is certain to be a burst of cold wind, known as the Lantern Festival wind", "the northwest wind blows at the beginning of the winter, and is called the trade wind of the beginning of the winter".

(3) Wind speed, wind ranges, wind in relation to clouds and rain

Yu Yu of the Yuan dynasty in his book *Tan Xuan* believed that: "the Qi (energy) in the north is cold, in the east it is like wind. If the wind blows from the north to the east, the clouds will become rain. If the wind blows higher, it travels further. The wind that does not rattle tree branches blows at a speed of 40 li, the one that breaks branches blows at 400 li and the one that bends large trees blows at 5,000 li [83]."

(4) Wind vanes

There were three kinds of surveying instruments used to gather and measure the wind direction: *xiangfengwu* (wind-indicating bird), *zhanfengqi* (wind-indicating flag), and *wuliang* ("five ounces" named after the weight of its materials).

The *xiangfengwu* were usually placed on roofs, the tops of towers and on other high buildings. The oldest surviving iron *xiangfengwu* in China is the phoenix on the brick tower at Yuanjue Temple in Hunyuan, Shanxi. This pagoda was built in the 3rd year of the Zhenglong reign in the Jin dynasty (1158) and was renovated in the Ming dynasty. The iron phoenix on top of the pagoda can turn with the wind [84].

Li Zhi of the Song dynasty gave an account of the *zhanfeng qi*: "When he was a canal officer, he set up a wind flag in front of his office and sent people to observe and record it in a booklet every day. At the same time, he also asked the transport officials in each region to record the direction of the ships every day and to check on them in detail after the ships had entered the port, and he would scold the officials who did not follow the regulations or were sloppy."

In Volume 34 of the *Xuanhe fengshi gaoli tujing* by Xu Jing, it is written that "a feather-woven instrument for measuring wind direction and wind power stands outside, which is called a *wuliang*".

7.6.5 Observation of Other Meteorological Phenomena

(1) Causes of rainbows

In the Northern Song dynasty, Sun Sigong's thought on the formation of rainbows was close to today's, namely: "The rainbow is the shadow of the sun in the rain, which happens when the sun lights up the rain [85]." His idea predated that of the Englishman Bacon (Roger Bacon, 1214–1294) by 200 years; the latter discovered that rainbows are formed by the sun shining on raindrops reflected in the sky. It was not until the seventeenth century that the rainbow was explained in a more precise manner. Shen Kuo further proposed that the positions of the rainbow and the sun were relative, and the observer needed to stand facing in a certain direction to see one [85].

(2) The cause of *bao guang*

The *bao guang* (lit. precious light, also known as "Buddha's light") from Mount Emei is a glory that appears between valleys after the rain, a reflection of sunlight through the water vapour. It resembles the light of the halo on the head of Buddha, hence the name "Buddha's light". In China, the most detailed account of its formation process was given by Fan Chengda, who saw the Buddha's light for himself at Mount Emei, and pointed out its conditions and relationship with the changes in the weather at that time: "whenever the Buddha's light is about to appear, there will always be a swathe of thick clouds, just like the cotton trees of Douluo (a type of tree translated from Sanskrit). The light comes out of the clouds." Fan Chengda also noticed the differences in the intensity of the illumination, describing it as "weak", "strong" and "clear", and he emphasised that "clear" glories were the rarest to be seen. His insights were all correct and they are the earliest and most vivid accounts of glories in the literature [57].

(3) Records of mirages

In Volume 21 of the *Mengxi bitan* by Shen Kuo, the phenomenon of a mirage is described: "In the sea of Dengzhou (near Shandong peninsula), there are sometimes clouds of air, looking like palace rooms, a pavilion, people, chariots and horses, which are so vivid. This is called *hai shi* ("seaside city"). Some people say that it is formed by the exhalation of the dragon. Ouyang Wenzhong once went on a mission to Heshuo, passing through Gaotang County, Liaocheng, Shandong, and at night in the post house, he heard ghosts and gods travelling through the sky, and the sound of carriages, horses and animals could be distinguished one by one. He spoke about it in great detail, so I will not repeat everything here. When I questioned an old local man, he said: 'Twenty years ago, I passed by this county in the daytime, and I could see people and things clearly. The locals also called this (kind of sight) the *hai shi*.' It is virtually the same as what we saw in Dengzhou."[1]

7.6.6 Weather Forecasting

(1) Based on experience

In Volume 25 of the *Mengxi bitan*, Shen Kuo commented: "Be wary of the wind when travelling on rivers and lakes. In winter, the strength of the wind builds up, so people can get prepared; in summer, strong winds gather quickly, leaving people unprepared for the attack. I once heard that businessmen have a way of avoiding such disasters. Generally, in summer, high winds blow in the afternoon, so people who want to travel must set off early in the morning, as the moon and stars are bright and clear without any clouds. Thus, travellers will avoid encountering gales."

Proverbs were often used to forecast the weather in ancient China. In the *Tianjia wuxing*, there is an abundance of relevant information. The book is three volumes in length. The first part describes methods of divination for agriculture for each month in the year. The second part focuses on the seasonal phenomena in astronomy, geography, and in the plant and animal kingdoms. The last part discusses several methods of divination, such as *sanxun* [three phases of each month, each phase is about ten days], *liujia* [the six combinations beginning with the first Heavenly Stem of the sixty combinations of the ten Heavenly Stems and the twelve Earthly Branches] and *qihou* (climate). Weather forecasts are divided into two types: short-term and medium-term ones. Short-term predictions include "when stars flicker, wind is near", "prepare a raincoat for the east wind" and so on. Medium-term forecasts include "only after the spring wind does the summer rain come" and "it is always cold and rainy in spring." These proverbs were applicable to the lower Yangzi River region and some of them are still useful today.

(2) Based on relevant theories

Once there was a long drought in the capital (present-day Kaifeng) and people were praying for rain. After several cloudy days in a row, there was still no rain. One day, the sky suddenly cleared and the sun shone. On that day, Shen Kuo went to the palace to report to the emperor, and the emperor asked Shen Kuo when it would rain,

[1] Cited from *The Dream Pool Essays*.

and Shen Kuo answered: "It will rain tomorrow." Even after so many cloudy days, everyone thought that it would not rain, and as the sun was so bright that day, could it rain the following day? But it turned out to be just as Shen Kuo said, and it rained heavily the next day. Shen Kuo had used the theory of the Five Evolutive Phases and Six Climatic Factors to forecast the weather and made a correct prediction. From the perspective of modern meteorology, when it is cloudy for several days, the air is rich in moisture, but the right thermal conditions are missing, so it fails to rain. Once it is sunny again, the heat is available to produce an upward airflow, which will cause convective instability and rainfall [86].

(3) By measuring air humidity

Su Shi said in the *Gewu cutan* that "charcoal and soil are placed on each side of the scale and hung in the chamber. When it is about to rain, the charcoal side is heavy, and when it is sunny, the charcoal side is light". The hygroscopicity of charcoal is much greater than soil. When the humidity of the air changes, the weight of charcoal also changes due to the absorption of moisture: the weight of charcoal increases more when the humidity is high and decreases more when the humidity is low. Humidity is directly related to the weather.

7.7 Geographical Literature on Borderlands and Extraterritoriality

During the Song and Yuan dynasties, due to active seafaring and frequent foreign contacts, together with the travel records of diplomats and religious activists and the writings of historians, much foreign geographical knowledge was acquired. As a result, several works on frontier and extra-territorial geography emerged.

7.7.1 Geographical Information Gathered from Foreign Exchange Trading

(1) *Pingzhou ketan*

The *Pingzhou ketan* ("Matters Worth Discussing from Pingzhou") was a "brush notes" style book written by the scholar Zhu Yu. The book narrated the story of his father, who had been entrusted with a diplomatic mission to the court of the Liao Empire and was appointed commander of Guangzhou. A small part of the three-volume book also touched upon the geography of Korea and the Samboja Kingdom. For the first time, this book recorded the fact that China used the compass for navigation.

(2) *Lingwai daida*

The 10-volume *Lingwai daida* ("Representative Answers from the Region beyond the Mountains [in the South]") is a geographical treatise by Zhou Qufei, who was a local magistrate in Guilin. This book is divided into 20 sections: the "Geography" section focuses on the geography of Guangxi and the "Foreign Countries" section introduces the geography, transport routes, produce and customs of the indigenous people of the countries in Southeast Asia, West Asia and East Africa. This only accounts for one fifth of the entire content. There are also sections relating to treasures and commodities. The aim of the author was to answer queries regarding matters of China's extreme southwest, hence the title *Linwai daida.*

(3) *Zhufan zhi*

The *Zhufan zhi* ("Gazetteer of Foreign Lands") is a 13th-century Song dynasty work by Zhao Rushi, who was a supervisor of the maritime trade in Quanzhou, Fujian. The author spent his spare time reading foreign maps and visiting foreign merchants, and learned from them the names, customs and products of those countries and how far away from China they were. He recorded what he saw and heard in this book, with the first volume detailing the geography of 57 Asian and African countries and their ties with China, and the second volume 47 kinds of traded products such as borneol, frankincense, myrrh, pearls, parrots, black pepper, etc.

(4) *Dade nanhai tu*

Chen Dazhen's *Dade Nanhai zhi* ("Records of the Southern Regions during the Reign of Dade") was finished in 1304. The original version contained 20 volumes and its catalogue can only be seen in the *Wenyuange shumu* ("Catalogue of the Hall of Literary Profundity"). The National Museum of China now holds the remnants of five volumes (Volumes 6–10). Only Volume 7 listed various barbarian nations and imported products, which were considered valuable materials for overseas shipping in the Yuan dynasty [87]. The book included a total of 124 countries that traded with China and 71 kinds of imported goods, a greater number than in the Tang and Song dynasties [88]: (1) nine types of treasure—ivories, rhino horns, crane tops, pearls, coral, sapphire, kingfisher feathers, tortoiseshells, hawksbills; (2) five types of cloth; (3) 14 types of incense goods; (4) 19 types of drugs, and mineral drugs including sulfur and androgen; (5) four types of wood; (6) six types of leather goods; and (7) 14 types of sundries. These commodities came from Indochina, Southeast Asia, India, West Asia, Africa, Europe, etc.

7.7.2 *Geographical Information Recorded from Travel and Expeditions*

(1) *Xiyu xingcheng ji*

In the 2nd year of the Qiande reign (964), the Song Emperor Zhao Kuangyin sent more than 300 shamans to Tianzhu to bring sarlra and Bedouin books back to China, and Wang Jiye was one of them. Twelve years later, in 976, he returned to China. By then, the Song Emperor Taizu was dead and Emperor Taizong Zhao Gui was on the throne. Wang Jiye came to the capital to present Sanskrit sutra scrolls and Buddhist relics to Emperor Taizong. Then Wang Jiye was granted the opportunity to choose a famous mountain where he could practise—Mount Emei in Sichuan. There, he concealed 42 volumes of the Nirvana Sutra from the West in the Niuxin Temple. At the back of each volume of the Nirvana Sutra, he added notes of his journey to the West, which was then named *Xiyu xingcheng ji* ("Record of a journey to the Western Regions"). Although this book was not very detailed, it was still a very rare example worldwide. On the first day of the seventh month of the fourth year of the Southern Song dynasty (1177), when Fan Chengda visited Niuxin Temple, he found these notes and transcribed them in full in the *Wuchuan lu* ("Record of a Boat Trip to Wu"), which has survived to this day. The main content covers the geography, temples and cities in Central and South India, and it is the only extant work of the Song dynasty that records the geography of India.

(2) *Xizhou chengji*

In May of the 6th year of the Xingguo reign of the Northern Song dynasty (981), Emperor Taizong sent two officials: Wang Yande and Bai Xun to visit Gaochang (present-day Turpan). In April of the first year of the Yongxi reign (984), Wang Yande and others returned and reported to Emperor Taizong on the progress of the diplomatic mission and what they saw along the way, which was put into the book *Xizhou shicheng ji* ("Travels to Xizhou"), shortened to *Xizhou chengji*, also known as *Shi gaochang ji* and *Gaochang xingji.* The full text contains about 1,800 words with rich geographical content. The account begins with his departure from Xia Prefecture (present-day Jingbian County, Shaanxi), across the desert of Inner Mongolia westwards via the Yellow River to Hami at Mazong Mountain. There were four deserts along the route, and Wang Yande mentions the names of two of them: Liuke Sha ("六窠沙", literally Six Nests Desert) and Dahuan guimei qi ("大患鬼魅碛", literally Broad Desert of Nightmare Creatures). The former belongs to the present-day Tengger Desert and Ulan Buh Desert region. The description of desert landscapes in the *Xizhou Chengji* has its unique characteristics and is an excellent example of documentation recording desert landscapes in ancient China. It also records the products, ethnic groups, customs, hydrology, history, and religions of the countries they learned about during the journey. The full text of the *Songshi: Gaochang zhuan* ("History of Song: Record of Gaochang") compiled by Tuo Tuo of the Yuan dynasty and others includes the *Xizhou Chengji.*

(3) *Xining shiliao tuchao*

The *Xining shiliao tuchao* ("Maps Made on the Mission to Khitan") was written by Shen Kuo. He was sent to the Liao court to negotiate regarding controversial territories. On his way back, he recorded his travels, including the details of the geography, folk-customs, and climate, among other things. Since the book was a first-hand account, it is of significant historical value. It also produced descriptions of the winding, directions and mileage of the post roads along the route, which people were able to convert into a map later on. It can be said that it was an accurate work on the history and geography of the Liao dynasty [89].

(4) *Beishi ji*

In the fourth year of the Xingding reign in the Jin period (1220), the emperor sent the Minister of Rites, Ugu Sun Zhongduan, to visit Mongolia in July, and meet with Genghis Khan on his Western expedition in December. After his return, he described the experience of his journey orally, and Liu Qi recorded it in the book entitled *Beishi ji* ("Record of a Journey to the North-West"). According to the *Posthumous Papers of Mr Wang Jing'an of Haining*, the full text was about 1000 words long and outlines the author's journey to the West, the topography, ethnic groups, history, climate, products, and the customs and traditions of various places along the way. It was one of the earliest books in Chinese literature to detail a Mediterranean type of climate.

(5) *Xiyou lu*

Yelü Chucai (1190–1244) was a scholar in the early Yuan dynasty. In 1218 Genghis Khan summoned him to Mongolia. The following year, he accompanied Genghis Khan's army on its western expedition. On his way back, in 1228, he wrote the *Xiyou lu* ("Record of a Journey to the West") which was about 4,800 words in length. It was divided into two parts. The first part was dedicated to the author's journey from Beijing, through Juyong Pass, Wuchuan (now Xuanhua, Hebei), Yunzhong (now Datong in Shanxi), across Tianshan Mountain (now Yin Mountain) and the desert, to his final destination—Genghis Khan's camp on the bank of the Kulun River. The following year, he accompanied Genghis Khan on his western expedition, crossed Jinshan (now the Altai Mountains), Bechbaliq (now Jimsar, Xinjiang), Yinshan (now Tianshan Mountain), Arima city (in present-day Huocheng, Xinjiang), the Yilie River (present-day Ili River), to Tacisi city (today's Jambul in the Republic of Kazakhstan), Kuzhan city (today's Leninabad in the Republic of Tajikistan), Bapu city (today's Lapu) and many other places. The geographical descriptions of these places are very reliable, meaning that the text is an important resource for the study of the historical geography of the Chu, Syr Darya, and Amu Darya regions in the thirteenth century. The second part of the book criticised Daoist patriarch Qiu Chuji without any geographical content.

(6) *Changchun zhenren xiyou ji*

The *Changchun zhenren xiyou ji* ("The Record of the Perfected Master Changchun's Journey to the West") is the report of the Daoist patriarch Qiu Chuji's journey to the court of the Mongol ruler Genghis Khan (c. 1206–1227). It was written by the Daoist monk Li Zhichang of the Jin period (1115–1234) and is two volumes in length.

In 1219, Qiu was invited to come to the Mongol empire to teach the art of longevity. Qiu set out from Shandong province, travelled through Beijing to Xuande (modern Xuanhua, Hebei) and then all the way to the khan's camp in the Hindukush. Qiu's disciple, Li Zhichang, accompanied Qiu on his journey west and recorded what he saw and heard along the way, as well as Qiu's poems and speeches. When Li returned to Beijing, he quickly compiled the draft into a book, which was divided into two volumes of more than 18,000 words. After the book was completed, it remained unpublished and went unnoticed in the Yuan and Ming dynasties. It was not until 1795 that Qian Daxin found the text in the Xuanmiao Temple in Suzhou and copied it. In the Daoguang reign of the Qing dynasty, it was included in the *Lianjunyi congshu* and *Huangchao fanshu yudi congshu.* The text was later translated into Russian, French and English, and attracted the attention of scholars worldwide. It is richer in geographic content than the *Xiyou lu,* and so it is of greater value.

(7) *Xishi ji*

Chang De was appointed by the emperor to go to Persia in 1259 to pay an official visit to Hulagu Khan. He set out from He Lin, crossed through the Mongolian Plateau and the Junggar Basin, crossed the Ili River, Syr Darya River and Amu Darya River, and arrived in the south of the Caspian Sea. He travelled beyond Yelu Chucai and Qiu Diji and reached the northwestern part of the Iranian plateau. In March of the 4th year of the Zhongtong reign (1263), Chang De dictated and Liu Yu wrote the book *Xishi ji.* The entire book consisted of about 2,500 words and was more informative than the *Xiyou lu.* It reported more than ten kinds of medicines in Samahan in Uttar Pradesh, India, "none of which existed in China". The method of pearl hunting in the Mediterranean Sea was similar to that in China, and the geography and products of India were also introduced.

(8) *Yajiang xingbu zhi* and *Liaodong xingbu zhi*

The *Yajiang xingbu zhi* ("Record of Travels Through the Jurisdictions Near the Yalu River") and *Liaodong xingbu zhi* ("Record of Travels Through Liaodong") were both written by Wang Ji of the Jin period (1127–1193). The former was completed in the second and third months of the 2nd year of the Mingchang reign (1191), but only the first half remains. The latter was written from February to April of the first year of the Jinchang reign (1190). At that time, the author was an official in charge of the Liaodong jurisdiction, penalties and prisons, and he inspected various ministries to assess their work. The books introduced the geography of the three *lu* (circuits): Shangjing, Dongjing and Beijing with many anecdotes, which supplemented the "Geography" section in the *History of the Jin Dynasty.*

(9) *Zhenla fengtu ji*

The *Zhenla fengtu ji* ("Land and People of Cambodia") is a description of Chenla, the ancient kingdom of Cambodia. The author Zhou Daguan of the Yuan dynasty had visited the country as a Chinese ambassador during the years 1295–1297. He started out from Wenzhou in March, and it took him five months to arrive in Angkor, the capital of Cambodia. He stayed there for a year and began his return to China in July 1297. Within a month he was back in Ningbo, Zhejiang province. Zhou described the travel route, the city of Angkor, the royal palace, the officials, the religion, the people and their customs and habits, the language, their ways of living, festivities, agriculture, trade, the landscape, animals, and much more. It is an important text for the study of the history and geography of Chenla and its interaction with the Yuan dynasty. In 1819, the book was translated into French.

(10) *Daoyi zhilue*

The *Daoyi Zhilue* ("Description of the Barbarians of the Isles") was written by the Yuan dynasty traveller Wang Dayuan in 1349. After the author turned 20 years old, he joined the merchant navy and travelled to the coast of the Indian Ocean and the islands of Southeast Asia on two separate voyages. He recorded what he saw in 99 articles and synthesised excerpts from the old records of his predecessors into one article. The book contained 220 names of foreign places and the accounts of some places were more detailed than in the than in The Travels of Marco Polo. The book provided information on mountains and rivers, climate, production, products, customs, people, trade, currency, culture, education, law, and so on. It is highly appreciated by scholars at home and abroad for its high historical value. There have been numerous translations into English, French and Japanese since 1888.

(11) *Dali xingji*

The *Dali xingji* ("Record of Travels in Dali"), also called *Nanzhao xingji*, was written by Guo Songnian in the Yuan dynasty. It was said he had travelled to Baiya, Zhaozhou, Longwei Guan and many other places, recording the mountains, rivers, local products and customs, and so on. He had proved that the five major surnames in the Dali area were Meng, Zheng, Zhao, Yang and Duan. Although the people had been moving to Dali from the mainland for hundreds of years, their palaces, buildings, languages, books, marriages, rituals, fights and battles still contained the legacy of the Han dynasty. This was seen as an example of the impact of the Yuan dynasty on the geography of Yunnan, especially on the history and on ethnic relations after Fan Chao's *Yunnan zhi* written in the Tang dynasty.

7.7.3 Geographical Literature on Borderlands and Extraterritoriality

(1) Frontier and extraterritorial geography in official history

The *Songshi* ("History of the Song"), compiled by Tuo Tuo and others, included a special section on frontier and extraterritorial geography—*Waiguo liezhuan* ("Biography of Foreign Countries") and *Manyi liezhuan* ("Biography of Barbarian Regions").

The eight-volume *Waiguo liezhuan* provided information about Dali, Ü-Tian, Gaochang and many other kingdoms, which are now the minority regions of China's borders. In addition, the geography of 17 countries, including Goryeo, and the history of their relations with China were also recorded. In the case of coastal countries, their sailing routes to China were also mentioned. The 4-volume *Manyi liezhuan* gave an account of the history of China's ethnic minorities and the geography of their settlements.

In the *Liaoshi* ("History of the Liao") compiled by Tuo Tuo and others, the *Gaoli-deng liezhuan* ("Biography of Goryeo and Other Countries") covered the history of interactions between Liao and Goryeo, though with little geographical content. The *Xixia liezhuan* ("Biography of Xixia") contained the history, customs and products of the Xi (Western) Xia Kingdom, and the relationship between Xi Xia and Song and Liao, etc.

In the *Yuanshi* ("History of the Yuan") compiled by Song Lian and others, there were three volumes of the *Waiyi liezhuan*, which described the history and geography of Goryeo, Japan, Annam, Burma, Java and other countries, as well as their interactions with the Yuan dynasty.

(2) Borderlands and extraterritoriality in other historical literature

The *Tongzhi* ("Comprehensive Treatises") by Zheng Qiao includes the seven volumes of the *Barbarian Tribes on the Four Borders*: the one-volume *Eastern Barbarians*, two-volume *Xi (Western) Rong Tribe*, two-volume *Southern Barbarians* and two-volume *Bei (Northern) Di Tribe.*

The *Wenxian tongkao* ("Comprehensive Investigations Based on Literary and Documentary Sources") by Ma Duanlin also contains a total of 25 volumes of the *Siyi kao*, which includes four volumes on the eastern countries, three volumes on the western countries, four volumes on the Western Regions, nine volumes on the northern countries and five volumes on the southern countries.

7.8 Views of the Shape of the Earth

During the Song and Yuan dynasties, the concept of geomorphic forms was further developed and the spherical Earth model was more clearly stated.

Cheng Hao and Cheng Yi seem to have touched on the concept of a round earth. At the end of the Southern Song dynasty, Li Ruchi in the *Dongyuan congshuo* refuted an old theory which compared the sky to an eggshell, the earth to the egg yolk. Here the so-called "old theory" used a metaphor of the shape of the sky and the terrain which was fundamentally different from the previous belief of the relative position of heaven and earth. Li Ruchi himself strongly advocated that the ground was as flat as a cake. As such, there is no doubt that the old theory quoted by Li Ruchi advanced the spherical Earth view, although it was once again in a position of being criticised and denied.

In the Yuan dynasty, the idea of the spherical shape of the earth was reintroduced from the Western world. In the 4th year of the Zhiyuan reign (1267), Jamāl al Din presented a wooden globe to the emperor, explaining that "the green part represents the ocean, which accounts for 70%, and the white part represents the land, which occupies 30%. Moreover, the rivers and lakes are marked across it. Small squares [meridians and parallels] are also drawn to calculate space and distance" [90]. The tellurion was more evocative and directly displayed the figure of the Earth. Even though it was only conserved in the imperial court, its significance and influence cannot be underestimated.

Li Zhi (1192–1279), in his book *Jingzhai gujin tou*, explicitly pointed out that heaven and earth were both spherical, "if the earth is square, then heavens and the land will end up crashing".

Yin Shizhen, in the *Langhuan ji* ("Record from Langhuan"), commented that the sky was shaped like an eggshell, the earth was like the egg yolk, and the people were compared to the egg white attached to the yolk (referring to the earth). This was definitely a clear statement of a spherical earth, and also articulated the important idea that there were people living all around the earth.

Huang Zhencheng (1287–1362), in the *Shangshu tongkao* ("Comprehensive Studies on the Book of Documents"), used two concentric circles to represent the sky and the land. The accompanying notes said the round earth stayed still in the middle of the universe to confirm the right south, which meant he had explicitly and correctly depicted the earth as a globe.

7.9 Advances in Geology

7.9.1 Geological Theories

(1) Changes in sea levels

Shen Kuo of the Northern Song dynasty took the ancient Chinese idea of sea level changes to a new stage. By observing fossils and rounded pebbles, he argued that the North China Plain used to be sea. He said, "I was appointed to inspect Hebei and travelled along the Taihang Mountains. I found that the rocks of the cliffs were often interspersed with snail and mussel shells and pebbles. This place should be a former seashore, but now the east is nearly a thousand li from the sea. It seems that the continent is actually the result of the accumulation of sediment. The Yellow River, Zhang River, Hutuo River, Zhuo River, Sanggan River, and the other rivers all contain turbid water. Now, in the west of Shaanxi, the water flows through the canyon below ground, over a hundred feet deep, and the sediments carried by the current to the east every year becomes the soil which has built the continent [91]."

Zhu Xi of the Southern Song dynasty believed that soft silt deposited in the ocean, buried underground for a long time, would solidify into hard rock and then be lifted up to form a high mountain. In the Yuan dynasty, Yu Qin used the example of the mussel shell fossils on the cliff face of Dayunding Mountain in Jinan, Shandong province, to further prove Shen Kuo's point [92].

(2) Theories of fossilisation

Shen Kuo suggested for the first time that the fossilised "bamboo shoots" in Yanzhou (modern Yan'an, Shaanxi) had been bamboos in ancient times [85]. However, modern palaeontologists have confirmed that the "bamboo shoots" mentioned by Shen Kuo are neocalamite or fern fossils in the Mesozoic strata [93]. Nonethe, his idea that wild plants buried underground would fossilise was very meaningful.

In the Northern Song dynasty, Zhang Shizheng proposed a very innovative theory on the formation of fish fossils, saying: "This was previously a lake or a river, the fish that swam here were sealed in the volcanic ash because of a volcanic eruption and turned into fossils with the passage of time" [94], which is consistent with modern geology.

(3) Theories of mineral deposits

The *Zaohua zhinan* ("Guide to the Creation of Things") specified that mutual transformation took place among natural copper, copper oxide and copper sulfide. In this book it was written that "in a copper mine, copper undergoes a series of chemical reactions that give the shiny metal a pale green outer layer called a patina, and after another series of reactions, it turns into chalcopyrite" [95]. This was a legacy of the mineral genesis theory of Liu An from the Han dynasty.

In the Southern Song dynasty, Du Wan came up with the new theory that ore is formed as a result of fumigation. He argued that "of the patina excavated, there are three kinds: one kind in darker green which can be carved to make objects, one kind in green and blue, and one kind in a mediocre colour…The reason for this may be due to exposure to copper fumes" [96].

(4) Theories of mountain formation

Most scholars in the Song and Yuan dynasties believed mountains were formed because of water erosion. For example, the *Xixi yucong* by Yao Kuan from the late Song dynasty mentions that the author once read a book by an unknown author which claimed that the mountains in the southeast of China were formed by water erosion. This is consistent with Shen Kuo's explanation of how Yandao Mountain was formed.

In the Northern Song dynasty, Zhu Xi expressed his views on mountain formation: "when earth was first separated from heaven, only water and fire existed, and water was at the origin of the mountains. Today we climb high and gaze far afield, discovering that the layers of the mountains look like the waves of the sea, a result of water crystallisation, we are just not sure when [97]." His idea that mountains were actually the crystallised dregs of water was 600 years ahead of the theory of Neptunism proposed in Germany in the eighteenth century.

7.9.2 Knowledge of Fossils

During the Song and Yuan dynasties, only two types of fossils were known, namely, fossilised remains and chemical fossils.

(1) Fossilised remains

Fossilised remains refer to fossils formed by biological remains, including animal bones, teeth, shells, plant stems, flowers and leaves, seeds, etc.

① Dragon bones

"Dragon bones" are fossilised mammal bones and teeth from the late Cenozoic. They were given this name even before the Spring and Autumn and Warring States periods. The Song dynasty scholars Su Song and Kou Zongshi doubted the dragon moulting theory, but they did not conduct any further investigation. During the Song dynasty, dragon bones were sold as commodities and there was quite a considerable amount of exploitation. According to the *Maoting kehua* by Huang Xiufu, in Sichuan province, from January to March in the Chinese lunar calendar, the elderly, with their children or grandchildren, would sell dragon stones in some rural markets, and these would sell out quite quickly. When asked where they dug, they answered Fendong Mountain in Lingchi County (today in the southeast of Chengdu, Sichuan province). This shows that there was a great demand at that time. Some of the stones were as large as ten or more *zhang*, and the smaller ones were between three to five *zhang*.

② *Shiyan*

A *Shiyan* (lit. stone swallow), or a spirifer, is a fossilised ocean-dwelling brachiopod shell and it was given this name because both sides of the shell are wide and shaped like a flying swallow with its wings spread. In the Song dynasty, Xie Ming and Du Wan explained in a scientific manner that the reason why sometimes stone swallows could fly was because of thermal expansion and contraction. In Xie Ming's explanation: "Stones were exposed to the sun and, after sudden rain, a stone might fall off because of thermal expansion and contraction, not because it could actually fly [98]." Foreign records of brachiopod fossils began in the Middle Ages with Gessner who mapped small-mouthed shellfish brachiopods from Voldenberg, Germany, in 1565. The paleontological determination of the phylum Brachiopoda was made by Davidson in 1853 when he studied the Devonian Brachiopods [99].

③ *Shiyu*

The *Shiyu ouji* by Yang Jian, *Juanyou zalu* by Zhang Shizheng, and the *Songchao shishi leiyuan* ("Categorised Garden of Historical Facts of the Song Dynasty") by Jiang Shaoyu respectively recorded the existence of *shiyu* (literally stone fish) in Cixi (now Ningbo, Zhejiang province) and in Longxi County, Gansu province. Zhang Shizheng even gave a reasonable explanation for their formation. Du Guan also commented that the *shiyu* looked quite vivid, just like a real fish. In the *Shilin shipu*, it is written that "if you scratch off a little bit and burn it, there is a smell of fish" [100].

④ Fossil molluscs

When Shen Kuo was travelling in the Taihang Mountains in Hebei, he saw that there were snail and mussel shells and stones like bird eggs on the cliffs, and thought they were the fossils of marine molluscs. In fact, from the stratigraphy and fossils he described, they may have been certain river and lake formations (such as the Sanmen Formation) from the late Cenozoic, and the fossils contained in them would have been bivalves and gastropods, making them freshwater mollusc shell fossils.

⑤ *Baotashi*

The *baotashi* (literally pagoda stone) is also known as the bamboo shoot stone and Taiji stone. It is an ancient nautilus shell fossil, whose shell is right-angled, cone-shaped or cylindrical, with many diaphragmatic walls and a funnel-shaped body tube in the centre of the diaphragm or slightly off centre, and its longitudinal section is shaped like a pagoda, so it is named "pagoda stone". It is mostly distributed in the Ordovician and Triassic strata and is the standard fossil of the Middle Ordovician. It is also called Taiji stone because its horizontal cross-section looks like a Taiji diagram. The earliest extant pagoda stone specimen in China (held in the Nanjing Institute of Vertebrate Zoology) is the so-called bamboo shoot fossil described in a poem by Huang Tingjian, a poet of the Northern Song dynasty. It is 19 cm long, 11.4 cm wide and 2.5 cm thick, with a branching bamboo shoot on the front, formed of clean white calcite and which is set in a greenish-grey surrounding rock.

⑥ Crab, prawn and silkworm fossils

Crab fossils, mostly found in Hainan Island and Guangxi, are buried in the strata. In the *Kaibao bencao* by Ma Zhi of the Northern Song dynasty, it is written that "crab fossils are often seen in the South China Sea. It used to be just a normal crab which gradually transformed into stone" [95]. Su Song in the *Bencao tujing* also recorded that crab fossils could be commonly found in coastal prefectures. "They are made of stone and resemble crabs, with mud and debris attached [95]." The *Kaibao bencao* also documents silkworm fossils: "Close to the reef are stones looking like silkworms [95]."

⑦ *Shishe*

The *shishe* (literally snake stone), was considered by some to be a fossilised ramshorn snail or fossilised scale wood. Further research is needed. According to the *Bencao tujing*, *shishe* appeared in rocks which were found beside the rivers flowing into the southern sea, and their shape was like a coiled snake with no head or tail-tip. No one knew what they were made of but guessed that they were similar to crab fossils [95]. Kou Zongshi thought differently from Su Song. He described that the colour of *shishe* was like earth on old walls, and their coils were as big as hawthorn, with a hollow centre and two thin ends, therefore not of the same kind as crab fossils [95].

⑧ *Shilong*

The *shilong* (lit. dragon stone) recorded in ancient literature may be fossilised scale wood. Scale wood was a kind of tall tree growing in the Carboniferous and Permian

Periods. After the leaves covering the branches fell off, the tree showed clear scale-like marks. In Volume 21 of the *Mengxi bitan* by Shen Kuo, it was said that "once in Ze Prefecture (in modern Jincheng, Shanxi province) when a household was drilling a well, they saw a thing winding like a dragon and snake. Afraid to touch it, they waited a long time and it did not move. Then they tried touching it, and it turned out to be stone. (I guess) it was similar to stone crabs".

⑨ *Neocalamites* and fossilised ferns

The bamboo fossils recorded in ancient literature may be *Neocalamites* or fossilised ferns in Mesozoic strata. Volume 21 of the *Mengxi bitan* says that "near Yan Prefecture (now in Yanchuan, Shaanxi province), the bank of a river collapsed and sank several dozens *chi* deep into the earth. There, a forest of bamboo shoots, amounting to hundreds of stems connected to roots, was entirely exposed in stone form".

⑩ Silicified wood

Silicified wood is one of the common plant fossils. The silicification of wood not only preserves the annual rings, but also the fine structure of the plant. The *Yudi guangji* ("Vast Records of the Empire") in the Song dynasty recorded that "between 1127 and 1130, in Suining Prefecture (Suining, Sichuan province), there were pine stones which looked like a pine, but they were formed of stone inside. In Yongchuan County, Chongqing Prefecture, there existed stone pines and Songhua stones; they are about two to three *chi* high and one can just get one's arms around them. Several mountains have these, and they look as if they have been burned by lightening".

⑪ Human fossils

In the Yuan dynasty, Xianyu Shu, in the *Kunxuezhai zalu* ("Collection of Poetry Critique and Miscellaneous Themes"), described the fact that "a fossil of a human form had been discovered in a green stone looking as if a human being had been fossilised alive. Cut by knives, they were actually human bones" [101].

⑫ Pseudo-fossils

In ancient China, it was discovered that after iron-manganese aqueous solutions filled in rock cracks, pine branch and pine needle shaped pseudo-fossils would be made. This had been well documented in the *Dongtian qinglu* ("Pure Records of the Cave Heaven"), *Yunlin shipu, and Shanju xinyu* of the Yuan dynasty. Pine stone is mostly found in sandstone, or in shale layers and fissures, with many black branches in irregular shapes (looking like aquatic plants, pine branches, snapdragon algae and other forms) clustering together. Chrysanthemum stones are also false fossils.

(2) Chemical fossils

Chemical fossils, or chemofossils, are chemicals found in rocks and fossil fuels (petroleum, coal, and natural gas) that provide an organic signature of ancient life. The amber unearthed in Fushun, Liaoning province, was formed together with the Fushun coal seam during the Tertiary period (40 million years ago). Fossil remains are the preserved remains of living organisms, such as animal bones, teeth, shells, plant stems, flowers, leaves and seeds, which include the following types.

Chemical fossils refer to organisms and organic matter preserved in rocks of various geological eras. For example, amber is the resin of gymnosperms flowing into the ground and formed after tens of millions of years of geological activity. Amber from Fushun, Liaoning, China, was developed with the Fushun coal seam during the Tertiary Period (40 million years ago). These ambers are the resin fossils of ancient pine, cypress, juniper, maple and other resinous trees. In colour they are yellow, orange, red, brown, green, milky white, etc. [102] Chen Cheng from the Song dynasty said that the *kun*-ants found in amber became stuck there before they got buried in the soil [103].

7.9.3 Knowledge of Volcanoes and Thermal Springs

(1) Volcanoes

The earliest Chinese work that recorded volcanoes abroad was Zhao Rushi's *Zhufan zhi* ("Gazetteer of Foreign Lands") in the Southern Song dynasty. When it described Sicily (in southern Italy, the largest and one of the most densely populated islands in the Mediterranean Sea), it said, "In this country there is a deep cave on fire all year round. The people throw large stones into it, and they soon explode and shatter into broken pieces. The mountain erupts every five years."

(2) Hot springs

During the Song and Yuan dynasties, there were only scattered descriptions of hot springs in some local gazetteers and notes. Wang Xiangzhi recorded 18 springs in his work *Yudi jisheng*. The features of the springs included the water temperature, whether they were warm, hot or boiling; the smell, as some smell like sulfur; the colour, as some turn red; the creatures, as fish may live in some springs; the function, as it was believed that some could cure certain diseases and others could process food and irrigate the fields; and their distribution. The springs recorded were located in Anhui, Jiangxi, Guanggong, Hainan, Hubei and Hunan.

There were three schools of thought on the causes of hot springs in the Song: one represented by Su Shi; one represented by Tang Geng; and one represented by Zhu Xi, Zhu Yi and Zhou Mi. Of these three schools, Su Shi's view was the closest to reality, linking the formation of hot springs with volcanoes. Tang Geng believed that a hot spring was not heated by sulfur but rather was hot by nature, which made some sense. The thought of Zhu Xi and others was derived from Chen Zangqi's in the Tang dynasty, which argued that not only sulfur but also alum stone can turn cold springs into hot springs, which later proved to be incorrect.

7.9.4 Knowledge of Minerals and Rocks

(1) Knowledge of minerals and rocks reflected in the mining and archaeological data

① Knowledge of minerals and rocks reflected in mining

The *Taiping huanyu ji* listed the names of 12 ores exploited at the Longpei *jian* (industrial unit) at the beginning of the Song dynasty.

In the volume "Dili zhi" of the *History of the Song*, 29 species of minerals and rocks were involved as tributes from different places, such as Yangqi stone, purple quartz, stalactite, Changli stone, mica and others from Shandong; powered gold and inkstone from Shaanxi; jade grinding sand and magnetite from Hebei; gravel, alum stone, white quartz, limonite, jade grinding sand, gypsum, dragon bones, salt and others from Shanxi; halloysitum album from Jiangsu; powdered gold, crystalware, mica and others from Jiangxi; silver and manganese from Hubei; cinnabar, mercury, powered gold, *shiyan* (stone swallow), silver and others from Hunan; powdered gold, *zengqing* [copper sulfate], *kongqing* [azurite], iron and steel from Sichuan; cinnabar and gold from Guizhou; stalactite, silver, lead, tin and inkstone from Guangdong; silver, cinnabar, gold, white quartz, amomum and others from Guangxi.

② Knowledge of mineral and rocks reflected in archaeological data

The authors have summarised the minerals and rocks in the Song, Liao, Jin and Yuan dynasties mentioned in the *Wenwu* ("Chinese Cultural Relics") and *Kaogu* ("Archaeology") journals published before 1990. There are many types: gold, silver, copper, iron, tin, lead, mercury, cinnabar, azurite, malachite green, kaolinite, China stone, cobalt ore, copperas, arsenic, coal, talc, sulfur, jade, crystal, agate, ochre, Duan stone [a kind of high-quality ink stone from Guangdong province], She stone [a type of stone used to make ink in Anhui province], She inkstone (chrismatite), Shoushan stone, Black inkstone, Mo stone [mainly exploited in Sichuan province], Qingtian stone [mainly found in Qingtian, Zhejiang], marble, limestone, granite, red sandstone, argillite, quartz sandstone, green sandstone, sandstone, slate, marl, sandy shale, mudstone, black shale, chalk, amber, calcium carbonate, Dengni inkstone, turquoise, malachite, sapphire. Their main uses were as follows:

Gold was used in jewellery, pigments, gilding, gold wares, etc. In the Song dynasty, gold medals and gold bars were used as savings, not as a circulating currency. They were mainly used in payments and as stored savings, and sometimes as a measure of value.

Silver: silverware, such as bowls, plates, cups, boxes, bottles, shoes, saucers, cups, jars, dishes, chopsticks, basins, spoons, lamps and stoves; silver jewellery and tools, such as hairpins, ear spoons and bracelets. There were also silver ingots, silver bars, silver nuggets, silver collars, silver coffins, silver towers, silver rafters, etc. More than 51 pieces of gold and silver in six batches have been discovered in Song dynasty tombs, 391 pieces in six batches in cellars, and two batches of five pieces at the base of a tower, plus ornaments: the total number may amount to 500 pieces.

Copper: coins, bronze utensils—kettles, bowls, small bowls, large cups, three-legged plates, bottles, spoons, chopsticks, potholders, pen holders, hanging pots, mirrors, tables, basins, lamps, tea trays, tea sets, buckets, plates, candleholders, percussion instruments (such as cymbals and gongs), Buddha statues, statues of heavenly kings, copper beasts, copper seals, copper weighing apparatus, etc.

Iron: handicraft tools, agricultural tools, weapons, firearms, furnaces, iron towers, iron oxen from tombs, iron pigs, iron land coupons, iron tombs, etc. The Liaocheng Tower in Shandong of the Northern Song dynasty, 15.5 m high, was originally 13 levels high, and now has 12 levels. The Dangyang Eight-sided Iron Tower in Hubei is 21 m high, with 13 floors, weighing more than 100,000 *jin*. The two iron lions of the Tongbai Huaidu Temple in Henan were cast in 1328–1329. There were many types of talismans in the Yuan dynasty, among which there were round cards made of iron.

Tin: this was used in alloys, such as in a bronze drum from the Song dynasty in Guangdong, containing 10.92% tin. Tin alloy-containing vessels included furnaces, long necked bottles, jugs, pots, boxes, pails, horses, pounding mortars, spittoons, powder boxes, trays, beans, cups, bowls, pans, saucers, etc. It was used for tin bottles, tin rings, tin dings, tin lamp holders, tin basins, etc.

Lead: used in alloys, such as in a bronze drum in Guangdong made in the Song dynasty, containing 13.17% lead.

Mercury: mercury was placed in the tombs of senior officials to prevent decomposition and was common practice in the Song dynasty. In the Southern Song tomb of Zhang Tongzhi and his wife in Huangyueling, Pukou District, Nanjing, a large amount of mercury was unearthed, and in addition to the amount lost, 5 *jin* of mercury were still obtained from the relic [104]. In the Song dynasty, the annual production of mercury was higher than that of gold: for example, in the first year of the Yuanfeng (1078), the annual production of gold was 10,710 taels, while the figure for mercury was 53,696 [104]. In the Liao dynasty tomb in Jianping, Liaoning province, a great deal of mercury was found in the coffins. The book *Xinshi* ("History of the Heart") written by Zheng Suonan recorded the use of mercury in the process of embalming corpses, saying when the corpse was being prepared for burial, they would wash the intestines and put mercury and salt inside.

Cinnabar: mainly used in traditional Chinese medicine, alchemy, pigments and mercury refining.

Azurite: a product of hematite, this is copper carbonate, with the chemical composition of $Cu_3[CO_3]_2(OH)_2$, and was mostly used in pigments, Chinese medicine, and copper refining.

Malachite green: produced in copper ore, this is copper carbonate, with the chemical composition of $Cu_2[CO_3](OH)_2$, and was mainly used as pigments, materials used for making crafts and in refining copper.

Kaolinite (*gaolingtu*): the raw material for manufacturing porcelain, this is a white or slightly grey or yellow clay, which contains kaolinite, quartz, mica, silicon dioxide, aluminium trioxide, ferric oxide, calcium oxide, magnesium oxide, potassium oxide, sodium oxide, etc. It is a mixture of kaolinite and greater or lesser amounts of quartz, mica and other impurities.

Kaolinite (*gaolingshi*): an aluminous silicate mineral with the chemical composition of $Al_4[Si_4O_{10}](OH)_8$. Its primary use was in ceramics and then it was used in paper making and construction.

Cishi (porcelain stone or China stone): a kind of rock composed of quartz and sericite minerals, it is white with a slight yellow tinge, or a green, grey or reddish rock. It can be used alone as a raw material for porcelain, without adding any other clay minerals. Some porcelain rocks can be used in glazes, so are also called glaze stones.

Cobalt ore: with the chemical composition m (Co, Ni) $O \cdot MnO_2 \cdot nH_2O$, this was used as the raw material for blue pigments and in cobalt refining, and as a blue colouring agent in the porcelain industry.

Copperas: a blue colouring agent used in the porcelain industry.

Arsenic: mainly used to make carbide, and also used in medicine and insecticides.

Coal: during the Song and Yuan dynasties, this was also known as charcoal stone, which was mainly used as fuel. Two coal mining sites from the Song dynasty have been found, one in Hebi, Henan province, and the other in Yu County, Henan province. The sites may have revealed the technical level of coal mining at that time.

Talc: mainly used in ancient burial objects, including figurines and bottle caps, etc. Sometimes it was also used to make seals.

Sulfur: mainly utilised for making gunpowder and traditional Chinese medicine, in alchemy, and a little has been seen spread on the bottom of coffins.

Jade: jade was still in vogue during the Song and Yuan dynasties. Emperor Gaozong of the Song had nearly a hundred volumes of the *Guyu tupu* ("Ancient Jade Atlas"), which outlined in detail the ancient jade and replica products in his collection. The standard of jade seals in the Yuan dynasty was superior to that of gold seals.

Crystal: crystal products unearthed from tombs in the Song and Yuan dynasties include rabbits, crystal ornaments, pendants, beads, crystal stone blocks, crystal rings, balls, lions, boxes, cups, strings of beads, necklaces, etc.

Agate: Agate products unearthed in tombs and towers from the Song and Yuan dynasties include beads and products such as earrings, bowls, goblets, cups, tubes, lion-shaped ornaments, child-shaped ornaments, peach-shaped ornaments, ball ornaments, beaded ornaments, belt ornaments, etc.

Ochre: ochre is a product of hematite and was mainly used as a pigment.

Duan stone: this was named after Duanzhou (now Zhaoqing, Guangdong), where it was produced. Duan Inkstone has been famous since the Tang dynasty, and seven pieces of Duan inkstones have been excavated from the tombs of the Song and Yuan dynasties. Three types of rocks were often used to make Duan inkstones, namely, argillaceous slate, mica argillaceous rock, and tuff silty mudstone. They are bluish or greyish purple in colour. Their mineral composition is mainly: mica 70–80%; chlorite 5–10%; quartz 2–3%; hematite 1–2%. The composition varies depending on the origin of the Duan inkstone. Rocks produced in the area of Kefu Mountain on the east bank of the Lingyang Gorge were known more specifically as Duanxi inkstones. Their rock type is mica mudstone or slate, and the texture is softer. Rocks produced at the Song pit, Meihua pit and other inkstone pits were known as Duanxian inkstones from the Beiling Mountains. Their rock type is hematite-bearing argillaceous rock, with a reddish colour and a slightly dry texture.

She stone: this was named after Shezhou (now She County) in Anhui province, and because its rock was produced in Longwei Mountain, She stone is also called Longwei inkstone, the poet Su Dongpo's favourite type. It was discovered during the Kaiyuan reign of the Tang dynasty (713–741) and became an imperial item during the Southern Tang dynasty. She inkstone is produced in the shallow metamorphic rocks of marine mud and sandy sedimentation of the Sinian system in the eastern section of the ancient land of Jiangnan; it is a chlorite-mica aggregate in slate, with a metacrystalline structure, mainly composed of phengite and pyrochlore, with a small amount of quartz, carbon, pyrite, etc. It is marked by a strong texture, excellent ink performance, hardness of 3–4 on the Mohs scale, and a specific gravity of 2.89–2.94. Its main pits include the Longwei Mountain Pit, Meizi Pit, Shuixian Pit, Jinxing Pit and others on Luowen Mountain, Xitou Pit, Yejiu Pit and others on Zhuchi Mountain, and the Bihei Pit on Ji Mountain.

Depending on its texture, She inkstone can be classified into five types: *meizi*, *luowen*, *jinxing*, *yinxing* and *jinyun*. Meizi (eyebrow) inkstones have markings like human eyebrows and can be further divided into 13 kinds based on their shape. Luowen (subtle stripes) inkstones have thick, thin and hidden markings. Jinxing (lit. gold star) inkstones, which display markings like gold sprinkles, have over ten different kinds. The formation of a variety of natural ornamentation is related to the geological environment of their raw ores, to the co-occurring metal sulfides, and also to the later magmatic hydrothermal activity and secondary oxidation.

Chrismatite: a paperweight made of chrismatite in the Southern Song dynasty was excavated in Dayi County, Sichuan province.

Shoushan stone: this was named after Shoushan Town in the northern suburbs of Fuzhou, Fujian province, and is the earliest seal stone mined and utilised in China. Shoushan stone is found in the acidic volcanic rocks of the Upper Jurassic, the product of volcanic hydrothermal alteration. Its hardness on the Mohs scale is about 2–2.5. Its colours are white, yellow-white, grey-white, yellow, yellow-purple, purple-red, light green, etc. It can be classified into *tian* (farmland)-*kengshi*, *shui* (water)-*kengshi* and *shan* (mountain)-*kengshi* according to its occurrence and the causes of its formation [105]. In the tombs of the Song dynasties in Fuzhou and Youxi, Fujian, 33 Shoushan stone figurines, including one bird and one fish-shaped stone, were unearthed. All the materials had been sourced locally.

Qingtian stone: this derived its name from its place of origin in Qingtian County, Zhejiang province. Qingtian stone is found in Late Jurassic rhyolite and rhyolitic welded tuff. The main mineral components are chlorite, kaolinite, quartz, mica, montmorillonite, corundum and silica. The colours are green, light yellow, orange, brick red, purple red. Varieties are categorised by colour, with light yellow being the best. Its hardness on the Mohs scale is about 2.3–2.6. Qingtian stone can be classified by colour, among which light yellow is the most desirable. The light green, Shoushan yellow and Changhua chicken bloodline are called the Three Treasures of Yin stone [106].

Marble: in ancient times, marble was classified as jade, but now it is primarily regarded as a kind of coloured stone mostly used for stone carving and as stone for architectural decoration, occasionally used as a jade material. Its main minerals are calcite and dolomite, sometimes containing traces of quartz, sericite, serpentine and others. The colour is diverse: white, grey, pink, green, yellow, brown, etc., with a variety of colour pattern stripes. Transparency ranges from translucent to opaque, depending on the impurities and the degree of alignment of the calcite. It measures between 3 and 3.5 on the Mohs scale of hardness.

Limestone: with a chemical composition of $CaCO_3$, limestone is a versatile material mainly used for construction, sculpture and so on. In many tombs, limestone products have been found. For example, in Fangcheng County, Henan province, a stone dragon, a chicken, a horse, tables and chairs were found in a tomb from the Song dynasty. In some places, limestones were used to make tombstones.

Granite: in the Song and Yuan dynasties, granite was mainly used for construction, such as Anping Bridge, Luoyang Bridge, and the stone tower in Kaiyuan Temple. Granite was also used to build tombs, and the colourful granite gravel was used to make tombstones. In a tomb from the Liao dynasty, a chessboard made of granite was excavated. In the Northern Song dynasty tomb in Chaozhou, Guangdong, one granite stone rafter was unearthed.

Red sandstone: a tombstone and a land deed (ground certificate) were found in a tomb from the Song dynasty in Linchuan, Jiangxi province.

Mud slate: in the Yuan dynasty tomb of Lingyuan in Liaoning, the burial chamber was built with mud slate. In the Liao dynasty tomb in Jinzhou, one piece of clay slate was unearthed.

Quartz sandstone: in Mingzhou Port, the three carved stones of the Southern Song dynasty are all made of quartz sandstone.

Green sandstone: one green sandstone epitaph was unearthed from the tomb of the Bei Dawang (Great King of North) of the Liao dynasty. In the tomb of the Zhao clan of the Liao dynasty in Chaoyang, Liaoning, two pairs of green sandstone coffins and three epitaphs were unearthed, and the sarcophagus had been carved from six green sandstone slabs.

Sandstone: in the Liao Tomb of Jianchang in Liaoning, one pair of sandstone coffins engraved with Khitan script were unearthed. They are yellowish white in colour, with carvings on all sides. In the Jin dynasty tombs in Yonghe County, Shanxi, one pair of greyish yellow sandstone coffins was unearthed.

Slate: in the tomb of the Northern Song dynasty in Macheng, Hubei province, an epitaph made of cyan slate was unearthed.

Marl: in the Northern Song dynasty tomb in Macheng, Hubei province, one pair of cyan marl stones was unearthed.

Sandy shale: in the Song tomb in Shanghai, one pair of sandy shale inkstones was unearthed.

Mud stone: in the Southern Song dynasty tomb in Wuchang, a single red mud inkstone was unearthed.

Black shale: in the Song tomb in Nan'an, Fujian, one epitaph made of black shale was unearthed.

Chalk: a white clay powder, chemical composition $CaCO_3$, mainly used as pigments, building materials and traditional Chinese medicine.

Amber: during the Song and Yuan dynasties, amber was mainly used for accessories for adornment, such as for wearing, beads, handles, leaf-shaped ornaments, earrings, double fish-shaped boxes, mandarin ducks, vials, etc.

White marble: this was mainly used in construction, large items included coffins, rafters, bearing stone under the door, relief railings, etc. Small utensils included cups, incense burners, lions, pillows, stone mortars, belt plates with carved dragons, etc.

Dengni Inkstone: the very fine mud under the bed of the Fen River was filtered, mixed with bonding material, dried and carved into inkstones, then fired in a kiln. Because of the different minerals in the mud as well as the different temperatures of the firing, the final products are in different colours.

(2) Knowledge of mineral properties

① Colour

One of the most noticeable properties of minerals is the colour, and many names for minerals are derived from their colours. Pseudochromatic minerals were recorded by Kou Zongshi, who had found that "with a beam of sunlight, the stone can disperse five-coloured light". In the *Bencao tujing* by Su Song, it is said that a "stalactite is white with reddish markings; vitriols have five colours: white, green, yellow, black, and red…". The *Yunlin shipu* by Du Wan also gave a description of the colours of minerals, such as the purple Duan stone, and dark black jade. Pseudochromatic minerals were recorded by Kou Zongshi in his book *Bencao yanyi* where he noted his findings that "with a beam of sunlight, the stone can disperse five-coloured light" [107]. The five-coloured light here is a kind of false colour, and this discovery was 500 years earlier than Issac Newton's [108].

② Streak

The streak of a mineral is the colour of the powder produced when it is dragged across a white unglazed porcelain plate. In ancient China, people used to scratch or grind on a stone and then observe the colour of the powder left behind. Shen Kuo mentioned using streaks to identify bezoar stone and pyrolusite, saying that "though there are five stone colours of bezoar stone, if you grind it and mix the powder with water, the one with the red fluid is genuine. For pyrolusite, the real one will produce a milky mixture" [109].

③ Diaphaneity

Du Wan used words such as "diaphanous" and "slightly transparent" in his book *Yunlin shipu* ("Stone Catalogue of Cloudy Forest"). He mentioned that "the *fuguang* [glowing] stone (calcite) looks diaphanous".

④ Cleavage

In the *Bencao tujing*, it was said that "mica can easily be split into extremely thin elastic plates". The *Kaibao bencao* by Ma Zhi also recorded that when calcite was

Table 7.1 A comparison between Du Wan's description of hardness and the Mohs scale

Du Wan's description	Extremely soft	Moderately soft	Slightly hard	Not very hard	Hard	Quite hard	Cannot be split with an axe
Mohs scale	1	2	3	3.5	4	5	Above 6

struck, it would break into lumps of small stones. The *Bencao yani* documented that fibrous gypsum had quite a regular organisation in parallel sheets.

⑤ Hardness

In the *Yunlin shipu,* Du Wan first detailed the hardness of various minerals. For example, he believed agalmatolite is "extremely soft"; mica is "moderately soft"; stalactite is "slightly hard"; malachite is "not very hard"; guard stones [stalagmite] is "hard"; Xiangzhou stone is "quite hard"; and Yongkang stone "cannot be split with an axe". To understand the relationship between the seven levels of hardness described by Du Wan and the Mohs scale of mineral hardness, we can refer to Table 7.1 for a comparison:

It was certainly a great achievement to have such a finely graduated hardness scale more than 800 years ago, and one which predated the Mohs hardness scale by almost 700 years.

⑥ Magnetism

Su Song recorded in the *Bencao tujing* that "lodestones can attract dozens of iron needles and can even pull a couple of pounds of bladed weapons". The *Bencao Yanyi* made a similar statement saying that "a magnet can attract needles and iron".

⑦ Crystal habit

In the *Yunlin shipu,* the crystalline shape of minerals is mentioned many times: "Hang stone has angular corners", Ying stone has "multiple edges and corners", and the Pusa [bodhisattva] stone is "hexagonal". This crystal belongs to the trigonal crystal system, and it is a hexagonal columnar crystal. Quartz is a relatively pure carbonate rock, that is, calcite, which belongs to the trigonal crystal system; the crystal is rhombohedral, and the aggregate is crystal clustered, so Du Wei said that it was "angular".

⑧ Mineral associations

Volume Five of the *Bencao yanyi* ("Extension of the Pharmacopoeia") mentioned the association between gold and *fenzi* stone (a kind of brownish-black mineral mainly composed of ferrous oxide and copper oxide) in that "if about a hundred *chi* away from the cave you find brown stones with a single charcoal-coloured head, then gold ore can surely be found [110]." In the *Bencao tujing,* Su Song first mentioned the relationship between asbestos and talc. He also said that "silver is mostly mixed with copper in the mine" [111].

(3) Knowledge of minerals and rocks in materia medica texts

① Types of minerals and rocks in works on materia medica

The *Zhenglei bencao* ("Collected Classified Materia Medica") is a book on pharmacology by the Song period scholar Tang Shenwei (1056–1093 AD). In the Jin empire, Zhang Cunhui revised and published it under the title *Chongxiu zhenghe zhenglei bencao.* In the 30-volume long book, only chapters three to five recorded mineral drugs, which were classified as being of "superior" (*shangpin*), "mediocre" (*zhongpin*) and "inferior" (*xiapin*) quality. The medicinal properties, functions and origins of each drug were described, without any description of the properties of the mineral rocks. Each drug was also illustrated with diagrams and texts, and the main treatments with the drug and its processing methods were outlined.

The *Bencao yanyi* is a book on drugs of various characteristics written by the Song period (960–1279) scholar Kou Zongshi (early twelfth century). It recorded 63 kinds of mineral drugs, and under each one, it described its shape, nature, colour, method of identification, medicinal properties, therapeutic effects, method of extraction, and corrections of previous mistakes, etc. It is much richer in mineral information than the *Chongxiu zhenghe zhenglei bencao.* For example, when talking about cinnabar, the author referred to the places of origin, the size of the stones, the colour, transparency, cleavage, quality and so on. In the item on asbestos, it was documented: "It is mainly mined in Shanggang (now in Changzhi, Shanxi), and is possibly a type of rock. Its colour is bluish white. It has the name 'no ashes stone' in Chinese on account of its incombustibility. Wherever there is talc, there is asbestos." For the first time, Su Song pointed out that asbestos and talc are often found in close proximity in the earth's soil [112].

② Identification of minerals and rocks in the works on materia medica

The *Bencao tujing* by Su Song recorded how to identify melanterite, saying: "Put it on an iron plate and heat it with charcoal. If red fluid flows out, it is melanterite."

There are four methods for identifying minerals and rocks described in the *Bencao yanyi* by Kou Zongshi.

First, to distinguish between gold and silver using chemical changes: "If there is a mixture of gold and silver, melt it in a pot and then add salammoniac (NH_4Cl), and the silver will disappear." When NH_4Cl is heated, hydrogen chloride gas (HCI) will be produced. HCl can react with a variety of metals or alloys, some of which produce volatile metal chlorides, but it reacts with silver to produce a white precipitate of silver chloride and does not react with gold. This is the theory behind the method [113].

Second, to identify minerals on the basis of their crystal form, cleavage and colour: "When distinguishing gypsum from calcite, the one with well-organised stripes is gypsum." This is because gypsum belongs to the monoclinic crystal system, and sometimes longitudinal stripes can be seen on the crystal's surface; calcite belongs to the tripartite crystal system, with complex crystal shapes, including rhombohedra, complex tripartite deltahedra, hexagonal columns and parallel bifaces.

Third, using colour to evaluate the level of mineral content: when talking about lithargyrum, it was said that "lithargyrum which is gold in colour is of a better quality". This is because the main component of lithargyrum is lead oxide (PbO), whose crystals are yellow or yellow–red in colour and can absorb carbon dioxide from the air, forming a white outer film of lead carbonate.

Fourth, the use of magnetic properties. Magnetite can attract pins or iron, so it can be used to identify magnetic ores.

(4) Knowledge of minerals and rocks in stone and inkstone collections

In the Song dynasty, a number of *shipu* (stone catalogues) and *yanpu* (inkstone catalogues) emerged, which were not available before the Song and were a major cultural feature of this period. This reflects quite rich knowledge of minerals and rocks and provide an empirical summary of the knowledge of minerals and rocks before the Song dynasty.

① *Shipu*

The *Yuyanggong shipu* and *Xuanhe Shipu* mainly recorded the stones suitable for making rockeries at that time. The names of the stones vary according to their shape. Because both only noted the names of the stones without any description of their features and quality, they are of little significance in ancient Chinese geology and mineralogy.

The *Yunlin shipu*, which was written in 1118–1133 by Du Wan, is the most complete and detailed *shipu* in ancient China. It is about 14,000 words long, and 116 kinds of stones are described. The origins, characteristics (shape, colour, texture, sound, hardness, lustre, diaphaneity, hygroscopicity), uses, causes and the processing methods of the stones are described in detail. Among the 116 kinds of stones there are relatively pure limestone, stalactite, sandstone, manganese-bearing or iron-bearing limestone or sandstone, relatively pure quartzite, agate, crystal, pyrophyllite, mica, talc, shale, relatively pure metal minerals and jade, and fossils. A wide range of production areas for the stones was recorded in the book including 19 present-day provinces, 83 counties, cities and districts [114].

The *Taihu shizhi* was compiled by Fan Chengda in the Southern Song dynasty. Fifteen kinds of rocks are recorded, but only three kinds are accompanied by descriptions of their properties, and these were mainly used for the construction of gardens, rockeries, decoration, inscriptions and pillar bases.

The *Dongtian qinglu* ("Pure Records of the Cave Heaven") is a book on antiques written by the Song period (960–1279) scholar Zhao Xihu. This book gives an account of eight kinds of rocks, two of which have detailed descriptions of their properties, such as the colour, sound and hardness, and with hardness used to help identify the genuine ones. The stones introduced were mostly named after the places where they were originally extracted, such as Taihu stone, Guichuan stone, Shao stone, etc. When describing Lingbi stone, the author said the sound when one knocked on the Lingbi stone was the same as when knocking on jade, and it was difficult to cut it, even with knives.

② *Yanpu*

During the Song and Yuan dynasties, educated people attached great importance to inkstones (*yan*), one of the four treasures of the literary room, and there were a number of famous *yanpu* (inkstone catalogues). The study of inkstones and materials had reached its apogee, which greatly promoted the study of Chinese minerals and rocks. Joseph Needham once commented that there seemed to be no comparable research in the West. Below are works that have survived to the present day [115].

a. *Yanpu* ("Notes on Inkstones") by Su Yijian

This book starts with the history of inkstones and then talks about the materials used to make inkstones such as jade, stone, iron, lacquer, wood, clay, mussel shells, etc. Then it describes the manufacturing method, and the excellence of several inkstones such as Duan, She, Guwa and Dengni. Finally, it refers to the anecdotes, poems, inscriptions and odes, etc.

b. *Yanshi* ("A History of Inkstones") by Mi Fu

Although the title of this book suggests a focus on historical inkstones, it in fact discusses the quality of ink stones from various pits and especially analyses Duan and She inkstones. Apart from these two, 24 other types of inkstones are mentioned, including Gexiangong stone from Fangcheng County in Tangzhou, Huayanni Temple stone from Wenzhou, Ni inkstone from Tongyuan, Qing stone from Qingzhou, Liting stone from Chengzhou, Gushan stone from Tanzhou, Green inkstone from Guizhou, brownish yellow inkstone from Suzhou, dim stone from Jianxi, Tao inkstone, Zizhou inkstone, Goryeo inkstone, jade-like stone from Qingzhou, Hongsi stone and so on. Mi Fu concluded that "an inkstone which consistently produces ink must be soft. When struck, it conveys a deep and rhythmic sound, and it develops a sunken hollow over time. One which is slow to make ink is hard and sounds solid when knocked. It will become slippery after use."

c. *Shezhou yanpu* ("Notes on Inkstones from Shezhou") by Tang Ji

This short text focuses on the inkstone produced in the city of Shezhou (modern Xiuning, Anhui province). It introduces 36 types of inkstone and explains how stones were detected, their place of origin, how they were processed and rated, but also, which types of inkstones were common, and any problems the natural material might be subject to.

d. *Sheyan shuo* ("On She Inkstone") by an unknown author

The book opens with the history of She inkstone and explains the different characteristics of She inkstone from different stone pits. Some of the pits had stones with markings of gold flowers or waves, others with coarse curves or patterns like tortoiseshell. She inkstone was first exploited in the Tang dynasty and was then regarded as precious from the era of Li Yu, the last ruler of the Southern Tang dynasty in China. Then, in the Song dynasty, She inkstone became equally as appreciated as Duan inkstone.

e. *Bian sheyan shuo* by an unknown author

This book describes the characteristics, merits and drawbacks of 25 varieties of She inkstones, for example that the "She inkstone with fine grain markings (known as

Luowen inkstone) is of the finest quality. It is hard and heavy, and its colour is green and clear without any impurities". Although only two out of the 25 types lacked illustrations, the author concluded that it was hard to distinguish between top quality Luowen inkstones and the low-grade mudstone-like Luowen stones produced in Qimen County.

f. *Duanxi yanpu* by an unknown author from the Southern Song dynasty

This book solely focuses on Duan inkstone, though its title suggests it includes various types. It starts with the location of the pits and the differences between the inkstones from these pits. The place where Duan stones were collected was usually three to four li away from the riverside, gradually rising to the quarrying places such as Xiayan (lower hill pit), Zhongyan (middle hill pit), Shangyan (higher hill pit), Longyan and Bangkeng [lit. clam pit]. Among these, the stones from Xiayan were the best quality ones, and next were those from Zhongyan and Longyan, followed by Shangyan, with the Bangkeng stone being the worst. "The purple and lustre rocks are the most valuable. Rough ones look greyish while lustrous ones are blue or purple. The emerald ones which have markings of perfect circles ('animal eyes') are precious, especially those with 'pupils' in the 'eye'. Stones with 'eyes' can be ranked easily, as red and yellowish 'eyes' are all low-grade." "Stones with eyes are gentle and clean; people from Duanxi comment that younger stones tend to have more eyes, and the older ones have fewer, because younger ones are finer and can produce more ink. Those with blue veins definitely have eyes, thus blue veins can be found on stones quarried from the middle and lower pits. The upper rocks are likely to be pure and bright. The shape of the eye is compared to that of a myna, parrot, sparrow, cock or cat and named after these animals. The top grade one is green and the lowest grade is yellowish." The price also varied depending on the pit, with the differences ranging from ten to several hundred times as much. The author continued with discussion of the shape and the blemishes stones may have when found in nature.

g. *Yanpu* by an unknown author

This is a collection of statements about inkstones, especially the famous ones, including Duan inkstone, Hongsi stone, Longwei stone, Zhuzhou inkstone (such as Jinque [gold sparrow] stone, Qingjin [green and gold] stone, Zijin [purple and gold] stone, Datuo stone, Cidong [magnetite cave] stone, Tao green stone, Laoshan red stone, guwa [old rooftile] inkstone, Dengni stone and others), copper-like inkstone, clam inkstone, crystal-like inkstone, jade-like inkstone, iron-like inkstone and so on. The author cited the source when quoting from earlier sources.

h. *Yanqian* by Gao Sisun

The *Siku quanshu zongmu tiyao* provides an overview and review of the book: "The first volume describes Duan inkstone with 19 sub-items. The second scroll gives a description of She inkstone with 20 sub-items. Then the third scroll records the other 65 types of inkstones. The last scroll contains poems and prose related to inkstones. There are four or five books on inkstones from the Song dynasty, most of which only provide details about the quality of the materials but rarely cover the allusions. This book was written relatively late, so it was able to draw on the strengths of others, while at the same time quoting from a wide range of sources." In the third volume, the 64 names of inkstones reflected the rich variety of materials used for inkstones during that period. Apart from metal, bricks, tiles, pottery, clay, lacquer, clams, and crocks, the rest were all rocks of 49 different kinds.

(5) Rocks and minerals in notes

① Mineral and rocks

a. In *Mengxi bitan and Bu bitan*

The *Mengqi bitan* ("Dream Pool Essays") is a large collection of essays in the so-called "brush notes" genre written by the Song period (960–1279) scholar Shen Kuo (1031–1094). The book is 26 volumes long and is enriched by the appendix B*u bitan* which is two volumes long (three volumes in some editions). Within these, there are seven items relating to minerals and rocks.

"Dicui pearl" [moonstone] in Volume 21, Item 371: Zhang Hongzhao believed that "there is a fluid and watery movement to the stone. Colour and light shift within the gem" [116]. The *Yunyan guoyan lu* ("Clouds and Mist Passing Before One's Eyes") also records the existence of enhydro agate: "It is empty inside. There is a leaf-shaped void within which water flows."

"Bamboo shoots" in Volume 21, Item 373: although Shen Kuo was not yet able to correctly verify the fossil, his comment that it was a prehistoric life form. Based on this, it was also reasonable for him to speculate on the paleoclimate. Modern palaeontologists believe that the "bamboo shoots" Shen Kuo referred to were Triassic fern neoreed and pseudo-fern fossils [117].

Petroleum in Volume 24, Item 421: the original text reads that "there is a kind of 'stone oil' in the territories of Fu Prefecture and Yan Prefecture which was referred to as 'greasy water' in the past. It is found at the water's edge, mixed with gravel and spring water, and it flows out slowly. Local people dip the tail of a pheasant in it and collect it in a tile jar. The oil is very much like pure lacquer and burns like burning sesame sticks, only with a very thick smoke, and the tents in which it has been burned turn black. I suspected that this smoke could be used, so tried sweeping up its soot to make ink, and the lustre of the ink is like black lacquer, and (the fact is) pine ink is no better than it. So, people make it in large quantities and name it 'Yanchuan *shiye*' (lit. stone liquid). This will probably be widely used in the world in the future. The oil is so abundant since it is produced under the earth, unlike pine wood which will be used up by a certain time."

This is an early reference to petroleum in China. The author described how people collected and used petroleum in Yanzhou (modern Yan'an, Shaanxi) and predicted that it would be of great use in later ages. He pointed out that the use of oil contributed to the protection of vegetation and the ecological environment. The term *shiyou* (petroleum) was first coined by him and is still used today.

"Bitter spring" in Volume 25, Item 455: it is said that chalcanthite, which is a kind of copper ore, can be boiled in water from a hot spring (the main component is $CuSO_4$–$5H_2O$) in the Yanshan Mountains in Xin Prefecture, and copper replaces the iron, thus refining the copper. This detailed approach was also mentioned in the *Songshi.*

Taiyin xuanjing stone [gypsum] in Volume 25, Item 455: "Taiyin xuanjing stone, created in the brine of Xiezhou (modern Yuncheng, Shanxi) salt pond with its very high salt content, can be found in the soil of the ditch. The large Taiyin xuanjing stones are like apricot leaves, while the small ones are like fish scales, both are hexagonal and squared off like turtle nails ... The colour is green and crystal clear. When struck, it will break along the straight plane, and the fractured area is also hexagonal, like a willow leaf. When it is heated with fire, it will split into pieces which are separated from each other, thin like willow leaves, white like frost and snow, smooth and polished, very lovely." The crystal shape, colour, lustre, transparency, cleavage, as well as the changes to this stone after heating were made clear, which made a major contribution to ancient Chinese mineralogy.

Bezoar stone in Volume 26, Item 496: the Kingdom of Dupo (around the island of Java in Indonesia) sent a tribute of two bezoar stones which were as big as Chinese dates, yellow and slightly resembling the stamen of a flower. There was also a piece of pyrolusite which looked like a lotus. They were stored in gold foil. The envoy was asked, "How do you verify whether the stone is genuine?" He answered: "Although there are five colours of bezoar stone, if you grind it and mix the powder with water, the one producing the red fluid is real. For pyrolusite, the authentic ones will produce a milky mixture." Bezoar stone is a mineral containing sulfur and is used in Chinese medicine as an antidote to detoxify after ingesting poison. Manganese oxide (MnO_2) is the main raw material required for the refining of manganese and is also used as a traditional Chinese medicine for external use to relieve pain and regenerate muscle. Bezoar stones are produced in Guilin and Yizhou of Guangxi, Guangzhou, Sichuan, Liaoning, Jilin and Shandong. In this very short text, Shen clearly explained the form, size, colour, characteristics, function, origin and identification of these two minerals.

b. *Yangwengong tanyuan* by Yang Yi

In his book, Yang Yi of the Northern Song dynasty gave a detailed description of the Pusa (Bodhisattva) stones (crystals) saying that "the stones are bright and spotless, looking like the kind of Langya stone and quartz crystal in Shangrao. When sunlight passes through the stones, it disperses and forms a five-colour light around it, like the halo behind a Buddha's head". The author clearly indicates the colour, diaphaneity, crystalline structure, and false colour of the Pusa stones.

c. *Juanyou zalu* ("Miscellaneous Records of Casual Travels") by Zhang Shizheng

In the "Mi Stone" section, the author mentioned an example of a Mi stone being used as a whetstone in Tongyuan. "It measures several *chi* in length and weighs one to two kilos; it is bright and blue, and blades sharpened by it are unlikely to turn rusty. A marvellous item indeed."

In the "Cinnabar" section, the author described the origin, production, mining method, crystalline form, colour, lustre, transparency, streak colour, and the value of cinnabar. It is well documented that "the sites for mining cinnabar are often very deep. In order to extract it, workers have to fill the well with firewood, burn it until the stone cracks at a high temperature. Then the cinnabar can be collected from on a jade-like layer in small niches once it condenses. The size of the cinnabar may be as small as an arrowhead, or as big as a hibiscus. It has the lustre of fine jade and turns red once ground."

In the "Shiyu" (lit. stone fish) section, the book also gave an account of the origin, production, genesis and identification, especially the genesis. Zhang Shizheng may have been the first ever to put forward this hypothesis, which was in line with the modern geological view. He posited that "*Shiyu* exist at Yulong. The stones are extracted from rock pieces which are dug up from underground. Most of the patterns on the stones are crucian carps or loaches, with a few others following them. If painted, they would look like real fish; if burned, they smell of fish. Yulong, where *shiyu* are excavated, used to be lakes and rivers which contained fish. Over a long time, the mountains receded and earth turned into stone, thus fish fossils can be found. Today's *shiyu* found in Hengzhou are no different from Longzhou's."

In the "Shiyan" section (lit. stone swallow), a kind of Palaeozoic marine brachiopod shell fossil is introduced. The name *shiyan* was derived from its particularly wide shell sides, like a swallow spreading its wings. Zhang Shizheng pointed out its origin and the scientific experiments conducted by his colleague Xie Hong (or Xie Ming) in order to correct an old belief. He said that "people in Lingling think that *shiyan* may fly when it rains. Doctor Xie recalled that 'when I studied in a mountain temple, I noticed some swallow-like stones high in the mountain. They were mostly exposed to the sunlight, and with sudden rain, a few would drop off probably due to the unexpected change in temperature, not because the stones could fly.' I was given a stone slate with a swallow pattern on it when I travelled across Yongzhou, and the locals called it a swallow nest."

d. *Youhuan jiwen* ("Memories of My Travels in Search of an Office") by Zhang Shinan

In the fifth volume, Zhang Shinan provided information about the origin, colour and quality of jade and the tool *yu dengzi* (literally jade grader) against which the quality of jade was measured. Zhang commented that "jade comes from Lantian [Modern Lantian County, Xi'an, Shaanxi] and Kungang [Kunlun Mountains]. In the *Bencao*, it is also written that 'the top jade is mostly produced in Lantian, Nanyang, Rinan and Lurongshui, while the best foreign jade comes from Khotan and Kashgar (present-day Xinjiang area)'. Yet today we no longer find jade in Lantian, Nanyang or Rinan. The jade used in sacrificial vessels, parts of imperial chariots and emperors' belongings is mainly from Khotan … probably today's treasure mainly comes from the Northwestern tribes: Xixia [Kingdom], Wutai Mountain and Yutian Kingdom. Jade has five colours: white as fat, yellow as steamed chestnuts, black as lacquer, red as a cockle or as rouge, and the colour green which has the greatest range of differences in quality."

In the ninth volume, Zhang recorded the jade and beautiful stones seen in various places, such as "in Jiezhou, the quality of the stone varies. It is delicate when first carved, but over time it loses its lustre. Therefore, it is mainly used for seals in the imperial court. Jie stone in several colours is abundant and easy to excavate, so it is cheap. In contrast, the stone from Bazhou and Jiading Prefecture is hard and difficult to cut, and it does not seem to differ from jade (known as Ba *pu* and Jia *pu* respectively) in quality, thus its price is several times that of Jie stone. In Xuanhua County, Xuzhou, there is also a jade called Xuanhua *pu*. Jade which is sourced from around the area of the Dadu River is of the top quality…".

e. *Xixi congyu* ("Miscellaneous Words from the West Stream") by Yao Kuan

In this book, Yao Kuan quoted from the *Linanfu renhexian tujing* and introduced a special kind of sand called *tuoyuesha* for making casting moulds which would not burst or crack at high temperatures and generally contained a significant amount of quartz. It was from Renhe County and could be used for polishing jade.

Yao Kuan also quoted from the *Taiping huanyu ji*: "In the Samboja Kingdom, there is a hill which gleams with various colours in the South China Sea. The hill stones seem to contain flames. Whenever a ship passes nearby, the water gushes out. People may strike the hill with a knife, and the stone that bounces up into the ship is a bezoar stone. When burned, it smells like sulfur."

As for nitre, Yao referred to the *Luhuo bencao* by Cui Fang and stated in the *Xixi congyu* that "nitre is not a stone but rather an element obtained from the nitrate-containing clay found in Shangcheng of Hebei and residential areas along the Huai and Wei rivers (today's Bin and Hui counties in the north of Henan). Burned together with impure mirabilite and salt, the mixture can produce lead and copper. Mirabilite, or *mangxiao* is officinal; the Xiao stones are derived from the 'frost' of the earth, which are often found in the mountains, rivers and lakes. The 'frost' is collected at night in winter, dissolved in water, evaporated and concentrated, and precipitated into saltpetre. Probably due to its ability to dissolve rocks, it is called Xiao stone ['xiao' means 'to dissolve'), and is different from mangxiao or puxiao (types of mirabilite)." Yao Kuan's investigation is correct.

f. *Jilei bian* ("Chicken Rib Chronicles") by Zhuang Chuo

In the *Jilei bian*, Zhuang Chuo talked about the benefits of stone charcoal (coal). As the ruling centre of the Southern Song dynasty was in Southern China, and the lack of coal led to the widespread cutting down of trees and to the surrounding mountains and forests becoming bare land, Zhuang Chuo was very far-sighted and concerned about the environment, a view which is still valuable today. He said: "The Su Shi in his introduction to the poem *Coal* mentioned that 'in the past, there was no coal in Pengcheng [nowadays Xuzhou in Jiangsu province]. In December of the first year of the Yuanfeng reign, they started looking for it and found it north of the town of Baitu, southwest of Xuzhou, where it was used to make weapons that were sharper than ordinary weapons.' Millions of households in the capital of the Northern Song dynasty used to rely on coal to live instead of firewood. However, nowadays even the abundant forest resources in the Southern Song are not able to satisfy the need for firewood so that most trees have been cleared. It is such a great pity!"

g. *Guihai yuheng zhi* ("Well-Balanced Records of Guihai") by Fan Chengda

In the volume on gold and stones, Fan Chengda recorded 11 kinds of minerals, including cinnabar, mercury, stalactites, *wumingyi* (pyrolusite) and copper. He also described in detail the place of origin, production conditions, cleavage, colour, quality, use and other features. Fan stated that "*wumingyi* is a small black stone, and many of these stones can be found in the mountains in Guilin".

h. *Lingwai daida* ("Representative Answers from the Region Beyond the Mountains") by Zhou Qufei

This book records seven kinds of minerals and five kinds of fossils, namely raw gold, cinnabar, mercury, copper, verdigris, stalactite, talc, stone swallow, stone crab, stone shrimp, stone plum, and stone cypress, etc. The raw gold mainly refers to the alluvial gold originating from Guangxi. Zhou Qufei compared the cinnabar from Chen Prefecture (now in Hunan province) with that from Yi County in Guangxi and thought these two kinds were similar. He also wrote about the origin of verdigris and its nature and function, saying that this mineral could be found in Hunan, Guangdong and Guangxi and that some types might be used for painting or decorating but some were like a crust, and too fragile to be used.

In his book *Zhizhengzhi ji,* Kong Qi also mentioned some kinds of minerals and rocks. In the item on sardonyx, he said that "among the agates, sardonyx is ranked highest. It displays a particularly beautiful striped appearance in hues of red, yellow, white and blue, and those which have one of these colours penetrating throughout are the best. The value of the sardonyx depends on the number of stripes, and too many or too few of them will decrease its value. Zhexi [a historical and cultural region of Zhejiang, referring to the three prefectures of Hangzhou, Huzhou and Jiaxing in the former Hangjiahu Plain in northern Zhejiang] collectors often vie with each other to possess agates. Whether it is for use or simply to admire them, they conserve agate wine cups, waist ornaments and hilts, and the price is not less than tens of silver ingots, which is higher than for jade. I think it is because it is forbidden to trade jade privately, so they have developed a passion for agates. Lü Zihou, governor of Jinling [modern Nanjing in Jiangsu province], has an agate bowl which was inherited from his grandfather. It can hold a litre of water and it is light-coloure. Only three scarlet grape-sized dots and one or two waxy colour dots appear on the gem, such an elegant piece. I then argued with these agate aficionados: 'Among gold, silver, copper, iron, and tin, gold artifacts are the most desirable. Pearls, though small in size, are of course a treasure. In terms of warmth, texture and colour, jade is more precious than gold, but gold is solid and less likely to be damaged. The colour of crystal is too plain, and agate too mottled. Therefore, crystals and agate are not worth collecting." Now the pursuit of agates is merely temporary and not widely acknowledged throughout history. Most officials in Yanjing [modern Beijing] do not appreciate agate, and only insignificant singers and dancers wear its products. As the saying goes: 'The price of gold and jade objects depends on the offer made by the buyer', which is especially true.

There is an item on "Lingbi Stone", a carbonate rock mined in Lingbi (an ancient county), Anhui province. While large specimens, due to their beautiful shapes, have become ideal ornaments for landscape gardens, small and extraordinary ones resemble mountain peaks and ravines, and thus are ideal for indoor displays. It is said that "the Lingbi stone can be appreciated in three ways: whether it is thin, wrinkled, and permeable. The one which has a clear resonance and is dark in colour is the Lingbi stone, and the one which only has a clear sound when struck is the Taihu Lake stone.

② Inkstone

a. *Chouchi biji* by Su Shi

Su Shi mentioned Fengzhou (lit. beak of the phoenix) inkstone in the *Chouchi biji* ("Brush records of Chouchi"), saying that "although Fengzhou inkstones attract much scholarly attention and the descriptions of them vary, it is hard to identify and track the genuine ones. They are the top of the range, better than Hongsi inkstones" [118].

In the *Shengshui yantan* ("Banquet Talks from River Sheng") by Tang Yanyou, it was written that he once obtained a piece of Hongsi inkstone from the Black Mountain (in modern Qingzhou, Shandong) and carved it into an inkstone. Red patterns were entangled with yellow ones, which looked like a lunar aureole, or mountains, or clouds and flowers. When it was returned to the box after use, it took several days to dry. Modern gemologists have pointed out that the Hongsi inkstone is found in the Ordovician Majiagou Formation and is a purplish-red microcrystalline chert with fine patterns. It was already famous in the Tang dynasty, but production ceased in the Song dynasty because the stone veins were exhausted, and very few physical objects were passed down. Now, a new vein of Hongsi inkstone has been found, and its stone quality and colour are the same as those described in ancient texts, leading to the renaissance of these famous, long-lost and extinct inkstones [119].

b. *Jiashi jiuwen* by Lu You

In this book, Lu You revealed that throughout his lifetime he had had a passion for inkstones, especially Hongsi inkstone, which Duan inkstone and She inkstone could not compare with: "The Hongsi type with a yellow base is considered the best. It has moderate hardness with a fine texture. The silk pattern is delicate and tends to be in parallel arrays of ripples or concentric circles. The Hongsi inkstone feels like cream when touched, is able to disperse ink like oil, keeps the ink as black as lacquer, and is capable of holding water without wasting it" and he added "I only got about twenty. Later the mountain collapsed, so there was no way to access it. The so-called red silk stones are all fake nowadays [120]."

c. *Chunzhu jiwen* ("Hearsay from the Spring Islet") by He Wei

The ninth volume of this book provides information about 24 kinds of inkstones, such as Duan inkstone, She inkstone and Dengni instone, but it made more references to miscellaneous stories than to the properties of the inkstones. Therefore, it has a lower status in the history of mineral petrology, but is indispensable as historical material.

d. *Youhuan jiwen* by Zhang Shinan

Volume Five of the *Youhuan jiwen* ("Memories of My Travels in Search of an Office") discusses the characteristics of the Duan inkstone: "everyone highly values the Duan inkstone amongst all kinds of inkstones. It has been named in multiple inkstone catalogues with many of its varieties. Some people cherish the attribute of the 'eye' [a natural inclusion that exists in the form of concentric circles], while some desire those without. Stones with green veins surely have 'eyes', and the tenderer ones have more, while the harder ones have fewer. The tender and fine inkstones produce more ink, so those with eyes are more desirable…a green 'eye' receives the highest praise, while a yellow or red one is considered inferior."

The book quotes Emperor Gaozong of Song: "Duan inkstones of the best quality are purple in colour and often come from Xiayan (lower hill pit), and a Duan inkstone looks firm, closely grained and smooth. Pouring water on it can produce ink and it does not make a noise when ground. The inferior Duan inkstones are all mixed with sand and earth. Apart from having rough textures, they look dry and reddish. Those from the new pits are entirely inappropriate because of their mediocre appearance and slippery nature. I personally dislike stones with 'eyes', since most of them are fake and of a lower grade which decreases the appeal of the inkstone. I have collected a piece of purple inkstone without any patterns." From his criticism, it can be concluded that "Duan inkstone with fewer 'eyes' and in a delicate colour is of a better quality".

Volume Seven describes Lixi inkstone: "Lixi inkstone is produced in the area of Yuanzhi (modern Yuanling and Zhijiang in Hunan province). The purple ones are similar to Duan inkstones but without 'eyes'. Some have markings of eyebrow lines, and some have patterns of golden belts. When grinding, this type of inkstone will easily produce ink, but over time will become slippery. Polishing with small stones will make it new again. Some of these inkstones are green with patterns like waves, and some are black with golden sprays, and some have been found to contain copper. Since the Qingyuan reign of the Southern Song dynasty, people have started to carve inkstones which has grown in popularity since then, but they are still lower quality than Duan and She inkstones."

e. *Xixi congyu* ("Miscellaneous Words from the West Stream") by Yao Kuan

The second volume of the *Xixi congyu* mentions that "the Hongsi inkstones are too smooth to sustain ink, but the ink produced is excellent for writing on paper".

With regard to Duan inkstone, Yao Kuan commented that "the inkstone from Xiayan (lower hill pit) is purple like pork liver; it is firm, dense, moisturised and sonorous when struck. However, it also breaks easily due to its high level of hardness and contains many defects. Qiufeng (lit. autumn maples) stones in lighter colours are not as firm and lubricous as stones from Xiayan. Meigen (lit. plum root) stones from Zhongyan (middle rock) and Taohua (lit. peach flower) stones from Shangyan (upper rock) are both light in colour, slightly sparse in texture and contain a mixture of sand and earth. Stones from the new pits are purplish red, dense, and larger in size without impurities, yet they are merely mountain rocks lacking solidity and moisture. Houli stones are similar to the stones from the new pits but are of a lower grade. Stones from Xi pit (west pit) are cyan and slightly black in colour, with coarse patterns and tend to produce a great deal of ink, and the best quality ones can be compared to She inkstone. Those which have patterns of multiple concentric circles and show a mixture of the colours blue, white, yellow and black are the most desirable. The larger the circles are, the worthier the stones. The locals viewed the circles which look bright, round, watery and unblemished as 'living eyes', the circles in similar shapes with blurry boundaries as 'tearful eyes', and the dim and irregular circles that have a whitish colour within and around them as 'lifeless eyes'." All these comments were based on experience.

f. *Dongtian qinglu* by Zhao Xihu

The *Dongtian qinglu* ("Pure Records of the Cave Heaven") records 12 items of "inkstone recognition", and mentions the "old pit in the low stratum of Duanxi", the "relatively new pit in the middle stratum of Duanxi", the "new pit in the upper stratum of Duanxi", etc.

With regard to Tao River green inkstone, Zhao Xihu said, "Apart from the Duan inkstone and She inkstone, Taohe inkstone, which is as green as jade, is the most precious. However, it is mined from the bottom of the Tao River (in modern Gansu province) and is difficult for people to access." Taohe inkstone, one of the four famous inkstones in China, is so named because it is produced along the Tao River in Gansu. It is extremely smooth with delicate colours, which allows for easier grinding without damaging the brushes. It was found during the Xining period of the Song dynasty (1068–1078) and was a tribute during the Song dynasty, so it is unlikely that it was seen by the general public. From the petrological point of view, it is a lime green slate, formed in the Late Carboniferous Period. The Taohe inkstone has two main species: the so-called Hong Tao (reddish Taohe inkstone) and Lü Tao (greenish Taohe inkstone). The former is cinnabar-coloured, fine and pure, and extremely rare. The latter has duck-head green and parrot green forms. In addition, there is a small amount of black Taohe inkstone [106].

As regards *moyu* (dark jade) inkstone, Zhao Xihu commented that "in E'zhou of Hubei, there is a piece of black jade gemstone, similar to the black stones in the low Duanxi pit. It is hard enough to serve as inkstone. Although it is not as bright as jade ware, often there is a white part (can be as wide as one *cun*) in the middle of the black. If only the black part is used to make inkstones, it will be above Longwei inkstone and inferior to Duan inkstone".

g. *Fuxuan yelu* by Chen Qiao

Chen Qiao, in the second volume of the *Fuxuan yelu* ("Miscellaneous Notes by the Rustic while Warming Himself under the Sun"), compared the advantages and disadvantages of inkstones made of various materials such as stones, pottery, giant clams, mud, etc. He commented that: "The top-class inkstone is Duan inkstone, then Taohe inkstone, followed by She inkstone and Jianxi inkstone. Duan and She inkstones mined from old pits are of better quality. The 'eye' in Duan inkstone is not precious in itself but rather it helps determine the authenticity, and the yellowish and reddish shades with a mynah's eye are the most desirable. As for the natural texture of She inkstone, there are four types of veins: straight lines, eyebrows, subtle stripes and gold sprinkles. The ones with straight lines are the most valuable, and the ones with gold sprinkles are the hardest, so are difficult to use. Tao inkstone is now extremely rare, and not to be confused with the kind of stone from Qimen in She County; it is light green in colour with a fine texture, and is considered to be false Tao inkstone by the locals. As it cannot work well with ink, paper and brush, it is cheap. The stones in Jianxi are extraordinary. As for the Fengwei (tail of phoenix) inkstone praised by Su Shi, it was believed that it came from Feng Mountain in Beiyuan [present-day Fujian province], but now there is no fine stone there and little is known about it." All the views of Chen Qiao are correct.

h. *Zhi zhengzhi ji* by Kong Qi

In the second volume of the *Zhi zhengzhi ji*, Kong Qi discussed the decline of Lowei [lit. dragon tail] inkstone: "It became difficult to access (the) high-quality Longwei inkstone from the beginning of the Yuan dynasty. After the Renchen War, even the low-quality ones became rare."

(6) Exchanges of minerals in the song and Yuan dynasties
① The Northern Song dynasty

In the Japan section of the *History of the Song,* it is written that "in the first year of the Duangong reign (988), the tribute included amber, green, red and white crystals, gold inkstone, gilt copper water bottles, iron swords, dragon bones and sulfur, totalling 700 catties". "In the 5th year of the Xining reign (1072), the Japanese monk Cheng Xun offered silver incense burners, white glaze, crystal, amber rosaries, etc." "In the first year of the Yuanfeng reign (1078), the Japanese monk Zhong Hui paid a tribute of 5,000 *liang* of mercury".

In the section on Goryeo in *the History of the Song*, it is written that there was a "tribute of gold, copper and sulfur in the 8th year of the Tiansheng reign (1030)".

In the section on Srivijaya in *the History of the Song*, it is recorded that "in the 4th year of the Kaibao reign (971), the Srivijaya paid a tribute of crystal and kerosene. In the 7th year (974), ivory, crystal rings, glazed vases and coral trees were also paid as tribute".

In the section on Champa in the *History of the Song*, it is written that "in the 7th year of the Kaibao reign (974), 40 *jin* of the Xitian [the West] iron were paid as tribute. In the 2nd year of the Tianxi reign (1018), 72 ivories and a thousand pieces of tortoiseshell were paid as tribute. In the 8th year of the Tianshing reign (1030), the tribute comprised tortoiseshell, rhinoceros horns and ivories. In the 5th year of the Xining reign (1072), the tribute included glazes, coral wine wares, and purple ore".

In the chapter on Dayi [an Islamic empire established by the Arabs in the Middle Ages], the description of the tribute paid by Dayi included 700 *jin* of Bin iron [a type of wrought iron known for its hardness) and a piece of pyrolusite.

In Volume Three of the *Mengxi bitan*, it is written that "in the Xining period of the Song dynasty, Dupo Kingdom (present-day Java or Sumatra, Indonesia) submitted two pieces of *mosuo* stone [a yellowish gem the size of a date] … and a piece of pyrolusite".

In the *Mozhuang manlu* ("Notes of Mo Zhuang") by Zhang Bangji, it is written that "During the Xuanhe reign, some barbarian countries paid a tribute of cat-eye stone, as round as the dragon's eye, as green as grapes, and which was able to extinguish fire".

② The Southern Song dynasty

According to the *Yunlu manchao* ("Loose Notes of the Cloud-covered Foothill") by Zhao Yanwei, the maritime trade bureau Shibo Si in Fujian often docked ships from various countries, with varied goods. For example, there were real pearls, ivory, coral, glaze, agate, tortoiseshell, etc. from Srivijaya (only those related to minerals are listed, as is the case below). From Goryeo there were silver, copper, mercury, etc. The minerals on board reflected the kinds of minerals being imported into China by foreign countries at that time.

The sixth volume of the *Baoqing siming zhi* [a local gazetteer on Ningbo, Zhejiang, written during the Baoqing reign] recorded the products obtained from Japan, such as exquisite bronze ware, gold, mercury and especially sulfur.

③ The Yuan dynasty

In the seventh volume of the *Nancun chuogeng lu* ("Retirement to the Countryside") by Tao Zongyi, the author outlined five types of gems: red stones of four kinds, green stones of three kinds, corundum of seven kinds, cat's eye of two kinds (cat's eye with a beam of light and Zoushui stone) and turquoise of three kinds.

In the *Baoshi shuo* ("Notes on Gems") written by Zhang Hongzhao, the author identified the stones above. He pointed out that the names of these stones were mostly phonetic. The red stones included ruby, garnet, ruby spinel and the last kind (Kumulan) remained unknown. The green stones were varieties of beryl of different shades of green. The seven kinds of *yagu* stones were actually garnets of various colours, such as red, purple-red, dark red, orange, yellow, green, blue, brown, and black. The places of origin were South Africa, India, Sri Lanka, Brazil, China, Afghanistan, Myanmar, etc. Zhang also argued in his *Baoshi shuo* that of the two kinds of cat's eye recorded by Tao, Zoushui stone was not real chrysoberyl despite displaying the cat's eye effect. The *dianzi* stones were turquoise from ancient Persia and Xiangyang, Hubei. Zhang Hongzhao believed that turquoise was a phosphate compound of copper and aluminium, with the colours of cyan, azure, green, yellow-green, apple green, etc. The azure colours were the most expensive. They were prone to fading in the sunlight after long exposure. They had a [Mohs] hardness of 6 and they were opaque.

In the *Zhenla fengtu ji* ("Land and people of Cambodia") by Zhou Daguan, goods from China are recorded: "People here value the gold from the Tang dynasty, since no gold is produced here. Then there are pewter, mercury, silver beads, sulfur, nitre, iron pots, copper plates, glass beads etc. from Zhen Prefecture (modern Yizheng, Jiangsu)."

The *Daoyi zhilue* ("Description of the Barbarians of the Isles") by Wang Dayuan gave an account of the minerals and rocks in some countries in Southeast Asia, South Asia and West Asia, such as tin ingots and lead from present-day Malacca, alluvial gold, silver, copper, tin and lead from Cochin (modern Vietnam), oil from Sampang (now Madura Island, Indonesia), and tin ingots from Long Ya Men or Dragon's Teeth Gate (modern Singapore).

References

1. Huang D黄鼎 *Qiandao siming tujing xu*乾道四明图经·序 [A Preface to *Qiandao Local Gazeteer of Siming]*
2. Liu W刘纬毅 (1986) Songdai fangzhi shulue 宋代方志述略 [A brief account of local gazetteers in the Song dynasty]. *Wenxian*文献 [Documentation] (4)
3. Wang W王文楚, Wei S魏嵩山 (1984) *Yuanfeng jiuyu zhi dianjiao* 元丰九域志点校·前言 [A revised and annotated version of the *Treatise on the Nine Regions during the Yuanfeng Reign*: Preface]. Chinese Publishing House 中华书局
4. *Yudi jisheng xu* 舆地纪胜·序 [Preface to the *Record of the Landscape of the Imperial Domain*]
5. Xu Y 许有壬 (Yuan dynasty) Dayitongzhi xu 大一统志·序 [Preface to the *Compilation of the Gazetteer of the Grand Unification*]. Recorded in the *Zhizheng ji juan sanshiwu* 至正集·卷三十五 [Responsibility, vol 35]
6. Zhao W 赵万里 (1966) *Yuan yitongzhi qianyan*元一统志前言 [Preface to the *Compilation of the Gazetteer of the Yuan Unification*]. Chinese Publishing House 中华书局, Beijing
7. Zhu S 朱思本 *Yuditu zixu* 舆地图·自序 [Author's preface to the *Terrestrial Maps*], recorded in *Zhen Yizhai zazhu* 贞一斋杂注 [Jottings from the Zhenyi Study]
8. *Songshi: Songzhun zhuan* 宋史·宋准传 [History of the Song: Biography of Song Zhun]
9. Wang X 王晓岩 (1992) *Fangzhi yanbian gailun* 方志演变概论 [A brief introduction to the evolution of local gazetteers]. Liaoshen shushe 辽沈书社, Shenyang
10. Huang D 黄鼎 *Qiandao siming tujing xu* 乾道四明图经·序 [Preface to the *Qiandao siming map guides*]
11. Wang Y 王燕玉 (1984) Fangzhi chuyi 方志刍议 [A preliminary study of local gazetteers] In: *Zhongguo defang shizhi luncun* 中国地方史志论从 [Collected essays on Chinese local histories and gazetteers]. Chinese Publishing House 中华书局, Beijing
12. Zhu S 朱士嘉 (1958) *Zhongguo difang zhi zonglu zengding ben* 中国地方志综录: 增订本 [A comprehensive record of Chinese local chronicles: revised version]. The Commercial Press 商务印书馆, Beijing
13. Fu Z傅振伦 (1984) *Cong Dunhuang faxian de tujing tan difang zhi de qiyuan* 从敦煌发现的图经谈方志的起源 [A discussion of the origin of local chronicles based on the *tujing* discovered in Dunhuang]. Chinese Publishing House 中华书局, Beijing
14. Xue H薛虹 (1984) *Zhongguo fangzhi xue gailun*中国方志学概论 [Introduction to the studies of Chinese local chronicles]. Heilongjiang People's Publishing House 黑龙江人民出版社, Harbin
15. Zhu C 朱长文 *Wujun tujing xuji xu* 吴郡图经续记·序 [Preface to the *Supplement to the Map Guides of Wu Prefecture*]
16. Wang X 王象之 *Yudi jisheng xu* 舆地纪胜·序 [Preface to the *Record of the Landscape of the Imperial Domain*]
17. *Guanglingzhi xu* 广陵志·序 [Preface to the local gazetteer of Guangling]
18. Li Z 李宗谔 *Xiangfu zhouxian tujing xu* 祥符州县图经·序 [Preface to the map guides of prefectures and counties in the Xiangfu reign]
19. Liang K 梁克家 *Sanshanzhi xu* 三山志·序 [Preface to the *Records of Three Mountains*]
20. Zhang J 章鉴 *Yubuzhi xu* 余不志·序 [Preface to the *Local Chronicle of Yubu*]; Kangxi deqing xianzhi yuanxu 康熙德清县志原·序 [Preface to the *Local Chronicle of Deqing County in the Kangxi reign*]
21. Yue S 乐史 Jin taiping huanyu jibiao 进太平寰宇记表 [Universal geography of the Taiping Era (976–983)]

22. Wang X王晓岩 (1992) *Fangzhi yanbian gailun*方志演变概论 [Introduction to the revolution of local gazetteers]. Liaoshen Publishing House 辽沈书社, Shenyang, pp 115–116
23. Wang X王晓岩 (1992) *Fangzhi yanbian gailun*方志演变概论 [Introduction to the revolution of local gazetteers]. Liaoshen Publishing House 辽沈书社, Shenyang, pp 116–118
24. *Siku quanshu zongmu juan sishiqi* 四库全书总目·卷四十七 [Complete library of the four treasuries, vol 47]
25. Feng L 冯立升 (1995) *Zhongguo gudai celiangxue shi* 中国古代测量学史 [A history of ancient Chinese surveying]. Inner Mongolia University Press, Huhhot, p 183
26. Office of Historical Studies on the Song dynasty at Hangzhou University杭州大学宋史研究室 (1985) *Shen Kuo yanjiu* 沈括研究 [Research on Shen Kuo]. Zhejiang People's Publishing House 浙江人民出版社, Hangzhou, p 98
27. Qi L 齐履谦 *Zhitaishi yuanshi guogong xingzhuang* 知太史院事郭公行状 [Biography of the Assistant Court Historian Guo Shoujing], recorded in *Guochao wenlei* 国朝文类 [Categorised literature from our dynasty]
28. *Yuanwen lei*: *Guo shoujing zhuan*元文类·郭守敬传 [Collected works from the Yuan dynasty: Biography of Guo Shoujing]
29. *Mengxi bitan*梦溪笔谈 [The dream pool essays]
30. Hu D胡道静 (1965) *Gudai ditu cehui jishu shang de qifa wenti* 古代地图测绘技术上的"七法"问题 [The "seven methods" of Chinese ancient mapping technology]. Beijing: Chinese Publishing House 中华书局; Cao, Wanru曹婉如 (1980) *Lun Shen Kuo zai ditu xue fangmian de gongxian* 论沈括在地图学方面的贡献 [A research on Shen Kuo's contribution to cartography]. Shanghai Science and Technology Press上海科学技术出版社, Shanghai
31. Luo H 罗洪先 *Guangyutu xu* 广舆图序 [Preface to the *Enlarged Terrestrial Atlas*]
32. *Songshi: Yiwen zhi* 宋史·艺文志 [History of the Song: Treatise on literature]
33. *Mengxi bitan: Bu bitan juansan* 梦溪笔谈·补笔谈卷三 [Dream pool essays: Addendum to the essays, vol 3]
34. *Yuhai juan shsii*玉海·卷十四 [Jade ocean, vol 14]
35. *Yuhai juan jiushiyii*玉海·卷九十一 [Jade ocean, vol 91]
36. Chen Y 陈玉龙 (1982) Shuoxue zhan hongtu yimo ning zhenzhi 硕学展宏图, 遗墨凝真知 [The boundless knowledge in written texts] *Shehui kexue zhanxian*社会科学战线 [Frontier of social science] (1)
37. Hu B胡邦波 (1988) Jingding jiankangzhi he zhizheng jinling xinzhi zhong de ditu chutan 景定《建康志》和至正《金陵新志》中的地图初探 [Research on the maps in the *Local Gazetteers of Nanjing in the Jingding Reign* and the *New Local Gazetteer of Nanjing in the Zhizheng Reign*]. *Ziran kexueshi yanjiu*自然科学史研究 [Studies on the history of natural science] (1)
38. Yang W杨文衡 (1990) *Chang'an zhitu de tedian he shuiping*《长安志图》的特点和水平 [Characteristics and quality of the *Illustrated Records of Chang'an*]. Cultural Relics Press文物出版社, Beijing
39. Zhao Y 赵彦卫 *Yunlu manchao: juanba* 云麓漫钞·卷八 [Miscellaneous notes of historical events, vol 8]
40. Wang Q 汪前进 (1986) Dui pingjiangtu, jingjiang fucheng tu, chang'an chengtu de yanjiu 对"平江图""静江府城图""长安成图"的研究 [A study of the *Pingjiang Map*, *Jingjiang Prefectural Map* and *Chang'an Map*], Master's thesis
41. Cao W 曹婉如 (1982) Xiancun zuizhao de yibu lishi dituji—lidai dili zhizhang tu 现存最早的一部历史地图集—历代地理指掌图 [The earliest extant historical atlas – easy-to-use geographical maps through the ages]. Kexueshi jikan 科学史集刊 [Collected papers on the history of science] (10)
42. Chen Z陈正祥 (1979) *Zhongguo ditu xue shi*中国地图学史 [The history of Chinese cartography]. The Commercial Press, Hong Kong Branch商务印书馆(香港)有限公司, Hong Kong, p 4
43. Yang W杨文衡 (1990) Chang'an zhitu de tedian he shuiping 长安志图的特定和水平 [Features and quality of *the Illustrated Records of Chang'an*]. In: *Zhongguo gudai dituji: Diyiji* 中国古代地图集: 第一辑 [Ancient Chinese atlas I]. Cultural Relics Press文物出版社, Beijing

44. Zheng X郑锡煌 (1982) Beisong shike jiuyu shouling tu 北宋石刻《九域守令图》[The stone engraved with the *Map of Administrative Units in the Nine Regions of the Northern Song Dynasty*]. *Ziran kexueshi yanjiu*自然科学史研究 [Stud Hist Nat Sci] (2)
45. Cao W曹婉如 (1963) Huayi tu he Yuji tu de jige wenti 华夷图和禹迹图的几个问题 [Several questions on the *Map of China and Barbarian Countries* and the *Map of Tracks of Yu*]. *Kexueshi jikan*科学史集刊 [Collected papers on the history of science] (6)
46. Qian Y钱镛 (1959) Pingjia tu bei 平江图碑 [The stele engraved with the *Map of Pingjiang Prefecture*]. *Wenwu*文物 [Relics] (2)
47. Yu S俞绳方 (1980) Woguo gudai chengshi guihua de yige jiezuo——song pingjiang tu 我国古代城市规划的一个杰作——宋《平江图》[A masterpiece of urban planning in ancient China—*Map of the Pingjiang Prefecture* in the Song dynasty]. *Jianzhu xuebao*建筑学报 [Journal of architecture] (1)
48. *Shiji: Qinshihuang benji* 史记·秦始皇本纪 [Records of the grand historian: Imperial biography of Qin Shi Huang]
49. *Songshu: Xiezhuang zhuan* 宋书·谢庄传 [Book of the Song: Biography of Xie Zhuang]
50. *Mengxi bitan juan ershiwu*梦溪笔谈·卷二十五 [The dream pool essays, vol 25]
51. Joseph N 李约瑟 (1975) *Zhongguo kexue jishushi (zhongyi ben): diywu juan diyifence*中国科学技术史 (中译本): 第五卷第一分册 [Science and civilisation in China (trans), Vol 5, Book 1]. Science Press科学出版社, Beijing, p 220
52. Zhu Xi 朱熹 *Zhuzi wenji: da lijizhang shu* 朱子文集·答李季章书 [Collected works of Zhuzi: Responding to Li Jizhang]
53. Zhi W支伟成 (1986) *Qingdai puxue dashi liezhuan: xiace* 清代朴学大师列传: 下册 [Biographies of masters of evidential scholarship in the Qing dynasty II]. Yuelu Publishing House, 岳麓书社, Changsha, p 461
54. *Songshi: Hequ zhi* 宋史·河渠志 [History of the Song: Treatise on rivers and canals]
55. Huang X 黄象曦 (Qing dynasty) *Wujiang shuikao zengji: juan'er* 吴江水考增辑·卷二 [Supplement to the *Research on the Water of the Wu River*, vol 2]
56. Wang X 王象之 *Yudi jisheng: juan yibai qishisi:Fuzhou* 舆地纪胜·卷一百七十四: 福州 [Record of the landscape of the imperial domain, vol 174, Fuzhou]
57. Yang W杨文衡 (1988) Fang Chengda de dilixue chengjiu 范成大的地理学成就 [Fan Chengda's great achievements in geography]. *Ziran kexueshi yanjiu*自然科学史研究 [Stud Hist Nat Sci] (2)
58. Dao G 道光 *Kuaiji xianzhi gao: juanliu*会稽县志稿·卷六 [Local chronicle of Kuaiji County, vol 6]
59. *Mengxi bitan: juan ershisi*梦溪笔谈·卷二十四 [The dream pool essays, vol 24]
60. Studies of Geoscience History at the Research Centre of Natural Science History, China Academy of Science 中国科学院自然科学史研究所地学史组 (1984) *Zhongguo gudai dili xueshi*中国古代地理学史 [A history of ancient Chinese geography]. Beijing: China Science Publishing & Media Ltd 科学出版社, pp 137–138
61. *Yuanshi: Dilizhi: Heyuan fulu* 元史·地理志·河源附录 [History of the Yuan: Treatise of geography: River origins appendix]
62. Huang S黄盛璋 (1956) Zailun huanghe heyuan wenti 再论黄河河源问题 [Further study on the origin of the Yellow River]. *Dili xuebao*地理学报 [Acta Geographica Sinica] 22 (1)
63. *Liangshimen ji: juanyi*梁石门集·卷一 [Collected essays of Liang Shimen, vol 1]
64. *Congshu jicheng chubian*丛书集成初编 [A preliminary compilation of the integrated series]
65. *Yishan wenji: juanba*一山文集·卷八 [Collected works of Yishan, vol 8]
66. *Mengxi bitan· juan ershisi*梦溪笔谈·卷二十四 [The dream pool essays, vol 24]
67. Zhu K 竺可桢 (1926) Beisong Shen Kuo duiyu dixue zhi gongxian yu jishu 北宋沈括对于地学之贡献与记述 [The contribution of Shen Kuo to geoscience and his accounts]. *Kexue* 科学 [Science] (11)
68. Gao Y高泳源 (1954) Woguo gudai dui yixie ziran dili xianxiang de renshi. 我国古代对一些自然地理现象的认识 [The understanding of some natural geographical phenomena in ancient China]. *Dili zhishi*地理知识 [Knowledge of geography] (7)

69. *Mengxi bitan: Bu bitan: juan'er* 梦溪笔谈·补笔谈·卷二 [Dream pool essays: Addendum to the essays, vol 2]
70. *Haichao jishuo juanshang*海潮辑说·卷上 [Collected writings on tides]
71. Zhongguo gudai chaoxi shiliao zhengli yanjiuzu 中国古代潮汐史料整理研究组 [Research group of the historical documentation on tides in ancient China] (1980) *Zhongguo gudai chaoxi lunzhu xuanyi* 中国古代潮汐论著选译 [Selected translation of ancient Chinese literature on tides]. Science Press 北京科学出版社, Beijing
72. Yao K 姚宽 *Xixi congyu: juanshang* 西溪丛语·卷上 [Miscellaneous notes from the West Stream I]
73. *Haitang lu: juan jiushi* 海塘录·卷九十 [Records of seas and rivers, vol 90]
74. Joseph N 李约瑟 (1975) *Zhongguo kexue jishu shi (zhongyiben): disijuan*中国科学技术史(中译本): 第四卷 [Science and Civilisation in China (trans), vol 4] China Science Publishing & Media科学出版社, Beijing, pp 757–786
75. Hu G 胡广 (Ming dyansty) *Xingli daquan shu: juan ershiqi*性理大全书·卷二十七 [A great compendium of hman nature and principles, vol 27]
76. *Zhuzi quanshu: juan wushi* 朱子全书·卷五十 [Completed works of Zhu Xi, vol 50]
77. *Shushu jiuzhang: juan'er*数书九章·卷二 [The mathematical treatise in nine sections, vol 2]
78. *Mengxi bitan: juan ershliui*梦溪笔谈·卷二十六 [The dream pool essays, vol 26]
79. *Zhuzi yulei: juan'er* 朱子语类·卷二 [A collection of Master Zhu's discourses, vol 2]
80. *Wuchuan lu* 吴船录 [Record of a boat trip to Wu]
81. *Mengxi bitan: juan ershiyi* 梦溪笔谈·卷二十一 [The dream pool essays, Vol 21]. According to the *Xu zizhi tongjian changbian* 续资治通鉴长编 [Sequel to the *Comprehensive Mirror in Aid of Governance*], the time should be the 10th year of the Xining reign in the Song dynasty (1077)
82. *Su Dongpo quanji: juan ershiyi*苏东坡全集·卷二十一 [Complete works of Su Dongpo, vol 21]
83. *Shuofu: juanwu* 说郛·卷五 [Persuasion of the suburbs, vol 5]
84. Wang Q王其亨 (1987) "Hunyun yuanjue si ta ji gudai houfengwu shiwu" 浑源圆觉寺塔及古代候风乌实物 [Yuanjue Temple Pagoda and the ancient wind-finding instrument]. *Wenwu* 文物 [Relics] (11)
85. *Mengxi bitan: juan ershiyi* 梦溪笔谈·卷二十一 [The dream pool essays, vol 21]
86. Office of Historical Studies on the Song Dynasty at Hangzhou University杭州大学宋史研究室 (1985) *Shen Kuo yanjiu* 沈括研究 [Research on Shen Kuo]. Zhejiang People's Publishing House浙江人民出版社, Hangzhou, p 101
87. Yao N姚楠 (1981) *Daoyi zhilue jiaoshi*岛夷志略校释 [A revised and annotated version of the *Description of the Barbarians of the Isles*]. Chinese Publishing House 中华书局, Beijing
88. Chen L陈连庆 (1985) *Dade nanhai zhi yanjiu*《大德南海志》研究 [Research on *the Records of the South China Sea in the Dade reign*]. Fujian People's Publishing House福建人民出版社, Fuzhou
89. Office of Historical Studies on the Song Dynasty at Hangzhou University杭州大学宋史研究室 (1985) *Shen Kuo yanjiu* 沈括研究 [Research on Shen Kuo]. Zhejiang People's Publishing House浙江人民出版社, Hangzhou, p 168
90. *Yuanshi: Tianwen zhi yi* 元史: 天文志一 [History of the Yuan: Records of astronomy I]
91. *Mengxi bitan: juan ershisi* 梦溪笔谈·卷二十四 [The dream pool essays, vol 24]
92. *Zhuzi yulei: juan jiushisi*朱子语类·卷九十四 [A collection of Master Zhu's discourses, vol 94]
93. 尹赞勋 (1984) *Zhongguo gushengwu xue zhi genmiao*中国古生物学之根苗 [The origin and budding of Chinese paleontology] Beijing: China Science Publishing & Media科学出版社, p 112
94. *Juanyou zalu* 倦游杂录 [Miscellaneous records of casual travels], recorded in *Shuofu: juan sanshisan, wanweishan tangben*说郛·卷三十三, 宛委山堂本 [Persuasion of the suburbs, Vol 33], the edition of Wanweishan Hall
95. *Bencao gangmu: juanshi*本草纲目·卷十 [Compendium of materia medica, vol 10]
96. *Yunlin shipu: juanshang*云林石谱·卷上 [Stone catalogue of the cloudy forest I]

97. *Zhuzi quanshu: juan sishijiu* 朱子全书·卷四十九 [Completed works of Zhu Xi, vol 49]
98. Zhang Shizheng 张师正 (Song dynasty) *Juanyou zalu* 倦游杂录 [Miscellaneous records of casual travels]
99. *Huashi*化石 [Fossils] 1978 (2)
100. *Yunlin shipu: juanzhong*云林石谱·卷中 [Stone catalogue of the cloudy forest II]
101. *Zhibuzu zhai congshu di ershijiu han*知不足斋丛书·第二十九函 [Book series from the Zhibuzu Study, Slipcase 29]
102. *Kexue dazhong*科学大众 [Popular science] 1957(2)
103. *Bencao gangmu juan sanshiqii*本草纲目·卷三十七 [Compendium of materia medica, vol 37]
104. *Wenwu*文物 [Relics] (1973), 4:59
105. Wu H 吴惠群 (1994) *Shiyong baoyu shixue*实用宝玉石学 [Practical gemology]. Higher Education Press高等教育出版社, Beijing, p 183
106. Wu H 吴惠群 (1994) *Shiyong baoyu shixue*实用宝玉石学 [Practical gemology]. Higher Education Press高等教育出版社, Beijing, p 194
107. *Bencao yanyi: juansi*本草衍义·卷四 [Extension of the pharmacopoeia, vol 4]
108. Lu X陆学善 (1984) *Zhongguo jingtu xue shiliao duoshi*中国晶体学史料掇拾 [Gleanings from the historical records on Chinese crystallography]. Shanghai Science and Technology Press上海科技出版社, Shanghai
109. *Mengxi bitan: Bu bitan: juansan* 梦溪笔谈·补笔谈·卷三 [The dream pool essays: Addendum to the essays, vol 3]
110. *Chongxiu zhenghe zhenglei bencao: juawu* 重修政和证类本草·卷五 [Revised Zhenghe materia medica, vol 5]
111. *Bencao gangmu: juanba*本草纲目·卷八 [Compendium of materia medica, vol 8]
112. Su L苏良赫, Li Z李仲均 (1982) "Zhongguo guji Zhong youguan shimian de jizai" 中国古籍中有关石棉的记载 [Records of asbestos in ancient Chinese literature]. *Diqiu kexue*地球科学 [Earth Sci] (1)
113. Zhou S周始民 (1977) Woguo gudai de yixie kuangwu jianding zhishi 我国古代的一些矿物鉴定知识 [Some knowledge of mineral identification in ancient China]. *Huaxue tongbao*化学通报 [Chem Bull] (2)
114. Yang W杨文衡 (1985) Shishu Yunlin shipu de kexue jiazhi 试述《云林石谱》的科学价值 [A preliminary exploration of the scientific value of the *Yunlin Stone Catalogue*]. In: *Kejishi wenji: di shisi ji* 科技史文集: 第14辑 [Collected papers on the history of science, vol 14]. Shanghai Science and Technology Press上海科技出版社, Shanghai, pp 169–178
115. Joseph N 李约瑟 (1976) *Zhongguo kexue jishu shi zhongyi ben: diwujuan di'er fence*中国科学技术史 (中译本) 第五卷第二分册 [Science and Civilisation in China (trans), vol 5, Book 2]. China Science Publishing & Media科学出版社, Beijing, p 390
116. *Shiya: juanshang: Shuijing* 石雅·卷上: 水精 [Elegant stones I: Crystals]
117. Zhen S甄朔南 (1982) Woguo gudai dui huashi de renshi 我国古代对化石的认识 [Knowledge of fossils in ancient China]. *Zhongguo keji shiliao*中国科技史料 [Hist Data Chin Sci Technol] (3)
118. *Chouchi biji: juanxia* 仇池笔记·卷下 [Brush records of Chouchi II], p 244
119. Wu H 吴惠群 (1994) *Shiyong baoyu shixue*实用宝玉石学 [Practical gemology]. Higher Education Press高等教育出版社, Beijing, p 195
120. *Jiashi jiuwen xia* 家世旧闻·下 [Family history and old anecdotes II]. Beijing: Zhonghua Book Company中华书局, p 225

Chapter 8
Geoscience in the Ming Dynasty

8.1 The Impact of the Social Environment on Geoscience

In the Ming dynasty, social environment influenced geoscience in the following aspects.

8.1.1 Political Control of Scientific Thought

Ever since the founding of the Ming dynasty, Emperor Zhu Yuanzhang had taken measures to establish ideological unity. He advocated Neo-Confucianism (*li xue*) and introduced an imperial civil examination system to select government officials. The scope of the examination was set out in the Confucian "Four Books and Five Classics" with Zhu Xi's annotations. The examination candidates had to write essays according to the designated annotations and they were required to write in the style of the ancients. By the middle period of the Ming dynasty, this kind of essay with its specific format had gained the name of *bagu wen* ("eight-part essay"). This required examinees' total submission to the rules and regulations prescribed by the imperial family. No independent and up-to-date thoughts were allowed. People's thinking was hedged in by Neo-Confucianism. To enact stricter ideological control, Zhu Yuanzhang also implemented many literary inquisitions, which meant that if certain words in a piece of writing appeared suspect, the writer would be executed. When Zhu Di (Emperor Chengzu of the Ming) ascended to the throne, he carried forward the ideological control of Zhu Yuanzhang and ordered Hu Guang and other officials to compile the 121 volumes of the *Wujing daquan* ("Great Compendium of Five Classics"), the 30 volumes of the *Sishu daquan* ("Great Compendium of Four Books") and the 70 volumes of the *Sishu daquan* ("Great Compendium of Human Nature and Principle"). He supported the Cheng-Zhu school of Neo-Confucianism with an enormous amount of propaganda and scholars were required to "uphold the

W. Yang and Q. Yang, *A History of Ancient Chinese Geoscience*, Understanding China,
https://doi.org/10.1007/978-981-97-4457-2_8

traditional ideas of the preceding Confucianists with no revision at all", which gave rise to the dominant position of the Cheng-Zhu school all over the empire. This kind of ideological control and academic atmosphere was highly disadvantageous to the development of science and that of geoscience in particular. Without innovation and new thoughts, there could be no advancement of science. We should also be aware that the Cheng-Zhu school of Neo-Confucianism was later challenged by Wang Yangming's school of Philosophy of Mind and the school of Pragmatism which came into being after the Song dynasty. In particular, under the influence of Pragmatism, many intellectuals abandoned the Cheng-Zhu school of Neo-Confucianism and stepped outside their study and home to seek out nature and the mysteries of the universe. Some of them achieved universally acknowledged successes in science, such as Wang Shixing, Xu Guangqi, Li Shizhen, Song Yingxing, and Xu Xiake, etc.

8.1.2 Zheng He's Voyages

From the second year of Emperor Yongle's reign (1405) to the eighth year of Emperor Xuande's reign (1433), Zheng He led his fleet to the Western Seas seven times under the instruction of the imperial government, which is not only a remarkable feat in China's history of marine navigation, but also a major event in the world history of marine navigation. Zheng He's voyages mainly influenced four aspects of geoscience. First, Zhen He opened up new sea routes and made contributions to China's maritime transport. Second, the *Zhenghe hanghai tu* ("Chart of Zheng He's Voyages"), which was drawn based Zheng He's navigation routes, enriched the body of knowledge in the fields of cartography, toponymy and geomorphology in the Ming dynasty. Third, four books written by members of Zheng He's entourage enlarged China's range of vision and knowledge in the field of geography. Finally, Zheng He's contact with other countries promoted exchanges of geographical and geological knowledge and goods and materials between China and other Afro-Asian countries. Meanwhile, it also widened the scope of Chinese people's geographical and geological knowledge and improved the material life of people during the Ming dynasty.

8.1.3 Resistance to Foreign Invasions Leading to Developments in Borderland Geography

From the 2nd year of Emperor Hongwu's reign (1369), Japanese pirates kept on looting, burning and killing all over the coastal areas of Shandong, Zhejiang, Fujian and Guangzhou, and they even stormed and captured many cities during an entire year. Invasions by the Japanese were most prevalent under Emperor Jiajing's reign, so the Ming imperial government had to launch a number of anti-Japanese battles. During wartime, books on strategies of war like the *Chouhai tubian* ("Illustrated

Book for Coastal Defence") appeared. This book, written about military geography and military maps, served as a significant basis for the defence of coastal areas and added rich content to borderland geography.

Apart from coastal defence, frontier defence was also a challenging mission for the Ming. In the north, there were border raids carried out by Wala and Tatar Mongolians. In the southwest, there was Si Renfa's rebellion on the China-Myanmar border of Yunnan province. In the northeast, there was the rise of the Jurchen tribes. They established the Later Jin dynasty, which eventually led to the downfall of the Ming dynasty. Through these frontier wars and frontier defence actions, a series of works on frontier geography and many frontier maps were produced, such as the *Jiubian tushuo* ("Records and Maps of the Nine Garrison Zones") and the *Xinan yifeng tuji* ("Records of Southwestern Social Customs"), which added to the understanding of geography in the Ming dynasty.

8.1.4 The Rise of Daoism

The Ming dynasty had had a close relationship with Taoism ever since Zhu Yuanzhang became emperor. Emperor Chengzu of Ming (Zhu Di) was a strong supporter of large-scale construction in the Wudang Mountains and made them a Taoist shrine to Taoism. And he also instructed officials to compile the Taoist Canon, so the practice of Taoism was vigorously implemented for a while. Emperor Shizong of Ming (Zhu Houcong) was a strong believer in the art of immortality as explained by the celestial beings and the Taoist priests, and his death was ultimately caused by consuming alchemists' elixirs. The Prince of Ning (Zhu Quan) (1390–1448) wrote books concerning "*Waidan* (a school of Daoist alchemy)", such as *Gengxin yuce* ("A Jade Book of Gengxin [Gold]"), *Zaowu qianchui* ("Nature as the Cultivating Force"), *Qiankun miwen* ("Rare Skills between Heaven and Earth") and *Qiankun shengyi* ("Essence of Living Things between Heaven and Earth"), and one relating to geography named the *Yiyu* tuzhi ("The Illustrated Record of Strange Countries"), but all these books are now lost. Only a few pieces of writing about minerals in *Genxin yuce* have been handed down as a part of the *Bencao gangmu* ("Compendium of Materia Medica"). The *Yiyu tuzhi* recorded the geography of foreign countries. Chen Lian wrote the *Luofu zhi* ("Records of Luofu Mountain"). Zha Zhilong wrote the *Taishi* ("History of Mount Tai"). All the Taoists mentioned contributed to the literature on geoscience.

8.1.5 Management of the Yellow River and the Grand Canal

The Ming dynasty harnessed the Yellow River and the Grand Canal by employing a large amount of manpower, money and material resources. And they achieved satisfactory results. Leading figures included Xu Youzhen, Bai Ang, Liu Daxia, Liu

Tianhe, Pan Jixun, and Wan Gong, etc. They had their own methods of harnessing rivers and had each written works on water conservancy. For example, Liu Tianhe wrote six volumes of the *Wenshui ji* ("Water Conservancy of the Yellow River and the Grand Canal"). Wan Gong produced two volumes of the *Zhishui quandi* ("Ways to Deal with the Yellow River and the Grand Canal"). Pan Jixun authored 14 volumes of the *Hefang yilan* ("A Survey of the Workings of the River Defences"). These theories and texts would have a great impact on water conservancy in the Qing dynasty.

8.1.6 Works Listing the Chinese Materia Medica

There were not many official compilations of *materia medica* during the Ming dynasty. One exception was the work of Liu Wentai, the deputy director of the imperial hospital, who wrote 42 volumes of the *Bencao pinhui jingya* ("Essentials of Materia Medica Distinctions"), during the reign of Emperor Hongzhi (Zhu Youcheng). This book classifies herbal medicines into three grades—superior, intermediate and inferior, and ten parts—jade, grass, wood and so on. It contains 1,815 types of herbal medicines with coloured illustrations but the original was not formally published. The handwritten text was not found until a fire broke out in the Palace Museum in 1923. And it finally began to circulate when the publisher Commercial Press printed it in 1936. Most works in this field were written by individuals, such as Zhu Su's *Jiuhuang bencao* ("Materia Medica for Famine Relief"), Xu Yanchun's *Bencao fahui* ("Elaboration on Materia Medica"), Wang Lun's *Bencao jiyao* ("Summarisation of Materia Medica"), Chen Jiamo's *Bencao mengquan* ("Enlightening Primer of Materia Medica"), Lao Mao's *Diannan bencao* ("Materia Medica of South Yunnan"), Li Shizhen's *Bencao gangmu* ("Compendium of Material Medica"), etc. Among them, the *Bencao gangmu* is the most relevant to geoscience for it contains descriptions of medicines made of metals and stones that pertain to the study of minerals and rocks. It is a repository of much vital information about minerals and rocks across the ages.

8.1.7 Development of the Mining Industry

In the Ming dynasty, except for Emperor Hongwu's suppression of the mining industry, other emperors' reigns witnessed different degrees of development in the mining industry. During Emperor Wanli's reign in the late Ming dynasty in particular, the development reached a point where "all places without exception were used for mining". The reason behind the promotion of the mining industry was to "excavate and smelt extensively in order to mint coins and forge tools". Gold, silver and copper coins were all indispensable as the currency used in transactions. Moreover, steel was the raw material of production tools. As a result, the excavation of metallic minerals was a priority. With the prosperity of the mining industry, people acquired a deep understanding of minerals and rocks, which promoted the development of mineralogy. The *Tiangong kaiwu* ("Exploitation of the Works of Nature") is representative of the writing in this field.

8.1.8 Visits of Western Missionaries

At the end of the sixteenth century, there existed a movement called "Learning from the West" as the Western missionaries from the Society of Jesus arrived in China. These missionaries attracted the attention of Chinese scholar-officials and intellectuals through their knowledge of Western science and technology, and thereby achieved their goal of preaching. Matteo Ricci is the most renowned of these missionaries. He arrived in Macao in 1582, went to Zhaoqin in Guangdong province in 1583, left for Beijing in 1598, settled down and did missionary work there from 1601 and died there in 1610. He incorporated Western science and technology into his preaching and his efforts promoted the spread of Western mathematics, astronomy and geography across China. The introduction of the world map, the spherical Earth theory and knowledge about world geography were especially useful, as this improved the Chinese people's understanding of geography, broadened their horizons and changed their thoughts. To a certain extent, it changed the direction of geoscience in late Ming and early Qing China.

8.1.9 Landscape Engineering and Geoscience

There were not only imperial gardens, but also the many private gardens of senior officers, eunuchs in power and members of the imperial family during the Ming dynasty. When it came to the construction of royal gardens, artistic beauty had to be taken into consideration in the design. Normally, man-made scenes were created which represented Nature, Heaven and Earth and the Universe. In the actual construction process, the norm was that rocks should be used to imitate mountains, wood to imitate various forms of houses, ditches to imitate lakes and flowers and trees to imitate landscapes. These were all indebted to geoscience. In turn, building gardens would also boost people's exploration of and accumulation of geoscientific knowledge. For example, Ji Cheng's *Yuanzhi* ("Craft of Gardens") includes 15 types of rocks for garden construction in the "Rock Selection" part. It not only describes the origins and properties of the rocks, but also the effects of using them to make gardens. The Section "Liuhe Rock" says:

> Liuhe Rock is a kind of small and irregular agate produced in sandy soil or alongside a river. White varieties exist which have fine stripes in five colours and which are as large as a fist. There also exist pure-coloured ones of five colours. This rock is the epitome of smoothness and translucence, and the colourful ones are preferred. Their appearance resembles brocade

when used in paved roads; and they seem natural and flawless when placed in man-made ravines.

8.2 The Seven Voyages of Zheng He and Texts on Overseas Geography

8.2.1 Significance of Zheng He's Expeditions

In the early fifteenth century, Zheng He made seven voyages to the Western Seas, which was not only a "great event in the early Ming dynasty" [1], but also a notable feat in the history of worldwide ocean exploration. Zheng He's voyages began 87 years before Christopher Columbus' arrival in the Americas after sailing across the Atlantic Ocean (1492), and 92 years earlier than Vasco da Gama's arrival at Calicut in India after sailing around the Cape of Good Hope (1497), which proves Zheng He was a pioneer in the history of worldwide ocean exploration. Zheng He's ocean explorations continued for 29 years, which was unmatched by any other navigator.

Zheng He (1371–1433), originally surnamed "Ma" and born in Kunyang (now called Jinning), Yunnan, into a Hui family, was later given the surname "Zheng" by Emperor Yongle. His family was not very well-off and his father died of a disease when he was 11 years old. His hometown was in turmoil and enduring the chaos of war, so he was forced to join Zhu Yuanzhang's army in Yunnan. Afterwards, he left his hometown for Peiping (now called Beijing) with the army and served as a servant to the Prince of Yan (Zhu Di). Zheng He was trusted and highly regarded by Zhu Di for his intelligence and loyalty and became his trusted deputy. In order to accede to the throne, Zhu Di launched the Jingnan Campaign in which Zheng He rendered outstanding service. After Zhu Di secured the throne, Zheng He, at the age of 34, was appointed chief eunuch of the Directorate of Palace Eunuchs which was a Standard Class, Rank Four position (in the nine-rank system of ancient China) [1]. In May 1405 (the third year of Emperor Yongle's rule), Zheng He was appointed as the eunuch commander-in-chief by Emperor Yongle and led a large fleet to the Western Seas. The so-called "Western Seas" at that time signified the areas west of the South China Sea and the Indian Ocean today. In the 29-year period from the 3rd year of Emperor Yongle's reign to the 8th year of Emperor Xuande's reign (1433), Zheng He made a total of seven epic voyages to the Western Seas and reached as far as the region of the east coast of Africa south of the equator. The dates and locations of Zheng He's seven voyages are described below.

The first voyage: from the winter of 1405 to September 1407, after setting off from Changle, Fujian, Zheng He's fleet visited Champa (in today's South Vietnam), Siam (now called Thailand), Java, Palembang, Malacca, Aru (in the northwest of the island of Sumatra), Sumatra, Nakur (on North Sumatra Island), Lide (on Northwest Sumatra Island), Lambri (on Northwest Sumatra Island), Bengal (Bangladesh), Ceylon (now called Sri Lanka), Liu-shan (the Maldive and Laccadive Islands), Kochi (in today's

India), Calicut (in today's India), Hormuz, Dhofar, Mecca and Aden [2], and then returned home.

The second voyage: from October 1407 to the summer of 1409, Zheng He's fleet visited Champa, Siam, Java, Malacca, Sumatra, Aru, Lambri, Ceylon, Cail (in the southwest of the Indian Peninsula), Cape Comorin (in today's India), Kollam (in today's India), Kochi, Calicut, Ahmedabad (in today's India), etc. [3].

The third voyage: from December 1409 to June 1411, Zheng He's fleet visited Champa, the Vulture Peak, Condore, Panduranga (in today's South Vietnam), Chenla (now called Cambodia), Siam, Karimata (now between Kalimantan and Benga), Gelam (on the west coast of Kalimantan), Java, Palembang, Jungala (now called Surabaya, a city on the island of Java), Timor, Malacca, Belitung (now in Indonesia), Pahang (now in Malaysia), Aur Island (on the southeast coast of the Malay Peninsula), Langkasuka (on the west coast of Sumatra Island), Sembilan (on the coast of Sumatra Island), Aru, Tamiang (now in Sumatra), Sumatra, Nakur (Nagur), Longyu Island ([probably Rondo/Bras Island] in the northwest of the island of Sumatra), the Cui-lan Islands (now a group of islands which form part of the North Nicobar Islands), Ceylon, the Maldives, Kollam, Kochi, Bengal, Brava, Juba, Mogadishu, Aden, Cisa ([probably Ras Sharwein or Ras Isa in Yemen] on the coast of the Red Sea), Dhofar, Hormuz,, Mecca, Ryukyu, the Batan Islands (in today's Philippine Archipelago), Brunei (on the island of Kalimantan), and Sulu (in today's Sulu Archipelago) [4]. According to the *Xingcha shenglan* ("Overall Survey of Star Raft"), apart from the regions mentioned above, Calicut and Dragon's Teeth Gate (on the southeast of the Strait of Malacca) were also visited by the fleet [5].

The fourth voyage: from the winter of 1413 to July 1415, Zheng He's fleet visited Champa, Pahang, Kelantan, Java, Palembang, Wu-yu [the Five Islands, or arguably the Babuyan Islands], Malacca, Sumatra, Aru, Lambri, Cail, Kochi, Calicut, Hormuz, Ceylon, the Maldives, Mogadishu, Brava, Malindi, Bilad (referring to the Island of Mozambique), Slan (referring to Sofala on the Island of Mozambique) [6]. This was the longest of Zheng He's seven voyages, and it opened up a new sea route from the Maldives to the east coast of Africa across the Indian Ocean and considerably shortened the sailing distance between China and all the ports of South Somalia. This marked the start of a new era in China's ocean exploration.

The fifth voyage: from the winter of 1417 to July 1419, Zheng He's fleet visited Champa, Pahang, Java, Palembang, Malacca, Sumatra, Lambri, the Cui-lan Islands, Ceylon, the Maldives, Kochi, Calicut, Hormuz, Ra's Sharwayn, Dhofar, Aden, Cisa, Mogadishu, Brava, Malindi, Juba and Mombasa.

The sixth voyage: from the spring of 1421 to August 1422, Zheng He's fleet visited Champa, Chenla, Siam, Pahang, Malacca, Sumatra, Lambri, the Cui-lan Islands, Chattogram, Bengal, Soli (now around Madras in India), Cape Comorin, Ahmedabad, Kochi, Calicut, Dhofar, Aden, Mogadishu, Brava, Juba, Malindi and Mombasa.

The seventh voyage: from the winter of 1431 to July 1433, Zheng He's fleet visited Champa, Surabaya (in Java), Palembang, Malacca, Lambri, Ceylon, Calicut, and Hormuz, etc.

At its largest, Zheng He's fleet consisted of 63 treasure ships and ships for transporting horses, food and the workforce, and warships of different sizes [7]. There were two sizes of treasure ships -arge and medium. Large ones were 44.4 *zhang* (about 148 m) in length, and 18 *zhang* (about 60 m) in width; medium ones were 37 *zhang* (about 123.33 m) in length, and 15 *zhang* (about 50 m) in width. The large treasure ship had nine masts. The interior design was sumptuous while the exterior was incomparably colossal meaning that the sails could not be hoisted and the anchor and rudder could not be controlled unless two or three hundred people worked together. The fleet had a crew of over 27,000 men at most [1]. Among them were commanders, interpreters, foremen (*huozhang*), quartermasters, rudder operators (*duogong*), riggers, anchormen, seamen, helmsmen, ironsmiths, caulkers, artisans, compradors, pursers, errand boys, assistants, reservists, veteran sailors, grooms, servants, cooks, family members, doctors, etc. [8]. Only a fleet with such a clear division of labour, tight organisation and strict discipline could have so successfully completed the Seven Treasure Voyages to the Western Seas.

8.2.2 Navigation Logs and the Geography of the Outside World

(1) Navigation logs

Although the original records of Zheng He's voyages have been destroyed [9], the contents were collected and cited in later literature and published as the *Zi Baohuanchang kaichuan cong Longjiangguan chushui zhidi waiguo zhufan tu* ("Map of Voyages from the Big Ship Factory in Longjiangguan to Foreign Countries"), later shortened to the *Zhenghe hanghai tu* ("Zheng He's Nautical Chart"). Nowadays, the *Zhenghe hanghai tu* is included in the *Wubei zhi* ("A Record of Military Affairs") and there is a separate edition authored by Xiang Da, which was published by the Zhonghua Book Company in 1982. Huang Xingzeng from the Ming dynasty listed the book as having the same status as the *Xingcha shenglan* ("The Overall Survey of the Star Raft") and *Yingya shenglan* ("Overall Survey of the Ocean Shores") in his author's preface to his work *Xiyang chaogong dianlu* ("Tributes from the Western Countries").

The *Zhenghe hanghai tu* (Fig. 8.1) is the earliest extant and most detailed nautical chart in China, showing the sea routes, directions and sailing distances of the voyages. There are 109 sea routes overall, with indications of place names, city walls, government offices, temples, towers, bridges, bays, reefs and beaches. The charts include more than 500 place names, of which nearly three-fifths are foreign place names, considerably more than those included in the *Lingwai daida* ("Notes from the Land Beyond the Passes"), *Zhufan zhi* ("Description of the Barbarian Nations") and *Daoyi zhilue* ("Descriptions of the Barbarians of the Isles") [10]. Mao Yuanyi, a scholar of the Ming dynasty, praised the *Hanghai tu* for being "detailed without exaggeration" [11]. Joseph Needham, a historian from England, also commended the charts for being "a true nautical chart" [12].

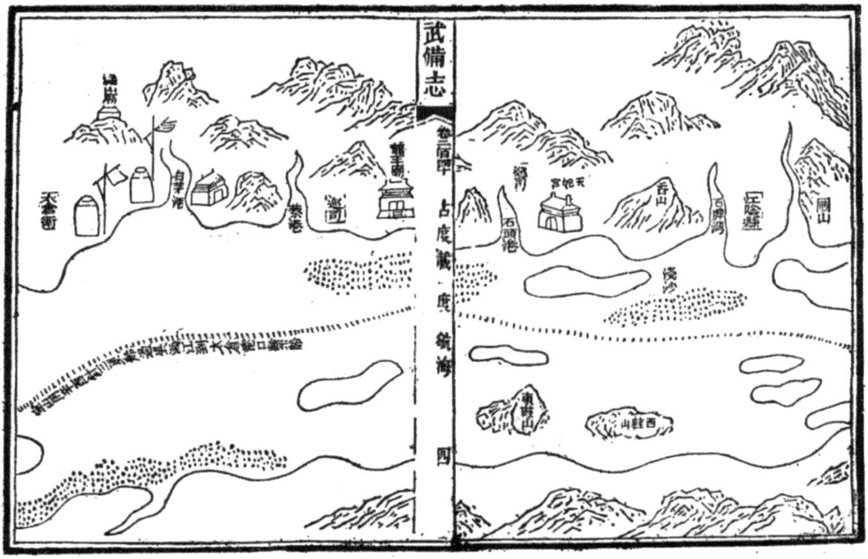

Fig. 8.1 Partial chart of Zheng He's voyages. Xiang [129]

(2) Geography of the outside world

To date, four geographical works about foreign places written by Zheng He's crew members are known of, i.e. Ma Huan's *Yingya shenglan*, Fei Xin's *Xingcha shenglan*, Gong Zhen's *Xiyang fanguo zhi* ("Records of the Foreign Countries in the Western Ocean") and Kuang Yu's *Huayi shenglan* ("Overall Survey of Chinese and Foreign Encounters"). The first three have survived until now while the last one has been lost.

The *Yingya shenglan* was written in 1416 and completed in a single volume. The author joined Zheng He's voyages on three occasions and served as an interpreter for the fleet. The book "encompasses both the positive and negative aspects of people with different origins, compares the differences and similarities between local customs and presents the knowledge accumulated regarding different local products and territorial structures, compiling all the above into one book in the correct fashion" [13]. The book is divided into different parts according to countries, and the schedule of sailing and the course of the ship from one country to another are usually recorded. A total of 20 countries are recorded, each one with a different treatment of the details about the location, climate, nationality, religion, customs, resources, costume, housing, commodity transactions, currency, culture, criminal law, calendric system, mythology, Zheng He's diplomatic activities and the situation of overseas Chinese people, etc.

The *Xingcha shenglan* was written in 1436, originally in two volumes, and some of the language was incoherent and offensive. Then it was adapted and recompiled by later scholars into four volumes with elegant and pithy language. There is another version, divided into two parts, which was revised and annotated by Feng Chengjun and published by the Commercial Press in 1936. The author joined four of Zheng He's voyages and, consequently, the book has a rich content, which includes sea routes, schedules, locations, social customs, natural resources, climate, calendric systems, buildings, languages, religions, nationalities, currencies, trade, mythologies, and Zheng He's diplomatic activities, etc.

The *Xiyang fanguo zhi* was written in 1434 and completed in a single volume. The author had experience of sailing with Zheng He. A total of 20 countries are in the record, and the structure and scope of the writing are basically identical to those of the *Yingya shenglan.* The book begins with three imperial edicts issued by Emperor Yongle and Emperor Xuande, which are an important resource for studying Zheng He's voyages to the Western Seas. The preface written by the author is also valuable in that it mentions important facts: crossing the oceans by means of the *Qianxing shu* (astronomical techniques); navigating with reference to a floating compass; shipmasters managing sailing according to sailing manuals and maps; the huge sails and rudders of treasure ships; the experience of saving fresh water. All these accounts are important materials for research into the history of ocean exploration in fifteenth century China.

The *Huayi shenglan* has been lost. The author was a doctor who joined three of Zheng He's voyages. In his spare time, he carefully observed the beautiful mountains and rivers of Western countries, recorded them one after another and included them in this atlas. However, only the preface written by Zhang Hong was handed down. According to the preface, the book recorded the locations, landscapes, social customs, natural resources and personages of Cochinchina, Champa, Java, Samboja, Malacca, Sumatra, Ceylon, Siam, Borneo and other countries, which gave the readers a sense of the "endless beauty of countless mountains and rivers" [14].

The works above broadened the geographical horizons of the Chinese people and enriched their knowledge of extraterritorial geography.

After Zheng He's voyages, some other geographical works about foreign places appeared, such as the *Xiyang chaogong dianlu* ("Tributes from the Western Countries"), *Haiyu* ("Talks about the Sea"), *Dongxiyang kao* ("On the [countries] in the Eastern and Western Oceans"), etc.

Huang Xingzeng's *Xiyang chaogong dianlu,* finished around the year 1520, consists of three volumes. It records the situation of 23 countries and regions in the South China Sea when Zheng He was voyaging to the Western Seas. Huang Xingzeng drew on a wide range of literature including the *Xingcha, Yingya* and *Zhenwei,* which he summarised in accordance with the norms, refined with standardised language and checked against the emperor's edicts [15]. The knowledge of *zhenwei* (the direction shown by the compass needle) recorded in this book is possibly cited from the book *Zhenwei bian.* The weakness of the book is that the author was not familiar with extraterritorial geography and, therefore, some place names and materials are often mistaken and misleading [16].

Huang Zhong's *Haiyu* comprises three volumes and was completed in the early years of Emperor Jiajing's rule. "In his late years, the author resigned from his official post and returned home. Instead, he devoted himself to researching the oceans and foreign merchant ships. Through investigation, he came to understand the landscapes, natural conditions and social customs of foreign places and compiled what he learnt in this collection [17]." The first volume "Social Customs" includes two items—Siam and Malacca, each containing the illustrated shipping routes towards the destination; the second volume "Natural Resources" is composed of 29 items which introduce the natural life and petroleum of foreign states; the third volume "Dangerous Road" records the perilous South China Sea area, which was then referred to as "ten thousand li [unit of length, equal to 500 m] of reefs" or "ten thousand li of shoals". Although the book is a collection of what the author heard from others, "the merchant seamen are in possession of timely news, so the facts shared by them seem reliable". Therefore, this work can be used to revise the mistakes in other historical records.

Zhang Xie's *Dongxiyang kao* consists of 12 volumes and was completed in 1616. This book was written as a guide to foreign trade in the late years of the Ming dynasty. The author agreed to write the book at the request of Tao Rong, the magistrate of Haicheng County. Later, he stopped his work halfway through because of personal circumstances. After that, Wang Qizong, the assistant governor of Zhangzhou Prefecture, asked him to resume writing. The materials in the book are either from historical and contemporary documents or are based on anecdotes. The author did not experience the events in the record himself. As a result, a few mistakes or incomprehensible accounts are hardly to be avoided in the book. But what is valuable about the book is that it records a relatively detailed history of the countries of Southeast Asia in the sixteenth century, especially the history of the western colonisers' plundering and enslavement of the Southeast Asian people. It describes the local specialties and the trading of overseas countries, the Western merchant ships entering China, and the types, quantities, specifications, tax systems and standards of import cargos. It also records navigational techniques and geographical knowledge, including information about the shipping range, shipping routes, navigation charts, depths of water, weather and tides, as well as about the environment, the people's living conditions, and the economic conditions and social customs of Taiwan, China. Therefore, the book does not merely focus on extraterritorial geography, but also concerns itself with frontier geography.

Moreover, Luo Yuejiong's *Xianbin lu* ("Records of Tributary Guests"), a work in eight volumes, also includes accounts of both extraterritorial and frontier geography. It documents the ethnic minorities in the frontier areas of China, as well as the history and geography of East Asian, Southeast Asian, Middle Asian and South Asian countries. The material in the work was taken from historical literature. The number of works cited amounts to 345 texts, some of which have been lost. Thanks to this work, part of their content has been preserved. But due to the author's unfamiliarity with extraterritorial geography, some mistakes are also included in the work.

In order to fight against Japanese pirates, Chinese scholars recorded the geography of Japan, such as Xue Jun's *Riben kaolue* ("A Brief Survey of Japan") and Zheng Ruozeng's *Riben tuzuan* ("Illustrated Treatise on Japan") during Emperor Jiajing's reign; Li Gongyan and Hao Jie's *Riben kao* ("A Survey of Japan"), Hou Jigao's *Riben fengtu ji* ("Geography and Customs of Japan") and Zheng Shungong's *Riben yijian* ("A Mirror on Japan") during Emperor Wanli's reign. Among them, the *Riben kaolue* was the first monograph to study Japan [18]. The most valuable part of this book can be found in the chapter entitled "A Brief on Translation" in which the author established a system for noting down Japanese pronunciation with Chinese characters. The system was used as a guide by frontier guards and the practice was then imitated by later scholars.

The *Riben yijian* comprises 16 volumes and was written in the 35th year of Emperor Jiajing's reign (1556). When the author went on a foreign mission from Guangzhou to Japan by order, he "enquired about all the landscapes and products along the trip and noted them all down" and compiled the notes into this book. The book is divided into three parts: the first part consists of the nine volumes of the "Qionghe huahai" (a discussion of the seas from an insignificant river), recording Japan's location, neighbouring countries, history, relations with China, terrains, headwaters, seasons, ethnic groups, names and surnames, monarch, officials, territory, cities, strategic passes and ferry crossings, bridges, roads, buildings, personages, natural resources, tools, laws, rites and music, costumes, social customs, religion, medicine, written language, books, transportation, etc., providing a very comprehensive geographical record of Japan. The second part comprises the three volumes of "Fuhai tujing", focusing on the illustrated shipping and transport routes between China and Japan. The third part consists of the four volumes of "Juedao xinbian" (the island retold), primarily presenting the maps and illustrated treatises of Japan's administrative divisions, city proper, dwelling houses, markets and islands, and secondarily recording Japan's landscapes, terrains, place names and natural resources. According to Zheng Shungong himself, the purpose of writing this book was to "obtain facts, seek lasting peace and stability in Southeastern Asia, and thereby pay back to the imperial court". Since the last tribute by Japan recorded in the book was in the 37th year of Emperor Jiajing's reign (1558), the book may also have been completed in 1558.

8.3 New Developments in Fangzhi

The further development of *fangzhi* (local gazetteers) in the Ming dynasty is reflected in three aspects: first, the number increased compared with in the Song and Yuan dynasties. Although there were 981 examples of *fangzhi* in the Song dynasty, most of them failed to survive. The number of *fangzhi* in the Ming dynasty, however, reached 990 [19]. Second, the scope of the local records broadened. Provincial annals, or *shengzhi*, of larger administrative divisions appeared. Even so, *shengzhi* did not really exist in the Ming dynasty as the administrative division equivalent to the provincial level was called the "Buzhengshi si": the provincial administration commission. Other names used for records of larger administrative divisions included *tujing* (illustrated classic), *shu* (book), *zhi* (record), *zongzhi* and *tongzhi* (general chronicle). Among these names, *tongzhi* was the most widely used. Third, the genres of *fangzhi* increased: the illustrated treatise of border affairs emerged. Several well-known Ming dyansty *fangzhi* are detailed below.

8.3.1 *Unification Chronicles*

In the third year of Emperor Hongwu's reign (1370), Emperor Taizu (Emperor Hongwu/Zhu Yuanzhang) considered that his territory was so vast that a record of it was necessary. Therefore, he ordered the Confucian scholars in the Imperial Academy to compile a unification chronicle of the Ming following the format of the *Yuan yitong zhi* ("Unification Chronicle of the Yuan"); however, the book was not finished. In September 1394, the *Huanyu tongqu* ("Major Routes of the Realm") in one volume compiled by the imperial government was published. This book records the names of all the posthouses and all the rules and principles of the local people starting from Liaodong to Guangxi, covering eight barbarian areas. The book does not have a preface or contents page, but not a single page is missing from the total of 145 pages. Nowadays there exists a Hongwu period version, copied by Mr Qu from the Changshu area, which is kept in Shanghai Library [20]. Under the reign of Emperor Jingtai (Zhu Qiyu) (1450–1457), the imperial court issued an edict to compile another book called the *Huanyu tongzhi* ("Comprehensive Records of the Universe"), which was completed in 119 volumes. After Emperor Yingzong (Zhu Qizhen) was restored to his throne, he ordered Li Xian and others to write another book called the *Daming yitong zhi* ("Unification Chronicle of the Great Yuan") in the second year of his reign (1458), based on the *Huanyu tongzhi*. In 1461, the book was finished comprising 90 volumes. There were some additions to the orginal during the reign of Emperor Wanli.

(1) Huanyu tongzhi

Chen Xun and others wrote *Huanyu tongzhi* in the seventh year of Emperor Jingtai's reign (1456). A Ming dynasty block-printed edition of it has been handed down. As a rarely seen local record, it is included in the *Xuanxiantang congshu xuji* ("Continuation of Collectanea from the Hall of Mysterious Contemplation"). It follows the narrative sequence of Beijing, Nanjing, and thirteen provincial administration commissions, recording the evolution of the administrative systems, the names of prefectures, the mountains and rivers, terrains, social customs, local products, cities, sacrificial canons, imperial mountains, palaces, temples, altars, political libraries, gardens, grand residences, government offices, the National Imperial Academy, schools, academies of classical learning, pavilions, roadside inns, halls, pools and residences, elevated pavilions, bridges, wells, passes, Buddhist and Taoist temples, ancestral temples, mausoleums, graves, historical sites, famous court officials, demotions, sojourns, personages, imperial examinations and poems of appreciation, reaching a total of 38 items. The book covers a wide range of topics, and the structure is quite concordant with existing rules. However, after the *Daming yitong zhi* was issued, this book became out of print; therefore, it is now a very rare text [21].

(2) *Daming yitong zhi*

The structure of the *Daming yitong zhi* followed the example of the *Dayuan yitong zhi*. It is divided into different parts in the narrative sequence of three capitals, Beijing,

Nanjing and Fengyang, and thirteen provincial administration commissions. Under the title of each prefecture, sub-prefecture and county are recorded more than ten items, including the evolution of the organisational system, the names of prefectures, the favourable terrains and landscapes, social customs, historical sites, and personages. And the last part contains accounts of foreign countries. The book is rather concise compared with *Dayuan yitong zhi*, for its total number of volumes amounts to less than a third of the number in that book. The main weakness of the book is that it contains many mistakes relating to historical events, and some sentences do not read smoothly. Even so, the book helps to preserve many of the resources of the Ming dynasty, and that is why it is still valuable. For example, the book records 372 caves in 95 prefectures, sub-prefectures or guard posts, among which 131 are accompanied by a detailed description. The descriptions of the caves include the geographical location, size, hydrology (subterranean rivers), water quality, utilisation of cave water, climate, fossils, minerals, stalactites, travertines, stalacto-stalagmites, stalagmites, stone curtains, uncultivable land, stone beds, cave corals, perlites, types of caves, animals in the caves and the utilisation of the caves. The account of caves in the book was unprecedentedly detailed. Therefore, Xu Xiake took it with him when he went on excursions and regarded it as a travel guidebook. Thus, the book was of great geographical value.

8.3.2 Local Chronicles

(1) The *Shanxi tongzhi* ("Comprehensive Gazetteer of Shanxi"), in 17 volumes, was written by Hu Mi during the reign of Emperor Chenghua (Zhu Jianshen) and revised by Li Kan. The existing edition is the block-printed edition of 1475 (the 11th year of Emperor Chenghua's reign). Hu Mi took four years to browse widely and collect a wide range of materials, and then he carefully selected and edited the useful ones as the content of the book. It gives a detailed account of the evolution of the organisational system, and the territory, landscape, strategic positions, local products, and number of households within the administrative division of Shanxi. Under the "number of households" item, the number of households in every prefecture, sub-prefecture and county, throughout the dynasties from the Han to the Ming, are all recorded in great detail, which makes the book a very valuable resource for the study of historical demography.

(2) The *Jiangxi tongzhi* ("Comprehensive Gazetteer of Jiangxi"), in 37 volumes, was written by Zhou Guang during the reign of Emperor Jiajing (Zhu Houcong) and revised by Lin Ting. Volumes One to Three contain the records of provincial administration commissions; these cover 13 items, namely, the evolution of the organisational system, strategic positions, cities, number of households, land tax, vassal states, the administration of military services, commissioners' offices, examination centres, temples, permanent officials, famous court official and lawbreakers. Volumes Four to Thirty-Seven contain records of all the prefectures, sub-prefectures and counties, covering 30 items. Among these are the 13

items previously listed, and some other items which include the names of prefectures, landscapes, social customs, funeral rites, schools, palaces, local products, posthouses, passes and bridges, good and bad omens, imperial tombs, historical sites, subject states, personages, upright and principled women, etc. The book has the following features: first, it created the new item of "lawbreaker", recounting the wrongdoings or malicious acts of local lawbreakers as a warning to posterity, just like "the biographies of wicked or oppressive officials" in the works recounting dynastic histories. In other words, it initiates a new practice of including both praise and denigration in local gazetteers. Secondly, it omitted the item about "art and literature" and relocated each tablet inscription to the item where it correctly belonged, which increased the practicability of district annals. Third, historical figures were included according to the chronology of the eras, rather than the category to which he or she belonged, which was a new method of ordering the historical accounts of famous figures [22]. There is much to be learned from the innovative approach of the author.

(3) The *Dianlue* ("A Survey of Yunnan"), in ten volumes, was written by Xie Zhaozhe during the reign of Emperor Wanli (Zhu Yijun). The book consists of ten treatises, and each treatise comprises one volume. Volume One "Geography" records the territory and the evolution of the organisational system of Yunnan; Volume Two "Topography" introduces the mountains, rivers and famous scenic spots; Volume Three focuses on "Local Products"; Volume Four "Customs and Habits" documents local features and customs; Volume Five "Eminent Officials" records famous court officials; Volume Six "Eminent Personalities" discusses capable villagers; Volume Seven "History and Administration" records the history of administrative systems; Volume Eight "Written Works" addresses art and literature; Volume Nine "Native Tribes" lists ethnic minorities; Volume Ten "Miscellaneous Matters" reviews trivial anecdotes and various conversations. Each treatise is preceded by a preface, which contains the gist of the part. The *Catalogue of the Complete Library of the Four Treasuries* commends the book, because "its citations are reliable, the narration is skilful, and the style is elegant and concise compared with other local chronicles". To conclude, it is a very characteristic provincial gazetteer of Yunnan.

(4) The *Songjiang fuzhi* ("Gazetteer of Songjiang Prefecture"), in 32 volumes, was written by Gu Qing during the reign of Emperor Zhengde (Zhu Houzhao) and revised by Chen Wei. The existing edition is the block-printed edition of 1512 (the 7th year of Emperor Zhengde's reign). Songjiang Prefecture was in charge of three counties—Huating, Shanghai and Qingpu—and the frontispiece features the *Map of the Songjiang Prefecture Jurisdiction*. The text is divided into 31 general items which include the evolution of the organisational system, dividing lines, territory, bodies of water, social customs, local products, number of households, corvée, land tax, cities, lanes and alleys, constables at village level, towns, bridges, government offices, schools, military fortifications, granaries, posthouses, altars and temples, thatched cottages, tombs, Buddhist and Taoist temples, historical sites, prefects and county magistrates, autographs, the achievements of eunuchs, the imperial civil examination system,

personages, incidents in past ages, good and bad omens, etc., and the items are further subdivided into 70 specific items. The accounts in the book start from the Spring and Autumn period and the Warring States period, and continue until Emperor Zhengde's reign in the Ming dynasty, covering both the past and the present in great detail [21].

(5) The *Yaozhou zhi* ("Yao Prefecture Gazetteer"), in 11 volumes, was written by Qiao Shijing during Emperor Jiajing's rule and revised by Li Tingyu. The existing edition is the block-printed edition of the 26th year of Emperor Jiajing's reign (1547). Volumes One and Two contain "*dili zhi*", recording the course of change in administrative geography, and the mountains, rivers, historical sites and tombs of all the dynasties included; Volume Three is about "the record of organisational systems" and "the record of ceremonies and rites of sacrificing"; Volume Four contains "the record of land taxes"; Volumes Five and Six contain the "records of officials"; Volumes Seven and Eight contain "the record of personages"; Volume Nine contains "the record of selection systems"; Volume Ten contains "the record of significant events"; Volume Eleven contains "the record of art and literature". This gazetteer is representative of the *jianti zhi* (brief annals) of the Ming dynasty, featuring a pithy style. The works cited are well-grounded without any exaggeration or anomalies. Nine records are arranged horizontally, and there are no subdivisions within these records. To illustrate, the Yao Prefecture was described as a single entity without further geographical divisions; things that once belonged to the Yao Prefecture but no longer did were excised from the book; the out-going officials were not included in the "officials" part; the deceased were not included in the "personages" part; only major events were included.

(6) The *Renhexian zhi* ("Renhe County Gazetteer"), in 14 volumes, was written and revised by Shen Chaoxuan during Emperor Jiajing's reign. The book was finished between 1542 (the 21st year of Emperor Jiajing's reign) and 1549 (the 28th year of Emperor Jiajing's reign) but was not published then. Nevertheless, a handwritten copy by Zheng Gui existed during Emperor Wanli's reign. Someone surnamed Ding from Qiantang County issued a revised version in 1893 (the 19th year of Emperor Guangxu's rule), which has been handed down to the present day. The author of the book referred to some prefecture gazetteers and works in other categories, and personally conducted some on-the-spot investigations and interviews, so the book had reliable sources. At the very beginning of the book are the preface, the table of contents, and the title. Then the text of the book is divided into 17 items which include the territory, bridges, mountains and rivers, government offices, social customs, schools, water conservancy, charitable policies, altars and temples, famous officials, the imperial civil examination system, personages, tombs, Buddhist and Taoist temples, books, inscriptions on tablets, notes on specific fields, etc., and the items are further subdivided into 86 specific items. This book was the first county gazetteer of Renhe County.

(7) The *Shouning daizhi* ("Shouning Gazetteer-in-Waiting"), in 2 volumes, was written by Feng Menglong during the reign of Emperor Chongzhen (Zhu

Youjian). Since its author considered the book deficient and waited for latecomers to provide supplementary material, the book was given the name "Gazetteer-in-Waiting", but it is still an outstanding gazetteer with a distinctive style. The gazetteer is divided into 28 items, which include the border territories, city wall and passes, county seat, schools, number of households, taxes, regulations on the salt industry, local products, social customs, seasons, Buddhist temples, unusual natural phenomena, etc. The book abandons the corrupt custom of using flattering remarks to present a false picture of peace and prosperity. Instead, it gives a faithful account of peoples' suffering, which is the most distinctive feature of the book. The book even lays bare Shouning's backward culture. Moreover, the whole book is written in the first person, which is unique in the history of China's local records.

(8) The *Jinshan weizhi* ("Jinshan County Gazetteer"), in 6 volumes, was written by Xia Youwen during Emperor Zhengde's reign and revised by Zhang Kui. It was published in 1517 (the 12th year of Emperor Zhengde's reign). The book mainly introduces the military systems of the Ming dynasty, giving a rather detailed account of the regulations of the military administration within the framework of the Weisuo System (military garrison units to maintain peace) and other military laws. The whole gazetteer is divided into two chronicles, ten chapters, and sixty items. In the first chronicle, Volume One is about territory, boundaries and construction; Volume Two is about buildings and temples; Volume Three is about high-ranking military officers and military policies. In the second chronicle, Volume One is about strategic and impregnable sites, and schools; Volume Two is about ceremonies of sacrificing, and personages; Volume Three is about local products. The book is based on concrete facts, so it is convincing. In addition, there is no specific item for "art and literature", so writings about military defence matters are attached to various chapters.

(9) The *Sizhen sanguan zhi* ("Annals of Four Military Towns and Three Passes"), in 10 volumes, was written by Liu Xiaozu during the reign of Emperor Wanli. It was completed between 1574 (the 2nd year of Emperor Wanli's reign) and 1576 (the 4th year of Emperor Wanli's reign). "*Sizhen* (Four Military Towns)" refers to Jizhou, Liaodong, Baoding and Changping while "*Sanguan* (Three Passes)" refers to Zijing, Juyong and Shanhai. There were dozens of annals or gazetteers about border defence affairs in the Ming dynasty, but this was the most detailed one.

From the gazetteers above, it can be seen that the gazetteers, whether at the level of the provincial administration commission, prefecture, sub-prefecture, county or guard post or whether about border defence affairs, all vary in their content and structure and have different foci.

8.4 Travellers' Contributions to Geoscience

8.4.1 *A Community of Travellers in the Late Ming Dynasty*

Field trips are a basic component of any geoscientific research work, and a major source of first-hand data.

In the early Ming dynasty, Chen Cheng (also styled as Zi Shi) from Jishui (now called Jishui County, Jiangxi province) was sent on a diplomatic mission abroad, and thereby became a traveller. He became a Jinshi (a successful candidate of the highest imperial examination) during Emperor Hongwu's reign and was given an official position as Xingren (envoy). He was once dispatched to Nan'an. During Emperor Yongle's reign, Herat paid tribute to the Ming. In response, acting on orders, Chen Cheng, a high-ranking clerk of the Ministry of Official Personnel Affairs, led eunuch Li Da and other envoys and went on another diplomatic mission through 17 countries including Herat and Samarkand. He returned in 1415, the 13th year of Emperor Yongle's reign, and wrote the *Shi xiyu ji* ("Record of a Journey to the Western Regions"), giving a detailed account of the landscapes, personages and social customs of those countries [23]. Apart from the geography of East Asian countries, the book also records the geography of the Turpan Region of China. The book adopts a pithy style, but still gives much emphasis to the geographical features of the countries. For example, according to a description in the book, "Herat is located in the southwest of Samarkand. The country is surrounded by mountains and a river flows westwards at its centre. The houses, which have a ground floor built on a high platform in the palaces, are built of piled stones instead of ridgepoles, beams, and clay tiles. In the house are several dozens of rooms, with windows and doors all decorated with the glitter of gold and Jade. Carpets are laid down on the floor, on which people sit when they are gathered, regardless of gender and status [24]."

In the middle and late Ming dynasty, under the influence of Pragmatism, many intellectuals developed a distaste for the imperial civil examination system featuring the use of *Bagu wen*, for it encouraged students to copy blindly from books without any innovative ideas. Therefore, intellectuals shifted their interest from the Cheng-Zhu school of Neo-Confucianism and Wang Yangming's school of Philosophy of Mind to the school of Pragmatism, which signified a transformation from vain talk about human nature and principles to the pursuit of practical applications. They asserted: "Do not pretend; do not go against nature; do not be unconscionable; do not curb your will; just follow your own inclinations [25]." They abandoned themselves to nature and considered travelling to be equally as important as reading. Some even argued that: "Reading the books unseen and reaching the places unvisited is akin to finding a valuable treasure or tasting a rare delicacy. This peculiar sensation of pleasure cannot be expressed in words." A crowd of these intellectuals ventured out of their studies and lost themselves in natural beauty, thus becoming a group of travellers in the late Ming dynasty. The dates of the births and deaths of 12 of these people have been verified.

Liu Song (1321–1381), was the author of the *Dongyou lu* ("Journey to the East").

Du Mu (1459–1525), a Jinshi of 1499, the 12th year of Emperor Hongzhi's reign, was the author of the six volumes of *You mingshan ji* ("Travels to Famous Mountains") and the two volumes of *Runzhou youshan ji* ("Travels to Mountains in Runzhou"). In 1576, He Tang compiled and enlarged the books into the *Introduction to the Famous Mountains* (48 volumes). In the preface, Du Mu claims that "I loathe to stay in one place. On the contrary, I like to immerse myself in the beauty of nature. Each time I hear of any place of natural beauty, I do my utmost to find it, no matter how far afield the place is. Even if I must clamber over vines or cleave the waves, I would still be more than delighted to go". It is thus evident that he was a huge fan of travelling.

Wang Jiusi (1468–1551), a Jinshi of 1496, the ninth year of Emperor Hongzhi's reign, was the author of the *Xiyou suibi* ("Essays on the Journey to the West").

Wang Shimao (1536–1588), a Jinshi of 1549, the 28th year of Emperor Jiajing's reign, was the author of *Mingshan youji* ("Travelogue of the Famous Mountains") (one volume) and *Guanluo jiyou* "Travels to Guanzhong and Luoyang" (one volume).

Tu Long (1542–1605), a Jinshi of 1577, the fifth year of Emperor Wanli's reign, was the author of *Mingliaozi's Travel Records* (1 volume). In 1593 (the 12th year of Emperor Wanli's reign), he shared his own views on and experiences of travelling in the preface to Wang Shixing's *Travels to the Five Sacred Mountains*. The preface reads:

> ...My friend, Wang Hengshu (courtesy name Wang Shixing) from Tiantai, is a man of magnanimous gestures. He transcended the worldly and travelled all over. He ventured to ascend high mountains to tour splendid scenic spots. He left tracks in all the beautiful places and made appraisals or comments about all of them in his own works. Among these places, the five great mountains are the most magnificent. Even though travellers throughout the centuries have always thought it impossible to visit them all, Hengshu managed to achieve this. He travelled through all the "Caverns of Heaven and Places of Blessing" in Taoism, hardly missing any of them. In light of his travel experiences, he wrote copious and fluent essays in which he compares himself to the deity of mountains, gains insights and practises alchemy at Buddhist temples, expounds the fundamental principles and clarifies any doubts in his mind. Most of his writings are about his understanding of the truth of Buddhism and Taoism. In other words, his writing possesses both elaborate rhetoric and natural instincts. Consequently, this proves that Hengshu is particularly favoured by nature. I am talented as well. However, I was only a young scholar who was confined to the back streets and I had no opportunity to travel around. I left the court not long after I started my official career. Therefore, I crossed barely one-tenth of the entire world. When I returned home, I had to cater to my centenarian family, so I did not dare travel afar. Recently, I have been engaged in Buddhist services. Although I cannot travel far afield, I truly yearn for distant journeys. I view small hills as eminent mountains, and narrow and hard to access cottages as palaces of the Celestial Ruler. I take pleasure and rest in my humble abode, feeling more than satisfied when I look around. Hengshu heard of my thoughts and laughed heartily at them. Despite this, I must wait until the conditions allow. Hengshu takes advantage of his official post and leaves footprints everywhere, and one day my straw sandals will be my carriage and take me to reach the abodes of the immortals or unchartered territories - yet none of these destinations shall halt my footsteps and I will keep on exploring. [26]

Wang Shixing (1547–1598), a Jinshi of 1577, the fifth year of Emperor Wanli's reign, was the author of the *Record of Extensive Wanderings* (2 volumes), *Travels to the Five Sacred Mountains* (10 volumes) and *Observations in the Realm* (6 volumes, but only 5 volumes can be traced; Volume Six is missing).

Hu Yinglin (1551–1602), who was a Juren [successful candidate in the provincial imperial examination] during Emperor Wanli's reign, but failed to pass the highest-level examination for a long time, was the author of *A Supplement to Liu Jun's Record of Jin Mountain* (16 volumes).

Chen Jiru (1558–1639) was a Zhusheng [successful candidate in the imperial examination at the county level] before he was 29 years old and a hermit after the age of 29. He wrote the *Record of Xu Zichang's Garden, Record of Wandering among the Peach Blossoms, Preface to Wang Jizhong's "The Call to Travel", Preface to "Travel Sketch of Min", Travels to Kongling Shoal*, etc. and revised *Travels to Famous Mountains* (4 volumes).

Yuan Hongdao (1568–1610), a Jinshi of 1592, the 20th year of Emperor Wanli's reign, acted successively as a county magistrate, instructor at the Confucian school of Shuntian Superior Prefecture, instructor of the Directorate of Education, secretary of the Bureau of Ceremonies of the Ministry of Rites, and as director of the Bureau of Records of the Ministry of Personnel, etc. He once travelled throughout the Wu-Yue area, and ascended Mountain Lu, Mount Song, Mount Hua and other mountains. He also wrote some travel notes, such as *Manjing youji* ("Diary of the Tour to Manjing"). Later scholars selected the travel notes from the *Complete Works of Yuan Zhonglang* and recompiled them into the *Travels of Yuan Zhonglang*, which was published in 1935 by the National Library of China Publishing House.

Cao Xuequan (1574–1646), a Jinshi of 1595, the 23rd year of Emperor Wanli's reign, was the author of the *Diary of An Official's Travels in Shuzhong* (4 volumes) and *Descriptions of Famous Sites of the Great Ming* (96 volumes).

Wang Siren (1576-?), a Jinshi of 1595, the 23rd year of Emperor Wanli's reign, was the author of the *You manjing ji* ("Travels to Manjing") and other travel writing texts, and the author of other books, including *The Call to Travel* (1 volume) and *Travel to Mountain Lu* (1 volume).

Xu Xiake (1587–1641), a commoner and learned person, was the author of *Xu Xiake's Travels* (10 volumes).

There were another 27 people without definite dates of birth and death, but it is still possible to deduce approximately when they lived.

Chen Yi, a Jinshi of 1517, the 12th year of Emperor Zhengde's reign, was the author of the *Travels to Famous Mountains* (4 volumes).

Tian Rucheng, a Jinshi of 1526, the 5th year of Emperor Jiajing's reign, was the author of the *Diary of an Official's Travels in the West of Yue* (1 volume).

Xu Qian, a Jinshi of 1529, the 8th year of Emperor Jiajing's reign, was the author of the *Travels in the Peach Blossom Spring.*

He Tang, a Jinshi of 1547, the 26th year of Emperor Jiajing's reign, wrote the *Travels to Famous Mountains* (17 volumes) and compiled the *Records about Travelling to Famous Mountains from Antiquity to the Present Day* (76 volumes).

Shen Meng, a Jinshi of 1553, the 32nd year of Emperor Jiajing's reign, wrote the *Introduction to the Famous Mountains* and compiled *A Survey of Famous Mountains and Scenic Spots* (15 volumes).

Shen Mingchen, secretary to Hu Zongxian during Emperor Jiajing's reign, was the author of the *Travel Notes of the Siming Mountains* (1 volume).

Pan Jingsheng, a secretariat drafter during Emperor Jiajing's reign, was the author of the *Travel Notes of the Yellow Sea.*

Li Tianzhi, a Jinshi of 1571, the 5th year of Emperor Longqing (Zhu Zaihou)'s reign, was the author of the *Travel Notes of Nine Mountains.*

Tang Hezheng, a Jinshi of 1571, the 5th year of Emperor Longqing's reign, was the author of the *Journey to the South* (3 volumes).

Wang Yu, a Jinshi of 1577, the 5th year of Emperor Wanli's reign, was the author of *A Dialogue of a Visitor to Chu.*

Zhou Yingbin, a Jinshi of 1583, the 11th year of Emperor Wanli's reign, was the author of the *Travel Notes of Mountains.*

Wang Jiamo, a Jinshi of 1586, the 14th year of Emperor Wanli's reign, was the author of the *Collection of Jiqiu* (47 volumes). In the book can be found the *Travel Notes of Northern Mountains*, which systematically describes the geographical landscape of the northwestern mountainous areas of Beijing.

Yang Boke, a Jinshi of 1586, the 14th year of Emperor Wanli's reign, was the author of *A Dialogue on the Eastern Expedition* and *Travels to Shu.*

Zeng Weifang, a Jinshi of 1589, the 17th year of Emperor Wanli's reign, was the author of the *Jottings on the Tour to Yan.*

Zhu Huafu, a Jinshi of 1592, the 20th year of Emperor Wanli's reign, was the author of the *Records of Scenic Spots across Yangcheng.*

Peng Zongmeng, a Jinshi of 1601, the 29th year of Emperor Wanli's reign, was the author of the *Miscellaneous Records on the River* (1 volume).

Yao Lusu, a Jinshi of 1601, the 29th year of Emperor Wanli's reign, was the author of the *Records of Scenic Spots across Chu.*

Fu Zhenshang, a Jinshi of 1607, the 35th year of Emperor Wanli's reign, was the author of the *Records of Secluded Scenic Spots of Shu.*

Qiu Zhaolin, a Jinshi of 1610, the 38th year of Emperor Wanli's reign, and the grand coordinator in Henan early in Chongzhen's reign, was the author of the *Record of the Tour to Mount Heng.*

Xu Rulan, a Jinshi of 1616, the 44th year of Emperor Wanli's reign, was the author of *A Brief Record of the Tour to Quzhou* (1 volume).

An Guoxian, living during Emperor Wanli's reign, was the author of the *Travel Notes on Nan'ao Island* (12 volumes) and *Travel Notes on Nanri Island* (10 volumes).

Lin Pei, the censor-in-chief of Nanjing during Emperor Wanli's reign, was the author of the *Travels to Jiannan.*

Cao Fan, a Juren who successfully passed the imperial examination at the county level in Shuntian Superior Prefecture in 1597, the 25th year of Emperor Wanli's reign, was the author of the *Jottings on the Tour to Yan* (1 volume) and *Travel Notes* (or *Travel Sketches*) (1 volume).

Zhang Zhende, the county magistrate of both Xingwen and Changning counties in Sichuan, died at his official post during the reign of Emperor Tianqi (Zhu Youjiao). He wrote the *Travels to E'mei Mountain.*

Huang Yisheng, an Imperial College student during Emperor Tianqi's reign, was the author of the *Travels to Famous Mountains* (6 volumes).

Ma Yuantiao, a Zhusheng [students who are enrolled in schools affiliated to different levels of governments after passing the Imperial Exam) in the late Ming dynasty, who died in a battle against the Qing government, was the author of the *Travels to Heng Mountain.*

Yu Zhanbai, born in the late Ming dynasty, was the author of the *Wuyue woyou* ("Travels to the Five Sacred Mountains") for which Chen Jiru wrote the preface.

Based on the estimated lives of the 39 people mentioned above, it is evident that they lived in the late Ming dynasty, and most of them were active from Wanli's reign to the end of the Ming dynasty. Over two-thirds of them were Jinshi while the others were Juren or Zhusheng or officials. Thus, it can be inferred that travelling was very popular in the late Ming dynasty and writing travel notes had also become a general trend. This trend fostered a group of travellers, among whom Wang Shixing and Xu Xiake were especially outstanding. They had an awareness of the need for field studies. Moreover, they were sharp-eyed and broadminded, for they not only took notice of things unnoticed by others but also set out to analogise and analyse things and phenomena from the perspective of geography as well. That is why they were able individually to create the two exceptional geographical works, *Xu Xiake's Travels* and *Guangzhi yi* ("A Broad Description of Human Geography"). Xu and Wang mutually spurred on each other's brilliance, which marked a great leap forward in China's geography. But the two varied in their emphasis—Wang Shixing focused on human geography while Xu Xiake focused on physical geography [27]. The contribution made by Xu and Wang was by no means an occasional, sporadic or unexpected occurrence, but rather was the inexorable outcome of the prevailing trend for travelling among the group of travellers in the late Ming dynasty.

8.4.2 Wang Shixing's Contributions to Human Geography

Wang Shixing, with the courtesy name Hengshu, the art name Taichu and the alias Yuanbai Daoren, was born in Linhai in Zhejiang. He became a Jinshi in 1577, the fifth year of Emperor Wanli's reign. He served successively as eunuch director of Langling, county magistrate, supervising secretary of the Office of Scrutiny for Rites, counsellor of the Sichuan Region, surveillance vice commissioner in charge of a military defence circuit in Lancang, Yunnan, education-intendant censor of Henan, administration vice commissioner of Shandong, vice minister of the Court of the Imperial Stud, chief minister of the Nanjing Court of State Ceremonial, and other posts. He was born into a poor family, but he was given to studying, so his poetry and prose became renowned all over the country. His temperament made him fond of travelling. Ever since he was young, he "has the same aspirations as Xiang Ziping, bursting to travel around the Five Sacred Mountains". Most of his tours, however, were so-called "official travel", which meant travelling when being transferred to another official post. Only a few of his tours were carried out for the sake of travel.

According to the *Gazetteer of Linhai County: Biography of Wang Shixing* (finished during Emperor Kangxi's reign): "(Wang Shixing) travelled at any time, to any place whilst in any official post, thereby becoming a prolific writer of travel notes. Taking advantages of his official posts, he toured almost all the country except for Fujian. Starting from the Five Sacred Mountains, he travelled to a great many famous mountains no matter how far-flung they were, and he would not fail to notice the finest details of the place." He served as an official in places which included Beijing, Nanjing, Henan, Sichuan, Guangxi, Guizhou, Yunnan, and Shandong, etc. He crossed the Five Sacred Mountains, as well as E'mei Mountain, Taihe Mountain, Mount Bai (now called Qiyun Mountain), Diancang Mountain, Jizu Mountain and many other famous mountains. His footprints covered the two capitals and twelve provincial administration commissions which equates to 17 provinces, cities or autonomous regions according to today's administrative divisions. He realised his aspirations as a youngster and became one of ancient China's outstanding travellers. Among his surviving works are the *Travels to the Five Sacred Mountains* (10 volumes), *Record of Extensive Wanderings* (2 volumes) and *Observations in the Realm* (only five volumes can be traced; Volume Six is missing) which were compiled and revised by Zhou Zhenhe and are all included in the *Three Geographic Records by Wang Shixing* [28]. It also includes seven scattered articles, of which four were selected from local gazetteers while the other three epitaphs were selected from the *Epitaphs Unearthed in Linhai Region.* The *Collected Essays on Wang Shixing* by Xu Jianchun and Liang Guangjun includes six pieces of poetry or prose, three epistles and one encyclopaedic article on the government (ten articles in total) [29]. Another work *Records of the East Lake* remains undiscovered to the present day.

Wang Shixing's achievements in geography mainly concentrated on human geography, but he also delivered some brilliant commentaries on physical geography. The achievements can be broadly classified into the nine areas below.

(1) Extraordinary contributions on the man-earth relationship

After conducting regional research in Zhejiang, Wang Shixing argued that: "Hangzhou, Jiaxing and Huzhou are plains that abound in rivers and lakes, so the people living there are 'People of the Water Town'; Jinhua, Quzhou, Yanzhou and Chuzhou are full of steep hills, so the people living there are 'People of the Valleys'; Ningbo, Shaoxing, Taizhou and Wenzhou are mountainous areas by the sea, so the people living there are 'People of the Coast'. The people in these three regions have varied customs. The 'People of the Water Town' make a living from boats. Therefore, they are capable of amassing a wide assortment of goods, so ordinary people can easily become rich and, therefore, lead a luxurious life. The officials are pompous while the commoners have little power. The 'People of the Valleys' live in the valleys where mists form. Therefore, they are of a fierce and headstrong nature, and so they despise the penal code and lead a frugal life. However, the rich commoners are so full of self-confidence that they gather around and regard the officials with disdain. The 'People of the Coast' eat outdoors and sleep by the water and struggle to survive all the perils. They are not very poor thanks to the sea, yet neither are they very rich due to their isolation from traders. As a result, the officials and commoners live at peace with each other, and they lead a semi-luxurious and semi-frugal life [30]." In other words, people living in different geographical environments have different production modes, lifestyles, customs, values and official-commoner relationships. It can also be inferred from this quote that Wang Shixing's thought on man-land relationship actually encompasses two mechanisms as to how the geographical environment has an impact on culture: one is the indirect mechanism of "geography-economy-culture", while the other is the direct mechanism of "geography-culture". It is admirable that Wang Shixing realised this 400 years ago.

Based on geographic factors, Wang Shixing divided the Zhejiang region into a water town cultural area, valley cultural area and coastal cultural area. This division was unprecedented.

In Wang Shixing's view, the distribution of talents was closely related to geography: "the regions north of the Yangzi River have vast landscapes, so the development of rites and music is fairly balanced across the land; the regions south of the Yangzi River have deep and complex landscapes, so the successful candidates of the imperial civil examinations converge in mountainous and watery areas: to name but a few, Yuyao and Cixi in Zhejiang, Quanzhou in Min (Fujian), Huangzhou in Chu (Huguang, referring to today's Hubei and Hu'nan), Neijiang and Fushun in Shu (Sichuan), and Quanzhou and Maping in Yue (Guangdong) [31]. These areas boast of more talents than other prefectures." Geographical environment is a major inhibiting factor in the training of talent, but the correlation is not absolute. Therefore, following on from the previous quotation, Wang Shixing further explained: "However, the literati and scholars are not confined to favourable landscapes. It is natural for some of them to adapt to different natural conditions." He then provided a number of examples to substantiate his view.

Wang Shixing also drew a theoretical conclusion that when it comes to the transformation and utilisation of the natural environment, human beings cannot avoid having an impact but must not do anything blindly. Instead, we should always abide by the basic principles of nature. In his words, "One cannot be so cowardly as to fall into inaction, whilst craving great achievements is also inadvisable [32]."

(2) Profound understanding of regional geography

Wang Shixing had a deep understanding of geographical regionality and he was especially good at spotting the geographic differences in different regions. Essays on this aspect are mostly included in his geographical monographs *Guangyou zhi* ("Record of Extensive Travels") and *Guangzhi yi* ("A Broad Description of Human Geography") written in his later years.

In the *Guangyou zhi,* the entire country falls within the scope of his regional studies. It analyses the disparities between different regions in terms of the natural environment (such as the terrains, strategic positions, social customs) and of the human factors (such as ethnic minorities, regions and dialects). In the "Terrain" section, the book elaborates on the mountains range extending along three main lines, namely, the northern "dragon", the central "dragon" and the southern "dragon"; in the chapter on "Strategic Position", it analyses the natural geographic conditions of two capitals and thirteen provincial administration commissions and concludes that the division is largely reasonable. No one had ever carried out analysis of this kind before. In the last paragraph, Wang Shixing comprehensively explores the differences between regional dialects: "Dialects vary from one place to another. That is to say, people from different regions may find it difficult to understand each other due to the nonstandard mode of vocalisation. The dialect may vary among prefectures, or on different sides of the Yangzi River … For example, people from Qi and Lu (Shandong) speak loudly and clearly; the Yangzhou locals like to stress the middle of an utterance; and natives of Hui and She (Huizhou and Shezhou) tend to prolong the last syllable." Judging from this, Wang Shixing had shown great interest in dialect geography.

After ceasing to travel in his later years, Wang Shixing found the argumentation in the *Guangyou zhi* too simplistic, so he wrote the five volumes of the *Guangzhi yi* to give a more in-depth account of his geographical thoughts. He wrote a preface to the book himself in 1597, the 25th year of Emperor Wanli's reign, but he did not survive until the book was published. Afterwards, Yang Tiyuan first block-printed the book in 1644, the first year of Emperor Shunzhi's reign in the Qing dynasty, and another edition was reprinted in 1676, the 15th year of Emperor Kangxi's reign. Very few copies of these two editions have survived. In 1817, the 22nd year of Emperor Jiaqing's reign, Song Shiluo from Linhai reprinted it with reference to a copy of Yang's edition, and his edition was included in the *Taizhou Series*. In 1981, the Zhonghua Book Company published a punctuated and revised edition by Lu Jinglin. The book is composed of six volumes in name, but, in reality, there are only five volumes, for Volume Six "Barbarians of Four Directions" is missing. The titles of these five volumes are: "A Compendium of the Land", "Two Capitals", "Four Provinces North of the Yangzi River", "Various Provinces South of the Yangzi River" and "Various Provinces in the Southwest".

"A Compendium of the Land" expounds the geographic situations all over the country, such as the course of change and development in the territories of past dynasties, the differences in taxation across the country, differences in local products, regional differences in national reserves, regional differences in talents, the differences in and causes of the flow of water in rivers, the distribution of frontier passes, the frontier force of the Ming dynasty, etc. The volume covers many subdisciplines of geography including evolution geography, economic geography, cultural geography, military geography, etc.

(3) Views of economic geography

Wang Shixing's insights regarding economic geography were brilliant and covered natural resources, the distribution of agriculture and the handicraft industry, the division of economic zones, commercial trade, land and water transportation, etc. With regard to typical local products in different places, he says in the book: "The harbours have always been a place for assembling goods, such as coins from Suzhou and Hangzhou, grain from Huaiyin, salt from Weiyang [Yangzhou], cargo from Linqing and Jining, chariots from Xuzhou, town gods from the capital, antiques from lantern fairs, rice from Wuxi, block prints from Jianyang, porcelain from Fuliang, dried fish from Ningbo and Taizhou, foreign merchant ships from the Fragrant Hills, female entertainers from Guangling and lacquerware from Wenzhou."

Wang Shixing also compared the differences between the north and the south in terms of local products, modes of transport and eating habits, and divided the whole country into economic zones. He wrote: "The southeast is abundant in fish, salt and round-grained non-glutinous rice; Zhongzhou (Henan) and Chu (Sichuan) are abundant in seafood; the southwest is abundant in gold and silver mines, gemstones, patterned shells, amber, cinnabar and mercury; the south is abundant in rhinoceros, elephants, Chinese prickly ash, perilla, foreign currencies; the north is abundant in cattle, goats, horses, mules and woollen blankets; Chuan (Sichuan), Gui (Guizhou), Qian (Yunnan) and Yue (Guangdong) are abundant in *huangpianmu* (an ancient wood) and *nanmu* (a precious wood unique to China and South Asia). The regions south of the Yangzi River are richly endowed with firewood for fire making while the regions north of the Yangzi River are richly endowed with coal for the same purpose. The northwest features high mountains, so people tend to travel by land instead of taking boats while the Southeast features vast bodies of water, so people tend to travel by boat instead of travelling by horses or horse-drawn carriages. People from Hainan eat fish and shrimps while those from the North dislike them because of their fishy smell. People from the regions north of the Great Wall eat cheese while those from the South hate it due to its disgusting taste. People from Hebei eat shallots, garlic, and longstamen onion bulbs while those from the regions south of the Yangzi River unconsciously reveal their dislike of the spicy flavour. These are habits born of the geographic conditions, so it is impossible to insist on a unified pattern."

Wang Shixing also mentioned the mining industry in Yunnan. The book reads: "Dian [Yunnan] is the most proficient in mining. People there have been excavating around the mine caves since the foundation of the Ming dynasty to fill in the gap in taxes, and they have never ceased excavating since then. Outcrops of mines can be spotted all over Dian. Even if the scale of some mines is not as large as a cave, the commoners usually excavate them by themselves. Each day, they only dig out the necessary amount from the mine to support one day's living, so their behaviour is permitted by the law. In the case of formal mine caves, the mine leaders need to report to the officials and then acquire permission. Then the leader will be allowed to recruit several miners depending on the size of the mine cave. Miners are under the control of the leader. If the mine has not yet become a formal mine cave, the expenditure on both public and private affairs is entirely the responsibility of the leader. The expenditure may be as high as the value of thousands of gold coins, if the mine cave is large enough. If the mine cave is established, however, the mines may be supervised by the specific official department. Each day, several miners enter the cave and work until sunset. The material excavated from the mines by them is piled up. Any profit can be divided into four parts: the first part is the tax given to the local authorities. The supervising officials take it and distribute it to the provincial administration commissioners. The second part is public expense. The mine leaders take the share, allocate it to public or private spending and keep a record of it. The third part belongs to the mine leaders. The last part is equally divided amongst the miners … Mining can be reasonably recommended as a way of satisfying the unexpected needs of the folk and administering relief to famine-striven refugees." In this section, the author described the history of Yunnan's mining industry, the mode of mining, the form of organisation, the meagre income of the workers, and the social benefits, etc.

(4) Views of cultural geography

Based on field investigations, Wang Shixing divided Zhejiang regions into two cultural areas: east Zhejiang and west Zhejiang. Volume Four of the *Guangzhi yi* reads: "Bordered by the river, the east and west have distinct customs. West Zhejiang features prosperity, fine and sensitive people, and has a great respect for cultural relics … It is abundant iin famous clans which have thousands of young servants. Therefore, they lead a life of luxury, much better than that of ordinary townspeople. The east features simplicity and sincerity, thrifty and unsophisticated people, and a great respect for honesty and frugality. Rich merchants are rarely seen here."

Considering geographic factors, Wang Shixing further divided the east Zhejiang cultural area into three subzones—Ningshao (Ningbo and Shaoxing), Jinqu (Jinhua and Quzhou) and Taiwen (Taizhou and Wenzhou), and summarised their cultural features, respectively: "There are three subdivisions of eastern Zhengjiang's customs: in Ningbo and Shaoxing, the imperial civil examination system is prevalent. Locals [who fail the exam] are accustomed to earning a living away from home. While some of them excel at rewriting books and creating literary works, others become well-off businessmen who will not give up faced with meagre profits. In Jinhua and Quzhou, people are sturdy, high-spirited and eloquent. The reserve military officers of the six prefectures mostly come from this subzone. In Taizhou and Wenzhou, people live besides the mountains and the seas, which means they make a living for themselves through hunting in the mountains, fishing on the sea and cultivating the fields. Since the merchants do not travel far afield to do business, they regard west Zhejiang as an utterly different land." The divisions and summaries correspond to the historical facts, so they serve as valuable references to the present day.

Wang Shixing divided Zhejiang into the east Zhejiang cultural area and the west Zhejiang cultural area. The division was not only on the grounds of the separation formed by the Qiantang River and the differences in the geographical environment, but it was also closely related to the political geography. He pointed out that "The dividing line between the east and west partly overlaps with the line between the Wu and Yue regions, so the landscapes and specialties of the east and west are quite different from one another. Western Zhejiang is a watery land without any mountains, so the customs there are elaborate and ingenious, which is similar to Suzhou and Changzhou. This is because the region can be traced back to being part of Wu. The east is close to mountains and rivers, so the customs there are simple and unadorned. This is because the region roughly overlaps with the historic land of Ouyue [33]."

Unlike the geographical environment of the Zhejiang region, the Central Plains region has a flat terrain and barely any mountains, so the customs within the region are basically identical. Wang Shixing pointed out in the *Guangyou zhi*: *Landscapes* that the "Central Plains is the only region with vast lands. There are no mountains for several thousand li, so people must draw man-made lines to divide the different areas."

In addition, Wang Shixing presented a number of descriptions of the religions, wedding and funeral customs, and the basic necessities of life, etc. of the different places. For example, "Yongzhou is near Yue [Guangdong], so the customs are slightly influenced by those of the ethnic minorities in the southwest. Men wear floor-length clothing and robes while women wear knee-length skirts and shorts and lay bare their shanks and feet, with no fear of the filth. Those who can wear straw sandals enjoy high status. They comb their hair into a high knot, wear large rings in their ears, and cast tin to make flower-like accessories which they wear all over their hair. The jobs of bailiffs in the post house, carter, stablehands, and the like are all undertaken by women instead of men. Thus, actors and actresses appear simultaneously in plays and the officials cannot prohibit it [34]."

(5) Extraordinary regional philosophy and regional planning

Wang Shixing possessed an extraordinary philosophy of regional planning and designed concrete regional planning schemes. The *Guangyou zhi: Miscellanies* divides southeast China as it was at that time into 14 natural areas and summarises the fundamental features of each area and the relationship between each natural area and its corresponding administrative area.

① Jinzhong (central Shanxi): "the Taihang Mountains run several thousand li along its eastern edge; the Hong River surrounds it on the west; deserts determine its northern edge. Therefore, it naturally became an administrative unit."
② Guanzhong (centre of the Four Passes): "Rivers and Tongguan determine its eastern boundary; Liangshan Temple in Jiange County fixes its southern boundary; the Qinling mountains define its northwestern limit; the Wei River is located on its left while the Han River is on its right; the Zhongnan Mountains can boast of being the location with the highest mountain. Therefore, it also naturally became an administrative unit."
③ Shuzhong (central Sichuan): "Around the region, peaks rise one after another and there is a vast expanse of fertile lands. The Min River cuts across and many other rivers run vertically, all flowing out from the Three Gorges. Therefore, it also naturally became an administrative region."
④ Chu [mostly in Hubei and Hunan]: "The Yangzi River runs through it horizontally. Nine rivers south of the Yangzi River converge in Dongting Lake while various rivers north of the Yangzi River merge to form the Han River, and all the rivers above finally flow together into the Yangzi River. It is encircled by numerous mountains in Hunan, Guangxi and Jiangxi at the front while the Jing Mountains surround it to the north. Therefore, it also naturally became an administrative region."
⑤ Jiangyou [Jiangxi]: "Mount Huang is on its left while Mount Lu is on its right, both starting from the south, meandering and surrounding it on the east, the west and the south. All the rivers rise in Jiangyou, running through Pengli [Boyang Lake] and flowing together into the Yangzi River. The mountains and

rivers both align with good features in *fengshui* [Chinese geomancy], especially as the Yangzi River flows behind it. Therefore, it also naturally became an administrative region."

⑥ Liangguang (Guangdong and Guangxi): "Liangguang is the region outside the Five Great Mountains, but Guangyou [Guangxi] also has its own administrative body. Three rivers meet at the north of Cangwu and another branch forms east of Meiling Mountain (or Dayu Mountain), and together they surround the northern edge of Liangguang.

⑦ Min (Fujian): "It is enclosed by seas at the front, so it also naturally became an administrative region."

⑧ Dian (Yunnan): "Central Dian in the southwest stretches for thousands of li, but Dian was once an ancient kingdom."

⑨ Guizhu (Guizhou): "At first, it was the gateway to Dian. Later, it had no choice but to establish its own provincial administration commission, because there are several sporadic rivers running through it, [which hindered it being a gateway to Dian], and Hubei and Eastern Sichuan could not reach a consensus on the area of jurisdiction."

⑩ Zhongyuan (Central Plains): "It covers a vast expanse of land. There are no mountains for several thousand li, so people had to draw man-made lines to divide the different areas. The Wei River (the largest tributary of the Yellow River) and the Han River flow across the region from the centre while the Huai River and Wei River (a river in Henan) also run through. With the Taihang Mountains behind it, the Jing Mountains in front, the Qin Mountains to the west and Mount Song standing tall in the middle, it also naturally became an administrative region."

⑪ Shandong: "Mount Tai, as the highest mountain in Shandong, has no clear boundaries with other provincial administration commissions like splendid mountains or rivers. However, it unites Qi, the ancient territory of the Grand Duke Jiang, and Lu, the ancient territory of the Duke of Zhou, into one whole.

⑫ Liangzhe (east and west Zhenjiang): "The dividing line between the east and west partly overlaps with the line between the Wu and Yue regions, so the landscapes and specialties of the east and west are quite different from one another. Western Zhejiang is a watery land without any mountains, so the customs there are elaborate and ingenious, which is similar to Suzhou and Changzhou. This is because the region can be traced back to being part of Wu. The east is close to mountains and rivers, so the customs there are simple and unadorned. This is because the region roughly overlaps with the historic land of Ouyue."

⑬ Southern Capital (Nanjing): "It features easy grain transhipment, splendid cultural relics and unified systems on cultural relics, which is similar to the Western Capital of the Zhou dynasty."

⑭ Northern Capital (Beijing): "the Taihang Mountains form a natural barrier, while the vast sea is fed by other branches of water. It is a key link to many barbarian places and a military stronghold, which is similar to the Eastern Capital of the Zhou dynasty [33]."

Wang Shixing analysed how the administrative divisions (two capitals and thirteen provincial administration commissions) corresponded to the natural divisions, the historical course of development and local conditions and customs. However, in the meantime, he realised that due to human and historical factors, the administrative divisions and the natural divisions were certainly not completely identical in the Ming dynasty. The Central Plains and Shandong are good counterexamples. As far as he was concerned, "Chu" was a reasonable natural division while "Huguang" was less reasonable as an administrative division, for the latter was so large that the division led to many weaknesses. He explained: "In the Spring and Autumn period and the Warring States period, Huguang was referred to as "the Great Chu Covering Six Thousand Li", spreading across the Huai River and Ru River, and its northern edge was once near the Yellow River. Nowadays, the Ming dynasty divides it into several provincial administration commissions, but Chu is still the largest one. It has 15 prefectures under its jurisdiction … People must travel several thousand li and it takes more than a month to arrive in Chu from other provinces. This also poses great difficulties for the receipt and handling of official documents and barbarian affairs. Chenzhou and Yongzhou sometimes share an education intendant, but, in other cases, they appoint different education intendants. Moreover, there are many ethnic minorities in Guizhu and Yuexi. Military institutions play the main role in their management while native chieftains have an auxiliary role. But these two regions do not stand out as independent provinces." To rationalise the administrative division in Huguang region, he suggested that "Chenzhou [a prefecture-level city in today's southeastern Hunan], Yuanzhou and Jingzhou should be administered by Guizhou while Yongzhou, Baoqing and Chenzhou should be placed under Yuexi's administration. In so doing, the thirteen provinces will be comparable sizes in terms of coverage and any problems in public life will be easily solved [35]." His analyses were correct and his suggestions were also valuable. In light of this, Huguang was further divided into two provinces, Hunan and Hubei, in the Qing dynasty.

(6) Completing the system of mountain ranges

The concept of a "mountain range" was expounded in the *Tribute of Yu* and the *Classic of Mountains and Seas*, dating back to the Warring States period. After further development in the Western Han, Eastern Han, Tang and Song dynasties, the Silie Shuo (Four-Range Theory"), Santiao Shuo (Three-River Theory), Shanhe Liangrong Shuo (Theory of Two Mountainous and Watery Regions) and Sandalong Shuo (Three-Grand-Dragon Theory) came into being. On the basis of the ancient Chinese theories on mountain ranges, Wang Shixing put forward a more detailed version of the "Three-Grand-Dragon Theory", and thereby promoted the totalisation and systematisation of theories on mountain ranges in ancient China. He said in *Guangyouzhi: Dizhi: Dimai* (Terrains):

> Since the past, *fengshui* masters have always claimed that all the mountains and rivers in the world originated in the Kunlun Mountain range which was later further divided into the "three dragons (three mountainous terrains)" and meandered throughout China. However, the reasons for the rise and fall of the "three dragons" had never been mentioned before. The dragons were stopped by waters. How can a person travel though a land with so many high

> mountains and deep valleys? The only way to find the mountains is to find the corresponding rivers. The Kunlun Mountains occupy the central position and are surrounded by mountains stretching into the wilderness in all directions. They form the southeastern line that stretches into China. This line can be further divided into three ranges in areas beyond the Great Wall: the range on the left surrounds the Yin Mountains and the Helan Mountains of the barbarians and continues into Shanxi. This range starts from the Taihang Mountains, meanders for several thousand li, passes Yiwulu Mountain and finally ends at the Liao Sea - this is called the "North Dragon". The middle range develops along Turpan, the Min Mountains and areas on the left and right of the Min River. The range on the left does not end until it surrounds Xuzhou while the right one stretches northwards towards Guanzhong and along the Sanguan Pass with the Wei River on its left and the Han River on its right. In the middle, however, the range passes through the Zhongnan Mountains, Taihua Mountains, Qin Mountains and Mount Song with Jingshan and the Huai River on the right and extensive plains on the left, and finally it joins the sea past Mount Tai - this is called the "Middle Dragon". The range on the right starts from the area west of Turpan, passes Lijiang towards Yunnan, surrounds Zhanyi and Guanling in Guizhu and stretches eastwards towards Yuanling. From there, the first subdivision stretches westwards from Wugang to the Xiang River and goes further westwards until Wuling; a second subdivision starts from the Haiyang Mountains of Guilin to the Jiuyi Mountains, Mount Heng, and the Xiang River and stretches eastwards until it draws near Mount Lu; a third subdivision starts from Yu Ling (also Dayu Mountain or Meiling Mountain), stretches towards Mount Huang and Mount Tianmu across grasslands, and ends at the lower reaches of the Yangzi River. The subdivision that goes along Yu Ling has a branch which passes through the Xianxia Pass towards Min and another branch which passes through Dapan Mountain, further towards the right and downwards to Cang Mountain, further to the left to Tiantai Mountain and Siming Mountain and finally stops at the sea - altogether, these branches are called the "South Dragon".

This is the most detailed theory on mountain ranges in ancient China, marking a new stage in the theories on mountain ranges. But admittedly, Wang Shixing's theory on mountain ranges is not totally aligned with the concepts concerning mountain ranges in modern geography. Wang Shixing's division of mountain ranges was based on the beliefs that "the dragons are stopped by waters" and that "the only way to find the mountains is to find the corresponding rivers", which was inherited from theories in Feng Shui. Modern geography, however, divides mountain ranges according to geological structure and age. Only mountains of identical structure and age can form part of the same mountain range [36]. Nevertheless, the theory regarding this in ancient China still had its positive effects and specific value, for it systematised and standardised the complicated distribution of the mountains and made it much easier for people to understand. Some division patterns of mountain ranges even coincide with those of our era, which is even more valuable.

(7) Advanced thoughts of on the transformation of terrains over time

Wang Shixing had grasped the mechanism of "sea-land transformation" to a higher degree than any of his predecessors [37]. It is even possible that his understanding of this mechanism attained the highest level of anyone throughout the ancient history of China. In the *Guangzhi yi: Various Provinces South of the Yangzi River*, Wang said: "In Shaoxing city, each street is accompanied by a river. The same is true in the countryside where the waterways are arranged like the chess pieces on a chessboard. Is this not created by nature? Some people argue that it cost a fortune of several hundreds of thousands from the national treasury to carry out the dredging of the Yue River (a branch of the Huai River). How could so much money be spent on dredging when there was still sufficient manpower? My answer is: Not exactly. This has always been a land dotted with rivers and lakes. At the very beginning, water overflowed here and there and then a few sandbanks came into being, so people gradually gathered around those places. That is why they are referred to as residents among wild rice and gourds. As time went by, the residents transported soil to level up the ground, build dykes and ditches for planting, or dredged harbours for sailing. With the passage of time and concerted efforts, the river courses were gradually established; the brick walls were gradually built; and a growing number of bridges and markets appeared. The situations in Jiaxing and Huzhou are certain to be similar. Now we have Qingcao Lake and Shaobo Lake in Huaiyang, but who knows if this will continue to be true in the future? And who can tell whether Yuezhong (now called Shaoxing) will remain a valley in the future? In the past, Lake Taihu dried up and relics of stone streets and houses were exposed. A conclusion can be drawn based on this analogy." In other words, under the dual effect of natural forces and human activities, the water town was gradually turned into land. This view conforms with the explanation of the shaping of Shaoxing, Jiaxing and Huzhou in geology, Quaternary geology, paleogeography, historical geography and geomorphology.

In the above quotation, Wang Shixing also pointed out the fact that the transformation of terrains is bidirectional rather than unidirectional. That is, seas may morph into fields while fields may also morph into seas or lakes. "Lake Taihu dried up and relics of stone streets and houses were exposed" is proof of this. Judging from this, Wang Shixing's research methods and conclusions quite closely resemble those of modern science. He took the development of theories on the transformation of terrains in ancient China to a new level.

(8) Contributions on toponymy

In his geological works, Wang Shixing not only recorded a large number of toponyms, but also provided explanations for some of them, elaborated on their origins, outlined their differences, corrected any mistakes and investigated their geological locations. Sometimes, he even used the toponyms when studying the changes in the natural and human environments [38].

China boasts of a vast land, a large population and various dialects. As a result, there may be a huge variety between generic names within the same category. Taking the "countryside fairs" in Wang Shixing's accounts as an example: "Wuning has the toponym 'Changzhouhai', which was confusing to me at first. Then I learned that it is the name of a fair. The ancients did business in 'shi' [the place to do business] at noon. Adopting this expression, people in Wu and Yuezhong refer to their fair as 'shi'. In Henan, people refer to the fair as 'ji' [gather around], for it is the place where the masses gather around. In regions south of the Five Mountains, people refer to it as 'xu' [leisure], for they do not hold fairs very often and enjoy having plenty of leisure time. In Xishu, people refer to it as *jie* ('痎', malaria), for it occurs as frequently as the onset of malaria. In regions south of the Yangzi River, people consider the names of diseases taboo, so they use a humorous expression *hai* instead [*hai* '亥'

and *jie* '痎' have similar forms] [35]." Thus, Wang Shixing not only commented on the differences between the terms for "fair" in different regions, but also explained the underlying reasons. *Xu Xiake's Travels* also includes different terms for "fair" in different regions: "Guizhou refers to it as *chang* ('场'); Yunnan as *jie* ('街'); and Guangxi as *xu* ('圩')" However, Xu did not explain why these names were adopted.

The toponyms in Wang Shixing's records can be divided into eight categories according to their connotations.

① Toponyms based on colours

Volume Seven of the *Travels to the Five Sacred Mountains* reads: "Diecai Hill (or Folded Brocade Hill) was behind Bagui Hall. Bagui Hall has now become untraceable, but heading northwards from the county and passing through multiple passes and mountains, you will find the mountain stretching eastwards with densely packed, colourful stones. That is why it is dubbed Diecai Hill in the *Tujing* ("Illustrated Classic"). According to what can be read in Volume Four of the *Travels to the Five Sacred Mountains*, Chicheng Mountain (or Red Walls Mountain) in Tiantai County "only has red stones with sawtooth-shaped walls, so it is named Chicheng Mountain". The *Observations in the Realm: Various Provinces of the Southwest* reads: "Kongming [Zhuge Liang] crosses the Lu River in May. Although it is not located in Luzhou, the Lu River flows northward for over a thousand li and passes through today's Huichuan, and is referred to as the Jinsha River or Black River, for its water looks black."

② Toponyms based on pictographs

Volume Seven of *Travels to the Five Sacred Mountains* reads: "Mount Jizu (Chicken-foot Mountain) towers high above numerous mountains in Binchuan, facing the south. With three mountain ranges to the front and one range to the rear, it is like a chicken's foot." The *Guangzhi yi: Various Provinces South of the Yangzi River* reads: "Jiuyi Mountains (Nine Puzzling Mountains) are where the "South Dragon" lies. There are nine jagged peaks set against each other, so people find it puzzling to tell them apart. That is how the mountain range acquired its name."

③ Toponyms based on the climate

The *Guangzhi yi: Four Provinces North of the Yangzi River* reads: "Liangzhou, literally 'Cold Prefecture', gets its name because it is famous for the coldest northwest winds. Even in May and June, there is frost in the daytime, which looks like heavy snow." The *Guangzhi yi: Various Provinces South of the Yangzi River* reads: "Leizhou, literally 'Thunder Prefecture', obtains its name from the frequent thunderstorms or because it is located south of the Lei River (literally 'Thunder River'). The Lei River flows under Qinglei Mountain (literally 'Lifting Thunder's Mountains'). Rising in Tonggu Village in Haikang County, it flows southwards for seventy li, then turns eastwards and converges with the sea. It was created in the wake of several thunderstorms. Because it is close to the South Sea, the thunder sounds just around the eaves of the houses."

④ Toponyms based on the terrain

The *Guangzhi yi: Four Provinces North of the Yangzi River* reads: "The name Hequ (literally 'River Bend') indicates that it is situated in a bend of the Yellow River. There was an army called Huoshan Jun (Volcano Army), for the place they belonged to has volcanoes and the cracks in the stones were full of smoke." The book also reads: "Wuding River (literally 'Mobile Water') abounds with quicksand, so carriage drivers must rush forward rapidly to cross, or they will sink into the sand. That is how it acquired its name."

⑤ Toponyms based on directions

The *Guangzhi yi: Four Provinces North of the Yangzi River* reads: "Chang'an has an alternative name which is Guanzhong (literally 'Central Pass'), for there is Hangu Pass in the east, Sanguan Pass in the west, Wuguan Pass in the south and Xiaoguan Pass in the north while Chang'an is right in the centre of the passes."

⑥ Toponyms based on numbers

The *Guangzhi yi: Various Provinces South of the Yangzi River* reads: "There are different versions of the Five Mountains. Pei Yuan's *Records of Guangzhou* lists the following mountains: 'Dayu Mountain, Shi'an Mountain, Linhe Mountain, Guiyang Mountain and Jieyang Mountain.' Deng Deming's *Records of Nankang* reads: 'The Five Mountains include: first, Tailing Mountain (Dayu Mountain) in Dayu; second, Qitian Mountain (Guiyang Mountain) in Guiyang; third, Dupang Mountain in Jiuzhen; fourth, Mengzhu Mountain (Linhe Mountain) in Linhe; fifth, Yuecheng Mountain (Shi'an Mountain) in Shi'an.' Thus, apart from the disputed Jiuzhen Mountain and Jieyang Mountain, the other four mountains are the same in both versions." *Guangzhi yu: Two Capitals* reads: "(Taihu Lake/Tai Lake) is referred to as Zhenze in *Yugong* and as Ju'ou in *Zhouli* and *Erya.* Some people regard Wuhu (Five Lake) as its alternative name, for it is a hydro-junction for five waterways. According to Yu Fan, it is connected to the Song River which flows across Changzhou in the east, to the Zha River flowing across Anji in the south, to the Jing River flowing across Yixing in the west, to Ge Lake flowing across Jinling in the north and to Jiu River flowing across Jiaxing in the southwest. According to Zhang Bo's *Records of Wu Kingdom*, Wuhu received its name because the distance around it measures five hundred li. According to the *Records of Yixing*, the name Five Lakes refers to Tai Lake, She Lake, Gui Lake, Yang Lake and Yao Lake. In Wei Zhao's view, Five Lakes refers to Xu Lake, Li Lake, Yao Lake, Ge Lake and Tai Lake. In the *Waterways Classic*, Five Lakes refers to Lake Changtang, Shegui Lake, Shang Lake, Ge Lake and Tai Lake. In the *Tujing* ("Illustrated Classic"), Five Lakes refers to Gong Lake, You Lake, Xu Lake, Meiliang Lake and Jinding Lake. In the *Shiji zhengyi* ("Annotation to Historical Records"), Five Lakes refers to the five bays on the eastern coast of Taihu Lake, that is, Jiao Lake, You Lake, Mo Lake, Huang Lake and Xu Lake. But all the above are based on pure conjecture."

⑦ The same toponyms for different places

Wang Shixing collected examples of the use of the same toponyms for different places in the counties and prefectures all over the country in *Observations in the Realm: A Compendium of the Land*: "Jianchang County in Jiangxi is administered by Nankang Prefecture rather than Jianchang Prefecture, while Nankang County is administered by Nan'an Prefecture rather than Nankang Prefecture. Both Ji'an and Guangxin have a place called 'Yongfeng'. As for Anren, Chongren, Anyi, Chongyi, Nanchang, Xinchang, Duchang, Ruichang, Guangchang, Jianchang, Huichang, Wannian, Wanzai, and Wan'an, etc., all these names were widely used by prefectures of different levels when they were first established. Taiping County in the Southern Metropolitan Region does not belong to the Taiping Prefecture, but to Ningguo Prefecture, while Jianning County in Fujian does not belong to Fujian Prefecture, but to Shaowu Prefecture. With regard to the toponym 'Taiping', the Southern Metropolitan Region has Taiping Prefecture; Guangxi has Taiping Prefecture and Taiping County; Taizhou Prefecture, Ningguo Prefecture and Pingyang Prefecture all have a Taiping County as well. As for the toponym 'Yongning', Yunnan has Yongning Prefecture; Guizhou has Yongning Sub-Prefecture; Ji'an Prefecture, Henan Prefecture and Longqing Sub-Prefecture all have a Yongning County as well." From Wang Shixing's perspective, this phenomenon of the "same toponyms for many different places" had multiple disadvantages and was "especially inconvenient when selecting officials and checking their political achievements".

Wang Shixing also collected some examples of the same toponyms being used for different places at different times in *Guangzhi yi: Four Provinces North of the Yangzi River*: "Today Liangzhe is known as Kuaiji Prefecture, but Kuaiji Prefecture used to be located near Dunhuang in the Northern Wei dynasty. People know Guazhou city in Weiyang, but they are unaware of the fact that a Guazhou city was located beside Dunhuang in the Tang dynasty. People know Shouchang County in Yanzhou, but it is not clear to them that Shouchang County was set a hundred and fifty li south of Shazhou in the Tang dynasty." These accounts were significant references for his contemporaries when reading ancient books as well as when compiling records of toponyms and historical maps.

⑧ Multiple toponyms for the same place

Wang Shixing also collected many examples of "multiple toponyms being used for the same place". For example, the *Guangzhi yi: Two Capitals* reads: "Mount Mao was first named Juqu Mountain. It was regarded as the Eighth Heavenly Abode and the First Blessed Spot in the *Daoshu* ("Daoist Books"). Later, the three Perfect Lords surnamed Mao ascended to heaven here and each occupies one peak. That is why the mountain is also called the 'Three Mao' Mountain." The East Lake in Ningbo, Zhejiang is another example. "It is thirty li from Yindong, fed by water from seventy-two streams, then irrigates seven villages, so it has another name which is Wanjin Lake [39]." In Zhejiang, there is a "Jinhua Mountain as high as a thousand *zhang* (unit of length equal to 1/3 metre) with one alternative name which is Long Mountain and another alternative which is North Mountain [39]."

(9) Accounts of physical geographical phenomena

Wang Shixing's accounts of physical geographical phenomena are detailed and concrete and, indeed, some are quite brilliant. For example, he discussed the causes of the differences between the Yangzi River and the Yellow River in the flow capacity at the river mouth. His discussion related to the drainage area, the terrains of the river valley, the number and size of branches, rainfall and its seasonal variations, groundwater levels, the properties of the soil, sources of river recharge, the watershed and other natural geographical factors. His arguments were completely in line with the theories of integrated physical geography, and they shared the method of analysis of modern integrated physical geography. He wrote:

> Only the two major rivers of China - the Yangzi River and the Yellow River - flow across the whole country. The Yellow River is fed by water from Shanxi, Shaanxi, Henan and Southern *Zhili* (known as the Southern Metropolitan Region) while the Yangzi River is fed by water from Sichuan, Huguang, Jiangxi and the Southern Metropolitan Region. The Yellow River rises in areas beyond the Great Wall and flows for five thousand li into the Central Plains, which is rather long distance. The Yangzi River, originating in the Min Mountains, develops over a much shorter distance. When the stream draws near the river mouth, the Yellow River enters the waterway of the Huai River and thereby empties into the sea. However, the river mouth of the Yangzi River covers a wide area of several dozen li. The Jing Mountains are located in between the waters which drain into the Yangzi River and the waters which drain into the Yellow River. In the areas north of the Jing Mountains, the water is hot and shallow. The water veins extend as deep as scores of *zhang* into the ground, hardly infiltrated by any soil. Only the Fen River, Wei River and Luo River feed fast flowing water into the Yellow River. The Su River, Huai River, Yi River and Si River all have a limited water flow. In the summer months, the rainfall increases so the water rises and runs rapidly, while in the other months, the water falls in the waterway so the river mouth is narrow. In the areas south of the Jing Mountains, the water flows through saline-alkali land and is virtually level with the ground, gushing regularly every now and then. Two rivers feed into the Yellow River from areas beyond the Great Wall, that is, the Dadu River and Li River; two other rivers meander for thousands of li from Lake Taihu, that is, Dongting Lake and Pengli Lake; many other bodies of water drain into it, including the Qize, Chao Lake, Huai River, Yangzhu Lake, and so on. When the snow melts and the spring water swells, the large stones in the upper reaches of the Yangzi River are submerged in the water and the water rises as high as twenty *zhang*. In the areas south of the Yangzi River, it rains continually all year round, so the flood never ends. That is why the water in this section flows slowly and the river mouth seems broad. The Yangzi River features slow flowing water, a broad channel and clay soil that is seldom dislodged. The Yellow River, however, features a rapid flow of water, a narrow channel and loose soil and sand so that it frequently bursts its banks [40]."

The above description is comprehensive and faithfully reflects the reality apart from the single mistake that the Yangzi River "originates in the Min Mountains".

In the *Guangzhi yi: Various Provinces of the Southwest*, Wang Shixing also provided a vivid picture of the physical geography of Guizhou: "Guizhou abounds with caves and gullies, as water passes through all the hollow caves in the mountains … Biyun Cave is in Pu'an. Each drop of water from Pu'an must flow through this cave. Since there are tunnels below the cave, the river can traverse the mountains. In the cave are 'fairylands' for planting, and the number of fields can amount to several score ... The place is characteristic of mountainous terrains with many snakes, dense fog and heavy rainfall. Sunny days appear at an interval of every three to five days. However, the sunny weather disappears quickly and the unpredictable rain follows close on its heels. Therefore, the locals equip themselves with felt cloaks, bamboo hats and bamboo sticks whenever they go out. The sticks are used for repelling snakes while the cloaks are in order to be prepared in case of sudden rains. As the saying goes, 'Sunny days never last more than three days, just as the ground never remains flat for more than three li.'"

Wang Shixing had some other brilliant arguments regarding the climate. According to the *Guangyouji: Social Customs*: "With the Yellow River as a dividing line, the North and the South actually do not vary significantly as regards their climate. However, the climate in the southeastern region is markedly different. For example, Qian (Guizhou) is usually overcast and rainy, while Dian (Yunnan) switches frequently between rainy and sunny weather; Yue (Guangdong) switches frequently between warm and cold weather, while Dian has a moderate temperature." This summary is correct. Then he provided further analysis of the cases of climate conditions in the southwestern region: "Qian tends to be overcast and rainy, for it is surrounded by mountains where clouds gather. Therefore, sunny days are rarely seen. As the proverb says: 'Sunny days never last more than three days, just as the ground never remains flat for more than three li [41].'" The climate features of Yunnan and their causes are as follows: "Dian has a unique climate ... It is neither too hot in summer, nor very cold in winter; people wear one or two layers of clothes in the daytime and cotton and linen-made clothes at night. This remains true all year round. Daylight neither lasts a very long time in summer, nor does it last a very short time in winter ... The southwestern winds blow all day round with no wind blowing from the northeast. Winter and spring winds kick up dust, which is the same as in the areas north of the Yangzi River. In other words, within a distance of merely two or three hundred li, the temperature and the timing for sowing differ greatly ... According to Li Yueshan, the place is far away from the Kunlun Mountains. The terrains are at an extremely high altitude, but since Dian is in the south, the climate is half-cold and half-hot. Due to the high terrains, the sun usually rises early and sets late. I am not sure whether this view is accurate [41]." Wang Shixing's conclusions on the climate features of Yunnan and Guizhou are correct. And the quote about Yunnan's climate from Li Yueshan in terms of latitude and vertical configuration is also correct.

In the *Guangyouzhi*, there is some discussion by Wang Shixing on the mechanism of climate formation, which is highly concordant with the principles in modern climatology:

The formation of cold or hot weather is partly affected by the heavens and partly by the earth. The blowing of the wind and the light of the sun are dependent on the heavens while the climate is dependent on the earth. If the earth is thin and porous, the vapour leaps up and it turns hot; if the earth is thick and dense, the vapour condenses and it turns cold. It is not simply a matter of location (south or north). I once read the *Report from Alien Regions* and learned that there is a place in regions north of the Yin Mountains where it is like spring all year round and the plants never wither.

I doubted whether the place really existed before, for the farther north a location is, the colder the climate becomes. Now I realise that this kind of place does exist, for the climate is not only related to the heavens and the location (south or north), but it is also closed related to the thickness and density of the earth. If the climate were completely dependent on the heavens, then since the high mountains are close to the heavens and near the earth, it should be hot rather than cold up there. But the fact is, the higher the mountain is, the colder it becomes. Therefore, this is possibly due to the thick soil and stones that draw the vapour down to the earth.

If expressed in modern climatological terms, the former paragraph means the temperature is not only related to the latitude (i.e. north or south), but also to the properties of the underlying surface of the Earth. The concept of "specific heat capacity" did not exist at that time, "the earth is thin and porous" and "the earth is thick and dense" were the possible explanations given the knowledge people had acquired by that time. In Wang Shixing's view, the factors affecting the coldness of the weather included both the heavens and the earth. The heavens manage "the blowing of the wind and the light of the sun". To illustrate, "the blowing of the wind" refers to atmospheric circulation; "the light of the sun" refers to solar radiation; "partly affected by the earth" refers to the influence of the underlying surface—the three factors mentioned by him are exactly the three major elements related to the formation of the climate. Wang Shixing drew the correct conclusion after climate observation and research in many locations around the country [42]. Taking all the ancient geographers in China into consideration, Wang Shixing is the only one we know of with such profound understanding.

Wang Shixing also wrote some remarkable expositions on karst landforms. Although he was outperformed by Xu Xiake who covered a wider range and delivered longer passages on this subject, the quality of what he expounded was on a par with that of Xu Xiake. The most representative examples are the 13 passages on karst landforms in Guilin in the *Travels to the Five Sacred Mountains*, almost all of which are extraordinary. The *Travels to the Five Sacred Mountains: Supplement to the Record of Guihai* reads: "Duxiu Mountain, located in the central county, covers a circular area with a radius as long as several hundred *bu*. Its height amounts to around a thousand *chi* [unit of length equal to 1/3 m]. It is a stony mountain, iron grey in colour, standing tall without jagged sides as if it were clearly cut by knife. There are no slopes or mounds in any direction. And it is not connected to any other mountain." "Seven-star Cave is located about one li east of the river. The mountainous caves there are as numerous as the Big Dipper (Beidou) in the sky. Among them, there is a cave named Qixia Cave which is halfway up the hill. It was the Mid-Autumn Festival when the Surveillance Vice Commissioner, Li Jun, and I entered the cave. We later moved on to Xingchun Cave. Li Jun was too fearful to enter, while I went straight inside. Entering the cave, I found the stones hanging up high, upside-down. I groped for the wall in the dark for more than a hundred *wu* [equal to half a step] then it became brighter, so I continued to climb up the cave without difficulty. It was even easier to climb down the cave. After descending ten steps, I found the view to be even brighter and broader. Raising my head, I caught sight of the 'carp jumping over the top of the cave'. The carp was so lifelike that I forgot it is not real … Having passed the gates, it became dark again so my eyesight could not perceive things at a distance. I can only estimate that the cave is around a hundred *chi* in width and around ten *xun* [unit of length equal to about eight *chi*] in height. With the help of a torch, I managed to see the stones in the shape of all things on earth on both sides and I asked the Taoist priests to point them out one by one: the 'elephant' curls its trunk and is lying down; the 'lion' is playing with the balls; the 'camel' has a long neck and humps; the 'Buddha of Xiang Mountain' is standing with his praying hands; the 'Cloth Bag Monk' is reclining on his side and laughing heartily with his mouth wide open. Half of the stones are stalactites, formed by drips falling for generation after generation … If you wonder who was brilliant enough to make them: it was miraculous Nature that managed to do so. The other stones are similar to beds, tables, nets, chessboards, fishes, birds and the hands and feet of Buddhas. There were too many things for my eyes to take in, nor can I remember them all or name them all. … There is also a Dragon Pond whose water was so cold and deep that I could not see the bottom. My heart throbbed with fear when I stood by it and looked down. Having covered seven or eight li, I caught sight of white waves again, in the middle of which stood a round mound about one *zhang* in length. This unique sight is colloquially named the 'Golden Hill Bathed in the Sea'. Moving along a divergent path for about one li, I encountered a landscape called the 'Buddhist Monastery' where a Bodhisattva sits halfway up the wall. The black stones that could be vaguely seen are about 7 *cun* [unit of length equal to 1/3 decimetre] in length. The air became even warmer inside. Going further eastwards from there, I caught sight of a white halo and it turned out to be an opening in the cave leading to the back of the mountain and Qinglin Temple." These two parts concerning Duxiu Mountain and the Seven-Star Cave are extremely precise, including the descriptions of the shape of Duxiu Mountain, the size of the caves, sediment of various kinds, stratification, the causes of the stalactites' formation, subterranean rivers, the climate in the caves, etc.

In addition to karst landforms, Wang Shixing also recorded other physical geographical phenomena, including the soil, animals, plants, products, the Buddhist light of E'mei Mountain, etc. All the examples mentioned above prove that Wang Shixing was not only a famous traveller in the Ming dynasty, but also a famous geographer of his era, for he accomplished outstanding achievements in geography by means of his field trips.

8.4.3 Xu Xiake's Contributions to Physical Geography

Another extraordinary traveller and geographer in the Ming dynasty was Xu Xiake (1587–1641), with the courtesy name Hongzu, the art name Zhenzhi and the alias Xiake, who was born in Jiangyin of the Southern Metropolitan Region (today's Jiangyin, Jiangsu province). He was born into a well-educated family of declining officials, so he "was extremely fond of marvellous books", especially those about history, geography or travels. And he aspired to tour throughout the Five Sacred

Mountains, or in his words "discover the mysteries of famous mountains and rivers" [43]. He was disgusted with the imperial civil examination system featuring the use of *Bagu wen*, so when he grew up, he never sat the imperial civil examination and never tried to become an official. After the age of 22, he firmly turned his back on the small world of his family and studies and embraced the vastness of nature instead by making trips and geographical investigations part of his schedule. During a 30-year period, his footprints covered more than half of China. He even visited some places more than once for the purpose of investigation. In the course of his trips, he "does not always follow the regular roads. Whenever there were places of interest, he takes a detour; he examines the general topographical features of mountains and rivers, then makes a more detailed study of them bit by bit. He ascends the mountains, no matter if there are paths or not. Even if the road is overgrown with weeds, he always manages to get through; he crosses the rivers, no matter if there are ferries or not. Even if the current is more than swift, he always manages to proceed. Sometimes a mountain is too high to climb, but he still finds ways to overcome any barriers and finally reach the summit. Sometimes a cave is too deep, but he still clings like an ape, creeps like a snake and finally travels to every single corner and every single cave. When he encounters a blind alley, he does not become anxious; when his journey is delayed, he has no regrets. He sleeps among the trees and stones and eats the fruits of the plants. He has no regard for winds and rains, is fearless of tigers and wolves, careless of travel itineraries and duration, and has no need of companions. He travels with his spirit and soul, with his body and life. From time immemorial, he is the only one to do so [44]." The above paragraph is a true portrayal of Xu Xiake's life of travel.

Apart from the harsh natural conditions, the social conditions were also extremely problematic. Given the multiple impacts of social unrest, peasants' revolts and the prevalence of thieves and invaders, travellers' personal safety could not be guaranteed. However, it was under such harsh social conditions that Xu Xiake persisted in his long-distance travel to the southwest. He resolutely left home and set foot on a journey of thousands of li. During his journey, he ran into robbers on four occasions, had goods stolen three times, and witnessed 31 cases of robbery and theft. Every time he ran out of money and food due to robbery or theft, he would sell his own belongings as a useful buffer in times of need. Since he was a commoner rather than an official, he had to travel at his own expense. Therefore, once he ran out of funds due to robbery or theft, he had to save money for travelling all over again, in a rather short period of time. One may well imagine how difficult that was. Despite the great difficulties and dangers on his journey, he always pressed forward indomitably.

Under the extremely harsh conditions, Xu Xiake not only insisted on travelling, but also insisted on writing travel notes. There was no fixed time or place for writing, which meant he wrote whenever he wanted. In hotels on the journey, in cottages, in temples, beside stones in the mountains, at roadsides, in the shade of trees, in spaces beside campfires—he could write in all these places. His works were not limited to journal-style travelogues, but also included monographs of geography, poems, epistles, and articles, etc. His manuscripts could "tower higher than the table" [45], if piled up. Later, at the turn of the Ming and Qing dynasties, a large number of his manuscripts went missing due to the wars. Even so, later generations sorted them out and the surviving work *Xu Xiake's Travels* still contains more than 600 thousand characters. The book touches upon geomorphology, geology, hydrology, climatology, biology, economic geography, transport geography, toponymy, epigraphy, literature, history, religions, buildings, products, environmental protection, nationalities, folklore, Feng Shui, etc. Among them, geomorphology, hydrology and biology account for the largest portions.

(1) Achievements in geomorphology

In 30 years of field trips, Xu Xiake had seen many different landforms. There are six types of landforms included in *Xu Xiake's Travels*—karst landforms, mountain landforms, red bed landforms, fluvial landforms, volcanic landforms and glacier landforms. There are up to 102 terms for landforms. Those in the karst landforms are especially systematic: (1) names for stony sprouts and karrens: *shigu* (stone bones), *shichi* (stone teeth*), shi'e* (calyx-like stones or blade-like stones), *hua'e* (calyx-like stones), *shiwen* (stone lines), etc.; (2) names for sinkholes: *longtan* (dragon pool) and *yuanjing*; (3) names for karst dolines: *yangfu* (up-looking cauldron), *fudi youjian* (gully below cauldron) and *jin* (well); (4) a name for shafts: *longjing* (dragon shaft); (5) names for solutional depressions: *panwa* (circular low-lying area), *dong* (cavern) and huanwa (round low-lying area); (6) a name for karst valleys: *changxing wu* (long depressed area); (7) names for karst basins: *panhe* (mountain-surrounded gully), *dian* (field) and *wu* (depressed area); (8) a name for a blind valley: *luoshui keng* (sinkhole); (9) a name for a dry valley: *kujian* (dry gully); (10) a name for a collapsed karst gorge: *xia* (gorge); (11) names for karst windows: *shixi* (stone crevice) and *shidou* (stone hole); (12) names for natural bridges: *xianren qiao* (immortal bridge), *tiansheng qiao* (natural bridge) and *shiliang* (stone bridge). The name *tiansheng qiao* first began circulating among the ordinary people. Xu Xiake introduced it into *Xu Xiake's Travels* and thence it has been handed down to the present day; (13) names for peak forest landforms: *shishan* (stone hill) and *shifeng* (stone peak); (14) a name for an isolated peak: *dushan* (isolated hill); (15) names for "light through caves": *chuanshan* (pierced hill), *chuandong* (pierced cave), *yueyan* (moon rock) and *yuliang shan* (moon hill). To summarise, with the series of 15 specific terms, Xu Xiake preliminarily established the system of proper nouns about karst landform.

Xu Xiaka explored a wide range of karst landforms, from Feilai Peak in Hangzhou in the east to the Baoshan region in southeastern Yunnan. Compared with all the other places, the range from southern Hunan to eastern Yunnan covers an area of 550 thousand square kilometres, which is much larger than the Dinaric Karst region in former Yugoslavia and the karst region of the southern Appalachian Mountains of North America. His accounts of karst landforms, which consist of about 100 thousand words, all demonstrate unprecedented scope and depth.

The karst landforms which feature most prominently are found in many provinces and cities of southeast China, south-central China and southwest China. For example, the stony sprouts, karrens, peak forests, karst dolines, sinkholes, swallet streams, natural bridges, subterranean streams, stalactites, stalagmites, and stone pillars are all well developed. Xu Xiake's descriptions of these landforms are not only vivid, but faithful to reality as well. His accounts of Yuhua Cave in Jiangyue, Fujian, and other karst landscapes nearby are generally in line with the current situation there. The place is mainly composed of karst basins and hills; subterranean streams are intermittently visible; and the karst caves can be divided into two or three layers. According to *Xu Xiake's Travels*, a group of sinkholes in Heqing, Yunnan, which has a total of 108 caves, drains all the water from the Yanggong River. "All the rivers converge into one in the middle of the mountain, flowing towards the southeastern foot of Longzhu Mountain."

Xu Xiake had acquired an in-depth understanding of the large-scale karst landform in the Southeast through field trips. He said: "Looking at the eastern border from far away, you can see high and steep mountains standing in great numbers similar to what can be seen in West Yue; apparently, the cluster of the mountain range starts from here (Luoping, Yunnan) in the west and ends at Daozhou in the northeast. The panoply of alpine mountains meanders for several thousand li, creating the fantastic scenery of the southwest. This place (Luoping, Yunnan) is the point furthest southwest [46]." Ranging from Daozhou in Hunan (now called Daoxian) to Luoping in Yunnan, most regions are formed of thick-layered limestone so the peak forest landform is well formed. Xu Xiake's arguments are highly concordant with the facts.

Xu Xiake paid a great deal of attention to the regional differences in karst landforms. The changes in karst landforms in Hunan from Qiyang to Linglin are described as follows: "Entering Qiyang, the texture of the stones becomes unusual; the colour of the stones looks glossy. After passing Qiyang, the mountains gradually become higher. Reaching Linglin, (karst terrains) proliferate everywhere [47]." Xu Xiake's observations and records are extremely detailed and accurate. He also compared the karst landforms of Liuzhou, Guilin and Yangshuo, pointing out the differences between them. Guilin and Yangshuo are regions of karst valleys and peak forests. In Xu Xiake's words, "Looking around, you can see nothing but stone pinnacles, without any earth-piled hills in between", "stone pinnacles stand in great numbers", "green, steep and towering mountains are as numerous as the trees in the forest". Liuzhou is a region of tablelands and isolated peaks where the peak forests are dispersed and the peaks are distributed in isolation on the tablelands. Isolated peaks are generally low in altitude, ranging from 50 to 100 m in height. Therefore, Xu Xiake commented that they resembled "an awl in a bag", which meant only the spires of the isolated peaks could be seen above the tablelands.

After arriving at Guiping in Guangxi, Xu Xiake compared the red bed landforms with their isolated peaks and hums in Guiping with the karst landforms with their peak forests in Guilin, Yangshuo, Liuzhou and Rongxian, and pointed out their differences: "The mountains in Guilin, Yangshuo, Liuzhou and Rongxian all tower like hairpins or bamboo shoots, but their stones are greenish-blue in colour and have a peculiar texture, arranged like layers of lotus petals. There are the impressions of footsteps and marks of climbing, so ascending the mountains is not very difficult; however, the colour of the stones in this place (Guiping) is red. There are boundless tall mountains in the shape of multi-layered cone-shaped pagodas or continuous columns towering side by side. It is only by hacking a way through the cracks that one can find a way through [48]."

After travelling through Guangxi, South Guizhou and South Yunnan, Xu Xiake also compared the karst landforms of these places, highlighting their differences: "Some mountains in West Yue are composed of a single substance while others are made of miscellaneous substances. The two types of mountains stand alone in different ways, never merging with one another. Mountains in South Dian are all surrounded by earth piled hills. Some of them are mixed with gravels, but these only account for a very small proportion. Most of them are depressions. The mountains in south Qian are a mixture of the first two, famous for their narrow and tall shape [49]." This argument is entirely accurate, for the features of karst are closely related to the characteristics of landforms. On top of the plateau, which is more than 2,000 or even 2,500 m high, small depressions, dolines and sinkholes are in the majority while some low peak forests can also be spotted. South Guizhou is a slope inclining towards the Guangxi Basin, so the groundwater mainly moves vertically. The peak forests are well-developed, featuring dense and tall peak forests, peak clusters and closed depressions. The peak forest landform typical of Guangxi is mainly distributed in limestone areas with pure lithology. Regions of impure limestones and limestones mixed with insoluble stones, however, are not that typical of peak forest landforms and even display gentle slopes and hills. Thus, Xu Xiake's arguments on the differences between karst landforms were of a highly scientific nature. His understanding in this field surpassed that of his predecessors and also that of the Western scientific community of his day.

Through investigations and comparative studies, Wang Shixing divided the stone pinnacles in Guangxi into five regions according to the differences in their peak forest landforms: (1) Li River Valley: mainly composed of peculiar pinnacles; (2) Liu River Valley: interspersed with stone pinnacles and earth-piled hills; (3) southwest Guilin: stone pinnacles in the shape of peak clusters; (4) southeast Guilin: stone pinnacles only to be found in a few regions such as Goulou Peak and Luocong Cave; (5) northwest Guilin: scattered in the shape of peak clusters. This division of karst landforms is concordant with the reality. He often made comparisons between different regions to study various kinds of geographical phenomena and came to many correct conclusions. This corresponds to the modern comparative regionalism approach.

Xu Xiake explored more than 250 caves in person, which demonstrated his keen interest in and selfless pursuit of caves. His exploration included 13 items about caves—the direction, size, type, morphostructure, speleothems, subterranean rivers (or lakes), waterfalls, living creatures, climate, acoustics, archaeology, utilisation, the colour of the stones and the cause of the formation. Taking the morphostructure of caves as an example, Xu Xiake recorded six basic forms: (1) Bottle Gourd Style: a cave system in which caves and channels of different sizes are connected. (2) Pavilion Style: the cave is divided into different layers and rooms inside. (3) Squatting Tiger Style: a pictographic name coined by Xu Xiake with the South Cave of the White Cliffs Fort in Sanlicheng in Guangxi as the most typical example. (4) Deep Well Style: as exemplified by the deep well cave of Huixian Mountain in Qingyuan of Guangxi. (5) Hall Style: normally seen in different sizes. (6) Conch Style: an unusual form, such as the upper chamber of Shanjuan Cave in Yixing, Jiangsu, and Meitian Cave in

Yongxin, Jiangxi. In Xu Xiake's era, nobody else was able to explore and study caves in as great detail as he did. His achievements in the study of caves were unparalleled in China and even throughout the world. He was the originator of speleology.

Xu Xiake was the first person to conduct on-site investigations of volcanic landforms in China and to describe their forms in vivid language. Although he neither proposed terms for volcanoes or volcanic landforms nor explained the origin of volcanic landforms, the landforms recorded by him, the pumice stones formed through volcanic eruptions and the local folklore all stand as evidence of Xu Xiake's careful investigation and descriptions concerning volcanic landforms. The studies of modern geology also prove that the places investigated and described by Xu Xiake actually experienced volcanic eruptions during the Pleistocene. The red, honeycomb-like pumice stones in Xu Xiake's accounts can still be found on Daying Mountain today. The "saddle-like hilltop of Daying Mountain" in fact refers to the form of a caldera; the hills "like two breasts on the chest" beside Daying Mountain are actually small parasitic cones.

(2) Achievements in hydrology

In *Xu Xiake's Travels* there is a wide coverage of types of waterbodies and hydrological features. In this book, there are descriptions of 551 rivers, 59 lakes, 131 ponds and pools, 8 mires and 2 seas. The hydrological characteristics of rivers in *Xu Xiake's Travels* include drainage areas, hydrological systems, the size of rivers, velocity of flow, sediment concentration, changes in water yield, water quality, watersheds, swallet streams, the relationship between topography and hydrology, regional differences in riverbeds and the relationship between the velocity of flow and the slope of the riverbed.

Examining and correcting the source of rivers was the main purpose of Xu Xiake's field trips. In *Panjiang kao* ("An Investigation of the Pan River"), he pointed out that the north branch of the Pan River flows eastwards to the Duni River (Hongshui River) rather than the You River, which corrected a mistake in the *Daming yitong zhi* ("Records of the Unity of the Great Ming"). In the *Sujiang jiyuan* ("Travel Note for Tracing the River Source"), he pointed out that the source of the Yangzi River was the Jinsha River, refuting and correcting the long-standing mistake that the Min River was the source of Yangzi River. After "travelling northwards through Three Qins [Qin (Shaanxi), Qinling, Qinchuan]", southwards to the Five Mountains and westwards to Shimen and the Jinsha River", Xu finally ascertained that the Yangzi River rises in the southern foot of the Kunlun Mountains: "It flows southwards through Shimen Pass and turns eastwards into Lijiang, thereby forming the Jinsha River; then it turns northwards into the large river of Xuzhou and finally merges with the branch from the Min River." The coverage of the Yangzi River makes it more than a thousand li longer than the "rising in the Min Mountains" version recorded in the *Yugong*.

As a result of investigations, Xu Xiake confirmed that the Lishe River is the upper branch of the Yuan River. The Lancang River and the Yuan River diverge from each other and empty into the sea. This was a correction of the mistake in the *Daming yitong zhi* which claimed that the Lishe River joined with the Lancang River and that the Lancang River fed into the Yuan River. It also corrected another mistake in the *Daming yitong zhi* and *Yongchang junzhi* ("Records of Yongchang Prefecture") which claimed that the Kuke River merged with the Lancang River.

Xu was also the first to set out the physical properties and hydrological characters of mires.

Xu Xiake had already realised that the velocity of flow was related to the slope of riverbed. Taking Jianxi Stream (today's Min River) and Ningyang Stream (today's Jiulong River) in Fujian as examples, he explained that "the shorter the distance, the faster the speed" [50], which means that when the water yields are similar, the steeper the slope of a riverbed is, the faster the velocity of the flow becomes. This view is completely correct. Xu was the first person to posit this view in China and throughout the whole world.

With regard to the relationship between topography and hydrology, Xu Xiake mentioned two aspects: first, topography affects the convergence of surface runoff; second, topography affects the flow direction of the waters.

In Wugong Mountain in Jiangxi, Xu Xiake discovered that hydrological situations varied significantly between Xianggong Ling (mountain) and the Lutai Terrace. On Xianggong Ling it is "hard to obtain a drip of water" while on the Lutai Terrace, although higher than Xianggong Ling, "it is surrounded by numerous mountains". "Miscellaneous streams converge here" and "form a great river". What is the reason behind this? Topography provided Xu with the answer: "The eastern branch of Wugong Mountain (where Xianggong Ling is located) extends along a ridge with several sub-ranges to the sides." This kind of landform offers no space for water reservation, so the surface runoff flows away very quickly and it is hard for water to converge. "The western branch of Wugong Mountain (where Lutai Terrace is located) is surrounded by numerous cliffs, high mountains and stones." Therefore, there is space for water reservation which boosts the convergence of surface runoff, so it significantly restores the groundwater. "Since the conditions of the soil and stones are quite different, the amount of wetness also differs significantly [51]."

As regards topography's influence on the direction of the water flow, Xu Xiake referred to Baotai Mountain in Yongping, Yunnan, as a typical example: "the mountain is like an annular hook while the river is like crossed arms … the Lancang River is the 'right arm' starting from Yunlong Sub-Prefecture, flowing around the southeastern foot of the mountain and elsewhere from behind the mountain after reaching the furthest point to the east. Shamu River is the 'left arm' starting from the eastern gorge of the southern mountains, flowing around the northwestern edge of the mountain through its col and flowing elsewhere from behind the mountain after reaching the furthest point to the west. The two rivers flow—one in and one out; one to and one fro, one clockwise and one anticlockwise, both wrapping around the foot of the mountain. Moreover, the southern branch of the mountain cuts through the middle, creating two tributaries of water—one flowing from the north to the south and from the east to the west, while the other flows from the south to the north, forming a natural barrier to Baotai Mountain. What a majestic scene of crisscrossing mountains and rivers! [52]".

(3) Achievements in biology

There are over 150 varieties of plants recorded in *Xu Xiake's Travels* (grain crops and vegetables are not included). Among them, there are 62 varieties from Yunnan, accounting for 41% of the total. On the one hand, "flowers and trees in Dian are all extraordinary", so Xu Xiake wrote more about these out of curiosity. On the other hand, Xu spent a long period of time travelling in Yunnan. He visited a number of places and left a great many records.

Xu Xiake remarked on the tropical and subtropical forest vegetation in southern Yunnan: "Tufted bamboos and jumbled vines leave no space to set foot in. No path can be found further downwards, for the path is bumpy and densely overgrown … Going southwards, I broke through brambles and thorns, hardly able to see the sky above my head or tread on the ground … The couch grasses are more than one *chi* thick and more than one *zhang* tall, so it is still hard to tell things apart whether in the sky or on the ground"; "thousands of trees form a luxuriant forest, covering the sky. The vines and mosses interlock with each other, making the road extremely serene and rugged … The leaves of the lofty greenery all sway in mid-air in large clusters, so not a single ray of light can penetrate through them. There are a good many betel peppers in the mountain here which are extremely large and long. Some could even be as long as six *zhang*. There is also a tree with a diameter of one *chi*. It is dotted with tiny sprouts on the exterior, too dense to find any gaps. In the centre exists a giant tree called a *mulong* (lit. wood dragon), the lower half of which is flat. It is five *chi* wide and stretches three *chi* into the soil [53]." These descriptions reveal the basic features of tropical and sub-tropical forests: first, they are evergreen forests; second, there are various kinds of plants and not much seasonal variation in the appearance of the plant community—however, its hierarchical structure is rather complex, with vines (especially woody vines), epiphytes and parasitic plants thriving there; third, the forest is well-developed; fourth, the root system is well-established with buttress roots.

Xu Xiake also acquired some knowledge of the vertical distribution and latitudinal distribution of flora. He often conducted investigations in precipitous mountains, so it was not challenging for him to discover the vertical variation in mountain flora. On Tiantai Mountain, he saw the mountain top "blasted by a bitterly cold wind and overgrown with weeds covered with frost as deep as around one *cun* [54]." On Qipan Mountain in Qunming, he saw "no tall trees on the summit and very few tufted weeds, for the altitude is too high and the climate is too cold [55]."

Furthermore, Xu Xiake discussed how latitude affects the distribution of plants. According to his observations, "longan trees could not be found" in Nandan, Guangxi, but they were present "in large numbers" [56] in Desheng, which is around 90 km southeast of Nandan. Xu Xiake acquired his knowledge of the vertical distribution and the latitudinal distribution of flora over 200 years earlier than the German geographer Alexander Von Humboldt (1769–1859).

Xu Xiake also mentioned the geographic boundaries and ranges of fauna: "Regions north of Heqing abound with yaks, while the regions south of Shunning abound with elephants." At the present day, the northern limit where elephants are found in China has already moved southwards to Xishuangbanna, and their number has dropped dramatically as well. Xu's accounts indicate that the geographical boundaries of elephant territory have moved southwards by more than 300 km in the past 350 years.

Xu Xiake's Travels provides many resources for the study of human geography, covering the handicraft industry, mining industry, agriculture, transport, business and trade, urban settlements, ethnic minorities, social customs, local products, etc., and reflecting the correlation between humans and the land.

(4) Xu Xiake's status in the history of science and technology worldwide

Xu Xiake's achievements in the fields of karst on the earth's surface and speleology were 150–200 years ahead of anybody in the West. Four specific areas are outlined below.

① He was foremost in terms of the scale of the study of karst landforms. Xu Xiake investigated the karst area of southwest China, covering an area of more than 500 thousand square kilometres. This is the largest area worldwide with continuous karst landforms and also the area with the most developed tropical and subtropical karst. This exceptional natural advantage provided him with an immense space and ample material resources in order to observe the karst phenomena, analyse the mechanism of their formation and propose inter-regional analogies. Early Western scholars' investigations and statements on karst landforms were generally focused on temperate karst areas in Europe and in the vicinity of the Mediterranean. The scale is much smaller. The well-known Dinaric Karst region in former Yugoslavia only stretches for 50 thousand square kilometres.

② He was ahead of other scholars in time: in the Western world, research on tropical karst did not appear until 1845, when the German traveller Franz Wilhelm Junghuhn described the landforms of *Goenoeng Sewoe* (Javanese, referring to "innumerable mountains"). Xu Xiake worked in this field 208 year earlier than he did. As regards the caves in the Dinaric Karst region, most western scholars regard *Le Monde Souterrain*, written by Jacques Gaffarel and published in 1654, as the earliest description of it. However, the book is no longer complete. Only a few fragments of text are preserved in the National Library of France in Paris, France. Moreover, its content has a strong flavour of mysticism. Despite the consideration given to this book, it was still published more than a decade later than *Xu Xiake's Travels*. The work *Die Ehre des Herzogsthums Krain* by Johann Weichard von Valvasor, published in 1689, records some caves, sinkholes, etc. in the author's hometown and its vicinity, but it was also influenced by mysticism and folklore. The spatial scale covered in the book is comparatively smaller and it was published half a century later than Xu's work.

③ He was the pioneer as regards content: before the nineteenth century, only a few researchers had tried to observe some karst areas and explain some karst phenomena. Knowledge about karst was still growing, but sporadically and gradually. Speleology was still at the stage of observing the phenomena and amassing materials. Various karst phenomena were not treated as an integral whole nor as a specific research object. There was no clear view on the formation mechanism and geographical distribution of karst landforms either. However, in the 1630s, Xu Xiake conducted multiple large-scale investigations, presented fairly systematic descriptions of tropical and sub-tropical karst phenomena, and proposed clear scientific views on the formation mechanism and geographical distribution of karst phenomena. No contemporary Western scholar was able to reflect on as many types of karst landform as Xu did in *Xu Xiake's Travels*. In fact, Xu's approach to research by hiking across such a vast area has rarely been seen, even today. More than 350 years ago, he explored more than 200 caves entirely by himself and noted precise and detailed accounts concerning almost all the branches of speleology. Lacking devices and equipment, Xu still managed to record the shapes and structural transformation of caves, as well as the directions, sizes, divergence and combination of paths. His astounding achievements were around two centuries in advance of the rest of the world. It was not until the 1850s onwards that better monographs on karstology were finally published, such as *Das Karstphänomen* by Jovan Cvijić published in 1893.

④ He was a leader in the field of philosophical materialism: although Xu Xiake lived in the sixteenth and seventeenth centuries, he had already abandoned the religious theological understanding of nature and epistemology. He recorded the original appearance of nature without any exaggeration, which gives his work great scientific value. He extensively collected the relevant literature, but at the same time he did not blindly follow the existing book knowledge. Instead, he built on the existing knowledge to add new discoveries and viewpoints. His thoughts displayed the pragmatism and the spirit of adventure that are both typical of modern scientists and they were not at all coloured by mysticism. In other words, he tried to explain all the natural phenomena he witnessed in a materialistic way rather than turn superstitiously to the deities. Therefore, it seems entirely justified to regard him as the earliest karstologist and speleologist as well as the most outstanding traveller and geographer in the world. Moreover, *Xu Xiake's Travels* is indeed the earliest work on karstology and speleology.

Xu Xiake's academic standing around the world has now been accepted by the global community in research on the history of science. The famous English science historian, Professor Joseph Needham (1900–1995) spoke highly of Xu Xiake's achievements: [back translated] "The extraordinary traveller Xu Xiake devoted his life to investigating the vast area of Western China and Southwestern China which were little known in his day. His most remarkable discovery was identifying the real headstreams of the Xi River and Yangzi River. Furthermore, he also discovered that the Lantsang and Nujiang are two different rivers [57]." "He was not interested in officialdom nor in any religion, but was particularly passionate about science and art … over the course of thirty years he trudged the most isolated and desolate areas in China and was exposed to all kinds of difficulties and hardships … His travelogue does not read like something written by a seventeenth century scholar, but more like the fieldnotes of a twentieth century surveyor. He was not only amazingly capable of analysing different landforms, but he also systematically used various technical terms, such as *ti* ('梯', terrace), *ping* ('坪', level ground), etc. These special terms expanded the meaning of the words for daily use. He used *bu* or li to carefully mark the size of anything he recorded and refrained from using vague descriptions [58]." Needham's comments are apt rather than fulsome compliments. He certainly realised that Xu's achievements were ahead of his time, indeed, not one century ahead, but three centuries. Professor Needham's remarks further prove that Xu Xiake was a world-class traveller and geographer in the seventeenth century.

8.5 Geographic Studies on Borderlands

8.5.1 Northern Frontier Defence

In the Ming dynasty, frontier defence in the north was significant, so many works on this topic were published. Some examples are listed below.

(1) *Jiubian tushuo* by Sun Yingyuan

The existing edition of the one-volume *Jiubian tushuo* ("Annotated Illustrations of the Nine Garrison Districts") is the block-printed version from 1569, the 3rd year of Emperor Longqing's reign. Later, Huo Ji added a memorial to the emperor and changed the authorship to himself. The book records the topographical features, armaments and land tax of the nine garrison zones in the north with their maps, making it a significant geographical work on border defence in the Ming dynasty.

(2) *Sanguan zhi* by Liao Xiyan

The one-volume book *Sanguan zhi* ("Annals of Three Passes") was finished in 1545, the 24th year of Emperor Jiajing's reign, but it can be divided into four sections according to its content: Section One contains maps with explanations. The section begins with a general map of three passes and is followed by separate maps of the passes—Yanmen, Ningwu and Piantou. Each map is accompanied by an elaboration of the general situation of the corresponding pass. Section Two is devoted to Yanmen Pass (now in the north of Dai County). Section Three is devoted to Ningwu Pass (now in Ningwu County). Section Four is devoted to Piantou Pass (now in Pianguan County). The descriptions of the passes all open with the evolution of their geographical features and of the historical systems, together with the events involved in appointing officers and stationing garrisons during past dynasties. The author then continues with an introduction to the mountains, rivers, stations and towns in each pass. The content is briefer than in past dynasties, but more detailed than others in the Ming dynasty. The work is a highly valuable resource in the field of military matters.

(3) *Juyongguan zhi* by Wang Shiqiao

The *Juyongguan zhi* ("Records of Junyong Pass") comprises ten volumes and the block-printed edition completed in 1548 (the 27th year of Emperor Jiajing's reign) has survived. Juyong Pass, which is located in today's northwestern area in Changping District, Beijing, is an important pass of the Great Wall. The frontispiece is a map of Juyong Pass and its description. Volume One introduces its history, dimensions, landscapes, divisions of land, mountains and rivers, passes and city walls; Volume Two discusses troops, lookout platforms, frontier situations and arrangements; Volume Three focuses on granaries, grasslands, storehouses, military training grounds, blockhouses, taxes and corvée, annual expenditure, government offices, schools and administrative bodies; and Volume Four addresses personages, social customs, local products and mines. Volumes Five to Ten are not directly related to geography, but the imperial edicts and memorials to the emperor record the local frontier situations. This book provides rich data with an emphasis on affairs concerning warfare and the defence of frontier passes.

(4) *Shanhaiguan zhi* by Zhan Rong

The *Shanhaiguan zhi* ("Records of Shanhai Pass") consists of a single volume which contains eight categories of maps. It was completed in 1534, the 13th year of Emperor Jiajing's reign. The first section has a general map of the area from Shanhai Pass to Huanghua Town (today's Huanghuacheng in Northwest Huairou) across 28 pages. On the following page is an individual map of Shanhai Pass, depicting the area from Shanhai Pass to the settlement of the garrison town of Huanghua. The statements about the number of troops and horses provide a great deal of detail. This book mainly focuses on defensive works, forces, and other military affairs.

(5) *Sizhen sanguan zhi* by Liu Xiaozu

The ten volumes of the *Sizhen sanguan zhi* ("Annals of Four Military Towns and Three Passes") was completed between 1574 and 1576 (from the 2nd year to the 4th year of Emperor Wanli's reign). The term "four military towns" refers to Jizhou Town (with a garrison in Santunying, today's northwest Qianxi, Hebei), Liaodong Town (with a garrison in Guangning, today's Beizhen in Liaoning; the garrison was moved to Liaoyang after Emperor Longqing rose to power), Baoding Town (with garrisons in Zijing Pass and Futuyu Valley) and Changping Town (with a garrison in Changping, today's Changping district of Beijing). "Three passes" refers to Zijing Pass (today's northwest Yi County, Hebei), Juyong Pass (today's northwest Changping District of Beijing) and Shanhai Pass (in today's Qinhuangdao, Hebei). The book contains records of institutional systems (maps, administrative divisions, the evolution of local governance), of the landscapes (territory, mountains and rivers, guards on the city wall), of the army (territories, military politics and discipline, armaments), of food (transport of civil goods, the state treasury, grain turned over as taxes, and salt policies in the appendix), of transportation, of strategies for governance (historical accounts of past dynasties, rules and regulations, defence), of edicts and memorials (imperial edicts, memorials to the emperor, eunuch conferences), of official posts (the arrangement, ranks of officials, ranks of officers), of talents, of merits (meritorious service, resourcefulness and bravery, integrity) and of previous experience and background (foreign states with tributary, subordinate tributary or invader status in the appendix). Although there were several scores of works on frontier politics in the Ming dynasty, this is the most detailed one.

(6) *Wubian dianze* by Xu Rijiu

The *Wubian dianze* ("Codes and Principles of the Five Garrisons") consists of 24 volumes compiled into five collections. Collection One contains 4 volumes about Jimen and Liaozuo; Collection Two contains six volumes about Xuanfu and Datong; Collection Three contains eight volumes about three garrison zones in Shaanxi; Collection Four contains five volumes about Dian, Yue, Qian and Shu; Collection Five contains one volume about resistance against Japanese pirates. The content of the book includes warfare and defence, flexible tactics, expeditions and frontier garrisons, soldiers' pay and walled villages. Therefore, it is a distinctive integrated record of frontier passes.

(7) *Jiubian tulun* by Xu Lun

The *Jiubian tulun* ("Illustrated Treatise of the Nine Garrison Districts") is a single volume consisting of nine articles. The author wrote the book by making reference to historical literature and archives and the situation at that time. He also drew a graphic with symbols representing the nine garrisons' landforms, the accessibility of mountains and rivers, the straightness of roads and strategic points. Studying the illustration, you can roughly understand the basic defence conditions at the nine garrisons; thanks to the treatise, you can briefly understand the whole story of frontier affairs. The first edition, block-printed in 1538, the 17th year of Emperor Jiajing's reign, survives to the present day.

8.5.2 Coastal Defence

Coastal defence in the Ming dynasty was strong during the reigns of Emperor Hongwu and Emperor Yongle. The Hongwu Emperor Zhu Yuanzhang established military institutions, built fortresses, set up the navy and had ships built all along the coastline, so quite a complete coastal defence system came into being. At that time, the coastal defence was solid. Despite raids by Japanese pirates, this did not lead to disaster ultimately. But after the middle period of Emperor Jiajing's reign, the harassment by Japanese pirates became rampant because of the lax coastal defences. There was no longer any peaceful land in the regions along the southeastern coast. The people suffered many atrocities. It was under such circumstances that the great work *Chouhai tubian* ("An Illustrated Compendium on Maritime Security") appeared in order to combat the aggressors.

The *Chouhai tubian,* written by Zheng Ruozeng, comprises 13 volumes. Based on the *Illustrated Treatise on Japan* and *Illustrated Treatise on the Ten Thousand Li Coastal Defence*, he expanded the archives and documentation at Hu Zongxian's supreme headquarters and completed this book. In 1562 (the 41st year of Emperor Jiajing's reign), the book was block-printed under the patronage of Hu Zongxian. There then existed the Longqing edition, Wanli edition, Tianqi edition and the reprinted version of 1692 (the 31st year of Emperor Kangxi's reign).

It is evident that the *Chouhai tubian* was an encyclopaedia of coastal defence for the purpose of combating invaders and promoting the defence of the coastline. The main content of the 13 volumes is as follows:

Volume One: Complete Map of the Land and Map of Coastal Mountains and Sands. These two maps are the maps of the regions mentioned in the book, from Guangdong in the south to Liaodong in the north. There are more than 70 maps in total.

Volume Two: Summary of Court Officials' Missions to Japan (appendix: Sailing Manuals to Japan), Summary of Japan's Tributes and Summary of Japanese Affairs (appendix: Map of Japan and Map of Japanese Pirates).

Volume Three: Map of Coastal Counties and Prefectures in Guangdong, Record of Upheavals of Japanese Pirates in Guangdong, Official Record of Military Defence in Guangdong (appendix: Military Organisation) and Matters Concerning Guangdong.

Volume Four: Map of Coastal Counties and Prefectures in Fujian, Record of the Overthrowing of Japanese Pirates in Fujian, Official Record of Military Defence in Fujian (appendix: Military Organisation) and Matters Concerning Fujian.

Volume Five: Map of Coastal Counties and Prefectures in Zhejiang, Record of the Overthrowing of Japanese Pirates in Zhejiang, Official Record of Military Defence in Zhejiang (appendix: Military Organisation) and Matters Concerning Zhejiang.

Volume Six: Map of Coastal Counties and Prefectures in the Metropolitan Region (Jiangsu), Record of the Overthrowing of Japanese Pirates in the Metropolitan Region, Official Record of Military Defence in the Metropolitan Region (appendix: Military Organisation) and Matters Concerning the Metropolitan Region.

Volume Seven: Map of Dengzhou, Liaozhou and Liaohai, Record of the Overthrowing of Japanese Pirates in Dengzhou, Liaozhou and Liaohai, Official Record of Military Defence in Dengzhou, Liaozhou and Liaohai (appendix: Military Organisation) and Matters Concerning Dengzhou, Liaozhou and Liaohai.

Volume Eight: General Chronological Table of the Perils Posed by Japanese Pirates and Graphic of Japanese Pirates' Whereabouts.

Volume Nine: Record of Great Victories.

Volume Ten: Record of Martyrdom.

Volume Eleven to Volume Thirteen: Strategies for Governance (Parts One to Three). Among them, in Volume Eleven and Volume Twelve are the opinions and suggestions extracted from the court as regards combating, resisting and eradicating the Japanese pirates. Volume Thirteen contains the illustrated record and illustrated treatise of armaments, serving as a template for the commanders in the combat against Japanese pirates. It contains some of the most widely used weapons and warships at that time, including the latest artillery, the Frankish cannons.

Another famous work by Zheng Ruozeng, the *Zhengkaiyang zazhu* ("Miscellaneous Theses of Zheng Kaiyang"), comprising 11 volumes, was also a very important work on coastal defence. The book was originally divided into *An Illustrated*

Compendium on Maritime Security, *Strategies of Governance in Regions South of the Yangzi River*, and an *Illustrated Treatise on Four Remote Areas*, etc., each one of which was a standalone volume. During the reign of Emperor Kangxi in the Qing dynasty, the author's great-great-great grandchild Zheng Qihong and his son Zheng Ding Yuan re-edited the book and integrated the volumes into one slipcase. Since then, the book has been divided into the *Defence Map of the Great Coast* (two volumes), one volume of the *Illustrated Record of Defence along the Yangzi River*, the *Illustrated Treatise on Japan* (one volume), the *Illustrated Treatise on Korea* (one volume), the *Illustrated Treatise on Vietnam* (one volume), the *Illustrated Treatise on Ryukyu* (one volume), the *General Map on Coastal Defence* (one volume), the *Complete Map of Maritime Transport* (one volume), the *Illustrated Treatise on the Yellow River* (one volume) and the *Impractical Tax Regime in Suzhou and Songjiang* (one volume). Of these, the *General Map on Coastal Defence* was the first draft of the *Illustrated Treatise on the Ten Thousand Li Coastal Defence.* Since these two maps differ in their level of detail, both versions have therefore been retained. The *Strategies of Governance in Regions South of the Yangzi River* is no longer included in this edition for reasons unknown. This book covers an even wider range of topics with some overlapping with the *Illustrated Book for Coastal Defence.*

8.5.3 Southern Frontiers

Works on the border guard in the south (including the southwest) in the Ming dynasty are represented by the *Dianxing jilue* ("A Brief Record of the Tour to Dian"), *Haicha yulu* ("Notes on Specialties of Hainan Island"), *Chiya* ("China") and *Xinanyi fengtu ji* ("Records of Southwestern Social Customs").

(1) *Dianxing jilue* by Feng Shike

The *Dianxing jilue* is a single volume in length. The author Feng Shike had the courtesy name Yuancheng and an art name Wensuo. He became a Jinshi during Emperor Longqing's reign. His highest official post was that of vice commissioner of Huguang. The book recounts his travels in Dian. Although it is not lengthy and is short of details, it is good at capturing the geographical features. The descriptions of the characteristics of the landforms and climate of Yunnan in this book far outperform those in previous works. It concludes with ten geographical features of Yunnan:

① South Dian is the best place, for the climate in June is just like that in the autumn, so people do not need to carry fans with them or wear clothes made of kudzu-hemp cloth.
② In the harsh winter, although the snow covers the tops of the mountains, the chill in the air does not penetrate through the skin, so people do not need to sit around a fire or cover themselves up with a quilt.
③ The climate is high and dry with no intermittent drizzle. The trees may be more than ten *zhang* high. Camellia blossoms are as large as bowls. The trunks of trees are so wide that you can put your arms around them.

④ There are several tens of thousands of cornus and pine trees. The floating clouds are as beautiful as brocades.
⑤ The sun, moon and stars look much larger and brighter than when observed from other places.
⑥ The flowers are mostly exotic.
⑦ The moon is full from the fifteenth to the twentieth day of the lunar month.
⑧ Winter does not last for a short period.
⑨ Hot springs can be found everywhere.
⑩ The caves are deep, remote and bewitchingly wonderful [59].

(2) *Haicha yulu* by Gu Jie.

The *Haicha yulu* was finished in 1527, the 6th year of Emperor Jiajing's reign. This book provided comprehensive accounts of the geography of Hainan Island. The book included "everything such as mountains, rivers, strategic points, local customs and culture, and all the strange living creatures you might ever have heard of". The original manuscript has been lost and only some fragments of text have survived to the present day, such as information about historical sites, plants, the slash-and-burn cultivation style, animals, social customs, climate, seafoods, the calendar of ethnic minorities, etc. [60].

(3) *Chiya* by Kuang Lu

The *Chiya* comprises three volumes. The author Kuang Lu, with the courtesy name Zhanruo, lived on the coast of the South Sea. He travelled to Guangxi and engaged with five different chieftains, with the surnames Cen, Lan, Hu, Hou and Pan respectively. For a period of time, he served as a secretary in charge of clerical work for the female chieftain, Yunduoniang, of the Yao people. Then he noted down what he had witnessed in the *Chiya*, including "the mountains, rivers and products with concise and elegant words. The order of passages is well founded. The quality of the book is as excellent as Fan Chengda's *Guihai yuheng zhi* ('Records of the Officer of the Land of the Cassia Sea')" [61]. Since the book describes the karst landforms in various places in Guangxi, it is an important document of karst landforms in ancient China as well as a good work on frontier geography.

(4) *Xinanyi fengtu ji* by Zhu Mengzhen

The *Xinanyi fengtu ji* was completed in a single volume. The author Zhu Mengzhen, with the courtesy name Bingqi, was born in Xingan (today's Xingan County, Jiangxi). He became a Jinshi during Emperor Longqing's reign. He once served as vice censor-in-chief and then as a circuit inspector in Shanxi. The book records the geography of Yunnan and its neighbourhood, including issues related to climate, vegetation, springs, animals, agriculture, nationalities, marriage, penalties, religions, fairs, products, armaments, etc.

In addition, some travelogues or notes from the Ming dynasty also include records of frontier geography, such as Chen Cheng's *Travelling Records of Western Regions*, Xu Xiake's *Xu Xiake's Travels* and Wang Shixing's *Travels to the Five*

Sacred Mountains, Record of Extensive Travels and *A Broad Description of Human Geography.*

8.6 Hydraulic Engineering and Hydrography

8.6.1 Water Conservancy Engineering

After Emperor Chengzu (Zhu Di) of the Ming dynasty moved the capital to Beijing, there was an urgent need to improve transport capacity in order to transport food and resources from the southeast to Beijing. Therefore, the Grand Canal had to be cleared of obstructions. Previously, the Huitong River built in the Yuan dynasty only had a "narrow course and shallow water", so it had already been "overburdened" [62]. Later, it fell into disuse as it had become silted up with sand. Emperor Chengzu took the advice of ministers and ordered the Minister of Works Song Li to take charge of constructing a new Huitong River. As soon as Song Li took office, he conducted a survey of local experts in water control. Among them, an elderly man from Wenshang named Bai Ying was passionate about hydraulic works and was familiar with the geographical environment of both the east and the west. He suggested that Song Li should build a new Dai Village Dam in Dongping, on the lower reaches of the Wen River, to stop the flow of the river and entirely divert it towards Nanwang, north of Jining. Nanwang had higher terrain than Jining and it was the dividing ridge between the south and the north of Jining. If the watershed were raised, there would be no barrier to sending water to the north. When the waters converged in Nanwang, 40% of the water should be directed to Xuzhou and Pizhou, and 60% northwards to Linqing. Without any hindrance, the problem of blocked waterways which dated from the Yuan dynasty could be solved. Song Li accepted Bai Ying's suggestion and the problem was indeed solved. Song Li also used sluice gates to control the waterflow and store water section by section, enabling the boats transporting grain to sail without any impediments. At that time, the "great economic artery" of the Grand Canal was completely unblocked, creating the conditions needed for the development of the economy, politics and the military in the Ming dynasty. Bai Ying made huge contributions to the openness of the Grand Canal thanks to his rich geographic knowledge.

Although the Grand Canal was free from obstacles, it still seemed overburdened and failed to satisfy the ever-growing economic demands of the North. To eliminate the need for the economy of the North to have to rely too much on the regions south of the Yangzi River, many men of insight and foresight suggested building field irrigation projects in the North in order to promote the economy there. The most famous work on this subject is the hydraulic expert Xu Zhenming's *Lushui ketan* ("Discussions of the Lu River"). The book primarily promoted the idea that we should use manpower to control droughts and floods. For this, Xu Zhenming went on field trips east of Beijing to investigate the distribution of terrains, soil types, rivers and springs and drew a corresponding map. In addition, he put forward concrete plans, conducted experiments and popularised his results. In his view, not only could the wastelands bordering the seas be cultivated, Miyun, Pinggu, Sanhe, Jizhou, Qian'an, Lulong, Funing, Fengrun, Yutian and other places with rich water sources for irrigation could also be cultivated. When he served as a commissioner in charge of reclamation, he recruited farmers from the South to plough paddy fields. In the course of a year, he reclaimed more than 39,000 *mu* [unit of area, equal to 1/6 acre] of wastelands. He continued to reconnoitre rivers in the territory and set about building several water conservancy projects to combine the drainage of floodwater with the irrigation of fields. He emphasised the role of water conservancy: "Flood disasters persist due to lack of water conservancy projects." Water "brings about disasters if it converges and benefits if it is drained; it brings about disasters if wasted and benefits if collected [63]." He argued that people should start building water conservancy projects in the upper reaches. "Farms are prevalent in prefectures like Shuntian, Zhending and Hejian. Half of these places have low and wet landforms. Water from the upper reaches of the fifteen rivers all empty into Mao'er Bay, so the water there is certain to overflow and become sanded up. On the upper reaches, it is advisable to immediately dredge the ditches and divert water to irrigate the fields in order to disperse the force of water. On the lower reaches, we can dig more tributaries to channel the excess water. The lowest lakes can be used to retain water. We can follow the Southerners' example of building dykes on higher terrains. In this way water conservancy can be boosted and flood disasters can then be eliminated [64]."

Xu Zhenming's thoughts on water control emphasised improving the geographical environment and combining water control with utilisation, demonstrating his pursuit of pragmatism. However, Xu Zhenming's scheme of water conservancy for land reclamation did not win the approval of the eunuchs and landlords in possession of vacant fields. They all raised objections and spoke ill of Xu Zhenming in the presence of the emperor. As a result, Xu Zhenming lost his official post, returned home and died of an illness soon afterwards in 1590 (the 18th year of Emperor Wanli's reign). Not all of his objectives were achieved. His contemporary Wang Shixing (died in 1598, the 26th year of Emperor Wanli's reign) disagreed with the "water conservancy for land reclamation" view, for he believed people should not forcibly alter the geographical environment: "The soil in regions south of the Yangzi River is muddy soil, while that in regions north of the Yangzi River is sandy soil. The former is wet, while the latter is dry. The former is suitable for sowing rice, while the latter is suitable for sowing sorghum, millet, wheat and beans. The rules were created by nature in the very beginning, so they should not be forcibly changed. Xu proposed building paddy fields in suburban Beijing in the *Lushui ketan*. There did exist a few paddy fields between Yutian and Fengrun, and it is true that the rice harvested there was very tasty, but this does not mean other places in Beijing were as suitable for growing rice. I doubted it from the very beginning, but I thought it was just casual talk by scholars. I did not expect that they would really take action. People forcibly excavated small reservoirs out of land with no water at all. It took the effort required to build ten *mu* of fields to build only one *mu* under such conditions. When the reservoirs were finished, they failed to retain water because they were made of sandy soil. Therefore, when it rained heavily, the water overflowed; when the rain stopped, the reservoirs dried up again. The northerners are lazy by nature and disregard supervision, so in fact the action almost stirred up trouble among the people. Fortunately, Xu was indicted, and the project was halted. I also heard that Shen Dacheng once built a hundred *qing* [unit of area, equal to 6.6667 ha] of fields on the estuary of Zhigu [Tianjin]. Years later, the fields were registered and ready for levies. However, they were swamped by the tide in a single night. It is thus clear that we must be neither so cowardly that we take no action nor unduly ambitious to accomplish achievements. Taking actions to respond to things when they happen is safer [65]." What Wang Shixing said was sensible to a certain extent. Therefore, apart from the objections by eunuchs and landlords, the flaws in Xu Zhenming's water control approaches also led to his indictment.

The breaching and rechannelling of the Yellow River happened most frequently in the Ming and Qing dynasties. In the Ming dynasty, it was breached every seven mouths on average. The river flooded frequently and the courses of the river, which stretched northwards or southwards intermittently, were very unstable and disorderly. Thus, managing the river was strenuous work. During the 138 years from Emperor Hongwu's rule to Emperor Hongzhi's rule, the Yellow River joined the Huai River and emptied into the Yellow Sea most of the time, or it flowed northeastwards through Shouzhang, passed through the Grand Canal and emptied into the Bohai Sea at other times. With regard to river management, the Ming people concentrated on the North rather than the South and the main focus of their work was to assure the river transport of grain. The pattern was "building dykes on the north bank and diverting waterflows on the south bank". During the 138 years from Emperor Zhengde's rule to Emperor Chongzhen's rule, the threat shifted to Shandong and the Southern Metropolitan Region. It was not until Pan Jixun joined in the river management work that the river courses were consolidated into one, flowing through the Huai River into the Yellow River. After Zhengde's rule, river-dredging activity was increased. Liu Tianhe, Zhuheng, Wangong, Pan Jixun and other famous figures involved in river dredging were all born during that period of time. In terms of strategies, "ensuring the safe transport of grain" had always been the top priority, and another task of "protecting the imperial tomb of Zhu Yuanzhang" was added to the agenda during Emperor Jiajing's reign, which made the river management work even more complicated.

In terms of theories on river management, Pan Jixun (1521–1595), who served as the superintendent of the Yellow River on four occasions inherited and developed the theories of "building dykes to restrain the water flow and washing away deposited sand with the water flow". He realised that "if the water converges, the force of water will become stronger, then the sand will be washed away and the river channel will become deeper. Build dykes to restrain the flow of the water and wash away the deposited sand with the water flow. Since the water no longer floods the banks, it will wash through the riverbed, which is the inevitable outcome. Therefore, consolidating the tributaries of the Yellow River actually helps divert the water flow" [66]. Because of this, Pan Jixun built dykes to consolidate the tributaries of the river according to the slope of the terrain. That is to say, he "made use of the clear water from the Huai

River to wash away the sand and silt of the Yellow River", and he "built the Gaoyan (the High Dam) to divert water from the Huai River to feed into the flow of clear water". "The estuary was thereby clear of blockages [67]." The impact of Pan Jixun's river management efforts was satisfying. The method not only increased the depth of the river channel, but also improved the sand deposition on the estuary through alluviation, which was an unprecedented achievement.

There were some drawbacks to Pan Jixun's river management: to ensure the openness of the canal and the security of Gaoyou and Baoying, he built the High Dam in the area west of Huai'an County in Jiangsu. But the dam blocked the channel from the Huai River to the Huliu River, causing deposition on the lower reaches of the Huai River, an increase in the water levels of the upper reaches and severe floods in Si County.

8.6.2 *Studies of Water Conservancy*

(1) *Lushui ketan* by Xu Zhenming

The *Lushui ketan* ("Discussions of the Lu River") was completed in a single volume. The author Xu Zhenming, with the courtesy name Rudong or Bocheng in another version, was born in Guixi (in today's Jiangxi). He became a Jinshi during Emperor Longqing's reign, and his highest official post was Vice Minister of the Seals Office. When he served as the supervising secretary of the Office of Scrutiny for Works, he submitted a memorial on water conservancy near the capital. He wrote this book on the way to Tongzhou. He elaborated on his opinions on water conservancy near the capital in the style of a dialogue between a host and a guest. His main thought was to use manpower to regulate floods and droughts caused by changes in rainfall: provided that manpower was employed to its full extent, natural disasters could be overcome. This was a legacy of the ancient belief that "man is bound to conquer nature", which has positive resonance. However, he did not realise that the relationship between human beings is sometimes even more important than productive forces. Therefore, ultimately, his ideals and aspirations were not realised due to obstruction from the ruling class of the feudal society [68].

(2) *Wenshui ji* by Liu Tianhe

The *Wenshui ji* ("Water Conservancy of the Yellow River and Canals") comprises six volumes. The author Liu Tianhe, with the courtesy name Yanghe, was born in Macheng, Hubei. His highest official post was as Minister of War. He was in charge of river management in 1535, the 14th year of Emperor Jiajing's reign. He argued that "defending the north bank of the Yellow River should be the priority" and concluded there were "six methods of planting willow trees". This work included pertinent analyses concerning the features of the water and sand of the Yellow River, as well as of the regular pattern of breaching along the Yellow River. The *Wenshui ji*, containing more than 60 thousand words, was completed in 1536, the 15th year of

Emperor Jiajing's reign. It includes 38 edicts, essays or reading notes. In the "General Argument on the Reasons for the Yellow River's Stagnation" and other relevant items, he presented the six reasons for deposition along the course of the river. There exists a block-printed edition dating from 1537, the 16th year of Emperor Jiajing's reign.

(3) *Zhishui quanti* by Wan Gong

The *Zhishui quanti* ("Means and Purposes of Water Control") comprises two volumes. The author Wan Gong, with the courtesy name Suqing, was born in Nanchang, Jiangxi. He became a Jinshi during Emperor Jiajing's reign. He served as secretary of the Bureau of Appointments of Nanjing, director of the Bureau of Evaluations and Vice Minister of the Court of Imperial Entertainments. He was appointed Left Vice Minister of War and Right Assistant Censor-in-chief, superintendent of the Yellow River and superintendent of military affairs in 1572, the 6th year of Emperor Longqing's reign. He was in charge of the governance of the Yellow River and the Grand Canal for over two years. He represented the school of thought: "build dykes to restrain the water flow and wash away the deposited sand with the water flow". The book was completed between 1577 and 1585 (the 5–13th years of Emperor Wanli's reign). It contains over 24,000 words without any preface or postscript. The narrative develops 148 items, and the content is divided into five areas: the renovation, protection and management systems of the Yellow River and the Grand Canal, a management system for grain transport, the river course of the Yellow River, the river course of the Grand Canal and theories of river governance. It was a very influential work on river governance and transport in the late Ming dynasty. This edition was reprinted during the reign of Emperor Wanli.

(4) *Hefang yilan* by Pan Jixun

The *Hefang yilan* ("A Survey of the Workings of River Defences") comprises 14 volumes. The author Pan Jixun (1521–1595), with the courtesy name Shiliang, was born in Wucheng (present-day Wuxing, Zhejiang). He became a Jinshi during Emperor Jiajing's reign. He served as superintendent of the Yellow River, Minister of Works and Right Censor-in-chief. During his career, he served as superintendent of the Yellow River four times, for a total of 27 years, and made great achievements. This book, which contains about 220 thousand characters, was finished in 1590, the 18th year of Emperor Wanli's reign. The last record in the book was finished in 1588, the 16th year of Emperor Wanli's reign. It is still a valuable reference today for it records the geographic correlation between the Yellow River, the Huai River and the Grand Canal, schemes of precautions against flood disasters, causes of past flood disasters and lessons learnt from them, comparative research on several water governance measures and the collection of literature on specific projects. There exists a block-printed edition dating from 1748, the 13th year of Emperor Qianlong's reign [69].

8.6.3 Knowledge in the Field of Hydrology

Hydrological knowledge could be found in various kinds of literature in the Ming dynasty. The content can be classified into the five aspects below.

(1) On the sources of the Yangzi River and the Yellow River

In 1378 (the 11th year of Emperor Hongwu's reign), Monk Zongle went on a pilgrimage to the Western Regions to collect Buddhist scriptures which had been commissioned. He returned via the source of the Yellow River in 1382 (the 15th year of Emperor Hongwu's reign). With the help of the locals, he observed the hydrological condition there and wrote the *Poem Attached to Preface on Gazing into the Source of the Yellow River*. The preface reads: "The source of the Yellow River lies in Mobilichiba Mountain, the barbarians refer to the Yellow River as "Mo" and to the Liniu River as "Bili", with the mountain "Chiba" in between as the watershed. Water from the southwest of the mountain joins the Liniu River, while water from the northeast is the source of the Yellow River [70]." The so-called "Mobilichiba Mountain" is actually the Bayan Har Mountains, and the source of the Yellow River is actually Kari Qu [71]. The Liniu River is the source the Yangzi River. Zongle made a new contribution to the history of the study of river sources by giving a definite account of the sources of the two longest rivers in China and their watershed, the Bayan Har Mountains.

Most scholars during the Ming dynasty still claimed that the "Min Mountains are the source of the Yangzi River" in their works as this was recorded in the *Yugong*, including the renowned traveller and geographer Wang Shixing. Some scholars' descriptions were unclear and ambiguous. For example, Wang Qi said that "the Yangzi River comes from the Min Mountains, but its source can be traced to the thousands of mountains in the Western barbarian regions" [72]. With regard to the source of the Yangzi River, the traveller and geographer in the late Ming dynasty, Xu Xiake, wrote the *Research on the Source of the Yangzi River* after long-term investigations and he explicitly pointed out that "when it comes to the source of the Yangzi River, the Jinsha River must come first". And he made a well-grounded argument to object to the mistake which said that "the Yangzi River comes from the Min Mountains". Tracing further upwards from the Jinsha River, Xu Xiake reckoned that the source was actually Liniushi (today's Tanggula Mountains) in the south of the Kunlun Mountains. He also asserted that people should judge the source of the Yangzi River by comparing the distance and length and he was opposed to the opinion which supported "seeking afar while forgoing what is close at hand" and "going for the tributaries instead of the main channel" [73]. All these statements are correct.

(2) Inhibiting factors of floods

Floods are influenced by rainfall: the more the rainfall, the higher the water levels. According to Pan Jixun, as regards the Yellow River, "in regions west of the upper reaches, if no rain falls, the muddy water recedes with ease and the water in the river channel flows rapidly; if it rains frequently, the muddy water struggles to recede and the water in the river flows sluggishly [74]."

Floods are influenced by the region and the season. According to Wang Fuzhi, "If the water levels of the Yellow River, Yangzi River, Huai River and streams all over the country rise at the same time, it must be raining excessively and simultaneously. However, in regions far from the source of the Yellow River, it is hard to tell whether it is rainy or sunny. The levels of the Yangzi River and Han River may rise because of melting snow. The Yellow River rises in June, as spring turns to summer. The Han River rises in summer; the Yangzi River rises in autumn; and most small streams rise in spring. The rivers rise here and subside there, never following the same schedule, all over China [75]."

Floods are influenced by the terrain. According to Lu Chen, "the Jin River flows through the gullies, which is similar in Minyue. However, even though people can cross the river on foot, the extremely strong and filthy river has strong and surging torrents. This is all because of the height of its source [76]." Xu Xiake went one step further by discussing the relationship between the slope of the riverbed and the velocity of flow. He took Ningyang Stream (today's Jiulong River) and Jianxi Stream (today's Min River) in Fujian as examples and elaborated that although the origins of these two rivers are at the same altitude, the Ningyang Stream flows a shorter distance to reach the sea, which he concluded meant "the shorter the distance, the faster the speed" [77].

Floods are influenced by lakes. According to Song Lian, "the Central Plains are vast and flat, so, without the convergence of Dongting Lake and Pengli Lake, the Yellow River often breaks through the dyke and causes damage [78]." According to Yu Ruwei, "Rivers keep flowing, while mires keep collecting water. Rivers run from their source, while mires retain water on the lower reaches. The harm caused by deposition in rivers is evident and everyone knows the significance of water control. The banks of vast lakes are breached, but few people have really devoted themselves to relevant studies. What is more, some people seek petty profits at the behest of despots and bullies, regardless of the fact that rivers and mires interact with each other. … The mires are both the gullies of the upper reaches and the sources of the lower reaches. If the mires are destroyed, the whole river disappears. Moreover, if there are large mires in the country, when the country suffers from floods, mires help to hold water preventing a sudden and drastic overflow of water; when the country suffers from droughts, mires help to store water, protecting the riverbed against drying up completely. All in all, water control is worth discussing [79]."

(3) On lakes and mires

Shen Qi discusses the erosion and deposition of lakes in the *Wujiangshui kao* ("Record of the Wujiang Waters"): "Our county (Wujiang County) has a 'collapsed lake' in the west and a 'newly raised sandbank' in the east. The soil added to the east specifically comes from the fields collapsed in the west." The term "collapsed lake" here refers to lake banks which collapsed due to the erosion effect of lake water; the expression "newly raised" here refers to the transportation and deposition of lake water. Through erosion and transportation, the lake water carried silt and soil from the west bank to accumulate on the east bank. That was how new sandbanks came into being. He also said: "I discovered shallow beaches in the shape of calipash. Although continually washed by the winds and waves, they seldom collapse or breach. To illustrate, if the bank is as straight as a wall and cuts across the trend of the waves, the bank is sure to fall; in contrast, if the slope is not steep and runs in the same direction as the waves, the bank will stand firm. Nowadays, if every household with fields were to fill their land with pebbles, rubble, soot or other kinds of small debris, then the terrain as a whole would form a slope. Consequently, with the addition of silt or plants such as water caltrops, reeds, poplars and willows, the water running in all directions would be under control and the embankments would not collapse as easily as in previous years [80]." Shen Qi's view was correct, for he not only ascertained the hydrological situation of the lakes, but also identified scientific approaches for protecting the banks from erosion.

Xu Xiake once saw many mires in his travels and recorded an especially detailed account of Ganhaizi Mire at Dazhai, Baoshan, Yunnan province: "I knew I had reached Ganhaizi when I saw the subsided gully in the west and round and flat ground in the middle like a besieged city. There were streams flowing by the residences and moving southwards into small ponds in the meadows. The ponds can be as large as a thousand *mu*. On the water grow thriving grasses and underneath them there is a mixture of floating grasses and soil. Streams also flow through in the middle, but the soil is not suitable for ploughing, for it does not retain water. If a walker touches the water with his foot, the water will be disturbed for several *zhang*. Cattle and horses can only drink water from the bank, or they will become stuck in the water if they stay in the middle for a while. Therefore, the residences are also constructed around the water's edge. People can only reclaim the slopes to sow wheat since there is no proper field in which to sow rice. In the southeast lies a gorge formed by two mountains side by side; water flows out from here and the road also stretches from here to Manao Mountain. Nevertheless, people cannot cross the middle of the sea, but they can go alongside the slopes and bays in the southwest. Therefore, I travelled south along the west cliff for more than one li and saw a round and clear pool with a diameter of over one *zhang* under the west cliff. Since it is as round as a mirror and is extremely clear, transparent and deep, it is also referred to as the Longtan (or Dragon Pool). The question is: what makes it stand out in this wilderness? [81]" This is an excerpt from a very brilliant thesis on mires, discussing the shape, size, living creatures, soil, hydrologic features, production, transportation, physical properties, etc. According to the description, Ganhaizi Mire is a peat mire with water flowing through from the middle [82].

(4) On course changes on the lower reaches of the Yellow River

In 1535, the 14th year of Emperor Jiajing's reign, the superintendent of the Yellow River Liu Tianhe put forward six reasons for rechannelling the Yellow River and engraved them on the *Illustrated Treatise of the Yellow River* tablet [83]. "The first reason: The water of the Yellow River is muddy now. Obstruction on the lower reaches, the divergent and slow water flow along the river course, the bends of the river course and the sudden rise in the water level in the hottest days of autumn all lead to deposition. The second reason: the water drops down from an extreme height in the northwest, so the dykes fail to stop the rapid current. The third reason: when deposition occurs frequently, the level of the riverbed is normally high … water flows on the surface of the ground, so the Yellow River can never be as deep as the Yangzi River. The fourth reason: there are no lakes or pools next to the Yellow River, so it cannot retain water. The fifth reason: flowing to regions downstream of Mengjin, the terrain is extremely flat, so the flow is free from the impediment of mountains. The sixth reason: in Henan … the soil is mixed with silt, so the river banks are prone to collapsing and the water is given to overflowing. As a result, every time it rains unusually in the northwest, the water from the river is certain to overflow without warning and breach the dykes. Each time the water breaches the dykes, it flows in all directions. As time goes by, deep water makes gullies and even rivers, while shallow water forms embankments through deposition. In the fortunate situation of the river course being straight, the lower reaches will also remain unimpeded for several decades. Otherwise, years later, the riverbed at both banks may rise or the water level of the river may suddenly rise, then it is impossible for the river to remain within its channel. That is why the Yellow River diverts its courses or breaches dykes so frequently." Pan Jixun also pointed out: "If the water diverts, the force of the water will weaken, then sand will accumulate and the river course will become silted up." On the contrary: "If the water converges, the force of the water will get stronger, then the sand will be washed away and the river will turn deeper." As a result, he proposed a method for rechannelling the Yellow River: "build dykes to restrain the water flow and wash away deposited sand with the water flow" and "retain limpid water to clean out the Yellow River" [84].

(5) On the water cycle

Ma Chutu, a Jinshi of 1556, the 35th year of Emperor Jiajing's reign, discussed the water cycle in *Introduction to the Study of Things*: "All the famous rivers under heaven flow down to the sea, but where does the sea eventually flow to? The answer is: at the end of the earth lies the water; at the end of the water lies the sky; the water flows to the skyline and cannot find its way to anywhere farther, so it eventually returns to the earth. And if the water from the sky eventually goes under the earth, then where does the underground water eventually flow to? The answer is: the water of the Yangzi River, the Huai River, the Yellow River and the Han River actually all spring up from underground. The underground water springs up and becomes the water of the rivers above; the water of the rivers above flows into the sea and becomes underground water again; then the underground water springs up all over again. This cycle proceeds forever without any impediments. It is just like the blood in our bodies, which flows from the feet to the head, then back to the feet from the head. The circulation seldom slows, but when it really pauses, abscesses appear." There is one mistake in this theory: sea water does not return to the earth by evaporating into clouds and falling down as rainfall. But instead, it returns to the earth via underground. However, generally speaking, his viewpoint that water circulates on the earth is correct, despite the misunderstanding of one link in the whole process.

8.7 Mapmaking and Overseas Collections

8.7.1 New Maps and Their Significance

In the early Ming dynasty, Emperor Zhu Yuanzhang ordered that all kinds of maps to be produced and submitted to the court: "In 1373, the 6th year of Emperor Hongwu's reign, all the sub-prefectures and prefectures were required to draw maps of strategic positions. In 1383, the 16th year of Emperor Hongwu's reign, regional military commissioners were required to draw maps of military institutions, geographical situations, mountains and rivers, passes and ferries, posts and forts, water and land routes, and warehouses, while the officials presented at court were required to draw maps of terrain and people. And in the summer of the 18th year of Emperor Hongwu's reign, the Terrestrial Atlas was presented to the Emperor [85]."

Maps of the Ming dynasty had a wide usage and there was a large variety of types. The number of atlases, which far exceeded the number in the Song and Yuan dynasties, included the *Guangyu tu* ("Enlarged Terrestrial Atlas"), *Huangming zhifang ditu* ("Administrative Maps of the Ming Dynasty"), *Guangyu kao* ("Enlarged Terrestrial Records"), *Yuditu kao* ("Investigating Geographical Maps of the Empires"), *Ditu zongyao* ("Compendium of Charts"), etc. Maps produced in this period are categorised as follows.

(1) Nautical charts

The *Zhenghe hanghai tu*, which is included in the *Wubei zhi* ("Treatise on Armament Technology"), was the first real nautical chart in ancient China. It was produced during Zheng He's voyage using authentic materials and is highly accurate. Following the example of the *Yangzi River Map*, the original map was drawn horizontally. The style was altered to resemble the form of a book when it was included in the *Treatise on Armament Technology* with 24 pages overall, including 1 page of preface, 20 pages of maps, 2 pages of maps on the astronomical technique, and 1 blank page. It unfolds from the right to the left. The orientation of the map is not fixed. From Nanjing to the mouth of the Yangzi River, "up" stands for "south" while "down" stands for "north". Traversing the mouth of the Yangzi River, the rule becomes "'up' for 'north' and 'down' for 'south'". Crossing the Bay of Bengal, it is basically "'up' for 'east' and 'down' for 'west'". In terms of the style of drawing, this map does not use the traditional "grid system" but uses the "view". The so-called "view" is a chart of the shape of the mountains. Provided that you match the sites in real life with those points on the chart, you can immediately find your location and at once tell the direction. Shipping routes are illustrated on the chart, as well as the directions and cruising ranges. The navigation courses and information about major sites indicated by the stars are marked out as well, such as "Calicut, 4 *zhi* [a unit measuring the height of the astrology board]", "Mogadishu, 2 *zhi* and 1 *jiao* [a unit measuring the angle between the horizontal plane and the 'eye-pole star line'] pointing at the polestar", etc. Since the scale is not shown on the chart, the distances and directions may not be accurate in some cases. Moreover, the chart varies a lot in the amount of

detail. For familiar places, the chart is quite detailed, while it can be fairly sketchy for unfamiliar ones. In order to present the coastal places and features on a narrow and limited layout, the vast Indian Ocean is forcefully condensed so that its upper half is the north coast of the Indian Ocean while the lower half becomes the narrow bay of East Africa, which leads to distortion of the graphic.

(2) Maps of administrative regions

Maps of administrative regions were abundant in the Ming dynasty and were mainly displayed in the form of atlases.

① *Yudi tu* with Yang Ziqi's postscript

This *Yudi tu* ("Terrestrial Atlas") is an atlas with Yang Ziqi's postscript immediately under the map, which is now preserved in Lushun Museum in Dalian, Liaoning province. It follows the practice of fair drawing with colour. Due to its antiquity, the colours on it are no longer clear. It was copied from the original chart between 1512 and 1513. The scale is around 1:1,760,000. The map is 164 cm long and 180 cm wide. It did not use the method of *ji li hua fang* (a plotted geometrical grid and graduated scale). The coastal lines are extremely accurate, even close to those in today's maps. The layout of the river systems, the directions of currents and the main curves are basically identical to those depicted in today's maps. Rivers are represented by double lines. The location of the source of the Yellow River is more accurate in this map than in the maps of the Song dynasty. Mountains are indicated by pictographic symbols. Administrative districts are represented by squares, circles, and rhombuses, from the highest to the lowest levels respectively. The names are noted inside these shapes, making it easier to perceive and recognise the levels of the administrative districts. The Great Wall, temples, tombs and bridges are all indicated with conspicuous symbols. The whole map includes over 20 kinds of map legend symbols, which is almost on a par with the *Guangyu tu*. Despite differences in the focus of the content and the map-making techniques, Yang's *Yudi tu* and *Guangyu tu* are quite similar to one another in many ways, which suggests the possibility that they were both based on Zhu Siben's *Yudi tu* [86].

② Li Mo's *Tianxia yudi tu* ("Terrestrial Atlas All Under Heaven")

In 1529, the 8th year of Emperor Jiajing's reign, Gui E, the Grand Academician (Daxueshi) presented his version of the *Yudi tu* to the emperor. It consisted of 17 maps with a general map on the first page, followed by one map of two capitals, one map of thirteen provincial administration commissions, and the maps of four barbarian areas affiliated to work. The maps were drawn by Li Mo. It may also be referred to as *Huangming yutu* or *Tianxia yuditu*. The maps are of a lesser quality, with the names of prefectures, sub-prefectures, counties and guard posts not listed in full [87].

③ Luo Hongxian's *Guangyu tu*

Luo Hongxian is a famous cartographer from the Ming dynasty, with the courtesy name Dafu and an art name Nian'an. He was born in Jishui, Jiangxi. In 1529, the 8th year of Emperor Jiajing's reign, he ranked first in the imperial examination for the selection of Jinshi and was given the post of Hanlin Senior Historiography Complier. In 1540, the 19th year of Emperor Jiajing's reign, he presented a memorial to the emperor with the Remonstrator Tang Shunzhi and Editing Clerk Zhao Shichun. This was against the will of the emperor, so he was dismissed from his post. After returning home, he "had few aspirations to achieve fame and wealth. Instead, Luo honed his martial arts skills and conducted further studies on cartography and history. His intensive research covered many fields, including astronomy, geographic records, rites and music, decrees and institutions, rivers and canals, frontier forts, the deployment of troops, yin-yang and arithmetic" [88]. In other words, he was fully committed to academic research. The *History of the Ming: Biography of Luo Hongxian* contained serious omissions in that it gave no account of the fact that he finished the *Guangyu tu* ("Enlarged Terrestrial Atlas"). According to the preface to the *Guangyu tu*, in the process of "studying cartography and history", he found that "although there are various maps, most of them do not avoid errors and omissions, and rectifying them is difficult". Therefore, he travelled around gathering geographical documentation in order to draw a map that was rich in content and high in quality. After three years of investigation, he obtained Zhu Siben's *Yudi tu* by chance. Through careful comparison with other atlases, he realised that Zhu's "*ji li hua fang* (a plotted geometrical grid and graduated scale) was a well-grounded method for mapping". Therefore, he redrew each map in keeping with Zhu's version and added the new materials he had collected. What he did "added to the incomplete parts of Zhu's maps". "The number of maps increased to several dozen. It took me more than ten years to include the historical transformations and development as completely as possible due to the difficulty of hiring scribes in the heart of the mountains." The atlas was finished around 1541 [89] and was named *Guangyu tu*.

It is certain that the *Guangyu tu* carried forward many of the positive aspects of the *Yudi tu* and overcame many of its drawbacks, thereby taking Zhu Siben's mapping method to a new level. In the 1555 edition (from the 34th year of Emperor Jiajing's reign), each grid is 18 mm long, which is equal to 500 li in reality (scale: around 1:15,500,000). In the map of the Northern Metropolitan Area, each grid is 17 mm long, which is equal to 100 li in reality (scale: around 1:3,200,000) [90]. The *Yudi tu* is "as long as 7 *chi*, so it is inconvenient to scroll it". To tackle this problem, Luo Hongxian adopted the format of an atlas which contains 45 maps in total. Among these, there is one general terrestrial map, 16 maps of two capitals and thirteen provincial administration commissions, 11 maps of nine garrison zones, 3 maps of the Yellow River, 3 maps of canals, 2 maps of ocean routes, 5 maps of the Tao, Songpan, Qianzhen and Mayang rivers and 4 maps of Korea, the Northern Regions, Vietnam and the Western Regions. The 16 provincial-level maps were redrawn from Zhu's *Yudi tu* while all the other maps were added by Luo Hongxian. Owing to the scarcity of documentation on extraterritorial regions, Luo Hongxian had to copy some maps without any changes from the *Shengijao guangbei tu* ("Extended Terrestrial Atlas with High Prestige") by Li Zemin from the Yuan dynasty, such as the *Dongnan haiyi zongtu* ("General Maps of Barbarians Areas on the Southeastern Sea") and *Xinan haiyi zongtu* ("General Maps of Barbarians Areas on the Southwestern Sea")

[91]. Before the *Guangyu tu* came into existence, maps already had legends, but they were not yet defined in explicit terms. Luo Hongxian applied 24 kinds of map legend symbols and "created a code to classify them". This was the first time that the names of symbols in map legends were explicitly recorded and widely used to symbolise terrains and surface features, much earlier than similar approaches by Western Europeans [92]. Moreover, all the maps of provincial-level regions have further explanations appended in charts or tables which provide additional remarks on the historical development and famous landscapes of the regions, the areas under the jurisdiction of each prefecture, sub-prefecture or county, the land tax collected, and so on.

Most Chinese maps in the sixteenth and seventeenth centuries were based on the *Guangyu tu*. It is thus clear that the atlas can boast of a vast coverage and long period of circulation and had a huge impact on the maps of later generations. For example, the *Guangyu kao* compiled by Wang Zuozhou (1604) includes a *Map of Northern Regions* in which the descriptions of Karakorum (the ancient capital of Mongolia) were all copied from those in the *Guangyu tu*; Chen Zushou's *Huangming zhifang tu* (1636) was an adaptation of the *Guangyu tu*, with the *Map of Northern Regions* in it completely identical to the one in the *Guangyu tu*. *Yudi tu* by Cheng Daosheng during Emperor Tianqi's reign, the *Ditu zongyao* by Wu Xueyan, Zhu Shaoben and others during Emperor Chongzhen's reign and the *Yutu beikao* by Pan Guangzu are all obviously drawn from the *Guangyu tu*, with minor additions or omissions. The grid maps of administrative districts in the Ming dynasty included in the *Dushi fangyu jiyao* ("Essentials of Historical Geography") by Gu Zuyu at the turn of the Ming and Qing dynasties were also all transferred from the *Guangyu tu*; even the two editions of the *Neifu yutu* ("Imperial Terrestrial Atlas") under Emperor Kangxi and Emperor Qianlong's reigns (Qing dynasty) also refer to the *Guangyu tu* and the map legends were faithfully copied from the original.

The *Guangyu tu* was reprinted seven times and hand-copied once in the Ming and Qing dynasties: first printed edition in 1555; second printed edition in 1558; hand copied edition between 1559 and 1560; printed edition by Hu Song in 1561; printed edition by Huo Ji and Han Jun'en in 1566; reprinted edition by Qian Dai in 1571; a block print of the Wanli period edition in the early Qing dynasty; a new block print by Zhang Xuelian of the Wanli period edition in 1799.

④ Chen Zushou's *Huangming zhifang ditu*

The *Huangming zhifang ditu* ("Administrative Maps of the Ming Dynasty") comprises three volumes and was completed between 1635 and 1637. There were 46 co-workers involved in the process of compiling and illustrating the map apart from Chen Zushou, so the atlas was the fruit of collective effort. When Chen Zushou led the compilation of the map, he was the Secretary of the Bureau of Operations of the Ministry of War, so he had access to confidential maps and the rare geographic books of the empire. The editorial team referred to the *Huanyu ji* ("Universal Geography of the Taiping Era [976–983]"), *Daming yitong zhi* ("Records of the Unity of the Great Ming"), *Daming guanzhi* ("Official Geography of the Great Ming"), *Guangyu kao* ("Enlarged Terrestrial Records"), *Luoji* ("Records of Luo"), Bianzhen tu ("Atlas of Frontier Districts"), *Chuanhai tu* ("Atlas of Rivers and Seas"), *Heyun* ("River Transport"), *Haiyun* ("Sea Transport"), *Jiangfang* ("River Defence"), *Haifang* ("Sea Defence") as well as to Gui E and Li Mo's *Tuxu* ("Maps with Commentaries"), Xu Lun's *Biantu* ("Frontier Charts"), Zheng Ruozeng's *Haitu* ("Nautical Charts") and other charts [93].

Since the *Huangming zhifang ditu* is an adaptation of the *Guangyu tu*, it is identical to the latter in many ways. But nonetheless, Chen Zushou's atlas is unique in the following four aspects.

First, Chen Zushou compensated for the deficiencies in the illustration of administrative divisions in Zhu Siben's atlas and in that of natural landscapes in Luo Hongxian's atlas. He treated both as (being) equally important.

Second, with regard to frontiers, Chen Zushou reckoned that "the most significant issue in relation to garrison zones lies in their defence against foreign barbarians. It is a priority to illustrate their locations in great detail." As a result, both frontier and interior factors are elaborately illustrated in his atlas [93].

Third, with regard to frontiers, Chen Zushou tried to encompass all the territories of the Ming dynasty in his charts, including areas that had temporarily fallen into the control of ethnic minorities. The aim was to motivate the generals' desire to recover the lost lands when they read the charts.

Fourth, in terms of rivers and waterway transport, Chen Zushou argued that the water system should be illustrated in great detail and that the diversion of river courses should also be marked on the charts.

Thus, the *Huangming zhifang ditu* is a map that incorporates the positive aspects and avoids the weaknesses of Zhu's and Luo's atlases. It places the emphasis on military matters. Compared with the *Enlarged Terrestrial Atlas*, it is evident that this map is characteristic of those used for military purposes, and it adds the maps of four garrison zones—Daning, Kaiping, Xinghe and Dongsheng as well as maps named *Jiangshan tu* ("Maps of Rivers and Mountains"), *Ruoshui tu* ("Maps of Waning Water"), *Heishui tu* ("Map of Amur River"), *Haifang tu* ("Map of Sea Defence"), *Taipu muma tu* ("Horse Herding for the Imperial Court"), etc.

In terms of the annotation of toponyms, the *Huangming zhifang ditu* followed the historical transformation of toponyms after Emperor Wanli's reign. The "Map of Japan" in the *Guangyu tu* was replaced by "Riben rukou tu" ("Map of Japanese Invasions"). As regards mapping techniques, special attention was devoted to the clarity and detail of the map and the design of symbols. The text of the annotations was reduced in size as much as possible to make the map neat and beautiful. All these features make it one of the most outstanding atlases of the Ming dynasty [94].

⑤ Wang Zuozhou's *Guangyu kao*

Guangyu kao is an atlas composed of two volumes. The National Library of China (previously named Beijing Library) has a fragmentary copy dating from Emperor Wanli's reign in its collection. The mapping methods and the choices made regarding significant factors are similar to those of *Guangyu tu*, so it belongs to the *Guangyu tu* system.

⑥ Cheng Daosheng's *Yuditu kao*

The *Yuditu kao* ("Investigating Imperial Maps") comprises 6 volumes. The author Cheng Daosheng, with the courtesy name Kesheng, was born in Haichang (present-day Gaozhou in Guangdong). The atlas was based on the *Records of the Unity of the Great Ming*, but it was revised to reflect the adjustment of administrative division in both Emperor Jiajing's and Emperor Longqing's reigns. It was completed during Emperor Tianqi's reign (1621–1627) and belongs to the *Guangyu tu* system. At the very beginning, there are the maps of two capitals and thirteen provincial administrative commissions. There are individual maps at the level of the provincial administrative commission, prefecture, sub-prefecture and county respectively, all annotated with the historical development of the area during the Ming dynasty [95]. The fourth volume, named "Jiubian tukao" (Illustrated Treatise of Nine Garrison Districts), contains a detailed account of the affairs of nine garrison zones and its overall format is similar to that of the *Guangyu tu*, *Huangming zhifang ditu* and *Ditu zongyao.*

⑦ *Ditu zongyao* by Wu Xueyan and others

The *Ditu zongyao* ("Compendium of Charts") comprises three volumes. It was compiled by Wu Xueyan, Zhu Guoda, Zhu Shaoben and Zhu Guogan. The first volume is the "Overview" which depicts the distances between different capitals and provincial administrative commissions, between frontier and interior areas and between China and barbarian areas. It also depicts the landscapes, the places of strategic significance and the layout of constellations. All these charts are annotated. The following volume illustrates the two capitals and thirteen provincial administrative commissions, all with annotations. The third volume illustrates the sources of the Yangzi River, the Yellow River and other rivers and canals, and the frontier areas, as well as their defence infrastructure; these are also all annotated [95]. In the second volume, the *Fenli tu* adopts the method of the *jili huafang* (a plotted geometrical grid and graduated scale) while *the Fenjie tu* does not. The latter is briefer than the former because only the boundary lines of prefectures and sub-prefectures are marked. The illustrated treatises of prefectures and sub-prefectures, on the other hand, include the famous mountains, jottings, major rivers, local products, famous eunuchs, travellers, ancient sites, passes, etc., which compensates for the deficiencies in the *Fenli tu* and *Fenjie tu.* As for the map legends, some are identical to those of the *Guangyu tu* while others are not. Some of the symbols for different levels of administrative divisions are not concise enough, such as the double-lined octagons for two capitals with toponyms inside. The provincial capitals are marked with double-lined hexagons, also with toponyms inside. Both these symbols are too complex to draw. Some obvious mistakes also exist in the atlas, such as mistaking the Min Mountains for the source of the Yangzi River, miscalculating that the Yellow River flows to the north of Vietnam and that the Nu River (the Salween River) flows to the south of Vietnam.

(3) Maps of frontier defence

During the Ming dynasty, "the foreign aggressors increased after the reign of Emperor Zhengtong (Zhu Qizhen), and when the Ming's rule came to an end, frontier defence had become a serious issue" [96]. There were nine garrison zones in the north, while the coastal areas had to tackle sea defence. Charts were essential to the major frontier defence events in both the north and the south.

① Jiubian tu

The term *jiubian* refers to the nine garrison districts, namely Liaodong, Xuanfu, Yansui, Ningxia, Gansu, Jizhou, Taiyuan and Guyuan in North China. These nine interconnected garrisons created a defence system in the north. Any map of these nine garrisons can be regarded as an example of *jiubian tu* (maps of nine garrison districts).

Almost all the military maps of the north in the Ming dynasty have *jiubian* in their names. Among the various kinds of "*jiubian tu*", the five major types are as follows: *Jiubian tulun*, *Jiubian tushuo*, *Jiubian tuzhi*, *Jiubian tu* and *Beifang biankou tu* ("Maps of the Northern Frontiers"). Each type includes several maps by different map makers at different times. For example, under the type of *Jiubian tulun*, there exist four atlases of the same name, made by Xu Lun, Zhang Guan, You Ying and Wang Guan respectively. Here, we only introduce the most representative works as examples.

Xu Lun's *Jiubian tulun* was first printed in 1538, the 17th year of Emperor Jiajing's reign. Each map has a brief commentary attached [97]. The content is factual and reliable.

Sun Yingyuan's *Jiubian tushuo* contains a general map, and separate maps and corresponding descriptions of each garrison town. The descriptions pointed out places which were hard to defend; remote and defendable places; places close to the enemy's barracks; places with weak military strength and favourable terrains, etc. They also introduced the reserves of military forces, horses, money and food. This atlas is similar to Luo Hongxian's *Jiubian tu*; therefore, it can be inferred that it had learned some lessons from the earlier work [98]. The Minister of War, Huo Ji, not only presented the *Jiubian tushuo* to the emperor, but also ordered the Grand Coordinators of Military Affairs of different towns to report the landscape and military forces of their jurisdiction to the Ministry of War in the form of *tushuo* (maps with commentaries). Based on these messages, the map maker then referred to the maps of past generations to compile *tushuo* for each town.

The author of the *Beifang biankou tu* is unknown. Unlike the standard format of *jiubian tu*, the map uses the perspective drawing method to include the scenery and landscape of garrison zones from Liaodong to Gansu in a single long scroll.

② Haifang tu

The *haifang tu* (charts of sea defence) were a type of map specific to the Ming dynasty for the purpose of fighting against Japanese pirates. In order to combat them, the Ming court "appointed Supreme Commanders, Grand Coordinators, Surveillance Vice

Commissioners, Regional Commanders, Assistant Regional Commanders, Mobile Corps Commanders and other officers in charge of military affairs" [99]. That is how the concept "sea defence" came into being. Apart from sea defence, there also existed river defence. Likewise, as well as the *haifang tu*, there were also *jiangfang tu* (maps of river defence).

Jiangfang tu can be found in Chen Zushou's *Huangming zhifang ditu* and Pan Guangzu's *Yudi tukao*.

Zheng Ruozeng's *Chouhai tubian: Haifang tu* (later included in *Wubei zhi*) comprises two major parts.

First, the *Riben rufan tu* ("Map of Japanese Aggression") gives a very detailed description of the route of Japanese aggression. In the map, "up" stands for the East and "down" stands for the West, and the shape of the coastal lines differs significantly from previous maps.

Second, the *Yanhai shansha tu* ("Map of Coastal Mountains and Islands") explicitly illustrates the location of barracks, command posts and lookouts.

The *jiangfang tu* already existed in the *Ditu zongyao* ("Compendium of Charts"). The most intact chart of river defence surviving nowadays is in the *Zheng Kaiyang zazhu* ("Miscellaneous Notes by Zheng Kaiyang") by Zheng Ruozeng. This map starts from present-day Ruichang, Jiangxi province, extending horizontally from the west to the east, to present-day Jinshan District of Shanghai. It contains 48 separate maps. In the maps, the Yangzi River is symbolised by the shape of waves, and the terrains and residential locations on both banks of the river are illustrated in detail.

③ Stone maps

As far as we know, all the maps carved on stones in the Ming dynasty were regional maps rather than national maps. For example, according to Qian Daxin in the *Zhuting riji* ("Diary of Zhuting"), he saw a map of Shaoxing Prefecture on a tablet in Hangzhou, which had been carved during the reign of Emperor Chenghua [100].

In the early 1960s, a tablet covered in maps was discovered in Huaiyuan Ancient Town, in Sanjiang County, Guangxi. Huaiyuan Ancient Town was established on an islet on the upper reaches of the Liu River during the Song dynasty. The rectangular-shaped town was 600 m long (from east to west) and 500 m wide (from north to south). A map was carved on a tablet on the left-hand wall of the archway chamber at the eastern city gate. The tablet is 83 cm wide and 132 cm long. Two maps of the county are carved on the tablet. The first map is the *Huaiyuanxian zongtu* ("General Map of Huaiyuan County") on the upper half of the tablet while the second one is the *Map of Huaiyuan Town* on the lower half of the tablet. The two maps are separated by a white line. The *General Map of Huaiyuan County* is 83 cm wide and 66 cm high. The *Map of Huaiyuan Town* is the same size as the general map. The general map contains inscriptions by Tian Heng, which indicates that the map was "renewed from old maps and carved on the tablet". The time of carving was not indicated on the tablet, but according to the county records, it was carved in the late Ming [101].

The *Map of Dengfeng County*, now preserved on a gable of Songyang Academy in Dengfeng, Henan, contains clear engravings of rivers, mountains, city walls, temples, places of interest, vegetation, etc. The map was carved in 1593, the 21st year of

Emperor Wanli's reign. Although the city wall and some buildings were gouged out from the site during "The Great Proletarian Cultural Revolution" [a political, social and cultural movement in China from 1966 to 1976], the tablet remained intact.

8.7.2 Overseas Collections of Chinese Maps

As evidenced by the extant materials, the number and location of disseminated maps from the Ming dynasty are as follows: three in Japan and one each in Spain, England, France and South Korea.

(1) Three maps of the Ming in Japan's collections

① *Huangming yitong dilizhi tu*

The *Huangming yitong dilizhi tu* ("Complete Geographical Map of the Ming") was reprinted by someone surnamed Wang in Qingquan. It was originally printed in 1536, the 15th year of Emperor Jiajing's reign, and is now held by the Museum Yamato Bankakan in Nara City in Japan. It is a woodblock-printed map, which is 103.5 cm long and 78 cm wide. There are 46 lines of text below the map, indicating the number of prefectures, sub-prefectures and counties in the two capitals and thirteen provincial administration commissions, along with a postscript by Yang Ziqi. Of the surviving maps, as a map of an administrative area, this is probably the earliest large woodcut national map from the Ming [102]. According to this map, the East, South and Southwest of China are all surrounded by a vast sea and Japan is only marked in the south of the East China Sea in tiny letters, without any outline of the boundaries. Some toponyms from the *Shanhai jing* are also marked on the southeastern seas. In addition, toponyms which include Java, Sulu, Dhofar and Malindi are marked on the west of the South China Sea and the southwestern seas. Like Yang Ziqi's *Yudi tu*, the author did not use the method of the *ji li hua fang* (a plotted geometrical grid and graduated scale).

② *Beijing cheng gongdian zhi tu*

The *Beijing cheng gongdian zhi tu* ("Map of Palaces in Beijing") was completed from 1531 (the 10th year of Emperor Jiajing's reign) to 1562 (the 41st year of Emperor Jiajing's reign), printed during Emperor Wanli's reign and is now conserved by Tohoku University in Miyagi Prefecture in Japan. It is 99.5 cm long and 49.5 cm wide. It covers the inner city of Beijing. Apart from the palaces, the government offices, altars, temples, city walls and main streets are also marked on the map. Above the map are 30 lines of text, elaborating on the names and durations of the reigns of previous emperors. The last reign name is "Wanli", which is referred to as the "current" reign name. Therefore, it can be inferred that the map was block-printed during Emperor Wanli's reign. According to the names of the palaces, the map was finished no later than 1562, the 41st year of Emperor Jiajing's reign. It is probably the earliest map of Beijing that still exists. The map does not indicate any scale; therefore, it does not display a high degree of accuracy and even contains some errors.

③ *Huangming yudi zhi tu*

The *Huangming yudi zhi tu* ("Terrestrial Map of the Ming") was revised and printed by Wu Ti in 1536 (the 15th year of Emperor Jiajing's reign), reprinted by Sun Qishu in 1631 (the 4th year of Emperor Chongzhen's reign), and is now in the collection of Kano Bunko (Kano Library) at Tohuku University and in Jingu Bunko (Jingu Library) in Japan. It is 135 cm long and 63.5 cm wide. The upper half is a complete map of China while the lower part contains words and tables, listing the number of prefectures, sub-prefectures and counties in two capitals and thirteen provincial administration commissions. Overall, the style of drafting is sketchy, and this is especially true for regions outside China. For example, "Japan" is simply marked by a rectangle with two characters "日本 (Japan)" inside.

(2) *Gujin xingsheng zhi tu* in a collection in Spain

The *Gujin xingsheng zhi tu* ("Modern and Ancient Topographic Map") was made by Yu Shi of the Ming dynasty. It was reprinted in 1555, the 34th year of Emperor Jiajing's reign, and is conserved in the General Archive of the Indies in Seville in Spain. It does not use the method of the *ji li hua fang*. It is much less accurate than the *Enlarged Terrestrial Atlas*, as significant errors exist in the mapping of the Yellow River, the Yangzi River and the coastlines. In the blank spaces on the map, brief texts are inserted to elaborate on the historical development and geographic terrains of the different regions.

(3) *Qiankun wanguo quantu gujin renwu shiji* in an English collection

The *Qiankun wanguo quantu gujin renwu shiji* ("Complete Map of Ten Thousand Nations between Heaven and Earth Illustrating Ancient and Modern People's Deeds") is a map printed in 1593, the 21st year of Emperor Wanli's reign, by Liang Zhou, the Assistant Instructor in the Confucian School of Wuxi County, Changzhou Prefecture. It is now owned by the English collector Philip Robinson. It is 171.5 cm long and 130.5 cm wide. It is a woodcut printed map. According to the introduction above the map, the map maker had referred to the *Guangyu tu*, *Kunyu wanguo quantu* ("Great Universal Geographic Map"), and Abraham Ortelius's *Theatre of the Orb of the World*, but he did not employ the advanced mapping methods in these atlases. The features of the map are as follows: first, it highlights many toponyms and provides many necessary explanations, which is similar to the *Gujin xingsheng zhi tu* ("Map of Advantageous Terrain Past and Present"). Rather than a pure atlas, it is more like a gazetteer with the necessary annotations. Second, in terms of its content, it focuses on China. Places outside China are simply marked with toponyms. Whether they are large or small, the nations are all drawn as islands on the oceans. This method of mapping was widely used in the Late Ming and Early Qing.

(4) *Wang Pan tishi yudi tu* in a collection in France

The *Wang Pan tishi yudi tu* ("Map of the Empire with Wan Pan's Inscription") was printed in 1594, the 22nd year of Emperor Wanli's reign. It was copied and added to from 1603 (the 31st year of Emperor Wanli's reign) to 1626 (the 6th year of Emperor Tianqi's reign). It is now in the collection of the National Library of France. It is 180 cm long and 190 cm wide. The extant version now is a coloured drawing on silk. The map covers as far east as Korea and Japan, as far west as the Tianshan Mountains and the Daliusha (Great Moving Sands), as far north as Mongolia and the Amur River and as far south as the South China Sea. The water system is marked in detail and the jurisdictional areas of the provinces are indicated by different colours. Symbols for toponyms are graded into different levels and used to signify over 5,000 toponyms. The author did not adopt the method of the *ji li hua fang*, but the level of accuracy is relatively high. Moreover, the orientations of the coastlines and rivers are generally correct. The bases of the mountains are marked in brown while the tops are in green; the ocean is symbolised by bluish-green waves. This map is not the original *Yudi tu* with Wang Pan's inscription, but rather a copy by an unknown Korean.

(5) Coloured map on a silk scroll in a collection in South Korea

On 10th May 1972, a map of seventeenth century China was discovered in the Central Library of Seoul National University. It is 3.62 m long and 2.5 m wide. It was drawn during the Late Ming dynasty. Several hundreds of Chinese cities are drawn in great detail on the silk scroll.

8.8 Jesuit Missionaries and the Introduction of Western Geoscience

In the Late Ming dynasty, Western missionaries arrived in China with the aim of colonial invasion. They followed in the footsteps of pirates and merchants and took their place. Missionaries had already acquired some degree of basic knowledge of the science and technology of the West at that time, so they made good use of this to win the trust of rulers in China's upper classes and this enabled them to carry out different activities. Matteo Ricci (1552–1610), the first Italian Jesuit to travel to China, was one of them. He arrived in Macao in 1580, the 8th year of Emperor Wanli's reign, and then travelled across Guangzhou, Zhaoqing, Nanchang, Nanjing and other places in China. In 1598, the 26th year of Emperor Wanli's reign, he went to Beijing for the first time and left soon afterwards. In 1600, the 28th year of Emperor Wanli's reign, he arrived in Beijing once again, and presented gifts to the emperor via the imperial eunuchs. Consequently, he was appreciated by the emperor and earned the privilege of buying land and building churches in Beijing. In the end, he died in Beijing.

Matteo Ricci took advantage of his knowledge of Western science and technology in his approach to preaching in China, introducing some subject knowledge of mathematics, astronomy and geoscience. With certain accomplishments in terms of mathematics and cartography, Matteo Ricci managed to translate and remodel traditional Chinese maps into "Western Style" maps. For example, in Alessandro Valignano's work *Huaguo qiguan*, or Wonders of the *Great Chinese Empire*, the attached Map of China was drawn by Matteo Ricci. During Matteo Ricci's stay in China, he measured the longitudes and latitudes of all the places he toured around, including Beijing, Nanchang, Hangzhou, Xi'an, Guangxi, Nanjing and other places. The level of accuracy is not significantly different from that of today's maps. According to his observations, the literati-officials were extremely interested in the World Map made by Westerners. Therefore, he decided to use the world map with explanations in Chinese as his tool for preaching [103]. This was even more evident in *Concerning the Christian Expedition to China Undertaken by the Society of Jesus* by Nicolas Trigault (1577–1628, of French nationality, who came to China in 1610): "Priest Ricci ... was quite expert in science subjects and, therefore, took on the mapping task ... He made good use of this, so many Chinese took the bait and were fascinated by the Society of Jesus." The World Map sketched by Matteo Ricci was regarded as a priceless treasure, widely reprinted and circulated and even voluntarily promoted by China's literati-official class. This map was reprinted or copied as many as 12 times in China (Table 8.1).

The *Shanhai yudi quantu* (Fig. 8.2) was augmented based on the *Wanguo quantu*. The longitudes and latitudes were all relabelled and annotations were added for the Chinese readers. This was the first world map block-printed in Chinese.

In 1595, the 23rd year of Emperor Wanli's reign, Matteo Ricci dedicated the *Wanguo tuzhi* to the Seigneur of Jian'an, Zhu Duojie. It was compiled and translated from an atlas in a language of the West brought by Matteo Ricci, but it was never printed and published, and it is now lost.

From 1595 to 1598 (from the 23rd year to the 26th year of Emperor Wanli's reign), the *Shanhai yudi quantu* was inscribed from the woodblock print by Wang Pan on the orders of Zhao Kehuai, the Grand Coordinator of Yingtian. It no longer exists.

In 1596, the 24th year of Emperor Wanli's reign, Matteo Ricci compiled the *Shijie tuji* for Wang Zuo in Nanchang. It no longer exists.

In 1599, the 27th year of Emperor Wanli's reign, at the request of Wu Zhongming, the Secretary of the Ministry of Personnel of Nanjing, Matteo Ricci revised the *Shanhai yudi quantu* in Nanjing. This book was longer in length and the annotations were also added. In the following year (1600), Wu Zhongming printed this edition.

In 1601, the 29th year of Emperor Wanli's reign, Feng Yingjing issued the *Yudi quantu* (a map of the eastern and western hemispheres). The original version of this edition is lost, but the copy of this map in Cheng Bai'er's *Fangyu shenglue: Waiyi zhuan* ("Brief on China's Geography: Biography of Barbarians") still exists todays.

In 1601, the 29th year of Emperor Wanli's reign, Matteo Ricci dedicated the entire volume of the *Wanguo tuzhi* to Emperor Shenzong. The map is now lost.

In 1602, the 30th year of Emperor Wanli's reign, Li Zhizao published the *Kunyu wanguo quantu* in Beijing. The whole atlas was divided into six maps and it was an enlarged version of Wu Zhongming's edition. In the same year (1602), a block print worker reprinted Li Zhizao's edition.

Table 8.1 The reprints and copies of Matteo Ricci's world map in China

Map name	Year	Copier	Place	Note
Shanhai yudi quantu	1584 (the 12th year of Emperor Wanli's reign)	Wang Pan (sent to the press)	Zhaoqing	
(*Shijie tuzhi*?)	1595 (the 23rd year of Emperor Wanli's reign)		Nanchang	Dedicated to Zhu Duojie (Seigneur of Jian'an)
Shanhai yudi tu	1595–1598 (from the 23rd year to the 26th year of Emperor Wanli's reign)	Zhao Kehuai (inscribed in stone)	Suzhou	Copied from Wang Pan's edition
(*Shijie tuji*?)	1596 (the 24th year of Emperor Wanli's reign)		Nanchang	Compiled for Wang Zuo
Shanhai yudi quantu	1600 (the 28th year of Emperor Wanli's reign)	Wu Zhongming (sent to the printing press)	Nanjing	Enlarged from Wang Pan's edition
Yudi quantu	1601 (the 29th year of Emperor Wanli's reign)	Feng Yingjing (sent to the printing press)	Beijing (?)	Maps of the eastern and western hemispheres
Wanguo tuzhi				One volume, presented to Emperor Shenzong (Emperor Wanli/Zhu Yijun) in 1601
Kunyu wanguo quantu	1602 (the 30th year of Emperor Wanli's reign)	Li Zhizao (sent to the press)	Beijing	Enlarged from Wu Zhongming's edition
Kunyu wanguo quantu	1602 (the 30th year of Emperor Wanli's reign)	An unknown woodblock print worker (woodblock print from the original)	Beijing	Copied from Li Zhizao's edition; the extant edition now is a facsimile by the Society of *Yugong*
Liangyi xuanlan tu	1603 (the 31st year of Emperor Wanli's reign)	Li Yingshi (sent to the printing press)	Beijing	Enlarged from Li Zhizao's edition; nowadays held in the collection of Liaoning Provincial Museum
Shanhai yudi quantu	1604 (the 32nd year of Emperor Wanli's reign)	Guo Zizhang (sent to the press)	Guizhou	Reduced size version from Wu Zhongming's edition
(*Kunyu wanguo quantu*?)	1608 (the 36th year of Emperor Wanli's reign)		Beijing	Copied by eunuchs from Li Zhizao's edition

Note "?" refers to uncertain times or places, or those translated from Western languages; "World Map" is a general name referring to maps without traceable Chinese names; "()" refers to maps which were completed, but not printed.

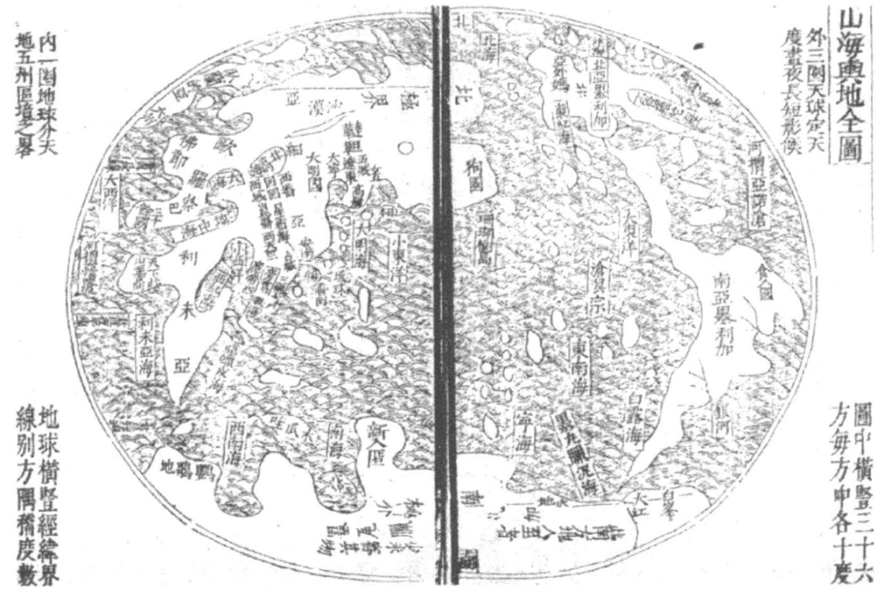

Fig. 8.2 *Shanhai yudi quantu* 山海舆地图. Jianqiu et al. [130]

In 1603, the 31st year of Emperor Wanli's reign, Li Yingshi published the *Liangyi xuanlan tu* in Beijing. The whole atlas was divided into eight maps and it was an enlarged version of Li Zhizao's edition.

In 1604, the 32nd year of Emperor Wanli's reign, Guo Zizhang reprinted Wu Zhongming's edition of *Shanhai yudi quanu* in a reduced format in Guizhou.

In 1608, the 36th year of Emperor Wanli's reign, eunuchs made several copies of Li Zhizao's edition of *Kunyu wanguo quantu* in Beijing.

In China, the extant world maps from Matteo Ricci's original are:

1) Colour-facsimile copy of the *Kunyu wanguo quantu* held in the collection of the Nanjing Museum
2) Ink-facsimile copy of the *Kunyu wanguo quantu* held in the collection of the National Museum of China
3) Block print copy of the *Liangyi xuanlan tu* held in the collection of Liaoning Provincial Museum
4) The *Kunyu wanguo quantu* facsimile copy conserved by the Society of *Yugong*
5) Facsimile copies of the *Map of the Eastern and Western Hemispheres* and *Yudi shanhai quantu* in Zhang Huang's *Tushu bian* ("Compilation of Books"), Cheng Bai'er's *Fangyu shenglue* ("A Brief Overview of the Entire Empire"), Wang Qi's *Sancai tuhui* ("Collected Illustrations of the Three Realms") and Pan Guangzu's *Yutu beikao* ("Reference Notes for a Terrestrial Atlas")
6) The *Kunyu wanguo quantu* copied by eunuchs, held in a collection in Taipei in Taiwan, China.

In countries overseas, there also exist copies of the *Kunyu wanguo quantu* (see Fig. 8.3 the copy preserved in the Nanjing Museum) in the collections held by the Vatican Apostolic Library in Rome and by Kyoto University Library in Japan and another one in the collection held by the Royal Geographical Society in England.

After the circulation of Matteo Ricci's world map, Chinese people had access to some new geographical knowledge, such as:

(1) Spherical Earth

In Matteo Ricci's world map, the Earth is illustrated as a sphere or hemispheres, which is a faithful representation of the real shape of the Earth. When elaborating on the Earth, Matteo Ricci also made great efforts to promote the theory of the Spherical Earth.

(2) Map projections

Map projections were not part of traditional Chinese mapping methods. When making maps, Matteo Ricci used three types of map projections. First, the Mollweide projection, also known as an equal-area projection. The latitudes are projected as a set of parallels. The central meridian is projected as a perpendicular straight line while other longitudes are symmetric curves concave toward the central meridian; the equator is perpendicular to the central meridian and it is double the length of the latter. The adjacent longitudes or latitudes are 10° apart from each other. The *Shanhai yudi quantu*, *Kunyu wanguo quantu* and *Liangyi xuanlan tu* all use this type of projection. Second, the Mollweide projection interrupted into two hemispheres, which is also an equal-area projection. Maps of the eastern and western hemispheres use this type

Fig. 8.3 *Kunyu wanguo quantu* 坤舆万国全图 [The great universal geographic map]. Chinese Academy of Surveying and Mapping (1998). *Zhonghua guditu zhenpin xuanji* 中华古地图珍品选集 [Selected collection of valuable maps of ancient China]. Harbin: Harbin Map Publishing House 哈尔滨地图出版社, pp 126–127

of projection. Third, the azimuthal equidistant projection. The centre of the graphic is the pole. The longitudes radiate from the pole in all directions as straight lines while the latitudes are concentric circles with the pole at their centre. Maps of the southern and northern hemispheres use this type of projection. Matteo Ricci brought map projection methods to China, enriched the Chinese people's knowledge in this regard and broadened their horizons.

(3) Division of five geographic zones on the Earth

In Matteo Ricci's map, the equator acts as the central axis dividing the northern and the southern hemispheres. Divided by the Tropic of Cancer, the Tropic of Capricorn, the Arctic Circle and the Antarctic Circle, the earth is categorised into five geographic zones, namely: the Torrid Zone, the North and South Temperate Zones, and the North and South Frigid Zones. The division into five geographic zones reflects the close relationship between latitude and climate.

(4) Distribution of water and land

Before the introduction of Matteo Ricci's world map into China, Jamāl al-Dīn al-Afghānī's terrestrial globe had already included the distribution of water and land, but it did not offer concrete names for the major continents and oceans. At that time, the Chinese did not have a comprehensive view of the world. Essentially, they only knew Asia, and parts of Europe and Africa. As for the Americas, Antarctica, and Oceania (also absent from Matteo Ricci's map), they had barely any understanding of these. However, in Matteo Ricci's world map, apart from Oceania which remained unknown at that time, the other six continents (Asia, Africa, North America, South America, Antarctica, Europe) are all set in the right locations and tagged with names. The scope of four major oceans (the Atlantic Ocean, Pacific Ocean, Indian Ocean, and Arctic Ocean) are also delineated. In summary, the general layout of water bodies and landmasses was already displayed on the map.

(5) Well-known mountains and rivers

In Matteo Ricci's world map, apart from the well-known mountains and rivers of China, famous rivers around the globe are also depicted, such as the Euphrates, the Nile, the Ganges, the Indus, the Ob, the Don, the Volga and the Amazon.

(6) Toponyms for countries and places

Some Chinese names of countries and places in Matteo Ricci's world map have already changed nowadays, such as Portugal (Fulangji → Putaoya), France (Fulangcha → Faguo), England (An'eliya → Yingguo), the Canary Islands (Fu Dao → Kaneili Qundao) and the Cape of Good Hope (Dalang Shan → Haowang Jiao). Other names have remained in use until the present day, such as Asia (Yaxiya), Europe (Ouluoba), America (Yamolijia/Yameilijia), Romania (Luomaniya), Canada (Jianada), Cuba (Guba), Chile (Zhili), Jamaica (Yamaijia), the Atlantic Ocean (Daxi Yang), the Mediterranean Sea (Dizhong Hai), the equator (Chidao), etc. [104].

In the wake of Matteo Ricci, some other westerners also began to draw world maps in China: Diego de Pantoja (Spanish, 1571–1618) drew the *Haiwai yutu* ("Map of Outside China") (not published); Giulio Aleni (Italian, 1582–1649) sketched the *Wanguo quantu* ("Complete Map of the Myriad Countries") [105]; Ferdinand Verbiest (Belgian, 1623–1688) and Michel Benoist (French, 1715–1774) both made their own versions of *Kunyu quantu*. These maps followed the same mapping methods and theories as Matteo Ricci's map, but they were not as influential. Some other contemporary maps introduced into China included: Giulio Aleni's *Zhifang waiji* ("Record of Foreign Lands") (5 volumes), *Dizhen jie* ("The Study of Earthquakes") (1 volume) by Nicholas Longobardi (Italian, 1559–1654) and *Taixi shuifa* ("Hydromethods of the Great West") (6 volumes) by Sabatino de Ursis (1575–1620).

The *Zhifang waiji* was originally translated by Diego de Pantoja from Matteo Ricci's *Wanguo quantu*, and then supplemented and polished by Giulio Aleni [106]. It "records the social customs and landscapes of exotic lands that were not included in terrestrial maps of ancient times and is hence entitled *Zhifang waiji*" [107]. The book was completed in 1623, the 3rd year of Emperor Tianqi's reign. It begins with a complete map showing myriad countries and a general map of the five major continents, and is then followed by content concerning Asia, Europe, Africa, the Americas and Antarctica respectively. As for each continent, first of all it presents a map and subsequently it provides corresponding records both in general and specifically related to each state. With regard to the states, the book gives accounts of directions, landscapes, climate, social customs, places of interest, local products, etc.—these aspects were not recorded separately. The last section "General Records of Four Major Oceans" briefs on the names of seas, islands, marine life, produce from the seas, the outline of seas, seagoing vessels, sea routes, etc. This was the first great work on world geography to be introduced to China from abroad.

The *Dizhen jie* (1 volume) was completed in 1624, the 4th year of Emperor Tianqi's reign, block printed in 1626 (the 6th year of Emperor Tianqi's reign) and reprinted in 1679 (the 18th year of Emperor Kangxi's reign). The book adopts the format of questions and answers, proposing 9 explanations for issues concerning earthquakes: (1) causes; (2) magnitudes; (3) locations; (4) sound; (5) severity; (6) timing of occurrences; (7) duration; (8) warning signs; (9) features. This was the first Western seismological work to be introduced to China, so some viewpoints contradict those of today's seismology.

The *Taixi shuifa* (6 volumes) was translated by the collective efforts of Xu Guangqi and Sabatino de Ursis in the summer of 1611, the 39th year of Emperor Wanli's reign. Their translation differs from traditional literal translation, for they only translated the applicable or advanced parts of the book, taking into consideration the hydrology tools that China already possessed. When they made tools using the methods indicated by the book, doing the corresponding tests while translating, they then took the methods of making the tools and the results of the tests into account. In the spring of 1612, the 40th year of Emperor Wanli's reign, the book was completed, making it the first drainage and irrigation monograph to be introduced into China. Apart from elaborating on drainage, irrigation and relevant tools, the book also discussed the factors affecting water quality and the methods for testing water quality.

8.9 Mineralogy and Petrology

8.9.1 *Applications of Mineral-Petrology in Mines, Buildings and Cultural Relics*

The development of mineralogy and petrology in the Ming dynasty can be described by focusing on three aspects: (1) applications of mineral-petrology in mines, buildings and cultural relics; (2) key studies on minerals and rocks; and (3) knowledge of minerals and rocks recorded in local gazetteers and notes.

(1) Knowledge of mineral-petrology in mines and buildings

In the early Ming dynasty, Emperor Zhu Yuanzhang drew lessons from the failure of the government-run mining industry of the Yuan dynasty and ordered that only the existing public mines were to be sustained; anyone who appealed for additional mining of this kind would be deprived of his official post. It was not until the 10th year of the Yongle reign (1412 AD) that the policy was changed. For example, "in 1414 (the 12th year of the Yongle reign, superintendents were sent to purchase and tax gold and silver in Chenzhou, Huguang, and Tongren, Guizhou"; "in 1421 (the 19th year of the Yongle reign), censors and National University students were sent to purchase and tax silver in Zhejiang and Fujian" [108]. Thereafter, the government-run mining industry of the Ming finally made some progress.

According to the statistics, the minerals and rocks used in the mining industry (both government-run and privately-run) included: gold, silver, copper, iron, tin, lead, zinc, alums (green, red, yellow, chalcanthite, *jishi* (lit. chicken-droppings)], cinnabar, mercury, coal, salt, kaolinite, clay, ochre, *wu ming yi* (a mineral with pyrolusite as its main component), carmine, *shiqing* (azurite), *shilü* (malachite), *shihuang* (pyrite), limestone, granite, sandstone, inkstone (with more than 10 varieties), sulfur, white arsenic (arsenopyrite), gypsum, saltpetre, mirabilite, red halloysite, orpiment, realgar, scorodite, touchstone, borax, calamine, sal ammoniac, natural gas, petroleum, jades (including Hetian jade, Lantian jade, Dushan jade, Xiuyan jade and other varieties of nephrite, as well as various kinds of jadeite), gemstones (ruby, sapphire, beryl, etc.), crystal, chrysoberyl (cat's eye), agate, amber, emerald, turquoise, etc. There were 48 kinds in total.

In the late Ming dynasty, Song Yingxing's *Tiangong kaiwu* ("Exploitation of the Work of Nature"), which was finished in 1637 (the 10th year of the Chongzhen reign), was a summary of the mining technology of the Ming dynasty. It covers not only mining technology, but also knowledge about minerals, including the names, types, properties, shapes, origins, colours, streaks, mining techniques, etc. Below are a few examples.

Volume Five "Salt" introduces six main types of salt in China, among which the records of the exploitation of sea salt and well salt are the most detailed. It also provides an example of boiling out salt using natural gas in Sichuan: "A highly amazing phenomenon is the fire wells of western Szechuan. These wells actually contain cold water, without the least sign of fire. But, split open a long bamboo stalk and take out its inner section partitions, then put the stalk together again and securely wrap it with varnished cloth. Next, place one end of this bamboo into the well, while the exposed end is connected to a curved section [of bamboo] which reaches into the bottom of a [salt]pan filled with brine. It will be seen that heat radiates [from the mouth of the bamboo] to boil the brine in the pan. Yet if the bamboo stalk is opened and examined, no sign at all of charring or evidence of burning can be seen."

Volume Seven "Ceramics: Brick" focuses on the properties of clay for making bricks: "Clay for making bricks is also obtained from the ground, in colour it is either blue, white, red or yellow (that found in Fukien and Kwangtung is mostly red; blue clay is produced chiefly in Kiangsu and Chekiang and is called "good clay"), and the best is the kind that is adhesive, fine-textured and sandless."

Volume Eleven "Calcination of Stone: Coal" outlines the geographical distribution, types, functions and exploitation techniques of coal. People used bamboo tubes to provide ventilation to ensure the safety of production activity in the mines. This was a very advanced practice at that time. The book reads: "Coal is reached in pits about fifty *chi* deep. The first appearance of the coal seam is accompanied by strong poisonous gas; therefore, some people erect a thick hollow bamboo pipe on the coal, thus extracting the poisonous gas and enabling the men to dig out and shovel up the coal underneath."

Volume Fourteen "The Metals: Gold" describes the shape, origin, hardness, colour, specific gravity, streak, ductility, etc. of natural gold. "In China, gold is produced in more than one hundred places, which we cannot enumerate here. Of the kind found in mountains, the large pieces are called "horse-shoe gold," the medium ones "olive" or "pendant gold," and the small ones "melon-seed gold." Of the gold obtained from water and sand, the large pieces are called "dog-head gold," and the small ones "wheat-husk gold" or "chaff gold." The gold obtained by underground mining is called "sand-powder gold," but the larger size is known as "bean gold." All these varieties of run-of-mine gold need to be washed and smelted before they can be made into ingots." The above excerpt elaborates on the origin, category (mountain gold and sand-powder gold) and shape of gold. "Most gold is produced in the southwest. The miners dig tunnels of some dozen *chang* into the hillside. Gold is usually found by the detection of its particular gangue … Placer gold is found mostly in the Jinsha River (formerly called Li River) in Yunnan. With its source in Tubo, this river flows through the Lijiang prefecture and reaches Beisheng district after a meandering course of more than 500 li. Gold is taken from several parts of it. In addition, gold is also panned from the river sand in places such as Tongchuan in northern Sichuan and Yuanling and Xupu in Huguang province. A piece of 'dog-head gold', which is called *jin mu* (mother gold), is obtained perhaps once in every hundred or thousand cases, and the rest [of the gold pannings] all look like wheat chaff. When smelted, the crude gold bullion will turn light yellow and it will become reddish golden after it is refined … at first it is soft to the bite." The excerpt introduces the mountain and sand-powder gold originating in Yunnan, elaborating on its shape, colour and hardness. "Gold is a heavy metal … and if a square inch of silver weighs one ounce, a piece of gold of the same dimensions weighs 1.2 [sic] ounces. Gold is also soft and can be bent like a willow twig. The grades of gold are classified in terms of 'seventy per cent green', 'eighty per cent yellow', 'ninety per cent purple' and 'pure red' gold, which can be easily determined by placing [the gold] upon the 'gold-testing stone' [touchstone]. (This stone is found in abundance in rivers in Kuangxin Prefecture, ranging from the size of a man's fist to that of a bushel, and it becomes shining black like lacquer when boiled in goose stock.)" The preceding excerpt elaborates on the specific gravity, hardness, colour and the method of judging the purity of gold from its streaks. All the excerpts concerning the properties of gold are correct, which is an illustration of the knowledge about minerals acquired by the Ming people.

As regards silver mines in Yunnan, Song Yingxing also gave a rather detailed account: "Most of the Yunnan silver is produced in Chuxiong, Yongchang, and Dali, with Qujing and Yao'an ranking second, followed by Zhenyuan. The presence of silver ore in rocks and caves is indicated by heaps of surface stones that are slightly brown in colour and scattered in such a way that they look like forked paths. Much time is required to dig the mine shafts, which are from ten to twenty *zhang* deep, and the ore can be reached when veins are found on the ground. The ore deposit is buried deep in the earth, branching off in different directions, so that each man should follow an individual vein and find the ore by tunnelling … Its proximity is indicated by the presence of small yellow stones in the vein or of thread-like marks on the stones. An ore deposit which contains native silver is called *jiao* (reef or lode); that which exists in small fragments is called *sha* [gravel, i.e. alluvial or glacial deposits]; and that which has branchlike markings on its surface is called *kuang* and is usually surrounded by valueless boulders." This excerpt introduces the origin, yield, veins and ore of silver mines in Yunnan.

According to Zhao Kuanghua, calamine ($ZnCO_3$) had already been used in the calcination of brass since the Liang period of the Southern dynasties [109]. It is thus evident that the Chinese were already familiar with this kind of mineral but did not know how to smelt zinc at that time. As of the mid-Ming period, zinc was successfully smelted in China and the earliest extant account on zinc smelting is *Tiangong kaiwu*. The so-called *wo qian* (lit. Japanese lead) is exactly metal zinc. In 1428 (the 3rd year of Emperor Xuande's reign), a large batch of brass *Ding* (cooking vessel with two loop handles and three or four legs) was cast, using as much as 13,600 *jin* (unit of weight, equal to 0.5 kg) of zinc. It is conceivable that there was rather a high level of zinc production in China at the beginning of the fifteenth century while zinc production did not begin in Europe until the eighteenth century. Therefore, China must have been producing zinc at least about 400 years earlier than Europe [110].

Volume Fourteen "The Metals: Tin" introduces the origin, category, shape and colour of tin. "Tin is produced in many places in southwestern China, but in very few in the northeastern parts of the country. Tin is called *he* in ancient books, because it was produced most abundantly in Linhe commandery (modern Hezhou, Guangxi). Eight-tenths of today's supply of tin comes from Nandan and Hechi in Guangxi, followed by Hengzhou and Yongzhou; large quantities are also produced in Dali and Chuxiong, but these places are too remote and not easily accessible. There are two kinds of tin ore: mountain tin (tin stone deposits on hillsides) and stream tin (alluvial deposits). Mountain tin includes two varieties, 'melon tin' and 'granular tin'. Melon tin is about the size of a small gourd, while granular tin is the size of beans. Both kinds are found close to the earth's surface and can therefore be accessed without mining deep underground. Sometimes a tin ore vein in the ground becomes full, causing the hillside to collapse, so that people can freely pick up the exposed tin ore. Stream tin is produced in the rivers of Hengzhou, Yongzhou, and in the river at Nandan, Guangxi. Stream tin is black and powdery, with the consistency of double-bolted flour."

Volume Fourteen "The Metals: Lead" clarifies the origin, category and smelting method of lead. "There are more lead-producing mines than there are copper or tin ones. Three kinds of lead can be distinguished: the first is derived from the ore from silver mines, with the lead enclosing the silver. A preliminary smelting will produce a lead-silver mass; following further smelting the lead is isolated from the silver and sinks to the bottom of the furnace. This is called 'silver-mine lead' and is found most abundantly in Yunnan. The second kind is derived from the ore from copper mines. When smelted in a furnace, the lead flows out first, followed by the copper. This is called 'copper-mine lead' and is produced mostly in Guizhou. The third kind is produced from the ore from lead mines. The miners dig tunnels into the hills, and with the aid of oil lamps, search for the ore veins. The path is usually twisted, as in silver mines. When the ore is washed and smelted, the lead obtained is called 'grass-joint lead', and it is produced in the greatest quantity in places such as Jiazhou and Lizhou in Sichuan. Apart from these three kinds, a pod-shaped or tadpole-shaped ore, called 'sinker lead', which is found in mountain streams, is produced in Yazhou. Furthermore, many other kinds of lead ore are produced elsewhere, such as the 'copper-mixed lead' of Shangshao in Guanxin Prefecture and of Leping in Jao Prefecture, and the *yinping* lead of Jianzhou."

Volume Eighteen "Pearls and Gems: Gems" provides information on the origin, yield, category and colour of gems. "Gemstones are produced in [underground] pits. Most gemstones occur in the various Western Barbarian regions, while in China they are produced only in Jinchiwei and Lijiang in Yunnan province. Similar to jade enclosed in its gangue crust, all large and small gemstones are encased in rock beds. Gold and silver are formed underneath a cover of accumulated earth … The stones belonging to the category of red-yellow gems are: *maojing* ('cat's-eye') [chrysoberyl], *mojieya* [carnelian], *xinghansha* [probably aventurine, sometimes called goldstone], amber, *munan* [probably beryl], *jiuhuang* [topaz], and *lazi* [ruby or rubellite]. Cat's-eye is yellow with a reddish tinge. The highest priced amber is called *yi* (worth five times the price of gold) and is a dark shade of red. Seen in daylight it is black, but under artificial light it is bright red. Beryl is pure yellow in colour, and ruby is pure red … Within the category of blue-green gems there are *pipizhu* [sapphire], *zumulü* [emerald], *yahushi* [possibly oriental bluish topaz], *kongqing* [malachite], and so on. As for the *mei gui* or 'round' gems [probably garnet or mica] which are the size of beans or green lentils, they exist in all colours—red, green, blue and yellow."

Some mistakes in previous works were also corrected in the *Tiangong kaiwu*. "The Metals: Tin" points out that "the notion that arsenic is the outcropping of tin is also erroneous". "The Metals: Iron" acknowledges that "when iron occurs on the shady side (yin) of a mountain, lodestone can be found on the sunny side (yang)", but it also adds "although there are several places where this is not so". The section "Calcination of Stone: Sulfur" indicates three misunderstandings in its predecessors: the first is that *fenshi* (lit. burned rock) was mistaken for aluminite; the second is that the so called "alum liquid" explanation is incorrect; and the third is that a solfatara is not necessarily located in the areas of thermal springs. Song Yingxing argued that "sulfur is produced by the condensation of the vapours arising from the heat treatment of rocks. Some authors wrongly believe that the heated rocks are alum rocks, and such confusion arises from the fact that sulfur is made partly from a special white variety of stones and partly from stones from coal mines which are (also) used to make alum, hence the term 'alum liquid'. It has also been maintained that sulfur is always present where there are hot springs in China, yet along the eastern seacoast and in Kwangtung in the south, sulfur is produced in places where there are no hot springs. The reason for this fallacy is that the similarity between the vapours of the hot springs and sulfur has caused people to speculate about their connection." However, Song Yingxing was incorrect in denying the existence of "native sulfur" and "liquid sulfur.

In the late Ming dynasty, Ji Cheng wrote the *Yuanzhi* ("The Craft of Gardens"), a monograph in three volumes dedicated to garden architecture. Ji Cheng, with the courtesy name Wupi and the art name Pidaoren, was born in 1582 (the 10th year of the Wanli reign) in Wujiang (today's Wujiang County, Jiangsu). In his youth, he was an artist and a poet. Approaching middle age, he started to make rockeries for manmade gardens, including the Dongdi Garden, Wu Garden, Ying Garden and other famous gardens in areas south of the Yangzi River. The *Yuanzhi* is a summary of his experience in garden construction and also the first systematic monograph on the theories of garden construction. It was first published in 1635, the 8th year of the Chongzhen reign. The original name of the book was *Yuanmu* (lit. management of gardens). Cao Yuanfu, a prominent figure at that time, thought highly of the book after reading it, and suggested that it should be renamed *Yuanye* (lit. design and construction of gardens). Since the book had some connections to Ruan Dacheng [a "treacherous court official"], it was banned during the Qing dynasty. It was preserved in the contents of the *Xianqing ouji* ("Sketches of Idle Pleasure") by Li Liweng. The original edition from Emperor Chongzhen's reign can only be found in full in the Cabinet Library of the National Archives of Japan, while no copy has yet been found in China [111]. In 1988, China Architecture and Building Press published the *Yuanzhi zhushi* ("Annotated Craft of Gardens") with Chen Zhi's annotations.

In Volume Three, there is a section named "Selection of Rocks" which selects 15 kinds of rocks that are suitable for building gardens, describing their origin, properties, category, colour, sound when struck, hardness, as well as the method of extraction, processing and utilisation. This book reflects the level of rock knowledge in the Ming dynasty. The book reads:

> Taihu rocks. The best rocks from the water's edge beside Dongting Hill in Suzhou Prefecture are those from Xiaoxia bay. They are naturally firm and glossy and within them are shapes such as "deep hollows", "eyeholes", "twists", and "strange grooves". They are of various colours: white, greyish-black and blackish-grey. The striations of the rock formation run criss-cross over them or swirl around them, rising and falling. There are pits and hollows all over the surface of the stone. These have been created by the assaults of the wind and waves and are known as "bullet holes". When tapped the rocks produce a faint sound. The workmen

who gather these rocks take their mallets and drills and go into the deep water; they select a rock of an interesting shape and chisel it away, and then bind it with a thick rope and suspend it from a large boat; then they arrange a wooden framework and hoist the rock up out of the water. These rocks are valued for their size and height and are particularly suitable for setting up individually in front of a gallery or hall, or for placing below a lofty pine-tree or strangely shaped plant. If formed into artificial mountains or positioned throughout a park or among scattered pavilions, they make a particularly fine spectacle. They have been collected for a very long time, from ancient times until the present day, so they are not easy to come by nowadays.

Kunshan rocks. The rocks in the ground near Ma'anshan in Kunshan County are covered by red soil. Once they have been dug out of the ground, you must wait for an inordinate amount of time while they are picked free of soil and are washed down. They are knobbly in shape and jut up into the air, without the sloping shoulders of a more rounded mountain peak, and when you tap them, they produce no sound. They are pure white in colour. They can be positioned amidst small trees, or have irises planted in their interstices, or else they can be set in troughs to make the focal point of a miniature landscape. They cannot be put to any major use.

Yixing rocks. The rocks from the mountains in the area around Master Zhang's Cave and Shanquan's Temple in Yixing County can be conveniently transported by water from their bamboo groves. They are hard, with eyeholes and strange grooves like those from the Great Lake. One kind is black, coarse-grained, with some yellow in it; another kind is white and smooth-grained. If used to form mountains, they cannot be made into overhanging cliffs as they may be too weak to hold firm.

Longtan rocks. The area from Longtan (or Dragon Pool), about seventy li downstream from Nanjing along the banks of the Yangzi, where there is a place called Seven Stars Convent, down to Hillmouth and Granary Head, produces several types of rock, some of which appear above ground, while others are half-buried. One type is grey and hard, with cracks and holes going right through it, like those from the Great Lake. Another type is pale grey and hard and seems rather clumsy, but it can be used to form either the foot or the summit of a mountain. Another type is mottled and rough, without any holes, and is suited to standing on its own. Another type is grey, with many markings like the wrinkles of a walnut.

Qinglong Hill rocks. There are rocks from Qinglong Hill (or Green Dragon Hill) in Nanjing which have large, twisting holes in them, but these are all enhanced by the craftsman's chisel. When used as peak stones, they only look good from one side. Most people in the past have used these as a main peak, as if they were Great Lake rocks, whereas 'patterned rocks' have generally been known as 'foot rocks', forming the foot of a mountain. When the mountain is laid out symmetrically like a censer between two vases on an altar, and then has one of these Qinglong Hill peaks added on top, it really does look like a 'hill of knives' or a 'tree of swords'.

Lingbi rocks. There is a place called Stone Chime Hill in Lingbi County, Suzhou, where there are rocks underground. These have been quarried for so many years that there is now a large pit there several fathoms deep. In their natural state, these rocks are completely covered with red mud, and the local people usually scrape them all over with an iron blade. This must be done three times, and once the natural colour of the stone comes through, they use a wire or bamboo brush together with powdered porcelain to brush it clean and make it glossy. When these stones are struck, they make a ringing, bell-like sound. These may be shaped and carved with an adze and chisel to bring out their beauty. There is a type of stone which is flat and plain-looking or has 'cloud-vapour' markings: this can be hung up in a room to act as a musical chime.

Xianshan (or Xian Mountain) rocks. The whole area around Daxian Mountain to the south of the prefectural city of Zhenjiang produces rocks. They are yellow in colour, clear, glossy and firm, and when struck they emit a sound. Some are grey-brown in colour. Most of the rocks have interconnecting eyeholes, and they are good for constructing artificial mountains.

Xuan rocks. Xuan rocks come from an area belonging to Ningguo County. They are pure white in colour, but they are mostly covered in red earth and you need to brush and wash them before you can see their natural appearance. These rocks, in particular, ought to be kept for a long time, and the older they are, the whiter they become so that they become just like a snowy mountain.

Hukou rocks. There are several types of rock from Hukou in Jiangzhou, some from underwater and some from the edge of the water. One type is grey in colour and appears naturally in the shapes of pointed and rounded peaks, cliffs and gullies, or looking like all kinds of natural objects. Another type is flat and thin with deep hollows and eyeholes running right through the stone, very like a plank of wood planed down with a sharp blade. The striations in the stone are like the bristles of a brush, and the colour is also slightly glossy. When struck, they emit a sound.

Ying rocks. Between Hanguang and Zhenyang counties in Yingzhou, there are several types of rocks, which are found in stream beds. One type is light grey with white veins winding through it. Another type is slightly greyish black, and another is a pale green, and they all have the shapes of pointed or rounded peaks with concavities and eyeholes that twist around and interconnect. The texture of this rock is slightly glossy, and a faint sound echoes, when it is struck. These rocks can be set out as table ornaments, or used as a feature of interest with a potted plant, or put together to form a miniature landscape. There is another type which is white in colour, with peaks which thrust up on all sides into many projecting points. It is slightly translucent, with reflective surfaces in which objects can be mirrored. When struck it does not resonate.

Sanbing (or "Routed Soldiers") rocks. They are located to the south of Chao Lake, and the area has stones of all sizes and every possible shape poking up all over the hills. The stone is firm and blue-black in colour.

Huang rocks. Huang rocks can still be obtained. Their texture is firm and cannot be carved with an adze or chisel, and the striations in the stone are rough and coarse. Places such as Huangshan near Changzhou, Yaofengshan near Suzhou, Tushan near Zhenjiang, and all along the Yangzi to above Caishi all produce these rocks.

Jinchuan Rocks. There are five-coloured ones, and pure green ones, and ones with markings like pine tree bark in a painting.

Patterned rock group. The patterned rocks from the Song dynasty appear all over the area of Henan which borders on Shandong.

Pebbles from Luhe. Among the sand and by the water's edge at Lingju Cliff (or Spiritual Dwelling Cliff) in Luhe County can be found agate pebbles which are very fine and small. Some are as large as a fist, pure white with five-coloured markings. Others only have the five-coloured markings. They are very warm-looking, glossy and translucent. If you pick out the ones with fine colours and patterns you can use them to lay a floor resembling brocade. Or if you place them in the bed of a stream or anywhere with flowing water, they will naturally appear bright and shining.

(2) Knowledge of mineral-petrology in cultural relics

The author examined the journals of *Wenwu* ("Cultural Relics") and *Kaogu* ("Archaeology") from before 1990 and collected the records of minerals and rocks in them. Overall, there were 36 kinds of minerals and rocks reflected in cultural relics from the Ming dynasty. Among them are 25 kinds of minerals, namely gold, silver, copper, iron, tin, lead, *wumingyi*, mercury, mica, jade, ruby, sapphire, alabaster, emerald, agate, crystal, gemstone, chrysoberyl (cat's eye), turquoise, talc, pyrophyllite (Shoushan stone), marble, white marble, limestone, and purple clay; 6 kinds of rocks, namely Duan stone, Songhua stone, alunite, red and green sandstone, biopelite and granite; 2 kinds of fossils, namely amber and beeswax; 3 kinds of organic gemstones, namely coral, pearl and ivory. A brief description of the uses of the 36 minerals and rocks above follows.

Gold: gold and silver were banned in the early Ming, but the people continued to trade with gold and silver throughout. Gold ware from the ancient tombs includes crowns, ornaments with jadeite inlays, ornaments with gemstone inlays, jewellery with gemstone inlays, rings with gemstone inlays, coins, sachets, earrings, buckles, belt hooks, ornaments with flower designs, sceptres, headpieces, necklaces, tweezers, toothpicks, earpicks, knives, bracelets, hairpins, ear pendants, a statue of Zhen Wu, bars, ingots, embroidery needles, hairpins with phoenix designs, ornaments with flower designs, flakes, ornaments with crane designs, ornaments with cow designs, buttons with decorations, beads, butterflies, dragons, buttons, pendants, dressing cases, chains, and sheets, etc.

Silver: during the Yingzong reign, silver was officially acknowledged as currency and was forged into silver ingots of different sizes. In 1955, 128 pieces of silver ingots were unearthed in a Ming dynasty tomb on Jiusheng Mountain in Hongya, Sichuan. These ingots date from different periods, such as the from the 8th, 9th and 16th years of the Zhengde reign, as well as the 1st and 2nd years of the Jiajing reign. During the Wanli reign, since silver mines had been authorised, there were two sizes of official silver coins and the smaller one has the characters "*kuang yin* [silver from official mines]" on its back around the hole in the centre [112]. Five pieces of "*jinhua yin* [substitute for grain tax]" have also been unearthed in suburban Shanghai. From the time of Emperor Yingzong's reign, land tax could be paid in silver. The regulation was that each teal [unit of weight for silver and gold, equal to about 31 g] of silver was equivalent to four *dan* [unit of dry measure for grain in China, equal to about 1 hectolitre] of rice. The Palace Treasury collected a total of more than four million *dan* of rice and wheat from Jiangxi, Zhejiang, Huguang, Fujian, Guangdong and Guangxi, which was equal to more than one million teal of silver, or Jinhua silver, in value [113]. The silverware excavated from the tombs includes hairpins (one tine), hair nets, hairpins (two tines), hairpins with phoenix designs, ingots, rings, hooks, coins, *jue* [a wine vessel with an ovoid body supported by three splayed triangular legs], earrings, coffin screws, flowers, bracelets, plates, bowls, small spoons, chopsticks, large spoons, shovels, hair bun ornaments with flower designs, pendants, buckles, buttons with decorations, joss money, foreign coins (from Venice, in Italy, and Bangladesh), etc.

Bronze: the bronze ware excavated from Ming tombs includes braziers, basins, incensories, stoves, trumpets, horns, mirrors, knives, incense burners, candle holders, small pots, square stoves, candlesticks, vases, lids, spoons, keys, stirrups, plates, scissors, gongs, ornaments for boxes or cases, wine vessels, locks, pipes, coins, *dui* [a flat-bottomed sheath for the blade of a spear], ingots, handwarmers, *gu* [a tall and slender wine vessel with a trumpet-like mouth], pitch-pots, hairpins, belt ornaments, etc.

Iron: the ironware excavated from Ming tombs includes *yinshan* (or silver mountains, but made of iron), ploughs, strings of cash, daggers, locks, suits of armour, stoves, skewers, helmets, halberds, swords, pieces of armour, heads of ornamental sceptres, knives, spears, arrowheads, bows, etc.

Tin: the tinware unearthed from Ming tombs includes ingots, sacrificial vessels (incense holders, incense burners, candlesticks, cups, flagons, pots, lamps, jars, ingots, boxes, plates), stationery products (inkstones, penholders), burial objects (*ding*, cooking pots, tea strainers, cup sets, lampstands, *bangu* [small drum used for marking time], *jue*, bowls, chopsticks, spoons, bottles, teacups, candlesticks, basins, jars with a lid, plates, dishes, saucers, boxes, pots with loop handles, oil lamps, goblets, pots), burial objects for daily use (tall covered *ding*, wine vessels, round basins, penholders, inkstones, oil lamps on high feet, creels, wine vessels with loop handles, small plates, amphoras, flagons, plates in the shape of the flower of the water chestnut), etc. In the Ming dynasty, tin vessels dominated as the vessels of the "ordinary people", as well as the prescribed raw material for burial objects. In the Ming dynasty, it was common for people to use tin vessels as burial objects. Tin is soft and pliable, so it is mostly used in alloys, but rarely used alone in the making of vessels. Before the Ming dynasty, tinware burial objects were hardly ever seen. Although tin vessels were occasionally spotted among the burial objects in the Song and Yuan dynasties, they amounted to just a small number.

Lead: the lead ware excavated from Ming tombs includes dog figurines, chicken figurines, inkstones, human figurines in tombs, small jars, candlesticks, incense burners, scissors, mirrors, land certificates (for the dead), lead coins (the Qianheng Chongbao of the Southern Han dynasty), tomb guardian jars (for the dead), bowls, pots, etc.

Wumingyi: some people regard *Wuminhyi* as "a mineral belonging to the family of ferromanganese ore, usually associated with psilomelane, pyrolusite and limonite. It forms in the mines of manganese ore or at the bottom of swamps. Its main component is MnO_2, but it is also mingled with iron [114]." Others hold that it is "extremely similar to pyrolusite". However, according to Volume One of Zhu Hang's *Detailed Records of Jinjiang*, "*Wumingzi* (another name for *wumingyi*) from Tianze Ridge in Shanggao is used to paint porcelain in Jingdezhen, Raozhou." Therefore, it could be composed of "*wanhua* (also *guhua*)" [115]. In the Ming dynasty, the *wumingyii* used for making blue-and-white porcelains did not belong to ferromanganese ore, but to asbolite whose main component is CoO (the so-called *guhua*). *wumingyi* was mainly produced in Yunnan, Fujian and Jiangxi. Among Huize, Yiliang and Xuanwei in Yunnan, the highest quality ones and those with the highest cobalt content were the varieties found in Xuanwei. According to the farmers, "*jinpian* (gold sheet)" and "*zhumi* (pearl)" were the highest quality types (cobalt content: 4.5–11%); "*lingjiao* (water chestnut)" was medium quality (cobalt content: 1–3%); "*wuya hei* (crow-like black)" was of the lowest quality (cobalt content: 0.2–0.8%). The highest quality types could be directly used in ceramics without further processing. Cobalt was used as a photographic coupler on porcelains. It had already been used in glazed ceramic beads, which were popular before the Warring States period. The "blue colour" in Tang-sancai (Tang tri-coloured glazed pottery) was also generated by using cobalt.

The discovery of blue-and-white porcelains (from both the Tang and Song dynasties) also proves that cobalt had long been used to produce the colour blue in underglazed decoration. The *Xuande ding yipu* ("Treatise on Ding during the Xuande Reign"), completed in 1428 (the 3rd year of the Xuande reign), was the first document to mention blue paint on porcelains in China. The book states: "*Wumingyi*...is used to tint *ding* blue". Volume Two states that *wumingyi* is a pigment used in porcelainising blue-and-white porcelains. Volume Three records that this batch of *wumingyi* was received from the Imperial Academy of Medicine. It is thus clear that *wumingyi* could be used as both medicine and pigment at that time. In the *Xuande ding yipu*, all imported materials were labelled with the country of origin. However, *wumingyi*, malachite and white wax were not tagged with such labels, so they were probably domestically produced. It can be inferred that the Imperial Academy of Medicine had domestic *wumingyi* in reserve [116].

Mercury: in Ming tombs, some of the dead were injected with mercury; some high-ranking officials were buried with mercury at the bottom of their coffins. A large number of mercury artefacts were unearthed in the tomb of a Ming couple in Tongliang, Sichuan [117]. This was a funerary custom inherited from the past.

Mica: mica has rarely been found in Ming dynasty tombs. A sachet of sheet mica to prevent decomposition was unearthed in the tomb of Wang Xijue in Huqiu, Suzhou (Wang was a Prime Minister and was buried in 1613). The antiseptic was very effective in the tomb, because, during the period of more than 360 years from the time of burial to 1975, the body had remained as if it had just been buried [118].

Jade: large quantities of jade ware were unearthed in Ming tombs or at the bases of towers, including hairpins, ornaments, hairpieces, rings, cicada figurines, *gangmao* (rectangular jade accessory with a hole in the centre, worn to ward off evil spirits), round flat jade pieces (with a hole in the centre, used for ceremonial purposes), *huanfang* ("magic squares", worn by Muslims as amulets), child figurines, seals (white jade seals and black jade seals are exceptional), jade belt decoration pieces, jade belts, pendants, drop pendants, beads, pig figurines, mandarin duck figurines, white pendants, *gui* (elongated tablet of jade with a triangular top held by ancient rulers during ceremonies), pipes, loops, flower ornaments attached to ribbons, human figurines, sheep figurines, fish figurines, loach figurines, drop beads, Buddhas, leather belts with jade ornaments, waist belts with jade ornaments, nose plugs [for the dead], hexagon-shaped beads, belt hooks, etc.

Ruby: many rubies were unearthed in Ming tombs. They are inlaid in gold and silver jewellery, for example, in gold flowers, in gold crowns, in gold earrings, in gold hairpins or in gold hair ornaments.

Sapphire: the sapphires unearthed in Ming tombs are similar to rubies and are also inlaid in various types of gold and silver jewellery.

Alabaster, emerald and other gemstones: those unearthed in Ming tombs are also used in ways similar to rubies.

Agate: agate ornaments have not been unearthed in large numbers in Ming tombs; they are inlaid in jewellery including seven-ridged crowns, pendants and hair nets.

Crystal: the crystal ornaments unearthed in Ming tombs include pendants, seals, and cups, etc.

Chrysoberyl (cat's eye): there was not a large amount of cat's eye unearthed in Ming tombs. It was only found in a pair of gold hairpins (two tines) from the tomb of the Xu Fu couple in Nanjing.

Turquoise: only a small amount of turquoise was unearthed in Ming tombs as well. However, a gold hairpin (one tine) inlaid with turquoise, a pair of gold hairpins (two tines) inlaid with turquoise, and a piece of gold hair ornament inlaid with turquoise were also unearthed in the tomb of the Xu Fu couple in Nanjing.

Talc: 112 pieces of talc were unearthed from the Ming tomb in Changyi, Shandong, including guardian figurines, attendant figurines, object holder figurines, music player figurines, architectural features and daily necessities. The talc was produced in Laizhou and was commonly known as Laizhou jade. The talc burial objects were also found in nearby counties or districts. The scale of the burial objects workshops was quite impressive, because the craftsmen had accumulated experience from generation to generation.

Shoushan stone (pyrophyllite): the history of Shoushan stone seals can be traced back to the Jin dynasty. Not many Shoushan stone products have been handed down from the Ming dynasty, and the only two extant seals are those of Li Zhuowu, which were carved from the "old" Shoushan stone. One is in the Quanzhou Committee of Cultural Relics while the other is in the National Museum of China in Beijing. No Shoushan stone products have been unearthed in Ming tombs so far. Shoushan stone is a hydrated aluminosilicate mineral, scientifically known as pyrophyllite, with a fine texture, magnificent colours, and a waxy or pearl lustre. Under the microscope, it appears as a schistose clay mineral. Minerals among its components, including Fe_2O_3, MgO and K_2O, all serve as pigments. It comes from Shoushan Town in the northern suburb of Fuzhou, Fujian.

Marble: a tomb memorial tablet made of marble was unearthed from a Ming tomb in Dali, Yunnan. Dianchang Mountain in Dali abounds in marble, which is recorded in detail in *Xu Xiake's Travels*.

White marble: Beijing white marble is a kind of marble that came into being as a result of contact metamorphism in the Sinian System. A tomb memorial tablet made of white marble was unearthed from the Ming tomb of the Zhaoyong General Dai Xian and his wife. There exists another Ming tomb in Fucheng, Hebei, where the coffin is composed of eight white marble slabs.

Limestone: limestone was widely used in the Ming dynasty. Apart from in burning lime, it was also used in construction. For example, some limestone slabs were used to cap tombs, some were used for tombstones, and others were used for buildings, roads, dams, etc.

Purple clay: this is the raw material for making purple clay ware. There are several varieties, such as *nen ni* (tender clay), *shihuang ni* (yellow clay), *tianqing ni* (lit. sky-blue clay), *lao ni* (old clay), *bai ni* (white clay), etc. Purple clay pottery emerged in the Northern Song dynasty and flourished after the middle of the Ming dynasty. During Emperor Zhengde's reign, there was a famous craftsman called Gong Chun, who developed his own unique style of purple clay ware and was consequently respected as its originator. During the Wanli reign, the famous craftsmen Shi Dabin, Li Zhongfang and Xu Youquan were sometimes honoured as the "Three Great Masters of Purple Clay Pots". Works by Shi Dabin unearthed from Ming tombs include a hexagonal purple clay pot from a Ming tomb in Jiangdu, Jiangsu, and another purple clay pot from a Ming tomb in Wuxi, Jiangsu.

Duan stone: produced in Duanzhou (now known as Zhaoqing), Guangdong, Duan stone is the raw material for the famous Duan inkstones. Duan inkstones have been famous since the Tang dynasty and were loved by the literati. Duan stone was also unearthed from Ming tombs, such as a piece from a Ming tomb in Wuxi, Jiangsu. There are also heirlooms, such as the Xuande Xiayan (lower hill pit) Inkstone in Huang Ren's collection. Xuande stones are located halfway up Pingfeng Mountain in Duanzhou, for it was excavated during the Xuande reign. The quality of Xuande stones is better than that of the stones from the old quarry, while those from the lower hills are especially good examples of Xuande stones. The Xuande inkstones produced from the Xiayan with its fine and smooth texture are among the top-quality inkstones [119].

Songhua stone (also known as Songhua jade): this is the material for making Songhua inkstones. Songhua inkstones had already been produced in the early Ming dynasty. The scientific name for Songhua stone is calcipulverite. It can be dark green or light green in colour, and some are mixed with yellow. Its grain is like the bristles of brushes, with a fine, smooth, hard and dense texture. It is composed of minerals which include calcite, microcrystalline quartz, and a few clay minerals with microcrystalline and microlayer structures. It has a hardness of 4 on the Mohs scale. Songhua stone is found in the central Nanlao Formation of the Sinian System, originating from the source of the Songhua River—the foot of the Changbai Mountains [120]. It cannot produce liquid ink as easily as a Duan inkstone or a She inkstone.

Alunite: three alunite tomb memorial tablets were unearthed in the Ming imperial tombs in Nancheng, Jiangxi.

Sandstone (the red and green types): sandstone was used to build coffin chambers. For example, the coffin chambers in a Ming family tomb in Pingwu, Sichuan, were built with sandstone slabs and sheets. It was also used to build the foundations of house walls and to carve stone men (officials and officers), stone horses, stone tigers, stone sheep and baluster columns in front of the tombs.

Biopelite: the tablets erected in the tomb of a Ming eunuch in Dongshan, Guangzhou, were (made) of biopelite.

Granite: granite was widely used in the Ming dynasty for building houses, bridges, dams, tower bases and many other various kinds of foundations. For example, a variety of bionic stone carvings in front of the Ming tombs in Anshan, Liaoning, have granite bases.

Amber: amber products excavated from Ming tombs include beads, a string of 108 rosaries, headpieces, etc.

Beeswax: beeswax was a fossilised resin that was used to make seals in the Ming dynasty. A pair of beeswax seals from the late Ming dynasty was found in Jiangsu [121].

Coral: people in the Ming dynasty collected coral as a kind of biological gem. A remnant of red coral was unearthed in the tomb of a Ming eunuch in Dongshan, Guangzhou. Some people also processed coral into ornaments.

Pearl: pearl is both a valuable Chinese medicine and a precious living material. It is as cherished as various gemstones because of its indispensable role in ornaments. For example, the phoenix crown excavated from the Ming tomb in Lanzhou was inlaid with more than 2,000 pearls, which were used together with gemstones.

Ivory: ivory, also from a treasured living material, was used to make valuable crafts and ornaments. For example, a piece of ivory tablet (used by officials when received in audience by the emperor) was excavated from a Ming tomb in Nanjing.

8.9.2 Applications of Mineral-Petrology in Chinese Alchemy and Medicine

(1) Knowledge of mineral-petrology reflected in Taoist elixirs of immortality

Taoism was revitalised by Emperor Chengzu of the Ming dynasty, who constructed the Taoist palaces and temples in the Wudang Mountains and also ordered officials to compile the *Taoist Canon.* Therefore, Taoist events prevailed for a while. Emperor Xianzong (Zhu Jianshen) placed Taoist priests in important positions. Emperor Jiajing was keen on alchemy. He pursued immortality, advocated Taoism and trusted Taoist priests. During his 45-year reign, he kept making and collecting Taoist elixirs of immortality. Ultimately, he died of poisoning from taking elixirs, and the imperial court put the alchemists to death, so the art of *waidan* (outer alchemy) suffered a great blow. There were not many works about alchemy and elixirs in the Ming dynasty. However, the Prince of Ning (Zhu Quan) was an exception, for he wrote many books on *waidan*, including *A Jade Book of Gengxin (Gold)*, *The Processing of Nature*, *Rare Skills between Heaven and Earth* and *Essence of Living Things between Heaven and Earth*, but they have all been lost. According to the *History of the Ming: Princes II*, "the Prince of Ning (Zhu Quan) is the 17th son (according to the tomb memorial tablet [122], he should be the 16th son) of Emperor Taizu. He was given the land of Ning in the 24th year of the Hongwu Reign. Two years later (according to the tomb memorial tablet, it should be three years later), he proceeded to his post in the Great Ning … In February of the first year of the Yongle reign (according to the tomb memorial tablet, it should be the 2nd day of March), he was reassigned to the land of Nanchang … From then, he hid his capacities and bided his time. He played *guqin* [a seven-stringed plucked instrument] and read books all day long, experiencing no more trouble through the rest of the Yongle reign … During the reign of Emperor Renzong (Zhu Gaochi), he associated with literati all day long and became obsessed with the art of 'Taoist ascension'. Thus, he styled himself as 'Quxian'. He died in the 13th year of the Zhengtong reign." A few parts of the content of *A Jade Book of Gengxin (Gold)* were preserved in the *Compendium of Materia Medica.* A collection of its quotes are as follows:

> *Qinglanggan*: formed in the cliffs of the South China Sea, it absorbs qi (vital force) from both yin (the feminine and negative principle of nature) and yang (the masculine and positive principle of nature). It resembles beads, but looks redder in colour.

Dansha (cinnabar): the best types of *dansha* are those from the caves in the Wuxi Mountains, for they acquire the qi from due south. Those from the mountains in Mayang, next to Wuxi, are ranked second best in quality. Those from Yunnan, Persia and the Western Regions are also clear and smooth. A kind of sandstone from Liuzhou is similar to *chensha* (another name for *dansha*), but it is as round as honey locust seeds in shape and it cannot be used as medicine. Local *dansha* from Shangzhou and Qianzhou, along with *dansha* from Xuanzhou and Xinzhou all have poisonous components, including gold, silver, bronze and lead inside, so they cannot be consumed either.

Lingsha (artificial cinnabar): *lingsha* is of the utmost divinity. It is formed by combining sulfur and mercury, called *danji*. It draws its power from Nature, absorbs the essence of yin and yang, transfers from the five elements (metal, wood, water, fire and earth, held by the ancient Chinese to compose the physical universe) and produces *jiuhai* (also called *jiuzhuan*) gold elixir of immortality through Taoist alchemy. The types which have not been smelted in *ding* are called *qingjin dantou*, while those which have been smelted in *ding* are called *lingsha*. There are three kinds of *lingsha*: those smelted in the fire for 24 hours are called *jinding* (gold *ding*) *lingsha*; those smelted in the fire for 24 hours, with nine rounds of feeding and extinguishing the flames, are called *jiuzhuan* (nine-round) *lingsha*; those smelted for 30 days are called *yijia laohuo* (doctors' slow-fired) *lingsha*. It is advisable to take *lingsha* with the ashes of mulberry stems which have been sprinkled with vinegar after cooking.

Buhuimu: *buhuimu* is a stone medicine loaded with yin qi (negative energy). It is produced in the barbarian areas of Southwest China. The types from Lizhou and Maozhou are especially high in quality. It is shaped like a needle. This stone is the so-called "wood-like", "smokeless when burnt" medicine.

Yangqi stone (actinolite): Yangqi stone is a stone loaded with yang qi (positive energy). The stones from Jianjin Mountain in Qizhou are the best. Those which are shaped like arrowheads are very potent, while those shaped like dogs' teeth are not very powerful. Genuine Yangqi stone should be covered by heavy snow.

Taiyi yuliang (limonitum): *taiyi yuliang* (or *yuyuliang*) is a stone loaded with yin qi. It can be found everywhere. It takes the form of overlapping sheets which are dark purple in colour. It contains pyrite and has the "hottest energy", so wherever *taiyi yuliang* exists, the snow melts first.

Kongqing (a kind of azurite): *kongqing* is a stone loaded with qi. It comes from Shangrao. The types which take the shape of stalactites are of a high quality. It is found in large purple sheets which reflect light. It is constantly found in the water from the Shu Roads (the paths linking Chang'an in Shaanxi and Chengdu in Sichuan), the Yan Roads (the path in today's Ya'an in Sichuan) and Beidai Mountain and is used to make elixirs. Some pieces of this stone are as large as fists and shaped like eggs. Hollow with oil-like water inside, it is considered to be effective in curing blindness. Those from copper mines also make a great raw material for paints. *Yangmeiqing* and *shiqing* belong to the same category as *kongqing*, but they have different levels of coarseness. When used to "transform people into immortals", *zengqing* is the best, *kongqing* is ranked second, while *yangmeiqing* is ranked third.

Tesheng Yu stone: Yu stone is a stone loaded with yang qi. It is found in valleys. When it is washed out from the water, it is similar to alum. Stones with a transverse grain are the best. Heat the stone medicine until it gasifies to make cinnabar and mercury. The appearance of the stone is quite similar to calcite, but genuine *tesheng* Yu stone does not freeze in water. Those excavated from gold mines are called *woxue* Yu stone.

Posuo stone (bezoar stone, also called *mosuo* stone): found in Samboja, *posuo* stone is a stone loaded with yang qi. There are tall and colourful mountains south of the sea, where stones are vitreous. The stones roll over underwater like arrows in flight. People on the passing boats use knives and axes to dig out the stone. Then they burn it to obtain gasified sulfur. The stones are shaped like yellow dragon's teeth [the fossils of ancient mammals'

> teeth] and the types which have great hardness and weight are the best. This stone is used to collect the "Five Jin" [five minerals: gold, silver, copper, iron and tin], to smelt "Three Huang" [three kinds of traditional Chinese medicine: Huangqin, Huanglian and Huangbo] and to make lead and mercury.
>
> Huaru stone (dolomite): Huaru stone is a stone medicine loaded with yin qi. Found in the valleys of Daizhou, it exists in various colours and it may be used as a medicine in place of cinnabar. It can also be found in Wenshan and Pengxian in Sichuan.
>
> *Shiliuhuang* (mineral-laden sulfur): there are two kinds of sulfurs: *shiliuhuang* [mineral-laden sulfur] from Liuqiu Mountain by the South China Sea and *tuliuhuang* (native sulfur) from Guangnan. Those which are noiseless when chewed and those imported from Japan are both of high quality. Nowadays, people use it together with saltpetre as beacon fire, which plays an important role in military activity.

The quotes only introduce 11 minerals, mentioning the traits, forms, colours, hardness and origins of the minerals, with special emphasis on the role of minerals in alchemy. Unfortunately, the book has been lost, so we cannot get the full picture. However, the elaboration on the 11 minerals cited suffices to illustrate the close relationship between alchemy and minerals. Alchemy certainly led to the accumulation and development of mineral knowledge in China.

(2) Knowledge of mineral-petrology reflected in traditional Chinese medicine

There were not many texts on traditional Chinese medicine in the Ming dynasty. The only two extant works are the *Bencao pinhui jingyao* ("Collection of the Essential Medical Herbs of Materia Medica") and *Bencao gangmu* ("Compendium of Materia Medica").

The *Bencao pinhui jingyao*, in 42 volumes, was completed in 1505 (the 18th year of the Hongzhi reign) by the Administrative Assistant of the Imperial Academy of Medicine Liu Wentai and his colleagues on the edict of the emperor. The script was kept in the Inner Court and was not printed at that time. It was not until 1700, the 39th year of Emperor Kangxi's reign, that the script was rediscovered. At that time, the Medical Assistant of the Imperial Academy of Medicine Wang Daojun and his colleagues re-edited the book and added around 490 items from the *Bencao gangmu* and other books, composing 10 volumes as an addendum. By then, there were two editions of the book: one was the illustrated edition from the Ming dynasty, while the other was the copy from the Qing dynasty. In 1937, the Commercial Press typeset and printed the book without any illustrations, marking the first publication of the book. In 1956, the book was reprinted by the Commercial Press. In 1963, People's Medical Publishing House republished the book based on the 1956 edition by the Commercial Press. This book firstly presents ten items including jade and stones, followed by three sections of medicines. Each medicine quotes content about materia medica from the *Shennong bencao jing*, *Mingyi bielu* ("Supplementary Records of Famous Doctors") and *Bencao shiyi* ("A Supplement to Materia Medica"), as well as relevant accounts from other works from the Tang and Song dynasties. It provides a detailed record of the main effects of the medicine, as well as the alias, origin and method of distinguishing the real substance from fake versions. The annotations to all the items are written according to past works on materia medica. The annotations quoted from other works are either marked with their sources or included in *bie lu* (quotations as supplementary records). On the other hand, the annotations which have been revised or any additions are marked as *jin an* (quotations as grounds for argument). Thus, this book is mainly composed of the arguments of past generations, while only the few *jin an* are really reflective of the viewpoints of the Ming dynasty.

For example, a *jin an* in Volume Two "Bianqing: Earth" states: "According to Su Gong, Bianqing (literally flat and green) is identical to Lüqing (literally green and blue). Other works from the Tang dynasty contain the same view. However, Lüqing is deep green in fist-shaped blocks, while Bianqing is light green in thin layers. The ancestors named them according to their appearance. Moreover, their properties and effects are also different as well. It is thus clear that they are probably not the same substance." As it turns out, the argument in the *jin an* is correct. Lüqing, whose molecular formula is $CuCO_3 \cdot Cu(OH)_2$, is also known as *shilü*, *dalü* or *kongque* (Peacock) stone". *Bianqing* is another name for *shiqing*, whose molecular formula is $2CuCO_3 \cdot Cu(OH)_2$, and which is also known as *daqing*, and the ones with paler colours are called *baiqing* or *biqing*.

A *jin an* in Volume Two "Lugan Stone: Earth" states: "This (Lugan stone) comes from the valleys of Sichuan, Huguang and Chizhou. It is smooth and soft, and takes the form of blocks stacked in layers. Lugan stones vary in size. There are pink ones like the petals of plum blossoms; there also exist greenish white ones with interleaved lines in relief. The pure white ones with a smooth texture are top notch, while those in other colours with a coarse texture are poor quality. Lugan stone refers to brass smelted from miscellaneous copper alloys in a 'crab shell furnace'."

A *jin an* in Volume Two "Eguan Stone: Earth" states: "This stone comes from Sichuan and regions south of the Five Mountains. It can also be found in the area now known as Licheng County in Ji'nan. Three *cun* in length, it is round with schistose layers, white and brittle. It acquired its name 'Eguan stone (gooseneck stone)' since it is hollow like a gooseneck."

A *jin an* in Volume Three "Yinnie: Earth" states: "*Konggongnie* and *yinnie* are both formed from stalactites. The hollow ones with fewer branches are *konggongnie*, while the solid ones with more branches like *jiangshi* (literally ginger stone) are *yinnie*. If it is light and is hollow in the centre, it is a *shizhongru* (stalactite); if water drips accumulate and spread like a plate, it is a *shichuang* (stone bed); if there are spikes on the surface of the stone bed like antlers, it is a *shihua* (cave flower). Although the three kinds of *yinnie* are the same in essence, they differ in quality and effect."

A *jin an* in Volume Three "Shichuang: Earth" states: "The four kinds of stalactite are the same in essence. But as opinions vary, no unanimous conclusion can be drawn. Since the lower end of a stalactite resembles the breast (known as *ru* in Chinese), it is named *zhongru* (also called *shizhongru* and *zhongrushi*, meaning stalactite). The two kinds of *nie* (literally tender shoot) (*konggongnie* and *yinnie*), or '*niezi*' according to Mencius, also develop from a *zhongru* (stalactite). A *shichuang* (literally stone bed) is the bed-like coagulum on the ground formed by the drops from the suspended stalactite. And a *shihua* (literally cave flower) refers to the flower-like spikes on the *shichuang*. Behold! There are grounds for all the names given by our ancestors."

A *jin an* in Volume Three "Chang Stone: Earth" states: "This book has already elaborated on Li stone (fibrous gypsum) and Chang stone (feldspar) in two separate items. They have different properties and effects. How could they possibly be the same substance? Nowadays Chang stone exists on Liaoban Mountain in Junzhou, although it is mistaken for Li stone by the local people."

A *jin an* in Volume Six "Meng Stone: Earth" reads: "Earlier works did not record the origins of the stone. Nowadays, it can be found in the Qi and Lu mountains (both in today's Shandong). It is green with a sporadic golden gleam, and hard with a fine texture. The fastest way to produce it is to use an apparatus to grind it, and the powder produced is very fine.

A *jin an* not only details the properties and sources of minerals, but also some concrete methods for extracting medicine from them. The properties, sources and other information recorded in the book are generally correct, which reflects the level of mineral knowledge acquired by people in the Ming dynasty.

The book contains a total of 110 kinds of minerals and rocks, i.e. *dansha* (cinnabar), mica, nephrite jade shavings, alum, green alum, willow catkin alum (extracted from white alum), saltpetre, mirabilite (in Chinese, *mangxiao*, *puxiao* or *mayaxiao*), *xuanming* powder (air-dried mirabilite), calcrete, talc, chalcanthite, *kongqing* (a kind of azurite), *zengqing* (another kind of azurite), yellow alum, Persian white alum, gold shavings, *jinya* gold deposits, silver shavings, raw silver, iron, raw iron, copper ore, raw copper, xanthoconite, pyrite, petroleum, burnt stone, *yuyuliang* or *taiyi yuliang* (limonitum), white quartz, purple quartz, gedanite of five colours, *baiqing, lüqing, bianqing* (the three minerals are all from azurite), yellow limonite powders, *wumingyi* (a mineral with pyrolusite as its main component), Pusa stone (literally bodhisattva stone), Posuo stone (meaning bezoar stone), Lugan stone, Eguan stone (literally gooseneck stone), Yun stone (saltwater crystal), *shiliuhuang* (mineral-laden sulfur), Xuanhuang stone, *shizhongru* or *shisui* (stalactite), realgar, orpiment, gypsum, calcite, calcitum, *yinnie*, *kongongnie*, shihua (cave flower), shichuang (stone bed), Chang stone (feldspar), Li stone (fibrous gypsum), magnetite, Xuan stone (black stone), Yangqi stone (actinolite), grindstone, rhodonite, *shinao* (a kind of stone mingled with talc), *shixie* (crab fossil), mercury, mercury powder, *lingsha* (artificial cinnabar), lithargite, coral, agate, *Tridacna*, salt, halitum, sea salt, halide salt, halite, atacamite, Taiyin xuanjing stone (selenite gypsum), Heiyang stone (literally black sheep stone), Baiyang stone (literally white sheep stone), *fuqin* (literally skin green), *shishe* (stone snake), limestone, Yu stone (arsenopyrite), white arsenic, sal ammoniac, miniumite, lead, lead powder, stannic oxide, hematite, *shiyan* (stone swallow), chalk, *qinglanggan* (a kind of halogenide), Jinxing stone (schistose copper), *tesheng* Yu stone (unassociated arsenopyrite), *woxue* Yu stone (yellowish white arsenopyrite), urinary stone, meng stone (schistose chlorite), Qiang stone, Maifan stone (a kind of subvolcanic rock), Cang stone, Huaru stone (dolomite), *shican* (silkworm fossil), naphtha, *buhuimu* (incombustible stone), borax, etc.

In the late Ming dynasty, Li Shizhen (1518–1593) wrote the 52 volumes of the *Bencao gangmu*, of which Volume Eight to Volume Eleven constitute the Gold and Stone Section, containing 97 kinds of minerals and rocks. It summarises the research results from the *Shennong baicao jing* and *Zhenglei bencao* ("Classified Materia

Medica from Historical Classics for Emergency"). It is especial innovative in its classification. Before it, inorganic medicines were all listed in the Jade and Stone Section. Li Shizhen further divided inorganic drugs into six categories—water, earth, gold, jade, stone and salt—in the Gold and Stone Section. The Water Category contains 43 species of aqueous solutions, instead of a single kind of mineral. In the Earth Category, there are 61 species, including various kinds of soil and burned clay. In the Gold Category, there are 28 species, including metals, alloys, metal compounds and metal products. In the Jade Category, there are 14 species, most of which are the purer silicides. In the Stone Category, there are 72 species, most of which are silicate rocks and insoluble natural salts. In the Salt Category, there are 20 species, most of which are soluble natural salts.

Li Shizhen gave a more meticulous explanation of the difference between pure metals and alloys than his predecessors. In the descriptions of various kinds of metals, nearly all the descriptions of metal items after the label "Shizhen said" include this differentiation. For example, in the Gold Category, Shizhen said: "There are two kinds of gold—mountain gold and sand gold. Its colour is aligned with its purity: gold which is 70% pure is green, 80% is yellow, 90% is purple and pure gold is red. If it is mixed with silver, the alloy is soft and appears green if it is struck with another stone. If it is mixed with copper, the alloy is hard and produces a sound when struck with another stone." This part introduces the method for identifying pure gold, gold-silver alloys and gold-copper alloys. Under the item "Red Copper", Li Shizhen clearly and correctly discusses the differences between pure copper and copper alloy: "Copper can be classified into red copper, white copper and yellow copper. Red copper comes from the mountains in Sichuan, Huguang, Yunnan and Guizhou. The indigenous people excavate it from cave mines and refine it. People use Lugan stone to make yellow copper, which is similar to gold in colour. Pi stone is refined to make 'white copper'. Miscellaneous tin is refined to make 'loud copper (copper-lead–tin alloy)'." Lugan stone here refers to siderite. Pi stone here refers to maucherite. Yellow copper (bronze) is copper-zinc alloy. White copper is copper-nickel alloy. Green copper is copper-tin alloy.

It can be concluded that Li Shizhen has made contributions to the knowledge of minerals and rocks from the three following aspects: first, the preservation of ancient knowledge of minerals and rocks; second, a more sensible classification of minerals; third, more accurate identification of minerals, metals and alloys, inclusion of another 13 kinds of new inorganic medicines as well as clarification of "one term for many things" or "several names for the same thing with different appearances" [123].

8.9.3 *Studies of Mineralogy and Petrology*

In the Ming dynasty, there were no monographs on minerals but there were monographs on rocks: for example, Lin Youlin's four-volume *Suyuan Stone Catalogue*, Du Jun's two-volume *Stone Appreciation* and Jiang Zhen's three-volume *Record of She Inkstone*. These three monographs on rocks also record minerals and fossils. This shows that there was no strict division between minerals and rocks in ancient China. Beginning with the *Yunlin Stone Collection* from the Song dynasty, stone catalogues actually contained a decent amount of information on minerals and fossils. Works from the Ming dynasty also continued this tradition.

Lin Youlin, with the courtesy name Renfu, was a native of Huating, Songjiang. His father, Lin Jingyi, a Jinshi during Emperor Longqing's reign, was the Chief Minister of the Nanjing Court of the Imperial Stud. Thanks to his father, Lin Youlin was appointed to the official post of Governor of Long'an. He was expert at drawing landscapes and fond of travelling. "With luxurious clothing and accessories, I went on my dreamy journeys to the Five Lakes and other magnificent places around China every year." He compiled the *Records of Guqin on the Qinglian Boat* and wrote the *Suyuan Stone Catalogue* and *Indoctrination of Buddhist Pearl*.

The *Suyuan Stone Catalogue* was completed in 1613 (the 41st year of the Wanli reign). The "Introductory Remarks" reads: "This book is a review of strange stones in ancient and modern books. And each time I encounter one that echoes in my heart, I describe its appearance and insert the inscriptions afterwards. I am deeply passionate about landscapes and especially interested in strange stones; although I do not revere rocks as Mi Fu did. But I draw whatever bizarre stones that I see with my own eyes. Then my amassed drawings form a whole book as time goes by. Every time I open the book, the stones seem to be right in front of my eyes. Most of the strange stones are found in those splendid mountains, and the ones listed in my book are only a few of those I have seen. But somehow everyone is able to access the panoramic view by examining a part of it. The selected stone must be exquisite. If it is a clumsy lump of stone, even though it is an ancient stone, it is not worth recording, such as the Bian Stone of the Tomb of Yu and the Lian Stone of Yulin. Most of the best stones have already been described in the works of many well-known literati, so it is impossible to include them all. But there are some that have not been spotted among ordinary ones, which have never been introduced. It is no bad thing that the people who have the insights identify them and include them in their works. Stones have both form and spirit. The book can only portray their forms. As for the beauty of their spirit, which is unpredictably fascinating, only people with the right sense of aesthetics can appreciate it. Illustration is always one-sided. Some stones are admirable on three sides or even four sides … All the stones included in this book are small, which is adequate for readers to be able to appreciate them. As for mountain ranges and the gigantic stones pointing to the sky, though equally admirable, they have to be omitted from the book since the paper is too small to accommodate them. Mi Yuanzhang's Yanshan Inkstone was taken by a friend, so I could not see it. Consequently, I merely drew it based on my imagination. I have now gathered all the bizarre stones under heaven in my book. Therefore, all of them, including the Yanshan Inkstone, can reappear vividly in front of the readers. From now on, the book will be kept in studies so that people no longer regret not being able to see the bizarre stones with their own eyes. There are too many strange stones of various shapes. Since I cannot record them all, it seems that I am bragging about them. However, I must have concrete evidence before I include them in my book. How dare I boast of my work in the world of art, if it is merely grounded in conjecture and speculation!" These "Introductory Remarks" give a clear explanation of the writer's aims.

The book contains 101 kinds of minerals and rocks. Each item has either a short text containing a description, or inscriptions written by earlier scholars. They are all painted as illustrations, to express the real spirit of the stone together with its form. The descriptions include not only the aesthetic value of the rocks, but also the size, properties, colour, sound when struck, and coarseness. Therefore, the book is a very important reference book regarding minerals and rocks. For example, in Volume One "Yongning Stone" can be read: "Stones come from both Jianghua and Yongning in Daozhou. Some are found in dispersed mountains, while others are from flat lands. Overlapping with one another, they vary significantly in size. There is extruded greyish black stone in Jianghua, which stands erect, prominent among the other stones. It is coarse and dry, and it produces a sound when struck. Those formed in Yongning, on the other hand, vary from over one *chi*, or two or three *chi* to more than ten *chi* in size. They can be either as large as fists or small but densely spread. In any case, all these stones excavated from mountains have a strange appearance. In terms of colour, they can be deep blue, light blue or slightly black. The stone is hard and smooth and produces a sound when struck. Some stones have a lot of small pits on the surface, resembling the circular holes washed out by Taihu Lake. They are steep with perforated structures on all sides, as high as several *chi*, presenting an imposing vista of magnificent mountains. These stones are crossed with gullies and surrounded by peaks. Each one resembles a different object, but these are too many to enumerate. For example, one kind of inkstone takes the form of cloud-capped peaks with white bumps and 'a belt' surrounding the mountainside. It has 'Haiyue' in small seal characters. This is a one-in-a-million rare treasure." This explanatory text, which talks about the origin, appearance, colour, occurrence, surface coarseness, sound produced when struck, hardness, size and other features of rocks, provides a significant record of the knowledge of minerals and rocks in the Ming dynasty. Unfortunately, there are not many such explanatory texts, and most of them are inscriptions or stories about scholars' love for stones.

Du Jun's *Shipin* ("Stone Appreciation") was completed in 1617, the 45th year of the Wanli reign. It is a monograph on minerals and rocks, which includes more than 500 kinds of minerals and rocks. The *Siku quanshu zongmu* ("Catalogue of the Complete Library of the Four Treasuries") reads: "Du Jun, with the courtesy name 'Kaiyuan', was native to Songjiang. The book was completed in 1617, recording the names of stones from ancient times in quite a disorderly manner. It quotes excessively from other works of the same kind. It even alludes to Tu Long and Chen Jiru's words. Therefore, to some extent, it can be read for reference."

Jiang Zhen's *Record of She Inkstones* was in 3 volumes, but the *Siku quanshu zongmu* has only retained its catalogue. The *Siku quanshu zongmu* reads: "Jiang Zhen, with the courtesy name Jifu, was native to Wuyuan. He was the Instructor of Shaoxing Prefecture. The book was compiled according to the *Yanzhi* ("Record of Inkstones") by the Grand Commandant Ye Lianggui in Raozhou and his brother Ye Liangqi, the Grand Commandant in Dongchang. It is generally based on Song Zhiping's *Sheyan pu ("She Inkstone Manual")* and Hong Shi's *Yanshuo* ("On Inkstones"), but it has a few additions."

8.9.4 Mineral-Petrological References in Chronicles, Encyclopaedias and Annotations

(1) Knowledge of mineral-petrology recorded in local chronicles

The *Daming yitong zhi* ("Records of the Unity of the Great Ming") by Li Xian and others, which was finished in 1461 (the 5th year of the Tianshun reign), records some material about minerals and rocks. For example:

> Jubao Mountain (literally treasure-gathering mountain) is located on the side of the Terrace of the Rain Flowers outside Jubao Gate in Yingtian Prefecture (today's Nanjing). It acquired its name because it abounds with fine agates.
>
> Dongting Mountain is located in Taihu Lake, which is 130 li west of the prefectural city of Suzhou (in today's Wuzhong District and Xiangcheng District of Suzhou). It is the origin of Taihu stone. The stones formed in the water are the best. The stones with many holes inside are smooth and humid, and they produce a sound when struck. However, those found in the mountains lack moisture and lustre.
>
> Qingshi Mountain (literally chime stone mountain) is located in Pizhou of Huai'an Prefecture (today's Pizhou city). It is 80 li southwest of the downtown area and close to the river. Stones can be found on the mountain, and they produce a loud and clear sound when struck, so can be used as chimes. The so-called "floating chime of Sibin" in the *Yugong is* related to the stones which come from this mountain.
>
> Shigao Mountain (literally gypsum mountain) is located in Fengyang Prefecture (today's Fengyang County). It is 50 li southwest of the prefectural city. Gypsum is found in the mountain and can be collected at any time. Yunmu Mountain (mica mountain) is located 40 li southwest of the old prefectural city. Allegedly, mica can be found on the mountain. Zhongru Mountain (stalactite mountain) is located 60 li southwest of the old prefectural city. The caves in the mountain produce stalactites. Jing Mountain in Huaiyuan County is in the southwest of the county's jurisdiction. There is a jade mine to the northwest of the mountain where Bian He once excavated jade. The white stones there are extraordinary, for the stones in other places are all greenish black. To the east exists Bianhe Cave, which is also known as Baopu Rock. Kun Mountain in Lujiang County in Luzhou Prefecture is 40 li south of the county's jurisdiction. Since the mountain produces alums, it is also known as Fan Mountain (alum mountain).
>
> Tongling County is a source of copper, lead, iron and tin. Tongguan Mountain, also known as Liguo Mountain, is 10 li south of the county's jurisdiction. There is a spring that never dries up on the mountain, which can be used to leach iron in order to smelt copper from it. It was once chosen as the official copper smelting site.
>
> Mo Mountain (graphite mountain) in Yi County of Huizhou Prefecture is 16 li southwest of the county's jurisdiction. The mountain produces graphite. The locals have long been excavating graphite from here, so it has become a graphite-excavating mine over time.
>
> Longwei Mountain in Wuyuan County is 100 li southeast of the county boundary and extends westwards to Wuxi River. During the reign of Emperor Kaiyuan (Li Longji) (713–741) in the Tang dynasty, someone saw the smooth and crystal-clear stones here, and took them to make inkstones. That is how She stone began to spread around. There are five categories of

She stone: first, Meizi stones (7 kinds); second, Waishan Luowen (13 kinds); third, Lishan Luowen (1 kind); fourth, Jinxing (3 kinds); fifth, Lükeng (1 kind). All these stones are jointly known as Longwei inkstone. Generally speaking, the best examples of She stone are the green ones with green halos and many star-like flecks.

Minqing County and Fuqing County in Fuzhou Prefecture are sources of iron. Jixia in Huai'an County is where Shoushan stone is produced. This kind of stone is as clear as jade. With a soft texture, it is easy to carve.

Nanping in Fujiang is the origin of Huawen stone. This kind of stone is green with simple inclusions. Since these inclusions may be in the form of mountains, water bodies, fowl and fish, they are used to make inkstones or screens.

Mixiong Mountain in Songmeng Sub-prefecture (today's Songmeng County) is the origin of *wuse huashi* (five-coloured stone). This kind of stone is similar to agate in appearance and can be used for making tools. The stones are as large as beans and come in a mixture of five colours. They can be crafted as part of a miniscape. Diancang Mountain in Dali Prefecture (today's Dali County) is the origin of Diancang stone. The stone is white with green patterns that may be in the form of mountains, water and plants, so people usually carve it to make screens. Laiyu Mountain (Lai jade mountain) in Shiping Sub-prefecture (today's Shiping County) is the origin of Lai jade. Zhongxiu Mountain is the origin of *zishi* (purple stone) which can be carved into inkstones. Malong Sub-prefecture (today's Malong County) is the origin of the *shiyan* (stone swallow). This is a swallow-shaped stone with patterns. The large ones are referred to as "male", while the small ones are referred to as "female". It is a medicine used for eye diseases.

Tuoji Island in the sea in Penglai County is the origin of Luowen Jinxing inkstone, also known as Xuelang inkstone.

Xingtai County is the origin of Jieyu sandstone.

Qingyu Mountain in Yaozhou (today's Yaozhou district of Tongchuan) is the origin of *qingshi* (green stone). During the reign of Emperor Tianbao (Li Longji) (742–756) in the Tang dynasty, the stone was excavated to make chimes. Afterwards, the ritual music of suburban temples appeared; therefore, people ceased to use Sibin chimes.

Taozhou in Gansu is the origin of Tao inkstone. Huashi Gorge in Jiezhou is the origin of *shiqing* (azurite), which is a green stone with black patterns. These patterns may be in the form of pine trees, cypress trees, figures, streams, bridges, rivers, stones, mountain forests, buildings, the sun and the moon. It can be used to make screens.

Ningguta (today's Ning'an) is the origin of Songhua stone which may be used to make inkstones.

Volume Eighty records the petroleum and gas in Nanxiong, Guangdong: "The oil mountain, which is as high as several thousand *ren* [equal to seven or eight *chi*], is 120 li east of the prefectural city. It stands tall and erect, and beside it a small cave produces oil, which people tend to collect for the economic profits." This is an important record of how people prospected for minerals in ancient times which provides pointers for present-day prospecting for petroleum and gas in this area. There are too many similar accounts to list them all.

The *Prefectural Record of Hangzhou* during the reign of Emperor Wanli reads: "Baosheng Temple in the Manao Slope in Qiantang County of Hangzhou Prefecture has gravels which are similar to agates. That is how the Manao Slope (agate slope) acquired its name. There was once a Manao Court (agate court) as well."

The *Sub-prefectural Record of Anji* during the reign of Emperor Jiajing reads: "Tongxian Mountain of Anji Sub-Prefecture is 30 li east of the boundary of the sub-prefecture, and natural copper can be found there.

The *Prefectural Record of Ningbo* during the reign of Emperor Jiajing reads: "Xi Mountain (tin mountain) in the Yin County in Ningbo Prefecture is 50 li southwest of the county border. It acquired its name because it produces lots of tin."

The *County Record of Changshan* during the reign of Emperor Wanli reads: "Yan Mountain (inkstone mountain) in Changshan County is the origin of Jinxing stone which may be used to make inkstones. People from all over come to trade for them. The mountain is also the origin of *zishi* (purple stone). *Zishi* with a white belt in the middle is called *zipao yudai* (purple gown with a jade belt). This is a precious variety, but we no longer have access to it."

(2) Knowledge of mineral-petrology in reference books

The *Assembled Pictures of the Three Realms* compiled by Wang Qi and Wang Siyi has a volume called "Treasures", which records 55 kinds of minerals, rocks and fossils, namely, south pearl and north pearl, raw gold, raw silver, coral, raw iron, Baiyang stone (lit. white sheep stone), lead, tin, *qinglanggan* (a kind of halide), *dansha* (cinnabar), *kongqing* (a kind of azurite), *zengqing* (another kind of azurite), mercury, natural copper, jade, *shilü* (malachite), Chang stone (feldspar), realgar, orpiment, mica, stalactite, alum stone, Heiyang stone (literally black sheep stone), saltpetre, talc, chalcanthite, *yuyuliang* (limonitum), white quartz, purple quartz, red halloysite, white halloysite, *wumingyi* (a mineral with pyrolusite as its main component), yellow powders in limonite, *shiliuhuang* (mineral-laden sulfur), gypsum, Xuan stone (black stone), Yangqi stone (actinolite), calcitum, *shixie* (crab fossil), lithargite, rhodonite, Xuanjing stone (gypsum selenite), *shishe* (literally stone snake), Yu stone (arsenopyrite), white arsenic, sal ammoniac, *shiyan* (stone swallow), Jinya gold deposits, Jinxing stone (literally gold star stone), Yinxing stone (literally silver star stone), Qiang stone, raw yellow stone, Huaru stone (dolomite), *buhuimu* (literally incombustible stone), etc. Attached to each item are descriptions of properties, origin, occurrences, crystal size, colour, fracture, cleavage, quality and usage, reflecting the level of mineral and rock knowledge in the Ming dynasty. For example:

> *Dansha* (cinnabar): *dansha* occurs in Chenzhou, Yizhou and Jiehzhou. The kind from Chenzhou, which is the best, is called *chensha*. It occurs in in deep mountains and stone cliffs. The local people must dig underneath the ground for several dozens of *chi* until they are finally able to spot the veins containing it. These stones are "white stones", for *dansha* develops on them, so the white stones are also referred to as "the bed of *dansha*". As for *dansha*, the large crystals are like the testes of roosters while the small ones are like pomegranate seeds. They take the shape of lotus seed pods or arrowheads. On the stone "bed", they are dark purple in colour, similar to the colour of iron, but bright and clear at the same time. Smashed ones can be used to build a wall. And they are dispersible like mica. All the above are real *chensha*. If they are not mingled with gravels, the quality is even better. In the past, they extruded from earth's rocks, rather than formed on stone beds. According to Tao Yinju's annotations, this kind of stone usually comes from several barbarian areas in Xichuan, Wuling. Today's Chenzhou is exactly the Wuling of the past. Although the stone is dubbed *chensha*, it is seldom found in Chenzhou. Instead, it is usually excavated in Xizhou, Xuzhou and Jinzhou. Therefore, maybe he mistook other places for Xichuan in Wuling. *Yisha* is also found as a

large chunk of cinnabar and it can also be used to build walls after being smashed, but there is hardly any *yisha* that is as red as *chensha* or resembles it in appearance. Yisha is worse quality than *dansha*, for it is entirely found in the gravelly earth rather than on a bed of white stones. However, *jiesha* from Jiezhou is even worse. So instead of being used as medicine, *jiesha* may only be used as pigments. Among all the *dansha* of different origins, the best ones are called *guangmingsha* (literally bright sandstone), the second-class ones are *kekuai*, the third-class ones are *lusu* and the worst ones are *mosha* (literally the worst sandstone)".

Kongqing: *kongqing* comes from the valleys of Yizhou and the copper-producing areas of Yuejun Mountain. *Kongqing* is a by-product of exquisitely fumigated copper. Now it can also be occasionally found in Xinzhou. Normally, the stone is hollow. Those with slurry inside are extremely rare. There also exist some like the testes of roosters and smaller ones like beans. Excavation of the stone is possible all year round.

Zengqing: *zengqing* occurs in the same mountain as *kongqing*. The two kinds of stones are similar in shape and have no apparent differences in colour and texture. But *zengqing* usually appear as strings of pearls. It is rarely spotted nowadays.

Hua Stone (talc): there are two kinds of Hua stone: stones from Daozhou and Yongzhou are white and smooth like congealed fat, while those from Laizhou and Haozhou are green stones with white and black flecks and a coarse texture and are also known as Ban stone (fleck stone). Both types can be used to make exquisite tools. Hua stone is as soft as mud when first dug out, but it becomes tougher later. Therefore, everyone processes it into tools immediately in the cave where it was excavated.

(3) Knowledge of mineral-petrology in annotations

① Cao Zhao's *Gegu yaolun*

The *Gegu yaolun* ("Study on Cultural Relic Identification"), written by Cao Zhao of the Ming dynasty in 3 volumes, was finished in 1387 (the 20th year of the Hongwu reign). Cao Zhao was native to Songjiang and had the courtesy name Mingzhong. His father "was learned and accomplished, and was ardent about collecting antiques, including copybooks for calligraphy, famous drawings, *guqin*, antique inkstones, ritual vessels and wine vessels. He restored them as treasures in his study and pavilions. He frequently associated with the sellers of those antiques." Cao Zhao stated that "I have had a passion for antiques since I was young, so I used to stand by my father to admire those antiques. Each time I found something unfamiliar, I would search through all the reference books in his possession to trace its origin and judge its authenticity. I have maintained this habit until old age and I am still worried that my research is not extensive and profound enough. In recent years, some profligate sons of rich parents have also delved into this kind of elegant and refined preoccupation. However, they lack the skill to tell the true worth of treasures, despite their interest in them. Therefore, I have compiled this book, which covers the skills involved in judging the authenticity and quality of ancient bronze vessels, painting and calligraphy and peculiar objects, which are organised in different sections. I wish to present this book *Gegu yaolun* to the true lovers of this field [124]." In the book, the sections "Essay on Ancient Inkstones", "Essay on Treasures" and "Essay on Bizarre Stones" are the most relevant to minerals and rocks.

The second volume "Essay on Ancient Inkstones" records 13 kinds of inkstones, that is, the inkstones from the upper, middle and lower reaches of the Duan River;

"quasi" Duan stone; 7 kinds of She stone, including Longwei, Luowen, Shuasi, Jinyinjian Shuasi, Meizi, Jinxing, Yinxing; Jinxing stone from Wanzhou; Taohe stone (stone from rivers in Lintao). Moreover, there also exist two kinds of tile inkstones, namely Tongque inkstone and Weiyang inkstone. According to the book, stones from the old pits on the lower reaches of the Duan River are "pebbles, as black as paint and as fine and smooth as jade. The stone has small holes with halos inside it. Six or seven holes connect to one another, forming a 'string of stars'... It is soundless when struck and when ground on for ink." Stones from the old pits on the middle reaches of the Duan River are "also pebbles as purple as tender liver and as fine and smooth as jade. The stone has small holes like green beans or has green or white streaks on its surface. Vertical and round streaks are referred to as 'holes' while horizontal and long streaks are referred to as 'stripes'. It is soundless when struck and when ground on for ink. Its sharpness and glare will not fade away through long-term use." Stones from the new pits on the middle reaches of the Duan River are "light purple. The stone has small holes like the eyes of mynahs with halos inside. The tender ones are soundless when struck, but they produce subtle sounds when used for grinding ink. The sharpness and glare will fade away with long term use. The stones may be dry or humid, but humid ones are rare. This kind of stone is three grades poorer in quality compared with stones from the lower reaches of the Duan River." Stones from the upper reaches of the Duan River are "all grey and coarse. Their holes are as large as the eyes of roosters. The stone produces sounds when struck and ground on for ink. It remains as clear as a mirror even after long term use ... Only Duan stone has 'eyes'. As the proverb goes: 'There is no Duan stone without eyes.' The eyes in Duan stone can be further categorised into 'live eyes', 'tearful eyes' and 'lifeless eyes', and the first kind is the best. Another saying is that 'stones with more eyes are a good medicine for diseases'." Cao Zhao gave a clear explanation of the properties of Duan stone.

In the case of She stone, Cao Zhao said: "It is also a pebble of a light greenish black colour with no patterns. It is as fine and smooth as jade. When it is moistened, it turns slightly purple with discernible white patterns forming pictures of mountains, rivers, the sun and the moon. Then the patterns all disappear when it dries up. The large stones are four or five *cun* in length, usually used as the base for carving moon-shaped inkstones out of its original shape. Pure black ones also exist, which are as precious as the stones from the lower reaches of the Duan River."

In terms of Taohe stone (Lintao River stone), Cao said: "The green stones from the Tao River are bluish green, as smooth as jade. Their ability to produce ink is worse than that of stones from the lower reaches of the Duan River. Since it is excavated from the deep riverbeds of the great rivers in Lintao, it is extremely rare. Nowadays, these so-called green Taohe stones are mostly Li stones or valley stones from Changsha. Li stone is smooth and bright, so ink is not prone to staying on it."

The second volume "Essay on Treasures" records four categories of treasures, namely, jade, gemstones, fossils and biological treasures. These are further divided into 38 kinds. For example, Cao Zhen commented on jade as follows: "Jade which comes from Tian (today's Hetian in the Xinjiang Uygar Autonomous Region has five colours. Jade cannot be cleaved by sharp knives. It is smooth and humid, so when you stroke it, you will feel the spirituality of it." "Regarding white jade, the best examples are those in creamy colours. The second-class ones are those which are the colour of soup, oil or snowflakes. The best kinds of yellow jade are those which resemble chestnuts, called *ganhuangyu*; light brown ones are ranked second. As regards *biyu* (jasper), bluish cyan or blackish cyan kinds are the best ones. Light blue ones with tiny black flecks are ranked second. Black jades are as black as paint, but they are not expensive and can also be found in West Sichuan. As regards red jade, those as red as cockscombs are the best, but are rarely seen. As regards *lüyu* (green jade), deep green kinds are the best, while the quality of light green ones is not very good. *Ganqingyu* (opaque cyan jade) is turquoise-yellow. *Caiqingyu* (lit. vegetable jade) is neither cyan nor green, for its colour is more similar to the colour of vegetables. It is the lowest quality of all the types of jade above." The origin, colour, hardness, surface fineness, type and quality of jade are all mentioned here, providing a basis for people to be able to identify the authenticity of jade.

As regards agate, Cao said: "Agates mostly occur in the North, but they can also be spotted in western or southern barbarian areas. Agate is neither stone nor jade. It is hard and brittle. Even sharp knives cannot cut it … The ones with a brocade texture and vermilion patterns are called Jinhong agates; the ones in pure black with one crossing white line are called Hezi agates; the ones in mixed colours of black and white are called Jiezi agates; the ones with red and white silk-like stripes are called *chansi* agates. All these types are valuable. The ones in light watery red are called Jiangshui agates; the ones with purplish red flowers are called Jiangban agates; there are also some types the colour of jellyfish or which have the shape of a rabbit's face, but neither are expensive. As the proverb goes: 'Agates without the colour red are always cheap.'"

This paragraph makes clear the origin, colour, hardness, type and quality of agate.

As regards crystal, Cao said: "It is hard and brittle, and knives cannot chip it. It is as white as ice with great transparency and brightness. Stones without any scratches are the best … Crystals can be found in many places, but those from Japan are the best; those from the South are white; those from the North are black; and those from Xinzhou are opaque." The descriptions here on the properties of crystals are correct, but it is not necessarily the case that "those from the South are white; those from the North are black".

As regards amber, Cao said: "Amber occurs in southern and western barbarian areas. It is the essence of maple trees after years of crystallisation. It is clear and humid, in shades of yellow. Those similar to rosin and yellowish red in colour are called Ming amber (bright amber); those with an aroma are called Xiang amber (fragrant amber); those light yellow in colour are called La amber (wax amber). None of these types are expensive. Examples which are deep red in colour are called Xie amber (blood amber). This type comes from Korea and Japan and has valuable bees, ants or pine twigs inside. If you rub amber on the skin and then put some scraps of paper on it, you will find that they dance in the air. Some people pass off dyed ram's horns as genuine amber." This paragraph clearly explains the origin, formation, colours, properties, types, and identification method of amber.

The third volume "Essay on Bizarre Stones" records the stones in different places, as well as a few minerals. There are 17 kinds of stones recorded, including Lingbi stone (a kind of calcite), Ying stone (another kind of calcite), Guichuan stone, Kunshan stone (Kun Mountain stone), Taihu stone (Taihu Lake stone), Zhuye Manao stone (bamboo agate stone), native agate, Hongsi stone (red stripe stone), Nanyang stone, Yong stone, Chuan stone, Hushan stone, Xia stone, Wu stone, Guiwen stone (turtle-pattern stone), touchstone and mica stone. Overall, the book introduces the origin, colour, hardness, fineness, patterns, processing methods, usages, identification methods, etc. of the stones. For example, Cao Zhao introduced Lingbi stone as follows: "It occurs deep in the mountains in Lingbi County. You will see it very soon after you start excavating. It is as black as paint, alternating with white stripes. It can be buried in the sand like the jades. The stone cannot form mountains. The good types include Woniu (crouched ox), Handan (bud) and Panchi (a kind of hornless dragon). When you knock on it, you will find that the sound resonates as clearly as with jade. Even sharp knives cannot cleave it. Due to the fact that this stone can retain fragrance, people tend to keep it in their studies. In this way, a fragrance can linger for a whole day. People may pass off dyed Taihu stone as genuine Lingbi stone, but fake examples will turn into fragments if they are scratched with a knife."

As regards Nanyang stone, Cao said: "Those with pure green patterns are the best. Those with light green patterns or five-coloured cloud patterns are both ranked second in quality. The stone is extremely hard with a fine and smooth texture. Slates cut from the stone can be processed into tabletops, inkstones or screens. It turns bright when you illuminate it with a light or in the rays of the light from outside the window. Large stones are rarely seen. People also know it as *liuhuang* stone (sulfur stone)."

② Gao Lian's *Zunsheng Bajian*

Gao Lian, with the courtesy name Shenfu and the art name Ruinan Daoren, was native to Qiantang (today's Hangzhou) in Zhejiang. His date of birth and death are not clearly known, but he lived around the reign of Emperor Wanli. Tu Long, a famous dramatist of the Ming dynasty, acclaimed him as "erudite and good at identifying talented people and with a wide collection of books in his family for generations". Gao Lian was a poet and dramatist in the Ming dynasty. He was versed in ballads and composed two pieces of *nanqu* [a local opera in southern China), *Yuzan ji* ("The Story of Jade Hairpin") and *Jiexiao ji* ("The Story of Morality and Fidelity"), as well as a poetry collection *Yashangzhai shicao* ("Poem in the Yashang Study"). The *Zunsheng bajian* ("Eight Commentaries on Health Care") is a collection of his miscellaneous writings.

The *Zunsheng bajian* is in 19 volumes with more than 400 thousand words. Among them, Volume Fifteen "Commentary on Leisurely Appreciation" contains a section "On Inkstones", which is a narrative on 38 types of stones as the raw material for making inkstones, including Duan stone, She stone, black Duan stone, Lixi stone, green stone from the Tao River, *jinxing* (gold star) stone, Quzhou stone, *heijiao* (black corner) stone, *hongsi* (red silk) stone, *huangyu* (yellow jade) stone, purple-gold stone, *quejin-moyu* (magpie gold and dark jade) stone, crystal stone, white stone, celestial stone, cinnabar stone, Tang stone, Suzhou stone, purple stone, gold stone, *jinque* (canary) stone, Qingzhou stone, green stone, jade-inlaid stone, *rong* stone, crimson stone, Huai stone, Ning stone, Xuancheng stone, Ji stone, Kui stone, bright stone, *cidong* (magnetic travertine) stone, *liu* (willow) stone, *cheng* stone, *le* stone, Tuoji Island stone, Datuo stone, etc. Apart from the stones mentioned above, there were also inkstones made of other materials, such as Dengni inkstone, silver inkstone, copper inkstone, brick inkstone, *qi* (paint) inkstone, clamshell inkstone, magnetic inkstone, etc.

As regards Duan stone, Gao Lian said: "Those from the old pit are greenish-black stones as smooth as jade. They have holes on the surface with five or six green halos around. The centre of the holes, however, is usually slightly yellow with black flecks inside. Since the holes resemble the eyes of mynahs, these inkstones are also named after mynahs. The holes can be further divided into three categories: those with many transparent halos are 'live eyes'; those with opaque halos are 'tearful eyes'; and the yellowish ones without any halos are 'lifeless eyes'. According to the commentaries, 'Each preceding one is better.'... Duan Stone from the pits on northern hills generally have eyes, while those from other pits may or may not have eyes. As for the arrangement of eyes, some stones have multiple ones connected to each other like strings of stars; some have around a dozen well-proportioned holes, with four other eyes around the sides; some others have white flecks like millet seeds, which become nearly invisible in water. If the stone produces no sound both when struck and when ground on to make ink, it must be a stone from the lower hills. Stones from the lower hills are now extremely rare. Stones from the upper and middle hills, on the other hand, are both greyish purple like pork livers. Nowadays, Duan stone primarily features one huge rooster eye with few halos. It produces a sound both when struck and ground on. It also has a coarse texture." The text makes the nature and types of Duan stone very clear. The other inkstones are described briefly, or even just named.

③ Zhang Yingwen's *Qingmi cang*

Zhang Yingwen, with the courtesy name Maoshi, the art name Yifu and the alias Pihe Xiansheng, was native to Kunshan. He was erudite about both the past and the present and was on friendly terms with Wang Shizhen. He lived during the reigns of Emperor Jiajing and Emperor Wanli. Later he moved from Jiading to Changzhou to collect valued pieces of calligraphy or paintings created by masters. He wrote the *Qingmi cang* ("Writings on the Connoisseurship of Ming Calligraphies"). In the book are two relevant sections, "On Bizarre Stones" and "On Pearls and Gemstones" [125]. In "On Bizarre Stones" can be read:

> Recently, there are masses of exquisite stones made of *chensha* (cinnabar, also known as *zhusha*) and *shiqing* (azurite). I have seen several dozen pieces of *zhusha* resembling arrowheads, which were as heavy as three or four *jin*. Shiqing as heavy as two or three *jin* is also found in great amounts. I once saw a piece of arrowhead-like *zhusha* in Baixia, which was as heavy as four *jin* and two *liang* [unit of weight, 1/10 *jin*, equal to 50 grams]. The stone was surrounded with arrowhead-like bulges, each as large as a finger. Another time, I saw a piece of *shiqing* in Wuzhong, which was as heavy as two *jin* and twelve *liang* [around

1,600 grams]. Its colour was like the feathers of a kingfisher. Nevertheless, both stones were demanding a high price and failed to sell for years.

Lingbi stone comes from Lingbi County in Hongzhou. It is as black as paint with slender white patterns on it like jade. It produces a clear sound when struck. The pieces which cannot be damaged by sharp knives are genuine. Once I acquired one which is only as large as a fist. It has an undulating surface and shows no sign of artificial artistry by mallet and chisel. This extremely exquisite artifact originally belonged to Mi Fu, then it was transferred to the Songxue Study and finally it joined my collection. I also got another one which is three *cun* and three *fen* [unit of length, 1/10 *cun* equal to 1/3 centimetre] in length and three *cun* and six *fen* [around 12 centimetres] in height. It was chiselled from the Sword Pond on Tiger Hill. I sold it to someone interested in it, but still hold it in my heart. The pure black ones with no white patterns are Ying stones. Those which turn into fragments when scraped are counterfeits made of Taihu stone.

In terms of marble, the pieces which are as white as jade or as black as ink are of high quality. The grey ones (mixed with white and slight green) are not worth any consideration. However, the one I got, which presented a landscape of mountains, rivers and clouds just like in Mi Fu's drawing, was an unparalleled treasure for making screens. I keep several screens made of marbles in my study. Each time I see the beautiful landscapes on them, it feels as if I am really visiting the places.

As regards Kunshan stone, the larger the piece, the more precious it is. There are two types of Kunshan stone, *jigupian* (lit. chicken bone slice) and *hutaokuai* (lit. walnut piece). *Jigupian* is the better quality. During the reign of Emperor Jiajing, I saw one as tall as over one *zhang* and as long as seven or eight *chi*. The lower half of that stone was of the *hutaokuai* type while the upper half was of the *jigupian* type. As white as jade, it was extremely exquisite. An illustrious family in Songjiang spent eighty strings of a thousand cash coins on buying the stone to build a large library.

The origin, shape, colour, hardness, texture, sound when struck, identification methods, and aesthetic value of rocks are all mentioned above, reflecting the level of knowledge of rocks in the Ming dynasty. In "On Pearls and Gemstones" can be read:

Among pearls, the best ones are luminous pearls and clam pearls. Those from Western countries are pink with strong glazes. Those from freshwater are second best. However, those from the seas north, south or east of China are not worth collecting. I once saw a clam pearl as large as a magpie's egg. If placed under the sunshine, it would appear like cream-coloured drips that refracted a colourful glare. I heard that the trader sold it for 300 strings of a thousand cash coins, which indicates how amazing it is.

Diamond is similar to amethyst in appearance and can be carved further to make jade. If smashed against iron, it will be damaged, while the iron remains intact. Diamond can be further divided into "north diamonds" and "south diamonds". The latter is useful, while the former is not. The best pieces are the wine yellow ones, while pea green, lake blue and purple ones are second best.

As for chrysoberyl (cat's eye), the wine yellow ones are also the best, while pea green, lake blue dragonfly head (greenish blue) and black ones are second best. There is a way to recognise the genuine ones: "A line above, an area below". "A line above" refers to the fact that the stone reflects a plume of bright light while "an area below" means that its base colour is close to silver.

Zhubabi (referred to as "green stone" or "green gemstone" in the *Nancun chuogeng lu* ("Retirement to the Countryside Nancun") is dark green; *zumulü* (emerald), also known as *zhushuilü* (another kind of green gemstone) is bright green; both kinds of green gemstone emit a glare like that of dragonflies' wings. *Taohuaci* (red gemstone) is light red and extremely tender; *hongyagu* (another kind of red gemstone) is a similar colour to *taohuaci*, but with white waterdrops on it. *Bizheda* (garnet) is dark red and tender, shaped like thin squares. *Yahuqing* (another kind of garnet), also known as *qingyagu*, is dark green. *Nilan* (another kind of garnet) is light green and slightly bright. Among all the gemstones mentioned above, those as soft as an autumn stream are the best, while the minutely flawed ones are less desirable. I once bought one piece of *taohua(ci)* and one piece of *yahu(qing)*, each one with twenty strings of a thousand cash coins. I tried heating them over a big fire for a whole day, but they remained almost intact.

The names of the gemstones mentioned here are basically the same as those in the *Nancun chuogeng lu* despite a few differences. Obviously, the names of the gemstones presented in this book are from the *Nancun chuogeng lu*.

④ Cao Xuequan's *Shuzhonguang ji*

Cao Xuequan (1574–1647), with the courtesy name Nengqi and the art name Shicang, was native to Houguan (today's Fuzhou) in Fujian. He became a Jinshi in 1595 (the 23rd year of the Wanli reign) and was once the Secretary of the Ministry of Revenue. Then he was transferred to Nanjing and had another post as the Director of the Left Court of the Judicial Review after passing an imperial official test. Later, he was free of any official duties for seven years and completely devoted himself to learning. Then he successively served as the Director of the Nanjing Bureau of the Ministry of Revenue, the Right Administration Vice Commissioner of Sichuan province and the Surveillance Commissioner of Sichuan province. Then he was selected to sit an imperial official test again and was considered for another change of post. In 1622, the 2nd year of Emperor Tianqi's reign, he was assigned to the post of the Right Assistant Commissioner of Guangxi province. In the early Chongzhen period, he was promoted to the post of Vice Commissioner of Guangxi province, but he refused to take the position. Instead, he stayed at home for twenty years and dedicated himself to writing the book *Selected Poems of Shicang in the Past Twelve Dynasties* in Shicang Garden where he lived. The book later became a runaway success. When the Prince of Tang established his rule in Min (Fujian), Cao Xuequan was appointed Chief Minister of the Court of Imperial Sacrifices. Then he was transferred to the post of Right Vice Minister of the Ministry of Rites and the Hanlin Academician Expositor-in-waiting. Then he was promoted once again to the post of Personnel Minister as well as Grand Guardian of the Heir Apparent. However, the rebellion of the Prince of Tang ended in failure. The prince had to escape into the remote mountains and finally chose to commit suicide by hanging himself at the age of 74. Cao composed quite a number of poems which were all included in the *Shicang ji* ("Collections from Shicang") [126]. The *Shuzhong guangji* ("Broad Records of Shu") was completed when he was about to change posts from Right Administration Vice Commissioner of Sichuan province to Surveillance Commissioner of Sichuan province. The book is divided into twenty items, among which "Gas Wells and Oil Wells" in Volume 66 provides records of how the salt wells in Sichuan produced petroleum. In this volume it is written: "According to the *General Treatises*, salt wells were exploited in Yizhou in 1521 (the last year of Emperor Zhengde's reign). Oil was occasionally found in the wells and appeared even brighter at night. If it was irrigated with water, the blaze burned even more fiercely; if ashes were used, the blaze would be extinguished. Realgar sulfur gas would be emitted during the process, which is called realgar oil or sulfur oil by native people. A few more wells have been exploited recently. This exploitation was overseen by imperial officials. The substance people obtained was indeed petroleum, but it was excavated from the wells. The wells can produce petroleum because they are connected to the sources of realgar, sulfur and halloysite. According to the *Yibu tanzi* ("Topics of Conversations in Yizhou"), counties including Jiamei, Qingshen, Jingyan, Hongya and Jianwei all have wells. All the people use the oil from the wells to light lamps. When the official in charge went out walking during the night, he used a thick bamboo tube with oil inside as a torch. With the help of a single tube, he could walk for several li. The price of this kind of oil is only half that of normal oil, but the illumination strength is definitely not any worse. I once passed by Jianwei in 1611 with a tube like this and obtained this kind of oil by chance. I asked for some in order to give it a try, and it really worked. However, due to the fierce properties of the oil, people cannot get too close to it. When you carry the flames forward, if you do not dodge in time, your hands are likely to get burned. Later on, monks from Mount Lu took the oil-laden tube away to broaden the horizons of those living deep in the mountains." This passage is a compelling account, for it is his own personal experience of using petroleum.

"Stones as Medicine" in Volume 64 talks about the minerals and rocks produced in Shuzhong, such as *dansha*, *kongqing*, *bianqing*, mica, *taiyi xuanjing* stone (gypsum selenite), stalactite, dragon bone, mirabilite (in Chinese, *mangxiao*, *puxiao* or *mayaxiao*), sulfur, etc. "Jade and Five Jin [five minerals: gold, silver, copper, iron and tin]" in Volume 67, talks about minerals and rocks produced in Shuzhong, such as jade, black jade, crystal, *pusa* stone (bodhisattva stone), green pearl, gold, silver, iron, tin, lead, copper and others. However, they are all quoted from historical materials and few of these are from the Ming dynasty.

⑤ Lu Rong's *Shuyan zaji*

The *Shuyuan zaji* ("Jottings from the Shu Garden") in 15 volumes was written by Lu Rong. Lu Rong (1436–1496), with the courtesy name Wenliang and the art name Shizhai, was native to Taicang. He became a Jinshi in 1466, the 2nd year of the Chenghua reign. At one time he was appointed to the post of Secretary of Nanjing province, then transferred to Director of the Bureau of Operations in the Ministry of War and, ultimately, became Administration Commissioner of Zhejiang province. Volume Fourteen of this book talks about the exploring, mining, smelting and other techniques of processing copper and silver ores, and also discusses the purity of the copper and silver in the ores. The volume reads: "Five Jin (five minerals) are found in the high mountains and lofty hills. Initially, some minute lodes of ore protruded from hard rocks. Those who are well versed in identifying ores spot them, chisel minerals from them and try to heat the minerals. The ores vary in terms of colour, shape and fineness. Silver is commonly found in these, but the amount is not fixed. Of a basket of minerals as heavy as 25 *jin*, there may be as much as two or three liang of silver or as little as three or four *qian* [unit of weight, equal to 5 g]. It is hard to judge whether the lode is deep or not. Some lodes terminate shortly after they extend from the surface of the ground; some may penetrate as deep as several *zhang*; some may be narrow at the beginning but widen over time; some terminate once but as people keep excavating, they come across them again, and these are, therefore, dubbed *guobi*, or 'wall-traversing' lodes; some others start in one place, disappear suddenly and reappear around eight *chi* to one *zhang* away, and are, therefore, dubbed *xiamatiao*, or 'toad jump' lodes. In general, the miners are like woodpeckers, for the depth could be several *zhang*, several dozen *zhang* or even several hundred *zhang*. Miners will not stop excavating until the ore runs out." This paragraph makes clear the situation of silver lodes as well as the methods of prospection and mining approaches. The purity of the silver in the minerals, according to the figures recorded, is projected to range from 0.075% to 0.75% [127]. As for the purity of the copper in the minerals, the book reads: "The amount of copper found in the minerals is not fixed, but basically a basket of minerals [more than 30 *jin*] can yield one *jin* of copper." Based on this, it can be judged that the purity of the copper was around 3.3% [128].

⑥ Wang Linheng's *Yuejian bian*

In 1600 (the 28th year of Emperor Wanli's reign), Wang Linheng, with the courtesy name Zhizhi, held a position in Guangdong. He recorded what he had seen and heard during his term there and collated his records into the *Yuejian bian* ("Compilation of Experiences in Yue [Guangdong]") in four volumes in 1601 (the 29th year of the Wanli reign). Volume Three "Record of Local Products" talks about minerals and rocks including pearls, gold, inkstones, Ying stone (a kind of calcite), *shixie* (crab fossil), etc. The book reads:

> Pearls have their origins in the seas southeast of Lianzhou. Their exploitation takes place in winter and spring and ceases in summer and autumn. There are mountains all around the north coast of the sea, so the north wind blows strongly in winter and spring. That is why it is safer to go by boat to collect pearls in winter and spring. In the Warring States period, Wei state, where pearls that can illuminate the path for chariots were collected, was located on the northern frontier quite far away from Lianzhou. During the Sui dynasty, candles were not used in the palace. Instead, a great many pearls were hung up there to bring "bright daylight" into the whole palace. Nowadays, however, even the imperial treasury has hardly any luminous pearls, let alone the ordinary people. After all, it was the distant past before the Sui and Tang dynasties, when people were unable to understand the profits to be gained from fishing. That is why clams from thousands of years ago may still survive in the seas until today to give out bright light. Today's coastal residents, however, learn to fish in their own ponds. Rich households even fence in the fishing lakes. Despite repeated prohibitions, fishing remains popular among them, and they go with nets that can easily capture clams with pearls inside. How can we expect pearls to shine for long under such circumstances? Recently, a mining official showed me two pearls: one, which was as heavy as seven *qian*, was in the shape of a lion; the other, which was as heavy as three *qian*, was round on the surface and flat on the bottom. They were both shell-clinging pearls, which are not valuable. Nonetheless, the mining official thought them something precious and was eager to offer them to the court as a tribute.

Gold can be mined from the sandbanks in Dianbai County and Shicheng County. The mining official heard of this and ordered someone to come and obtain gold. They usually sift through the sand to discover the precious gold. Large pieces are like grains, while small ones are like bran, and they could be gathered without requiring to be melted. Later, I was stationed at Qingyuan, purported to be another site where gold can be found. I suppose there are many locations, but fearing that the news may cause disturbances among the folk, the officials and bigwigs do not dare to make the information public.

Duan inkstone occurs in Lingyangxia (lit. antelope canyon) Mountain. The canyon is over 30 li away from the centre of Duanzhou, which was precisely the site of the battle over the south sea [under the governance of Jiaozhou] between Wu Buzhi and Qian Bo. For those who want to enter Duanzhou from Guangxi, Lingyangxia is a strategic point that they must pass through. Recently, since the tax officials ordered people to excavate the mountains to collect inkstones, the reserves of *yumuqing* (a kind of Duan inkstone resembling mynahs' eyes) are almost exhausted.

Ying stone is collected by those who love it in my home village. The stone is generally valued for being black, hard and fine. As I observe the middle-sized stones from Yingde city nowadays, I find them to be green but not smooth, coarse but not hard, and they have few ridges and peaks and tend to break. A mountain that used to produce high-quality Ying stones is almost depleted of any Ying stone now. The Ying stones we can find now are all collected from other mountains, and so they are much less valuable.

Shixie (crab fossils) are found in the waters of Yazhou, Linchaun. Each one is quite soft when first collected, but it becomes as tough as a stone when taken out of the water. According to the *Haicha yulu* ("Notes on Specialties of Hainan Island"): "*Shixie* come from the Port of Yulin in Yazhou. Around a half li into the port is the place with the finest and coldest soil. Once a crab climbs inside, it will be frozen and transformed into a fossil after a while. *Shixie* can be used as a medical substance to improve eyesight."

References

1. *Mingshi: Zhenghe zhuan* 明史·郑和传 [History of the Ming dynasty: Biography of Zheng He]
2. Tan Q谈迁*Guoque: juan shisan*国榷 卷十三 [Discussions about the state, vol 13]
3. *Ming chengzu shilu: juan wushijiu*明成祖实录·卷五十九 [Authentic records of Emperor Chengzu of the Ming dynasty, vol 59].
4. Lu R 陆容*Shuyuan zaj: juansani*菽园杂记·卷三 [Jottings from the Shu Garden, vol 3]
5. The textual research about toponyms refers to the *Nautical Chart of Zheng He*郑和航海图compiled by Xiang Da 向达, *Zheng He* 郑和by Zhu Xie 朱偰, *Tributes from the Western countries* 西洋朝贡典录collated and annotated by Xie Fang 谢方, etc.
6. *Ming chengzu shilu: juan yibai sanshisi*明成祖实录 卷一百三十四 [Authentic records of Emperor Chengzu of the Ming dynasty, vol 134]
7. Tan Q谈迁*Guoque: juan shisan* 国榷·卷十三 [Discussions about the state, vol 13]
8. Zhu Y 祝允明*Qianwen ji* 前闻记 [Accounts of events before the Ming dynasty]
9. Yan C严从简*Shuyu zhouzi lu: juan ba* 殊域周咨录·卷八 [Historical and geographical works of peripheral countries in the Ming dynasty, vol 8]
10. Niu Z钮仲勋 (1985) *Zhenghe hanghai tu de chubu yanjiu*《郑和航海图》的初步研究 [Preliminary Research on the *Nautical Chart of Zheng He*] In: *Zhenghe xia xiyang lunwenji: diyiji*

郑和下西洋论文集: 第一集 [Collected papers on Zheng He's voyages to the West, vol 1]. China Communications Press 北京人民交通出版社, Beijing
11. *Wubeizhi: juan erbaisishi, "Hanghai"* 武备志·卷二百四十"航海"条 [Treatise on armament technology, vol 240, "Navigation"]
12. Joseph N 李约瑟 (1976) *Zhongguo kexue jishushi (zhongyi ben): diywu juan diyifence*中国科学技术史 (中译本): 第五卷第一分册 [Science and civilisation in China (trans), vol 5, Book 1. Science Press科学出版社, Beijing, p 169
13. *Yingya shenglan: Xu*瀛涯胜览·序 [Extensive survey of the ocean shores: Preface]
14. Zhang H张洪*Huayi shenglan: Xu,* 华夷胜览·序 [Extensive survey of China and foreign countries: Preface], cited from *Guitiangao: juansan*归田稿·卷三 [Script of resigning from the office and returning home, vol 3], 常熟市图书馆藏抄本 [Version: the manuscript reserved in the Changshu Library]
15. *Xiyang chaogong dianlu: Zixu*西洋朝贡典录·自序 [Tributes from the Western countries: Author's preface]
16. Xie F谢方 (1982) *Xiyang chaogong dianlu jiaozhu: qianyan*西洋朝贡典录校注·前言 [Annotation of the *Tributes from the Western Countries*: Preface]. Zhonghua Book Company中华书局, Beijing
17. Ji Y纪昀 (1965) (Qing dynasty) *Sikuquanshu zongmu tiyao* 四库全书总目提要 [Catalogue of the complete library of the four treasuries]. Zhonghua Book Company中华书局, Beijing
18. Wang X汪向荣 (1985) *Zhongri guanxishi wenxian lunkao* 中日关系史文献论考 [Research on the historical literature about Sino-Japanese relations]. Yuelu Press岳麓书社, Changsha
19. Wang Y王燕玉*Fangzhi chuyi* 方志刍议 (1984) [On local chronicles]. In: Chinese Local Records Association (CLRA) 中国地方志史协会 (ed) *Zhongguo difangzhishi luncong*中国地方史志论丛 [Collected essays on the history of Chinese local chronicles]. Zhonghua Book Company中华书局, Beijing
20. Chen G陈光贻 (1987) *Xijian difangzhi tiyao: shangce*稀见地方志提要: 上册 [Abstracts of rare local chronicles I]. Shandong Qilu Press齐鲁书社, Jinan, p 18
21. Huang W黄苇 (1987) *Zhongguo difangzhi cidian*中国地方志辞典 [Dictionary of China's local chronicles]. Huangshan Publishing House黄山书社, Hefei
22. Liu W刘纬毅 (1991) *Zhongguo difangzhi*中国地方志 [China's local chronicles]. Xinhua Publishing House新华出版社, Beijing, p 108
23. *Mingshi·Xiyu liezhuan si*明史·西域列传四 [History of the Ming dynasty: Biographies of the Western Regions IV]
24. *Shi xiyu ji*使西域记 [Travels to the Western Regions], 载《学海类编》第26函 [the 26th Slipcase of the *Xuehai leibian* (Classified compilation of learning materials)]
25. Li Zhi李贽*Fen shu: juan er*焚书·卷二 [A book to be burned, vol 2]
26. Zhou Z周振鹤 (1993) *Wangshixing dilishu sanzhong* 王士性地理书三种 [Three geographic records by Wang Shixing]. Shanghai Classics Publishing House上海古籍出版社, Shanghai, pp 25–26
27. Zhou Z周振鹤 (1997) *Xuxiake yu mingdai houqi lvxingjia qunti*徐霞客与明代后期旅行家群体 [Xu Xiake and the community of travellers from the Late Ming dynasty]. In *Xu xiake yanjiu: diyiji* 徐霞客研究: 第1辑 [Research on Xu Xiake, Vol 1]. Xueyuan Publishing House 学苑出版社, Beijing, pp 52–61
28. Zhou Z周振鹤 (1993) *Wangshixing dilishu sanzhong* 王士性地理书三种 [Three geographic records by Wang Shixing]. Shanghai Classics Publishing House上海古籍出版社, Shanghai
29. Xu J徐建春, Liang G梁光军 (1994) *Wangshixing lungao*王士性论稿 [Collected essays on Wang Shixing]. Hangzhou University Press杭州大学出版社, Hangzhou
30. Wang S王士性 (Ming dynasty) *Guanzhiyi: juansi*广志绎: 卷四 [Observations in the realm, vol 4]. In: Zhou Zhenhe 周振鹤 (1993) (ed) *Wangshixing dilishu sanzhong*王士性地理书三种 [Three geographic records by Wang Shixing].: Shanghai Classics Publishing House上海古籍出版社, Shanghai, p 324
31. Wang S王士性 (Ming dynasty) *Guanzhiyi: juanyi*广志绎: 卷一 [Observations in the realm, vol 1]. In: Zhou Zhenhe 周振鹤 (1993) (ed) *Wangshixing dilishu sanzhong*王士性地理书三种 [Three geographic records by Wang Shixing]. Shanghai Classics Publishing House上海古籍出版社, Shanghai, p 243

32. Wang S王士性 (Ming dyansty) *Guanzhiyi: juan'er*广志绎: 卷二 [Observations in the realm, Vol 2] In Zhou Zhenhe 周振鹤 (1993) (ed) *Wangshixing dilishu sanzhong*王士性地理书三种 [Three geographic records by Wang Shixing]. Shanghai Ancient Books Publishing House 上海古籍出版社, Shanghai
33. *Guangyouzhi: Xingsheng*广游志·形胜 [Records of extensive wanderings: Strategic positions]
34. *Guanzhiyi: jiangnan zhusheng: huguang*广志绎·江南诸省·湖广 [Observations in the realm: Various provinces south of the Yangzi River: Huguang]
35. *Guanzhiyi: jiangnan zhusheng*广志绎·江南诸省 [Observations in the realm: Various provinces in south of the Yangzi River]
36. Weng W翁文灏 (1925) *Zhongguo shanmai kao*中国山脉考 [Investigations on China's mountain ranges]. *Kexue*科学 [Science] 9(10)
37. Xu J徐建春, Liang G 梁光军 (1994) *Wangshixing lungao*王士性论稿 [Collected essays on Wang Shixing]. Hangzhou University Press杭州大学出版社, Hangzhou, p 52
38. Xu J徐建春, Liang G 梁光军 (1994) *Wangshixing lungao* 王士性论稿 [Collected essays on Wang Shixing]. Hangzhou University Press杭州大学出版社, Hangzhou, p 56
39. *Wuyue youcao: juan si*五岳游草·卷四 [Travels to the Five Sacred Mountains, vol 4]
40. *Guanzhiyi: fangyu yalue*广志绎·方舆崖略 [Observations in the realm: A compendium of the land]
41. *Guangyouzhi: fengtu*广游志·风土 [Records of extensive wanderings: Social customs]
42. Xu J徐建春, Guangjun Liang 梁光军 (1994) *Wangshixing lungao*王士性论稿 [Collected essays on Wang Shixing]. Hangzhou University Press杭州大学出版社, Hangzhou, p 29
43. Chen H陈函辉 (1993) *Xuxiake muzhiming*徐霞客墓志铭 [Epitaph of Xu Xiake]. In: *Xuxiake youji*徐霞客游记 [Xu Xiake's travels]. Shanghai Classics Publishing House上海古籍出版社, Shanghai
44. Pan L潘耒 "Xuxiake youji: Xu" 徐霞客游记·序 [Xu Xiake's travels: Preface]. In *Suichu tang ji: juanqi*遂初堂集·卷七 [Collected works from Suichu Library, vol 7], also published in the *Xuxiake youji* 徐霞客游记 [Xu Xiake's travels]. Shanghai Ancient Books Publishing House 上海古籍出版社, Shanghai
45. Qian Q钱谦益 "Xuxiake zhuan" 徐霞客传 [Biography of Xu Xiake]. In: *Muzhai quanji: juan qishiyi*牧斋全集·卷七十一 [Complete works of Muzhai, vol 71], also published in the *Xuxiake youji* 徐霞客游记 [Xu Xiake's travels]. Shanghai Ancient Books Publishing House 上海古籍出版社, Shanghai
46. Xu Hongzu (Xu Xiake) 徐弘祖 (霞客) (Ming dynasty) *Xuxiake youji*徐霞客游记 [Xu Xiake's travels]. Shanghai Ancient Books Publishing House 上海古籍出版社, 1993, Shanghai, p 697
47. Xu H (Xu Xiake) 徐弘祖 (霞客) (Ming dynasty) *Xuxiake youji*徐霞客游记 [Xu Xiake's travels]. Shanghai Ancient Books Publishing House 上海古籍出版社, 1993, Shanghai, p 267
48. Xu H (Xu Xiake) 徐弘祖 (霞客) (Ming dynasty) *Xuxiake youji*徐霞客游记 [Xu Xiake's travels]. Shanghai Ancient Books Publishing House 上海古籍出版社, 1993, Shanghai, p 408
49. Xu H (Xu Xiake) 徐弘祖 (霞客) (Ming dynasty) *Xuxiake youji*徐霞客游记 [Xu Xiake's travels]. Shanghai Ancient Books Publishing House 上海古籍出版社, 1993, Shanghai, p 711
50. Xu H (Xu Xiake) 徐弘祖 (霞客) (Ming dynasty) *Xuxiake youji*徐霞客游记 [Xu Xiake's travels]. Shanghai Ancient Books Publishing House 上海古籍出版社, 1993, Shanghai, p 60
51. Xu H (Xu Xiake) 徐弘祖 (霞客) (Ming dynasty) *Xuxiake youji*徐霞客游记 [Xu Xiake's travels]. Shanghai Ancient Books Publishing House 上海古籍出版社, 1993, Shanghai, p 165
52. Xu H (Xu Xiake) 徐弘祖 (霞客) (Ming dynasty) *Xuxiake youji*徐霞客游记 [Xu Xiake's travels]. Shanghai Ancient Books Publishing House 上海古籍出版社, 1993, Shanghai, pp 949–950
53. Xu H (Xu Xiake) 徐弘祖 (霞客) (Ming dynasty) *Xuxiake youji*徐霞客游记 [Xu Xiake's travels]. Shanghai Ancient Books Publishing House上海古籍出版社, 1993, Shanghai, pp 1054–1055

54. Xu H (Xu Xiake) 徐弘祖 (霞客) (Ming dynasty) *Xuxiake youji*徐霞客游记 [Xu Xiake's travels]. Shanghai Ancient Books Publishing House 上海古籍出版社, 1993, Shanghai, p 2
55. Xu H (Xu Xiake) 徐弘祖 (霞客) (Ming dynasty) *Xuxiake youji*徐霞客游记 [Xu Xiake's travels]. Shanghai Ancient Books Publishing House 上海古籍出版社, 1993, Shanghai, p 783
56. Xu H (Xu Xiake) 徐弘祖 (霞客) (Ming dynasty) *Xuxiake youji*徐霞客游记 [Xu Xiake's travels]. Shanghai Ancient Books Publishing House 上海古籍出版社, 1993, Shanghai, p 612
57. Joseph N李约瑟 (1976) *Zhongguo kexue jishushi (zhongyi ben):diyi juan*中国科学技术史 (中译本): 第一卷 [Science and civilisation in China (trans), vol 1], Science Press科学出版社, Beijing, p 150
58. Joseph N李约瑟 (1976) *Zhongguo kexue jishushi (zhongyi ben):diwu juan diyifence*中国科学技术史 (中译本): 第五卷第一分册 [Science and civilisation in China (trans), vol 5, Book 1], Science Press科学出版社, Beijing, pp 61–63
59. Quoted from the 22nd Slipcase of the *Xu shuofu* 续说郛 [Supplement to the *Persuasion of the Suburbs*]
60. Gu J顾岕 *Haicha yulu* 海槎余录 [Notes on the specialities of Hainan Island], also published in the *Jilu huibian* 纪录汇编 [Collection of records] compiled by Shen Jiefu沈节甫
61. *Sikuquanshu zongmu tiyao: shibu: dili lei*四库全书总目提要·史部·地理类 [Catalogue of the complete library of the four treasuries: History section: Geography]
62. *Mingshi: Songli zhuan*明史·宋礼传 [History of the Ming dynasty: Biography of Song Li]
63. *Lushui ketan*潞水客谈 [Discussions on the irrigation and water conservancy]
64. *Mingshi: Xuzhenming zhuan*明史·徐贞明传 [History of the Ming dynasty: Biography of Xu Zhenming]
65. Wang Shixing王士性*Guanzhiyi: Liangdu*广志绎·两都 [Observations in the realm: Two capitals]
66. Pan J潘季驯*Hefang yilan: Heyi bianhuo*河防一览·河议辨惑 [A survey of the river defence work: Clarification of confusing issues regarding river governance]
67. *Mingshi: Panjixun zhuan*明史·潘季驯传 [History of the Ming dynasty: Biography of Pan Jixun]
68. Hou R侯仁之 (1959) *Zhongguo dilixue jianshi*中国地理学简史 [A brief history of Chinese geography]. Former Department of Geology and Geography, Peking University北京大学地质地理系, Beijing, p 95
69. Wu F吴枫 (1991) *Zhonghua guwenxian da cidian:Dili juan*中华古文献大辞典·地理卷 [The dictionary of ancient Chinese literature: Geography]. Jilin Literature and History Press吉林文史出版社, Changchun, p 216
70. Quoted in the *Yugong zhuizhi: juan shisan* 禹贡锥指·卷十三 [An analysis of the *Tribute of Yu*, vol 13]
71. Geoscience History Group, Institute of the History of Natural Sciences, Chinese Academy of Sciences中国科学院自然科学史研究所地学史组 (1984) *Zhongguo gudai dilixue shi*中国古代地理学史 [A history of ancient Chinese geography]. Science Press科学出版社, Beijing, p 145
72. *Sancai tuhui: Dili juan si*三才图会·地理卷四 [Assembled pictures of the three realms: Geography, vol 4]
73. *Jiangyuan kao*江源考 [Research on the source of the Yangzi River], also published in the *Xuxiake youji* 徐霞客游记 [Xu Xiake's travels]. Shanghai Ancient Books Publishing House 上海古籍出版社, 1982, Shanghai, pp 1128–1130
74. *Hefang yilan: juan shi'er*河防一览·卷十二 [A survey of the river defence work, vol 12]
75. *Chuanshan yishu: juan yi*船山遗书·卷一 [Posthumous works of Chuanshan: vol 1]
76. Quoted in the *Yugong zhuizhi: juan shiba* 禹贡锥指·卷十八 [An analysis of the *Tribute of Yu*, vol 18]
77. *Xuxiake youji*徐霞客游记 [Xu Xiake's travels], (1982), p 60
78. *Xingshui jinjian: juan ershiyi*行水金鉴·卷二十一 [Golden mirror of waterways, vol 21]
79. *Nongzheng quanshu: juan shi'er*农政全书·卷十二 [Complete treatise on agriculture, vol 12]

80. Huang Xiangxi黄象曦*Wujiang kao zengji: juan er*吴江考增辑·卷二 [Supplement to *Research on the Wu River*, vol 2]
81. *Xuxiake youji*徐霞客游记 [Xu Xiake's travels], (1982), pp 1040–1041
82. Zhao D赵德祥 (1982) Woguo lishi shang zhaoze de mingcheng fenlei ji miaoshu" 我国历史上沼泽的名称、分类及描述 [Terms, classifications and descriptions of marshes in Chinese history]. *Dili kexue*地理科学 [Scientia geographica sinica] 2(1)
83. The tablet is now preserved in the Forest of Stone Steles Museum in Xi'an (aka. Xi'an Beilin Museum), according to Zhou Weizhou's周伟州 (1975) article published in *Wenwu*文物 [Relics] (3)
84. *Hefang yilan: juan er*河防一览·卷二 [A survey of the river defence work, vol 2]
85. Zheng X郑晓*Jinyan leibian: shang*今言类编 (上) [Classified compilation of contemporary talks I]. Published in the 3rd Slipcase of *Shengchao yishi* 胜朝遗事 [Anecdotes of the defunct dynasty]
86. Geoscience History Group, Institute of History of the Natural Sciences, Chinese Academy of Sciences中国科学院自然科学史研究所地学史组 (1984) *Zhongguo gudai dilixue shi*中国古代地理学史 [A history of ancient Chinese geography]. Science Press科学出版社, Beijing, pp 315–317
87. Wang Y王庸 (1934) Gui'e de yudi zhizhanagtu he limo de tianxia yuditu" 桂萼的《舆地指掌图》和李默的《天下舆地图》 [Gui E's *"Finger-and-Palm" Terrestrial Atlas* and Li Mo's *Terrestrial Atlas of the Empire*]. *Yugong*禹贡 [Tribute of Yu] 1(11)
88. *Mingshi: Luohongxian zhuan* 明史·罗洪先传 [History of the Ming dynasty: Biography of Luo Hongxian]
89. Geoscience History Group, Institute of History of the Natural Sciences, Chinese Academy of Sciences中国科学院自然科学史研究所地学史组 (1984) *Zhongguo gudai dilixue shi*中国古代地理学史 [A history of ancient Chinese geography]. Beijing: Science Press科学出版社, p 318, Note 1
90. Gao J高俊 (1962) Mingqing liangdai quanguo he shengqu dituji bianzhi gaikuang 明清两代全国和省区地图集编制概况 [An overview of the production of the national provincial atlas in the Ming and Qing dynasties (14–19th centuries)]. *Cehuixue bao*测绘学报 [Acta Geodaetica et Cartographica Sinica] 5(4)
91. Liu C刘彩玉 (1962) Luohongxian yu guangyutu 罗洪先与《广舆图》 [Luo Hongxian and the *Enlarged Terrestrial Atlas*]. *Guangming ribao*光明日报 [Guangming daily] 03–13
92. Chen Z陈正祥 (1979) *Zhongguo dituxue shi*中国地图学史 [A history of China's cartography]. The Commercial Press (Hong Kong) 商务印书馆香港分馆, Hong Kong, p 34
93. Chen Z陈祖绶, *Huangming zhifang ditu daxu*皇明职方地图大序 [Administrative maps of the Ming dynasty: Preface]
94. Lu L卢良志 (1984) *Zhongguo dituxue shi*中国地图学史 [A history of China's cartography]. Surveying and Mapping Press测绘出版社, Beijing, pp 105–107
95. Chen G陈光贻 (1987) *Xijian difangzhi tiyao: shangce*稀见地方志提要: 上册 [Abstracts of rare local chronicles I]. Shandong Qilu Press山东齐鲁书社, Jinan
96. *Mingshi: Bingzhi san* 明史·兵志三 [History of the Ming dynasty: Military records III]
97. Views vary in the number of volumes of Xu Lun's 许论*Jiubian tulun* 九边图论 [Illustrated treatise of nine garrisons]. The *Qianqing tang shumu* 千顷堂书目 [Catalogue of the Qianqing Library] and the *Mingshi: yiwen zhi* 明史·艺文志 [History of the Ming dynasty: Literature records] claim that it consists of three volumes while the *Tianyi ge shumu* 天一阁书目 [Catalogue of the Tianyi Pavillion] claims that it consists of one volume
98. Lu L卢良志 (1984) *Zhongguo dituxue shi*中国地图学史 [A history of China's cartography]. Surveying and Mapping Press测绘出版社, Beijing, p 114
99. *Mingshi: Bingzhi san* 明史·兵志三 [History of the Ming dynasty: Military Treatise III]
100. Published in the 2nd Slipcase of the *Ouxiang lingshi* 藕香零拾 [Lotus fragrances gathered together]
101. *Kaogu*考古 [Archaeology] (1965) 8:429
102. Ren J任金城 (1987) Liushi zai guowai de yixie zhongguo mingdai ditu" 流失在国外的一些中国明代地图 [Maps of China from the Ming dynasty that are scattered abroad]. *Zhongguo*

*keji shiliao*中国科技史料 [Historical data of Chinese science and technology] (1). The maps introduced below are referenced in this article
103. Quoted from the article by Hong Weilian洪煨莲 (1936) in the *Yugong* (banyue kan) 禹贡 (半月刊) [Tribute of Yu (biweekly)] 5(3)
104. Cao W曹婉如, Bo S薄树人, Zheng X郑锡煌 (1983) Zhongguo xiancun limadou shijie ditu de yanjiu 中国现存利玛窦世界地图的研究 [A study of the existing world maps in China by Matteo Ricci]. *Wenwu*文物 [Relics] (12)
105. Quoted from the *Zhifang waiji* 职方外纪 [Records of foreign lands]
106. Giulio Aleni艾儒略*Zhifang waiji: Xu*职方外纪·序 [Records of foreign lands: Preface]
107. *Siku quanshu zongmu: juan qishiyi*四库全书总目·卷七十一 [Catalogue of the complete library of the four treasuries, vol 71]
108. Ji H et al 嵇璜等*Xu wenxian tongkao: Kengye*续文献通考·坑冶 [Supplement to the *Comprehensive study of historical documents*: Smelting of mineral deposits]
109. Zhao K赵匡华, Zhou J周嘉华 (1998) Zhongguo kexue jishu shi: Huaxue juan 中国科学技术史·化学卷 [Science and civilisation in China: Chemistry]. Science Press科学出版社, Beijing, p 185
110. Zhang Z张子高 (1964) *Zhongguo huaxue shigao (gudai zhi bu)* 中国化学史稿 (古代之部) [Historical records of Chinese chemistry: Ancient times]. Science Press科学出版社, Beijing, pp 108–112
111. Sun J孙剑 "Jicheng" 计成 [Ji Cheng]. In: Du Shiran杜石然 (1993) *Zhongguo gudai kexuejia zhuanji: xiaji* 中国古代科学家传记: 下集 [Biographies of Chinese ancient scientists II]. Science Press科学出版社, Beijing, pp 935–938
112. *Wenwu*文物 [Relics] (1983) 7:96
113. *Wenwu*文物 [Relics] (1987) 3:50
114. Liu Y刘友樑 (1962) *Kuangwuyao yu danyao*矿物药与丹药 [Mineral medicine and elixir of immortality]. Shanghai Scientific and Technical Publishers上海科学技术出版社, Shanghai
115. Wang J王嘉荫 (1957) *Bencao gangmu de kuangwu shiliao*本草纲目的矿物史料 [Historical documents of minerals in the *Compendium of Materia Medica*]. Science Press科学出版社, Beijing, p 33
116. Wang Q汪庆正 (1982) *Qinghualiao kao*青花料考 [Research on blue and white porcelain material]. *Wenwu*文物 [Relics] (8):59
117. *Wenwu*文物 [Relics] (1989), 7:43
118. *Wenwu*文物 [Relics] (1975), 3:51
119. *Wenwu*文物 [Relics] (1985), 3:36
120. Wu H吴惠群 (1994) *Shiyong baoyushixue*实用宝玉石学 [Practical study of China's gems and jades]. Higher Education Press高等教育出版社, p 196. Also published in *Wenwu*文物 [Relics] 1980, 1:86
121. *Wenwu cankao ziliao*文物参考资料 [References of cultural relics] (1955)(7):156
122. *Kaogu*考古 [Archaeology] (1962) 4:204
123. Li Z李仲均 (1995) Woguo bencaoxue Zhong jizai de yaoyong kuangwu shilue 我国本草学中记载的药用矿物史略 [A brief history of medicinal minerals in Chinese herbology]. In: *Dizhixue shi luncong (san)*地质学史论丛 (三) [Collected essays on the history of geology, vol 3]. China University of Geosciences Press中国地质大学出版社, Beijing
124. *Gegu yaolun: Yuanxu*格古要论·原序 [Study on cultural relic identification: Original preface]
125. Sang X桑行之, Wang F王福康 (1993) *Shuoshi*说石 [On stones]. Shanghai Scientific and Technological Education Publishing House上海科技教育出版社, Shanghai, p 638
126. *Mingshi: Caoxuequan zhuan* 明史·曹学佺传 [History of the Ming dynasty: Biography of Cao Xuequan]
127. Xia X夏湘蓉, Li Z李仲均, Wang G王根元 (1980) *Zhongguo gudai kuangye kaifa shi*中国古代矿业开发史 [History of mining development in ancient China]. Geological Publishing House地质出版社, Beijing, p 289
128. Xia X夏湘蓉, Li Z李仲均, Wang G王根元 (1980) *Zhongguo gudai kuangye kaifa shi*中国古代矿业开发史 [History of mining development in ancient China]. Geological Publishing House地质出版社, Beijing, p 252

129. Xiang D向达 (1982) *Zheng He's Nautical Chart* [郑和航海图]. Zhonghua Book Company 中华书局, Beijing, pp 26–27
130. Jianqiu Z朱鉴秋, Jiarong C陈佳荣, Jiang Q钱江 (2017) Zhongwai jiaotong gudituji 中外交通古地图集 [The ancient atlas of communications between China and overseas territories]. Zhongxi Book Company上海中西书局, Shanghai, p 158

Chapter 9
Geoscience in the Qing Dynasty

9.1 The Impact of the Social Environment on Geoscience

First of all, with Manchu nobles as its core, the Qing regime of rulers from multiple ethnic groups had been diligently cultivating relations with ethnic minorities in the border areas which consolidated the development of frontier geography. The outstanding geographical works include Yang Bin's *Liubian jilue* ("Brief Records of Willow Fence Borders"), Wu Yanchen's *Ningguta jilue* ("Brief Records of Ningguta"), Xi Qing's *Heilongjiang waiji* ("Local Record of Heilongjiang"), Chunyuan's *Xiyu wenjian lu* ("Records of Things Seen and Heard in the Western Regions"), Fu Heng's *Xiyu tongwen zhi* ("Dictionary of Languages in the Western Regions"), Qi Yunshi's *Xiyu shidi* ("Interpretation of the Geography in the Western Regions") and *Xichui zongtong shilue* ("General Records of the Western Regions"), Xu Song's *Xinjiang shilue* ("Brief Records of Xinjiang"), and Song Jun's *Xizhao tulue* ("Illustrated Records of Tibet").

Second, the development of mining and the handicraft industry in the Qing dynasty boosted mineralogy and petrology. Representative works include Sun Tingquan's *Yanshan zaji* ("Records of Mount Yan"), Wu Qi's *Lingnan fengwu ji* ("Records of the Customs of Guangdong"), Min Linsi's *Huangshan songshi pu* ("Catalogues of Huangshan Turquoise"), Yu Minzhong and Liang Guozhi's *Xiqing yanpu* ("Records of Xiqing Inkstone"), Song Luo's *Guaishi zan* ("Tribute to Unusual Stones"), Zhu Jiuding's *Shipu* ("Records of Stone"), Gao Zhao's *Duanxi yanshi kao* ("Records of Duanxi Inkstone") and *Guanshi lu* ("Records of Viewing Stones"), Cheng Xing's *Xuanshi ji* ("Records of Selecting Stones"), Wu Lanxiu's *Duanxi yanshi* ("History of Duanxi Inkstone"), Mao Qiling's *Hou guanshi lu* ("Later Records of Viewing Stones"), Gu Yingtai's *Bowu yaolan* ("Key Reviews of Myriad Things"), Shen Xin's *Guaishi lu* ("Records of Strange Stones"), and Qu Dajun's *Guangdong xinyu* ("New Perspectives of Guangdong").

Third, the advancement of water conservancy fostered works on water conservancy and water systems, with Qi Zhaonan's *Shuidao tigang* ("Guide to the Network

W. Yang and Q. Yang, *A History of Ancient Chinese Geoscience*, Understanding China,
https://doi.org/10.1007/978-981-97-4457-2_9

of Waterways"), Xu Song's *Xiyu shuidao ji* ("Records of the Waterways in the Western Regions"), Jin Fu's *Zhihe fanglue* ("General Strategies for Treating Rivers"), Zhai Junlian's *Haitang lu* ("Records of Seas and Rivers"), and Fu Zehong and Zheng Yuanqing's *Xingshui jinjian* ("Golden Mirror of Channelling Water") as representative works.

Fourth, the harsh literary inquisition and other cultural policies of the Qing dynasty severely restricted the thinking of intellectuals. In order to survive, they had no alternative but to enter the political field through taking the imperial examinations and throwing themselves into the words of the ancient sages and textual criticism in order to make a living. The fact that scholars steered away from real-life problems and academic debates nurtured a generation of outstanding talents in textual research to a certain extent. Representative works of ancient geographical document research include Hu Wei's *Yugong zhuizhi* ("An Analysis of the *Tribute of Yu*"), Quan Zuwang's *Quanjiao shuijing zhu* ("Quan's Correction of the Commentary on the Water Classic"), Zhao Yiqing's *Shuijing zhushi* ("Interpretation of the Commentary on the Water Classic"), Dai Zhen's *Dianben shuijing zhu* ("Imperial Edition of the Commentary on the Water Classic"), Bi Yuan's *Shanhaijing xinjiaozheng* ("New Correction of the Classic of Mountains and Seas"), and Hao Yixing's *Shanhaijing jianshu* ("Commentaries on the Classic of Mountains and Seas").

Fifth, in order to provide the rulers with relevant information, the Qing dynasty attached great importance to the compilation of *fangzhi*, meaning the development of chronicles of ancient China was on course to to reach new heights. The annals and records can roughly be divided into three types: *zongzhi* (general chronicles), *tongzhi* (provincial chronicles) and *difangzhi* (chronicles about the prefecture, district, county, landscape, lake embankment, water conservancy and customs, etc.). In terms of quantity, there are more than 8,100 examples of *difangzhi* in China, including about 5,600 from the Qing dynasty, accounting for 70% of the total number [1]. Through the efforts of Zhang Xuecheng and others, a relatively complete and systematic theory of local chronicles was established.

Sixth, the introduction and assimilation of western cartography took place mainly during the reigns of Kangxi, Yongzheng and Qianlong, resulting in the completion of *Huangyu quanlan tu* and *Qianlong neifu tu*.

Seventh, a group of geographers and their works were influenced by the "statecraft" ideology (also known as "humanistic pragmatism"), such as Gu Yanwu, Gu Zuyu, Liu Xianting, Sun Lan, and Liang Fen.

9.2 Geographical Thought During the Qing Dynasty

Geographical thoughts during this period can be summarised according to five aspects.

First, the ideology of "statecraft" emphasised that geographical research should serve society and be used by society, mentioned Guan Yanwu. He believed that a gentleman should engage in scholarship to understand the truth and save the world.

Insignificant skills of poetry and prose produced no benefit [2]. Liu Xianting maintained that scholars must endeavour to analyse the successes and failures in the past and present to broaden their minds, which could ensure that society would prosper and that people would live in peace [3].

Second, geography should explore the laws of nature. This philosophy was represented by Liu Xianting and Sun Lan.

Liu Xianting first criticised the shortcomings of the previous geographical references. He said that the geographical chronicles merely recorded human affairs, such as territory, construction history, mountains and rivers, historic sites, cities, situations, customs, officials and people, and rarely touched on the relationship and rules of change between various factors in the geographical environment. As regards a specific method for improvement, he suggested that "before outlining the geographical territories, the longitude and latitude of Polaris should be derived from the height of Polaris and the angle of the horizon, and then the natural laws such as the morning and evening of the solar terms are to be calculated according to the differences in longitude and latitude. The climate has been changing, so we should record more phenological phenomena as the basis for future climate change research [4]."

Sun Lan, who styled himself as Liuting, made the point regarding the differences between *zhi*, *ji*, and *shuo* very explicitly in his *Liuting yudi yu shuo* ("Discussion on Geography by Liuting"): "a *zhi* is the record of authentic works, and a *ji* documents what happened. Yet a *shuo* is different. It explains why something is and how it should be. A study that depicts what it is like before and after Creation is called a *yudishuo* (study of geography). It focuses on what mountains and plains are, how they were formed, why human beings live, the origins of states, which mountains are divided into different cordillera, how water bodies relieve floods and keep flowing, how history bears witness to its evolution and development, and how heroes rule their countries. The land of China lies at the south-east corner of the world, and its people only account for a small portion of those in the whole world. Yet there are differences and similarities among different parts of the country. If we do not observe them in a comprehensive manner, how can we understand the universe? Moreover, the orbital periods of the moon and the sun are different from each other, and so is the terrain. Most of China's territory lies north of 23° north latitude, and the climate, the length of day and night, people's characteristics and living habits as well as people's understanding of the world remain unchanged. Is it not true that it is only when we know almost everything that we appreciate the immensity of the earth? If we never see or hear the universe, are we aware that everything is constantly changing?"

In fact, the idea that geography should explore the natural laws of the Earth has a long history. Confucius clearly explained the method for exploring heaven and earth in the *Zhouyi: Xici shangzhuan* ("Book of Changes: Treatise on the Appended Remarks I") and *Xici xiazhuan* ("Book of Changes: Treatise on the Appended Remarks II") saying that the *Yiching* was written in accordance with the principles of heaven and earth. The sage, in accordance with the *Yiching*, looking up, contemplates the brilliant phenomena of the heavens and, looking down, examines the definite arrangements of the earth, thus he knows the causes of darkness (or, what is obscure) and light (or, what is bright). He also added in the *Xici xiazhuan* that "in ancient times, when Bao-xi had come to rule over everything under heaven, looking up, he contemplated the brilliant forms exhibited in the sky, and looking down he surveyed the patterns shown on the earth. He contemplated the ornamental appearances of birds and beasts and the diverse suitability of the soil. Near at hand, in his own person, he found things for consideration, and likewise at a distance, in things in general. Based on this he devised the eight trigrams, to show fully the attributes of the spirit-like and intelligent (operations working secretly), and to classify the qualities of the myriads of things." Although there is not enough evidence to prove that the method was used by the Fuxi, the first of the Three Primordial Sovereigns, it undoubtedly stretches back at least to the Spring and Autumn period of Confucius.

Sima Qian of the Western Han dynasty inherited this tradition and advocated "exploring the relative relationship between natural phenomena and human society and being familiar with the development and evolution of society from ancient times to the present." In the *Records of the Grand Historian: Taishigong Autobiography*, Sima Qian believed that the idea of "respecting natural disciplines" had been initiated by the Yellow Emperor. In the *Records of the Grand Historian: Treatises on Astronomy*, Sima Qian elaborated on the content which related to "respecting natural disciplines", namely that since ancient times, rulers had tried to find the laws of the sun, the moon and other stars in the sky. During the regime of the five ancient rulers of China and the Xia, Shang and Zhou dynasties, people gradually learned about the differences between indigenous races and outsiders and divided the country into 12 states. The rulers looked up to observe brilliant forms exhibited on the sky and looked down for the patterns of the earth. They found that the sun and moon were in the sky while yin and yang were on the earth, that there were five stars [Venus, Jupiter, Mercury, Mars, and Saturn] in the sky while five elements on the earth, that there were constellations in the sky and states on the earth.

The third aspect is the determinism of the geographical environment. In Volume One "Topography" of the *Tianxia junguo libing shu* ("On Benefits and Faults of the Empire's Local Administration"), Gu Yanwu explained the reasons for the economic and cultural underdevelopment of Guizhou and western Guangdong with the theory of the dragon-shaped mountain range in geomancy, illustrating his environmental determinism. "The mountains in western Guangdong and Guizhou range from west to east. There are good bodies of water but no open ones in western Guangdong, while in Guizhou there are neither mountains nor closed waters. Thus, the Dragons cannot stay here and cities in these areas could only lie between rivers and mountains. As such, western Guangdong and Guizhou, although part of China, do not prosper with this great nation." Moreover, Liu Xianting and Wei Yuan also supported a similar idea.

Fourth, the thinking on the correlation between human beings and the earth emphasised the interaction between people and the geographical environment, rather than a relationship in which one determined the other [5], which was relatively scientific. Gu Zuyu's exposition in the early Qing dynasty reflected this thought: "Besides, what is so enduring about the advantages of geography? The Hangu Pass and Jiange are among the most important strategic fortresses in China. At the beginning, the state of Qin used the inaccessibility of the Hangu Pass to repel the attacks by the other six states in the east, but until the end of the Qin dynasty, it was not enough to defend against the rebels nationwide. Zhuge Wuhou sent troops to Jiange and surprised Qinlong, seeking to conquer the Sanfu area; while Liu Chan, although his state was based around Jiange, could barely even keep Chengdu as part of his territory. So, if there is no appropriate person to guard the iron walls and the rolling moat, their function can hardly be compared to low hills and shallow water on the ground. With the right person in charge, even dead wood and rotten plants can be used to cause difficulties for the enemy. As such, curved slopes and narrow and winding mountain roads are not necessarily located in the steep Qionglai Road and cliffs of the Taihang Mountains; deep valleys where the sun does not shine and ravines of thousands of feet deep are not necessarily located in the dark Min River gorge and deep Dongting Lake. Low walls that reach the shoulders sometimes play a bigger role than walls of a hundred feet high. A small ditch that can only soak the wheels of a chariot can be more useful than fast-flowing rivers. If a person only knows how to take advantage of high mountains, deep valleys and famous cities, without knowing that the advantage lies in being familiar with the terrain and making good use of it, what is the point of him talking about geographical advantages?" [6] He realised from historical events that the geographical settings of mountains and rivers had a significant impact on attack and defence in war, but it was human beings and their subjective initiative that played the decisive role.

Gu Zuyu also realised that in addition to factors related to humans and the terrain, opportunity was also important. He said, "Shaanxi is located in the upper reaches, which determines the rise and fall of the whole country. Therefore, although its strength is weak, troops dispatched from Shaanxi should not be underestimated. Even if they cannot dominate the country, they will certainly cause unrest."

In the *Liuting yudi yu shuo*, Sun Lan also talked about human-earth correlation. He believed that the impact of the relationship between the geographical environment and human society changed with the times and situations. He said that "the Yellow Emperor pacified the vassals and settled the world; the First Emperor of Qin (or Qin Shi Huang) abolished feudalism and set up prefectures and counties," which were the inevitable results of historical development, and the abolition of the square-fields system was also imperative [7]. He further argued that cultural differences were affected by a variety of factors, including the geographical environment, the era, and the people's temperament and habits [8].

Fifth, Sun Lan proposed the theory of "full water flowing down" on the formation of fluvial geomorphology. This originally came out of the *Zhouyi: Qiangua* ("Book of Changes: Hexagram of Modesty"), referring to the fact that the law of the Earth was to become full and flow downwards, which served as a summary of erosion by and sedimentation of flowing water. It is particularly noteworthy that the *Zhouyi* emphasised "the law of the Earth", which shows that the ancients regarded the erosion and accumulation of the earth's surface due to external forces as a law of nature [9]. In this way, Sun Lan's theory inherited and further developed the tradition of the *Zhouyi: Qiangua*. He regarded erosion and accumulation as two interrelated aspects of the process of geomorphic development which together shape and change the surface. Moreover, the causes of fluvial geomorphic evolution can be divided into three types: the gradual change factor, the abrupt change factor and the human factor, which was a creative approach in the history of Chinese geography. He wrote in the *Liuting yudi yu shuo*: "The reasons behind the phenomenon of 'full water flowing down' vary significantly. It might be the result of a lapse of time, human activities, or factors of change. In the case of a lapse of time, a prolonged downpour of rain washes away the mountain, causing flash floods, and over time, the location of the waters will change. As for human activities, the topography changes over time as mountains are excavated to build roads, flood waters are drained into rivers, and river water is diverted to irrigate farmland. Other factors that might cause full water flowing down include landslides, earthquakes, and sudden landscape changes, etc."

9.3 Studies of Hydraulic Engineering and Hydrography

9.3.1 River Management Experts and Their Texts

In the early stages of the Qing dynasty, the famous experts in Yellow River control were Jin Fu, Chen Huang, Kang Jitian, Zhang Boxing, Wang Quanyi, Guo Dachang and others, most of whom left a legacy of water conservancy works.

(1) Jin Fu

Jin Fu (1632–1692) served as Governor of the River for 15 years from the 16th year of the Kangxi reign (1677) until his death. With the help of Chen Huang, Jin Fu had achieved great successes in river management in his post. After his death, Kangxi judged him to be "a great contributor to the country" for saving millions of people from flood disasters and promoting social stability. Because of his major reconstruction, the canal was unblocked for a long time. Jin Fu wrote the *Zhihe fanglue* ("General Strategies for Governing the River") and *Jinwen xianggong zoushu* ("Memorials of Official Jin Fu"), reflecting on his river management thought and experience.

The *Zhihe fanglue*, former title *Zhihe shu* ("Book of Treating Rivers"), is composed of 10 volumes. Volume One is a summary of the author's memoirs and treatises on river management. In the 32nd year of Qianlong's reign (1767), it was reorganised by Cui Yingjie and the title changed to its current name. The record ends in the 21st year of the Kangxi reign (1682), with a total of about 1.81 million words. At the beginning of the volume, there are seven maps, including a map of the Yellow River, a map of previous flood-preventing facilities built on the Yellow River, a map of new flood-preventing facilities built on the Yellow River, a map of the confluence of the Huai River, a map of canals, a map of lakes in Huainan, and a map of water conveyance on the five rivers of Wen, Si, Yi, Guang and Ji, with illustrations attached. The text is divided into nine parts: river management, rivers and ditches, springs, lakes and pools, water transport, floods, river courses, memoirs, and famous treatises.

(2) Chen Huang

Chen Huang (1637–1688) was born in Qiantang (now Hangzhou), Zhejiang province, with the courtesy name of Tianyi and the literary name of Shengzhai. He helped Jin Fu manage the river for 10 years, and the major planning measures were all Chen Huang's initiatives. Jin Fu also acknowledged frankly that "everything I planned to build was proposed by Chen Huang." However, such a talented water conservancy expert was wrongly executed and died unfairly at the age of 52. Most of his works on river management have been lost, and the original manuscript of the *Tianyi yishu* ("Posthumous works of *Tianyi*") collected by later generations has not been handed down. Only Yang Xiangji's copy of *Tianyi yishu* is now in the National Library. According to Chen Huang's *Hefang zhaiyao* ("Abstract of River Management"), Zhang Aisheng, Chen Huang's fellow countryman, compiled 12 of the articles in the *Hefang shuyan* ("On River Management"). Both *Hefang zhaiyao* and *Hefang shuyan* were appended to Jin Fu's *Zhihe fanglue* allowing them to be preserved for later ages. Chen Huang also wrote more than 1,600 articles in the 28 volumes of the *Compilation of River Defence in Past Dynasties* as well as the *Zhihe ce* ("Strategies on River Management"), which has been lost. The *Tianyi yishu* and *Hefang shuyan*, important documents of water conservancy, have preserved some of Chen Huang's river management theories and engineering know-how and are extremely rich in content and are still valuable references today.

(3) Kang Jitian

Kang Jitian (1728–1813), who was born in Xingxian county, Shanxi province, was a scholar in the Qianlong period, with the courtesy name of Maoyuan. He once served as the prime minister of Jiangsu and was ordered to control the Dong River. Later, he was promoted to the post of governor of the Jiangnan river course and advocated narrowing the river channel, so as to manage dredging and flood control. Throughout life his highest post was the rank of Sanpinqing, that is, an intermediate official of the Qing dynasty. Kang collected numerous books and finished the 30-volume *Hequ jiwen* ("Chronicles of Rivers and Canals") of more than 560,000 words, in the 2nd year of the Jiaqing reign (1797), which focused on the evolution and control of the Huangyun river course and also dealt with local water conservancy, structured in chronological order. With age as the timeline and events as the object, the work described the governance of the Yangzi River, the Bohai Sea, the Yellow River, the Hutuo River, the Zhang River and other rivers and ditches in the past dynasties from Yu the Great to the Jiaqing period of the Qing dynasty. Kang also assessed the advantages and disadvantages, the contexts and origins, and the gains and losses of various water systems. He commented in particular on the famous water control masters from the time of the Ming and Qing dynasties.

(4) Zhang Boxing

Zhang Boxing (1652–1725), a scholar of the Kangxi period, Minister of Rites and Governor of River Management, was born in Yifeng (now Lankao in Henan province). With the courtesy name of Xiaoxian, he also styled himself as Jing'an. He summarised his experience during his tenure as Governor of River Management with reference to ancient books, and successfully completed the 8 volumes of the *Juji yide* ("Experience in Managing the Shandong Canal"), of about 150,000 words, in the 43rd year of the Kangxi reign (1704). The book consisted of 234 articles including a general overview of the canal, the canal source, discussion of river management, the Huang-Huai water conservancy project, and a compilation of water transport, etc., with the author's practical experience of river management as the main content, while, at the same time, several water conservancy engineering theories were recommended. Zhang paid great attention to the successful river control methods of his predecessors and quoted the important experience and comments of water control experts after the Song and Yuan dynasties which were of great historical value. He was also the author of the *Zhengyitang wenji* (Collected Essays from Zhengyi Hall).

(5) Wang Quanyi

Wang Quanyi, a senior river worker during Qianlong's reign and the father-in-law of Guo Dachang, was so skilful in engineering that he helped Guo Dachang achieve great advances in the field of river management. Wang Quanyi recorded his decades of experience as a river worker. Later, Wang's records were compiled by the river governor, Xu Duan, into the *Anlan jiyao* ("Summary of Quiet Waves") and *Huilan jiyao* ("Summary of Surging Waves"), which were signed with Xu Duan's name. Both books were included in the sixth and seventh volumes of the edition of *Zhihe huilan* ("Collection of Treating Rivers") from the 11th year of the Guangxu (1885), consisting of several sections, and they are now in the library of the Ministry of Water Resources.

(6) Guo Dachang

Guo Dachang, a senior river worker from the end of the Qianlong to the Jiaqing period, was Wang Quanyi's son-in-law. According to Bao Shichen's record in the *Zhongqu yishao: Guojun zhuan* ("Thoughts on Efficient Water Transport: Biography of Guo Jun"), Guo Dachang was born into an ordinary peasant family in Shanyang (now Huai'an), in northern Jiangsu province. At the age of 16, he worked as a government clerk assistant. He was so hardworking and fond of learning that he became proficient in accounting and surpassed his teacher after only three years. He was born in a region of rivers and lakes which made him familiar with water life, and skilled in constructing dams, he was known as the "senior dam worker" at that time. In his life, he had always been "cautious and plain-spoken", upright, unswerving and conscientious, insisting on saving money and materials during projects. However, he had been rejected and suppressed by the river officials and had not been appointed to any important position. Although Guo Dachang was just an ordinary river worker, he cared about the overall situation and actively offered suggestions in the middle of the Qing dynasty, when river administration was corrupt and the river courses were ruined. Finally, in the well-deserved post of river control master, Guo Dachang made the Yellow River and the Huai River empty into the sea together, easing the pressure in the lower reaches of the Yellow River.

(7) Fu Zehong

Fu Zehong presided over the compilation of the 175 volumes of the *Xingshui jinjian* ("Golden Mirror of Channeling Water"), which was completed in the 3rd year of the Yongzheng (1725) and published by the Commercial Press in 1937, including more than 370 kinds of water conservancy documents and regulations from the pre-Qin period to the 60th year of the Kangxi (1721), containing about 1.6 million words, and focusing on the period from the late Ming dynasty to the Kangxi reign of the Qing dynasty. Among the 175 volumes of the text, there were 60 volumes on the Yellow River, 10 volumes on the Huai River, 10 volumes on the Yangzi River and the Han River, 5 volumes on the Ji River, 70 volumes on the canal, 8 volumes of general statements on the Yellow River and the canal, and 12 volumes on the officials, the conscripted labourers, and the transport regulations within the chronological narrative structure, making it an important water conservancy document.

(8) Li Shixu

Li Shixu and others compiled the 156 volumes of the *Xu xingshui jinjian* ("Sequel to the Golden Mirror of Channelling Water"), which was completed in the 11th year of the Daoguang reign (1831) and published by the Commercial Press in 1937. The content included dated from as early as the 25th year of the Jiaqing reign (1820). Unlike the *Xingshui jinjian*, it contained a large amount of original engineering archives, which had been collected into the 121 volumes of the "Zhangdu" (official documents). This book also contains 13 volumes on the Yongding River.

(9) Zhai Junlian

Zhai Junlian (date of birth and death unknown), also known as Chunzhi, was born in Renhe (today's Hangzhou), Zhejiang province, and once served as a Juren [a successful candidate in the imperial examination] and as a government secretary during the Qianlong period. In addition to the *Haitang lu* ("Records of Seas and the Rivers"), he also wrote the *Zhouyi zhangju zhengyi* ("Comments on the Differences between Ancient and Modern Versions of the *Book of Changes*"). Since he grew up in Renhe, Zhai Junlian was aware of the importance of building seawalls. After extensive reading and detailed textual research, he completed the 26 volumes of the *Haitang lu* during the reign of Qianlong. The book was not published at that time, and minor additions were made when it was included in the *Siku quanshu* in the 46th year of the Qianlong reign (1781). In 1934, when the Shanghai Commercial Press printed the *Siku quanshu*, it was officially published.

The *Haitang lu* detailed the process and construction of seawalls built in previous dynasties, as well as the regional distribution and evolution of seawalls before the 29th year of the Qianlong reign (1764). The first part of the volume recorded the legal system promulgated by the emperor, and then it was divided into nine chapters, covering illustrations, territory, architecture, scenic spots, historic sites, shrines, memorials, arts and miscellaneous issues.

The Chapter "Illustrations" included 15 items, such as drawings of seawalls, riverwalls, river diversions, seawall repairs, walls made of stones and firewood, 18-layer walls, water carrying tools, buttresses, a half-moon dam, a fish-scale stone pond from the Qing dynasty, all with text to explain the length, materials, specifications, construction essentials and other relevant information.

"Territory" described the evolution of seawalls and experience(s) of surveying seawalls during the Qianlong period.

"Architecture" recorded the history of seawall construction in past dynasties.

"Scenic Spots" and "Historic Sites" recorded the mountains, rivers, lakes, ponds, riversides, stones, bridges, pavilions and other features, as well as poems and prose texts.

"Shrines" documented the sea god temples, sacrificial texts and poems.

"Memorials" included communication from ministers of the Ministry of Works to the throne.

"Arts" included hundreds of examples of descriptive prose writing, statements, scripts, articles, memorials, arguments, records, and poems from the Northern and Southern dynasties to the Qianlong period.

"Miscellaneous" gave an account of the causes of tides and tidal trends.

This book, like the *Zhihe fanglue*, was a work of water conservancy engineering, but it contained a lot of hydrogeological knowledge. It covered the past and its contemporary era, and was rich in data, which are valuable historical geoscience data.

9.3.2 Studies of Hydrography

A number of river system works appeared in the Qing dynasty; these were inspired by the focus on water conservancy projects at that time. The following is a chronological introduction to these river system works and their authors.

(1) Huang Zongxi's *Jinshui jing* ("The Present Water Classic")

Huang Zongxi (1610–1695), a thinker and historian during the Ming and Qing dynasties, with the courtesy name of Taichong, who styled himself as Nanlei and was also known as Mr. Lizhou, was born in Yuyao, Zhejiang province. His father was a celebrated scholar in Donglin, learning from Liu Zongzhou. Huang Zongxi called the restoration society together to struggle with the eunuchs and was persecuted several times. When the army of the Qing dynasty went south, he recruited righteous soldiers and set up the "Shizhong Camp" (or Camp of Perpetual Loyalty) to fight against the Qing government, and then he was engaged by the King of the Lu to be his state censor. After the decline of the Ming dynasty, he lived in seclusion and concentrated on writing. He was very knowledgeable of the past and the present and proficient in astronomy, geography, arithmetic, music, the classics, and history, as well as learned in Buddhist and Daoist literature. Huang Zongxi's works include the *Ming ruxue an* ("Scholarly Annals of Ming Confucians"), *Erchengxue an* ("Scholarly Annals of Cheng Yi and Cheng Hao"), *Mingshi an* ("Cases of Ming History"), *Lidai jiazi kao* ("Records of Jiazi in the Past Dynasties"), *Mingyi daifang lu* ("An Obscured Paragon of Virtue Awaiting a Royal Visit"), *Nanlei wending* ("Collected Works of Nanlei"),

Wenyue ("Documents and Deeds"), *Simingshan zhi* ("A Book of Mount Siming"), *Jin shuijing* and others. In 1986, Zhejiang Ancient Books Publishing House published the *Complete Works of Huang Zongxi*, edited by Shen Shanhong.

The *Jin shuijing* is a monograph describing the origin and flow of watercourses throughout the country, in the early Qing dynasty, which is not divided into volumes and contains about 20,000 words. After the author's preface, the book opens with the outline "Jin shuijing biao", followed by the sections of "Northern Water" and "Southern Water", recording the rivers and lakes according to the key linkages between them in their courses to the seas. The intention of Huang Zongxi's *Jin shuijing* was to correct the mistakes in the *Shuijing zhu* [10]. Huang divided the national water system into two categories: north and south. Water in the north includes the Yellow River and its tributaries, the rivers in the northeast of China, the rivers in Hebei and Shandong provinces, and the Huai River; while water in the south includes the Yangzi River and its tributaries, the rivers in Zhejiang and Fujian provinces, the Pearl River water system, and the rivers in Yunnan. His partitioning method of the water system was more logical than that of the *Shuijing zhu*. However, there are still many omissions, such as the lack of inner rivers in the classification of the water system. In addition, it also made some errors regarding specific issues. For example, the Heilongjiang River was mistakenly believed to enter the Songhua River; the Cao'e River was mistakenly believed to join the Qiantang River; the Lu River (now Nu River) was mistakenly believed to feed into the Daying River, and so on.

(2) Qi Zhaonan's *Shuidao tigang*

Qi Zhaonan (1703–1768), who was born into an official family in Tiantai, Zhejiang province, had the courtesy name of Cifeng, and also styled himself as Qiongtai and Xiyuan in his old age. In the 7th year of the Yongzheng reign (1729), he became a deputy nominee after passing the provincial examination. In the first year of the Qianlong reign (1736), he took the final imperial examination, winning the 8th place of second rank, and was awarded the status of the position of Hanlin Academician. He had successively served as You Zhongyun with responsibility for assisting the prince of internal affairs, as Rijiang Qijuzhu Guan [lecturer of historical classics and scriptures], as Shidu Xueshi with responsibility for reviewing and compiling literature and historical documentation, as Shangshufang Xingzou [assistant lecturer for the princes], and as Ceremonial Master. He was knowledgeable and had an amazingly retentive memory. However, in the 14th year of the Qianlong reign (1749), his memory was greatly reduced due to a severe brain injury, so he returned to his hometown to preside over the academy of classical learning and engage in writing. He participated in the compilation of the *Qing Dynasty National Chorography* and the *Compendium of the Ming Dynasty*. Among his 11 works, the most important one is the 28 volumes of the *Shuidao tigang* ("Guide to the Network of Waterways") of about 300,000 words, acclaimed as the "Water Classic of the Qing Dynasty", which was completed in the 26th year of the Qianlong reign (1761).

This book was based on the imperial map of the Kangxi period, documenting the actual mountains and rivers, as well as the city sites with the names of the day, and partially recording the historical sites. When writing about certain water systems, Qi Zhaonan started from the source, taking the main channel as the key structural link, and its tributaries as the specific items; in terms of the national water system, Qi Zhaonan used the large rivers as the leading subject and the small rivers as secondary to it. The book recorded a wide range of water systems (18° to 56° N, 73° to 145° E), which surpassed any other work on water systems in history. The number of rivers recorded was 5,980, four times as many as those in the *Shuijing zhu*. Moreover, the method of describing water systems and the approach to classification had made obvious progress; these were more scientific and practical than those of his predecessors.

According to the author's classification, the ocean should be the broad heading, so the first volume provided detailed statistics and descriptions of the inlets of large and small rivers from north to south, including the cities, islands and ports, and especially more thorough descriptions of Taiwan and Hainan. The second volume referred to rivers structured with the major coastal rivers first, which was consistent with the order of the first volume, starting with the Yalu River and the Liao River along the Bohai Sea, and then describing the Hai River, the North Canal, the Yellow River and its tributaries, the Huai River and its tributaries, the South Canal, the Yangzi River and its tributaries, the Jiangnan Canal, the Taihu Lake system, the Qiantang River, the Min River, the Pearl River and its tributaries. The following volumes described the water bodies in various provinces, such as Yunnan, Tibet, Northeast China, Inner Mongolia, and the Western Regions. In terms of writing style, the writer completely broke away from the boundaries of the administrative divisions, and was guided by the natural water system, describing each river from its upper reaches to the river mouth within the clear framework of the water system. When recording the confluence points, bends and canyons of rivers, he not only noted the place names, but also the degrees of longitude and latitude, an approach initiated by Qi Zhaonan.

Before Qi Zhaonan, the works on water systems focused on the central part of China and touched upon little or even none of the water systems in the southeast, southwest, northeast and in Xinjiang province. The *Shuidao tigang* met the needs of society and made up for the deficiencies of previous works on water systems. The author's spirit of seeking truth based on evidence is also commendable. For example, when he speculated about the directions of waterways, he would add expressions like "is suspected to be," and he was very frank about the matters that were unclear. Although the author had not personally visited the watercourses, he still managed to surpass the achievements of his predecessors, mainly for two reasons. First, the author was diligent and tenacious. It took him ten years of continual research to publish the book. Second, the author had a superior working environment, which gave him the opportunity to gain access to the *Huangyu quantu* ("Complete Atlas of the Empire") and other precious secret books hidden in the treasury of the imperial palace.

The book was finalised in 1761, but it was not published during the author's lifetime. Eight years after the author's death, in 1776, it was published thanks to the efforts of his son Qi Shiqian [11].

(3) Xu Song's *Xiyu shuidao ji* ("Records of the Waterways in the Western Regions")

Xu Song (1781–1848), whose courtesy name was Xingbo, was born in Shuntianfu, Daxing (now Beijing), and his ancestral hometown was located in Shangyu, Zhejiang province. He studied under Wu Xilin at an early age, and "at the age of nine, he took the preliminary-level imperial examination and was admitted to the school in the county" [12]. In the tenth year of the Jiaqing (1805), he was awarded the title of Jinshi [a nominee having passed the highest-level imperial exam] and worked as a Bianxiu [officer responsible for compiling and editing histories]. In the 13th year of the Jiaqing reign, he was on duty in the south study of the Forbidden City. In the 17th year of the Jiaqing reign, he was impeached by Censor Zhao Shen, and was banished to Ili until the 24th year of the Jiaqing reign. The following year, he completed the book *Zongtong shilue* ("Records of General Matters") based on his investigations in Xinjiang province, which was later named *Xinjiang shilue* ("Brief Records of Xinjiang") by Emperor Daoguang. Later, he successively served as the Neige Zhongshu [officer responsible for court paperwork], the Zhuyinju Yuanwailang [head of the National Printing Bureau], the Jiangxidao Jiancha Yushi (supervising censor of Jiangxi), and the Yulin Zhifu (mayor of Yulin). During his life, he enjoyed studying the classics, and he was especially proficient in historical events and geography. In addition, his works included the *Tang liangjingcheng fangkao* ("Records of Chang'an and Luoyang in the Tang Dynasty"), *Song huiyao* ("Song Dynasty Manuscript Compendium"), *Tang dengke jikao* ("Research on Examination Graduates of the Tang Period"), *Song sansi tiaoli kao* ("Records of the Regulations of Three Financial Bureaux in the Song Dynasty"), *Houhanshu xiyuzhuan buzhu* ("Supplement Notes to the Western Regions Biography of the Book of Later Han"), *Changchun zhenren xiyou jikao kao* ("The Record of the Perfected Master Changchun's Journey to the West"), *Xin Jiaozhu dilizhiji jie* ("Interpretation of the New Annotated Geography Records"), *Yuanshi xibei dili kao* ("Geographical Records of the Northwest in the Yuan Dynasty"), *Xixia dili kao* ("Geographical Records of the Western Xia Regime"), etc.

During Xu Song's exile in Ili, the General of Ili, Song Jun, heard that he was proficient in geography and invited him to revise the *Ili Zongtong shilue* ("General Records of Ili") compiled by Wang Tingkai and Qi Yunshi. Xu Song did not rigidly adhere to their original views and judgments, but rather carried out field investigations and travelled all over the north and south of the Tianshan mountain range [13]. During the investigations, he recorded the topographic features of mountains and rivers at all times, asked the local people for evidence, investigated various documents related to geography and, over the years [13], compiled the five volumes of the *Xiyu shuidao ji*, which was an unprecedented masterpiece recording the waterways of the Western Regions. Deng Tingzhen said that it had five virtues: the first was that it filled in gaps; the second was its practicality; the third, that it was based on personal experience; the fourth, that it contained many textual annotations; the fifth, that the book clearly distinguished between the different kinds of water bodies. "It is a rare, fine book with these five advantages" [14]. Liang Qichao called it "a meticulous masterpiece." [15].

The *Xiyu shuidao ji* was written in the first year of the Daoguang (1821) in the style of the *Shuijing zhu*. There are three kinds of text in the book: records, annotations and explanatory notes. The records are brief; the annotations are detailed; and the interpretation serves to explain.

Four characteristics are observed in the *Xiyu shuidao ji*.

First, structured around eleven lakes, such as Lop Nur, Barkol Lake, Balkhash Lake, Sayram Lake, Zaysan Lake, and Issyk Lake, the book describes the water system in the west of Jiayuguan in Gansu and Xinjiang, which displays the properties of inland rivers. The principle of dividing inland water systems by lakes was introduced by Xu Song, and this principle was not used for a long time after him.

Second, within the framework of the lakes, the book relates the rivers, and lists the cities, settlements, tributaries, mountains, and the longitude and latitude of some places along the rivers, as well as the history, products, ethnic minorities, landforms, historical documents, dialects, hydrology, water conservancy, garrisons, etc., making it not only a comprehensive work on water systems, but also an outstanding geographical reference.

Third, as an integral part of the book, maps are included in four volumes; these were drawn according to the method of *ji li hua fang* and to scale, and the length of each square represents one hundred or fifty li. However, in the *Xiaofanghuzhai yudi congchao* ("Geography Series from Wang Xiqi's Study"), the maps were deleted, which was a pity. The glaciers and glacial landforms recorded in the book are more advanced than those in the *Xiyu wenjian lu* ("Record of Things Seen and Heard in the Western Regions").

Fourth, using longitude and latitude to mark the sources and intersections of rivers was a very scientific approach by Xu Song.

Due to the limitations of the times, there are inevitably some shortcomings in the *Xiyu shuidao ji*. For example:

The records are not comprehensive vis-à-vis the watercourses in the Western Regions, given that the rivers were categorised by lake water systems. Rivers that did not belong to the systems of the eleven lakes listed, such as the Niya River and rivers to the east of the Tarim Basin, were not included.

Some of the information regarding the rivers listed in the records is not comprehensive or accurate. For example, there was no record of the tributaries to the south of the source of the Kashi River, which is on the upper stream of the Ili River. Moreover, only a few branches, such as the Maoman River in the upper reaches of the Kones River, were recorded. The Chelsen River flowing into the Lop Nur was a big river, but there was no record of it in the book.

Despite the comparatively correct understanding of rivers since the Tang dynasty, the book still supported the theory that the undercurrent of the Yellow River had more than one source. Moreover, the author devoted a lot of space to criticising previous correct understanding of river sources, which was quite backward-looking [16].

In addition to the above-mentioned works on water systems, there were Huang Maocai's *Xijiao shuiji* ("Waterways on Western Border"), He Qiutao's *Beijiao shuidao kao* ("Records of Waterways on Western Border"), *Selenggehe yuanliu kao* ("Source and Flow of Selenga River") and *E'erqisihe yuanliu kao* ("Source and Flow of Irtysh River"), and Miao Husun's *Eluosi shuidaoji* ("Records of Russian Waterways"), which expanded the vision and scope of water system research.

9.4 Reviewing and Annotating the Extant Geographical Literature

The first section of this chapter has already mentioned that, in order to control the thoughts of the intellectual class, the rulers of the Qing dynasty adopted a high-pressure policy and implemented the literary inquisition; on the other hand, they implemented the controlling policy of directing the intellectual class towards the path of the imperial examination. Such policies made academia oppressive and divorced academic activities from reality and production. Many famous scholars, unable to explore the laws of natural and social development, could only undertake some collation and interpretation of the classics, that is, organising ancient geographical documents. Hu Wei, Bi Yuan, Hao Yixing, Quan Zuwang, Zhao Yiqing, Dai Zhen and others made outstanding contributions in this regard.

9.4.1 Hu Wei

Hu Wei (1632–1714), formerly known as Hu Weisheng, was born into a family of generations of successful imperial examination candidates in Deqing county, Zhejiang province. With the courtesy name of Feiming, he also styled himself as Dongqiao in his later years and had been eager to learn since childhood. His father died when he was aged 12 and he followed his mother to avoid the chaos in the valley. At the age of 15, he became a scholar's student, hoping to pass the imperial examination. But he was so unlucky that he always failed the examination. Since he did not have much success until the age of 40, he "devoted himself to the study of geography" [17]. At the age of 58, he participated in the revision of the *Daqing yitong zhi*, and worked alongside Huang Yi, Gu Zuyu, Yan Ruoqu, Zha Shenxing and others day and night. This opportunity allowed him to verify a large number of historical materials, so he copied the content related to the *Yugong* and then compared these documents with Li Daoyuan's *Shuijing zhu* [18]. In the 33rd year of Kangxi's reign (1694), Hu Wei returned to his hometown to recuperate from an illness. "He declined all personal entertainment and buried himself in the classics. Through continuous researching, he grasped the key points, interpreted *Yugong* according to the classics, and classified it in an orderly manner depending on the content of the chapters and sentences. It took three years to complete this geography work, the 20 volumes of the *Yugong zhuizhi*." The book was completed in the 36th year of the Kangxi reign (1697), when Hu Wei was 65 years old. In the 41st year of the Kangxi reign (1702), forty-seven hand-drawn copies of the *Yugong Map* were appended to the book, and the woodblock was carved in Wumen the following year. In the 44th year of the Kangxi reign (1705), it was presented to Emperor Kangxi who awarded a plaque with the words "a senior man devoted himself to learning," which made Hu Wei famous at the age of 73.

The *Yugong zhuizhi,* which focused on interpreting and annotating the *Yugong*, is divided into five parts, namely, “Illustration”, “Jiuzhou”, “Daoshan”, “Daoshui” and “Wufu”. Liang Qichao posited that this book “opened up the specialised knowledge for the study of the evolution of geography [15].” Gu Jiegang believed that among the scholars who had interpreted the *Yugong*, “Hu Wei exerted the most force, and his *Yugong zhuizhi* was a great summary [19].” Zou Yilin maintained that Hu Wei’s *Yugong zhuizhi* made at least three contributions to the history of geographical development:

(1) Hu Wei’s annotation of the *Yugong* rejected the stereotype of the former scholars who only worked hard on textual research and it analysed various geographical phenomena in the *Yugong* through the eyes of geographers, thereby further explicating the implications of the *Yugong*, the earliest geographical work in China.
(2) Hu Wei also explored the content of the *Yugong* more deeply from the perspective of historical geography, not only reflecting the geographical features of the era of the *Yugong*, but also extending his exploration to the rise and decline of past dynasties after the Warring States period, which helped readers gain a deeper understanding of the formation of geographical phenomena in the early Qing dynasty.
(3) The *Yugong* is a work on regional economic geography, which was fully developed by Hu Wei. Through the annotation of the *Yugong*, he expounded his views on the advantages and disadvantages of regional developments and social-economic developments in past dynasties, some of which undoubtedly have important reference value for contemporary politics [20].

9.4.2 Bi Yuan

Bi Yuan (1730–1797), with the courtesy name of Xiangheng and nickname of Qiu Fan, who also styled himself as Lingyan Shanren, was born in Zhenyang (now Taicang, Jiangsu), Jiangnan. He was a Jinshi in Qianlong’s reign, serving as the governor of Huguang. He was familiar with the classics, history, epigraphy and geography, and loved to write books, which included the *Chuanjing biao* (“Transmission of Confucian Classics”), *Jingdian bianzheng* (“Identification of Confucian Classics”), *Xu Zizhi tongjian* (“Sequel to a Comprehensive Mirror to Aid in Government”), *Shanhaijing xin jiaozheng* (“New Correction of the Classic of Mountains and Seas”), *Annotation of the Book of Jin: Geography*, *Guanzhong shengji tuji* (“Records of Scenic Spots in Guanzhong”), *Xi’ansheng zhi* (“Records of Xi’an Province”), *Guanzhong zhongzhou shanzuo jinshi zhuji* (“Records of Metal and Stones in Shandong Province of the Central Plains”), *Lingyanshan ren shi ji* (“An Anthology of the Poems of Lingyan Mountain Hermit”).

The 18 volumes of the *Shanhaijing xin jiaozheng* contain textual research on the titles, characters, and the names of mountains and waterways in the *Shanhai jing*, which has strengthened the geographical value of the *Shanhai jing* [21].

9.4.3 Hao Yixing

Hao Yixing (1755–1823), who was born in Qixia, Shandong province, had the courtesy name of Xunjiu, and also styled himself as Langao. He had the title Jiaqing Jinshi, serving as the doctor of the Ministry of Household. He was deeply engaged in annotation or textual research on words, and wrote the *Erya yishu* ("Comments on *Erya*"), *Shanhaijing ding'e* ("Correction of *Classic of Mountains and Seas*"), *Shanhaijing jianshu* ("Commentaries on Classic of the Mountains and Seas"), etc.

The *Shanhaijing ding'e* was completed in the 9th year of the Jiaqing reign (1804), with more than 14,000 words. Following the order of the original book, all the edited scriptures were in large characters, and the original notes and the corrected parts were in small characters below. By comparing the differences in the notes from past dynasties, more than 300 mistakes were corrected based on the relatively common version. Although there was a basis for the revision, the author still retained the original text, imitating Zheng Xuan's annotation of the scriptures without changing the words.

The 18 volumes of the *Shanhaijing jianshu* form one of the best examples of all syntheses: the result of searching through and citing from all the classics and works of history, philosophy and literature with detailed notes. The book was completed in the 9th year of the Jiaqing reign, and the citation of historical events ended in that year, with more than 200,000 words of notes, structured in the order of the 18 chapters of Guo Pu's notes. The large characters are the main text, and the small characters are double-line notes with Guo's notes at the start, indicating "Yixing's notes" before the commentaries and providing references for the quotations. The record was still based on the geography of mountains and seas, and the notes were widely quoted from documents dating from after the Qin dynasty. The mountains and rivers, animals and plants, tribes, ancient countries and people in the pre-Qin period were all supported by documents conveying several different opinions. The mountains and rivers were considered in terms of their orientation, trends, sources, water levels, climate, vegetation, scenic spots, historic sites, production, mineral springs, caves, and undercurrents; the real names, aliases, shapes, characteristics and functions of all kinds of animals and plants were examined in detail; the structure, hardness, colour, symbiosis and the identification method of minerals were also considered. This book supplements and introduces new information on the basis of the original text, covering a wide range of cultural and geographic knowledge, such as nature, history, nationality, transportation, animals, plants, ancient culture, and ancient dialects. Moreover, this book is rich in content relating to language, art, religion, medicine, literature and other historical materials which was widely used by subsequent generations [22].

9.4.4 *Quan Zuwang*

Quan Zuwang (1705–1755), who was born in Yinxian county, Zhejiang province (now Ningbo city), had the courtesy name Shaoyi and the nickname Xieshan. At the age of 14, he became a student of a government school, and in the 7th year of the Yongzheng's reign (1729), he passed the imperial exam and was sent to the capital. In the first year of Qianlong's reign (1736), he gained the title of Jinshi and was then awarded the status of Hanlin Academician, as a magistrate in the county. His 40 volumes of the *Qijiao shuijing zhu* ("Seven Proofreads of *Commentary on the Water Classic*") constitute an important achievement in research on the *Shuijing zhu.*

Quan Zuwang's main contributions to the study of Li Daoyuan's work are in the following two aspects: the first is that he made a distinction between the classics and the notes. It had been very common practice to confuse the notes with the classics in the Song and Ming editions. Quan Zuwang, Zhao Yiqing and Dai Zhen made significant contributions by differentiating between the notes and the original text of the classics, but Quan Shi led the way. The second aspect was Quan Zuwang's creative approach in applying double-line notes and explaining the original notes. The *Qijiao Shuijing zhu* was not published during the author's lifetime. After his death, it was completed by Wang Zicai and Dong Pei, and published by Xue Fucheng in the 14th year of the Guangxu reign (1888).

9.4.5 *Zhao Yiqing*

Zhao Yiqing (1709–1764), with the courtesy name Chengfu and the nickname Zhengfu, and who also styled himself as Dongqian, was the son of the poet Zhao Gulin and a student of Quan Zuwang. Enlightened by Quan Zuwang, he carried out research on the *Shuijing zhu.* He agreed with Quan Zuwang's idea of applying double-line notes and explaining the original notes, so he worked hard to identify the meaning of the text. By using different font sizes to distinguish between the original notes and the interpretation of them, he made the text clear and coherent. Zhao Yiqing first delved into Zhu Mouwei's *Shuijing zhujian* ("Annotation on the *Commentary on the Water Classic*") and commented on its strengths and weaknesses in his writing of the 12 volumes of the *Shuijing zhujian kanwu* ("Correction of the *Notes on Commentary on the Water Classic*"). Then, based on the *Shuijing zhujian*, and referring to Quan Zuwang's *Wujiao shuijing zhu* ("Five Proofreadings of the *Commentary on the Water Classic*") written in the 3rd year of the Qianlong reign (1738) and many other versions, he completed the 40 volumes of the *Shuijing zhushi* ("Notes on the *Commentary on the Water Classic*") in the 19th year of the Qianlong reign (1754). Not only was the collation of the text accurate and its annotation detailed, but it also copied the long-lost preface by Li Daoyuan from Sun Qian's collated edition, adding 21 bodies of water such as the Fu River and the Hutuo River. Bi Yuan praised Zhao Yiqing's contribution to research on the *Shuijing zhu* in the preface to the book, and he said that this work compensated for Li Daoyuan's mistakes, which was indispensable. After the completion of the *Shuijing zhushi*, it only circulated as a copy for more than 30 years, and it was not published by his son, Zhao Zaiyuan, until the 51st year of the Qianlong reign (1786).

9.4.6 Dai Zhen

Dai Zhen (1723–1777), with the courtesy name Shenxiu and the nickname Dongyuan, was born in Xiuning, Anhui province. He was not able to speak until he was ten years old, but he was thoughtful and inquisitive when reading. Growing up in an impoverished family, he went with his father to Nanfeng, Jiangxi province, and set up a private school for children in order to make a living at the age of 18. In the 16th year of Qianlong's reign (1751), he became a scholar's student and studied under Jiang Yong. At the age of thirty-three, he went to the capital to sit the imperial examination, but failed it repeatedly, so he travelled around the country. He compiled the *Fenzhoufu zhi* ("Fenzhou Prefecture Annals") and *Fenyangxian zhi* ("Fenyang County Annals") on a visit to Shanxi and revised the *Zhili hequ shu* ("Book of Rivers and Canals in Zhili") after his tour of Zhili. Later, he lectured at Jinhua Academy in East Zhejiang. In the 38th year of the Qianlong reign (1773), he participated in the editing and compilation of the *Siku quanshu* as a nominee. Later, he was awarded the titles of Jinshi and Shujishi [a short-term post in the National Academy]. During his five years in the archive, he wrote a significant amount.

Dai Zhen had studied the *Shuijing zhu* in his early years, and he completed the Weiboxie Study version of the *Shuijing zhu* in the 37th year of Qianlong's reign (1772). After entering the Si Ku Archive, he took charge of editing the *Shuijing zhu*, and published the moveable type printed version of the *Shuijing zhu* in the 39th year of the Qianlong reign, that is, the *Daijiao shuijing zhu*. Once this book was published, none of the previous annotated versions could compete with it, and Dai Zhen reached an extremely high position in the field of textual research on Li Daoyuan [23].

Dai Zhen's main achievements from studying the *Shuijing zhu* are as follows: on the one hand, he studied the evidence in the original book; on the other hand, he formulated three principles for distinguishing the notes from the original text of the classics by comparing the maps and referring to the mapping method. First, annotators should pay attention to the occurrence of the names of the water bodies, since in the classics they are simple, but in the notes they are complex. Second, the classics often describe "flowing past somewhere", while the notes often use "flowing through somewhere." Third, the "counties" mentioned in the classics are often altered to the "former counties" in the notes. Based on these three principles, it is possible to differentiate the classics from the notes [24]. The *Daijiao shuijing zhu* ("Dai's Correction of the *Commentary on the Water Classic*") was based on Zhao Yiqin's *Shuijing zhushi*, incorporating the strengths of Quan Zuwang's edition and many other editions, and is an excellent version that represents all the achievements of the textual research on the work by Li Daoyuan in the Ming and Qing dynasties.

Dai Zhen summed up the achievements of the *Daijiao shuijing zhu* as follows: "This book has a new look, with 2,128 words added, 1,448 words deleted and 37,115 words revised [25]."

9.5 Introduction and Assimilation of Western Geography and Surveying in the Early Qing Dynasty

9.5.1 Introducing Western Surveying Theories

During the reigns of Emperor Kangxi and Emperor Qianlong of the Qing dynasty, the Chinese government hired ten Western missionaries with surveying and mapping expertise, such as Joachim Bouvet (French, 1656–1730), Jean-Baptise Régis (French, 1663–1738), and Pierre Jartoux (French, 1668–1720), to carry out geodetic surveying and mapping in China and to impart their knowledge in this field, thus introducing Western topography, which had a great impact on traditional Chinese surveying and cartography.

Emperor Kangxi Xuanye (1654–1722), who was one of the most influential emperors in Chinese history, had worked hard to learn Chinese and foreign knowledge since childhood, and to be prepared for his future role. In the 28th year of the Kangxi reign (1689), after the signing of the Treaty of Nerchinsk, he realised the significant role maps played in politics and diplomacy and decided to introduce advanced techniques of Western surveying and mapping. Therefore, he asked ministers to recommend experts and buy instruments. When he visited around the country, he also asked the missionaries in China to accompany him and measure the longitude and latitude. During the ten years from the 28th year to the 47th year of the Kangxi reign, all the missionaries hired were working on measuring the longitude and latitude of various places to prepare for the mapping. After the 47th year of Kangxi's reign (1708), the triangulation method was applied, and the work of surveying and mapping officially began [26]. The entirety of the work was presided over by Emperor Kangxi, who decided the major plans, policies and regulations, giving his personal attention to the specific candidates, and to the organisation and quality. Before the large-scale mapping work was carried out, Emperor Kangxi ordered the missionaries to conduct an experiment and, first of all, draw a map near Beijing. After personally collating the results and comparing with the old map, he found that the new map was far better than the old one [27], which made him more determined and confident about conducting geodetic surveys and drawing new maps throughout the country. In the 47th year of Kangxi's reign, large-scale surveying and mapping work was carried out nationwide. The survey team was composed of missionaries from various countries and Chinese scholars, totalling over 200 people [28], mainly missionaries, led by Joachim Bouvet, Jean-Baptise Régis and Pierre Jartoux, and they journeyed to various places in groups to take triangulation readings and longitude and latitude measurements, and determined 641 latitudinal and longitudinal points. After eight years, the measurement work was completed. This was followed by two years of collation and then, under the auspices of Pierre Jartoux, the famous *Huangyu quanlan tu* ("Complete Atlas of the Empire") was completed. "The atlas was completed in the 58th year of the Kangxi, and it included a complete map and 32 regional maps [29]."

The completion of the *Huangyu quanlan tu*, which was unprecedented in the history of world geodesy and cartography, made a tremendous contribution to China's cartography. The systematic mapping of Europe began in France in the eighteenth century and was completed in 1793 thanks to the efforts of Cassini and his son. Measuring the longitude and latitude of Britain did not begin until 1857, and the topographic map was completed in 1870. The mapping work conducted by other European countries under the auspices of their governments began in the early nineteenth century, almost 100 years later than that of China in the early Qing dynasty.

It is noteworthy that the ethnic groups in China had also made contributions to the *Huangyu quanlan tu*. For example, in Tibet, Emperor Kangxi sent people to consult with the Dalai Lama at that time and ask for a map of Tibet. However, Emperor Kangxi was not satisfied with the simple map, so he asked two lamas working in the Qing Astronomical Bureau to study mapping with Western missionaries, and then sent them to Tibet to carry out independent surveying and mapping; they completed the mapping work in the fifty-sixth year of the Kangxi (1717).

Emperor Kangxi presided over the mapping of the national map and accomplished two major achievements:

First, in order to unify the length used in the measurements, Emperor Kangxi himself stipulated that one degree of latitude on the meridian was equal to 200 li (1 li = 1800 ft) [30], and each foot was equivalent to 1/1000 s of latitude on the line of longitude. This practice of defining the scale by the earth's shape was decades in advance of the practice of defining the length of metres based on the distance from the equator at the end of the eighteenth century in France.

Second, in the 41st year of the Kangxi reign (1702), the linear distance from Bazhou to Jiaohe was measured on the "prime meridian" passing through Beijing. In the following eight years, the linear distance of each degree between 40° N and 47° N was measured in Northeast China, discovering that the higher the latitude, the longer the distance between each longitude. Although the significance of this discovery was not clear at that time, it was the first time that measurements in the field proved that the Earth is an oblate spheroid.

9.5.2 National Maps Produced During the Rule of Emperor Kangxi and of Emperor Qianlong

According to the records of the *History of the Qing Dynasty: Biography of He Guozong* and Shao Yichen's *Concise Catalogue of Si Ku*, "this Kangxi atlas included a general map of the whole country and 32 provincial and regional maps. The imperial map made by Quan Zuwang was scrutinised more than the method created by the

Yuan people [31]", which was the earliest version of the *Huangyu quanlan tu* ("The Complete Atlas of the Empire"); however, the part with the national general map has now been lost. Thereafter, two other versions of the *Huangyu quanlan tu* appeared. One of these is the version consisting of 41 copper plates found in the Imperial Palace in Shenyang, which was reprinted with the title of *Qingneifu yitong yudi mitu* ("The Secret Map of the Qing Dynasty's Imperial Palace") by Jin Liang; the other one comprises the maps in the *Tushu jicheng* ("Collection of Pictures and Writings"). This set of maps is very detailed, and it is divided into 227 pieces according to the provinces and prefectures, but it does not indicate longitude and latitude, nor does it include maps of areas beyond the borders [32]. According to the *Qingneifu yitong yudi mitu*, this map uses pseudo-cylindrical projection. The meridian through Beijing is the central longitude line, which is divided according to latitude, and the latitude line is perpendicular to the central longitude line. In terms of the language of the text notes, Chinese characters are used in the part focusing on Inner China and Manchu script is used in the part focusing on the frontier [33]. The scale is 1:1,400,000 to 1:5,000,000 [34].

Since the completion of the *Huangyu quanlan tu*, there have been many versions, and five main ones are commonly seen nowadays.

(1) An ink woodblock print version from the Kangxi period, measuring 210 cm high by 226 cm wide, which is not drawn to scale, is now conserved in the Palace Museum Library. In the map, the northeast extends to Sakhalin Island, the southeast to Taiwan Island, the west to Yelken City west of Aksu, the north to Baikal Lake, and the south to Hainan Island. The map uses the trapezoidal projection method to draw the longitude and latitude lines, which are straight lines, and takes the meridian passing through Beijing as the "prime meridian". Except for the central line of longitude, all the other longitude lines are inclined to the North Pole, and the latitude arc becomes increasingly shorter as it approaches the north. Both inside and outside Shanhaiguan Pass are annotated in Chinese characters, and the provincial capital and main places are marked with yellow paper. Tibet and the extreme west of Mongolia on the map are sketched, with many blank spaces. At the top of map is written in regular script "*Huangyu quanlan tu*" in ink, and the signature is not marked with the year [35].

(2) An ink woodblock print version from the Kangxi period, measuring 212 cm high by 340 cm wide, which is not drawn to scale, is now conserved in the Palace Museum Library. The four boundaries in this image are slightly different from those in the above map. Except for being the same in the north and southeast, the northwest extends to the Ili River, the west extends to the west of Talasaibila, and the southwest extends to the west of Leh district. In addition, the map of Tibet and the extreme west of Mongolia are very detailed. In Lhasa, the yellow paper is labelled as "Usizang", and Mount Kailash and the Yarlung Zangbo River are labelled with yellow notes, while on Mount Qomolangma, only the words "Zhumulangma Alin" are engraved in small characters. The map uses the trapezoidal projection method to draw longitude and latitude lines, and it takes the meridian passing through Beijing as the "prime meridian". The area

covered in this map is larger than in the one above. The place names inside and outside the Shanhaiguan Pass and in the northeast, northwest and southwest are marked with Chinese characters, and 93 yellow notes are pasted on the map. At the top of the map is written in regular script "*Huangyu quanlan tu*" in ink, and the signature is not marked with the year.

(3) The painted paper version of *Huangyu quanlan tu* from the Kangxi Period, which has no longitude and latitude lines and is composed of 16 pieces, is now stored in the First Historical Archives of China. The place names inside the Yumenguan Pass are annotated with Chinese characters, and the places outside the pass and in remote areas are annotated with Manchu characters, with similar boundaries to the two maps mentioned above.

(4) The *Der Jesuiten-Atlas der Kanghsi-Zeit* published by the German Walter Fuchs, at the University of Furen, in 1943, is actually another version of the *Huangyu quanlan tu* with 32 woodcut pictures. At the time of reproduction, Fuchs printed out four copies of the Heyuan Map published during the Kangxi reign, meaning the total number of copies was actually 36.

(5) During the period of the Republic of China (1912–1949), Jin Liang found 47 bronze plates in the Forbidden City of Shenyang: as 6 of these were blank, there were actually 41. With 5° of latitude per row, from 18° to 55° N, there are 8 rows represented in 41 bronze plates. Place names are noted in Chinese characters in Inner China, and Manchu characters are used in places outside the Shanhaiguan Pass, or in the Northeast, the Northwest and remote areas. There is a lithographic print copied by Jin Liang [35].

During the reign of Yongzheng, the Qing imperial government ordered the surveyors recommended by He Guozong to survey the canals, and the Wei River, Zhang River, Yellow River and other rivers and to continue the surveying and mapping work during the Kangxi period. Dominique Parrenin, Jean-Baptise Régis, Pierre Jartoux, Ehrenbert Xavier Fridelli, João Francisco Cardoso, Joseph-Anna-Marie de Moyria de Mailla (1669–1748, French, arrived in China in 1702 [1703]), Romanus Hinder and other western missionaries who had mapped China for Emperor Kangxi still served in the Qing court in the early years of the Yongzheng reign. During the reign of Yongzheng, the map from the Kangxi period was revised and printed into the *Huangyu quanlan tu* (10th edition) (see Fig. 9.1). This map combined the longitude and latitude lines with the traditional Chinese mapping method of the square grid. Each square represented 200 li, and the longitude and latitude lines were completely straight, so it was also called the *Huangyu fangge quantu* ("Complete Graticule Atlas of the Empire"). Such a method of mixing the longitude and latitude lines with the square grid was unsuccessful and retrograde. As a result, the accuracy of the map was reduced, and the higher the latitude, the greater the error. The only copies of this map are held by the collections of the First Historical Archives and the Library of the Chinese Academy of Sciences, and they were not widely circulated in China and had little impact.

The mapping work conducted during the Kangxi period was not complete. Therefore, during the reign of Qianlong (1736–1795), it was further boosted. Based on the

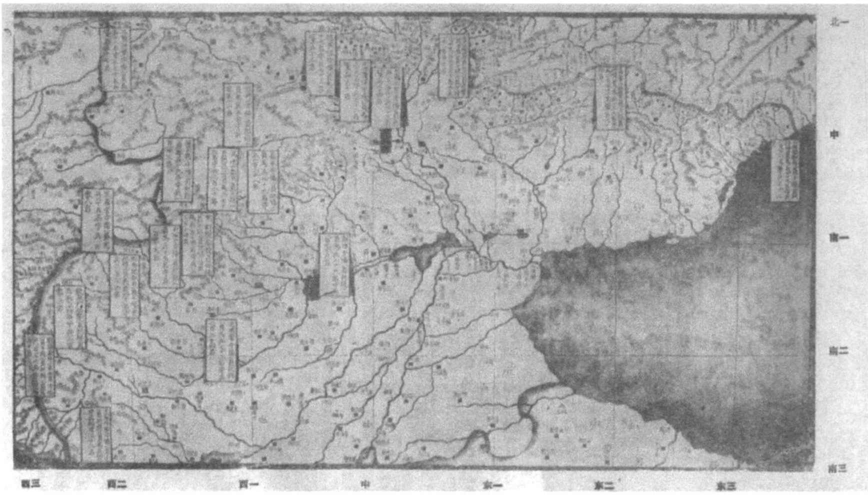

Fig. 9.1 *Huangyu tu* 皇舆图, *10th Edition.* Yu, Fushun 於福顺 (1983) "Qing yongzheng shipai huangyu tu de chubu yanjiu" 清雍正十排《皇舆图》的初步研究 [Preliminary research on the *Huangyu tu* from the Yongzheng period]. *Wenwu*文物 [Relics] (12):71–75

Huangyu quanlan tu from the Kangxi period, the map was expanded westwards and northwards, and some places were measured. For example, in Xinjiang, during the Kangxi period, the extent of the surveyed land only stretched as far as Hami. In the 21st year of the Qianlong reign (1756), He Guozong and Ming Antu were sent to the area west of Hami to measure the longitude and latitude and the *Western Regions Atlas* was drawn, which became a model for all subsequent maps of Xinjiang. By the 24th year of the Qianlong reign (1759), the Qing government had completely quashed the rebellion in the south of Xinjiang and sent Ming Antu's team to carry out geodesic surveys. After a year of hard work, they had successfully completed the task, laying the foundation for the completion of the national map. In the 25th year of the Qianlong reign (1760), the *Qianlong shisanpai tu* ("The Thirteen Edition of the Map of Qianlong") was completed. If it is compared to the *Huangyu quanlan tu*, most of its features are identical, except for the changes to one or two place names in the provinces within the Yumenguan Pass. However, being more than double the size of the *Hiangyu quanlan tu*, the scope of the *Qianlong shisanpai tu* is relatively large, including the Arctic Ocean in the north, the Indian Ocean in the south, and the Baltic Sea, the Mediterranean Sea, and the Red Sea in the west; this is obviously not just a picture of all of China, but rather a picture of the whole of Asia.

The results of the mapping during the reigns of Kangxi and Qianlong not only affected the entirety of the Qing dynasty, but also the early years of the Republic of China. Most maps published during these periods were based on the *Huangyu quanlan tu* which is a useful source of scientific data nowadays. However, after the map was completed, it was so secret that it was hidden in the inner government and only a few senior officials could see it. Thus, it was not popularised at the time, resulting in a great reduction in the practical value of the map. Popularising the surveying and mapping methods used at that time was difficult without written records, so for a period of time in the middle of the Qing dynasty, mapping went back to the old stereotype. This was the result of the restrictive policy by the rulers of the Qing dynasty, which should be taken as a warning.

9.5.3 Introducing Western Geography

(1) Ferdinand Verbiest and *Kun yutu shuo*

In the 13th year of the Kangxi reign (1674), Ferdinand Verbiest (1623–1688, Belgian, who had arrived in China in 1656 [1658]) wrote the two volumes of the *Kun yutu shuo* ("Illustrated Account of the World") and one volume of the *Kunyu waiji* ("Additional Illustrated Account of the World").

The first volume of the *Kun yutu shuo* covers the north and south poles, earthquakes, mountains, the movement of the sea, the tides of the sea, rivers, wind, clouds and rain, and people, etc., coverage which equates to physical geography. The second volume is divided into Asia and other countries and islands, Europe and its islands, Africa and its islands, America and its islands, and a general introduction to the world, tidal races, sea products and vessels. The content of the second volume is mostly from the *Zhifang waiji* ("Records of the Lands Beyond the Imperial Administration"), and the first volume contains information introduced by Matteo Ricci with a small amount of new data. At the end of the book, there is a map of 23 animals and a *Qiqi tu* ("Map of Seven Wonders"), depicting the seven major engineering projects in the world.

The *Kunyu waiji* records the world's unheard-of in a concise format. Most of the content was acquired from the *Kun yutu shuo* and *Zhifang waiji* ("Records of the Lands Beyond the Imperial Administration"), and a handful of accounts are about the latest inventions at that time. For example, there are tiny striking clocks that could be hidden in a ring and guns that could be shot 40 times within 30 min, ranking among the most advanced weapons at that time.

Due to the lack of new content in the two books, they do not have the same influence and value as the *Zhifang waiji.*

(2) Michel Benoist and *Kun yutu shuogao*

Jiang Youren (Michel Benoist, 1715–1774, French, arrived in China in 1744) had the courtesy name of Deyi. In the 26th year of the Qianlong reign (1761), he presented the hand-painted *Kunyu quantu,* and engraved it on wood in the 32nd year of the Qianlong reign (1767). The picture is 4.33–4.66 m wide and 2 m high, depicting the two hemispheres. Illustrations and textual descriptions of the astronomical content are arranged around this picture, which shows that the Copernican theory is the only correct one. It also introduces Kepler's three laws and the theory of the ellipsoid shape of the Earth, etc. Later, it was transcribed into a book, namely the

Kun yutu shuogao ("Transcription of Illustrated Account of the World"), containing the *Kunyu quantu shuo*, longitude and latitude lines, four continents, Yaxiya zhou (lit. the transliteration of "Asia"), Europe, Libya, America, stars, together with discussion of the four seasons, the differences in the radius of the earth, maps, eclipses, the sun, the five stars, and the celestial sphere, etc. Later, Qian Daxin renamed the *Kun yutu shuogao* as *Diqiu tushuo* ("World Geographic Atlas") and officially published it. However, at that time, Ruan Yuan and others stubbornly opposed the heliocentric theory, which seriously hindered the spread of Copernicus's theory and the theory of the oblate spheroid in China. Moreover, Jiang Youren was also the author of the *Kunyu quantu huiyi* ("Illustrated Complete Map of the World"), which included a map of the world and information about the armillary sphere, the horizon, longitude and latitude, and the reasons for the alterations in the nights and days and the four seasons [36].

(3) The introduction of western meteorology in the early Qing dynasty

Ferdinand Verbiest (commonly known as Nan Huai Ren) was the first to introduce western meteorological instruments into China; he brought the thermometer and hygrometer into China in the 16th year of the Shunzhi reign (1659). In the 9th year of the Kangxi reign (1670), he made thermometers and hygrometers himself. In the tenth year of Kangxi's reign, he published the *Yan qi shuo*, which explained the production and application of thermometers and was the first book to introduce European thermometers. The thermometer made by Nan Huai Ren was an early air thermometer with a flexible tip and an arbitrary thermometric scale [37], which could be used to measure the change in temperature at a certain period, as well as the temperature of soil and the human body. The hygrometer manufactured by him was used to measure the humidity in the air, and was a type of string hygroscopic hygrometer.

In the 8th year of the Qianlong reign (1743), the French missionary Antoine Gaubil (1689–1759, arrived in China in 1721) set up a weather station in Beijing for meteorological observation. In the 20–25th years of Qianlong's reign (1755–1760), the French missionary Jean Joseph Marie Amiot (1718–1793, arrived in China in 1750) also observed the temperature, pressure, cloud cover, rainfall and wind direction in Beijing twice a day, once at sunrise and once at 3 pm. The observations by these two people cannot be verified. However, Qian Deming's six-year long observation statistics were published by Messier in the *Journal of Mathematics and Geography* in Paris, which may be the earliest comprehensive modern meteorological record by any foreigner in China [38].

9.6 The Culmination of the Local Chronicles

The Qing dynasty was the peak period in the development of Chinese *fangzhi* (local chronicles), which was reflected in the following three aspects:

Firstly, in terms of quantity, there are more than 8,100 *fangzhi* in China, and about 5,600 date from the Qing dynasty, accounting for 70% [39].

Secondly, in terms of categories, there are records of national chorography, general records, prefectural and county records, township records, landscape records; records of lake embankments, water conservancy projects, salt wells and salt fields; records of palaces, temples, ancestral temples, tombs and customs; records of scenic spots and historic sites, etc.

Thirdly, from the perspective of local chronicles theory, the Qing dynasty established a relatively complete and systematic theory of *fangzhi* through the efforts of Zhang Xuecheng and others.

The Qing imperial government at all levels attached great importance to the compilation of *fangzhi* by actively promoting them and exercising strict control over them. The ten emperors of the Qing dynasty compiled and printed many *fangzhi*, especially in the Kangxi, Qianlong and Guangxu reigns. From a regional perspective, Hebei, Henan, Shandong and other provinces in the north compiled more chronicles than Jiangsu, Zhejiang and other places, breaking the pattern existing since the Song dynasty of there being more chronicles in the south than in the north. At the same time, the border provinces and regions also began to revise their records. The government strictly controlled the activity around the compilation of *fangzhi*. Almost all *fangzhi* above the prefecture and county levels were compiled by officials, and few were compiled by individuals. In terms of the method employed in the writing, a major drawback of the Qing dynasty's chronicles was the obligation to report only what was good while concealing what was unpleasant [40].

There were three editions of the National Chorography of the Qing dynasty: the compilation of the first one was ordered in the 24th year of the Kangxi reign (1685) and completed in the 8th year of the Qianlong reign (1743), totalling 342 volumes. The second one was compiled in the 29th year of the Qianlong reign (1764) and completed in the 49th year of the Qianlong reign (1784), totalling 500 volumes. The third one was begun in the 16th year of the Jiaqing reign (1811) and was completed in the 22nd year of the Daoguang reign (1842), totalling 560 volumes. Since the materials in the third edition cover the period up to the 15th year of the Jiaqing reign, the book is commonly known as the *Jiaqing zhi*, namely, *Jiaqing chongxiu yitong zhi* ("Revised Chorography During the Jiaqing Period").

The *Jiaqing chongxiu yitong zhi* collected information from all parts of the country and was divided into units relating to the provinces and special regions in the order of Beijing, Zhili, Jiangsu, Anhui, Shanxi and other provinces, Xinjiang, Mongolia and other regions, with an appendix containing countries with diplomatic ties with China. Each province's section contained an overview of the geographical environment, construction history, household registration and land tax, and the famous officials, etc. In addition, the book detailed 27 kinds of features of prefectures and provinces, such as maps, statistical tables, territory, construction history, customs, cities, schools, household registration, land tax, officials, mountains and rivers, historic sites, passes, dykes, tombs, temples, figures, martyrs, immortals, and local products. The geographical phenomena recorded was based on the *fangzhi* of many places in the entire nation, so this information became very rich, surpassing all previous chorography at a national level.

There were two ways of writing the *tongzhi* (general annals) of each province. One was to describe the various systems and the biographies of people organised into units at the level of the prefecture and municipality; the other was to take the outline of the biography and *fangzhi* as the structure, and then describe the situation in each prefecture and county. Taking the *Zhejiang tongzhi* ("General Annals of Zhejiang") compiled by Li Wei and others as an example, at the beginning of the work is a three-volume imperial decree, and then the text introduces 54 topics, such as illustrated maps, territory, administrative system, mountains and rivers, cities, schools, public offices, historic sites, water conservancy, seawalls, land taxes, household registration, storage, water transport, salt law, military systems, coastal defence, customs, products, vassals, officials, elections, figures, martyrs, shrines, temples, tombs, classics, steles, artistic works, miscellaneous notes, totalling 280 volumes. "The cited works were listed, and the most recent news that had not been recorded previously was also marked. As for any strange stories, the author would provide some evidence based on his research below [41]."

The style of the *fangzhi* in prefectures and counties was roughly the same. Taking the *Jingxian zhi* ("Chronicles of Jing County") compiled by Hong Liangji in Jiaqing as an example, its classification reflected that of most prefecture and county chronicles. The book has 32 volumes, including Volume One on administrative evolution (global vistas, territory, customs); Volume Two on cities (the old city, streets, township, bridge, port, mail); Volumes Three and Four on mountains and rivers; Volume Five on food and goods, taxes, household registration, corvees and properties, etc.; Volumes Six and Seven on schools; Volumes Eight on colleges; Volume Nine on temples; Volume Ten on government offices; Volume 11 on historic sites; Volume 12 on gold and stones; Volume 13 on the list of official posts, Volumes 14 and 15 on elections; Volume 16 on famous officials; Volumes 17 and 25 on figures, martyrs and temples; Volume 26 on art and literature; Volume 27 on miscellaneous information (chronicles, disasters or blessings, and anecdotes); Volume 28 on debates; Volume 29 on the origins of old records; Volumes 30 and 32 on poetry. The *Jingxian zhi* is one of the famous records of the Qing dynasty, with comprehensive content, rigorous textual research, and emphasis on authentic data rather than hearsay.

What is common to all the *fangzhi* is that they all adopt the styles of the records, biographies, pictures and tables which come from the official histories. The styles of the privately written chronicles of prefectures, counties, villages and towns vary, some are roughly the same as the official ones, but some are quite different from the official ones.

Before the Qing dynasty, scholars had always classified *fangzhi* into geographical categories. In the Qing dynasty, Zhang Xuecheng proposed several theories: first of all, local chronicles should be part of the local history. Second, in order to be practical, the content relating to modern times should be recorded in detail and the history should be summarised roughly. Third, the contents of local chronicles should be divided into three sections: *zhi*, *zhanggu* and *wenzheng* [42]. *Zhi* is the main body, which "imitates the style of the biographical history books." *Zhanggu* is like a compendium of government and social institutions, which collates the important local authority documents into a special book structured according to categories. *Wenzheng* is similar to a literature review, and its "main purpose is to prove history", and it selects poems and essays that reflect the local living conditions. Fourth, the local chronicles should advocate the "four genres". The first, *ji*, refers to the chronological chronicle of events. The second is *zhuan*, which signifies additional remarks relating to the figures or subject matter. The third is *kao*, or bibliography, which attaches importance to political and religious ceremonies and local customs. The fourth relates to tables and figures. The fifth theory was that, when compiling *fangzhi*, one should overcome the difficulties involved in field investigations, balance public opinions and search for documents, and avoid confusion, literary style and stereotypes to create concise, rigorous and elegant *fangzhi*. Sixth, one should distinguish between the styles and content of various kinds of *fangzhi* [43]. Zhang Xuecheng's theoretical views above were not comprehensive enough at the time, especially when the *fangzhi* were placed in the category of historiography, which weakened the geographical value of *fangzhi* and affected the development of geography. Still, his theory laid the foundation for the start of research on local chronicles in China, thus making an indispensable contribution to this field.

9.7 Studies of Borderland Geography

Studies of borderland geography works before the Opium War in the Qing dynasty were the result of the internal affairs of Qing governance. Some scholars studied frontier geography in order to expand their geographical vision on their own initiative, while others were banished or exiled to the frontier because they had offended the imperial court. In their adversity, these scholars did not yield to their suffering but rallied to study the geography of the frontier by writing books and promulgating theories, dedicating themselves to the service of their motherland. Some famous works of this kind include the following.

9.7.1 Yang Bin's Liubian jilue *(or* Quanliao beikao*)*

Yang Bin (1650–1740), with the courtesy name Keshi and the nickname Shanyin Gengfu, who also styled himself as Dapiao, was born in Zhejiang. His father was exiled to Ningguta (now Ning'an city, Heilongjiang province), so he asked to replace his father but his request was refused. Later, he went to the court to redress the grievances and was allowed to meet his father at the Wicker Palisade. People outside the Great Wall called him the dutiful son Yang. He had a talent for writing books and was proficient in engraving inscriptions on steles, producing the works *Jinshi yuanliu* ("Origins and Spread of Gold and Stone"), *Xifatang gao* ("Scripts from the Hall of Elegance") and *Liubian jilue* ("Brief Records of Willow Fence Borders").

The Liubian, that is, the Wicker Palisade in the Qing dynasty, was set up from Kaiyuan to the north of Jilin City to prevent inland people from entering the mountains to harvest ginseng in the early Qing. At the beginning of the Kangxi, Yang Bin's father was exiled to Ningguta. The records of the five volumes of the *Liubian jilue* are what Bin saw and heard on his way to visit his father in the 28th year of the Kangxi reign (1689). The book was written in the late Kangxi period, containing about 550,000 words, with an author's preface, describing the internal and external situation of the Liubian, the mountain and river settlements, the official system, the post stations and the mileage between the stations, the products, the local scenery, the language customs, the markets, the geographical evolution, etc., and is the earliest existing private book about the geographical conditions of Jilin and Liaoning in the Qing dynasty.

The book first describes the position of the twenty-one gateways and the extent of the Liubian, and then records forty-six stations and the mileage of the sections from Beijing to Ninggu Pagoda, which is two thousand eight hundred and seventy-eight li in total. The data, such as the names and locations of the 36 tribes under the jurisdiction of Ningguta, the tributes of eight Tatar tribes to the northeast of Ningguta, as well as their political and economic relations with Ningguta, are important for studying the history and geography of Northeast China. The book also details the specialities of the Northeast and the customs of marriage, hospitality, and the religious activities of all the ethnic groups. Seventy-five poems are attached at the end of the book in chronological order, which are mostly lyrical works. However, there are still some inaccuracies in the geographical records. For example, the Huanglong Fu (City) was once set up in Kaiyuan, but this was unknown to the author; the Hujiang River (Jin River) was mistaken for the Yalu River.

9.7.2 Wu Chenchen's **Ningguta jilue**

Wu Chenchen (1664–?), with the courtesy name Nanrong and the nickname Suhui, was the son of Wu Zhaoqian, who was exiled to the Ninggu Pagoda in the 14th year of the Shunzhi (1657) and became father to Wu Chenchen in the 3rd year of the Kangxi reign (1664). In the 20th year of the Kangxi reign (1681), Wu Zhaoqian was released, and Wu Chenchen entered the Shanhai Pass with his father after living in Ningguta for 17 years. Forty years later, based on his experience during his youth, he wrote the volume of *Ningguta jilue* ("Brief Records of Ningguta"), which vividly described the local climate, products, customs, vegetation, etc., and introduced the border defence measures, the agricultural system, the Manchu language, and the name and mileage of each station along the border defences at that time. Finally, according to his friends, the fact that a volcano erupted 50 li northeast of Pukui (now Qiqihar) in June or July of the 59th year of the Kangxi reign (1720) was recorded in his book, which is an example of rare and precious volcanic data in China.

9.7.3 *Zhang Jinyan's* Ningguta shanshui ji

Zhang Jinyan (1600?–1672?), a Chongzhen Jinshi of the Ming dynasty, with the courtesy name of Tangong, who was born in Xinxiang, served as magistrate of a county, commissioner, and Minister of War, etc. in the Ming, Dashun and Qing dynasties. In the early years of Kangxi's reign, he was exiled to Ningguta because of the literary inquisition, and died there. He collated the *Tianxia mingshansheng gaiji* ("Overview of Famous Mountains and Scenic Spots in the World"). Moreover, after visiting towns, villages, scenic spots, mountains and rivers in the area of present-day Mudanjiang, he wrote the *Ningguta shanshui ji* ("Records of the Ningguta Landscape") in the 9th year of the Kangxi reign (1670), of about 18,000 words, which was the first landscape chronicle in Heilongjiang province. The book includes three prefaces, recording 11 mountains, 5 bodies of water, 5 towns and villages, and 3 scenic spots, and the nationalities, customs, vegetation, birds, animals, insects and fish contained in the section "Zaji" (miscellaneous notes) are of great value. In 1984, Heilongjiang People's Publishing House printed a lithographic copy based on the Songshi Study block-printed edition from during the reign of Kangxi [44].

9.7.4 *Xiqing's* Heilongjiang waiji

Xiqing, with the courtesy name Yanzhai and the surname Sirin Gioro, the great-grandson of Ortai and the author of the *Jukui ji* ("Collection of Top Talents"), was born in the Bordered Blue Banner [the eastern part of Ulanqab League in Inner Mongolia] of Manchu, successively serving in posts as a junior official, such as the Guanku (responsible for treasury matters) and the Lixing Zhushi (responsible for criminal affairs). In his spare time away from political matters, Xiqing researched local customs extensively, widely read the official archives, consulted historical and geographical records, and collected the old stories from numerous books. In the 15th year of the Jiaqing reign (1810), Xiqing compiled the eight volumes of the *Heilongjiang waiji* ("Local Record of Heilongjiang"), about 61,000 words, with a map of the borderland attached. The text is divided into 28 categories, including territory, key towns, mountains and rivers, climate, situation, evolution, castles, stations, tribes, races, household registration, official system, military system, salaries and grain, income and expenditure, military equipment, tributes, criminal law, markets, exiles, dialects, food, customs, officials, and products. Covering a wide range of subjects, the book is the first *fangzhi* of Heilongjiang with detailed and accurate records and extremely systematic and fluent writing. The full text of the book includes the retranslated *Sino-Russian Treaty of Nebuchu*, which contains two more clauses than those recorded in the *Qinghui dian* ("The Judicial Corpus of the Qing Dynasty"), *Pingding luocha ffanglue* ("Military Annals of Pacification of the Rakshas") and *Shengjing tongzhi* ("General Record of Qing Dynasty Mukden") [45].

9.7.5 *Huang Shujing's* Taihai shicha lu

Huang Shujing, with the nickname Yupu, born in Daxing (now Beijing), Zhili, had the title of Kangxi Jinshi, and worked as an official overseeing the Changzhenyang-tong Dao [governments of Changzhou, Zhenjiang, Yangzhou and Tongzhou], in Changzhen Yang Tongdao. In the 61st year of the Kangxi reign (1722), when he was the imperial historian, he visited Taiwan and wrote the eight volumes of the *Taihai shicha lu* ("Record of a Tour Inspection in the Taiwan Strait"), about 80,000 words long. Volume One to Volume Five contain sketches and notes on Chikan (around Tainan, present-day Taiwan), with 22 items which include the primitive society, situation, ocean, tides, wind, climate, water courses, seagoing vessels, castles, military equipment, traders, and tributes. Volume Five to Volume Seven are about customs, detailing the residences of local minorities in Zhuluo in the north and Fengshan Mountain in the south and their diet, clothing, marriages, funerals, and uses of utensils. Volume Eight is divided into 8 items, such as the "Shengfan" (barbarians), "Shufan" (educated aborigines), "Sheshang" [the Han businessmen trading with minorities], "Shexiang" (public grain), "Deer Hunting", "Fanyi" [constables and procedures for arresting criminals] and "Tributes and Gifts". This book details the local customs, folk customs and products, as well as the dangerous mountains and rivers, the seaways and the wind, with reference to old documents and the author's experience, making it a significant geographic reference to Taiwan in the early years of China's history, and to the study of Taiwan's history and national customs [46].

9.7.6 *Chen Lunjiong's* Haiguo wenjian lu

Chen Lunjiong, with the courtesy name Zizhai, was born into a bureaucratic family in Tong'an County, Quanzhou Prefecture, Fujian province (now Tong'an District, Xiamen). His father, Chen Ang, followed Shi Lang to enter Taiwan in the 20th year of the Kangxi reign (1681), and later served as Vice Governor of Guangdong. Chen Lunjiong inherited his father's rank, and worked as Deputy General of Penghu, Chief Officer of Taipeng Town, and supervisor of the Ningbo navy in Zhejiang province, which were all located by the sea, so he was familiar with the situations of sea routes. When he was young, he travelled across oceans with his father, "paying attention to the customs of foreign people [47]." When he was in "Yinzhengao (now Gaozhou, Guangdong province), Lei (now Leizhou, Guangdong province), and Lian (now Hepu county, Guangxi province)", which shared borders with foreign countries, "he often deals with foreigners," so "he inquired about their national customs and examined their atlases" [48]. Later, based on his own experience and the information gathered overseas, he wrote the two volumes of the *Haiguo wenjian lu* ("Hearsay from Overseas Countries") in the 8th year of the Yongzheng reign (1730).

There are eight parts to the records in the first volume which are as follows.

(1) The "Tianxia yanhai xingshi lu" ("Records of the Coastal Situation in the World") describes the geographical situation of China's coastal areas. From Bohai Bay in the north to the Beibu Gulf in the south, including Taiwan and Hainan Island, it records the coastal landforms, hydrology, shipping, coastal defences and other cultural and geographical features along the coastline of more than 32,000 li [49].
(2) The "Dongyang ji" ("Records of Dongyang") describes the sea lanes, natural conditions, national systems, nationalities, customs, and products of Korea, Japan and the Ryukyu Islands. Since the author had visited Japan, its description is more detailed.
(3) The "Dongnan yang" ("The Southeast Seas") describes the physical and cultural geography of Taiwan, the Philippines, the Celebes (now Sulawesi), the Moluccas and Borneo, and records the details of the process of Zheng Chenggong's recapture of Taiwan.
(4) The "Nanyang ji" ("Records of Nanyang"), "Xiao xiyang ji" ("Records of India and the Middle East") and "Da xiyang ji" ("Records of the Atlantic Ocean") describe the geography of the Indochinese Peninsula, the Malay Peninsula, the Sunda Islands, South Asia, West Asia and Central Asia, Africa and Europe.
(5) The "Kunlun" ("The Kunlun Island") describes Kunlun Island in today's South Vietnam.
(6) The "Nanao qi" ("Records of the South China Sea") describes the islands in the South China Sea.

There are six maps in the second volume:

(1) the *Sihai zongtu* ("Map of the Eastern Hemisphere") is a map of the Eastern Hemisphere,
(2) the *Yanhai quantu* ("Coastal Map") is a coastal map of China,
(3) the *Taiwan tu* ("Taiwan Map") is a map of the west coast of Taiwan Island,
(4) the *Taiwan houshan tu* ("Map of the East Coast of Taiwan Island") is a picture of the east coast of Taiwan Island,
(5) the *Penghu tu* ("Penghu Map"),
(6) the *Qiongzhou tu* ("Hainan Island Map") is a map of Hainan Island.

These maps are more detailed and accurate than the *Zhenghe hanghai tu* ("Nautical Chart of Zheng He"), and Zheng Ruozeng's *Haifang tu* ("Coastal Defense Maps") and *Neifu yutu* ("Imperial Palace Maps"). The downside is that there is no map of the five continents, only the map of the Eastern Hemisphere, and the *Sihai zongtu* is not as detailed as Matteo Ricci's *Wanguo yutu* ("Great Map of Ten Thousand Countries").

The material in *Haiguo wenjian lu* is detailed and reliable, which had an impact on world geography treatises in China after the Opium War, such as Wei Yuan's *Haiguo tuzhi: Chouhai* ("Atlas and Description of the Countries Beyond the Seas: Coast Defence"), Cai Fangbing's *Guang zhiping lue: Haifang* ("The Strategies of Governing the Country and Benefiting the World: Coast Defence"), and Xu Jiyu's *Yinghuan zhilue: Riben* ("Concise Records of the World: Japan").

9.7.7 Chunyuan's Xiyu wenjian lu

Chunyuan, given name Qishiyi, surname Nimaca Hala, was born in the Zhenglan Banner of Manchu. In the 19th year of the Qianlong reign (1754), he was awarded the title of Jinshi. In the 42nd year of the Qianlong reign (1777), when he served in the Military Office in Kuche, he wrote the 8 volumes of the *Xiyu wwenjian lu* ("Records of Things Seen and Heard in the Western Regions"), which has several alternative titles, such as the *Xiyu zongzhi*, *Xinjiang zhilue*, *Xiyu ji*, *Xiyu suotan* and *Xiayu suotan*. The first volume is a map showing the territory of the Western Regions. And the rest of the content successively consists of 2 volumes of *Xinjiang jilue* ("Brief Records of Xinjiang"), which contains the geographical situation by region, and the recording officers and soldiers in more than ten cities, including Hami, Urumqi and Ili; 2 volumes of the *Waifan Liezhuan* ("Biographies of Vassal States") about over 20 tribes and vassal states, such as the Kazaks and Buruts [Kirghiz people], and their organisations and customs; 2 volumes of the *Xichui jishi benmo* ("Historical Events from the Western Frontier in Their Entirety"), recording the events that pacified Xinjiang several times in the early Qing dynasty; 1 volume of the *Hui jiang fengtu ji* ("Records of Natural Conditions and Social Customs in Xinjiang"), which includes Xinjiang's customs and glaciers, as well as the religions, laws and seasons; and 1 volume of the *Juntai daoli biao* ("Records of Postal Distance"), which was used as a basis for people to talk about mileage in the Western Regions.

9.7.8 Huangyu xiyu tuzhi *Under the Order of the Emperor*

In the 27th year of the Qianlong (1762), Fu Heng, Chu Tingzhang and others were ordered to compile the 48 volumes of the *Huangyu xiyu tuzhi* ("Illustrated Treatise on the Imperial Western Territories"), with 4 volumes of the emperor's poems at the beginning. There are 3 volumes of maps, including 21 new drawings and 12 old drawings by previous cartographers, followed by a textual description; 2 volumes of lists from the Qin and Han dynasties to the Yuan and Ming dynasties; 2 volumes on measuring longitude and latitude with ancient methods; 12 volumes on territories described according to four roads; 4 volumes on mountains; 5 volumes on water; 2 volumes on official systems; 1 volume on military defence; 2 volumes on garrison policies; 1 volume on tributes; 1 volume on finance; 1 volume on schools; 2 volumes on the nobility; 1 volume on customs; 1 volume on music; 2 volumes on clothing; 1 volume on local products; 3 volumes on vassal states; 2 volumes of miscellaneous records which include occurrences of quicksand. This was the first official general chronicle in Xinjiang to record "martial arts" in detail.

9.7.9 *Songyun's* Weizang tongzhi

Songyun (1754–1835), with the courtesy name Xiangpu and the family name Malat, was born in the Zhenglan Banner in Mongolia, and once served as Bitieshi (officer in charge of court paperwork, a position set especially for members of the royal family in the Qing dynasty), Daxueshi or the Grand Secretary, and as the minister stationed in Tibet. When he first became the minister in Tibet during Jiaqing's reign, based on his own experience, he wrote a volume of the *Xizhao tulue* ("Records and Maps of Tibet"), which described 28 items in Tibet, such as the mountains, rivers, and border passes, labelled with the mileage from Chengdu to the Tibetan courier stations attached. Then he wrote the 16 volumes of the *Weizang tongzhi* ("A Local Gazetteer of Tibet") of more than 238,000 words, concluding in the Qianlong period. The first part lists imperial poems. The main text is divided into seven sections: the world, monks and folk, defence and appeasement, brief chronicles, outlands, art and literature, and the classics. The main text covers 26 items, of which the "Wuxu" item (subsidies for entitled groups) is an original creation by the author. The territories, mountains and rivers, stations and lamas recorded in the book are more detailed than in the *Xizang tuzhi* ("Tibet Chronicle"), and the textual research was mostly based on ancient books and notes. The sources of the data are extensive, including many documents related to Tibetan affairs since the Han and Tang dynasties, old Tibetan chronicles, Tibetan archives and documents preserved in Tibet, official documents in the Qianlong period, and inscriptions on steles from past dynasties.

9.8 Geology, Mineralogy and Petrology

9.8.1 *Knowledge in the Field of Geology*

Three aspects of geological knowledge in the Qing dynasty are described: fossils, volcanic hot springs and earthquakes.

(1) Knowledge of fossils

The fifth volume of the *Qiuping xinyu* in the reign of Qianlong described in detail the *longgu* (or "dragon bones", mammalian skeleton and tooth fossils from the late Cenozoic): "In the mountains of Taizhou (now Taigu County of Shanxi province), peasants can often find dragon bones, about 3 m long and 4 *wei* [1 *wei* = 31.68 cm] around. The dragon bones are hollow and are heavier than stones. Craftsmen often use them to make artifacts. The colour of these bones varies: some of them are ivory-white, some red and purple, and some green. They feature fish-scale textures and scholars value them a lot. Once a man had a bowl-shaped dragon bone which weighed several kilos. But it was too hard and brittle, could not be made into artifacts and could only be used as an ingredient in medicines."

Tan Cui (1725–1801) wrote the *Dianhai yuheng zhi* ("Annals of Dianchi and Erhai Lakes") in the 4th year of the Jiaqing reign (1799), and the second volume mentioned the *shiyan* (shell fossil of a brachiopod) in Mayun County, Yunnan province: "Fossilia spiriferis and tanner crabs could be found everywhere and could be used to cure wounds. Fossilia spiriferis were found in Malong (now Malong County, Yunnan province). The big ones are male and the small ones are female. They were also excavated in Luquan."

The earliest record of the *bianfushi* (lit. bat stone) in China is found in Volume 20 of Wang Shizhen's *Chibei ou tan* from the 30th year of Kangxi's reign in the Qing dynasty (1691). The book recorded that in March of the 10th year of the Chongzhen reign (1637), Zhang Yandeng visited Mount Tai and stayed at Dawenkou [a well-known Neolithic site], and he found *bianfushi* on the riverside. A *bianfushi* is the fossil of a trilobite, and is the typical fossil from the late Precambrian, 500 million years ago. Rood, who referred to the trilobite as *Trinucleus* in 1698, was the first person outside China to recognise trilobites. It can thus be inferred that China discovered trilobite fossils 61 years earlier than other foreign countries.

The earliest document recording pagoda stone was the fourth volume of the *Guangyang zaji* ("Miscellaneous Notes of Guangyang") by Liu Xianting in the early Qing dynasty. Later, Volume 22 of the *Hubei tongzhi* compiled by Zhang Xuecheng in the 59th year of the Qianlong reign (1794) also recorded the occurrences of pagoda stones in Donghu Lake, Changyang and Xinshan. The *baotashi*, or "pagoda stone", also known as "bamboo shoot stone" and "Taiji stone", is the shell fossil of an ancient nautilus, whose shell is rectangular, conical or cylindrical with many diaphragms in the body. Since its longitudinal section resembles a pagoda, it is called pagoda stone. It is mostly found in Ordovician and Triassic strata and is the standard fossil of the Middle Ordovician Period. It is also called Taiji stone because, horizontally, it looks like a Taiji diagram.

The fourth volume of Li Xinheng's *Jinchuan suoji* ("Jinchuan Chronicles") during the Jiaqing period recorded the *shilong* (lit. stone dragon), which may be a *Lepidodendron* fossil in Li County, Sichuan province. A *Lepidodendron* was a tall tree, growing in the Carboniferous Period or Permian Period. After the leaves fall off, the branches show clear scaly marks. It is noted in the book that when local people were building houses on the hillside, they found fossilised bones in the soil, which were several metres long, and there was no way to remove them with a shovel. Therefore, people used water to wash away the sand on the surface of the fossil and found that it had scales, claws and teeth, as if it were a dragon, except that it had no eyes.

The earliest record of a *Psaronius* fossil (also known as Eight Diagrams stone) is Liu Luan's *Wu Shi Hu* in the Qing dynasty, which is included in the *Xieyu Congtan*: "During the reign of Emperor Chongzhen of the Ming dynasty, Tang Xuanzhi of Guizhi (now Guizhi, Anhui province) sent farmers into the mountains to pick rocks and burn them into ash to spread over the fields. After these people cracked open a stone with a hammer one day, they found that there were both yin and yang and *bagua* inside. But they did not think it was strange [50]." The *Psaronius*, which is a tall tree fern, grew during the Permian Period 200 million years ago. If the trunk of the tree is cut transversely and polished, it can be seen that the vascular bundle is composed of seven concentric rings, the innermost one is circular, and the rest are intermittent strips, some of which look like the Eight Diagrams, hence it is named *baguashi*, or "Eight Diagram" stone.

Amber is formed from the resin of gymnosperms flowing into the ground over tens of millions of years. In the Qing dynasty, Gu Yingtai's exposition of amber in the 8th volume of the *Bowu yaolan* ("Key Review of Myriad Things") made three new contributions compared with previous works: first, the places of origin—Ryukyu, Gaochang and Yongchang—were recorded; second, he declared that superior amber should be as red as chicken blood without any cracks or dirt inside; third, he advised that when seeking to distinguish true amber from fake examples, not only can we check whether it attracts grass after it is rubbed, but also whether it emits a fragrance after being rubbed and heated. The pieces which produce fragrance are genuine, and those which have no fragrance are fake.

(2) Knowledge of volcanoes and hot springs

In the Qing dynasty, many people recorded the volcanic activities in Taiwan. For example, Yu Yonghe said in the *Cailiu riji* ("Diary of Sulfur Extraction") in the 36th year of the Kangxi reign (1697) that "the ground near the sulfur pit of Mount Huangshan in Beitou is heated and barren, and the boulders around the mountain are eroded into powder by sulfur gas billowing from the ground." "Do not look directly at the toxic flame of the sulfur pit, and if you touch it carelessly, you will develop an appalling headache."

In the 61st year of the Kangxi reign (1722), Huang Shuyu also recorded the mud volcano in the southern part of Taiwan Island in his *Taihai shicha lu*. He described how "a mountain split in Fengshan County, creating a gap 24 m long and 12 m wide. Black mud gushed out of the crack, and a fire began to burn the following night. The fire was several metres high and very hot, and no one dared to approach".

The *Fengshan xianzhi* ("Chronicle of Fengshan County") recorded that "on June 26th of the Yongzheng period (1723), near Chishan, the sky glowed red, black mud erupted from the ground, and the surrounding vegetation was all turned to ashes."

Volume Two of the *Min zaji* ("Miscellaneous Notes of Minnan") recorded the mud volcano in Zhuluo County, Taiwan, also known as a pseudo volcano: "There is spring water flowing from the crevices of the stone, accompanied by fire that comes out of the spring and which can set things on fire. This natural fire comes out of the water, which is really strange." "There is a mountain slope where Fengshan County's Red Mountain and the Danshui Creek meet, which is about five kilometres long. The red fire can be seen in the soil of the slope, so it is also called Red Mud Bay." This is a mud dome continuously forced up by underground natural gas with water, mud, sand and rock debris. It may be the end of volcanic activity or the product of the eruption of natural gas, which can be understood as a sign of oil and gas seepage.

Xiqing's *Heilongjiang waiji* recorded the eruption of the southeast volcano in Merghen (now Nenjiang County, Heilongjiang province) in the 58th year of the Kangxi reign (1719): "The fire suddenly appeared inside the ground, stones flew out, and the sound was so loud that it could be heard everywhere. After a few days, the fire extinguished, and the place then became a swamp."

In the *Ningguta jilue*, Wu Chenchen recorded the formation of a volcano in the northeast of Pukui (Qiqihar) in June or July of the 59th year of the Kangxi reign (1720): "Suddenly this place is on fire, and there is a large amount of smoke. The sound is very loud, just like thunder. It goes on and on during the day and at night, and the sound can be heard more than 30 km away. Most of the stuff flying out of the fire consists of black stones and sulfur. This phenomenon lasted for many years and, later on, a mountain was formed."

In the two volcanic eruptions mentioned above in Wudalianchi (or Five Connected Lakes), in the City of Dedu, Heilongjiang province, the lava flow was 17 km long from the north to the south, covering an area of 68 km^2 and was 20–40 m thick. It covered the basalt platform that had erupted earlier, blocking the channel of the Bai River and forming the Wudalianchi.

The first volume of the *Kangxi jixia gewu bian* ("Kangxi's Scientific Works in His Leisure Time") recorded the Chahayan Volcano in Heilongjiang; Chen Ding's *Dian youji* ("Yunnan Travel Notes") recorded the Wuding Volcano in Yunnan: "There is a flame inside the cave day and night, and it has not been extinguished for many years. When people walk in, it smells like coal. There are also yellowish-white ashes, like the dung of cattle and horses, which crumble when pinched."

In the Qing dynasty, Qu Dajun's *Guangdong xinyu* recorded 88 hot springs distributed in 44 counties. It also described how the temperature of the Guangdong Dianbai Hot Spring changed over time: "The spring water has a slight smell of sulfur, and the temperature is so high that it can cook chicken and eggs. The temperature of the spring is highest from three to five o'clock in the afternoon, from eleven to one o'clock at midday and from five to seven o'clock in the evening, and not so hot at other times." Gu Zuyu recorded more than 90 hot springs across the country in the *Dushi fangyu jiyao* ("Geography Essentials for Understanding History").

(3) Knowledge of earthquakes

In the Qing dynasty, Dong Han's *Sangang zhilue* ("A Brief Introduction to Sankang") and Ren Shu's *Dizhen ji* ("Earthquake Record") [51] recorded the great earthquake that occurred in Beijing on 28th July, in the 18th year of the Kangxi reign (1679): "From the northwest, sandy soil was soaring into the sky and black clouds were blocking the sun. The people were lurching as if they were sitting on waves. After a while, there was suddenly a very loud noise, and the animals were frightened and fled everywhere. That night there were three earthquakes in a row, and the ground split open into cracks of several metres wide. There was a big ditch in Deshengmen from which water flowed out. There were many officials and civilians injured in the quake, and some even had their whole families die in the quake. Another earthquake occurred at noon on 29 July, and a large one at midnight on 1 August. Aftershocks occurred from time to time thereafter, and people stayed outdoors during the day and night, not daring to go indoors. Bodies piled up like a small mountain, making them unrecognisable. More houses collapsed in Tongzhou. There were cracks in the streets in Zhuozhou and Liangxiang, and black water gushed out from them, which was several metres deep. The ground in Shanhaiguan and Sanhe caved in and transformed into rivers. There were earthquakes around Beijing for an entire month."

"Suddenly there was a loud noise from the ground, and after that, the sound became louder and could be heard everywhere, as if 100,000 troops and horses were running. People could see the windows shaking as if pummelled by a strong wind and the waves of the sea. People hid under their beds or behind their doors, so they could hide from the earthquake. Afterwards, the ground subsided, and people ran to the top of the hill and climbed up into the trees to take shelter." The two works explained the environmental changes before the earthquake, the sounds of the earthquake, the fissures in the ground, the aftershocks, the damage, prevention measures and so on, providing a summary of the experience of the earthquake.

In May of the 7th year of Emperor Kangxi's reign in the Qing dynasty (1668), Wu Mingxuan, Deputy Supervisor of the Qing Bureau of Astronomy, proposed building a new seismograph made of a copper plate with a ball [52], which was quite different from the seismograph invented by Zhang Heng. However, this proposal was not implemented, and consequently its impact cannot be assessed.

There were many examples of predictions of earthquakes in the Qing dynasty, such as ground light, ground sound, foreshocks, aftershocks, groundwater anomalies, meteorological anomalies, and abnormalities in animal behaviour. In 1755, Wang Yichen wrote the *Yinchuan xiaozhi* ("Brief Records of Yinchuan") which summarised some common methods of making predictions among people. Using the Ningxia earthquake occurrence in spring and winter as an example, Wang mooted the idea of a relationship between well water changes, ground sound and abnormalities in animal behaviour and earthquakes, suggesting the theory of integrating various precursor phenomena to predict earthquakes.

9.8.2 Knowledge in the Field of Mineral-Petrology

(1) Mineral-petrology reflected in mining and cultural relics

① Mineral-petrology reflected in mining

In the Qing dynasty, the mining industry was very developed, and there was a relatively large range of minerals, which included iron, copper, tin, lead, zinc, gold, silver, mercury, coal, oil, natural gas, salt, sulfur, jade, the raw materials of ceramics, inkstone, and stone. Thus, this knowledge of minerals was the foundation for the development of mining.

During the Yongzheng, Qianlong and Jiaqing periods, Yunnan province abounded in copper. The average amount of Yunnan copper used to cast money each year was estimated to be more than 11 million *jin* (=1/2 kg). Given the proportion of materials used in casting money, this part of copper required about 91.1 million *jin* of zinc, 1.4 million *jin* of lead and 0.44 million *jin* of tin each year [53]. The highest annual production of mercury in Guizhou in the Qing dynasty reached more than 5,000 *jin*, over 1,000 *jin* more than the highest national production of mercury in the Northern Song dynasty. Such a large-scale mining output must have been based on an extensive knowledge of minerals and rocks.

According to archaeological data and documents, there were 106 names for mineral and rocks in the Qing dynasty, which was the largest number in any ancient dynasty, and they included clay, kaolin, copper, alluvial gold, tin, lead, zinc, crystal, turquoise, agalmatolite, agate, Qingtian stone, white marble, jasper, black jade, coal, gardenia oil, well salt, lake salt, sea salt, rock salt, nitrate, sulfur, sal ammoniac, silver skin, black and white sand, xanthoconite, raw silver, lead-silver, mineralised iron, magnetite, lodestone, perlimonite, talc, asbestos, graphite, lead and zinc ores, calamine, white copper, borax, sodium silicate, mirabilite, limestone, marble, mica, alum, bornite, natural copper, black lead, silver ore, turquoise, stalactite, gypsum, calcite, fluorite, jadeite, emerald, ruby, purple gemstone, white quartz, purple quartz, white gemstone, sapphire, chalk, etc.

Prospecting methods in the Qing dynasty included:

a. Using the outcroppings to find ore. For example, Zhang Shenshu described the outcropping of azurite as follows: "The outcroppings appear between the cliffs and caves, revealing a blue colour, shaped like wisps or ribbons [54]." Wang Chang also said that "where there is copper ore, there must be green outcrops of different widths between the rocks. After a long period of rain, the mountain collapses, and the copper will be exposed [55]."

In ancient times, not only were the outcrops used for prospecting, but also for calculating the size and depth of ore bodies. The detailed method for exploring the size and depth of an ore body based on the outcrops was described in the *Diannan kuangchang tulue* ("Illustrated Account on Mining in Southern Yunnan Province").

At the beginning of the Qing dynasty, Sun Tingquan described the methods for identifying outcrops for local mining in the 4th volume of the *Yanshan zazhi*. He said that weathered coal floats on the surface of the soil, with a dull colour, a slight smell of sulfur, and low thermal power; the raw coal in the rock stratum is buried deep underground following the run of rock and has a bright colour, abundant sulfur, high thermal power, and a strong smell when ignited.

b. Using the symbiotic relationship of metal minerals to find ore. Volume 15 of the *Guangdong xinyu* says: "There is crystal next to the tanzanite, and powder stone next to the gold."
c. Using spring water to find ore. It is recorded in *Guangdong Xinyu:* "In the mountain, if there is yellow water seeping out, then iron ore is present there. Follow this path, digging deep, you can find a lot of iron ore."
d. Using the odours of minerals to find ore. For example, Li Diaoyuan heard that there were silver ores on the mountains in Yingde, Qingyuan and other places in Guangdong. When the sun heats the silver ore on the mountains, white droplets form on the surface of the ore, and the smell is pungent. Guided by this special smell of minerals, the silver deposits can be found [56].
e. Using the relationship between rock strata and ore deposits in prospecting. For example, Sun Tingquan recorded a case of using the relationship between rock strata and ore deposits to find coal mines in the *Yanshan zaji*: people looking for coal must first observe the rock properties on the mountain. If there are layers

of shale on the mountain, there may be coal. If the mountain is full of limestone and sandstone, there is no coal. When people see a black coal seam outcropping on the land, they should carefully measure the layers of rocks above and below the coal seam to avoid water inrush. The underground conditions can be inferred from the overlying rock strata, and the distant conditions can be inferred from the near rock strata.

② Mineral-petrology reflected in cultural relics

In the Qing dynasty, the types of minerals and rocks reflected in the unearthed cultural relics include:

a. Gold and silver

The tomb of the Qing dynasty princess in Tongyu, Jilin, revealed four gold peacock ornaments inlaid with turquoise; 1 pair of golden dragon hairpins, inlaid with beads; 1 pair of gemstone and gold hairpins, inlaid with rubies; 1 gold crown, a variety of gold ornaments, 1 gold ring and 1 silver ring, each one with a ruby in the centre; 1 gold ingot, weighing 37.4 g; 26 silver hairpin rods; 1 silver ingot, weighing 103 g; 1 silver bracelet, weighing 44.5 g [57]. In No. 2 Tomb of the Qing dynasty in the western suburbs of Beijing, there were unearthed 2 pairs of gold bracelets, 2 gold phoenix ornaments, 5 gold hairpins inlaid with rubies; 1 gold hairpin inlaid with rubies; 2 gold hairpins, inlaid with fake crystals; 1 silver pipe, wrapped in gold [58]. In Bahrain Right Banner, Inner Mongolia, the total weight of gold and silver burial objects unearthed from the tombs of Mongolian nobles in the Qing dynasty is 5,825 g. They comprise 1 gold and silver box, weighing 1,950 g, with a gold outer layer and a silver inner layer; 1 large silver box, weighing 2,400 g; 1 small silver box, weighing 1,225 g; 15 pieces of silver ornaments, weighing 250 g in total; 1 hemispherical, translucent, yellow-green opal, with a shining band in the middle [59]. In the silver ingot cellar of the Qing dynasty in Guiyang, 31 pieces of silver ingots were unearthed, with a total weight of 41.2 *jin*. There are four kinds of silver ingots: horseshoe shaped, rectangular waist shaped, square bucket and round [60].

b. Copper

17 pieces of bronze ware and 14 pieces of copper coins were unearthed from the tomb of the Qing dynasty princess in Tongyu, Jilin province. Six bronze wares were unearthed from the Qing tombs in the western suburbs of Beijing.

c. Mineral-petrology reflected in other cultural relics

The cultural relics also reveal 22 inkstone materials, such as pottery inkstone and Hongsi inkstone (one of the most precious Lu inkstones, produced in Hongsi Cave in Heishan, Yidu County, and preserved in the structure of the Ordovician Majiagou Formation) [61], redware inkstone, Songhua inkstone (produced in the Tonghua area, Jilin province, at the foot of Changbai Mountain, the source of the Songhua River, which is a microcrystalline limestone of a light green colour, composed of calcite, microcrystalline quartz and a small amount of clay minerals) [61], Nishan inkstone (one of the Lu inkstones, produced in Nishan, Qufu, which is an earthy yellow argillaceous microcrystalline limestone, hard and moist, impermeable, and which is not stained by ink) [61], etc.; turquoise, pearl, ruby, Shoushan stone (produced in Shoushan Township, Fuzhou City, a Jurassic acidic altered volcanic rock, composed of dickite, kaolinite, quartz, talc, bauxite and pyrite), amber, agate, jade, sapphire, ivory, Qingtian stone (produced in Qingtian County, Zhejiang province, which is Late Jurassic rhyolite, and is mainly composed of pyrophyllite, kaolinite, quartz mica, montmorillonite, corundum and sillimanite), *furongshi* (rose quartz, a kind of Shoushan stone), white marble, crystal, jasper, *moyu* (dark jade), pseudocrystal, and *maojingshi* (or cat's eye, a golden emerald, beryllium, aluminium oxide, hardness 8–8.5, with a cat's eye optical effect, hence the name).

In short, in terms of minerals and rocks, and differing from previous dynasties, the cultural relics of the Qing dynasty reflect a decrease in metals and an increase in gemstones, suggesting a growing interest on the part of the upper class in pursuing luxury goods.

(2) Mineral-petrology recorded in local chronicles and notes

① Mineral-petrology recorded in local chronicles

There are numerous local chronicles from the Qing dynasty, including many containing knowledge of minerals and rocks: for example, pages 16–19 of the *Mianning xianzhi* ("Chronicle of Mianning County"). Volume Five, in the Xianfeng period, records the ore body outcrop on page 15: "At the Shizishan factory, for every 50 kg of black lead refined by workers, 10 kg and 2.4 kg must be paid in as taxes, respectively. The remaining 37.6 kg are generally sold to the government (at the rate of one tael and six coins of silver per fifty kilograms) and the other half to the manufacturers. Each month, 300 to 400 kg of lead can be produced, and once a year, a summary is made and taxed and sold separately, totalling about 5,000 kg, which may increase if it is a leap year. Every year, lead needs to be transported to Dashubao Lead Station, which is managed by the local government." The output, tax collection, share, use and other items relating to lead in Mianning, Sichuan, were also described.

Volume 22 of the *Hubei tongzhi* ("General Chronicle of Hubei") compiled by Wu Xiongguang and Chen Shi provides a record of the *baotashi* (lit. pagoda stone). There are many pagoda stones in the Ordovician reddish-brown limestone at the junction of Sichuan and Hubei in the Three Gorges of the Yangzi River, and there are also one to three feet of pagoda stones in the Ordovician limestone near Yichang. Pagoda stone is also found in the brown marl of Huxi, Xingshan County, Hubei province [62].

The *Guiyang zhili zhouzhi* ("The Chronicle of Guiyang Zhili Prefecture") from the Tongzhi period, in Volume 20 "Commodities", pages 17–23, records the gold, silver, copper, iron, lead, tin, crystal and charcoal mined in Guiyang. It notes that "Guiyang is known as the place where eight kinds of treasures gather, and the so-called eight kinds of treasures refer to gold, silver, copper, iron, lead, tin, crystal, and stone charcoal. The crystal is in the mountain of Baifu. Crystal stone has six points and five colours. But most of them are white crystals. Stone charcoal is found in many mountains, and people can collect 500 kg a day, which can make a profit of one thousand in coins. Lead ore produces the most profit, so many people buy mountains one after another in order to mine for lead ore. Raw silver is black lead, which is soft in texture and a higher grade of purity can be extracted by putting it on a fire. There are many kinds of sand, and the types found on stones are called old sand, which is white in colour, and they include three kinds: *sileng* (four-sided), *sugang* and *lianyubai* (chub white). The sand in the earth is called *daosha*, the best is *ziyi* (purple cloth), followed by *jiaoba*, *yinyi* (silver skin), *mawei* (horsetail), *yulin* (fish scales), all named according to their colour or shape. There are two types like mud: the *jiaoni* will float on the surface when put into water, and *louni* will turn into a paste when put into water. These two contain the highest amount of silver, which cannot be found unless you are an experienced craftsman. When mining, if earth mud is encountered, 5,000 kg can be extracted per day; if stone sand is encountered, only 5 kg can be extracted per day. The maximum amount of lead extracted from sand is 5,000 kg, and the minimum is only a few kilograms. Silver is usually found in the hard earth: this has many varieties, including *lanjia* (blue armour) *gu*, *jiaoni* (pepper clay) *gu*, *tongshi* (bronze arrow) *gu*, and *lüdou* (green bean) *gu*. "Gu" means silver mixed with sand."

Wu Zhenfang's *Lingnan zaji* ("Miscellaneous Notes of Lingnan") described how Japan imported zinc ingots (white lead) from Hubei, China, and possibly used it to produce silver. The lead produced after refining silver was commonly known as black lead. It recorded that the origin of white lead was in central Chu (now Hubei). Traders entered Chu through Lechang. The price of white lead was three taels per picul, and when it arrived in Guangdong it was sold to foreign merchant ships for six taels per picul. The merchant ships transported this white lead to Japan, where they could extract sixteen taels of silver for every one hundred *jin*. The remainder of the lead was *wuqian* [black lead], also called Japanese lead, although it did not originate in Japan. After the silver was extracted in Japan, the Japanese lead would be shipped to China at a price of six taels per hundred *jin*. As for the method of refining, sharing this knowledge with China was not allowed, so Chinese artisans did not know how it was refined either.

Huang Benji's *Hunan fangwu zhi* ("Records of the Customs of Hunan") records: "The coal which comes from Leiyang emits no smoke, while that found in other counties produces thick smoke, which can damage clothes. People used to call it charcoal. There are two kinds of charcoal, namely lump charcoal used in Xiangtan and grey charcoal used in Changsha. Sometimes it is necessary to excavate two or three li to mine the coal, so there are many local lawsuits on account of damage to tombs."

Cao Shuqiao's *Diannan zazhi* ("Miscellaneous Notes of Southern Yunnan") provides a record of the coal in Kunming Mountain, mostly distributed in the southeast of Tengyue Prefecture, which was in the form of columns and used to replace firewood for cooking. It is recorded in the book: "The mountains in Kunming were the main source of wood coal, which became popular during the reign of the Kangxi Emperor in the Qing dynasty. Qujing also produced wood coal. People dug very far into the ground to find the coal, lifted it to the surface with a winch and sold it to the local people for their fires and cooking. The wooden coals were shaped like house beams, doors, windows, railings, etc. Some also resembled large tree roots with a textured surface. The wood coal burns with a flame different from that of other coals. In the south-eastern part of Tengkuo, the fields erupt with lightning and the soil is black. Locals call it sea, and it looks like coal but is actually earth. People use it to replace firewood. For centuries this local soil has been inexhaustible. This phenomenon is very strange. Could this also be wood coal?" [63]

Yan Ruyu's *Sansheng bianfang beilan* ("Overview of Border Defence in Three Provinces"), in Volume 9 "Mountain Goods", describes how "the ancients said that Nanshan (Shaanxi) produced gold, silver and copper, but in fact there is only iron ore, which is red and shaped like stone."

According to Min Xu's *Yueshu* ("Guangdong and Guangxi Province Records"), "Guangxi is rich in gold and tin, and they can be found in many untouched caves. Gold ore is located in sand mixed with soil. To extract it requires panning and grinding. The tin which occurs in Fuhe is of excellent quality, and there is also tin in Qingyuan. Sodium nitrate is found in Nanning."

Zhang Shu's *Qianzhong jiwen* ("A Chronicle of Central Guizhou") mentions that "there are 17 kinds of silver, including the desirable *huangyin* (yellow silver) which occurs in Sichuan: its original form resembles a bundle of leaves; raw silver looks like hard tin; *musha* silver is red and black; *qianyin* (lead silver) has a special texture. These are all silver".

Volume 15 of the *Guizhou tongzhi* ("General Chronicle of Guizhou Province") from the Qianlong period provides a record of the cinnabar produced by Wanshanchang in Tongren, which was shaped like an arrowhead, and was also known as arrowhead sand. This cinnabar, which was produced in the crevices of rocks, was the most precious and difficult to mine.

Tan Cui's *Qiannang* ("An Encyclopedia of Guizhou") noted: "The main production area for *Jinxing* Stone (gold star stone) is the Xingshi Pond in Si Zhou, and it is very difficult to mine this material. The cave is filled with water and the path inside is very rugged. The cave is extremely dark. The bottom of the pool, however, is very bright. After being polished, the quarried gold star stone emits a lovely light inside, as if the dark clouds in the sky have dispersed and the sun is shining. There is a flickering light in the middle of the stone, which comes and goes. A stone with such an appearance is a real gold star stone. This stone is suitable for making ink and will not dry out when placed in water." It also records that *Puding* stone resembles jade and produces a red ink after polishing. "This stone in Sizhou has a textured surface and when polished can be used as an ink stone. This stone can also be made into a screen, the best quality of which is produced in Yuanzhou. This stone has a green surface and a yellow interior. The yellow part of the stone can be carved into branches and the green part into leaves, just like the old pine trees. This stone can also be made into a model of a landscape pavilion, which looks very beautiful. This is a really good thing in the south of Guizhou."

Zhang Hong's *Diannan Xinyu* ("A New Encyclopaedia of Southern Yunnan") mentions that "the colour of Lügong stone (green ore) resembles the green feathers of a parrot, and its lustre is very beautiful. The best ores can be added to green pine, and for every thousand *jin*, less than ten to twenty *jin* of copper can be extracted. Vessels made from this copper are very expensive ... The best green stone is called

Lvxila (green tin wax), which contains fifty to sixty per cent copper; the second class is called Baixila (white tin wax) or Lantou Xi La (rotten head tin wax); the third class is called Zhusha Xi La (vermilion tin wax), which contains thirty to forty per cent copper; and the worst is called Niu Ban Jin, which may only be sold for about the same price as coal".

The second volume of Tan Cui's *Dianhai yuheng zhi* ("Well-balanced Records of Yunnan") drew on Zhang Hong's *Diannan xinyu*. The book also recorded the author's theory on the formation of coal which had posited in 1804 that "there are many earthquakes in Yunnan, and the fallen trees are buried underground, so coal is formed after hundreds of millions of years," which was the earliest theory in the world to propose that coal is generated by plants, 71 years earlier than the theory advanced by the German geologist Messit in 1875.

According to Xia Xiangrong and others, the stone, natural copper and gold foil are considered to be chalcopyrite ($CuFeS_2$, 34.5% copper), a common copper sulfide mineral, often found in dense masses. It has a metallic lustre, is a brilliant brass or golden yellow in colour and is brittle. The green tin wax is a pyroxene (Cu_2S, 79.8% copper), often covered with a green layer of copper carbonate, with a light lead-grey colour on the newly cut surface. The red tin wax is bornite (Cu_5FeS_4, 63.3% copper), with a copper red or bronze colour on the newly cut surface, and often with a purple hue. It is also known as purple gold tin wax. White tin wax, or grey tin wax, is also an ore with pyroxene as the main component. There are many names for tin waxes, but in general there appear to be no more than two types: one is a copper ore with pyroxene as its main component; the other is a copper ore with bornite as its main component. There is also a gunpowder crisp, or tenorite (CuO, containing 79.8% copper), an oxidised mineral, often in a shiny scaly or black earthy form [64].

According to the fifth volume of Chunyuan's *Xinjiang jilue*, "Kuche City abounds in copper, nitrate, realgar, and ammonia. Ammonia comes from the mountains in the north of the city. There are many stone caves in the mountains. In spring, summer and autumn, the caves are on fire, and people cannot approach them. In winter, when the fire is extinguished by heavy snow, people can enter the cave to mine ammonia."

In the "Ziliujing (artesian well) shuo" of the *Fushun xianzhi*, compiled in the Tongzhi period, Wu Dingli not only introduced the drilling technology of Sichuan salt wells, but also introduced the bittern, natural gas and oil produced in the salt wells.

With regard to natural gas, the *Fushun xianzhi* ("Chronicle of Fushun County") says: "When natural gas meets fire, it burns. In the beginning, people used natural gas for cooking, leading it up the hill through bamboo tubes, so that one or two dozen stoves could be used for cooking. The bamboo tube used to draw the natural gas would not be damaged. Some of the fires fuelled by natural gas were large and some were small. If a gas well is shallow, only seven to eight or three to five stoves can burn, and in some cases only one stove can burn. These fires are white in colour and have a relatively gentle flame. If the gas well is deeper, then it is the same colour as black water. The fires in deep wells are very far below the surface. They do not sting the eyes and have no smell and can be used for hundreds of fireplaces."

As for petroleum, the *Fushun xianzhi* recorded its colours, including white, cyan, yellow and black, among which the white liquid was considered best. Oil could take the place of candles, and it could also set fire to rocks on the beach to repel ships.

In Zhang Hongzhao's *Gukuang lu* ("Records of Ancient Mines"), the *Qing yitong zhi* ("Records of the Unity of the Qing Dynasty") and the mineral resources recorded in the general annals of all the provinces, prefectures, and counties are included, from which we know the mineral resources in all parts of the Qing dynasty. However, most of the content deals with the geographical distribution of minerals, and only a small part involves mineral knowledge, as in the following examples.

The *Chenzhou fuzhi* ("Chronicle of Chenzhou Prefecture") mentions that "the Mingshan stone produced in Lixi Town, which is deep purple with blue tones, quite hard in texture ... It can be carved into screens, inkstones, ink tables, paperweights, pen containers, etc., with very delicate carvings of flowers, birds, figures, mountains, water and pavilions..." "In addition, Yongshunting (today's Yongshun county) produces crystal stones and *bailingshi* of different sizes..."

In the *Lingnan congshu* ("Collectanea of Lingnan") written by Guo Chun in 1830, 27 kinds of mineral rocks are recorded, including greenstone, *taohuashi* (lit. peach-blossom marble), *yinxingshi* (lit. silver-star stone [hydrous aluminium phosphate]), *jinxingshi* (lit. golden-star stone [aventurine]), slate (can be made into inkstones), siliceous slate, pumice, gypsum, limonite, opal, azurite, purple quartz, magnetite, sulfur, mica, alum, stone, talc, calamine, cinnabar, borax, chalk, etc.

② Mineral-petrology recorded in notes

a. Aisin Gioro, Xuanye: *Kangxi jixia gewu bian*

The book discusses *shiyan* (rock salt), *shiyu* (fish fossils), Hanhai pebbles, mirabilite, petrified wood, Aba-a [a Mongolian tribe] salt, etc. The original text is as follows:

> Rock salt is produced in places where the Hui people live, and it is found in almost every crack in the mountain. Local people often fire arrows to shoot it down from dangerous places. This salt is white and bright, somewhat like crystal with a mild texture, and it will not explode when put into the fire. It can cure diseases, which is completely different from the salt in Inner China. There is also red rock salt from shallow rivers, which is edible but not officinal.
>
> There is a kind of blue-white fish stone in Kerqin, which is usually fish-shaped, as if painted with estradiol [diarsenic disulfide (As_2S_2)]. I have read the *Shuijing zhu* and *Youyang zazu* which said that Hengyang has a fish stone mountain. The stones on the mountain are fish-shaped, just like those carved on it. In addition, the *Chibei outan* ("The Collection of Wang Shizhen's Writings) said that there is a fish stone ditch in Yuxian County. Fish stone has the effect of repelling moths.
>
> Agate stones are found in the Hanhai desert, with brilliant colours, and a pure and glossy texture, just like the bright red grains in pomegranate fruits, or like the smooth white flesh in a half-open litchi fruit.
>
> In the Ming dynasty, the export of mirabilite was strictly prohibited. There were many places producing mirabilite outside the gateways of the Great Wall, such as Karka and Erut (western Mongolia), which produced a better quality than that produced in Inner China.
>
> The water is extremely cool in places such as Heilongjiang and Jilin. In rivers there are fossils made of trees, the appearance and texture of which are the same as those of rocks. Even the grain of the wood and the traces of insects are still clearly visible. Some are not completely

petrified, and half of the wood is still preserved. It is the best stone for sharpening knives and arrows. Antlers and human bones can also be turned into fossils.

There are different kinds of salt. The sea salt and well salt used in the south must be boiled and refined. Like ploughing and sowing, as in the case of cultivated land, the salt pond in Xiezhou, Shanxi province, needs to be dredged to maintain its borders, and water is drawn from pools to fill it. When the south wind blows in summer and autumn, it crystallises. Therefore, the ancients believed that sea salt and well salt depended on artificial labour, and salt from Xiezhou relied on the weather. Only in Abagai and Zhangjiakou, where livestock are raised, is there a kind of salt that comes out of the water and forms naturally without being boiled or dried by the wind. The salt varies in size, is green and black in colour, and its taste is not inferior to the salt produced in Inner China.

b. Sun Tingquan's *Yanshan zaji* ("Miscellaneous Notes on Mount Yan")

Sun Tingquan (1613–1674), with the courtesy name Bodu or Meixian, who also styled himself as Zhiting, was born in Yidu (now Yidu, Shandong province). In the 13th year of the Chongzhen reign of the Ming dynasty (1640), he gained the title of Jinshi. During the Qing dynasty, he served successively as Tuiguan (local officer in charge of penalties) of the Henan government, Head of the Ministry of Officials, and Bachelor of the Inner Academy. In the 5th year of the Kangxi reign (1666), Sun Tingquan wrote this book after collecting old anecdotes and events. The book comprises four volumes, and the fourth volume records the local minerals, like charcoal, iron, yellow lead, white alum, copperas, Zishi inkstone, and coloured glaze, as well as products made from mineral rocks. The coal exploration and mining technology recorded is the most detailed and accurate information in ancient Chinese history, which is very valuable.

c. Wu Qi: *Lingnan fengwu ji* ("A Record of the Customs of Guangdong")

This book records ten kinds of rocks in Lingnan, the first of which is Duan inkstone, saying:

Duanxi is located in Lingyang Gorge, Zhaoqing. The inkstone is produced in the old pit of Duanxi. The rock in the middle of the old pit is soft and smooth and is an excellent product. The inkstone which is the colour of sheep's liver is the best, followed by the *qinghua* (blue flecks) pattern, starling-eye pattern, flame-burned pattern, plantain leaf pattern, water ripple pattern and ancient pagoda pattern. The inkstone looks as if there is sand on its surface, but it is quite smooth to touch, and the blue and white examples are the top-grade ones. There are also differences in the blue-and-white stripes. The blue-and-white pattern in the stone can only be seen by moistening it with water; if the dots are scattered, it is fake.

Yingde stone comes from Yingde County, Shaozhou Prefecture. When superior Yingde stone is struck, it will make a sound like gold and jade. Yingde stone has the following three characteristics: it is wrinkled and folded, with gaps on the stone, and thin and as hard as iron.

Xingyan stone comes from Qixingyan (or Seven Star Cave) in Zhaoqing Prefecture, and it has a white background and a cyan pattern, similar to marble, but not as hard. Xingyan stone with landscape patterns can be used to decorate several tables, screens and beds. White Xingyan stone can also be used to powder the face after being crushed, so its powders are commonly applied by ladies in eastern Guangdong.

Lashi (chrismatite) is as yellow as wax, with different sizes and a smooth texture, and is a specialty of Lingnan.

Nenshi (lit. tender stone) comes from Nanxiong Prefecture and can be carved into the gold inside.

Huameishi (graphite) comes from Huizhou Prefecture and can be used by women for painting their eyebrows.

Graphite comes out of the stream in Shixing County, Nanxiong Prefecture, and can be used for darkening eyebrows.

Amethyst is found on Baoshan Mountain in Dongguan County, and comes in different sizes, and it is crystal clear with a purple lustre. It is of decorative use and can also be applied medically for treating infertility in women.

Danzao (lit. furnace for making elixirs) mud comes out of Ge Hong's furnace on Luofu Mountain, and is like a small pellet, red and yellow in colour, and it can be used to treat heartache or acclimatisation. It is soluble in water and will smoke when put into water.

Linglongshi (exquisite stone) comes from a cave in Shixing County, Nanxiong Prefecture, and it is as white as snow and as big as a lion. If you put one lion-shaped stone and one elephant-shaped stone into a container with vinegar, the two will be drawn together even if they are a foot apart.

d. Ji Yun: *Yuewei caotang biji* ("Notes of the Thatched Abode of Close Observations")

In Volume 15 of this book can be read:

Zhangzhou, in Fujian province, produces crystal of all kinds of colours. However, a red one has never been seen, so purple is considered the most valuable. There is another kind called gold crystal, which is completely different from yellow crystal and is the most difficult to acquire; even when occasionally found, it is only as small as cowpeas and melon seeds. There is only one in Duke Haicheng's house, resembling a three-legged toad, which can be used as a fan pendant. It looks like pure gold, crystal clear, and is a rare treasure.

Therefore, the value of goods and people's preferences are not immutable. I remember when I was young, ginseng, coral and lapis lazuli were not expensive, but now the price increases day by day. The price of turquoise and tourmaline was expensive at that time, but now they are becoming cheaper and cheaper. At that time, no one thought that Yunnan jadeite was jade; now it is considered a treasure, and its price is far higher than true jade. In the past, people liked the bright red coral which looks like pomegranate flowers, but now people like the light red coral which resembles cherries. The examples of white coral are considered to be the most precious.

In the inspection yamen, there is a Taihu stone, which is higher than the eaves and has a natural and variegated texture. It is said to be a relic from the Liao and Jin dynasties. According to my investigations, people in the Jin dynasty once demolished a strange stone in the garden of Genyue and transported it to the north. Is this stone the so-called "Qingyun Wonderful Peak" of that time? However, Dading Prefecture was the capital and that is now Daning city. The Xingzhong Prefecture of the Liao dynasty was downgraded to a county in the Jin dynasty. Consequently, the stone would not have been placed in the county's yamen, so this is not clear. It is also said that the rocks of Tuer Mountain in the capital are all from Genyue, which I saw when I was young. My residence at Hufangqiao was the former residence of Duke Weixin. On the east side of the hall, there is a stone seven or eight *chi* high, which is said to have been gifted during the reign of Emperor Yongzheng and transported from Tuer Mountain. Of all the Taihu stones in Nancheng, this one ranks highest.

e. Qu Dajun: *Guangdong xinyu* ("New Perspectives on Guangdong")

Qu Dajun (1630–1696), originally known as Shaolong, with the courtesy names Wengshan and Jiezi, and who also styled himself as Panyu, was born in Fanyu, Guangdong province, and was a scholar in the Ming dynasty. After the Qing army overcame Guangzhou, Qu Dajun became a monk, and travelled north and south. Soon after that, he abandoned Buddhism and returned to Confucianism, associating with Wei Geng, Gu Yanwu, and Li Yindu. After the pacification of the Three Feudatories in the Qing dynasty, Qu Dajun lived in seclusion and wrote books, and his poems were particularly famous. This book was written in Qu Dajun's late years. There are 28 volumes in the book. Volume 14, Volume 15 and Volume 23 record 16 kinds of mineral rocks, including stalactite, mica, salt, gold, silver, copper, iron, jade, crystal, coral, amber, glass, etc. Volume Five records 24 kinds of rocks, such as chrismatite, *wuyangshi* (lit. five-goat stone), Nantai stone, *jiuyaoshi* (lit. nine-star stone), sea stone, Duan stone, and Duanxi inkstone. In the description of these minerals and rocks, there is some knowledge about the properties of minerals and rocks. Here are a few examples:

> Shao stone: it is found 40 li north of Shaozhou, towering into the clouds, surrounded by 36 strange rocks. The stone is hollow and makes a sound when the wind blows.
>
> *Dayingshi*: this means stones on Mount Yingde. Most of these stones are vertical in shape. Each stone has many faces. The stones are not connected to each other and are independent of each other. A single stone has one or many peaks. The smaller stones are not attached to the larger ones. The larger stones are also not attached to each other. The smaller stones have many bumps on their sides. The larger stones stand tall and have no extra angles.
>
> *Xiaoyingshi*: there are many strange stones in Yingzhou. Those which form clusters on foundations are called hill peaks, and those which spread without any foundations are called stones. *Xiaoyingshi* (small Ying stone) is an archetypal representation of the landform of Mount Yingde. A fine *xiaoyingshi* should be wrinkled and transparent, thin and as hard as iron. Such stones are divided into two categories: the yang stone is exposed on the ground and the yin stone is hidden in the earth.
>
> Chrismatite is found in Lingnan. Most of the streams in Conghua, Qingyuan, Yong'an and Enping contain chrismatite. High-quality chrismatite is as yellow as jade, with a pattern of caves and peaks on its surface.
>
> Duan stone: in Huanggang Village, along the northwest side of the Lingyangxia Valley, there were more than 500 households making a living from stones. Half of them carved purple stone to make inkstones; the other half carved white stone and Jin stone to make screens, tables, plates, and other objects, generating 10,000 *liang* of gold each year. The white stone is a resource which the people from Huanggang have been able to profit from since the Song dynasty, and many foreigners come to buy it.
>
> Jin stone is the raw material for screens and tables, with a white background and cyan pattern, similar to marble, but not as hard. Pure white Jin stone comes from Qixingyan, and its surface turns black when exposed to erosion by the wind. The whitest Jin stone is used to make face powder, which is one of women's cosmetics, like graphite in Huizhou and Shixing for darkening the eyebrows.
>
> Duanxi inkstone is dark in colour and found in the Lingyangxia Valley. Since the Tang and Song dynasties, talented scholars have used Duanxi stone as inkstone. The old pit is divided into three layers. The upper layer is close to the sand, and the texture is not pure enough;

the middle layer produces the best quality, with flame lines and leaf lines on the surface; the lower layer is also inferior in quality.

Silver: there were silver mines in the mountains of Guangdong. The book *Shixing ji* ("Records of Shixing County") records that there was light at the place where Xiaoshou Mountain crumbled, which was the original silver ore, and that silver could be obtained by refinement. Legend has it that in two places, Yingde and Qingyuan, where there are silver mines in the mountains, white gas would rise up and the grass and trees would turn white when they were covered with it. When the weather was very hot there would be silver droplets, white in colour and with a pungent smell. The silver ore here is either red, white or black, some with a textured surface, and this mineral is known as Longkou (dragon's mouth). As you continue to dig down along Longkou, you can see soil that looks like burned coal, which is the *kuangmiao* (ore seedling). Keep digging down and you will see the ore. At first, there is a very small quantity of this ore, which gradually becomes more abundant and then less. Some parts of it looks like beads, and other parts like vines. This ore becomes very rare and then gradually peters out, and when it disappears people look for new deposits, until eventually the land is mined out and becomes a wasteland. The best quality of this ore is when it is soft and relatively large. Harder ores are the second best. These two types are known as *mingkuang* (bright ores). Slightly inferior are the *jiashikuang* (ores with stones in them), of which the green ones are the best and the red, black, yellow and white ones are the second best. The worst is the *shatukuang* (sandy clay ore). Once extracted, the surface suspension is removed and the remaining material is burnt with fire. The best ones are those shaped like melons; those shaped like vines are of poorer quality.

Iron: iron ore is produced in the mountains in central Guangdong. Iron ore is found where yellow water seeps out. Following this path, more deposits of iron ore may be found.

Crystal: Wuzhi Mountain in Qiongzhou prefecture is rich in crystal, which is dazzling and looks as if the sun is shining after rain. Most of the silver and yellow-purple crystals come from Minzhang; red crystal can burn with a flame; white crystal can withstand water and fire. The place where crystal is buried will have red clouds at night.

(3) Mineral-petrological literature

The works about minerals and rocks in the Qing dynasty can be roughly divided into three categories, namely, works on ceramics, minerals and petrography.

① Mineral knowledge in ceramics literature

a. Zhu Yan: *Taoshuo*

Zhu Yan, with the courtesy name Tongchuan, also known as Li Ting, was born in Haiyan. In the 31st year of the Qianlong reign (1766), he was awarded the title of Jinshi and appointed as magistrate of Fuping, Zhili, and later he served as Wu Shaoshi's aide. As a writer who was good at poetry and prose, and a painter and a scholar of antiquity, he had completed numerous works. In Jiangxi, he concentrated on the porcelain industry and wrote the 6 volumes of the *Taoshuo* ("On Porcelain"), which was the first history of ceramics in China.

Volume One describes the porcelain stone (mainly containing quartz and sericite) occurring in the mountains of Pingli and Gukou in Qimen County, Huizhou, and records the raw materials and processing methods of porcelain.

The book also mentions that "the porcelain stone is produced from the two mountains in Pingli Gukou in Qimen County, Huizhou, Jiangnan province, 200 li from the kiln factory. The porcelain stones are taken from the cavern and cut open, and inside the stones are black patterns like the antlers of a deer. The local people set up machinery powered by a stream, and they pound and wash the porcelain stones and make them into the shape of clay bricks, given the name of Baidun. The stone is pure in colour and fine in texture, and is used to make stripped objects, filled with white, blue and white, and curved patterns. There are also varieties of porcelain stones from Raozhou, such as Gaoling, Yuxiang and Yutan, all of which are mined and prepared in the same way as Baidun. They can be used for producing coarse porcelain." Here the author describes the preparation and use of glaze and glaze ash. The glaze is made of glaze stone and glaze ash, which uses calcium oxide (CaO) as the main flux, so it is called calcium glaze, traditionally called grey glaze. The main composition of glaze ash is calcium carbonate [65].

"The raw materials for making a saggar include three types of earth: black, white and red, and they are found in Lichun Village, northeast of Jingdezhen. There are also black and yellow types found in Baoshi Mountain. They will be mixed into clay and heated in the fire."

"The raw material for the blue-and-white flower design on porcelain is *qingliao* (asbolite). The superior *qingliao* comes from Shaoxing and Jinhua, Zhejiang province, and is dark yellow in colour, large and round. If it is taken to Jingdezhen and buried in a kiln for three days, it can be sold after being washed. The *qingliao* produced in Jiangxi and Guangdong provinces is light and is not fire-resistant and can only be used to paint less delicate wares." The *qingliao* mentioned here refers to cobalt ore, with the chemical composition of SiO_2, Al_2O_3, Fe_2O_3, MnO, CoO, NiO, etc. [66].

In Volume Three, the raw materials and the origin of porcelain are discussed:

> Clay is sourced from Machang Mountain, Xinzhengdu, Fuliang County. Both the clay and the black soil have blue-black patterns and white spots on the surface. The superior clay is white, dotted with golden spots.
>
> *Shimo* [a kind of rubble as small as a green bean] is produced in Hutian, and it is mixed with high-quality porcelain clay to make hard ceramic containers.
>
> Glaze clay is produced in Xinzhengdu. Changling glaze clay is bluish yellow; Yikeng glaze clay is white, and both have cypress leaf patterns. (As clay contains impurities or organic matter, they form several markings. The types of clay minerals and aqueous oxides shaped in a dynamic environment are unreliable.)
>
> Scarlet clay is fired to make red porcelain, which was lost during the Zhengde and Jiajing years. Since then, the firing method is not as good as before, only the *fanhong* (iron red) colour is achievable.
>
> During the Xuande reign, the official kiln used *xihong* gem to make the red goblet with three fish patterns.
>
> During Xuande's reign, the official kiln used cinnabar to make small pots and large bowls which are as red as the sun.
>
> Cinnabar, used at Xuan klin for small pots and large bowls, is as red as the sun in colour.
>
> The blue colour is created from *potangqing* (cobalt) found in Leping. During the reign of Jiajing, the people of Leping and the people of Jingdezhen fought with each other, resulting in heavy casualties. Since that time, *potangqing* has never appeared in any historical documents relating to Jingdezhen pottery making. *Shiziqing* (also called *wumingzi* or *youzi*) was also used to produce the colour blue. It is a type of cobalt ore. It is a weathered residual deposit

of andesite tuff in Early Tertiary red rocks. The ore is mostly scattered in flakes, edges and strips in the upper part of the weathered crust of andesite tuff. It contains 0.2–5.8% cobalt and 20–23% manganese. It is commonly used for painting porcelain in Jingdezhen. In addition, the blue copper ore used for painting is also known as *shiqing* or *daqing*. *Shiziqing* is found in Ruizhou (Shanggao County, Jiangxi province).

During the Xuande reign, the blue and white wares were coloured with Su-Boni *qing*, which was lost during the reign of Chenghua. In the Ming dynasty, Kalimantan was known in China as Boni. And Su-Boni *qing* probably refers to the cobalt ore that was produced or re-exported from the Sumatra and Boni areas.

Huiqing with cinnabar spots (a cobalt ore of foreign origin imported from Southeast Asia and West Asia for use in fine blue and white porcelain) is a colouring matter of the finest quality, closely followed by the kind with silver spots. During the reign of Zhengde, the eunuch suppressed Yunnan and obtained Huiqing from foreign countries. During the reign of Jiajing, the official kiln adopted Huiqing, which was later lost.

Black ochre occurs in Xinjian, Luling (now Ji'an, Jiangxi province) and is used for painting porcelain. This signifies ores with a reddish-brown colour that can be used as a pigment, the red earthy form of which is known as a substitute for ochre mainly used as the colour red. Black ochre is a generic term for low-grade cobalt ores or manganese clays that are rich in limonite.

b. Lan Pu: *Jingdezhen taolu* ("Catalogue of Jingdezhen Porcelain")

Volume Three recorded the minerals used in various glazes, such as:

Purple-gold glaze: this is composed of *zijin* stone (a kind of micrite), water and ash refined from pot water. Emerald glaze: this is synthesised by heating copper, water and saltpetre. Golden glaze: this is made of black lead powder and crushed ochre. Red glaze: this is made of green alum and lead powder. Purple glaze: this is composed of black lead powder and *shiziqing* (azurite). *Jiaoqing* glaze: this is composed of ash refined from pot water and *shiziqing*. Jiaolü glaze: this is made of black lead powder and copper powder. *Douyou* glaze: this is made of ash refined from *douqing* oil (green-bean oil) and loess. Pure white glaze: this is synthesised from ash refined from glaze water. *Jiaohuang* glaze: this is made of mirabilite and ochre. Bright red glaze: this is made of red copper strips and purple quartz. Bright green glaze: this is made of asbolite with glaze. *Dongqing* glaze: this is synthesised with purple-gold glaze and mixed with a small amount of asbolite. *Longquan* glaze: this is made of purple gold glaze and a small amount of asbolite. *Lujun* glaze: this is synthesized from crystallised mirabilite and glaze. Cracked glaze: this is produced in Sanbaopeng.

The minerals mentioned in the "colourants needed for pottery colours" include: lead powder, nitre, alum, black lead, ochre, *shiziqing*, *zijin* stone, and five-colour quartz.

Volume Four deals with various materials of porcelain, such as:

The quality of Jingdezhen pottery varies, depending on the raw materials used. Some are made from official clay, some from fine clay, some from medium clay, some from talc, some from glaze stone and kaolin, some from talc with white stone, some from Yugan without kaolin, some from yellow clay, some from collected slag. Different clays are used depending on the object to be made.

The porcelain clay produced in Macang, which is part of Yidong County, is the most famous; this was exhausted at the end of the Ming dynasty and then reappeared. The porcelain clay is hard, fine, white, and sweaty, and is often used to form the matrix of porcelain. The porcelain clay from some places is hard, white and sometimes oily while the clay from other places is soft and white.

The finest porcelain is made of glaze stone. This forms half of the composition of the second finest. The rough quality kind is made of water glaze which is a mixture of water and glaze ash. In the 3rd year of the Jiaqing reign, the glaze stone was discovered in Leping, Linyi, and the people of Jiangxi built factories to mine glaze stone.

Gaolin (Kaolin), which was once the name of the east of Gaoling Mountain, is a raw material for porcelain which is processed into brick-like clay blocks. At first, Wang, He, Feng and Fang families were engaged in this industry, and later practitioners in Wu Town borrowed their names.

Pingli clay and Gekou clay are both found in Qimen County. Since the emergence of Yugan clay, the use of these two for porcelain making has declined. In addition, Xiaoli soil found near the south can be mixed for use, but it is not as good as Yugan clay.

c. Wu Qian: Yangxian mingtao lu

The two volumes of the *Yangxian mingtao lu* ("Catalogue of Famous Porcelain in Yang Xian") introduced the redware and its raw materials in Yixing, Jiangsu province. The chapter "Material Selection" says:

Soft, yellow clay, which is sticky and plastic, is found in Zhaozhuang Mountain and can be used to mix various coloured soils together.

Shihuang clay, found in Zhaozhuang Mountain, is a hard rock that has not been exposed to the wind and sun, and which turns vermilion after firing.

Tianqing clay is found in Lishu and turns a dark liver colour during firing. Mixed with the *tianqing* clay is *lipi* (lit. pear skin) clay, and it turns yellowish green during firing. Light red clay turns light yellow during firing, and light yellow becomes green when fired. *Mikou* clay turns light ochre during firing. *Lipi* clay and white sand become light charcoal in colour when fired.

Laoni clay is found in Tuan Mountain and white dots appear during firing. When mixed with azure and realgar it turns a dark colour.

White clay is found in Dachao Mountain and is used to make pottery bottles, vats and drinking vessels.

d. Ao Xuanbao: *Minghu tulu* ("Illustrated Catalogue of Tea Pots")

The section "Clay Colour" in the first volume says: The colours of purple clay are rich and varied. In simple terms, the colours are red and purple, which is the true colours of the purple clay teapot, in dark or light shades. In addition, there are varieties in white, black, yellow, pear skin colour, pine flower colour, iron colour, chestnut colour, light black colour, purplish red, dark liver colour, and with gold and silver spots. Moreover, there are also records of begonia red, vermilion purple, Dingyao-kiln white, cold golden, eaglewood colour, aquamarine, peach skin colour, sunflower yellow and other colours.

The raw ore of the purple clay is layered sedimentary rock. The white mud is often distributed over the mountainside, while purple clay is found in the flat place at the foot of the mountain, and is mainly composed of illite and kaolinite, with unequal amounts of montmorillonite, quartz, mica clasts and goethite, and a slight amount of rutile, ilmenite, tourmaline, zircon, etc. [67].

② Literature on minerology

There were not many books on minerals in the past dynasties, and the same was true of the Qing dynasty. The most famous is Wu Qijun's *Diannan kuangchang tulue* ("Illustrated Account of Mining Distribution in Southern Yunnan Province") written in the 24th to the 25th year of the Daoguang reign (1844–1845). The book is divided into two parts, one is the "Diannan kuangchang yucheng tulue" and the other is the "Yunnan kuangchang gongqi tulue" ("Illustrated Account on Mining Technology in Southern Yunnan Province"). The former deals with the geographical distribution of the mines and the distances between them; the latter deals with mining technology, ore types, smelting, tools, etc. and is a monograph on mining technology. The following is a brief introduction to the mineral knowledge contained in the "Yunnan kuangchang gongqi tulue".

First, the book discusses outcrops, namely various forms of outcrops of ore deposits. The book says:

> Minerals are found inside and outcrops are found outside. As the saying goes, if there are minerals in one mountain, there will be outcrops on thousands of mountains, just like the relationship between the melon and the vine. A new mine is relatively shallow, while an old mine is buried deep in the mountains.

The relationship between outcrops and deposits is further discussed:

a. Han outcrop: its colour is dry and light, and there are no minerals.
b. Pushan outcrop: it is scattered and rootless, with a small amount of the minerals beneath it.
c. Shusheng outcrop: it is straight up without any branches, with only a small amount of minerals.
d. Mopan outcrop: it twists and turns and spreads downwards. After a few years, it will surely cause floods.
e. Kuadao outcrop: it extends obliquely and intermittently. There are varying amounts of mineral resources buried below.
f. Big outcrop: it is wide and thick extending over several *chi* and has a hard texture, which cannot be mined speedily. It occurs in a string and is more sustainable.

Secondly, the book deals with the grade and type of ore. For example:

a. There are dozens of kinds of copper ores. The purple-golden copper ores are of the highest quality. The natural copper ores are the most durable and can be used to make utensils.
b. There are dozens of silver ores, and the blackish green silver ores are superior grade. The second best is grit salt, and it can produce 1–8 *liang* of silver.
c. Black lead ore can be further classified by its markings, and it only contains a small amount of silver.
d. Copper-covered silver ore: this black ore with a frosted or shiny surface contains silver. First, iron-like minerals are extracted from it in a large furnace; then it is put into a pusher furnace to remove any impurities; finally, it enters a small furnace where the silver wrapped in the copper is extracted.

e. Silver-covered copper ore: the ore with a green or jagged surface contains copper.
f. Tin ore is refined from clay pots. It is said that there is silver inside and only Vietnamese people know how to extract it.

In the section "Diannan changkuang tulue", there is the article *Caitong liantong ji* ("Records of Copper Mining and Refining") by Shen Shu, which also describes various copper ores and which says that:

> The superior copper ores are yellowish green, bean green and dark green. The *baixila* (lit. white tin wax) is of good quality, with a white colour, high density, and lines along the edges like clusters of needles; the *youxila* (lit. oily white tin wax) is brightly coloured; the *hongxila* is red in colour; the *zijin xila* (lit. purple gold tin wax) is of a dark purple colour; the *huoyaosu* is extremely valuable, and is dark coloured, with a crisp texture and a high degree of purity; the Yazi Mine produces good quality ore, and is tiered like hillsides, and mining it is like demolishing a brick wall; the Yansha Mine produces cyan black ore, but if it is blemished with yellowish green colour then it is less desirable; the *chuanhualü* is a kind of copper ore mixed in stone of average quality; the poorest quality ore is the *songlü* (lit. pine green), which is very low in purity and can only be used as a pigment.

③ Mineral-petrography in stone catalogues

Several works of *shipu* (stone catalogues) comprising lists of inkstones, fossils, jade, seal stones and various ornamental stones reflect the level of mineral and rock knowledge in the Qing dynasty. The following is an introduction to stone catalogues.

a. Song Luo: Guaishi zan

Song Luo wrote the *Guaishi zan* ("Tribute to Unusual Stones") in the 4th year of the Kangxi reign (1665), which describes 16 ornamental stones. They are named after their shapes, and no place of origin is recorded. There are some descriptions of the properties of the stones, for example:

Yichunsheng ("beautiful and pleasant spring"): this is a kind of aragonite, yellow and white in colour, with red patterns and sharp edges on the surface.

Ziyuan fuluan ("purple mandarin duck hatching eggs"): it is very small in size; the top is light ink coloured, and the bottom is transparent purple; the surface has white fine lines resembling a mandarin duck's eyebrows. On its side is an egg-shaped pattern, tucked under the duck's wing.

Hantan qiuzao ("cold pools with autumn algae"): it resembles a squashed jujube, is white in colour, and contains multiple roots of plants.

Hongshuijin ("red marshmallow"): it is the size of a chestnut, and its pattern and colour are like brocade.

Zhuxia longyue ("moonlight in the red lantern"): it is the size of a red bean, with a round shape and a faint glow.

Guimian ("the face of a ghost"): it is the size of a finger ring, and its colours are red and white.

Yubeiye ("white pattra leaf"): its exterior is light ink coloured, and its interior is dark purple; its surface has a white pattra leaf pattern.

b. Gao Zhao: *Guanshi lu*

Gao Zhao, with the courtesy name of Yunke, who was a famous scholar in the late Ming and early Qing dynasties, was born in Houguan County, Fujian province. In the early years of the Qing dynasty, Gao Zhao returned from Jiangzuo to his hometown, just when his friend Chen Yueshan was making a fortune by excavating Shoushan stones. He recorded more than 140 beautiful Shoushan stones he saw at the home of his friends', including Chen Yueshan and Lin Daoyi, and wrote the *Guanshi lu* ("Records of Stone Viewing"), a monograph on Shoushan stone, in the 7th year of the Kangxi reign (1668). The book contains records of rock properties and rock processing technology, totalling 1,700 words. The author classified Shoushan stones into divine, wonderful and above average examples, and described each one according to its shape and colour. Gao Zhao also spoke highly of the artistic skills of Pan Zihe, Xie Yi, Yang Yuxuan and other famous stone carving artists at that time, summarising their experienced discernment and polishing skills when carving Shoushan stone.

Gao wrote that: "According to Bian Erji's *Shoushanshi ji* ("The Story of Shoushan Stone"), in the past three or four years, the profiteers have exhausted every means to dig in the mountains and mine a large quantity of minerals, and have finally sourced many high-quality Shoushan stones, including Shoushan stones like jade, amber, glass, tortoiseshell, cinnabar, agate and rhinoceros' horn. The colours and textures of the stones are different. In addition to green, yellow, red, white and black, there are also light yellow, eagle brown, butterfly pink, partridge grey and wormwood green, and some are like silk, yarn, honey, sauce, and fish scales. The colour of wormwood green used to be considered best, but today, all kinds of Shoushan stones are very precious. Shoushan stone exhibits the shapes of peaks and waves in fine lines on its surface, which are indistinct and diverse." "Shoushan stone has patterns and markings resembling water. Before carving, you should first analyse its characteristics, and then observe its patterns, so as to remove any impurities, retain the water markings, and carve the finished product. Shoushan stone is warm and moist, which means it may heat up during cutting, and it will crack because of overheating, if sawed for a long time. The best way to cut a Shoushan stone is to pour water over it slowly while sawing.

c. Mao Qiling: *Hou guanshi lu*

Mao Qiling (1623–1761), with the courtesy name Dake, who also styled himself as Chuqing, was a well-known scholar at the end of the Ming dynasty and the beginning of the Qing dynasty, who was proficient in the classics and history, and was known as Mr. Xihe in the literary world. He was born in Xiaoshan County, Zhejiang province, and at one time served as the review officer of the Imperial Academy and the compilation officer of the Ming History Museum. He was talented in the writing of rhythmical prose, poetry and the theoretical criticism of poetry, and had written numerous poems and texts throughout his life. In the 26th year of the Kangxi reign (1687), he moved to live in Fuzhou and received 49 Shoushan stones. After reading Gao Zhaoyou's *Guanshi lu*, he wrote the *Hou guanshi lu* ("Later Records of Viewing

Stones"), containing a total of 3,600 words of detailed description and evaluation of the specifications, colours, quality, and the carving of the Shoushan stone seals he himself had collected.

> Two pieces of *aiyelü* (superior wormwood green stone): each is one *cun* long, one *cun* wide, one *cun* high, and one *cun* in diameter, with one *chiniu* [Chinese dragon without horns] figure as the seal knob. Overall, the body of the seal is bright and in a graduated green colour, and its bottom looks like a western beer bottle. Put together, the two *chiniu* seem to turn their heads and relax their feet, wriggling their way through the clear, green waves.
>
> One *yangzhi* suet jade: this two-and-a-half *cun* high, two *cun* wide, and one *cun* long seal has a pearly white and lustrous knob. The jade is gentle and pure white, as its name "suet" suggests.
>
> One *geyansha* (pigeon-eye stone): this is in a fine material from the old pit and is two *cun* high, one *cun* wide and one *cun* long. The seal knob is formed by an apotropaic item. The entire body of the seal is litchi red with sparkly white curves. In the white there are cinnabar beads standing out, hence the name "pigeon-eye".

d. Gao Zhao: *Duanxi yanshi kao* ("Records of Duanxi Inkstone")

This text describes the characteristics of the Duanxi inkstones from eleven quarries. Most of the content is consistent with Qu Dajun's *Guangdong xinyu.* In the winter of the 26th year of the Kangxi reign (1687), while travelling in Duanzhou, Gao Zhao consulted the stone workers about the difference in quarries thirty years ago compared with nowadays. Then, he met with his fellow scholars Liang Yaoting, Qu Dajun, Chen Yuanxiao and Wang Puyi in Songtai, one after another, to discuss how to distinguish between stones. Gao even made his way to visit three experts in the field of inkstones in order to verify his theories. He argued that once the stones were cracked out from the mountain, its vitality would gradually diminish and within less than a decade, this mountain would turn into a valley and be nowhere to be found.

e. Shen Xin: *Guaishi lu*

Shen Xin, a scholar in Yongzheng's reign with the courtesy name Fang Zhong, was born in Renhe, and was a follower of Zha Shenxing in his early years. Shen Xin's poems are in the style of Zha Shenxing, and he wrote the *Gushishanfang shiji* ("Collection of Poems from Shen Xin's Study in Mountain") and *Guaishi lu.*

The *Guaishi lu,* which was completed in the 14th year of the Qianlong reign (1749), describes the origin, properties, colours and uses of 21 kinds of stones, including *yunyu* (hidden jade) stone, Jinque (gold bird) stone, *zijin* (purple gold) stone, *hongsi* stone, peony stone, plum blossom stone, bamboo leaf stone, Laoshan stone, Tuoji stone, Tanziwo stone, Beihai stone, turquoise stone, mirror stone, fish stone, phoenix stone, coloured stone, sea stone, fine white stone, aragonite, stone powders, horse-teeth stone. For example:

> *Yunyu* stone: found in the Zi River, Yidu County, it is dark blue in colour and hard; it was once used to make inkstones.
>
> Jinque stone: found in Jinque Mountain, Yidu County, it is cyan in colour and hard, and is suitable for making inkstones.

Zijin stone: found in Yishan Mountain, Linqu County, it is purple in colour and hard, and is very good for inkstones.

Hongsi stone: found in Heishan Mountain, Linqu County, it is a very hard material. According to the *Inkstone Record* by Tang Xun, the colour of *hongsi* stone is red intermixed with yellow, which can be divided into red patterns with a yellow background and yellow patterns with a red background.

Peony stone: found in Yalü Mountain, Yexian County, it is harder than Lai jade, slightly red and extremely delicate; it can be used to make utensils.

Plum blossom stone: found in Yalü Mountain, Yexian County, it has a texture similar to that of peony stone; its colour is green intermixed with white.

Bamboo leaf stone: found in Daze Mountain, Yexian County, this stone is white in colour and scores high in hardness, with bamboo leaf veins on the surface. But its texture is rough and not deserving of appreciation.

Laoshan stone: found on the seashore at the foot of Lao Mountain, Jimo County, this stone is very hard, and its colour is like that of sacrificial vessels in the Qin and Han dynasties, with white lines. Smaller stones can be made into inkstones, and larger ones can be used to make handicrafts for indoor display.

Tuoji stone: found on Tuoji Island, Penglai County (now Changdao County), it is blue-black in colour and hard in texture, and is suitable for making inkstones. The Tuoji stone with golden spots and white wavy patterns is the rarest.

Tanziwo stone: found near Danya Mountain, Penglai County, it is as small as gorgon fruit, with five colours and a gentle texture.

Beihai stone: this is found in the islands in Penglai County, where the waves are wild and few people tread.

Turquoise stone: found on Dazhu Island in the sea of Penglai County, it is slightly yellow in colour, hard in texture but light in weight.

Mirror stone: found on Mo Island, Penglai County, it is dark yellow and its surface is bright like a mirror. Today, there is a stone tablet near Penglai Pavilion which is made of mirror stone.

Fish stone: found in the volcano in Laiyang County with patterns of fish and water plants, this stone has a colour like *Patrinia scaniosaefolia.* After polishing, it can be embedded in screens and tables.

Phoenix stone: this stone is found in Phoenix Mountain, Laiyang County. Because it looks like jade, people call it phoenix jade too. It is slightly rough, and its colour is light yellow or white; it can be used to make inkstones, paperweights, and seals, etc.

Coloured stone: found in Wulong Mountain, Laiyang County, this stone has a very fine texture, and its colours are diverse and beautiful, including blue, yellow and white. It can be used to make seals which are comparable to Qingtian stone.

Sea stone: found on the island in Laiyang County, it is dark coloured, with peaks and caves, making it suitable for creating a rockery or potted landscape.

Fine White stone: found in the sandy coastal soil of Wendeng County, it is white in colour and as small as a gorgon fruit.

Aragonite: found on Qinghua Island in Rongcheng County, it is hard, rich in colour, with a gorgeous pattern, and it looks like different kinds of things. When placed in a white porcelain basin with clear water, it is dazzling and thus a good interior decoration.

Stone powders: these have been found in Wei County since the Tang dynasty. People have been crushing and washing the stones in Wei River and burn them into inkstones.

> Horse-teeth stone: found in Black Mountain, Boshan County, it is calcined as a filamentous raw material to produce chess pieces, wires if burned with chemicals, and whistling toys, all belonging to glass wares.

f. Zhu Jiuding: *Shipu*

Zhu Jiuding, with the courtesy names Junnan and Tiean, was born in Qiantang, and was the author of the *Zhu tie'an ji* ("Collected Works by Zhu Tie'an") and *Yueqing ji* ("Collected Works from Yueqing"). He had loved stones since childhood, and then he acquired many strange stones in Zhenwu Lake, Zhongjiang County, so he wrote the *Shipu* ("Stone Catalogue") [68]. This work records 20 kinds of stones, which are named after their shape or place name, and it describes the forms, nature, colours and sizes of these stones. Here are some examples:

> Fishing Alone in Surging Waves: the shape of this stone is like a tangerine; it is partly white and partly black in colour, and the white part contains pine leaf patterns. It seems as if there is a person sitting alone fishing in a deep mountain, an impression which is very lifelike.
>
> Wave and Necklace: this stone is three *cun* long and shaped like a flute. The front of the stone is white at the top and slightly red at the bottom, like the beads on a necklace. The back of the stone looks like waves rising from the sea.
>
> Immortal Playing with a Dragon: this stone is 1.2 *cun* long, with white patterns on a black background. The pattern on the front is like lotus leaves and the pattern on the back is like an immortal surrounded by two dragons.
>
> Brocade: this stone is like brocade, colourful, and with fine lines in white and yellow, like silk threads. Its quality is comparable to jade and it has a circumference of more than half a *cun*.
>
> Pine Shadow: this stone is about 1.5 *cun* long and 0.9 cun wide. It is grey and white, and its texture is bright and gentle. The surface pattern looks like the shadow of pine trees beside a stream in late autumn and rippling waves of water.

In addition, there are Flower in the Mirror, Mountain God, Dragon Egg, Black Bamboo, Auspicious Cloud, Flute, etc.

g. Gu Yingtai: *Bowu yaolan*

Gu Yingtai, with the courtesy name Gengyu, who was born in Fengrun, was a Jinshi during the Shunzhi reign, and served as the Deputy Education Officer of Zhejiang province. He was fond of reading widely, and was competent at writing articles, with works such as the *Mingshi jishi benmo* ("Historical Events from the Ming Period in Their Entirety"), *Zhuyitang ji* ("Collected Works from Zhuyi Hall"), *Bowu yaolan,* and so on.

The *Bowu yaolan* was written from the first year to the 18th year of the Shunzhi reign (1644–1661). It contains 12 volumes, the third of which is about gold, including its origins, types and the identification method; Volume Four is about silver, and covers the origin, types, grades and the identification method; Volume Six is about gems, covering the origin, types and identification methods of gemstones; Volume Seven is about jade, including the origin, types and identification methods of jade; Volume Eight is about agate, coral, crystal, amber, and beeswax. In terms of amber, it discusses the origin, composition, colour, formation reasons, quality, and identification method of amber throughout the world, so it can be read as a monograph on amber.

In the sixth volume on gemstones, in addition to the origins of foreign gemstones, there are also records of rubies produced in Yunnan and Tibet in China. For example, "In the Yongle period of the Ming dynasty, there was a huge Yunnan ruby, weighing 155 g, dark red, and crystal clear. At that time, it was valued at 3,000 *liang* of silver. Later, the largest ruby only weighed 10 g. The price depends on the colour." "The ruby found in Baojing, Yunnan, is bright red with a yellow-black colour with broken lines inside." "The topaz found in Baojing, Yunnan, is as yellow as golden amber, crystal clear, and small in size." "The sapphire found in Baojing, Yunnan, is as green as turquoise, and a large stone weighs up to 50 g."

The tenth volume on stone describes 30 kinds of strange stones, including Lingbi stone, Taihu stone, Kunshan stone, Ying stone, Hukou stone, Yongkang stone, Yongzhou stone, Bianzhou stone, Quanzhou stone, Pingxiang stone, Xiukou stone, Songhua stone, Liuhe stone, Baizi agate, Baohua stone, Duanzhou stone, Wuyuan stone, Zhaoxin stone, Fenghua stone, marble, peach blossom stone, *hongsi* stone, *wuwei* stone, Bodhisattva stone, snow wave stone, *guo* stone, Qiuchi stone, Qingxi stone, Xiangyang stone, Jinling stone, etc. The volume not only details the origins, but also the shapes, colours, sizes, hardness, sounds when struck, grains, purposes, prices, counterfeit identification, etc.

h. Ruan Yuan: *Enping chakeng yanshi ji* ("Records of Inkstone in Chakeng Village, Enping County") [69]

This work explains that Chakeng stone (Enping stone) was found in a mountain more than 40 *li* south of Enping County, Lingnan. At the beginning of the Jiaqing, the residents on the mountain began to dig up Chakeng stone and brought it to Duanzhou, so the artisans in Duanzhou started to mine Chakeng stone, which was used to counterfeit Duan stone. The old quarries in Duanzhou became exhausted and closed, so Duan stones were being sourced from new quarries at the time of writing, including those marked by patterns of *yunao* (fish brain), *qinghua* (blue flecks), *huona* (iron-seared) and bird's eyes, similar to those from the old quarry. "Although Enping stone is inferior to the Duan stone mined in the old quarry, its ink-producing capacity is better than that of Duan stone from the new one. As Duan stone has no coating, inkstone workers must whittle it to form shapes. Enping stone, however, is naturally found in the shape of ovals, triangles and other forms and is one of the thousands of pieces of stone (mixed with soil) which form a large rock. Therefore, it is easier to obtain smaller pieces of Enping stone. Some weigh tens of *jin*, others weigh tens of *liang*. Unlike Duan stone, Enping stone has a coloured coating, on which different patterns may appear, such as yellow dragons, red pomegranate, green plantain leaves, cobwebs, pine bark and tiger skin spots. Once opened, the interior of an Enping stone may look entirely different from Duan stone. Consequently, inkstone workers have had to hold onto the similar ones in order to substitute them for Duan stone, so Enping stone has not been known about for more than 20 years."

i. Cao Rong: *Yanlu* ("Inkstone Catalogue") [70]

Cao Rong, with the courtesy names Qiuyue and Jiegong, who also styled himself as Duipu, was a native of Xiushui. He was awarded the title of Jinshi during the Chongzhen period of the Ming dynasty and served as an imperial censor. At the beginning of the Shunzhi reign, Cao Rong was granted an initial post, and later promoted to the position of Guangdong commissioner before being demoted to Yanghedao in Shanxi. He was talented in writing poems and had similarities to Gong Dingzhi, with works such as the *Chongzhen wushi zaixiang zhuan* ("Biography of the Fifty Prime Ministers of Chongzhen"), *Liuyu shiji* ("Liu Yu's Stories"), *Jinshi biao* ("Table of Metal and Stone"), *Juanpu shizhi ji* ("Records of Plants and Vegetables"), *Yueyou cao* ("Travel Notes of Guangdong Province"), *Jingcitang shiji* ("Collection of Jingcitang Poetry"), *Yanlu*, etc.

The *Yanlu*, which is devoted to Duanxi inkstone, is divided into seven parts: mountains and rivers, divinity, mining, categories, species, identification and caution. As regards the nature of Duanxi stone, he commented: "Duanxi stone occurs in the water and never sees the wind and sun throughout the year, so its texture is gentle and moist, similar to the feel of jade, and it makes the sound of gold when struck, and is very suitable for making inkstones. Pure purple Duanxi stone is very rare and hard to buy."

j. Chen Ling: *Duanshi ni* ("Records of Duan Stone")

Chen Ling, born in the middle years of the Kangxi, was originally from Fujian, and later settled in Haiyan, Zhejiang province. He disdained an official career, read widely, and was proficient in geography and medicine. In his later years, he was very interested in Taoism and claimed to be a Qingyang Taoist. In the 15th year of the Qianlong reign (1750), the *Duanshi ni* was completed. This paper not only discusses the rock structure of Duanxi stone, but also expounds on its types and properties. It can be read that:

> Sandong (three-cave) stone has three layers, each several *chi* deep. The upper layer is just under the sand on the mountain with traces of moth-eaten texture, slightly hard, pure purple, moist and bright, and this layer is called capstone; the middle layer is fine and in graduated shades of cyan, called waist stone; the lower layer is soft and tender, and slightly white in colour, and is called foot stone. The middle and lower layers have white plantain leaf and flame-burned patterns, which are crystal clear. The top layer is extremely coarse and dry, and cannot be used to make inkstone; the bottom layer is sand slate, too delicate to use to make the ink bright and thick.

A few examples are given illustrating the types and properties of Duanxi stone:

- Plantain leaf white mixed with *qinghua* (blue flecks): its texture is extremely delicate, and its colour is off-white.
- Jade-like purple mixed with *qinghua*: its texture is delicate and its colour is extremely purple.
- Light purple mixed with *qinghua*: its texture is fine and moist, and its colour is light purple.

– Black mixed with *qinghua*: its texture is very fine, gentle and jade-like, and it is dark blue and grey in colour.

k. Tang Bingjun: *Wenfang sikao tu shuo* ("Illustrated Investigations of the Study Room")

This book was completed in the 40th year of the Qianlong (1775). When referring to Duan stone in Volume Two, the author explained its rock structure:

> All stones have three layers: the upper layer is slightly rough, in the middle layer round spots appear, and the lower layer is fragile. The most precious Duanxi stone is the type with veins called *qinghua*, which is very exquisite and marked by minute and faint blue flecks. The next best example is the type with patches of larger flecks, but only a superfine texture reveals *qinghua*. Other superior types show cinnabar or emerald markings. The lowest layer and the top layer are relatively cheap, only the middle layer is of high purity, and is the very essence of the stone.

As regards the variety of Duan stone, the author commented:

> Stones mined from different quarries vary in quality, softness, dryness, thickness, colour and markings. Zi stone is black in colour, fine as jade, and has circles on the surface. Within a circle there are three to six layers of halo, and the black spot in its centre is like the pupil of an eye. At times, six or seven rings are interconnected, resembling irregular stars in the sky. When struck, the stone will make a clear sound, and there is no sound when grinding ink on it. Zi stone had become rare since the Southern Tang dynasty and supplies became exhausted during the Qingli reign of the Song dynasty.

In terms of lithology, he stated that: "If the stone is dry, the colour is towards yellow, brown, or grey, and if it is moist, the colour is blue or purple. The flame-burned markings suggest the hardest part of the stone, and it looks like coagulated blood and, therefore, it is red, purple or black in colour. Where the plantain leaf white colour appears signifies the gentlest part of the stone, and the white part looks pure without any mottling."

Apart from Duan stone, he further described inkstones produced in other places. For example:

– Beijing: in Meishan Mountain, Yanji, there is *wujin* inkstone which is black in colour mixed with golden stars (spots) on the surface, a very valuable type.
– She inkstone: this is found in Longwei Mountain, Wuyuan, Huizhou, and Lianwu River, Shanxi; it is hard and bright, and suitable for making inkstones. There are five grades, of which the one with golden spots is the most precious.
– Lingyan stone in Suzhou can also be used as inkstones. There are two colours: light green and eel yellow.
– Changshan County and Kaihua County in Quzhou, Zhejiang province, produce black stone, inferior to She inkstone in terms of hardness and brightness, but it is of a decent quality.
– Datuo stone is found in Guizhou County, Jingzhou Prefecture, Huguang. It is blue-black with spots and rough patterns. Locals called the river Tuo, so Datuo stone mostly appears in rivers and is only used by people near rivers.
– There is a kind of stone from Yuanzhou, Chenzhou prefecture, which is dark in colour, rough and dry in quality, and some examples have small circles on them.

- Lixi stone, found at the border of Changde and Chenzhou, has a light cyan surface, and is deep purple and red inside, and some stones have golden veins. A fine Lixi stone is as bright as a mirror when used for a long time.
- Changsha green inkstone, also known as Tao stone, is often the exterior of Li stone. It is moist and bright, but it does not produce ink.
- Guwa inkstone in Xiangzhou, Henan, is rough and dry, and every example produces ink.
- Zijin inkstone, produced in Qingzhou Prefecture, Shandong, is too rough to use to make the ink thick and glossy.
- Hongsi inkstone, also from Qingzhou, needs a large amount of water when used for grinding ink.
- Horn stone in Jiangzhou, Shanxi province, is as white as ox horns and has wavy patterns. But it is too smooth to grind ink on and is generally used to grind pills.
- Green stone in Tao River in Lintao Prefecture, Shaanxi province, is green in colour, as moist as jade, and comparable to Duan stone. This stone is produced in the deep water of the river and is difficult to obtain.
- The Wanzhou cliffs, in Qiongzhou Prefecture, Guangdong province, produce *jinxing* (golden star) stone, which is black and as fine as jade, and is used as inkstone. The texture is not as ideal as that of Duan stone. If it is wet with water, golden spots appear. *Jinxing* inkstones can produce ink sustainably and do not become dry, so they are very valuable.

The fourth volume introduces the varieties and properties of jade and other jade-like stones, saying:

> In the western region of Khotan, there is five-colour jade which cannot be cut by a sharp knife. It is warm and moist and feels like spring water. When considering jade wares, the white colour is the most valuable, and yellow and blue varieties are also expensive. Pearly white is the most desirable, whilst the shades of white like rice soup is cold. Items in snowflake or oily white colour are all inferior. *Xueqin*, or the blood-red colour, represents the highest quality.
>
> *Caiyu* (lit. vegetable-like jade), which is neither blue nor green, but like the colour of vegetable leaves, is the poorest quality jade.
>
> *Guanziyu* (lit. jar jade), which is crisp and not moist, is made by the northern people by burning stones with chemicals in jars. Other types made from burning include Qingzhou *liaodeng* and *liaojiaozi*.
>
> There are many sources of crystal, but Japan is the leading producer, and the best crystal is blemish free and has no striking marks. In addition, there exists a synthetic pseudo-crystal which is produced through the decoction of medicinal herbs and which contains bubbles. Also known as *xiaozi* (nitrate), it is dark blue in colour and not sufficiently transparent.
>
> There are scores of jade-like stones, and more than ten kinds which are inferior in quality to jade. For example, Mao Mountain stone from Jurong, Jiangsu province, is white and shiny. Some varieties have water veins or rice grain patterns. High-quality Maoshan stone is similar to genuine jade, which is hard and cannot be scraped with a knife. The stones found in Laizhou, Shandong province, are green or white. The white ones are better than the green ones, and the top-quality ones are glossy. White ones containing yellow impurities and green ones containing a light greyish colour are not so desirable.

> Lingyan Mountain gemstone: this comes from Lingyan Mountain in Liuhe County, south of the Yangzi River. When it rains heavily, the stones are washed into the stream from the mountain. Residents living nearby often pick them up after the rain. The ones which shine under the sun will be collected and sold to the rich. The bright red ones are the best quality. This kind of stone is harder than jade, glossier than pearl, clearer than crystal, and more dazzling than agate.

l. Ji Nan: *Shiyin yantan* ("Record of Inkstones in Shiyin")

Ji Nan, with the courtesy name Shouqiao, who also styled himself as Shiyin Daoren, was a native of Xiushui and served as an Anji Inspector. He indulged in writing and was a talented painter, and he did so well in depicting red plum blossom that people even called him Mr. Red Plum Blossom. He was the author of the *Yiyu caotang gao* ("Papers from Yiyu Cottage"), *Duanxi yankeng kao* ("Records of Duanxi Inkstone Pit"), and *Shiyin yantan*, etc. The *Shiyin yantan* was completed during the reign of Jiaqing.

The *Shiyin yantan* covers a total of 12 topics: tracing the sources, new and old pits, identification, colours, circular patterns, *qinghua* (blue flecks), *jiaoyebai* (white plantain leaf), stone markings, *sandie* (three-layered) stone, sounds, rubbing ink, and written records. Here, the nature of its discussion is presented by taking the "sound" as an example.

> On sound: a stone in the water with a tender texture has a clear and melodious sound, and a stone as tender as mud makes no sound. Su Dongpo called the former "chime stone", and Mi Shi believed that stones from lower rock are finer and sound clear when struck. Only old stone sounds sonorous and powerful, but it cannot be used to grind ink. Some posited that the sound of knocking and grinding is one of the criteria for judging the authenticity of stones in water, yet this may not be entirely correct.

m. Li Zhaoluo: *Duanxi yankeng ji* ("Records of Duanxi Inkstone Pit") (transcript)

Li Zhaoluo (1769–1841) was born in Yanghu (now Wujin), Jiangsu province. He had the courtesy name Shenqi and was later known as Yangyi Laoren. He was committed to the study of historiography, comprehensive statutes and to an extensive examination of documents. In the 10th year of the Jiaqing reign (1805), he was awarded the title of Jinshi, and was appointed as magistrate of Fengtai. Later, he worked in Guangzhou and Yangzhou, and lectured at Jiyang Academy in Jiangyin in his later years. His main works include: *Lidai yudi yange tu* ("Map of the Evolution of Administrative Geography in the Past Dynasties"), *Daqing yitong yudi quantu* ("Complete Map of the Unified Lands of the Great Qing"), *Lidai dilizhi yunbian jinshi* ("The Compilation and Interpretation of the Geographical Records in the Past Dynasties"), *Qing dili yunbian* ("Qing Dynasty Geographical Compilation"), *Fengtaixian zhi* ("Chronicle of Fengtai County"), *Yangyizhai wenji* ("Collected Works of the Yangyizhai Study"), and *Duanxi yankeng ji*, etc.

The *Duanxi yankeng ji* was completed in 1821 and records the various quarries and pits, the mining technology, mining history, sizes of quarries, and the characteristics and colours of Duanxi inkstone. He wrote:

> When quarrying, you must observe the bright and colourful stone veins and then dig down... The pits opened before the Song dynasty have no stones nowadays. Very occasionally one can find red and purple stones, but they do not produce ink any more. Stones in Daxi Cave have different qualities depending on the layer from which they are sourced. The ones from the top layer are all good except that the ink they produce becomes dry faster. The ones from the bottom layer tend to have water markings and they are porous with thorny bits mixed in. The middle layer is the very essence of the stone and the ones with patterns of *qinghua* and *jiaobai* stand out... In a nutshell, stones from Daxi Cave are of good quality, fine and transparent, suitable for making inkstones. When struck, they make a low sound, and when exposed in the sun, they cast no shadow. Stones from no other quarry outperform those from Daxi Cave.

n. Wu Lanxiu: *Duanxi yanshi*

Wu Lanxiu, with the courtesy name Shihua, was a native of Jiaying. He was awarded the title of Jinshi during the Jiaqing reign and served as Xinyi Inspector. Wu had many collections of books and he named his study the "Guardian of Scriptures". He was the author of the *Nanhan ji* ("Annals of the Southern Han Dynasty"), *Songshi: Dilizhi buzheng* ("History of the Song Dynasty: Supplement and Correction of Geographical Records"), *Duanxi yanshi* ("History of Duanxi Inkstone") and others.

The three volumes of *Duanxi yanshi* were completed in the 14th year of the Daoguang reign (1834).

Volume One is a picture of an inkstone quarry, which is unique to this book. The reader can understand the situation at a glance.

Volume Two is the main part of this book, and covers raw stones, stone products, stone defects, stone colours, stone sounds, inkstone workers, inkstone styles, the price of inkstones, inkstone use, inkstone maintenance, inkstone collections, inkstone inscription, etc. The mineral and rock knowledge associated with inkstones is also covered in this volume.

Volume Three contains tribute inkstones, the opening of pits, anecdotes, etc.

Some of the statements in the book are based on the views of others, such as *Baoyantang: Inkstone Evaluation* by He Chuanyao, *Yanlin shiyi* ("Supplement to Inkstones) by Shi Runzhang, *Duanxi yanpu ji* ("Records of Duanxi Inkstone Genealogy") by Yuan Shu, *Linghai jianwen* ("The Story of Guangdong and Guangxi Regions") by Qian Yikai, and other documents. The author respected the achievements and intellectual property rights of others, so he marked his own point of view with "from Lanxiu". Here are some examples of "fromhor Lanxiu" to illustrate the author's views.

"From Lanxiu" one: a stone that makes the sound of wood when struck is the highest grade, and a stone that makes the sound of gold and tiles is the lowest grade. The stones in the old pit all sound like wood. Therefore, if the stone is moist, it produces a deep sound, and if the stone is dry, it sounds ethereal.

"From Lanxiu" two: the stone containing jade is found on the side of the mountain, not on the flat ground. It should be mined from the broad, flat place, where the essence of the stone is located.

The above discussion reveals that scholars and officials in the Qing dynasty were very interested in the study of inkstones, and their works were more numerous and detailed than those of previous dynasties. Some views were more creative, and the recording of the origins of inkstones was wider than before. It can be said that the Qing people attained a new level in the study of inkstones building on the foundation laid by their predecessors.

9.9 Introduction and Assimilation of Modern Western Geoscience

After the Opium War, China became a semi-colonial and semi-feudal society and was oppressed by imperialist powers. As a means of resisting the invasion and oppression of imperialism, numerous insightful people advocated "learning from the advanced technologies in the West in order to fight against the invasion by the Western powers" (*Haiguo tuzhi* by Wei Yuan). Following this line of reasoning, Chinese people took the initiative of introducing Western science and technology. Modern geography and geology were introduced into China as a result of this trend.

9.9.1 *Modern Meteorology*

The Russians carried out intermittent meteorological observation over a long period in Beijing over a long period, from the 21st year of the Daoguang reign (1841) until the 9th year of the Guangxu reign (1883). In the 11th year of the Tongzhi reign (1872), the French Catholic Church set up an observatory in Xujiahui, Shanghai, to observe atmospheric pressure, temperature, humidity, wind, evaporation, rainfall, ground temperature, grass temperature, clouds and other topics, resulting in a total of 78 years of records. In the 24th year of the Guangxu reign (1898), Germany forcibly rented Jiaozhou Bay and established a weather station which took measurements three times a day until it was occupied by Japan in 1914. These meteorological observations served imperialist aggression against China.

During the Self-Strengthening Movement (from the 1960s to the 1990s), Chinese scholars cooperated with foreigners and translated some contemporary meteorological works. For example, in 1877, the Jiangnan Machinery Manufacturing Bureau published the four volumes of the *Cehou congtan* ("A Series of Discussions on Weather Measurement"), translated orally by the American Carl T. Kreyer (who had come to China in 1866) and transcribed by Hua Hengfang. This book was translated from "Meteorology" in the *Encyclopedia Britannica*, discussing general meteorological principles, meteorological elements and their formation, general calculation methods, air moisture content and atmospheric optical phenomena. In 1880, the *Cehou zhuqi shuo / Ce hou qi shuo* ("Records of Meteorological Instruments"), also known as *Cehou qishuo*, was orally translated by John Fryer (1839–1928, an

Englishman, arrived in 1861) and transcribed by Jiang Heng on the basis of *On Meteorological Instruments,* published in London in 1864, which discussed various meteorological measuring instruments and their principles and functions, divided into seven categories, with 74 illustrations. Hua Hengfang and Jin Kaili jointly translated the *Yufeng yaoshu* ("Important Methods for Avoiding Typhoons"), which introduced the general nature of typhoons and the rules governing them, as well as the methods navigators could use to avoid typhoons, and which had practical value. The first volume of the *Qixue congtan* ("A Series of Discussions on Pneumatic Instruments"), jointly translated by Hua Hengfang and John Fryer, deals with the manufacturing history, methods, principles, various structural forms and advantages and disadvantages of mercury barometers and thermometers; the second volume addresses the structure, method, principle, and measurements and calculations of empty box barometers. In addition, in 1880, the *Zaoshibiao shuo* ("Treatise on the Dampness Gauge") was jointly translated by John Fryer and Xu Shou and the *Feng yubiao shuo* ("Treatise on the Wind and Rain Gauge") was jointly translated by John Fryer and Hua Hengfang, but they were not published [71]. These works had practical value, but had not been popularised at that time, so had little impact on society.

9.9.2 Modern Cartography

From the 20th year of the Daoguang reign (1840) to the 2nd year of Xianfeng's reign (1852), the first atlas of the world employing the graticule system in China, *Haiguo tuzhi,* appeared incrementally. During the reign of Tongzhi (1862–1874), Hu Linyi drew the *Daqing yitong yutu* ("Map of Unified Realms of the Great Qing") based on the *Huangyu quanlan tu* ("Complete Atlas of the Empire"), which was in the collection of the treasury of the imperial palace, meaning the public could indirectly access the publication of the mapping achievements of the early Qing dynasty. However, Hu Linyi combined the latitude and longitude system with the square grid method and did not fully apply the mapping technology of the graticule system.

In the 24th year of the Daoguang reign (1844), Zou Boqi (1819–1869), the pioneer of modern Chinese science, drew the large volume of *Huangyu quantu,* using the system of latitude and longitude, which was published in the 13th year of the Tongzhi reign (1874). During the Tongzhi period, Zou Boqi also undertook the mapping of Guangdong. In terms of drawing method, the ellipse drawing method was proposed. He believed that the two poles of the earth were flat and the radius at the pole was smaller than the radius at the equator. Therefore, the influence of the elliptical shape of the Earth should be taken into account in the calculations. This view was very prescient. He also wrote the manuscript of the *Celiang beiyao* ("Measurement Essentials"), but it was not published. In addition to the description of measurement methods, this draft further documented the manufacturing of measuring instruments, including the barometric altimeter that was used to "gauge the height of a mountain" and the spirit level, mercury level and others [72].

In 1876, Jiangnan Machinery Manufacturing Bureau published the *Cedi huitu* ("Geodetic Mapping"), translated jointly by John Fryer and Xu Shou, and whose original author was Fuluma of the United Kingdom [73]. Xu Jianyin translated the *Cedi jiefa* ("Geodesic Methods"), which was presented in the 22nd year of the Guangxu reign (1896) and preserved from then by order of the emperor. As the original work was never published, the specific content is unknown.

9.9.3 Modern Geography

In the 18th year of the Daoguang reign (1838), Lin Zexu (1785–1850) organised the translation of the Encyclopaedia of Geography, a Description of the Earth, Physical, Statistical, Civil & Political (often abbreviated to the *Great Book of World Geography*) written by Hugh Murray, who was Scottish. The Chinese version *Sizhou zhi* ("Encyclopedia of Four Continents"), containing over 8700 words, is the earliest translation of a book of world geography in China. Later, Wei Yuan incorporated the content of the *Sizhou zhi* into the *Haiguo tuzhi.*

In the 27th year of the Daoguang reign (1847), Jose Martins-Marquez, who was Portuguese, published the ten volumes of the *Xinshi dili beikao* ("New Interpretation of the Encyclopedia of Geography"), also known as the *Waiguo dili beikao* ("Encyclopedia of Foreign Geography") in Chinese. Volume One to Volume Three cover general theories, equivalent to natural geography, and Volume Four and the subsequent volumes introduce the geographical locations and products of the continents and countries in the world.

In the 3rd year of the Xianfeng reign (1853), Dr William Muirhead (1822–1900), a Scottish geographer who had come to China around 1846, published the two sections of the *Dili quanzhi* ("Universal Geography") in Chinese. The first section is on world geography, containing volumes on different continents, and in the second section, apart from Volume One which focuses on geology, the rest of the volumes cover various branches of physical geography: Volume Two is on geomorphology, Volume Three is on hydrology, Volume Four and Volume Five are on meteorology and climatology, Volume Six is on plant geography, Volume Seven is on animal geography, Volume Eight is on population geography, Volume Nine is on mathematical geography, and Volume Ten is on the history of geography. The scope of the discussion is extensive. In addition, in 1857, Dr Muirhead launched the journal *Liuhe congtan* ("Shanghae Serial"), in which he published *On the Situation of the Earth, On Water and Land Boundaries, On Islands, On Mountain Plateaus, On Earthquakes and Volcanoes, On Plains, On the Tidal Advection of Waves, On Rivers and Lakes*, etc. He also edited *Dayingguo zhi, Tianwen dili, Zhishi wumen*, etc.

In 1900, the Jiangnan Machinery Manufacturing Bureau published the *Nongwu tuzhi lun,* translated orally by Edward Thomas Williams (1854–1944), who had arrived in China in 1887, and transcribed by Fan Xiyong. This translated book was based on the *Nongye turang xue* ("The Soil, its Nature, Relations, and Fundamental Principles of Management") which had been written by the famous American agronomist Franklin Hiram King (1848–1911). This was the earliest agricultural soil science book introduced into China, containing 12 chapters in total, and including information on the physical properties of soils, the origin of soils, conservation of soil moisture, soil temperature, irrigation, physical effects of tillage and fertilisers, etc.

9.9.4 Modern Geology

The first person to spread modern geological knowledge in China was the Scotsman William Muirhead, who elaborated on a number of geological phenomena and introduced geological knowledge in his book *Dili quanzhi*. He defined geology as one of the three major fields of geography. Aware of the problem inherent in the fact that the extant Chinese geographic literature only discussed "types of landscape, local conditions and customs", and not "the form and nature of the Earth," he specifically introduced some Western geological theories.

In the second section of the *Dili quanzhi*, the chapter "Geology" explains earthquakes, volcanoes, ground fissures, and the gradual erosion of the soil, which belong to the domain of tectonic geology; "Topography" expounds on landforms and their changes; "Water" addresses water quality, rivers and lakes, ocean waves, tides, springs, and water-induced terranes, which involves theories of hydrology, water erosion and sedimentation. Muirhead believed that the formation of bedrock was due to water erosion, transportation and sedimentation. He also said that fossils were buried in the ground and sank with the sand. His method of stratigraphic division was basically the same as Charles Lyell's, which divides the strata into the primary, secondary and tertiary layers from old to new (see Table 9.1 for more details).

It can be seen from Table 9.1 that the *Dili quanzhi* basically equates to the scientific levels of the Italian geologist Giovanni Arduino and of Abraham Gottlob Werner of the theory of Neptunism in Germany in the 1860s. It only uses Lyell's Tertiary system divisions from 1833, with no mention of the fact that the Primary, Secondary and Tertiary systems were renamed the Paleozoic, Mesozoic and Cenozoic by Phillips and Conibel in 1839. However, the Chinese terms in the geologic time scale not only take into account the habits of the Chinese people, but also follow the terms used by the British geological community. Compared with the *Dixue qianshi* ("A Brief Explanation of Geoscience"), this is more suited to the characteristics of the Chinese language. Therefore, it is more widespread than the *Dixue qianshi* [74]. Notably, the term *dizhi* (geology) was used for the first time in the *Dixue qianshi* and spread to Japan in 1859 [75].

The earliest modern geological masterpieces introduced into China were the *Elements of Geology* by Charles Lyell (1797–1875), translated into Chinese by Daniel Jerome Macgowan, an American Baptist missionary and physician (1814–1893, arrived 1843), and Chinese mathematician Hua Hengfang (1833–1902) (Fig. 9.2). The translation work began in 1869, was completed in 1871, and was published by the Jiangnan Manufacturing Bureau in 1873, with the title *Dixue qianshi*, comprising

Table 9.1 Comparison between the geologic time scale in the *Dixue quanzhi* and the geologic time scale of the nineteenth century

Surface stratum	Water and ice stratum	Cainozoic era	Quaternary period	Recent and Post-glacial age
Tertiary	Pliocene		Tertiary period	Pliocene
	Miocene			Miocene
	Eocene			Oligocene Eocene Paleocene
Secondary	Chalk	Mesozoic era	Secondary period	Cretaceous
	Oolite			Jurassic
	New Red Sandstone			Triassic; Permian
	Coal			Permian; Carboniferous
	Old Red Sandstone			Devonian
Primary	Silurian	Paleozoic era	Primary period	Silurian
				Ordovician
	Cambrian			Cambrian
Fossil	*Jinxing* stone [a kind of argillaceous slate in the Sinian system]	Precambrian era	Precambrian period	Precambrian
	Wenshi stone [shale] stratum			
Granite	Granite			

Wang, Zixian 王子贤, and Wang, Hengli 王恒礼 *Jianming dizhi xueshi* 简明地质学史 [A Concise History of Geology], p 206

38 volumes containing more than 230,000 words. This was the first of Lyell's works to be translated into Chinese. The *Dixue qianshi* deals with the composition of the Earth's crust from the new to the old according to the sequence of rock strata and geological history. This was the first time that Chinese readers had been introduced systematically, completely and in detail to the fundamentals of modern geology. The *Dixue qianshi* was also the first work to present the terms of Lyell's divisions of geological ages in China. After the publication of the *Dixue qianshi*, it was circulated in China as a textbook. Lu Xun once used it when he studied at the Mining Railway School affiliated to the Jiangnan Army Academy, in 1899.

The first modern crystal mineralogy masterpiece introduced into China was the *Manual of Mineralogy* (written in 1848) by James Dwight Dana (1813–1895), a famous American mineralogist, which was translated by Macgowan and Hua Hengfang [71]. The translation predated that of the *Dixue qianshi*. Work on the translation began in the 6th year of the Tongzhi reign (1867) and was completed in the 8th

Fig. 9.2 A Portrait of Hua Hengfang 华蘅芳像. Gao Qi 高奇 (2014) Zoujin zhongguo keji diantang 走进中国科技殿堂 [Entering the palace of Chinese science]. Jinan: Shandong University Press 山东大学出版社, p. 254

year of the Tongzhi reign (1869). It was published by the Jiangnan Manufacturing Bureau in the 11th year of the Tongzhi reign (1872) and entitled *Jinshi shibie*, in 12 volumes, with more than 170,000 words. This was Dana's first work to be translated into Chinese. The book systematically discusses the crystal form, the physical and chemical properties of minerals, and the classification methods of minerals. The purpose of this book was to be able to help identify minerals and this was also the aim of the translator. The *Jinshi shibie* ("Mineral Identification") revealed the theoretical system and testing methods of mineral crystals to Chinese readers for the first time, which was undoubtedly beneficial to the development of Chinese mineralogy.

In 1884, John Fryer (Fig. 9.3) published the *Kuangshi tushuo* ("Illustrated Account of Ores"), mainly covering the methods of identifying minerals and the classification of minerals, which was of practical value at that time.

Fig. 9.3 A Portrait of John Fryer傅兰雅像, cited from *Zhongguo kexue jishushi: tulujuan* 中国科学技术史-图录卷 [History of science and technology in China: Illustrations], p. 613

From 1902 to 1904, Yu Heqin and Yu Heyin were involved in translating the *Geologic Textbook* by the Japanese author Yokoyama Yojiro.

In 1903, Shen Hong translated and published the *Textbook of Mineralogy* from Japanese.

In 1907, Du Yaquan and Du Jutian translated and published Yokoyama Yojiro's *Preliminary Mineralogy Textbook*.

In addition to the above translated works, a number of translated works on geological and mineral engineering technology were also published during this period, such as the publication of the 12 volumes of the *Kaimei yaofa* ("Important Methods of Coal Mining") in 1871, interpreted by John Fryer and transcribed by Wang Dejun. In 1879, the *Jingkuang gongcheng* ("Well and Mine Engineering") was published, translated by John Fryer and Zhao Yuanyi, which introduced the technology of drilling wells and mines and gunpowder blasting technology. In 1896, the seven volumes of the *Qiukuang zhinan* ("Guide to Mining"), translated by John Fryer and Pan Song, were published. Moreover, the Jiangnan Manufacturing Bureau published Wang Ruran's translation *Xiangdi tanjinshi fa* and Wang Ruhuai's compilation of the *Kuangxue zhenquan* ("A Comprehensive Record of Mining Sciences").

9.9.5 Studies Conducted by Foreign Explorers and Scholars on Chinese Geology and Geography

After the Opium War [76], from the middle of the nineteenth century to the 1930s, more than 100 foreign explorers and scholars came to China to conduct geological and geographical investigations. Regardless of their different objectives, their work had dual significance. On the one hand, the vast majority of them were directly dispatched by their home imperialist powers with the intention of ultimately colonising China. Although a few were not motivated by political gains, their research results directly or indirectly served the imperialist invasion of China. On the other hand, these foreigners' visits to China transmitted knowledge of modern geology and geography, and proposed preliminary theories based on the first-hand data they gathered, which laid the groundwork for the development of modern geoscience and played a certain role in promoting it in China. Below are some significant research activities carried out by foreigners in China.

From 1861 to 1872, the German geologist and geographer Ferdinand von Richthofen (1833–1905) inspected China. In 1861, as a geologist, he came to China for the first time with the Prussian Elinborough Mission, but he failed to reach to the mainland due to the vigorous development of the Taiping Heavenly Kingdom revolutionary movement. From 1868 to 1872, with the financial support of the Bank of America, he returned to investigate the resources in various places. Later, sponsored by the British Chamber of Commerce in Shanghai, Ferdinand undertook seven research trips in mainland China, covering more than half of China, such as Shandong, Hebei, Shanxi, Sichuan, Central China, South China, Western China and Northeast

China, and collected a great deal of geological and geographical data. Upon returning home, he wrote the three-volume book *China*, the five-volume *Atlas of China*, and *Shandong and its Gateway - Jiaozhou*, which provided the basis for the German colonialists to seize Jiaozhou Bay and build the Jiaozhou-Jinan Railway in 1897. His research on the causes of the formation of the Loess Plateau in China opened up a pathway for future research by Chinese scholars.

In 1865, the American geologist Raphael Pumpelly (1837–1923) came to China to investigate the coal resources in various provinces along the upper reaches of the Yangzi River. Then he went to Beijing and Tianjin, and finally arrived at St. Petersburg by land via Mongolia and Siberia. After returning home, he wrote *Geological Surveys in China, Mongolia and Japan from 1862 to 1865* (1867) (the author was in Japan from 1861 to 1863) and *Crossing America and Asia*. In his two-volume *Memoirs* published in 1918, there were 13 chapters dedicated to China.

Between 1866 to 1889, the Russian geographer Nikolay Mikhaylovich Przhevalsky [Н. М. Пржевадьский], and geologist and geographer В. А. Обручев [Vladimir Afanasyevich Obruchev] went to Inner Mongolia, Xinjiang, Tibet, Qinghai, Gansu, Shaanxi and Hebei to conduct geological surveys, based on which they wrote a large number of works discussing the geological and geographical conditions of the above areas, providing information for the invasion of China by Imperial Russia.

In 1873, Catholic missionaries established Xujiahui Observatory in Shanghai, the Japanese set up weather stations in the south of the three northeastern provinces, and the Russians set up weather stations along the Middle East railway to capture China's meteorological and climatic data.

From 1877 to 1880, the Austrian Ludwig von Lóczy investigated the geology of the lower reaches of the Yangzi River, Gansu, Sichuan and Yunnan, and completed a number of geological reports (in German), discussing the landscape conditions of the southwest.

In 1890, the Japanese government sent Kyochibe [translated from Chinese], Director of the Department of Geographic Surveys, Ministry of Agriculture and Commerce, to visit China and conduct studies.

The Frenchman M. A. Leclere, in 1898, and Honoré Lantenois, in 1903, visited Yunnan for research purpose.

At the end of the Qing dynasty, the Japanese founded the East Asian Tongwen Academy in China to train researchers, who then conducted extensive investigations and collected data on the mainland of China. At the beginning of the Republic of China, the *China Economic Geography* and the *China Provincial Chronicles*, each containing several volumes, were published, which became indispensable sources of information for the Japanese aggressors.

From the end of the nineteenth century to the 1930s, Swedish geographer Sven Anders Hedin (1865–1952) led six expeditions in Tibet, Xinjiang, Mongolia and Central Asia (1885–1886, 1890–1891, 1893–1897, 1899–1902, 1905–1909, 1923–1930), over a total of more than 10 years. He organised the "Northwest Scientific Expedition Team" joined by many Chinese scholars. Sven Anders Hedin wrote numerous works, including the two-volume *Through Asia* (1898), the two-volume

Central Asia and Tibet (1903), the two [three]-volume *Trans-Himalaya: Discoveries and Adventures in Tibet* (1909–1913), *From Beijing to Moscow* (1924), *My Life as an Explorer* (1926) [1925], *Across the Gobi Desert* (1931), *Rehe Imperial Capital* (1931), *The Silk Road* (1938), etc.

From 1903 to 1904, the American geologist Bailey Willis undertook reconnaissance research in Beijing, Shandong, Shanxi, Shaanxi and the provinces and regions of the Yangzi River basin in China. In 1907, he published the *Geological Study of China* and attached a special geological map outlining the stratigraphic system of northern China. He also proposed a systemic view of the geological history and geography of China.

In 1903 [1904], David MacDonald, who was British, commanded a military expedition to investigate Tibet and gathered rich natural and economic geographical data. Later, Henry Hubert Hayden and Sidney Gerald Burrard co-authored *A Sketch of the Geography and Geology of the Himalaya Mountains and Tibet*.

In 1903, the Frenchman Honoré Lantenois surveyed Yunnan.

From 1904 to 1905, the American geologist Raphael Pumpelly returned to China with the geographer William Morris Davis (1850–1934) and his student Ellsworth Huntington (1876–1947). They reached the Tianshan Mountains and the Tarim Basin.

From 1907 to 1910, the British prospector J. C. Brown investigated Yunnan several times, wrote many geological reports, and produced a relatively detailed discussion on the geology of the southwest.

From 1909 to 1911, the French geologist Jacques Deprat (1880–1935) travelled to the eastern part of Yunnan from Annan (today's Vietnam) on a research expedition which lasted 15 months, and he wrote the *Étude Géologique du Yunnan Oriental* ("Geological Research in Eastern Yunnan").

In 1910, the Japanese geologist Bunjiro Koto [translated from Chinese] conducted a systematic survey of Northeast China and wrote the *Geological Summary of China and its Adjacent Areas.*

9.9.6 The Beginning of Modern Chinese Geoscience in the Late Qing Dynasty

(1) Surveying cartography

In the 21st year of the Guangxu reign (1895), the first school in China to train cartographic talents was established in Beijing—the Beijing Academy of Military Surveying and Mapping. In the 22nd year of the Guangxu reign (1896), Zou Daijun founded the Association of Cartography, raised funds to translate and print a world atlas, and also planned the compilation of provincial maps with the latest geographic data in China. In the 29th year of the Guangxu reign (1903), the *Zhongwai yudi quantu* ("Complete Maps of Chinese and Foreign Geographies") was published, including 3 general maps, 1 map of China, 24 maps of provinces, 38 maps of countries and continents, and 2 maps of islands. Zou applied the method of conic projection and hatching to draw mountains and classified the settlements according to the number of residents. This work was used by the Academy as a textbook for intermediate level students at that time. In the 33rd year of the Guangxu reign (1907), Zhou Shitang and Sun Haihuan, basing their work on Zou Daijun's atlas, compiled the hardback octavo edition of *The Great Maps of China and Foreign Countries in the Twentieth Century*. In addition to the common Chinese and foreign maps in this atlas, there are also special maps, such as terrain maps. It is the work which best represents the atlas in the late Qing dynasty [77].

In the 29th year of the Guangxu reign (1903), the Ministry of War became the Ministry of the Army, and the Fourth Department of the Military Advisory Office, the supreme advisory body in the Ministry of the Army, founded the Beijing Army Geodetic Bureau, with responsibility for surveying and mapping at national level. In this Geodetic Bureau, there were three units: Triangulation, Topography and Cartography, with officials specialised in the printing and publishing of maps. The Geodetic Bureau once formulated a plan to draft a national 1:25,000 topographic map, but it was never realised, except that a few military training fields were surveyed with such large-scale topographic maps, including 55 maps in the Hejian area near Baoding and 50 maps in Zhangde, Henan. In the 34th year of the Guangxu reign (1908), the Nanyang Land Survey Department drafted sixteen 1:25,000 topographic maps of parts of Anhui province. Jiangsu province completed dozens of 1:25,000 topographic maps near Nanjing. This was emblematic of the burgeoning industry of modern surveying and cartography in China.

(2) Geology

Since modern geology had been introduced to China, the question then was how to digest and absorb it. Three solutions were adopted at that time: first, sending students to study abroad; second, establishing domestic education programmes to train those with talent; and third, founding industrial and mining enterprises to cultivate apprentices through applied practice and to provide them with work opportunities. Zhang Hongzhao, Ding Wenjiang, Lu Xun, Weng Wenhao and Li Siguang, the pioneers of modern Chinese geology, were all once overseas students. After returning home, they wrote books and engaged in teaching or scientific research, making outstanding contributions to the establishment of modern Chinese geology.

A significant amount of work also went into the construction of schools, the preparation of geological departments and the training of talent in the field of geological science. For example, in 1897, the Tangshan Railway Mining School established the Department of Mining and Metallurgy; in 1898, a similar department was established in the Railway School affiliated to the Nanjing Army Academy, which was Lu Xun's Alma Mater; in 1903, the Imperial Peiyang University established a department of mining and metallurgy; and in 1909, the Imperial University of Peking launched a geology programme. These schools laid the foundations for modern geology in China and trained a number of experts in this field. At the beginning of the twentieth century, works on geology and minerals produced by Chinese scholars included Lu Xun's *A Brief Introduction to Chinese Geology* in 1903; *The Latest Textbook on Geology*, translated and edited by Zhang Xiangwen in 1905; Du Yaquan's *The*

Latest Middle School Textbook on Mineralogy, Chen Wenzhe and Chen Rongjing's *Textbook of Geology*, Gu Lang and Lu Xun's *Chinese Mineral Annals*, all in 1906; Zhang Xiangwen's *Physiography* (with some geological content) in 1908; and Kuang Rongguang's *Geological Maps of Zhili Province*, *Mineral Map of Zhili Province* and *Historical Stratum in Zhili Province* in 1910. These publications showed that modern Chinese geology was flourishing at that time.

(3) Geography

The Imperial University of Peking, founded in the 24th year of the Guangxu reign (1898), introduced geography courses, which was the beginning of the formal provision of degree programmes in geography in colleges and universities as required by the Chinese government. The Nanyang Public School, opened in Shanghai in 1897, also offered geography courses, which was the beginning of the teaching of geography courses in primary and secondary schools in China. In 1902, the Qing government issued the *Statute of the Academy Approved by the Emperor*, stipulating that the preparatory and political departments of universities must offer courses in Chinese and foreign geography, and the business department must offer courses in business geography. In 1903, the Qing government implemented the *Statute of the Academy Approved through Memoirs*, stipulating that geography courses were compulsory in the classics, literature, natural science, agriculture, commerce and other subject fields; that Chinese and foreign geography should be taught in the liberal arts, within a three-year academic programme. In 1906, the Qing government also issued the *Compulsory Courses for Excellent Teachers*, stipulating that courses in fields such as general geography, Chinese geography, continental geography, geology, physiography, and human geography should be provided in colleges for primary and middle-school teacher training. In 1905, Han Pucun, a geography teacher at the Jingshi Yixue Guan (Institute of Translation Studies, formerly the Jingshi Daxue Tang, or the Imperial University of Peking), compiled the *Yixue Guan Lectures on Geography*, which contained three subject areas: first, mathematical geography, namely, astrogeography; second, physical geography, namely physiography; third, political geography, namely human geography. These educational measures fostered the germination and development of modern Chinese geography [78].

Zhang Xiangwen (1866–1933), the pioneer of modern geography in China, with the courtesy name Weixi, who also styled himself as Zhuangu Jushi, was born in Siyang County, Jiangsu province. In 1899, he taught Chinese and geography at Nanyang Public School in Shanghai, and he himself wrote geography textbooks. In 1901, he compiled two of the earliest geography textbooks in China, namely, *Elementary Geography Textbook* and *Intermediate Geography Textbook*. These two books, which were reprinted many times, with a circulation of more than two million copies, were very popular and had a great impact on geographical education. In 1905, he compiled a secondary school textbook *Physiography* in Shanghai, which was the earliest physical geography textbook in China, and it also reflects the process of Chinese scholars' immersion in, and digestion and understanding of modern Western geography. *Physiography* is a compilation of material on the universe, land, water, air and biology. Except for the fact that soil was not listed as an independent area, but rather was attached to the section on land, it actually includes all the research objects of modern general physical geography, which was relatively advanced at that time [79]. Compared with traditional Chinese geography, this was a large stride forwards, although one weakness was that the book accepted Western environmental determinism.

In 1909, on the initiative of Zhang Xiangwen, the Chinese Geological Society was established in Tianjin. In addition to Zhang Xiangwen, who was the president, there were 27 other founding members, including Bai Yayu, Tao Maoli, Han Huaili, Zhang Boling, Wu Dingchang and Sun Shizheng. In the following year, the *Journal of Geoscience* was published. Although this journal was a preliminary publication, it played an important role in promoting the development of modern geography in China. It was an indication that modern geography in China was thriving and had shifted from the stage of dissemination to the stage of self-reliance.

References

1. Feng, E冯尔康 (1986) *Qingshi shiliaoxue chugao*清史史料学初稿 [The first draft on the historiography of the Qing dynasty], p 120
2. *Tinglin wenji: juansi yu renshu ershiwu*亭林文集·卷四, "与人书二十五" [Collected works of works of Tinglin, vol 4, "Correspondence with friends No 25"]
3. Liu X刘献廷 (1957) (Qing dynasty) *Guangyang zaji* 广阳杂记 [Miscellaneous notes on Guangyang]. Chinese Publishing House中华书局, Beijing, pp 198–199
4. Liu X刘献廷 (1957) (Qing dynasty) *Guangyang zaji* 广阳杂记 [Miscellaneous notes on Guangyang]. Chinese Publishing House中华书局, Beijing, pp 150–151
5. Zhao, R赵荣, Yang Z杨正泰 (1998) *Zhongguo dili xueshi (qing dai)* 中国地理学史(清代) [A History of Chinese geography (Qing dynasty)]. Commercial Press商务印书馆, Beijing, p 188
6. *Dushi fangyu jiyao: Zongxu*读史方舆纪要·总序 [General preface to the *Essentials of Geography for Reading History*]
7. *Liuting yudiyu shuo* 柳庭舆地隅说·卷上 [Liuting'study on geography I], 1885
8. *Liuting yudiyu shuo* 柳庭舆地隅说·卷下 [Liuting'study on geography II], 1885
9. Studies of Geoscience History at the Research Centre of Natural Science History, China Academy of Science 中国科学院自然科学史研究所地学史组 (1984) *Zhongguo gudai dili xueshi*中国古代地理学史 [A history of ancient Chinese geography]. China Science Publishing & Media Ltd. 科学出版社, Beijing, p 8
10. Huang Z黄宗羲 (1986) *Jin shuijing:Xu*今水经·序 [Modern *Water Classic*: Preface]. In: *Hunag zongxi quanji di'erce*黄宗羲全集第二册 [A complete collection of Huang Zongxi's works, Book 2]. Zhejiang Ancient Books Publishing House杭州: 浙江古籍出版社, Hangzhou
11. Chen, R陈瑞平, Qi Z齐召南 (1985) Shuidao tigang chutan《水道提纲》初探 [A preliminary study of the *Compendium of Waterways*]. *Keji wenshi ji di shisi ji*科技文史集第14辑 [Literature and history of science, vol 14] Shanghai Science Press上海科学出版社, Shanghai
12. *Xu beizhuanji: juan qishiba*续碑传集·卷七十八 [Supplement to the *Collection of Monuments and Biographies*]
13. *Xu beizhuangji: juan qishiba*续碑传集·卷七十八 [Supplement to the *Collection of Monuments and Biographies*]
14. *Xiyu shuidaoji: Deng Tingzhen xu*西域水道记·邓廷桢序 [Records of the water courses of the Western Regions: Preface by Deng Tingzhen]
15. Liang, Q梁启超, Zhongguo jin sanbainian xueshushi 中国近三百年学术史 [Three hundred years of the history of Chinese academic research]. In: *Yinbingshi wenji zhi qishiwu*饮冰室文集之七十五 [Collected papers from the Ice Drinking Room No 75]

16. Zhao R赵荣, Yang Z杨正泰 (1998) Zhongguo dilixue shi中国地理学史 清代 [A history of Chinese geography: Qing dynasty], The Commercial Press商务印书馆, Beijing, p 97
17. *Beizhuan ji: juan yibai sanshiyi*碑传集·卷一百三十一 [Collection of monuments and biographies, vol 131]
18. Hang S杭世骏, *Hudongqiao xiansheng muzhiming*胡东桥先生墓志铭 [Epitaph of Mr Hu Dongqiao], recorded in *Daogutang wenji juan sishi*道古堂文集·卷四十 [Collected papers from the Daogutang Hall, vol 40]
19. Gu J顾颉刚 (1959) *Yugong zhushi*《禹贡》注释 [Annotated *Yugong*]. Zhongguo gudai dili mingzhu xuandu中国古代地理名著选读 [Selected readings of masterworks about ancient Chinese geography]. Science Press科学出版社, Beijing
20. Zou Y邹逸麟, Hu W胡渭 (1993) *Zhongguo lidai dilixuejia pingzhuan:disan juan*中国历代地理学家评传: 第三卷 [Biography of Chinese geographers throughout history, vol 3], Shandong Education Press山东教育出版社, Jinan
21. Hou R侯仁之 (1962) *Zhongguo gudai dilixue jianshi*中国古代地理学简史 [A Brief history of ancient Chinese geography]. Science Press科学出版社, Beijing, p 79
22. Wu F吴枫, Wang Z王兆明, Fu L傅朗云 (eds) (1991) *Zhongguo guwenxian dacidian: dili juan*中国古文献大辞典·地理卷 [Dictionary of ancient Chinese documents: Geography]. Jilin Literature and History Press 吉林文史出版社, Changchun, pp 17–18
23. Refer to Chen Qiaoyi's 陈桥驿 (1987) Shuijing zhu yanjiu erji水经注研究二集 [Research on the *Commentary on the Water Classic, two volumes*]. Shanxi People's Press 山西人民出版社, pp 8–9. Zhai Z翟忠义 (1989) *Zhongguo dili xuejia* 中国地理学家 [Chinese geographers]. Shandong Education Press 山东教育出版社, pp 320–326
24. Zhi W支伟成 (1986) *Qingdai puxue dashi liezhuan xia* 清代朴学大师列传下 [Biographies of masters of evidential scholarship in the Qing dynasty II]. Yuelu Publishing House岳麓书社, Changsha
25. Chen Q陈桥驿 (1987) *Shuijing zhu yanjiu erji* 水经注研究二集 [Research on the *Commentary on the Water Classic, two volumes*]. Shanxi People's Publishing House山西人民出版社, Taiyuan, p 214
26. Weng W翁文灏 (1930) Qingchu cehui ditukao 清初测绘地图考 [Records of surveying and mapping in the early Qing dynasty]. *Dixue zazhi*地学杂志 [Geosci J] (3)
27. Ren J任金诚 (1982) Kangxi he Qianlong shiqi woguo ditu cehui shiye de chenjiu yiqi pingjia 康熙和乾隆时期我国地图测绘事业的成就以其评价 [Achievements of the cartographic undertakings in the Kangxi and Qianlong periods of the Qing dynasty and their value]. *Kexueshi jikan: dishiqi*科学史集刊: 第10期 [Collected papers on the history of science, vol 10]. Geology Press地质出版社, Beijing
28. Ren N忍侬 (1917) Zhongguo ditu zuozhi zhi yanjiu 中国地图作制之研究 [Research on mapmaking in China]. *Dongfan zazhi*东方杂志 [East J] 14(2)
29. *Qing shigao: He Guozong zhuan*清史稿·何国宗传 [A historical manuscript from the Qing dynasty: Biography of He Guozong]
30. *Daqing shengzu renhuangdi shilu: juan erbai sishiliu*大清圣祖仁皇帝实录: 卷二百四十六 [Authentic records of Emperor Shengzu (Kangxi) of the Qing dynasty, vol 246]
31. Shao Y邵懿辰 (1959) *Zending siku jianming mulu biaozhu*增订四库简明目录标注 [Additional annotations to the *Brief catalogue of the Siku quanshu*]. Zhonghua Book Company中华书局, Beijing
32. Shao Y邵懿辰 (1959) *Zending siku jianming mulu biaozhu*增订四库简明目录标注 [Additional annotations to the *Brief catalogue of the Siku quanshu*]. Zhonghua Book Company中华书局, Beijing, p 278
33. Lu L卢良志 (1984) *Zhongguo ditu xueshi*中国地图学史 [A history of Chinese cartography]. Surveying and Mapping Press测绘出版社, Beijing, p 183
34. Studies of Geoscience History at the Research Centre of Natural Science History, China Academy of Science中国科学院自然科学史研究所地学史组 (1984) *Zhongguo gudai dili xueshi*中国古代地理学史 [A history of ancient Chinese geography]. China Science Publishing & Media Ltd. 科学出版社, Beijing, p 324

35. Feng B玛宝琳 (1985) Kangxi huangyu quanlantu de cehui kaolue 康熙《皇舆全览图》的测绘考略 [A study on the mapping in the *Complete View of the Imperial Territories* in Kangxi period]. *Gugong buowuyuan yuankan*故宫博物院院刊 [Palace Mus J] (1)
36. Fang H方豪 (1987) *Zhongxi jiaotong shi xiace*中西交通史: 下册 [A history of communication in China and western countries II]. Yuelu Press岳麓书社, Changsha, p 880
37. Wang B王冰 (1986) Nan Huairen jieshao de wenduji he wenduji shixi" 南怀仁介绍的温度计和温度计试析 [Analysis of the thermometers introduced by Ferdinand Verbiest]. *Ziran kexue shi yanjiu* 自然科学史研究 [Stud Hist Nat Sci] 1
38. Hong S洪世年, Chen陈文言 W (1983) *Zhongguo qixiang shi*中国气象史 [Meteorological history of China]. China Agriculture Press农业出版社, Beijing, pp 88–89
39. Feng E冯尔康 (1986) *Qingshi shiliaoxue chugao*清史史料学初稿 [The first draft of the historiography of the Qing dynasty]. Nankai University Press南开大学出版社, Tianjin, p 120
40. Liu G刘光禄 (1982) Qingdai bianxiu fangzhi gaishu" 清代编休方志概述 [A summary of local chronicles in the Qing dynasty]. *Wenxian* 文献 [Documentation] 11
41. *Siku quanshu zongmu tiyao*四库全书总目提要 [Catalogue of the complete library of the four treasuries]
42. Zhang X章学诚 Fangzhi lisanshu yi 方志立三书议 [A discussion on setting three kinds of stylistic rules and regulations in local chronicles]. In: *Wenshi tongyi*文史通义 [General interpretation of historiography]
43. Cang X仓修良 (1984) *Zhang Xuecheng he wenshi tongyi*章学诚和文史通义 [Zhang Xuecheng and his *General interpretation of historiography*]. Zhonghua Book Company中华书局, Beijing
44. Wu F吴枫, Wang Z王兆明, Fu L傅朗云 (eds) (1991) Zhongguo guwenxian dacidian: dili juan中国古文献大辞典·地理卷 [Dictionary of ancient Chinese documents: Geography]. Jilin Literature and History Press 吉林文史出版社, Changchun, p 103
45. Wu F吴枫, Wang Z王兆明, Fu L傅朗云 (eds) (1991) *Zhongguo guwenxian dacidian: dili juan*中国古文献大辞典·地理卷 [Dictionary of ancient Chinese documents: Geography]. Jilin Literature and History Press 吉林文史出版社, Changchun, p 339
46. Wu F吴枫, Wang Z王兆明, Fu L傅朗云 (eds) (1991) *Zhongguo guwenxian dacidian: dili juan*中国古文献大辞典·地理卷 [Dictionary of ancient Chinese documents: Geography]. Jilin Literature and History Press 吉林文史出版社, Changchun, p 110
47. *Tonganxian zhi: Renwu:juan sanshi*同安县志·人物:卷三十(民国)[Chronicle of Tongan County: People, vol 30 (Republic of China)]
48. *Haiguo wenjian lu: Xu*海国闻见录·序 [Record of matters seen and heard in the maritime countries: Preface]
49. Chen D陈代光 (1984) Chen Lundong yu *haiguo wenjianlu* 陈伦炯与《海国闻见录》[Chen Lundong and his *Record of Matters Seen and Heard in the Maritime Countries*]. In: *Haiguo wenjianlu jiaozhu*《海国闻见录》校注 [Annotation of the *Record of Matters Seen and Heard in the Maritime Countries*]. Zhongzhou Ancient Texts Press中州古籍出版社, Zhengzhou
50. *Xieyu congtan sanji*屑玉从谈三集 [A series of discussions on fine jade, Book 3]
51. *Sanhe xianzhi qianlong kanben*三河县志·乾隆刊本 [Chronicle of Sanhe County, Qianlong edition]
52. *Shengzu shilu: jian ershiliu*圣祖实录· 卷二十六 [Authentic records of Emperor Shengzu (Kangxi), vol 26]
53. Xia X夏湘蓉, Li Z李仲均, Wang G王根元 (1980) *Zhongguo Gudai Kuangye Kaifashi*中国古代矿业开发史 [History of mining development in ancient China]. Geology Press北京地质出版社, Beijing, p 169
54. *Yunnan tongzhi gao: juan qishisi* 云南通志稿·卷七十四 [General chronicle of Yunnan, vol 74]
55. Wu Q吴其浚 *Diannan kuangchang tulue* 滇南矿厂图略 [Sketch maps of mineral deposits in south Yunnan]
56. Li D李调元 *Nanyue biji: juanwu* 南越笔记·卷五 [Notes of Nanyue, vol 5]
57. *Wenwu*文物 [Relics] (1984) 11:76
58. *Wenwu*文物 [Relics] (1963) 11:50

59. *Wenwu*文物 [Relics] (1985) 1:94
60. *Kaogu*考古 [Archaeology] (1985) 7:670
61. Wu H吴惠群 (1994) *Shiyong baoshixue* 实用宝石学 [Practical gemology]. Higher Education Press 高等教育出版社, Beijing, pp 195–196
62. Liu Z刘昭民 (1986) *Zhonghua dizhixue shi*中华地质学史 [A history of Chinese geology]. Taiwan Commercial Press台湾商务印书馆, Taibei
63. The Research Centre of Qing History, and the Teaching and Research Office of the Chinese Political System, School of Archiving, Renmin University of China中国人民大学清史研究所, 中国人民大学档案系中国政治制度教研室 (1983) *Qingdai de chanye: xiace*清代的产业: 下册 [Industries of the Qing dynasty II] Chinese Publishing House中华书局, Beijing, p 486
64. Xia X夏湘蓉, Li Z李仲均, Wang G王根元 (1980) *Zhongguo gudai kuangye kaifashi*中国古代矿业开发史 [History of mining development in ancient China]. Geology Press北京地质出版社, Beijing, pp 249–250
65. Lu J卢嘉锡, Li J李家治 (1998) *Zhongguo kexue jishu shi: taoci juan*中国科学技术史·陶瓷卷 [History of science and technology in China: Ceramics]. Science Press科学出版社, Beijing, p 322
66. Lu J卢嘉锡, Li J李家治 (1998) *Zhongguo kexue jishu shi: taoci juan*中国科学技术史·陶瓷卷 [History of science and technology in China: Porcelain]. Science Press科学出版社, Beijing, p 383
67. Li J李家治 (1998) *Zhongguo kexue jishu shi: taoci juan*中国科学技术史·陶瓷卷 [History of science and technology in China: Porcelain]. Science Press科学出版社, Beijing, p 457
68. *Tanji congshu juan sishisi: Shipu*檀几丛书: 卷四十四《石谱》[Tanji series, vol 44: Stone catalogue]
69. Deng J 邓经元 (1993) *Yanjing shiji: xiace* (by Ruan Yuan) 揅经室集: 下册 (作者: 阮元) [Collected works from the Yanjing Study by Ruan Yuan]. Chinese Publishing House 中华书局, Beijing, p 704
70. *Meishu congshu: chuji diliuji dierhan* 美术丛书: 初集第六辑第二函 [Fine art series, Book 1, vol 6, Slipcase 2]
71. Wang B王冰 (1986) Mingqing shiqi wulixue yizhu shumu kao 明清时期物理学译著书目考 [Study on the translated books of physics during the Ming and Qing dynasties]. *Zhongguo keji shilliao*中国科技史料 [Hist Data Chin Sci Technol] (5)
72. Li D李迪, Bai S白尚恕 (1984) Woguo jindai kexue xianqu Zou Boqi 我国近代科学先驱邹伯奇 [The pioneer of science in modern China: Zou Boqi]. *Ziran kexueshi yanjiu*自然科学史研究 [Res Hist Nat Sci] (4)
73. Xu Z徐振亚, Ruan S阮慎康 (1986) Xu Shou fuzi zusun yizhu jianjie 徐寿父子祖孙译著简介 [A brief introduction to the translated works of the three generations of Xu Shou]. *Zhongguo keji shilliao*中国科技史料 [Hist Data Chin Sci Technol] (1)
74. Wang Z王子贤, Wang H王恒礼 (1985) *Jianming dizhixue shi*简明地质学史 [A concise history of geology]. Henan Science and Technology Press河南科学技术出版社, Zhengzhou, pp 205–206
75. Li E李鄂荣 (1986) 'Dizhi' yi ci heshi chuxian yu woguo wenxian "地质"一词何时出现于我国文献 [An investigation of when the term *dizhi* first appeared in Chinese literature]. In: *Dizhixueshi luncong yi*地质学史论丛 (一) [Essays on the history of geology I]. Geology Press 地质出版社, Beijing; see also Li E李鄂荣 (1986)"Guanyu *dixue qianshi* de jige wenti 关于《地学浅释》的几个问题 [Several questions about the book *A Brief Explanation of Geology*]. In: *Dizhixue shilun cong yi* 地质学史论丛 (一) [Essays on the history of geology I]. Geology Press地质出版社, Beijing
76. This section refers to the following documents: 1. Yang, Wuyang 杨吾杨 (1984) *Dilixue sixiangshi gangyao* 地理学思想史纲要 [Outline of the history of geographic thought], printed by the Department of Geography of Henan University 河南大学地理系. 2. Wang, Zixian 王子贤, and Hengli Wang王恒礼 (1985) *Jianming dizhi xueshi* 简明地质学史 [A concise history of geology], Henan Science and Technology Press 河南科学技术出版社. 3. Ju, Jiwu 鞠继武 (1987) *Zhongguo dilixue fazhanshi* 中国地理学发展史 [History of the development of

Chinese geography], Jiangsu Education Press 江苏教育出版社. 4. Hou, Renzhi 候仁之 (1959) *Zhongguo dilixue jianzhi* 中国地理学简史 [A brief history of Chinese geography], printed by the Department of Geology and Geography of Beijing University. 5. Translation Office of the Institute of Modern History 近代史所翻译室 (1984) *Jindai laihua waiguo renming cidian* [Dictionary of foreign names in Modern China], China Social Sciences Press 中国社会科学出版社

77. Lu L卢良志 (1984) *Zhongguo dituxue shi*中国地图学史 [A history of Chinese cartography]. Surveying and Mapping Press 北京测绘出版社, Beijing; Jing Y金应春, Qiu F丘富科 (1984) *Zhongguo ditu shihua*中国地图史话 [Historical narratives of Chinese maps]. Science Press科学出版社, Beijing
78. Ju J鞠继武 (1987) *Zhongguo dilixue fazhanshi*中国地理学发展史 [History of the development of geography in China]. Jiangsu Education Press江苏教育出版社, Nanjing, p 211
79. Lin C林超 (1982) Zhongguo xiandai dilixue mengya shiqi de Zhang Xiangwen he zhongguo dixue hui 中国现代地理学萌芽时期的张相文和中国地学会 [Zhang Xiangwen and the Society of Chinese Geography in the infancy of modern geography in China]. *Ziran kexueshi yanjiu*自然科学史研究 [Studies in the history of natural science] (2)

MIX
Papier aus verantwortungsvollen Quellen
Paper from responsible sources
FSC® C105338

If you have any concerns about our products,
you can contact us on
ProductSafety@springernature.com

In case Publisher is established outside the EU,
the EU authorized representative is:
Springer Nature Customer Service Center GmbH
Europaplatz 3, 69115 Heidelberg, Germany

Printed by Libri Plureos GmbH
in Hamburg, Germany